PC UW
LLYFRGELL
LIBRARY
ABERYSTWYTH

Deformation Mechanisms, Rheology and Tectonics: Current Status and Future Perspectives

Special Publication reviewing procedures

The Society makes every effort to ensure that the scientific and production quality of its books matches that of its journals. Since 1997, all book proposals have been refereed by specialist reviewers as well as by the Society's Books Editorial Committee. If the referees identify weaknesses in the proposal, these must be addressed before the proposal is accepted.

Once the book is accepted, the Society has a team of Book Editors (listed above) who ensure that the volume editors follow strict guidelines on refereeing and quality control. We insist that individual papers can only be accepted after satisfactory review by two independent referees. The questions on the review forms are similar to those for *Journal of the Geological Society*. The referees' forms and comments must be available to the Society's Book Editors on request.

Although many of the books result from meetings, the editors are expected to commission papers that were not presented at the meeting to ensure that the book provides a balanced coverage of the subject. Being accepted for presentation at the meeting does not guarantee inclusion in the book.

Geological Society Special Publications are included in the ISI Index of Scientific Book Contents, but they do not have an impact factor, the latter being applicable only to journals.

More information about submitting a proposal and producing a Special Publication can be found on the Society's web site: www.geolsoc.org.uk.

It is recommended that reference to all or part of this book should be made in one of the following ways:

DE MEER, S., DRURY, M. R., DE BRESSER, J. H. P. & PENNOCK, G. M. (eds) 2002. *Deformation Mechanisms, Rheology and Tectonics: Current Status and Future Perspectives*. Geological Society, London, Special Publications, 200.

ZHU, W., MONTÉSI, L. G. J. & WONG, T.-F. 2002. Effects of stress on the anisotropic development of permeablity during mechanical compactions of porous sandstones. *In*: DE MEER, S., DRURY, M. R., DE BRESSER, J. H. P. & PENNOCK, G. M. (eds) *Deformation Mechanisms, Rheology and Tectonics: Current Status and Future Perspectives*. Geological Society, London, Special Publications, 200, 119–136.

GEOLOGICAL SOCIETY SPECIAL PUBLICATION NO. 200

Deformation Mechanisms, Rheology and Tectonics: Current Status and Future Perspectives

EDITED BY

S. DE MEER, M. R. DRURY, J. H. P. DE BRESSER & G. M. PENNOCK

Utrecht University, The Netherlands

2002

Published by

The Geological Society

London

THE GEOLOGICAL SOCIETY

The Geological Society of London (GSL) was founded in 1807. It is the oldest national geological society in the world and the largest in Europe. It was incorporated under Royal Charter in 1825 and is Registered Charity 210161.

The Society is the UK national learned and professional society for geology with a worldwide Fellowship (FGS) of 9000. The Society has the power to confer Chartered status on suitably qualified Fellows, and about 2000 of the Fellowship carry the title (CGeol). Chartered Geologists may also obtain the equivalent European title, European Geologist (EurGeol). One fifth of the Society's fellowship resides outside the UK. To find out more about the Society, log on to www.geolsoc.org.uk.

The Geological Society Publishing House (Bath, UK) produces the Society's international journals and books, and acts as European distributor for selected publications of the American Association of Petroleum Geologists (AAPG), the American Geological Institute (AGI), the Indonesian Petroleum Association (IPA), the Geological Society of America (GSA), the Society for Sedimentary Geology (SEPM) and the Geologists' Association (GA). Joint marketing agreements ensure that GSL Fellows may purchase these societies' publications at a discount. The Society's online bookshop (accessible from www.geolsoc.org.uk) offers secure book purchasing with your credit or debit card.

To find out about joining the Society and benefiting from substantial discounts on publications of GSL and other societies worldwide, consult www.geolsoc.org.uk, or contact the Fellowship Department at: The Geological Society, Burlington House, Piccadilly, London W1J 0BG: Tel. +44 (0)20 7434 9944; Fax +44 (0)20 7439 8975; E-mail: enquiries@geolsoc.org.uk.

For information about the Society's meetings, consult *Events* on www.geolsoc.org.uk. To find out more about the Society's Corporate Affiliates Scheme, write to enquiries@geolsoc.org.uk.

Published by The Geological Society from:
The Geological Society Publishing House
Unit 7, Brassmill Enterprise Centre
Brassmill Lane
Bath BA1 3JN, UK

(*Orders*: Tel. +44 (0)1225 445046
Fax +44 (0)1225 442836
Online bookshop: http://bookshop.geolsoc.org.uk

British Library Cataloguing in Publication Data
A catalogue record for this book is available from the British Library.

ISBN 1-86239-117-3

Typeset by Wyvern 21 Ltd
Printed by Cambrian Printers Ltd, Aberystwyth

Distributors
USA
AAPG Bookstore
PO Box 979
Tulsa
OK 74101-0979
USA
Orders: Tel. +1 918 584-2555
Fax +1 918 560-2652
E-mail bookstore@aapg.org

India
Affiliated East-West Press PVT Ltd
G-1/16 Ansari Road, Daryaganj,
New Delhi 110 002
India
Orders: Tel. +91 11 327-9113
Fax +91 11 326-0538
E-mail affiliat@nda.vsnl.net.in

Japan
Kanda Book Trading Co.
Cityhouse Tama 204
Tsurumaki 1-3-10
Tama-shi
Tokyo 206-0034
Japan
Orders: Tel. +81 (0)423 57-7650
Fax +81 (0)423 57-7651

Contents

Crust and lithosphere tectonics

Preface

This special volume is a collection of original papers and review articles based on work presented at the Deformation mechanisms, Rheology and Tectonics (DRT2001) conference, which was held in Noordwijkerhout, Netherlands in April 2001. DRT2001 was the twelfth in a series of international conferences that began in 1976 with a meeting organised by Henk Zwart, Richard Lisle, Gordon Lister and Paul Williams from the Geologisch en Mineralogisch Instituut der Rijksuniversiteit, Leiden. Special publications arising from these meetings are listed below. The Leiden meeting on Fabrics, Microtextures and Microtectonics 'was designed to bring together as many as possible of the people active in this field; not only geologists but also material scientists from other disciplines' (Lister *et al.* 1977). The topics considered in the conferences have evolved and changed but have always remained within the broad field of deformation processes. In 1999 a permanent name was adopted for the conference series, namely, 'Deformation mechanisms, Rheology and Tectonics' (Dresen & Handy 2001). The DRT meetings are devoted to the study of deformation behaviour and the rheology of rocks and minerals and to encourage dialogue between researchers working on all scales of field, experimental and theoretical studies of rock deformation. Recent DRT meetings have aimed to provide a forum where field geologists could get state-of-the-art information on experimental and theoretical studies and where theoreticians and experimentalists could debate the problems and questions posed by natural structures and microstructures (Schmid *et al.* 1999). The main focus of DRT2001 in Noordwijkerhout was on the progress made in the 25 years since the original Leiden meeting and the direction that our research should take in the new millennium. Professor Henk Zwart and Professor Paul Williams were special guests at the meeting. DRT2001 was organized by an informal group from the Faculty of Earth Sciences, Utrecht University including Siese de Meer, Martyn Drury, Magda Martens, Pat Trimby, Gill Pennock, Saskia ten Grotenhuis, and Jaap Liezenberg with support from Professor Chris Spiers and Professor Stan White.

We thank the sponsors of DRT2001 for financial support towards the conference and this special publication. We would also like to especially thank the referees for their important contribution: P. Bate, C. Beaumont, T. Blenkinsop , B. Bos, B. den Brok, P. Chopra, S. Covey-Crump, G. Dresen, M. Drury, B. Evans, D. Gapais, J. Ghoussoub, J.-P. Gratier , H. Green II, F. Gueydan, M. Handy, F. Heidelbach, R. Heilbronner, R. Holdsworth, K. Kanagawa, R. Kerrick, J. Kruhl, K. Kunze, B. Leiss, G. Lloyd, R. Lisle. I. Main, P. Mason, A. McGaig, J. Newman, D. Nieuwland, Y-D. Park, C. Peach, J. Raphanel, F. Renard, E. Rutter, C. Simpson, K. Schulmann, W. Skrotski, C. Spiers, P. Trimby, J. Urai , J. White, C. Wilson and six anonymous reviewers.

Martyn Drury, Siese de Meer, Hans de Bresser and Gill Pennock
Utrecht March 2002.

DRT conference volumes

Leiden, Netherlands 1976. LISTER, G.S., WILLIAMS, P.F., ZWART, H.J. & LISLE, R.J. (eds) 1977. Fabrics, microstructures and microtectonics. *Tectonophysics*, **39**, 1–487.

Göttingen, W. Germany, 1981. LISTER G.S., BEHR, H.-J., WEBER, K. & ZWART, H.J. (eds) 1981. The effect of deformation on rocks. *Tectonophysics*, **78**, 1–698.

Zürich, Switzerland, 1982. HANCOCK, P.L., KLAPER, E.M., MANCKTELOW, N.S. & RAMSAY, J.G. (eds) 1984. Planar and linear fabrics of deformed rocks. *Journal of Structural Geology*, **6**, 1–287.

Utrecht, Netherlands, 1985. ZWART, H.J. MARTENS, M., VAN DER MOLEN, I., PASSCHIER, C.W., SPIERS, C.J., & VISSERS, R.L.M. (eds) 1987. Tectonic and structural processes on a Macro-meso- and micro-scale. *Tectonophysics*, **135**, 1–251.

Uppsala, Sweden, 1987. TALBOT, C. (ed.) 1988. *Geological Kinematics and Dynamics: special volume in honour of the 70th birthday of Hans Ramberg*. Acta Universitatis Upsaliensis, Bulletin of the Geological Institutions of the University of Uppsala, New Series, **14**.

Leeds, Gt. Britain, 1989. KNIPE, R.J. & RUTTER, E.H. (eds) 1990. *Deformation Mechanisms, Rheology and Tectonics*. Geological Society, London, Special Publications, **54**.

Montpellier, France, 1991. BURG J-P, MAINPRICE, D. & PETIT, J.P. (eds) 1992. Special Issue-Mechanical instabilities in rocks and tectonics. *Journal of Structural Geology*, **14**, 893–1109.

Graz, Austria, 1993. WALLBRECHER, E., UNZOG, W. & BRANDMAYR, M. (eds) 1994. Structures and tectonics at different lithospheric levels – a selection of papers presented at the International conference on structures and tectonics at different lithopsheric levels, Graz, Austria. *Journal of Structural Geology*, **16**, 1495–1575.

Prague, Czechoslovakia 1995. SCHULMANN, K. (ed.) 1997. Thermal and mechanical interactions in deep-seated rocks. *Tectonophysics*, **280**, 1–197.

Basel, Switzerland, 1997. SCHMID, S.M., HEILBRONNER, R. & STÜNITZ, H. (eds) 1999. Deformation mechanisms in nature and experiment. *Tectonophysics*, **303**, 1–319.

Neustadt an der Weinstrasse, Germany, 1999. DRESEN, G. & HANDY, M. (eds) 2001. International conference on 'Deformation mechanisms, rheology and microstructures' Neustadt an der Weinstrasse, 22–26 March 1999. *International Journal of Earth Sciences (Geologische Rundschau)* **90**, 1–210.

List of Sponsors

KNAW: Royal Netherlands Academy of Arts and Sciences.
ISIS: Netherlands Research Centre for Integrated Solid Earth Sciences.
VMSG: Vening Meinesz Research School of Geodynamics.
GOI: Geodynamics Research Institute, Faculty of Earth Sciences, Utrecht University.
Stichting Electronen Microscopie Netherlands.
Nederlandse Aardolie Maatschappij bv., Netherlands.
Akzo Nobel Chemicals bv, Netherlands
Electronen Optik Service GmbH, Germany.
HKL Technology ApS, Denmark.
Corus, Netherlands.
NITG: Netherlands Institute for Applied Geosciences.
TSL/EDX, Netherlands.
FEI/ Philips Electron Optics, Netherlands.

Current issues and new developments in deformation mechanisms, rheology and tectonics

S. DE MEER, M. R. DRURY, J. H. P. DE BRESSER & G. M. PENNOCK

Vening Meinesz Research School of Geodynamics, Faculty of Earth Sciences, Utrecht University, P.O. Box 80.021, 3508 TA Utrecht, The Netherlands (e-mail: s.demeer@geo.uu.nl)

Abstract: We present a selective overview of current issues and outstanding problems in the field of deformation mechanisms, rheology and tectonics. A large part of present-day research activities can be grouped into four broad themes. First, the effect of fluids on deformation is the subject of many field and laboratory studies. Fundamental aspects of grain boundary structure and the diffusive properties of fluid-filled grain contacts are currently being investigated, applying modern techniques of light photomicrography, electrical conductivity measurement and Fourier Transform Infrared (FTIR) microanalysis. Second, the interpretation of microstructures and textures is a topic of continuous attention. An improved understanding of the evolution of recrystallization microstructures, boundary misorientations and crystallographic preferred orientations has resulted from the systematic application of new, quantitative analysis and modelling techniques. Third, investigation of the rheology of crust and mantle minerals remains an essential scientific goal. There is a focus on improving the accuracy of flow laws, in order to extrapolate these to nature. Aspects of strain and phase changes are now being taken into account. Fourth, crust and lithosphere tectonics form a subject of research focused on large-scale problems, where the use of analogue models has been particularly successful. However, there still exists a major lack of understanding regarding the microphysical basis of crust- and lithosphere-scale localization of deformation.

The motion and deformation of rocks are processes of fundamental importance in shaping the Earth, from the outer crustal layers to the deep mantle. Reconstructions of the evolution of the Earth therefore require detailed knowledge of the geometry of deformation structures and their relative timing, of the motions leading to deformation structures and of the mechanisms governing these motions. These problems concern structures on all scales, from grain scale or smaller to regional or global scale. Earth scientists in the early years of rock deformation studies focused strongly on extensive, detailed descriptions of structures. Since the 1960s, the emphasis has been more on the mechanisms behind structure development and on the role of the rheological or flow properties of rocks during deformation within the framework of large-scale tectonics. Integration of laboratory research, theoretical work on microphysical processes, microstructural and outcrop-scale studies, and modelling of tectonics has become more widespread, but at the same time the field has broadened enormously. Consequently, the need for dialogue between researchers from different disciplines is ever increasing.

The objective of this paper is to present a selective overview of some current issues and recent developments in the field of deformation mechanisms, rheology and tectonics. We have subdivided our review into four broad themes that reflect a large part of present-day research activities: (1) the effect of fluids on deformation; (2) the interpretation of microstructures and textures; (3) deformation mechanisms and rheology of crust and upper mantle minerals; and (4) crust and lithosphere tectonics. This introductory paper also serves to introduce the papers presented in this volume.

The effect of fluids on deformation

Fluids influence virtually all aspects of deformation mechanisms and rheology in the Earth on scales ranging from grain to plate boundaries (Carter *et al.* 1990). Deformation in turn has an important influence on fluid distributions in rocks (Daines & Kohlstedt 1997) and on rock transport properties (Fischer & Paterson 1989).

The involvement of water in deformation has been demonstrated in numerous field and laboratory studies. One of the principal effects on rheology results from the presence of water in grain boundaries. Grain boundary water supports a fast intergranular diffusion path, which allows stress-driven mass transport, resulting in permanent, time-dependent deformation

From: DE MEER, S., DRURY, M. R., DE BRESSER, J. H. P. & PENNOCK, G. M. (eds) 2002. *Deformation Mechanisms, Rheology and Tectonics: Current Status and Future Perspectives*. Geological Society, London, Special Publications, **200**, 1–27. 0305-8719/02/$15

(Paterson 1973, 1995; Rutter 1976, 1983; Green 1984; Lehner 1990, 1995; De Meer & Spiers 1995, 1999). This process of dissolution–precipitation creep (or pressure solution) is an important mechanism for: compaction in sedimentary rocks (Tada *et al.* 1987); healing, sealing and strength recovery in active fault zones (Sleep & Blanpied 1992; Hickman *et al.* 1995; Bos & Spiers 2000; Bos *et al.* 2000; Imber e*t al.* 2001); deformation under low temperature metamorphic conditions (Elliot 1973; Stöckhert *et al.* 1999); and evaporite flow (Spiers *et al.* 1990; Spiers & Carter 1998).

Despite the large amount of work already done on pressure-solution creep, many unresolved problems remain. At present, the elementary diffusive and interfacial processes remain poorly understood. In particular, the structure and diffusive properties of water-bearing grain boundaries are the subject of ongoing debate. Experimental studies of pressure-solution creep in crustal rocks have largely focused on compaction of granular quartz or quartz–phyllosilicate mixtures (Schutjens 1991; Mullis 1993; Dewers & Hajash 1995; Renard & Ortoleva 1997). However, compared with time scales accessible in the laboratory, pressure solution is slow in these materials, hampering reliable determination of bulk kinetics or the rate-controlling mechanism.

The involvement of fluids in faulting processes and shear zone development is widely recognized. A comprehensive review on the mechanical involvement of fluids in faulting is given by Hickman *et al.* (1995). Fluids are linked to a variety of faulting processes, including long-term structural and compositional evolution of fault zones, fault creep, and the propagation, arrest, and recurrence of earthquake ruptures. Besides the physical role of fluid pressures controlling rock strength in crustal faults, it is also clear that fluids can exert mechanical influence through a variety of chemical effects. In recent years, much attention has been focused on the role of pressure solution in the strength recovery, healing and sealing of faults and the role of phyllosilicates therein (e.g. Gratier *et al.* 1994; Bos & Spiers 2000; Bos *et al.* 2000). It is generally believed that pressure solution and subcritical crack growth have a significant weakening effect on fault zone rheology. However, Bos and co-workers found that in their experiments on halite–clay mixtures, pressure solution only resulted in weakening of fault gouges when clay was added. In the monomineralic halite fault gouge, pressure-solution compaction and healing effects dominated, leading to frictional behaviour. In halite–clay mixtures, the presence of phyllosilicates at grain boundaries prevented grain contact healing, leading to a mechanism of frictional sliding along clay foliae, accommodated by pressure solution of asperities.

Recent results on dissolution–precipitation creep, the properties of water-bearing grain boundaries, and the role of fluids in faulting (as well as vein formation) are discussed below.

Dissolution–precipitation creep

Dissolution–precipitation creep involves three serial steps (Fig. 1): (1) dissolution of material at grain boundaries under high normal stress;

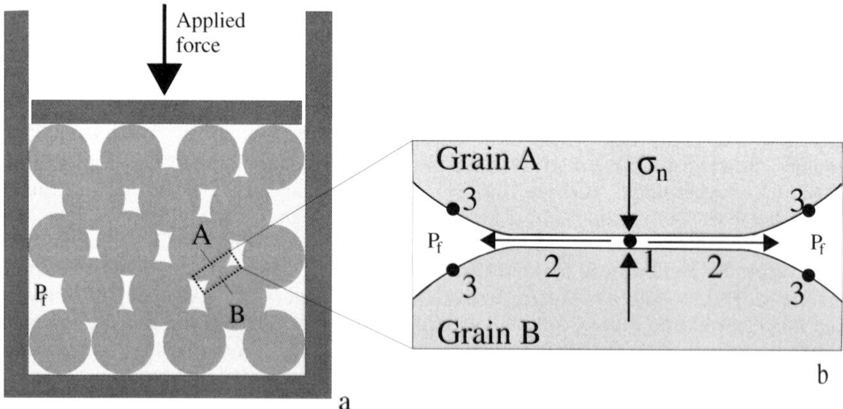

Fig. 1. Schematic illustration of pressure-solution creep. (**a**) Uniaxial compaction of a granular aggregate in the presence of saturated solution (saturated with respect to the stressed solid) at fluid pressure P_f. (**b**) Enlargement of grain contact area showing the three serial steps of pressure-solution creep: 1 – dissolution within the stressed grain boundary; 2 – diffusion through the grain boundary fluid; 3 – precipitation on the pore walls. σ_n is the effective mean normal stress across the contact.

(2) diffusion through the grain boundary fluid phase; and (3) precipitation at grain contacts under low normal stress or on free pore walls (Raj 1982; Rutter 1983; Lehner 1990; De Meer & Spiers 1997). Since interfacial reactions and diffusion occur serially, either the grain boundary diffusive properties or the interface reaction kinetics control the rate of the process.

Zhang *et al.* (2002) present the first systematic investigation into the effect of applied stress, grain size and Mg^{2+} content on the compaction of wet calcite powder at room temperature. Well-controlled starting aggregates were prepared by dry compacting granular samples before wet compaction. Dry compaction was carried out at applied stresses higher than those for wet compaction. This minimized grain rearrangement and sliding during the subsequent wet compaction stage of the tests. An acoustic emission transducer receiver was incorporated in the experimental set-up in order to detect any grain cracking or brittle deformation. As no acoustic emissions were detected and as the experimental conditions did not favour solid-state plastic deformation, intergranular pressure solution is proposed to be the most likely mechanism for compaction of the calcite aggregates. By systematically increasing the amount of Mg^{2+} added to the solution phase, compaction was drastically inhibited. This agrees with the crystal growth literature in which it is known that Mg^{2+} inhibits precipitation of calcite (Reddy & Wang 1980; Mucci & Morse 1983). It is therefore inferred that pressure solution in the calcite aggregates was rate limited by the precipitation step.

In nature, clastic sediments often contain a significant amount of clay, which can strongly influence the rate of pressure solution in a number of ways. For example, clay coatings on pore walls may have an inhibiting effect on precipitation, thus slowing down pressure solution creep (Baker *et al.* 1980; Tada & Siever 1989; Mullis 1991). On the other hand, small amounts of clay within grain boundaries may promote relatively rapid and ongoing pressure-solution creep (e.g. Weyl 1959; Dewers & Ortoleva 1991). Experimental evidence for such an effect has been reported by Hickman & Evans (1995). The accelerating effect of clay on intergranular pressure-solution creep has been attributed to different mechanisms. If diffusion is rate controlling, a clay film on grain boundaries may enhance grain boundary diffusion as it consists of a collection of clay platelets separated by thin water films that provide 'easy' diffusion paths (Weyl 1959; Hickman & Evans 1995). Alternatively, Renard & Ortoleva (1997) put forward the hypothesis that enhanced diffusion can be caused by a thick fluid film supported between clay minerals and, for example, quartz. Clay minerals have a relatively large surface charge, leading to large hydration forces. Therefore, clay–quartz boundaries are expected to have a thicker film than quartz–quartz boundaries (Israelachvili 1992; Heidug 1995), promoting water film diffusion. When dissolution or precipitation is rate controlling, cations in the solution released by the dissolution of clays cause changes in the solubility, dissolution and precipitation rates of quartz (e.g. Dove & Rimstidt 1994; Renard *et al.* 1997; Dove 1999), leading to acceleration or deceleration of pressure-solution creep rates. Furthermore, clay coating of available precipitation sites (pore walls) will also lead to a decrease in pressure-solution creep rates. Apparently, the presence of clays can result in both an increase and a decrease of intergranular pressure-solution creep rates in quartz.

In addition to experimental work, the effect of clay on pressure-solution creep has also been modelled in numerical studies. Gundersen *et al.* (2002) present the results of a numerical study on the effects of clay on pressure-solution creep in sandstone. They studied the effect of clay on the dissolution, grain boundary diffusion and precipitation steps as well as on global transport (centimetre to decimetre scale). More specifically, Gundersen *et al.* studied the effect of clays on: (1) the increase of the kinetic coefficient of dissolution or increased solubility; (2) the coating of pore walls that inhibits precipitation; and (3) the increase of the grain boundary water film thickness that enhances diffusion rates. The results of the modelling work show that dissolution is a purely local process that governs the amount of mass transport which dissolves at the grain-to-grain contact. The model also predicts that diffusion affects both local and global (cm to dm scale) processes as it governs the rate of mass transport into the pore volume. Finally, the model predicts that precipitation controls global mass transport by limiting fluid supersaturation.

The suggestion that clay has an accelerating effect on pressure-solution creep in sandstone (Gundersen *et al.*), can be assessed using the results of compaction creep experiments on quartz–muscovite mixtures from Niemeijer & Spiers (2002). These experiments (at 500 °C) show that the compaction rate is not accelerated by the addition of muscovite; rather, a modest decrease in compaction rate is observed. The results of previous work (Niemeijer *et al.* 2002) implied that, under similar experimental conditions in muscovite-free samples, pressure solution in quartz is rate limited by the dissolution step. The decrease in compaction rate by the addition of muscovite observed by Niemeijer & Spiers may be caused by dissolved Al^{3+} dominating any accelerating effects of alkali-metal cations. This is expected to decrease the solubility, dissolution and precipitation rates of quartz.

Structure and properties of water-bearing grain boundaries

Recently, attention has been focused on *in situ* observation of grain boundary structure and measurement of the diffusive properties of actively dissolving fluid-filled grain contacts using a variety of new techniques such as: reflected light interferometry and transmitted light photo-micrography (Hickman & Evans 1991, 1995); electrical conductivity (De Meer *et al.* 2002); and Fourier Transform Infrared (FTIR) microanalysis (De Meer pers. comm.). The aim of these experiments was to elucidate the structure and diffusive properties of wetted grain boundaries. Water may be present (Fig. 2) in the form of: (1) strongly adsorbed thin films (Rutter 1976, 1983; Hickman & Evans 1991, 1995; Renard & Ortoleva 1997); (2) non-equilibrium or dynamically stable island–channel networks or films (Raj & Chyung 1981; Raj 1982; Lehner 1990; Spiers & Schutjens 1990); or (3) isolated inclusions or connected crack arrays (Gratz 1991; Den Brok 1992). It is of major importance to determine the structure of actively 'pressure dissolving' grain boundaries,

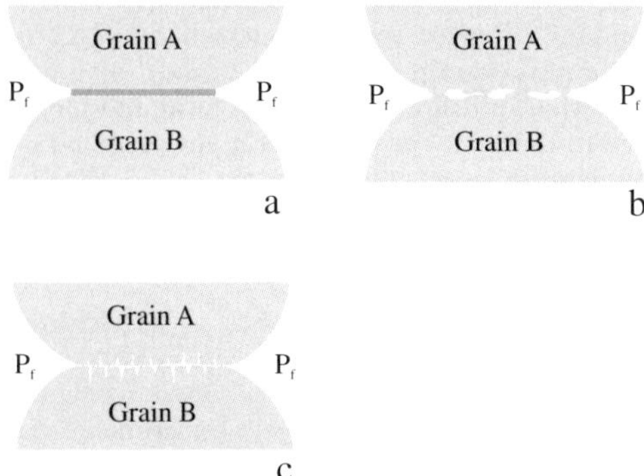

Fig. 2. Structure of water-bearing grain boundaries. (**a**) Adsorbed thin film (thickness up to *c.* 20 nm). (**b**) Dynamically stable island-channel structure with channel thickness up to ~250 nm. An adsorbed thin film might be present at grain-to-grain contacts points. (**c**) Array of connected cracks (Gratz' model; Gratz 1991). P_f is the pore fluid pressure.

as different grain boundary structures may lead to orders of magnitude difference in the predicted pressure-solution creep rates.

The observations by Den Brok *et al.* (2002), on the development of stress-induced solid/fluid interface roughness, are consistent with the dynamically stable island–channel structure above. From theoretical considerations, it is known that the flat surface of an elastically strained solid is morphologically unstable (e.g. Leroy & Heidug 1994). Furthermore, the elastic strain energy of a flat surface can be relaxed by the formation of a rough surface. This roughness can develop, for example, by diffusion through an aqueous solution. Den Brok *et al.* performed *in situ* experiments on (K-) alum single crystals in order to study this process. In the experiments, grooves developed in orientations perpendicular to the maximum compressive stress. These grooves were dynamic in nature, and moved as the local stress field changed.

Channelized and pervasive fluid flow

Regional to crustal scale shear zones are often described as channels of fluid flow (e.g. Etheridge *et al.* 1984; Pili *et al.* 1997). Le Hébel *et al.* (2002) investigated the role of fluids in deformation processes in a regional shear zone in Southern Brittany. They used geochemical methods to assess the amount of fluids involved and the scale of mass transfer. The microstructures preserved and extensive vein development were consistent with solution–precipitation creep as the predominant deformation mechanism. Dissolution of quartz and feldspars resulted in local enrichment of the rock in micas. Quartz and feldspar had been precipitated in numerous veins of which the composition reflected the local rock composition. Their geochemical data suggested that the extent of local volume loss of quartz and feldspar was in the range of 45–75%. Oxygen isotope data were consistent with local control of fluid compositions and Le Hébel *et al.* show that the amount of fluid involved during deformation was limited and fluid transfer occurred on a restricted scale with no significant fluid flow across lithological layers. Le Hébel *et al.* conclude that the regional shear zone they studied in Southern Brittany acted as a trap for early fluids rather than providing a channel for extensive fluid migration.

In nature, vein-growth mechanisms can be divided into two main classes: (1) vein formation dominated by advective fluid flow (e.g. McCaig *et al.* 1995); and (2) vein formation dominated by diffusional transport (e.g. Jamtveit & Yardly

1997; Bons 2000). In the first class of vein-growth mechanisms, material that precipitates in the vein (nutrients) can be transported over long distances to a vein-growth site. In the second class of vein-growth mechanisms, it is generally believed that nutrients are derived locally (cm to dm scale) due to the limited distance of diffusional transport. Diffusional transport along concentration gradients is normally cited as driving the nutrient transport towards growing antitaxial fibrous veins (e.g. Bons & Jessell 1997; Means & Li, 2001). This suggestion has been tested by Elburg *et al.* (2002), using constraints from geochemistry, on carbonaceous shale-hosted fibrous calcite veins from the northern Flinders Ranges, South Australia. From major and trace element data they infer that, besides locally-derived nutrients, material was transported over distances of at least decimetres to over 100 m. The fibrous texture of a vein is thus no proof of local derivation of nutrients. Furthermore, they found that fluid flow was pervasive, as there was no evidence for preferential channelized fluid flow through the veins.

In order to understand the effects of fluids in active tectonic environments, the fluid transport properties of rock need to be evaluated, particularly permeability. However, permeability is difficult to estimate in many geological processes because of its sensitivity to pressure, temperature and stress. Laboratory experiments on permeability development as a function of pressure, temperature and stress provide useful constraints on fluid transport properties of rock. Experimental observations have showed that permeability can be modified significantly under both hydrostatic and non-hydrostatic stresses. Moreover, permeability depends sensitively on the anisotropic development of damage. In most theoretical models to date, permeability is commonly prescribed as either a constant or a function of the effective mean stress (e.g. Ingebritsen & Sanford 1998), rather than a more complete second-rank tensor description. Because it is very difficult to measure permeability simultaneously in several different directions at elevated pressures, knowledge about permeability anisotropy is largely lacking. For this reason, Zhu *et al.* (2002) have developed a so-called 'hybrid triaxial compression test'. When combined with conventional triaxial compression experiments, this new type of testing makes it possible to measure permeability along both the minimum and maximum principal stress directions. The tests provide quantitative estimates of the development of permeability anisotropy as a function of effective mean and differential stress. In their experiments on sandstones, permeability showed negligible stress-induced anisotropy before the onset of shear-enhanced compaction. During the initiation of cataclastic flow, the permeability tensor showed significant anisotropy which diminished again with progressive development of cataclastic flow. Permeability anisotropy is thus transient in nature under the applied experimental conditions.

The interpretation of microstructures and textures

The development and application of new techniques of microstructure and texture characterization, such as automated electron back scattered diffraction (EBSD) and similar electron microscopy methods (Randle & Engler 2000; Kunze *et al.* 1994; Leiss *et al.* 2000; Prior *et al.* 1999), as well as computer-integrated polarization microscopy (CIP) and related light microscopy techniques (Panozzo-Heilbronner & Pauli 1993; Fueten & Goodchild 2001) have the potential to revolutionize our understanding of microstructural processes in materials. With EBSD, the complete orientation from regions as small as 0.5 μm can be measured and displayed as an orientation map. CIP involves computer-assisted collection and analysis of crystallographic orientations in optically uniaxial minerals using light microscopy. Maps showing the orientation, misorientations and orientation gradients of the *c*-axis can be produced with the CIP method.

Over the last twenty-five years a tremendous amount of work on microstructures in rocks and minerals has allowed a qualitative understanding of microstructural development to be established (Passchier & Trouw 1996; Blenkinsop 2000). New techniques offer the possibility of providing a complete characterization of microstructures. Combining these microstructural studies with experimental and theoretical studies should lead to an improved quantitative understanding of the microstructural development in minerals and rocks.

Dynamic recrystallization

Recrystallization during deformation has a strong effect on microstructure and texture development (Avé Lallemant 1975; Karato 1988; Wenk *et al.* 1997; Herwegh *et al.* 1997) and may have an important influence on rheology of all types of materials (White 1977; White *et al.* 1980; Urai *et al.* 1986; Peach *et al.* 2001).

The basic mechanisms of dynamic recrystallization are reasonably well understood from a qualitative viewpoint (e.g. Poirier & Nicolas 1975; White 1977; Poirier & Guillopé 1979), but quantitative general theories of recrystallization are lacking in areas that account for the development of microstructure with strain, the dependence of recrystallized grain size on deformation conditions, the variations of recrystallization mechanisms with deformation conditions, and the effect of recrystallization on texture.

Dynamic recrystallization has been studied extensively in metals. Metallurgists recognize three types of recrystallization, termed conventional dynamic recrystallization, continuous dynamic recrystallization and geometric dynamic recrystallization (Humphreys & Hatherly 1995; Doherty *et al.* 1997). The first type of recrystallization involves the formation of new dislocation-free grains in the deformed or recovered structure. The dislocation-free grains then grow at the expense of the old deformed grains. In continuous dynamic recrystallization a new grain structure evolves gradually and homogeneously during deformation with no distinct nucleation and growth of strain free grains (Humphreys & Hatherly 1995, 167–171; Doherty *et al.* 1997, 248). Mechanisms proposed in continuous recrystallization include subgrain growth, the development of new high-angle boundaries by merging of lower-angle boundaries, and the increase of boundary misorientation through the accumulation of dislocations into subgrain boundaries. Continuous recrystallization occurs in metal alloys in which grain boundary migration is inhibited by a high second phase and solute content (Doherty *et al.* 1997). Geometric dynamic recrystallization is important in metals which do not undergo conventional discontinuous dynamic recrystallization, like aluminium (Humphreys 1982). The mechanism (Fig. 3) involves the formation of new grains from the original high-angle grain boundaries once the original grains have flattened to approximately the diameter of the subgrain size, without involvement of rotation recrystallization. Kassner (in Doherty *et al.* 1997) suggests that geometric dynamic recrystallization may generally occur in high stacking fault materials at high strains, resulting in substantial grain size reduction (Humphreys & Hatherley 1995; Drury & Humphreys 1986).

The recrystallization terminology used by geologists concentrates on the role of grain boundary migration (migration recrystallization) versus the formation of new high-angle grain boundaries from sub-boundaries (rotation recrystallization) (White 1977; Poirier & Guillopé 1979; Drury *et al.* 1985; Urai *et al.* 1986; Drury & Urai 1990; Hirth & Tullis 1992). In many cases recrystallization in minerals involves hybrid mechanisms (Drury & Urai 1990) which occur by a combination of migration and rotation processes. The different terminology reflects differences between recrystallization in metals and minerals and also, in some cases, a different classification of similar mechanisms (Humphreys & Hatherly 1995; Drury & Urai 1990). Geometric dynamic recrystallization has not yet been recognized in minerals. However, high-strain microstructures of marble deformed in torsion (Pieri *et al.* 2001*a*, *b*) resemble geometric recrystallization structures (cf. Fig. 3), suggesting that this mechanism may be operative in geological materials.

In order to estimate temperature and strain rate conditions in naturally deformed rocks, a link is needed between experimentally and naturally deformed samples. One way of making such a link is to construct a recrystallization mechanism map for dynamically recrystallized rock covering laboratory as well as natural conditions. Stipp *et al.* (2002*b*) present such a map for dynamically recrystallized quartz. Natural data are obtained at the Tonale fault zone (Italian Alps)

Geometric dynamic recrystallisation

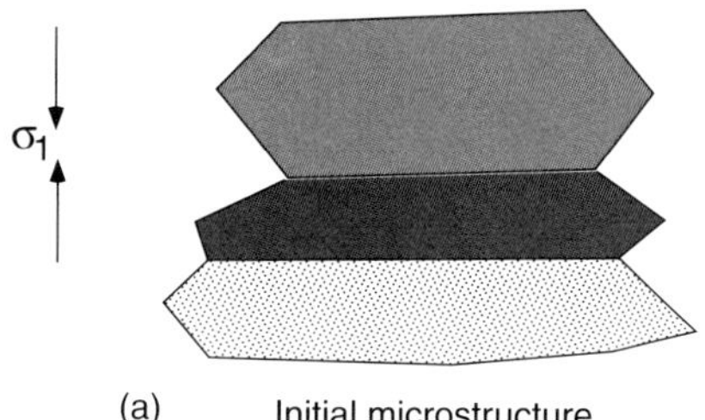

(a) Initial microstructure

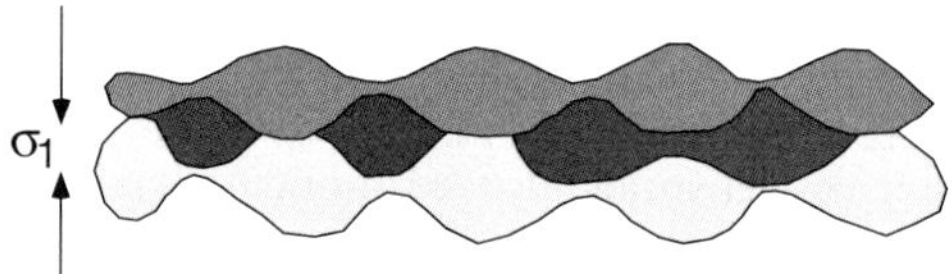

(b) High strain equi-dimensional microstructure

Fig. 3. Geometric dynamic recrystallization. (**a**) Elongated grains form at intermediate strains, showing cusps along grain boundaries. (**b**) At high strains, geometric dynamic recrystallization occurs; an equi-dimensional grain structure is produced by the impingement of irregular grain boundaries once the grains are strained to a width that is smaller than the amplitude of the grain boundary irregularities.

where a temperature gradient was determined across a single shear zone using mineral assemblages in the metasedimentary rocks in the shear zone (Stipp *et al.* 2002*a*). Three distinct dynamic recrystallization regimes were defined within increasing temperature ranges, namely, regimes of bulging recrystallization, subgrain rotation recrystallization and grain boundary migration recrystallization. Here, bulging is a combination of migration and rotation processes. This subdivision corresponds to regimes recognized in experimentally deformed quartzites (Hirth & Tullis 1992) of known deformation conditions. An estimate of the differential stress during deformation is obtained by applying a recrystallized grain size piezometric relation to the observed natural quartz microstructures. Stress estimation, however, relies on the theoretical approach of Twiss (1977), since experimental calibration of recrystallized grain size and stress is not only very limited for quartz (see Hacker *et al.* 1990), but also fails to take into account the proposed presence of different recrystallization regimes.

Boundary misorientations

During recrystallization, large misorientations are a necessary precursor to the development of new high-angle grain boundaries. Grain boundaries with certain misorientations have a higher mobility and may affect grain growth and texture evolution (Humphreys & Hatherly 1995). Formulating the development of misorientations, as a function of strain and temperature and in terms of microstructural processes, is therefore an important step towards improving our understanding of the role of grain boundaries in recrystallization and texture formation. The increasing volume of data from automated EBSD studies, coupled with statistical analysis, is improving our understanding of misorientation (Randle *et al.* 2001; Wheeler *et al.* 2001; Prior 1999; Lloyd 2000, Humphreys 2001). Fig. 4a shows deformed halite, mapped using EBSD. Misorientations, subgrain and grain scale textures can be related to the microstructure, and displayed as a misorientation angle distribution (Fig. 4b). Large-step scan mapping (step size greater than the average subgrain size, but less than the average grain size) is suitable for determining textures (Fig. 4c) and can give a reasonable estimate of average misorientations in materials with long-range gradients (Pennock *et al.* 2002). The information available from misorientation data from EBSD studies is a rapidly developing field and interpreting the data in terms of microstructural and textural changes is likely not only to improve our understanding of these processes, but may lead to routine use of new microstructural parameters.

An important development in misorientation studies of subgrains in materials deformed at elevated temperature is the observed power law relationship between the average misorientation and strain (Hughes *et al.* 1997; Pennock *et al.* 2002). The theoretical subgrain misorientations values predicted at higher strains by Mika & Dawson (1999) using finite element analysis modelling are in good agreement with the experimental values found by Hughes *et al.*, who also found that misorientation distributions from transmission electron microscopy studies could be scaled to a single curve. Misorientation distributions could, therefore, be useful in determining the amount of strain accommodated by dislocation creep in natural rocks.

The distributions of the minimum angle of misorientation (disorientation) from EBSD studies are often presented, especially for subgrain boundaries (Mainprice *et al.* 1993; Faul & Fitz Gerald 1999; Fliervoet *et al.* 1999). During rotation recrystallization, the misorientation angle distribution shifts to higher angles at higher strains (in aluminium, Hughes *et al.* 1997; in NaCl, Trimby *et al.* 2000; Pennock *et al.* 2002). Grain boundary migration recrystallization reduces the frequency of intermediate misorientation angles in halite and in quartz at low strains (Trimby *et al.* 2000, 1998), although these misorientations increase at higher strains in quartz (Neumann 2000; Trimby *et al.* 1998).

Problems in accurately defining the transition from a low-angle boundary to a high-angle boundary make it difficult to define the 'grain size' in rocks (White 1977; Drury & Urai 1990; Trimby *et al.* 1998). Trimby *et al.* (1998) introduced the 'grain boundary hierarchy' concept that accounts for the distribution of grain sizes and boundary misorientations. The grain boundary hierarchy is defined by the variation of domain size with the minimum misorientation angle (Fig. 4d). Such data can be readily obtained using the EBSD technique in the SEM. The information embodied in the grain boundary hierarchy is essential to the proper characterization of grain and subgrain size distributions. Usually an arbitrary angle is used to separate sub-boundaries from grain boundaries. Most geologists use an angle of 10° (White 1977) while material scientists use 15°. In materials with subgrains, the grain size will depend on which definition is used. This problem is illustrated in Figure 4d which shows the variation in measured grain size as a function

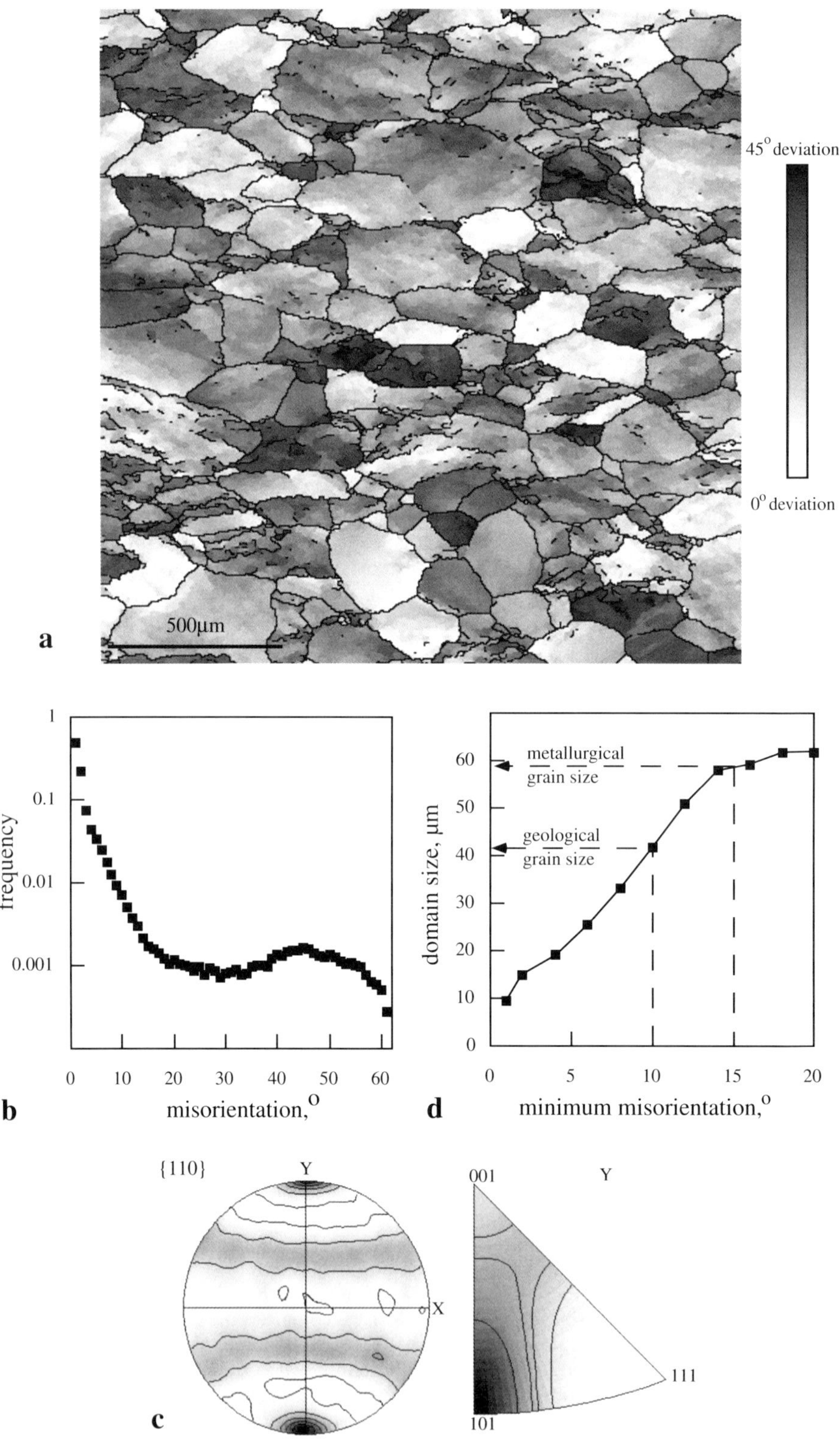

45° deviation
0° deviation
500μm
a
frequency
misorientation, °
b
domain size, μm
minimum misorientation, °
metallurgical grain size
geological grain size
d
{110}
Y
X
001
101
111
c

of minimum misorientation angle in a sample of halite deformed at high temperature. If a geological definition of a grain boundary is used, the grain size is 40 μm, while a grain size of 60 μm is obtained if the metallurgical definition of grain boundaries is used. This discrepancy arises because of the type of misorientation angle distribution developed during recovery and subgrain rotation (Fig. 4b). If the grain size is measured by light microscopy, then the distinction between subgrain boundaries and grain boundaries can only be made qualitatively.

While there may be a relatively sharp structural transition from subgrain to grain boundaries at an angle ranging between 10–25°, the properties of the boundaries may vary continuously with misorientation. Boundaries should be considered on the basis of their properties rather than structure (Lloyd & Freeman 1994; Lloyd *et al.* 1997). Consequently, the minimum misorientation used to define a grain boundary could be different when considering different processes. For recrystallization, boundary mobility is the key parameter and the transition from low mobility to high mobility may occur at low misorientations (Humphreys & Hatherly 1995). It is noted in this respect that the orientation of grain boundary might also be of importance (Randle 1998).

Textures

Crystallographic textures, or lattice preferred orientations, in metamorphic rocks can provide information on the mechanisms, kinematics and conditions of flow in the Earth (e.g. Law 1990, Wenk & Christie 1991, Schmid 1994, Bunge *et al.* 1994 and Leiss *et al.* 2000).

Leiss *et al.* (2002) used neutron diffraction to analyse the textures of polyphase amphibolites containing amphibole, plagioclase and occasional quartz. They found that care was needed in quantitative texture analysis and that more than three individual pole figures must be measured to obtain accurate results for plagioclase. Leiss *et al.* also found that two end-member types of texture developed in amphibole and plagioclase from amphibolites in the Ingales complex, Northern Cascades, USA. They suggest that the textures developed by crystal plastic slip, with texture reflecting different strain states in different units of the Ingalis nappe complex.

Neutron diffraction techniques were used by Zucali *et al.* (2002) to study a deformed hornblendite from the Sesia-Lanzo zone (Italian Alps). This eclogite facies rock consists almost entirely of amphiboles with grain sizes ranging from 0.1–0.8 mm. Conventional X-ray diffraction methods are not particularly suitable for quantitative texture analysis of this type of rock, since the technique only allows sampling of a relatively small part of the aggregate, resulting in poor statistics if the grain size is large (e.g. Kocks *et al.* 1998). Part of the problem can be solved if results of several parallel samples are summed (e.g. Schwerdtner *et al.* 1971). In an attempt to further improve quantitative analysis, Zucali *et al.* compared their results from neutron diffraction with spectra of X-ray diffraction summed from at least three slabs of the specimen. The analyses resulted in textures that agree with earlier work on amphibolite facies rocks (Gapais & Brun 1981), and showed that the X-ray technique alone provides enough information to obtain a reliable quantitative texture analysis.

The presence of a lattice preferred orientation in a rock usually indicates that dislocation creep processes were active during deformation; grain microstructures in the same rock are then often implicitly assumed to be associated with the deformation process. Heilbronner & Tullis (2002) performed texture analysis on experimentally deformed quartzites which were subsequently annealed at the deformation temperature. The analysis was confined to *c*-axis pole figures constructed from orientation images obtained by light microscopy methods. These images were calculated using the CIP method (Heilbronner 2000). Heilbronner & Tullis found that the microstructures of the samples changed substantially during annealing,

Fig. 4. Microstructural and textural information from EBSD mapping of halite, deformed (in compression) to a strain of 50%, containing 6 ppm water. (**a**) EBSD map: a dark line is drawn for differences in orientation >10° between neighbouring pixels. The majority of boundaries >10° surround grains but many also occur within grains. The difference in shading within the grains shows the deviation of ⟨110⟩ poles from the compression axis (vertical). 4 μm step scan map after replacing 25% unindexed pixels, mostly along etched boundaries. (**b**) Misorientation angle distribution, showing the frequency of misorientations between pixels; low angle misorientations dominate the distribution. (**c**) Boundary hierarchy, showing that the average domain size (the diameter of a circle, with the equivalent area as the domain) depends on the minimum misorientation angle used to define the domain. (**d**) {110} pole figure, where Y represents the compression axis, and inverse pole figure of compression axis. Equal area projection, upper hemisphere, contoured in 0.5 steps of mean uniform density, *c*. 3000 grains, 40 μm step scan size.

with a static recrystallized grain size that increased by a factor of 2 to 5. In contrast, the textures remained more or less unchanged. Grain microstructure and texture thus document different events from the history of the rock. Peridotites from the Ronda massif, southern Spain, form a natural example of this (Van der Wal & Vissers 1996). These upper mantle rocks show olivine crystallographic preferred orientations that developed during mylonitization, but the granular microstructure of the rock is attributed to static recrystallization. These findings pose problems on the interpretation of natural deformed rocks in general, and quartzites in particular (e.g. Stöckhert *et al.* 1999; Hirth *et al.* 2001). In such studies, grain size is used to constrain flow stress or strain rate, yet annealed recrystallized grains cannot be used to estimate palaeostress (Twiss 1977). The results of Heilbronner & Tullis show that crystallographic preferred orientations or textures cannot be used to distinguish annealed grains from deformed grains. However, grain-shape criteria are expected to be of use.

Computer simulation of microstructure development

Computer simulation modelling of microstructure development provides a means of improving our understanding of the interplay of the processes involved in deformation, including dynamic recrystallization. Furthermore, such models can be used to pinpoint key areas for future research and for predicting microstructures based on deformation and temperature regimes that are geologically important but which are not attainable experimentally. Jessell & Bons (2002) have reviewed the current status of numerical modelling of all scales of microstructure evolution in rocks, paying particular attention to simulation and prediction of texture development, grain boundary geometries, crystal growth and deformation in two-phase systems. The combination of EBSD studies of materials before and after deformation is a powerful technique for studying detailed changes in orientation, microstructure and texture, which can then be compared to numerical models (Bhattacharyya *et al.* 2001). Finite element models have been used to investigate inhomogeneous deformation and lattice rotations on the scale of individual grains (Zhang *et al.* 1994; Zhang & Wilson 1997; Mika & Dawson 1999).

Piazolo *et al.* (2002) used a numerical model (Jessell *et al.* 2001) to simulate dynamic recrystallization in quartz. The results can be compared with experimentally and naturally deformed quartzites (Stipp *et al.* 2002*b*; Hirth & Tullis 1992). The simulations show that the microstructure and grain size at high strain depends on the relative rates of rotation recrystallization, nucleation recrystallization and grain boundary migration, with the fastest process dominating a microstructure. The rates of these processes depend mainly on temperature and strain rate. Three recrystallization regimes have been recognized in experimentally deformed quartz (Hirth and Tullis 1992). The simulations of Piazolo *et al.* show that a transition from rotation dominated recrystallization (quartz regime II – Hirth & Tullis terminology) to migration dominated recrystallization (quartz regime III) can be associated with an increase in grain boundary mobility, which may be related to increasing temperature and/or water content. An increase of temperature, however, will also change the driving force for grain boundary migration and this effect needs to be incorporated in future recrystallization models.

Deformation mechanisms and rheology of crust and upper mantle minerals

Experimental deformation studies can provide direct information on the deformation mechanisms and rheology of minerals and rocks. The main limitation of such data is the problem of extrapolation to slow natural strain rate and from laboratory to larger scales (Paterson 1987, 2001). In contrast, studies of microstructures in naturally deformed minerals can potentially provide information on the deformation mechanisms that actually operate in the Earth. The limitations on studies of natural microstructures are that the deformation conditions are unknown, and the processes which are dominant in microstructure development may not be the processes which control the rheology (Bos & Spiers 2000).

Flow laws and extrapolation to nature

Creep laws for crust and upper mantle rocks are usually thought to follow the Dorn-type 'power law' equation. Renner & Evans (2002) show that data for low strain deformation of calcite do not fit this type of law, and that exponential creep laws (e.g. Goetze 1978; Tsenn & Carter 1987) are more appropriate. The strength of different calcite rocks in the dislocation creep regime varies with grain size, similar to the Hall-Petch relationship found in metals. Renner

& Evans suggest that an extra variable dependent on grain size – or subgrain size – is needed in calcite flow laws and they show that flow laws for several dislocation creep mechanisms can be modified to include this effect. An alternative approach is the internal state variable approach (e.g. Covey-Crump 1998). Renner & Evans show that the internal state variable model proposed by Stone (1991) provides a good description of calcite flow strength. In the Stone model, deformation occurs by a combination of subgrain-size dependent glide in addition to diffusion-controlled subgrain boundary migration, with subgrain size and distribution as the internal state variables.

The extrapolation of calcite flow laws to natural conditions is discussed by De Bresser *et al.* (2002) who note that there is a discrepancy between nature and experiment. Natural mylonitic microstructures suggest that dislocation creep is the dominant mechanism, while the extrapolation of experimental flow laws predict that grain-size-sensitive creep should be dominant. A variety of flow laws has been obtained for marbles, and extrapolation of these flow laws produces large variations in strength predictions for calcite under natural conditions. De Bresser *et al.* suggest that there may be a 'missing link' in current constitutive equations for creep in calcite (see also Renner & Evans). As an alternative approach, De Bresser *et al.* derive a hypothetical flow law based on estimates of deformation conditions in natural calcite mylonites.

Rheological and microstructural evolution towards high strain

Current flow laws for most minerals are based on low-strain experiments, yet natural deformation often involves enormous strain. The recent expansion of methods for high-strain experiments (direct shear in a saw cut assembly, Zhang *et al.* 2000; and torsion testing, Casey *et al.* 1998; Paterson & Olgaard 2000) may lead to a new generation of high-strain flow laws. Exciting results have been published on the mechanical, microstructural and textural evolution towards high strain of olivine (Bystricky *et al.* 2000), calcite (Pieri *et al.* 2001*a*, *b*) and anhydrite (Heidelbach *et al.* 2001).

In olivine, Bystricky *et al.* (2000) found that the stress decreased by 15–20% during large-strain torsion deformation of olivine up to shear strains of 5 (Fig. 5). This softening was associated with a grain size decrease from 20 μm to 3–6 μm. The development of a strong texture indicates that dislocation creep accommodated a substantial fraction of the deformation. These results agree well with observations made by Zhang *et al.* (2000) in simple shear deformation of synthetic olivine aggregates.

In calcite, current results show softening at high strains associated with dynamic recrystallization (Schmid *et al.* 1987; Rutter 1999; Pieri *et al.* 2001*a*, *b*) although softening is moderate in most studies (maximum 10% in the torsion tests of Pieri *et al.* 2001*a*, at $\gamma = 10$) and not associated with obvious shear localization. Grain size reduction was common in all tested calcite materials, but no evidence was found for a switch in mechanism from dislocation creep to grain-size-sensitive (diffusion) creep (cf. Rutter & Brodie 1988), although diffusion processes might have contributed to deformation at high strain according to Pieri *et al.* A detailed characterization of the change in grain size distribution with increasing strain in experimental, uniaxial deformation of Carrara marble is given by Ter Heege *et al.* (2002). In this work, a bimodal distribution at the start evolved into a grain size distribution close to log normal (cf. Ranalli 1984) at strains of ~35%. The median and average grain size decreased with increasing strain, but showed a complex dependence on temperature. Associated with the evolution of grain size with increasing strain, a general weakening of 10–25% was observed by Ter Heege *et al.* This weakening can be accounted for if the change in grain size distribution with increasing strain is included in composite flow laws comprising dislocation and diffusion processes. A gradual shift in distribution towards smaller grain sizes then results in an increased contribution to the deformation of the relatively weak diffusion-creep mechanisms.

In anhydrite, strain weakening up to 50% has been observed in torsion test to shear strains of ~8 (see Heidelbach *et al.* 2001). The weakening was accompanied by a reduction in grain size (from 12 to 6 μm). A switch in dominant deformation mechanism as a function of strain has been proposed for this material, from dislocation creep to diffusion creep, but full details have not yet been published. Textures of the deformed anhydrite were not weakened, but rather increased in strength towards higher strain (Heidelbach *et al.* 2001). This seems to contradict the general belief that diffusion creep does not result in a strong texture development.

All three materials discussed above showed grain size reduction (by dynamic recrystallization) and mechanical weakening towards high strain. However, weakening was limited (at least in case of calcite and olivine), and was substantially less than would be expected if a complete

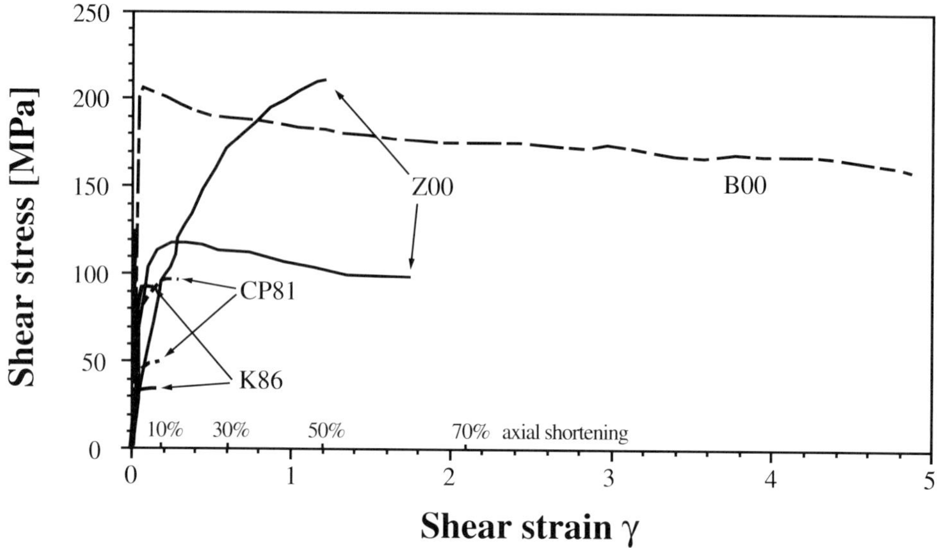

Fig. 5. Selection of stress–strain curves for olivine materials at T = 1200–1300 °C, P = 300 MPa. B00: synthetic aggregates (grain size D ≈ 20 μm) of hot-pressed San Carlos olivine powders deformed in torsion at T = 1200 °C and shear strain rate $6 \times 10^{-5}\ s^{-1}$ (Bystricky *et al.* 2000), Z00: synthetic aggregates (D ≈ 35 μm) of hot-pressed San Carlos olivine powders deformed in direct shear in a saw cut assembly at T = 1200°C (top curve) and 1300 °C (bottom curve), shear strain rate $10^{-5}\ s^{-1}$ (Zhang *et al.* 2000). CP81: natural Åheim dunite (D ≈ 900 μm; top curve) and Anita Bay dunite (D ≈ 100 μm; bottom curve) deformed in axial compression at T = 1200 °C, strain rate $10^{-5}\ s^{-1}$ (Chopra & Paterson 1981). K86: synthetic aggregates (grain size *c.* 65 μm) of hot-pressed San Carlos olivine powders deformed dry (top curve) and wet (bottom curve) in axial compression at T = 1300 °C, strain rate $10^{-5}\ s^{-1}$ (Karato *et al.* 1986). Note that the compressional stresses (σ) measured for CP81 and K86 have been converted into shear stresses (τ) applying $\tau = (1/\sqrt{3})\sigma$.

switch in deformation mechanism had taken place. De Bresser *et al.* (1998, 2001) have argued that rather than producing a switch in mechanism, dynamic recrystallization might lead to a balance between grain size reduction and grain growth processes set up in the neighbourhood of the boundary between the dislocation-creep field and the diffusion-creep field on a deformation mechanism map. If this model holds, only minor rheological weakening can result from dynamic recrystallization accompanying deformation. Further, the mechanical behaviour at high strain should be described by composite flow laws encompassing dislocation as well as diffusion creep rather than by a single constitutive rate equation. It seems worthwhile to test this model against the high-strain experimental data .

The lower crust

Experimental constraints on the rheology of the lower crust are limited compared to upper crustal and upper mantle rocks (Kohlstedt *et al.* 1995). The composition of the lower crust may vary considerably from mafic to intermediate (Rudnick 1992). Thus, depending on local composition, the strength of the lower crust may be influenced by several minerals including quartz, mica, feldspar, amphibole, pyroxene and garnet. Rutter & Brodie (1992) have provided a comprehensive review on the rheology of the lower crust based on experimental studies and naturally deformed rocks from exhumed lower crustal terrains. The strength of the lower crust depends on the timing of deformation and metamorphism (Rutter & Brodie 1992). The rheology of lower crust undergoing prograde metamorphism is strongly influenced by dehydration reactions which generate high pore fluid pressures. Deformation mechanisms in this case may be dominated by solution–transfer and brittle deformation at low effective stress (Etheridge *et al.* 1984; Rutter & Brodie 1992). Localized shear zones may develop by the concentration of deformation into transiently fine-grained zones produced by dehydration reactions (Brodie & Rutter 1987). The onset of melting may also result in significant weakening of the lower crust (Burg & Vigneresse 2002).

After metamorphism, lower crustal rocks are expected to be relatively dry. Subsequent deformation should be controlled by quartz + feldspar in intermediate lower crust (White & Bretan 1985) and feldspar + pyroxene in mafic lower crust. Structures in lower crustal terranes show that deformation can be concentrated into fine-grained shear zones formed by metamorphic reactions (Rutter & Brodie 1992; Kruse & Stünitz 1999). These lower crustal fine-grained shear zones may deform by grain-size-sensitive creep processes (Bouillier & Gueguen 1975).

Deep seismic reflection profiles often show a highly reflective lower crust. Seismic reflectors in the lower crust have been interpreted as shear zones formed during crustal extension (Reston 1990). The lower crust is often considered to be relatively weak and to act as a decoupling zone between the higher strength upper crust and upper mantle (Reston 1990). This view has been questioned by Schmid *et al.* (1996) and Handy *et al.* (2001), who point out that the geometry of the lower crust revealed in seismic profiles of the Alps is consistent with a strong lower crust and detachment of the upper crust from the lower crust and upper mantle. Considering likely variations in lower crust composition, a large variation in lower crust strength might be expected depending on grain size and water content.

Early work on the rheology of feldspar has been reviewed by Tullis (1983). Recent experimental studies on anorthite have established flow laws for diffusion creep (Wang *et al.* 1996; Dimanov *et al.* 1999) and dislocation creep (Rybacki & Dresen 2000) under both wet and dry conditions. As with quartz and olivine, the water content has a strong influence on the strength and flow parameters of anorthite. The effect of melt on anorthite strength has been investigated by Dimanov *et al.* (1998, 2000). The rheology of dry diabase has been studied by Mackwell *et al.* (1998) who show that dry lower crust may be much stronger than expected from earlier experimental studies (Shelton & Tullis 1981). In feldspar-poor regions of the lower crust the strength may be controlled by pyroxenes. Bystricky & Mackwell (2001) describe experimental flow laws for clinopyroxenite which indicate that a pyroxene-rich lower crust could be stronger than the upper mantle.

Discoveries of ultra-high pressure mineral assemblages (Chopin 1984) have shown that continental crust may be subducted or thickened, so that familiar crustal minerals are replaced by high pressure polymorphs. While there are large uncertainties concerning the rheology of normal lower crust (Handy *et al.* 2001), there is currently very little known about the rheology of ultra-high pressure crust made up of minerals such as omphacite, garnet, jadeite and coesite. The rheology of high pressure crust has an important influence on processes such as orogenesis, subduction and exhumation (Dewey *et al.* 1993; Austreim 1997).

Stöckhert (2002) reviews the information from field-based and experimental studies on deformation mechanisms and stress levels in high pressure (HP) and ultra-high pressure (UHP) metamorphic rocks with the aim of constraining the physical conditions along subduction zones to depths of 100 km. HP and UHP rocks are either undeformed or deform by dissolution–precipitation creep suggesting very low stress levels during (U)HP metamorphism in felsic rocks. Available flow laws for dislocation creep also provide an upper bound to stress levels along subduction zones within the uncertainties of the extrapolation of experimental flow laws to natural conditions. Stöckhert concludes that deformation along subduction zones with a subduction channel filled with crustal material is: (1) highly localized; and (2) occurs predominantly by dissolution–precipitation creep with a Newtonian rheology at very low stress levels.

Stöckhert notes that eclogites often show different behaviour, as many studies (Van Roermund & Boland 1981; Lardeaux *et al.* 1986; Piepenbreier & Stöckhert 2001) have found evidence for dislocation creep even at temperatures as low as 400–500 °C. Deformation by dislocation creep at such low temperatures is incompatible with extrapolated flow laws for diopside (Boland & Tullis 1986; Bystricky & Mackwell 2001) which implies that Na-pyroxenes jadeite and omphacite have a much lower dislocation-creep strength than diopside. Preliminary creep data on synthetic jadeite (Orzol *et al.* 2001) indeed suggest a much lower strength than diopisde. However, recent creep data on eclogites (Jin *et al.* 2001) indicate that dry eclogite has a similar flow strength to harzburgite. A low creep strength in naturally deformed eclogites may be explained by a strong water weakening effect in Na-pyroxenes (Buatier *et al.* 1991). Clearly, further experimental studies on these materials are needed to resolve the 'eclogite rheology problem'.

Effects of melts on rheology

Melting during deformation occurs in the continental crust during orogenesis and in a wide range of tectonic situations in the upper mantle. Intuitively, melts are expected to drastically weaken rocks, but the effects of melt on rheology can be quite variable and complex. Recent reviews on this topic include Kohlstedt *et al.* (2000) for mantle rocks, Nicolas & Idelfonse (1996) for oceanic crust and Rosenberg (2001) for granitic compositions. Burg & Vigneresse (2002) discuss the rheology of partially molten felsic rocks, extending the analysis presented by Vigneresse *et al.* (1996) and Vigneresse & Tikoff (1999) to include non-linear effects of melting and crystallization. The rheology of partially molten rocks is often considered in terms of a 'rheologically critical melt percentage' which marks the transition at which most of the viscosity drop from full solid to full liquid occurs (Arzi 1978; Van der Molen & Paterson 1979; Rosenberg 2001). Vigneresse *et al.* (1996) suggested that the rheological transition between melt and solid occurs at different melt fractions during heating or cooling. For the case of melting during heating, the solid loses cohesion at a melt content of between 20–25%. In contrast, during cooling and crystallization a magma gains cohesion once the melt content has decreased to 25–30% melt. The term 'melt escape threshold' is proposed to describe the loss of cohesion of a melting rock and the term 'particle locking threshold' to describe the onset of cohesion in crystallizing magma. Burg & Vigneresse discuss how positive feedback effects tend to localize deformation during melting and how negative feedback effects during crystallization result in a more distributed deformation. In consequence, during melting, the viscosity of the solid–melt mixture can be described by the geometric average of solid and melt viscosity at constant stress, while during crystallization the arithmetic average applies. Based on this approach Burg & Vigneresse derive temperature, melt content and viscosity relationships for partially molten rocks that are completely different for the cases of melting and crystallization.

Crust and lithosphere tectonics

The tectonic structure and evolution of the crust and lithosphere can be analysed from different viewpoints. Regional field studies generally provide insight into the geometry and kinematics of deformed parts of the crust and upper mantle (Schmid *et al.* 1987). Microstructural analysis focuses on key structures within the large-scale deformation zones, such as faults or ductile shear zones (e.g. Imber *et al.* 2001). Inferences can be made on the dynamics of deformation, given that reliable palaeostress indicators are present (Blenkinsop 2000). Further, microstructures provide constraints on the flow mechanisms operative during deformation and on the initiation of strain localization (Vissers *et al.* 1997; Jin *et al.* 1998). It has become clear that the rheology of the crust and upper mantle has an important influence on a wide range of tectonic processes, from the development of orogenic belts to the evolution of sedimentary basins.

The regional analysis of fault surface slip data is a powerful method of tectonic analysis (e.g. Bénard *et al.* 1990). Wiesmayr *et al.* (2002) apply this method to a part of the exhumed crustal slab of the High Himalaya in northwestern Bhutan. Two sets of faults were distinguished, which differed in strike and age. Older faults in one set were found to be more steeply dipping, while the second set of faults, with different strike, showed more shallow dips. This might mean that the principal stresses rotated with time or, alternatively, that the entire rock mass rotated while the stress field remained constant. In the latter case, deep ramp structures on a crustal-scale thrust system could be the cause of the rotation.

Softening and localization

One aspect of lithosphere tectonics which puzzles many researchers is the localization of strain in deformation (shear) zones. Although localized zones are ubiquitous in natural settings, their initiation remains a matter of debate (e.g. Braun *et al.* 1999; Rutter 1999). In pressure-insensitive viscous materials, persistent strain localization may only occur if weak zones are present or are created by some softening process (Bowden 1970; Cobbold 1977; Poirier *et al.* 1979; Rutter 1999). However, localization can also involve hardening if deformation is dilatant (Rudnicki & Rice 1975; Hobbs *et al.* 1990; Burg 1999). High-strain experiments on single-phase rocks show that some strain softening occurs in many materials at high strain (Rutter 1999; Bystricky *et al.* 2000; Pieri *et al.* 2001*a*, *b*) but the amount of softening is limited and shear localization has not been observed (see above). More significant strain softening has been reported in anhydrite (Heidelbach *et al.* 2001) and feldspar aggregates (Tullis *et al.* 1989), but the concentration of deformation has only been reported in feldspar.

Most current experimental data are consistent with the suggestion of De Bresser *et al.* (1998, 2001) that dynamic recrystallization does not lead to drastic softening by inducing a change to dominant grain size sensitive deformation, which implies that other softening processes may be important. Deformation mechanism maps constructed for olivine from low strain experimental data (Drury & Fitz Gerald 1998; Jin *et al.* 1998) imply that dynamic recrystallization may induce significant softening and localization by inducing a switch from normal dislocation creep to a hybrid mechanism (Fig. 6) involving deformation by a combination of dislocation creep on the weak slip system and grain boundary sliding (Hirth & Kohlstedt 1995). Dynamic recrystallization can result in other types of softening including, geometric softening (Ion *et al.* 1982; White *et al.* 1985) and structural softening (Urai *et al.* 1986; Peach *et al.* 2001).

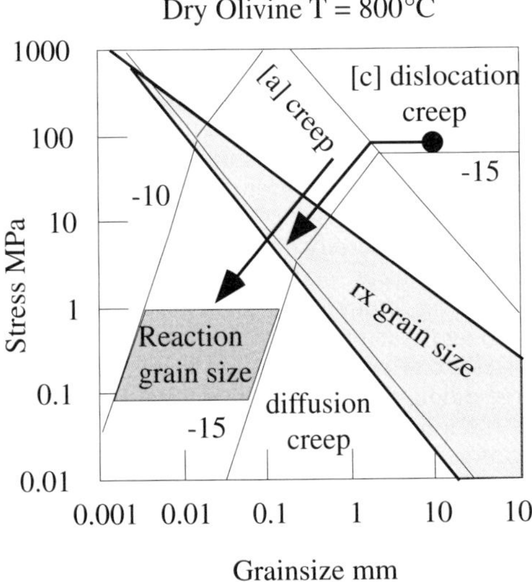

Fig. 6. Deformation mechanism map for dry olivine after Drury & Fitz Gerald (1998). Grain-size-insensitive dislocation creep controlled by slip on the hard [*c*] slip systems is dominant at high stress and large grain size. Diffusion creep and grain boundary sliding are dominant at low stress and small grain size. At intermediate stress and grain size the dominant mechanism is grain-size-sensitive dislocation creep where deformation occurs by a combination of slip on the weak slip system and grain boundary sliding (Hirth & Kohlstedt 1995). The shaded area marked 'rx grain size' shows the range of grain sizes expected for dynamic recrystallization in dry olivine. Under these conditions, grain size reduction by dynamic recrystallization may result in a switch from [*c*] dislocation creep to [*a*] dislocation creep. The shaded box marked as 'reaction grain size' shows typical grain sizes for polyphase bands that have been produced by metamorphic reaction. Grain size reduction by metamorphic reaction can result in change in deformation mechanism to diffusion creep in the fine-grained reaction productions (Newman *et al.* 1999; Handy & Stünitz 2002).

Microphysical and numerical models for dynamic recrystallization (Wenk *et al.* 1997; Shimizu 1998; Piazolo *et al.* 2002) are likely to play an important role in resolving how much softening can be produced by recrystallization.

Shear zones are commonly fine grained compared to the adjacent rocks, and thus grain size reduction is often suspected of acting as a softening mechanism. Cataclasis and metamorphic reactions are important processes of grain size reduction in natural shear zones (Stünitz & Fitz Gerald 1993; Vissers *et al.* 1997; Stünitz & Tullis 2001; Imber *et al.* 2001). Handy & Stünitz (2002) studied strain localization caused by fracturing and reaction weakening in spinel lherzolite from an exhumed passive continental margin. They observed two types of shear zones, with different mineral assemblages and deformation microstructures. In the first type of shear zone, ultra-fine grained products of fracturing and retrograde reactions under high temperature conditions (accompanied by fluid infiltration) allowed grain boundary sliding and diffusion processes to operate, probably lowering the strength of the rock by more than an order of magnitude (see Fig. 6). The second type of shear zone developed under low temperature conditions, and showed weakening associated with dilatancy and retrograde hydration, which was further enhanced once these type 2 zones coalesced subparallel to the lithosphere-scale extensional shearing plane. Extensional shear zones beneath rifted margins thus might nucleate as cracks in the initially strong upper mantle rock, subsequently evolving into trans-lithosphere weak zones as grain size is reduced due to retrograde reactions associated with fluid infiltration.

Tectonic models

Insights into crust and lithosphere tectonics have benefited substantially from a scale modelling approach in which the outer shell of the Earth is represented on laboratory scale by single-layer sandbox models or sand-silicone multilayers. Sandbox models usually consist of sand layers positioned above a rigid substratum, representing crust and lithospheric mantle, respectively. Sand layers and substratum are separated from each other by means of a plastic sheet. With this approach, characteristic structures of the crust have been reproduced and analysed in terms of geometry and evolution, such as: thrust wedge systems (Davis *et al.* 1983; Colletta *et al.*, 1991); horst and graben systems (Koopman *et al.* 1986); and strike slip fault zones (Schreurs 1994; Richard *et al.* 1995). In these models, the crust is considered as being made up of frictional, Mohr-Coulomb materials which are decoupled from the lithospheric mantle. In contrast, multilayer sand–silicone models incorporate frictional as well as viscous layers, which couple brittle and ductile rheologies. In this way, the strength profiles of a layered lithosphere (e.g. Goetze & Evans 1979; Kohlstedt *et al.* 1995) can be simulated. Essential in this approach is a correct scaling of relative strengths of the sand and silicone layers in the model with respect to brittle and ductile rheologies in nature.

Most analogue models are essentially two-dimensional, although three-dimensional models of, for example, continental indentation (e.g. Davy & Cobbold 1988) and oblique rifting (Tron & Brun 1991) have also been produced. The models almost invariably demonstrate that lithosphere-scale deformation patterns are highly dependent on the characteristics of the rheological layering as shown by Brun (2002). In particular, the rheology of the mantle and the brittle–ductile coupling determine whether deformation is distributed at lithosphere scale or localized in narrow zones. A high-strength upper mantle, represented by a sand layer in the analogue model, tends to localize deformation in zones of necking (in extension) or thrusting (in compression) which cross-cut the whole lithosphere (Davy *et al.* 1990, 1995). In contrast, models with a ductile upper mantle and lower crust below a brittle top layer, result in more-or-less homogeneous thinning or thickening, even if the ductile mantle is stronger than the ductile lower crust. Not only the relative strengths of the layers are of importance, but also the rates of deformation. At low strain rates, the lower crust may act as décollement between upper crust and mantle (Brun 2002) therefore favouring localization. At high strain rates, the increased strength of the lower crust couples upper crust and mantle, resulting in deformation to become more distributed. Patterns of distributed deformation appear to be in disagreement with structures in recent orogens (Butler 1986), but might apply to Archean or Proterozoic tectonics (e.g. Shackleton 1993; Brun 2002).

While analogue models provide important contributions to our understanding of lithosphere tectonics, temporal rheological changes, for example resulting from cooling or phase or grain size changes, cannot be incorporated. Also, measurement of stress or strain rate at any point in the model is not possible. This requires the application of numerical models (e.g. Beaumont *et al.* 1994; Braun *et al.* 1999; Gueydan *et al.* 2001).

Outstanding problems and goals for future work

We end this paper by outlining some of the outstanding problems related to topics reviewed in the previous sections.

Fluids and grain boundaries

Although during recent decades an enormous amount of field, laboratory and experimental work has been directed towards understanding the role of fluids in deformation processes, many questions remain unresolved. For example, the involvement of fluids in faulting remains uncertain. Questions that need to be answered include: what are fluid pressures at depth, what is the chemical role of fluids in fault and shear zones, and how can porosity and permeability be altered? The primary goal of future studies should be to identify and quantify the processes and parameters that are most important in controlling fault zone rheology (Hickman *et al.* 1995) and reactivation (Imber *et al.* 2001).

Despite extensive study and accumulation of a large amount of experimental data gathered on dissolution–precipitation creep (or pressure-solution creep), several important questions are still not fully answered. For example, the influence of clays on intergranular pressure solution in important rock-forming minerals such as quartz and calcite remains unclear from both the experimental and theoretical points of view. Furthermore, the structure and diffusive properties of grain boundaries during active pressure solution are still not well understood, although the different suggested grain boundary structures may lead to orders of magnitude difference in predicted strain rate (Den Brok 1998).

While the importance of fluids is widely recognized, current numerical models of geodynamic processes seldom include fluid effects. The rheology of the upper crust in the standard models is usually considered as a frictional or Mohr-Coulomb material (e.g. Beaumont *et al.* 1994). Current strength profiles for the crust only include the physical effect of fluids on effective pressure and frictional sliding (e.g. Kohlstedt *et al.* 1995; Evans & Kohlstedt 1995). Models which include the chemical effect of fluids, such as pressure solution are needed. For example, Bos & Spiers (2002) describe a new model for upper crustal deformation which involves frictional–viscous deformation by a combination of sliding along phyllosilicates accommodated by pressure solution of quartz.

Microstructure development

Quantitative methods of microstructure and texture characterization have provided extensive data on grain and subgrain sizes and shapes and on misorientation distributions. Future numerical models should investigate the full range of microstructural characteristics. The development of improved models for microstructure and texture development requires information on basic parameters as input for the models. This concerns information on diffusion coefficients, mobility and energies of subgrain and grain boundaries. Many of these parameters for minerals are unknown or badly constrained. There is a need for further experimental and microstructural studies of these basic properties (e.g. Duyster & Stöckert 2001).

Scaling laws between dynamic recrystallized grain size and stress are often used to estimate the flow stress during natural deformation (Mercier *et al.* 1977). If temperatures can be estimated from metamorphic mineral assemblages, then strain-rate estimates can be obtained from experimental flow laws, given that these flow laws are suitable for extrapolation to natural conditions. An assessment of environmental conditions during deformation is then attainable, thereby improving boundary conditions for geodynamic modelling.

Microphysical models for dynamic recrystallization (e.g. Shimizu 1998) suggest that the grain size should be dependent on temperature as well as stress. If this is generally valid, then large errors in stress estimates may occur (De Bresser *et al.* 2001). Another problem is the lack of reliable experimental data for dynamic recrystallization scaling laws. The application of early calibrations of scaling laws in olivine (Ross *et al.* 1980) and quartz (Mercier *et al.* 1977; Koch 1983) will result in overestimates of natural stresses (Green & Borch 1987). The calibration problems for quartz are discussed by Stipp *et al.* (2002*a*) who suggest that the theoretical model of Twiss (1977) is the best relationship to use. For olivine, Van der Wal *et al.* (1993) reported that the dynamically recrystallized grain size was independent of temperature and water content. However, later studies have shown that the grain size–stress relationship does vary with melt content (Hirth & Kohlstedt 1995), water content (Jung & Karato 2001) and flow behaviour (Van der Wal *et al.* 1993, Fliervoet *et al.* 1999, Zhang *et al.* 2000) (see Fig. 7). In consequence, parameters such as the water content and the rate-controlling deformation mechanism need to be identified, so that the correct scaling

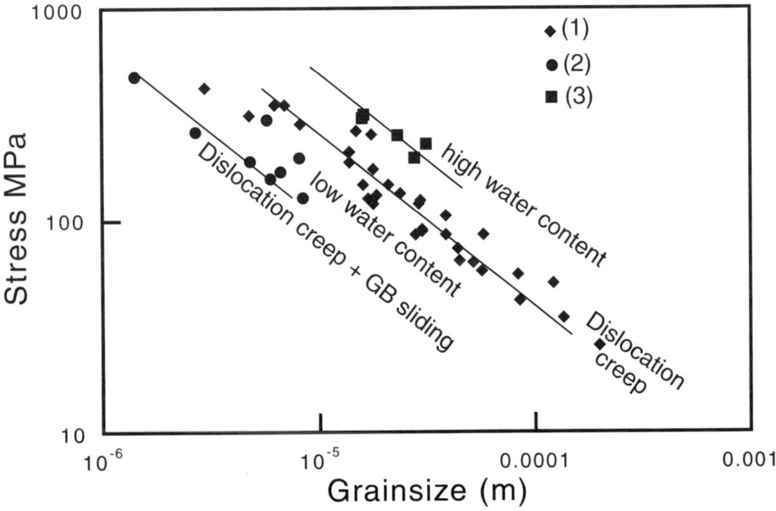

Fig. 7. Plot of dynamic recrystallized grain size versus differential stress in experimentally deformed olivine rocks. 1 – data on dry single crystals at 1650 °C and wet and dry dunites at 1200–1300 °C (Karato *et al.* 1982; Van der Wal *et al.* 1993). 2 – data from dry, fine grained synthetic olivine rocks (Zhang *et al.* 2000). 3 – data from olivine with high water content (Jung & Karato 2001). Plot shows that the scaling law between stress and grain size varies depending upon water content and which deformation mechanism dominates.

law can be used for stress estimates. A link between flow laws and scaling laws (Fig. 7) implies that self-consistent relationships must be used in the estimation of stresses and strain rates from natural microstructures.

The problems with calibration and application of grain size scaling laws impose limitations on the applicability of quantitative 'grain size palaeo-piezometery', although studies of grain size variation in naturally deformed rocks can give useful information on the qualitative variations of stress levels. Further studies on the controls of grain size during large-strain deformation are required.

High-strain flow laws

So far, published flow laws for geological materials (Kohlstedt *et al.* 1995; Evans & Kohlstedt 1995) have all been derived on the basis of low-strain deformation experiments. Recent high-strain experiments, however, have clearly shown that 'steady state' was probably not achieved at the strains for which the flow laws are calibrated. This is illustrated in Figure 5 for the case of olivine. Even if the amount of softening found under experimental conditions is relatively small, softening under natural conditions could be larger if the temperature and stress dependence of high-strain deformation are different to low-strain flow laws. Obviously, one primary goal of future experimental studies should be to establish high-strain flow laws. This may not be easy, because high-strain deformation of rocks might be governed by some combination of processes rather than a single mechanism. Special attention should thus be devoted to designing experiments that allow testing of end-member mechanisms, for example by fabricating well-controlled starting materials. Special emphasis must lie on grain size as being a fundamental parameter in the deformation behaviour of materials. Grain size is usually described by an average, one-dimensional value (i.e. the diameter) but other aspects may be important such as the distribution of grain sizes (Ter Heege *et al.* 2002) and the distribution of grain boundary misorientations (Trimby *et al.* 1998).

Considering the materials already studied in high-strain deformation, a new focus on lower crustal rocks is needed. For example, feldspars have been found to exhibit strong weakening at high strain (Tullis & Yund 1985) and this weakening can lead to shear localization (Tullis *et al.* 1989). Further information might allow more reliable interpretation of seismic reflectors in the lower crust.

Geodynamic models: significance of localization of deformation

It is important to obtain new experimental data and to develop new models for the grain size formed by dynamic phase transformations. Microstructures in natural shear zones (Stünitz & Fitz Gerald 1993; Fliervoet *et al.* 1997; Kruse & Stünitz 1999; Newman *et al.* 1999) show that the grain size of reaction products is generally a factor of 2–10 times smaller than the grain size in single-phase domains. Dynamic phase transformations might be of more importance in grain size reduction and associated rheological weakening (cf. Rutter & Brodie 1988) than conventional, mono-phase dynamic recrystallization.

Geodynamic modelling studies show that crustal and lithosphere scale localization has an important influence on the architecture and width of compressional orogenic belts (Ellis *et al.* 2001) and is a key process in continental break up (Regenauer-Lieb & Yuen 2000) and the initiation of subduction at passive margins (Branlund *et al.* 2000). Ductile shear localization may also be a crucial process that must be included in mantle convection models in order for the models to reproduce plate tectonics (e.g. Tackley 2000; Bercovici *et al.* 2000). To date, strain softening processes have usually been included in geodynamic models in an empirical way (Govers & Wortel 1995; Braun *et al.* 1999; Tackley 2000; Gueydan *et al.* 2001). There is an urgent need for materials based, microstructure and strain dependent rheological models that include the microphysics of the processes involved in deformation.

References

ARZI, A. A. 1978. Critical phenomena in the rheology of partially melted rocks. *Tectonophysics*, **44**, 173–184.

AUSTREIM, H. 1997. Influence of fluid and deformation on metamorphism in the deep crust and consequences for the geodynamics of collision zones. *In*: HACKER, B. R. & LIOU, J. G. (eds) *When Continents Collide: Geodynamics and Geochemistry of Ultrahigh Pressure Rocks. Petrology and Structural Geology*, **10**, 297–323, Kluwer, Dordrecht.

AVÉ LALLEMENT, H. G. 1975. Mechanism of preferred orientations of olivine in tectonic peridotite, *Geology*, **3**, 653–656.

BAKER, P. A., KASTNER, M., BYERLEE, J. D. & LOCKNER, D. A. 1980. Pressure solution and hydrothermal recrystallization of carbonate sediments – an experimental study. *Marine Geology*, **38**, 185–203.

BEAUMONT, C., FULLSACK, P. & HAMILTON, J. 1994. Styles of crustal deformation in compressional orogens caused by subduction of the underlying lithosphere. *Tectonophysics*, **232**, 119–132.

BÉNARD, F., MASCLE, A., LE GALL, B., DOLIGEZ, B. & ROSSI, T. 1990. Palaeostress fields in the Variscan foreland during Carboniferous, microstructural analysis in the British Isles. *Tectonophysics*, **177**, 1–13.

BERCOVICI, D., RICARD,Y. & RICHARDS, M. A. 2000. The relation between mantle dynamics and plate tectonics: a primer. *In*: RICHARDS, M. A., GORDON, R. G., VAN DER HILST, R. D. (eds) *The History and Dynamics of Global Plate Motions*. American Geophysical Union, Monographs, **121**, 5–46.

BHATTACHARYYA, A., EL-DANAF, E., KALIDINDI, S. R. & DOHERTY, R. D. 2001. Evolution of grain-scale microstructure during large strain simple compression of polycrystalline aluminium with quasi-columnar grains: OIM measurements and numerical simulations. *International Journal of Plasticity*, **17**, 861–883.

BLENKINSOP, T. 2000. *Deformation Microstructures and Mechanisms in Minerals and Rocks*. Kluwer Academic Publishers.

BOLAND, J. N. & TULLIS, T. E. 1986. Deformation Behavior of Wet and Dry Clinopyroxenite in the Brittle to Ductile Transition Region. *AGU Geophysical monograph*, **36**, 35–50.

BONS, P. D. 2000. The formation of veins and their microstructures. *In*: JESSELL, M. W. & URAI, J. L. (eds) *Stress, Strain and Structure, A Volume in Honour of W. D. Means*. Journal of the Virtual Explorer, **2**. World Wide Web Address: http://virtualexplorer.com.au/Vejournal/Volume2/www/contribs/bons

BONS, P. D. & JESSELL, M. W. 1997. Experimental simulation of the formation of fibrous veins by localised dissolution-precipitation creep. *Mineralogical Magazine*, **61**, 53–63.

BOS, B. & SPIERS, C. J. 2000. Effect of clays on fluid-assisted healing behaviour of gouge-bearing faults. *Earth and Planetary Science Letters*, **184**, 199–210.

BOS, B. & SPIERS, C. J. 2002. Frictional-viscous flow of phyllosilicate-bearing fault rock: Microphysical model and implications for crustal strength profile. *Journal of Geophysical Research*, in press.

BOS, B., PEACH, C. J. & SPIERS, C. J. 2000. Slip behavior of simulated gouge-bearing faults under conditions favoring pressure solution. *Journal of Geophysical Research B*, **105**, 16699–16717.

BOUILLER, A.-M. & GUEGUEN, Y. 1975. SP-mylonites: origin of some mylonites by superplastic flow. *Contributions to Mineralogy and Petrology*, **50**, 93–104.

BOWDEN, P. B. 1970. A criterion for inhomogeneous plastic deformation. *Philosopical Magazine*, **22**, 455–462.

BRANLUND, J., REGENAUER-LIEB, K. & YUEN, D. 2000. Fast ductile failure of passive margins from sediment loading. *Geophysical Research Letters*, **27**, 13, 1989–1992.

BRAUN, J., CHÉRY, J., POLIAKOV, A., MAINPRICE, D., VAUCHEZ, A., TOMASSI, A. & DAIGNIÈRES, M. 1999. A simple parameterization of strain localization in the ductile regime. *Journal of Geophysical Research B*, **104**, 25167–25181.

BRODIE, K. H. & RUTTER, E. H. 1987. The role of transiently fine-grained reaction products in syntectonic metamorphism: natural and experimental examples. *Canadian Journal of Earth Sciences*, **24**, 556–564.

BRUN, J.-P. 2002. Deformation of the continental lithosphere: insights from brittle-ductile models. *In*: DE MEER, S., DRURY, M. R., DE BRESSER, J. H. P. & PENNOCK, G. M. (eds) *Deformation Mechanisms, Rheology and Tectonics: Current Status and Future Perspectives*. Geological Society, London, Special Publications, **200**, 353–370.

BUATIER, M., VAN ROERMUND, H. L. M., DRURY, M. R. & LARDEAUX, J-M. 1991. Deformation and recrystallisation mechanisms in eclogites from the Sesia Lanzo zone, W. Alps. *Tectonophysics*, **195**, 11–27.

BUNGE, H. J., SIEGESMUND, S., SKROTZKI, W. & WEBER, K. (eds) 1994. *Textures of Geological Materials*. DGM Informationsgesellschaft, Oberursel.

BURG, J.-P. 1999. Ductile structures and instabilities: their implications for Variscan tectonics in the Ardennes. *Tectonophysics*, **309**, 1–25.

BURG, J.-P. & VIGNERESSE, J.-L. 2002. Non-linear effects on the rheology of partially molten felsic rocks. *In*: DE MEER, S., DRURY, M. R., DE BRESSER, J. H. P. & PENNOCK, G. M. (eds) *Deformation Mechanisms, Rheology and Tectonics: Current Status and Future Perspectives*. Geological Society, London, Special Publications, **200**, 275–292.

BUTLER, R. W. H. 1986. Thrust tectonics, deep structure and crustal subduction in the Alps and Himalayas. *Journal of the Geological Society, London*, **143**, 857–873.

BYSTRICKY, M. & MACKWELL, S. 2001. Creep of dry clinopyroxene aggregates. *Journal of Geophysical Research*, **106**, 13443–13454.

BYSTRICKY, M., KUNZE, K., BURLINE, L. & BURG, J.-P. 2000. High shear strain of olivine aggregates: rheological and seismic consequences. *Science*, **290**, 1564–1567.

CARTER, N. L., KRONENBERG, A. K., ROSS, J. V. & WILTSCHKO, D. V. 1990. Control of fluids on deformation of rocks. *In*: KNIPE, R. J. & RUTTER, E. H. (eds) *Deformation Mechanism, Rheology and Tectonics*. Geological Society, London Special Publications, **54**, 1–13.

CASEY, M., KUNZE, K. & OLGAARD, D. L. 1998. Texture of Solnhofen limestone deformed to high strains in torsion. *Journal of Structural Geology*, **20**, 255–267.

CHOPIN, C. 1984. Coesite and pure pyrope in high grade blueschists of the Western Alps. *Contributions to Mineralogy and Petrology*, **86**, 107–118.

CHOPRA, P. N. & PATERSON, M. S. 1981. The experimental deformation of dunite. *Tectonophysics*, **78**, 453–473.

COBBOLD, P. R. 1977. Description and origin of banded deformation structures. II Rheology and the growth of banded perturbations. *Canadian Journal of Earth Sciences*, **14**, 2510–2523.

COLLETTA, B., LETOUZEY, J., PINEDO, R., BALLARD, J. F. & BALÉ, P. 1991. Computerized X-ray tomography analysis of sandbox models: Examples of thin-skinned thrust systems. *Geology*, **19**, 1063–1067.

COVEY-CRUMP, S. J. 1998. Evolution of mechanical state in Carrara Marble during deformation at 400 °C to 700 °C. *Journal of Geophysical Research*, **103**, 29781–19794.

DAINES, M. J. & KOHLSTEDT, D. L. 1997. Influence of deformation on melt topology in peridotites. *Journal of Geophysical Research*, **102**, 10257–10271.

DAVIS, D., SUPPE, J. & DAHLEN, A. 1983. Mechanics of fold-and-thrust belts and accretionary wedges. *Journal of Geophysical Research*, **88**, 1153–1172.

DAVY, P. & COBBOLD, P. R. 1988. Indentation tectonics in nature and experiment. 1. Experiments scaled for gravity. *Bulletin of the Geological Institution of Uppsala*, **14**, 129–141.

DAVY, P., SORNETTE, A. & SORNETTE, D. 1990. Some consequences of a proposed fractal nature of continental faulting. *Nature*, **348**, 56–58.

DAVY, P., HANSEN, A., BONNET, E. & ZHANG, S. Z. 1995. Localisation and fault growth in layered brittle-ductile systems for deformations of the continental lithosphere. *Journal of Geophysical Research*, **100**, 6281–6294.

DE BRESSER, J. H. P., EVANS, B. & RENNER, J. 2002. On estimating the strength of calcite rocks under natural conditions. *In*: DE MEER, S., DRURY, M. R., DE BRESSER, J. H. P. & PENNOCK, G. M. (eds) *Deformation Mechanisms, Rheology and Tectonics: Current Status and Future Perspectives*. Geological Society, London, Special Publications, **200**, 309–331.

DE BRESSER, J. H. P., PEACH, C. J., REIJS, J. P. J. & SPIERS, C. J. 1998. On dynamic recrystallization during solid state flow: effects of stress and temperature. *Geophysical Research Letters*, **25**, 3459–3460.

DE BRESSER, J. H. P., TER HEEGE, J. H. & SPIERS, C. J. 2001. Grain size reduction by dynamic recrystallization: can it result in major rheological weakening? *International Journal of Earth Sciences (Geologische Rundschau)*, **90**, 28–45.

DE MEER, S. & SPIERS, C. J. 1995. Creep of wet gypsum aggregates under hydrostatic loading conditions. *Tectonophysics*, **245**, 171–184.

DE MEER, S. & SPIERS, C. J. 1997. Uniaxial compaction creep of wet gypsum aggregates. *Journal of Geophysical Research B*, **102**, 875–891.

DE MEER, S. & SPIERS, C. J. 1999. On mechanisms and kinetics of creep by intergranular pressure solution. *In*: JAMTVEIT, B. & MEAKIN, P. (eds) *Growth, Dissolution and Pattern Formation in Geosystems*. Kluwer Academic Publishers, Dordrecht, 345–366.

DE MEER, S., SPIERS, C. J., PEACH, C. J. & WATANABE, T. 2002. Diffusive properties of fluid-filled grain boundaries measured electrically during active pressure solution. *Earth and Planetary Science Letters*, **200**, 147–157.

DEN BROK, B., MOREL, J. & ZAHID, M. 2002. *In situ* experimental study of stress-induced solid/fluid interface roughness development. *In*: DE MEER, S., DRURY, M. R., DE BRESSER, J. H. P. & PENNOCK, G. M. (eds) *Deformation Mechanisms, Rheology and Tectonics: Current Status and Future Perspectives*. Geological Society, London, Special Publications, **200**, 73–84.

DEN BROK, S. J. W. 1992. An experimental investigation into the effect of water on the flow of quartzite. *Geologica Ultraiectina*, **95** (PhD thesis, Utrecht University).

DEN BROK, S. W. J. 1998. Effect of microcracking on pressure solution strain rate: the Gratz grain boundary model. *Geology*, **23**, 915–918.

DEWERS, T. & HAJASH, A. 1995. Rate laws for water-assisted compaction and stress-induced water-rock interaction in sandstone. *Journal of Geophysical Research B*, **100**, 13093–13112.

DEWERS, T. & ORTOLEVA, P. 1991. Influences of clay minerals on sandstone cementation and pressure solution. *Geology*, **19**, 1045–1048.

DEWEY, J. F., RYAN, P. D., & ANDERSEN, T. B. 1993. Orogenic uplift and collapse, crustal thickness, fabrics and metamorphic phase changs: the role of eclogites. *In*: PRICHARD, H. M., ALALBASTER, T., HARRIS, N. B. & NEARY, C. R. (eds) *Magmatic Processes and Plate tectonics*. Geological Society, London, Special Publications, **76**, 325–343.

DIMANOV, A., DRESEN, G. & WIRTH, R. 1998. High-temperature creep of partially molten plagioclase aggregates. *Journal of Geophysical Research*, **103**, 9651–9664.

DIMANOV, A., DRESEN, G., XIAO, X., & WIRTH, R. 1999. Grain boundary diffusion creep of synthetic anorthite aggregates: the effect of water. *Journal of Geophysical Research*, **104**, 10483–10497.

DIMANOV, A., WIRTH, R. & DRESEN, G. 2000. The effect of melt distribution on the rheology of plagioclase rocks. *Tectonophysics*, **328**, 307–327.

DOHERTY, R. D. & HUGHES, D. A. 1997. Current issues in recrystallization: a review. *Materials Science and Engineering A*, **238**, 219–274.

DOVE, P. M. 1999. The dissolution kinetics of quartz in aqueous mixed cation solutions. *Geochimica et Cosmochimica Acta*, **63**, 3715–3727.

DOVE, P. M. & RIMSTIDT, J. D. 1994. Silica-water interactions. *Reviews in Mineralogy*, **29**, 259–308.

DRURY, M. R. & FITZ GERALD, J. D. 1998. Mantle rheology: Insights from laboratory studies of deformation and phase transition. *In*: JACKSON, I. N. S. (ed) *The Earth's Mantle – Composition, Structure and Evolution*. Cambridge University Press, 503–559.

DRURY, M. R. & HUMPHREYS, F. J. 1986. The development of microstructure in Al-5% Mg during high temperature deformation. *Acta Metallurgica*, **34**, 2259–2271.

DRURY, M. R. & URAI, J. L. 1990. Deformation-related recrystallization processes. *Tectonophysics*, **172**, 235–253.

DRURY, M. R., HUMPHREYS, F. J. & WHITE, S. H. 1985. Large strain deformation studies using polycrystalline magnesium as a rock analogue. Part II: dynamic recrystallization mechanisms at high temperatures. *Physic of the Earth and Planetary Interiors*, **40**, 208–222.

DUYSTER, J. & STÖCKHERT, B. 2001. Grain boundary energies in olivine derived from natural microstructures. *Contributions to Mineralogy and Petrology*, **140**, 567–576.

ELBURG, M. A., BONS, P. D., FODEN, J. & PASSCHIER, C. W. 2002. The origin of fibrous veins: constraints from geochemistry. *In*: DE MEER, S., DRURY, M. R., DE BRESSER, J. H. P. & PENNOCK, G. M. (eds) *Deformation Mechanisms, Rheology and Tectonics: Current Status and Future Perspectives*. Geological Society, London, Special Publications, **200**, 103–118.

ELLIOT, D. 1973. Diffusion flow laws in metamorphic rocks. *Bulletin of the Geological Society of America*, **84**, 2645–2664.

ELLIS, S., WISSING, A. & PFIFFNER, A. 2001. Strain localization as a key to reconciling experimentally derived flow-law data with dynamic models of continental collision. *International Journal of Earth Sciences (Geologische Rundschau)*, **90**, 168–180.

ETHERIDGE, M. A., WALL, V. J. & VERNON, R. H. 1984. The role of the fluid phase during regional metamorphism and deformation. *Journal of Metamorphic Geology*, **1**, 205–226.

EVANS, B. & KOHLSTEDT, D. L. 1995. Rheology of rocks. *In*: AHRENS, T. J. (ed) *Handbook of Physical Constants Part 3 – Rock Physics and Phase Relations*. AGU, Washington DC, 148–165.

FAUL, U. H. & FITZ GERALD, J. D. 1999. Grain misorientations in partially molten olivine segregates: an electron backscatter diffraction study. *Physics and Chemistry of Minerals*, **26**, 187–197.

FISCHER, G. J. & PATERSON, M. S. 1989. Dilantancy during rock deformation at high-temperatures and pressures. *Journal of Geophysical Research*, **94**, 17607–17617.

FLIERVOET, T. F., WHITE, S. W. & DRURY, M. R. 1997. Evidence for dominant grain-boundary sliding deformation in greenschist and amphibole-grade polymineralic ultramylonites from the Redbank Deformed Zone, Central Australia. *Journal of Structural Geology*, **19**, 1495–1520.

FLIERVOET, T. F., DRURY, M. R., & CHOPRA, P. N. 1999. Crystallographic preferred orientations and misorientations in some olivine rocks deformed by diffusion or dislocation creep. *In*: SCHMID, S. M., HEILBRONNER, R., & STÜNITZ, H. (eds) *Deformation Mechanisms in Nature and Experiment. Tectonophysics*, **303**, 1–28.

FUETEN, F. & GOODCHILD, J. S. 2001. Quartz c-axes orientation determination using the rotating polarizer microscope. *Journal of Structural Geology*, **23**, 895–902.

GAPAIS, D. & BRUN, J.-P. 1981. A comparison of mineral grain fabrics and finite strain in amphibolites from eastern Finland. *Canadian Journal of Earth Sciences*, **18**, 995–1003.

GOETZE, C. 1978. The mechanisms of creep in olivine. *Philosophical Transactions of the Royal Society, London*, **288**, 99–119.

GOETZE, C. & EVANS, B. 1979. Stress and temperature in the bending lithosphere as constrained by experimental rock mechanics. *Geophysical Journal of the Royal Astronomic Society*, **59**, 463–478.

GOVERS, R. & WORTEL, M. J. R. 1995. Extension of stable continental lithosphere and the initiation of lithosphere scale faults. *Tectonics*, **14**, 1041–1055.

GRATIER, J. P., CHEN, T. & HELLMANN, R. 1994. Pressure solution as a mechanism for crack sealing around faults. *In*: HICKMAN, S., SIBSON, R. & BRUHN, R. (eds) *Proceedings USGS Red Book Conference on the Mechanical Involvement of Fluids in Faulting*. United States Geological Survey Open File Report, **94-228**, 279–300.

GRATZ, A. J. 1991. Solution transfer compaction of quartzites: progress towards a rate law. *Geology*, **19**, 901–904.

GREEN, H. W. 1984. Pressure solution creep: Some causes and mechanisms. *Journal of Geophysical Research B*, **89**, 4313–4318.

GREEN, H. W. & BORCH, R. S. 1987. A new molten salt cell for precision stress measurement at high pressure. *European Journal of Mineralogy*, **1**, 213–219.

GUEYDAN, F., LEROY, Y. M. & JOLIVET, L. 2001. Grain-size-sensitive flow and shear stress enhancement at the brittle-ductile transition of the continental crust. *International Journal of Earth Sciences (Geologische Rundschau)*, **90**, 181–196.

GUNDERSEN, E., DYSTHE, D., RENARD, F., BJØRLYKKE, K. & JAMVEIT, B. 2002. Numerical modelling of pressure solution in sandstone, rate-limiting processes and the effect of clays. *In*: DE MEER, S., DRURY, M. R., DE BRESSER, J. H. P. & PENNOCK, G. M. (eds) *Deformation Mechanisms, Rheology and Tectonics: Current Status and Future Perspectives*. Geological Society, London, Special Publications, **200**, 41–60.

HACKER, B. R., YIN, A., CHRISTIE, J. M. & SNOKE, A. W. 1990. Differential stress, strain rate, and temperatures of mylonitization in the Ruby mountains, Nevada: Implications for the rate and duration of uplift. *Journal of Geophysical Research B*, **95**, 8569–8580.

HANDY, M. R. & STÜNITZ, H. 2002. Strain localization by fracturing and reaction weakening – a mechanism for initiating exhumation of subcontinental mantle beneath rifted margins. *In*: DE MEER, S., DRURY, M. R., DE BRESSER, J. H. P. & PENNOCK, G. M. (eds) *Deformation Mechanisms, Rheology and Tectonics: Current Status and Future Perspectives*. Geological Society, London, Special Publications, **200**, 387–408.

HANDY, M., BRAUN, J. *ET AL*. 2001. Rheology and geodynamic modelling: the next step forward. *International Journal of Earth Sciences (Geologische Rundschau)*, **90**, 149–157.

HEIDELBACH, F., STRETTON, I. C. & KUNZE, K. 2001. Texture development of polycrystalline anhydrite experimentally deformed in torsion. *International Journal of Earth Sciences (Geologische Rundschau)*, **90**, 118–126.

HEIDUG, W. K. 1995. Intergranular solid-fluid phase transformations under stress: The effect of surface forces. *Journal of Geophysical Research B*, **100**, 5931–5940.

HEILBRONNER, R. 2000. Optical Orientation Imaging. *In*: JESSELL, M. W. & URAI, J. L. (eds) *Stress, Strain and Structure, A Volume in Honour of*

W. D. Means. Journal of the Virtual Explorer, **2**. World Wide Web Address: http://virtualexplorer.com. au/Vejournal/Volume2/www/contribs

HEILBRONNER, R. & TULLIS, J. 2002. The effect of static annealing on the microstructures and crystallographic preferred orientations of quartzites experimentally deformed in axial compression and shear. *In*: DE MEER, S., DRURY, M. R., DE BRESSER, J. H. P. & PENNOCK, G. M. (eds) *Deformation Mechanisms, Rheology and Tectonics: Current Status and Future Perspectives*. Geological Society, London, Special Publications, **200**, 191–218.

HERWEGH, M., HANDY, M. R. & HEILBRONNER, R. 1997. Temperature- and strain-rate-dependent microfabric evolution in monomineralic mylonite: evidence from in situ deformation of norcamphor. *Tectonophysics*, **280**, 83–106.

HICKMAN, S. H. & EVANS, B. 1991. Experimental pressure solution in halite: the effect of grain/interface boundary structure. *Journal of the Geological Society, London*, **148**, 549–560.

HICKMAN, S. H. & EVANS, B. 1995. Kinetics of pressure solution at halite-silica interfaces and intergranular clay films. *Journal of Geophysical Research B*, **100**, 13113–13132.

HICKMAN, S. H., SIBSON, R. & BRUHN, R. 1995. Introduction to special section: mechanical involvement of fluids in faulting. *Journal of Geophysical Research B*, **100**, 12831–12840.

HIRTH, G. & KOHLSTEDT, D. L. 1995. Experimental constraints on the dynamics of the partially molten upper mantle 2. Deformation in the dislocation creep regime. *Journal of Geophysical Research*, **100**, 15441–15449.

HIRTH, G. & TULLIS, J. 1992. Dislocation creep regimes in quartz aggregates. *Journal of Structural Geology*, **14**, 145–159.

HIRTH, G., TEYSSIER, C. & DUNLAP, W. J. 2001. An evaluation of quartzite flow laws based on comparisons between experimentally and naturally deformed rocks. *International Journal of Earth Sciences (Geologische Rundschau)*, **90**, 77–87.

HOBBS, B. E., MÜHLHUAS, H.-B., & ORD, A. 1990. Instability, softening and localization of deformation. *In*: KNIPE, R. J. & RUTTER, E. H. (eds) *Deformation Mechanism, Rheology and Tectonics*, Geological Society, London, Special Publications, **54**, 143–165.

HUGHES, D. A., LIU, Q., CHRZAN, D. C. & HANSEN, N. 1997. Scaling of microstructural parameters: misorientations of deformation induced boundaries. *Acta Materialia*, **45**, 105–112.

HUMPHREYS, F. J. 1982. Inhomogeneous deformation of some aluminium alloys. *In*: GIFKINS, R. C. (ed). *Proceedings of 6th International Conference On Strength of Metals and Alloys, Melbourne*, **1**, 625–630.

HUMPHREYS, F. J. 2001. Review: grain and subgrain characterization by electron diffraction. *Journal of Material Science*, **36**, 3833–3854.

HUMPHREYS, F. J. & HATHERLY, M. 1995. *Recrystallization and related annealing phenomena*. Elsevier Science, Amsterdam.

IMBER, J., HOLDSWORTH, R. E. & BUTLER, C. A. 2001. A reappraisal of the Sibson-Scholz fault zone model: the nature of the frictional to viscous ("brittle-ductile") transition along a long-lived, crustal-scale fault, Outer Hebrides, Scotland. *Tectonics*, **20**, 601–624.

INGEBRITSEN, S. E. & SANFORD, W. E. 1998. *Groundwater in Geological Processes*. Cambridge University Press, New York.

ION, S. E., HUMPHREYS, F. J. & WHITE, S. H. 1982. Dynamic recrystallisation and the development of microstucture during the high temperature deformation of magnesium. *Acta Metallurgica*, **30**, 1909–1919.

ISRAELACHVILI, J. N. 1992. Adhesion forces between surfaces in liquids and condensable vapours. *Surface Science Reports*, **14**, 109–159.

JAMTVEIT, B. & YARDLY, B. W. D. 1997. *Fluid Flow and Transport in Rocks – Mechanisms and Effects*. Chapman & Hall, London.

JESSELL, M. W. & BONS, P. D. 2002. The numerical simulation of microstructure. *In*: DE MEER, S., DRURY, M. R., DE BRESSER, J. H. P. & PENNOCK, G. M. (eds) *Deformation Mechanisms, Rheology and Tectonics: Current Status and Future Perspectives*. Geological Society, London, Special Publications, **200**, 137–148.

JESSELL, M., BONS, P., EVANS, L., BARR, T. & STUWE, K. 2001. Elle: the numerical simulation of metamorphic and deformation microstructures. *Computers and Geosciences*, **27**, 17–30.

JIN, J., KARATO, S.-I. & OBATA, M. 1998. Mechanisms of shear localization in the continental lithosphere: inference from the deformation microstructures of peridotites from the Ivrea zone, northwestern Italy. *Journal of Structural Geology*, **20**, 195–209.

JIN, Z. M., ZHANG, J., GREEN, H. W. & JIN, S. 2001. Eclogite rheology: Implications for subducted lithosphere. *Geology*, **29**, 667–670.

JUNG, H. & KARATO, S.-I. 2001. Effects of water on dynamically recrystallized grain size of olivine. *Journal of Structural Geology*, **23**, 1337–1344.

KARATO, S. I. 1988. The role of recrystallization in the preferred orientation of olivine. *Physics of the Earth and Planetary Interiors*, **51**, 107–122.

KARATO, S.-I., PATERSON, M. S., & FITZ GERALD, J. D. 1986. Rheology of synthetic olivine grain aggregates: influence of grain size and water. *Journal of Geophysical Research*, **91**, 8151–8176.

KARATO, S., TORIUMI, M. & FUJII, T. 1982. Dynamic recrystallization and high-temperature rheology of olivine. *In*: AKIMOTO, S. & MANGHNANI, M. H. (eds) *High-pressure Research in Geophysics*. Advances in Earth and Planetary Sciences, **12**, 171–189.

KOCH, P. S. 1983. *Rheology and microstructures of experimentally deformed quartz aggregates*. PhD thesis, University of California, LA, 1–464.

KOCKS, F., TOMÉ, C. & WENK, H.-R. 1998. *Texture and Anisotropy*. Cambridge University Press.

KOHLSTEDT, D. L., BAI, Q., WANG, Z.-C. & MEI, S. 2000. Rheology of partially molten rocks. *In*: BAGDASSAROV, N., LAPORTE, D. & THOMPSON, A. B. (eds) *Physics and Chemistry of partially molten rocks. Petrology and Structural Geology volume*, **11**, 3–28, Kluwer, Dordrecht.

KOHLSTEDT, D. L., EVANS, B. & MACKWELL, S. M. 1995. Strength of the lithosphere: constraints imposed by laboratory experiments. *Journal of Geophysical Research*, **100**, 17587–17603.

KOOPMAN, A., SPEKSNIJDER, A. & HORSFIELD, W. T. 1986. Sandbox model studies of inversion tectonics. *Tectonophysics*, **137**, 279–388.

KRUSE, R. & STÜNITZ, H. 1999. Deformation mechanisms and phase distribution in mafic, high temperature mylonites from the Jotun Nappe, Southern Norway. *In*: SCHMID, S. M., HEILBRONNER, R., & STÜNITZ, H. (eds) *Deformation Mechanisms in Nature and Experiment. Tectonophysics*, **303**, 223–249.

KUNZE, K., ADAMS, B. L., HEIDELBACH, F. & WENK, H.-R. 1994. Orientation imaging microscopy of calcite rocks. *In*: BUNGE, H. J., SIEGESMUND, S., SKROTZKI, W. & WEBER, K. (eds) *Textures of Geological Materials*. DGM Informationsgesellschaft, Oberursel, 127–146.

LARDEAUX, J. M., CARON, J. M., NISIO, P., PEQUIGNOT, G. & BOUDELLE, M. 1986. Microstructural criteria for reliable thermometry in low temperature eclogites. *Lithos*, **19**, 187–203.

LAW, R. D. 1990. Crystallographic fabrics: a selective review of their applications to research in structural geology. *In*: KNIPE, R. J. & RUTTER, E. H. (eds) *Deformation Mechanisms, Rheology and Tectonics*, Geological Society, London, Special Publications, **54**, 335–352.

LE HÉBEL, F., GAPAIS, D., FOURCADE, S. & CAPDEVILA, R. 2002. Fluid-assisted large stains in a crustal-scale décollement (Hercynian Belt of South Brittany, France). *In*: DE MEER, S., DRURY, M. R., DE BRESSER, J. H. P. & PENNOCK, G. M. (eds) *Deformation Mechanisms, Rheology and Tectonics: Current Status and Future Perspectives*. Geological Society, London, Special Publications, **200**, 85–102.

LEHNER, F. K. 1990. Thermodynamics of rock deformation by pressure solution. *In*: BARBER, D. & MEREDITH, P. D. (eds) *Deformation Processes in Minerals, Ceramics and Rocks*. Unwin Hyman, London, 296–333.

LEHNER, F. K. 1995. A model for intergranular pressure solution in open systems. *Tectonophysics*, **245**, 153–170.

LEISS, B., GRÖGER, H. R., ULLEMEYER, K & LEBIT, H. 2002. Textures and microstructures of naturally deformed amphibolites from the northern Cascades, NW USA. *In*: DE MEER, S., DRURY, M. R., DE BRESSER, J. H. P. & PENNOCK, G. M. (eds) *Deformation Mechanisms, Rheology and Tectonics: Current Status and Future Perspectives*. Geological Society, London, Special Publications, **200**, 219–238.

LEISS, B., ULLEMEYER, K. *ET AL*. 2000. Recent developments and goals in texture research of geological materials. *Journal of Structural Geology*, **22**, 1531–1540.

LEROY, Y. M. & HEIDUG, W. K. 1994. Geometrical evolution of stressed and curved solid-fluid phase boundaries, 2, Stability of cylindrical pores. *Journal of Geophysical Research B*, **99**, 517–530.

LLOYD, G. E. 2000. Grain boundary contact effects during faulting of quartzite: an SEM/EBSD analysis. *Journal of Structural Geology*, **22**, 1675–1693.

LLOYD, G. E. & FREEMAN, B. 1994. Dynamic recrystallization of quartz under greenschist conditions. *Journal of Structural Geology*, **16**, 867–881.

LLOYD, G. E., FARMER, A. B. & MAINPRICE, D. 1997. Misorientation analysis and the formation and orientation of subgrain and grain boundaries. *Tectonophysics*, **279**, 55–78.

MAINPRICE, D., LLOYD, G. E. & CASEY, M. 1993. Individual orientation measurements in quartz polycrystals – advantages and limitiations for texture and petrophysical property determinations. *Journal of Structural Geology*, **15**, 1169–1187.

MACKWELL, S. J., ZIMMERMAN, M. E, & KOHLSTEDT, D. L. 1998. High-temperature deformation of dry diabase with application to tectonics on Venus. *Journal of Geophysical Research*, **103**, 975–984.

MCCAIG, A. M., WAYNE, D. M., MARSHALL, J. D., BANKS, D. & HENDERSON, I. 1995. Isotopic and fluid inclusion studies of fluid movement along the Gavarnie Thrust, Central Pyrenees: reaction fronts in carbonate mylonites. *American Journal of Science*, **295**, 309–343.

MEANS, W. D. & LI, T. 2001. A laboratory simulation of fibrous veins: some first observations. *Journal of Structural Geology*, **23**, 857–863.

MERCIER, J.-C., ANDERSON, D. A. & CARTER, N. L. 1977. Stress in the lithosphere: inferences from steady-state flow of rocks. *Pageophysics*, **115**, 199–226.

MIKA, D. P. & DAWSON, P. R. 1999. Polycrystal plasticity modeling of intracrystalline boundary textures. *Acta Materialia*, **47**, 1355–1369.

MUCCI, F. & MORSE, J. W. 1983. The incorporation of Mg^{2+} and Sr^{2+} into calcite overgrowths: Influence of growth rate and solution composition. *Geochimica et Cosmochimica Acta*, **47**, 217–233.

MULLIS, A. M. 1991. The role of silica precipitation kinetics in determining the rate of quartz pressure solution. *Journal of Geophysical Research B*, **96**, 10007–10013.

MULLIS, A. M. 1993. Determination of the rate-limiting mechanism for quartz pressure solution. *Geochimica et Cosmochimica Acta*, **57**, 1499–1503.

NEUMANN, B. 2000. Texture development of recrystallized quartz polycrystals unraveled by orientation and misorientation characteristics. *Journal of Structural Geology*, **22**, 1695–1711.

NEWMAN, J., LAMB, W. M., DRURY, M. R. & VISSERS, R. L. M. 1999. Deformation processes in a peridotite shear zone: reaction-softening by a H_2O-deficient, continuous net transfer reaction. *Tectonophysics*, **303**, 193–222.

NICOLAS, A. & ILDEFONSE, B. 1996. Flow mechanism and viscosity in bastic magma chambers. *Geophysical Research Letters*, **23**, 2013–2016.

NIEMEIJER, A. R. & SPIERS, C. J. 2002. Compaction creep of quartz-muscovite mixtures at 500 °C: preliminary results on the influence of muscovite on pressure solution. *In*: DE MEER, S., DRURY, M. R., DE BRESSER, J. H. P. & PENNOCK, G. M. (eds) *Deformation Mechanisms, Rheology and*

Tectonics: Current Status and Future Perspectives. Geological Society, London, Special Publications, **200**, 61–72.

NIEMEIJER, A. R., SPIERS, C. J. & BOS, B. 2002. Compaction creep of quartz sand at 400–600 °C: Experimental evidence for dissolution controlled pressure solution. *Earth and Planetary Science Letters*, **195**, 261–275.

ORZOL, J., STÖCKHERT, B., & RUMMEL, F. 2001. Experimental Deformation of Synthetic Polycrystalline Jadeite Aggregates. *EOS, Transactions American Geophysical Union Fall* F1145.

PANNOZO-HEILBRONNER, R. & PAULI, C. 1993. Integrated spatial and orientation analysis of quartz c-axes by computer-aided microscopy. *Journal of Structural Geology*, **15**, 369–382.

PASSCHIER, C. W. & TROUW, R. A. J. 1996. *Microtectonics*. Springer, Berlin.

PATERSON, M. S. 1973. Thermodynamics and its geological application. *Reviews of Geophysics and Space Physics*, **11**, 355–389.

PATERSON, M. S. 1987. Problems in the extrapolation of laboratory data. *Tectonophysics*, **133**, 33–43.

PATERSON, M. S. 1995. A theory for granular flow accommodated by material transfer via an intergranular fluid. *Tectonophysics*, **245**, 135–152.

PATERSON, M. S. 2001. Relating experimental and geological rheology. *International Journal of Earth Sciences (Geologische Rundschau)*, **90**, 157–167.

PATERSON, M. S. & OLGAARD, D. L. 2000. Rock deformation tests to large shear strains in torsion. *Journal of Structural Geology*, **22**, 1341–1358.

PEACH, C. J., SPIERS, C. J. & TRIMBY, P. W. 2001. The effect of confining pressure on dilatation, recrystallization and flow behavior of rocksalt at 150 °C. *Journal of Geophysical Research*, **106**, 13315–13328.

PENNOCK, G. M., DRURY, M. R., TRIMBY, P. W. & SPIERS, C. J. 2002. Misorientations in hot deformed NaCl using EBSD. *Journal of Microscopy*, **205**, 285–294.

PIAZOLO, S., BONS, P. D., JESSEL, M. W., EVANS, L., & PASSCHEIR, C. W. 2002. Dominance of microstructural processes and their effect on microstructural development: insights from numerical modelling of dynamic recrystallization. *In*: DE MEER, S., DRURY, M. R., DE BRESSER, J. H. P. & PENNOCK, G. M. (eds) *Deformation Mechanisms, Rheology and Tectonics: Current Status and Future Perspectives*. Geological Society, London, Special Publications, **200**, 148–170.

PIEPENBREIER, D. & STÖCKHERT, B. 2001. Plastic flow of omphacite in eclogites at temperatures below 500 °C – implications for interplate coupling in subduction zones. *International Journal of Earth Sciences (Geologische Rundschau)*, **90**, 197–210.

PIERI, M., BURLINI, L., KUNZE, K., STRETTON, I. & OLGAARD, D. L. 2001*a*. Rheological and microstructural evolution of Carrara marble with high shear strain: results from high temperature torsion experiments. *Journal of Structural Geology*, **23**, 1393–1413.

PIERI, M., KUNZE, K., BURLINI, L., STRETTON, I., OLGAARD, D. L., BURG, J.-P. & WENK, H.-R. 2001*b*. Texture development of calcite by deformation and dynamic recrystallization at 1000K during torsion experiments of marble to large strains. *Tectonophysics, 330*, 119–140.

PILI, E., RICARD, Y., LARDEAUX, J. M., SHEPPARD, S. M. F. 1997. Lithospheric shear zones and mantle-crust connections. *Tectonophysics*, **280**, 15–29.

POIRIER, J. P. & GUILLOPÉ, M. 1979. Deformation-induced recrystallization of minerals. *Bulletin de Minéralogie*, **102**, 67–74.

POIRIER, J. P. & NICOLAS, A. 1975. Deformation-induced recrystallization by progressive misorientation of subgrain boundaries, with special reference to mantle peridotites. *Journal of Geology*, **83**, 707–720.

POIRIER, J. P., BOUCHEZ, J. L. & JONAS, J. J. 1979. A dynamic model for aseismic ductile shear zones. *Earth and Plantary Science Letters*, **43**, 441–453.

PRIOR, D. J. 1999. Problems in determining the orientation axis, for small angular orientations, using electron backscatter diffraction in the SEM. *Journal of Microscopy*, **195**, 217–225.

PRIOR, D. J., BOYLE, A. P. *ET AL*. 1999. The application of electron backscatter diffraction and orientation contrast imaging in the SEM to textural problems in rocks. *American Mineralogist*, **84**, 1741–1759.

RAJ, R. 1982. Creep in polycrystalline aggregates by matter transport through a liquid phase. *Journal of Geophysical Research B*, **87**, 4731–4739.

RAJ, R. & CHYUNG, C. K. 1981. Solution-precipitation creep in glass ceramics. *Acta Metallurgica*, **29**, 159–166.

RANALLI, G. 1984. Grain size distribution and flow stress in tectonites. *Journal of Structural Geology*, **6**, 443–447.

RANDLE, V. 1998. Overview No. 127. The role of the grain boundary plane in cubic polycrystals. *Acta Materialia*, **46**, 1459–1480.

RANDLE, V. & ENGLER, O. 2000. *Introduction to Texture Analysis: Macrotexture, Microtexture and Orientation Mapping*. Gordon and Breach Science Publishers, Amsterdam.

RANDLE, V., DAVIES, H. & CROSS, I. 2001. Grain boundary misorientation distributions. *Current Opinion in Solid State and Materials Science*, **5**, 3–8.

REDDY, M. M. & WANG, K. K. 1980. Crystallization of calcium carbonate in the presence of metal ions. I. Inhibition of magnesium ion at pH 8. 8 and 25 °C. *Journal of Crystal Growth*, **50**, 470–480.

REGENAUER-LIEB, K. & YUEN, D. A. 2000 Quasi-adiabatic instabilities associated with necking processes of an elasto-viscoplastic lithopshere. *Physics of the Earth and Planetary Interiors*, **118**, 89–102.

RENARD, F. & ORTOLEVA, P. 1997. Water films at grain-grain contacts: Debye-Hückel osmotic model of stress, salinity, and mineral dependence. *Geochimica et Cosmochimica Acta*, **61**, 1963–1970.

RENARD, F., ORTOLEVA, P. & GRATIER, J. P. 1997. Pressure solution in sandstones: Influence of clays and dependence on temperature and stress. *Tectonophysics*, **280**, 257–266.

RENNER, J. & EVANS, B. 2002. Do calcite rocks obey the power law creep equation? *In*: DE MEER, S., DRURY, M. R., DE BRESSER, J. H. P. & PENNOCK, G. M. (eds) *Deformation Mechanisms, Rheology and Tectonics: Current Status and Future Perspectives*. Geological Society, London, Special Publications, **200**, 293–308.

RESTON T. J. 1990. The lower crust and the extension of the continental lithsophere: kinematic analysis of BIRPS deep seismic data. *Tectonics*, **9**, 1235–1248.

RICHARD, P. D., NAYLOR, M. A. & KOOPMAN, A. 1995. Experimental models of strike-slip tectonics. *Petroleum Geoscience*, **1**, 71–80.

ROSENBERG, C. L. 2001. Deformation of partially molten granite: a review and comparison of experimental and natural case studies. *International Journal of Earth Sciences (Geologische Rundschau)*, **90**, 60–76.

ROSS, J. V., AVÉ LALLEMANT, H. G. & CARTER, N. L. 1980. Stress dependence of recrystallized grain and subgrain size in olivine. *Tectonophysics*, **70**, 39–61.

RUDNICK, R. J. 1992. Xenolith-samples of the lower crust. *In*: FOUNTAIN, D. M., ARCULUS, R. & KAY, R. W. (eds) *Continental Lower Crust*. Developments in Geotectonics, **23**, Elsevier, 269–316.

RUDNICKI, J. W. & RICE, J. R. 1975. Conditions for the localzation of deformation in pressure-sensitive dialant materials. *Journal of Mechanics and Physics of Solids*, **23**, 371–394.

RUTTER, E. H. 1976. The kinetics of rock deformation by pressure solution. *Philosophical Transactions of the Royal Society, London*, **283**, 203–219.

RUTTER, E. H. 1983. Pressure solution in nature, theory and experiment. *Journal of the Geological Society, London*, **140**, 725–740.

RUTTER, E. H. 1999. On the relationship between the formation of shear zones and the form of the flow law for rocks undergoing dynamic recrystallization. *In*: SCHMID, S. M., HEILBRONNER, R., & STÜNITZ, H. (eds) *Deformation Mechanisms in Nature And Experiment. Tectonophysics*, **303**, 147–158.

RUTTER, E. H. & BRODIE, K. H. 1988. The role of tectonic grain size reduction in the rheological stratification of the lithosphere. *Geologische Rundschau*, **77**, 295–308.

RUTTER, E. H. & BRODIE, K. H. 1992. Rheology of the lower crust. *In*: FOUNTAIN, D. M., ARCULUS, R. & KAY, R. W. (eds) *Continental Lower Crust*. Developments in Geotectonics, **23**, Elsevier, 202–265.

RYBACKI. E. & DRESEN, G. 2000. Dislocation and diffusion creep of synthetic anorthite aggregates. *Journal of Geophysical Research*, **105**, 26017–26036.

SCHMID, S. M. 1994. Textures of geological materials: computer model predictions versus empirical interpretations based on rock deformation experiments and field studies. *In*: BUNGE, H. J., SIEGESMUND, S., SKROTZKI, W. & WEBER, K. (eds) *Textures of Geological Materials*. DGM Informationsgesellschaft, Oberursel, 279–301.

SCHMID, S. M., PANOZZO, R. & BAUER, S. 1987. Simple shear experiments on calcite rocks: rheology and microfabric. *Journal of Structural Geology*, **9**, 747–778.

SCHMID, S. M., PFIFFNER, O. A., FROITZHEIM, N., SCHÖNBORN, G. & KISSLING, E. 1996. Geophysical-geological transect and tectonic evolution of the Swiss–Italian Alps. *Tectonics*, **15**, 1036–1064.

SCHREURS, G. 1994. Experiments on strike-slip faulting and rock rotation. *Geology*, **22**, 567–570.

SCHUTJENS, P. M. T. M. 1991. Experimental compaction of quartz sands at low effective stress and temperature conditions. *Journal of the Geological Society, London*, **148**, 527–539.

SCHWERDTNER, W. M. SHEEMAN, P. M. & RUCKLIDGE, J. C. 1971. Variation in degree of hornblende grain alignment within two boudinage structures. *Canadian Journal of Earth Sciences*, **8**, 144–149.

SHACKLETON, R. M. 1993. Tectonics of the lower crust – a view from the Usambara mountains, NE Tanzania. *Journal of Structural Geology*, **15**, 663–671.

SHELTON, G. & TULLIS, J. 1981. Experimental Flow Laws for Crustal Rocks. *EOS Transactions of the American Geophysical Union*, **62**, 396.

SHIMIZU, I. 1998. Stress and temperature dependence of recrystallized grain size: a subgrain misorientation model. *Geophysical Research Letters*, **25**, 4237–4240.

SLEEP, N. H. & BLANPIED, M. L. 1992. Creep, compaction and the weak rheology of major faults. *Nature*, **359**, 687–692.

SPIERS, C. J. & CARTER, N. L. 1998. Microphysics of rocksalt flow in nature. *In*: AUBERTIN, M. & HARDY, H. R. (eds) *The Mechanical Behaviour of Salt: Proceedings of the Fourth Conference*. Trans Technical Publications on Rock and Soil Mechanics, **22**, 115–128.

SPIERS, C. J. & SCHUTJENS, P. M. T. M. 1990. Densification of crystalline aggregates by fluid phase diffusional creep. *In*: BARBER, D. & MEREDITH, P. (eds) *Deformation Processes in Minerals, Ceramics and Rocks*. Unwin Hyman, London, 334–353.

SPIERS, C. J., SCHUTJENS, P. M. T. M., BRZESOWSKY, R. H., PEACH, C. J., LIEZENBERG, J. L. & ZWART, H. J. 1990. Experimental determination of constitutive parameters governing creep of rocksalt by pressure solution. *In*: KNIPE, R. J. & RUTTER, E. H. (eds) *Deformation Mechanism, Rheology and Tectonics*. Geological Society, London, Special Publications, **54**, 215–227.

STIPP, M., STÜNITZ, H., HEILBRONNER, R., & SCHMID, S. M. 2002*a*. The Eastern Tonale fault zone: A "natural laboratory" for crystal plastic deformation of quartz over a temperature range from 250 °C to 700 °C. *Journal of Structural Geology*, **24**, 1861–1884.

STIPP, M., STÜNITZ, H., HEILBRONNER, R. & SCHMID, S. M. 2002. Dynamic recrystallization of quartz: correlation between natural and experimental conditions. *In*: DE MEER, S., DRURY, M. R., DE BRESSER, J. H. P. & PENNOCK, G. M. (eds) *Deformation Mechanisms, Rheology and Tectonics: Current Status and Future Perspectives*. Geological Society, London, Special Publications, **200**, 171–190.

STÖCKHERT, B. 2002. Stress and deformation. *In*: DE MEER, S., DRURY, M. R., DE BRESSER, J. H. P. & PENNOCK, G. M. (eds) *Deformation Mechanisms, Rheology and Tectonics: Current Status and Future Perspectives*. Geological Society, London, Special Publications, **200**, 255–274.

STÖCKHERT, B., WACHMANN, M., KÜSTER, M. & BIMMERMAN, S. 1999. Low effective viscosity during high pressure metamorphism due to dissolution precipitation creep: the record of HP-LT metamorphic carbonates and siliciclastic rocks from Crete. *Tectonophysics*, **303**, 299–319.

STONE, D. S. 1991. Scaling laws in dislocation creep. *Acta Metallurgica Materialia*, **39**, 599–608.

STÜNITZ, H. & FITZ GERALD, J. D. 1993. Deformation of granitoids at low metamorphic grade. II: granular flow in albite-rich mylonites. *Tectonophysics*, **221**, 299–324.

STÜNITZ, H. & TULLIS, J. 2001. Weakening and strain localization produced by syn-deformational reaction of plagioclase. *International Journal of Earth Sciences*, **90**, 136–148.

TACKLEY, P. J. 2000. The quest for self-consistent generation of plate tectonics in mantle convection. *In*: RICHARDS, M. A., GORDON, R. G., VAN DER HILST, R. D. (eds) *The History and Dynamics of Global Plate Motions*. American Geophysical Union, Monograph, **121**, 47–72.

TADA, R. & SIEVER, R. 1989. Pressure solution during diagenesis. *Annual Reviews of Earth and Planetary Sciences*, **17**, 89–118.

TADA, R., MALIVA, R. & SIEVER, R. 1987. A new mechanism for pressure solution in porous quartzose sandstones. *Geochimica et Cosmochimica Acta*, **51**, 2295–2301.

TER HEEGE, J. H., DE BRESSER, J. H. P. & SPIERS, C. J. 2002. The influence of dynamic recrystallization on the grain size distribution and rheological behaviour of Carrara marble deformed in axial compression. *In*: DE MEER, S., DRURY, M. R., DE BRESSER, J. H. P. & PENNOCK, G. M. (eds) *Deformation Mechanisms, Rheology and Tectonics: Current Status and Future Perspectives*. Geological Society, London, Special Publications, **200**, 331–354.

TRIMBY, P. W., DRURY, M. R. & SPIERS, C. J. 2000. Recognising the crystallographic signature of recrystallization processes in deformed rocks: a study of experimentally deformed rocksalt. *Journal of Structural Geology*, **22**, 1609–1620.

TRIMBY, P. W., PRIOR, D. J. & WHEELER, J. 1998. Grain boundary hierarchy development in a quartz mylonite. *Journal of Structural Geology*, **20**, 917–935.

TRON, V. & BRUN, J.-P. 1991. Experiments on oblique rifting in brittle–ductile systems. *Tectonophysics*, **188**, 71–84.

TSENN, M. C. & CARTER, N. L. 1987. Upper limits of power law creep in rocks. *Tectonophysics*, **136**, 1–26.

TULLIS, J. 1983. Deformation of feldspars. *Reviews in Mineralogy*, **2**, 297–323.

TULLIS, J. & YUND, R. A. 1985. Dynamic recrystallization of felspar: a mechanism for ductile shear zone formation. *Geology*, **13**, 238–241.

TULLIS, J., DELL'ANGELO, L., & YUND, R. A. 1989. Ductile shear zones from brittle precursors in feldspathic rocks: The role of dynamic recrystallization. *In*: DUBA, A. G., DURHAM, W. B., HANDIN, J. W. & WANG, H. F. (eds) *The Brittle-ductile Transition in Rocks: The Heard Volume*. AGU Geophysical Monograph, **56**, 67–82.

TWISS, R. J. 1977. Theory and applicability of a recrystallized grain size paleopiezometer. *Pure and Applied Geophysics*, **115**, 227–244.

URAI, J. L., MEANS, W. D. & LISTER, G. S. 1986. Dynamic Recrystallization of Minerals. *AGU Geophysical Monograph*, **36**, 161–199.

VAN DER MOLEN, I. & PATERSON, M. S. 1979. Experimental deformation of partially-melted granite. *Contributions to Mineralogy and Petrology*, **70**, 299–318.

VAN DER WAL, D. & VISSERS, R. L. M. 1996. Structural petrology of the Ronda peridotite, SW Spain: deformation history. *Journal of Petrology*, **37**, 23–43.

VAN DER WAL, D., CHOPRA, P. N., DRURY, M. R., FITZ GERALD, J. D. 1993. Relationships between dynamically recrystallized grain size and deformation conditions in experimentally deformed olivine rocks. *Geophysical Research Letters*, **20**, 1479–1482.

VAN ROERMUND, H. L. M. & BOLAND, J. N. 1981. The dislocation substructures of naturally deformed omphacites. *Tectonophysics*, **78**, 403–418.

VIGNERESSE, J. L. & TIKOFF, B. 1999. Strain partitioning during partial melting and crystallizing felsic magmas. *Tectonophysics*, **312**, 117–132.

VIGNERESSE, J. L., BARBEY, P. & CUNEY, M. 1996. Rheological transitions during partial melting and crystallization with application to felsic magma segregation and transfer. *Journal of Petrology*, **37**, 1579–1600.

VISSERS, R. L. M., DRURY, M. R., NEWMAN, J. & FLIERVOET, T. F. 1997. Mylonitic deformation in upper mantle peridotites of the North Pyrenean Zone (France): implications for strength and strain localization in the lithopshere. *Tectonophysics*, **279**, 303–325.

WANG, Z., DRESEN, G. & WIRTH, R. 1996. Diffusion of fine-grained polycrystalline anorthite at high temperature. *Geophysical Research Letters*, **23**, 3111–3114.

WENK, H.-R. & CHRISTIE, J. M. 1991. Comments on the interpretation of deformation textures in rocks. *Journal of Structural Geology*, 13, 1091–1110.

WENK, H.-R., CANOVA, G., BRÉCHET, Y. & FLANDIN, L. 1997. A deformation-based model for recrystrallization of anisotropic materials. *Acta Materialia*, **45**, 3283–3296.

WEYL, P. K. 1959. Pressure solution and force of crystallization – a phenomenological theory. *Journal of Geophysical Research B*, **64**, 2001–2025.

WHEELER, J., PRIOR, D. J., JIANG, Z., SPIESS, R. & TRIMBY, P. W. 2001. The petrological significance of misorientations between grains. *Contributions to Mineralogy and Petrology*, **141**, 109–124.

WHITE, S. H. 1977. Geological significance of recovery and recrystallization processes in quartz. *Tectonophysics*, **39**, 143–170.

WHITE, S. H. & BRETAN, P. G. 1985. Rheological controls on the geometry of deep faults and the tectonic delamination of the continental crust. *Tectonics*, **4**, 303–309.

WHITE, S. H., BURROWS, S. E., CARRERAS, J., SHAW, N. D. & HUMPHREYS, F. J. 1980. On mylonites in ductile shear zones. *Journal of Structural Geology*, **2**, 175–187.

WHITE, S. H., DRURY, M. R., ION, S. E. & HUMPHREYS, F. J. 1985. Large strain deformation studies using polycrystalline magnesium as a rock analogue: Part I: grainsize palaeopiezometry in mylonite zones. *Physics of the Earth and Planetary Interiors*, **40**, 201–207.

WIESMAYR, U., EDWARDS, M. A., MEYER, M., KIDD, W. S. F., LEBER, D., HAUSLER, H. & WANGDA, D. 2002. Evidence for steady fault-accommodated strain in the High Himalaya: progressive fault rotation of the southern Tibet detachment system in NW-Bhutan. *In*: DE MEER, S., DRURY, M. R., DE BRESSER, J. H. P. & PENNOCK, G. M. (eds) *Deformation Mechanisms, Rheology and Tectonics: Current Status and Future Perspectives*. Geological Society, London, Special Publications, **200**, 371–386.

ZHANG, S., KARATO- S.-I., FITZ GERALD, J., FAUL, U. H. & ZHOU, Y. 2000. Simple shear deformation of olivine aggregates. *Tectonophysics*, **316**, 133–152.

ZHANG, Y., HOBBS, B. E. & JESSELL, M. W. 1994. The effect of grain boundary sliding on fabric development in polycrystalline aggregates. *Journal of Structural Geology*, **16**, 1315–1325.

ZHANG, Y. & WILSON, C. J. L. 1997. Lattice rotation in polycrystalline aggregates and single crystals with one slip system: a numerical and experimental approach. *Journal of Structural Geology*, **19**, 875–885.

ZHANG, X, SALESMANS, J., PEACH, C. J. & SPIERS, C. J. 2002. Compaction experiments on wet calcite powder at room temperature: evidence for operation of intergranular pressure solution. *In*: DE MEER, S., DRURY, M. R., DE BRESSER, J. H. P. & PENNOCK, G. M. (eds) *Deformation Mechanisms, Rheology and Tectonics: Current Status and Future Perspectives*. Geological Society, London, Special Publications, **200**, 29–40.

ZHU, W., MONTÉSI, L. G. J. & WONG, T.-F. 2002. Effects of stress on the anisotropic development of permeability during mechanical compaction of porous sandstones. *In*: DE MEER, S., DRURY, M. R., DE BRESSER, J. H. P. & PENNOCK, G. M. (eds) *Deformation Mechanisms, Rheology and Tectonics: Current Status and Future Perspectives*. Geological Society, London, Special Publications, **200**, 119–136.

ZUCALI, M., CHATEIGENER, D., DUGNANI, M., LUTTEROTTI, L. & OULADDIAF, B. 2002. Quantatitive texture analysis of glaucophanite deformed under eclogite facies conditions (Sesia-Lanzo zone, Western Alps): comparison between X-ray and neutron diffraction analysis. *In*: DE MEER, S., DRURY, M. R., DE BRESSER, J. H. P. & PENNOCK, G. M. (eds) *Deformation Mechanisms, Rheology and Tectonics: Current Status and Future Perspectives*. Geological Society, London, Special Publications, **200**, 239–254.

Compaction experiments on wet calcite powder at room temperature: evidence for operation of intergranular pressure solution

X. ZHANG, J. SALEMANS, C. J. PEACH & C. J. SPIERS

High Pressure and Temperature Laboratory, Faculty of Earth Sciences, Utrecht University, PO. Box 80.021, Utrecht, 3508 TA, The Netherlands (e-mail: zhangxm@geo.uu.nl)

Abstract: Dead weight uniaxial compaction creep experiments were carried out on fine-grained, super-pure calcite (<74 µm) at room temperature and applied effective stresses of 1–4 MPa. All samples were pre-compacted dry at a stress of 8 MPa, for 30 minutes, to obtain a well-controlled initial porosity. The samples were then wet-compacted under 'drained' conditions with pre-saturated solution as pore fluid. Control experiments, which were done either dry or with chemically inert pore fluid, showed negligible compaction. However, samples tested with saturated solution as pore fluid showed easily measurable compaction creep. The compaction strain rate decreased with increasing strain and increasing grain size, and increased with increasing applied stress. Addition of Mg^{2+} ions to the saturated solution dramatically inhibited compaction. From the literature, Mg^{2+} ions are known to inhibit calcite precipitation. By comparison with a theoretical model for intergranular pressure solution in calcite, the observed mechanical behaviour and the way that compaction responded to the pore fluid chemistry suggest that, under our experimental condition, intergranular pressure solution is the mechanism of the deformation and that precipitation is likely to be the rate-limiting step.

Intergranular pressure solution (IPS) is an effective mechanism of deformation and porosity and permeability reduction in both sedimentary and fault rocks under diagenetic and low-grade metamorphic conditions (Sorby 1863; Weyl 1959; Durney 1972; Rutter 1976, 1983; Tada & Siever 1989; De Meer & Spiers 1999). Both in silicate and carbonate rocks, evidence of intergranular pressure solution is widespread (Heald 1956; Bathurst 1958; Burger 1975; Cox & Withford 1987; Houseknecht 1987, 1988), and there is widespread interest in quantifying the rate of the process in these rock materials (Rutter 1976, 1983; Dewers & Hajash 1995; De Meer & Spiers 1999), notably in relation to hydrocarbons exploration and production research (Carozzi & Bergen 1987; Heydari 2000).

Much work has accordingly been done on the theoretical modelling of the intergranular pressure solution rate in two-component solid–liquid systems. In most of these models, grain boundaries are considered to contain fluid in some interconnected form which cannot be squeezed out, i.e. in a strongly adsorbed thin film (Weyl 1959; Rutter 1976, 1983; Robin 1978; Hickman & Evans 1991, 1995) or in a micro-scale island–channel network containing free fluid (Elliot 1973; Raj 1982; Lehner 1990, 1995; Spiers & Schutjens 1990). When there is no long-range transport of solid into or out of the system, the mechanism is believed to involve dissolution of material at grain contacts under high mean normal stress (high chemical potential), diffusion through the intergranular and pore fluid phase, and precipitation at interfaces under low mean normal stress (low chemical potential). As these processes operate in series under such conditions, the slowest step will be rate controlling at steady state (Raj 1982; Lehner 1990; De Meer & Spiers 1999). Despite the ambiguities of the details of the grain boundary structure, the main variables influencing intergranular pressure solution are well known. These are: grain size; effective stress; temperature; mineralogy; and pore fluid chemistry (Tada & Siever 1989; Rutter 1983; De Meer & Spiers 1999). While the existing models for creep by IPS can be reasonably applied to two-component systems such as NaCl plus saturated solution, the rate controlling parameters such as reaction rate or grain boundary diffusion coefficients are not well known. Moreover, the existing models have not been adapted for complex reaction mechanisms such as those involved in calcite dissolution/precipitation.

Numerous experimental investigations have been conducted in attempt to demonstrate the operation of IPS in the laboratory, to identify the detailed mechanism and to quantify the rate of the process. These experiments employ either single crystal contacts or fine-grained polycrystalline aggregate plus saturated solution. When single crystals are used, the cleaved or polished surfaces are usually indented with a knife-edge

From: DE MEER, S., DRURY, M. R., DE BRESSER, J. H. P. & PENNOCK, G. M. (eds) 2002. *Deformation Mechanisms, Rheology and Tectonics: Current Status and Future Perspectives*. Geological Society, London, Special Publications, **200**, 29–39. 0305-8719/02/$15 © The Geological Society of London.

piston (Tada & Siever 1986; Gratier 1993), or else the crystal is pressed against another crystal or glass (Hickman & Evans 1991, 1992, 1995; Schutjens & Spiers 1999; De Meer & Spiers 2001). In some single-crystal experiments, a regular hole is machined in the crystal to observe free surface dissolution occurring when the crystal is loaded (Sprunt & Nur 1977; Den Brok & Morel 2001). The main advantage of single-crystal experiments is that a well-controlled contact area is obtained and that the normal stress and displacement across the contact can be calculated or measured. In addition, because the contact is relatively big, it is possible to directly observe the dynamic evolution of the contacts under the optical microscope (Hickman & Evans 1991, 1992, 1995; Schutjens & Spiers 1999).

However, single crystal/contact experiments are only feasible for very soluble minerals such as NaCl, KCl and $NaNO_3$ (Hickman & Evans 1991; Gratier 1993; Schutjens & Spiers 1999). For the less soluble rock-forming minerals, the process is so slow that contact dissolution would not be expected to be observed within a reasonable time scale (Rutter 1976, 1983). In this situation, fine-grained polycrystalline aggregates are normally used and compaction experiments are employed, since these amplify the effect of the intergranular pressure solution process (Spiers & Schutjens 1990). Unfortunately, compaction creep experiments performed on wet quartz sand under effective stresses up to several tens of MPa and temperatures up to 400 °C have met with limited success in this respect. In such experiments, limited or ambiguous microstructural evidence for IPS has generally been found (De Boer *et al.* 1977) and the mechanical data are mostly inconsistent or insufficient to compare with models for IPS as the rate-controlling mechanism (Renton *et al.* 1969; De Boer *et al.* 1977; Cox & Paterson 1991; Schutjens 1991; Dewers & Hajash 1995; Rutter & Wanten 2000). Baker *et al.* (1980) carried out intergranular pressure solution experiments on deep sea carbonate (i.e. low magnesium calcite), Iceland spar and reagent-grade calcite powder at 25–100 MPa effective pressure and 22–180 °C, for 1 to 10 days. Significant porosity reduction to values as low as 9.9% occurred in 10 days. However, as high stresses and temperatures were used, the contribution from deformation mechanisms other than IPS is unknown. Recent work by Dewers & Muhari (2000) also suggests a mixture of cataclastic deformation and intergranular pressure solution in compaction experiments on Iceland spar. Thus the mechanism and kinetics of IPS in calcite remain poorly understood. Indeed, there is little or no proof for its operation under lab conditions.

The aim of this study is to determine if IPS operates in wet granular calcite compacted at low applied stress and at room temperature, placing emphasis on differentiating IPS from other possible deformation mechanisms and on trying to identify the rate-controlling process. As many authors have pointed out (Raj 1982; Tada & Siever 1989), one of the main problems of compaction experiments for IPS is to distinguish the compaction strain rate due to IPS from the effects of contact asperity removal, grain re-orientation, intergranular sliding, grain cracking and plastic deformation mechanisms. To minimize such effects in our experiments, all samples were subjected to initial dry loading (pre-compaction) at higher stress (8 MPa) than the test stresses (1–4 MPa). We used very fine-grained calcite powder ($<74\,\mu m$) in order to favour IPS and investigated the effects of Mg^{2+} ions in solution, since Mg^{2+} ions can inhibit precipitation of calcite (Reddy & Wang 1980; Davis *et al.* 2000).

Experimental method

The test materials used in this study were super-pure granular calcite (Merck 99.95% pure) with an average grain size of 15 μm (5–30 μm), plus high-purity calcite single crystals (Iceland spar) crushed and mechanically sieved into grain size fractions of 28–37, 37–45, 45–63, 63–74 μm. Chemical analysis showed that the crystal used contained impurities of magnitude of ppm. The super-pure Merck calcite grains have a rhombohedral shape. Pre-saturated solutions were prepared for each experimental run by adding excess test material (same grain size fraction as tested sample) to distilled water and stirring for 48 hours. Pore fluid Mg^{2+} concentration was varied by adding controlled amounts of $MgCl_26H_2O$ to the solution at the last stage of stirring.

The experiments were performed using the dead-weight compaction set-up, illustrated in Fig. 1 (see also De Meer & Spiers 1997). The detailed sample assembly is shown in Fig. 1b. A glass capillary tube with inner diameter of 2.1 mm, mounted in the brass base was used as a 'pressure vessel'. Two stainless steel pistons were used to load the samples. Paper filters were placed between the sample and pistons to prevent grains penetrating or jamming between the pistons and the glass tube. An acoustic emission receiver was installed in the brass base

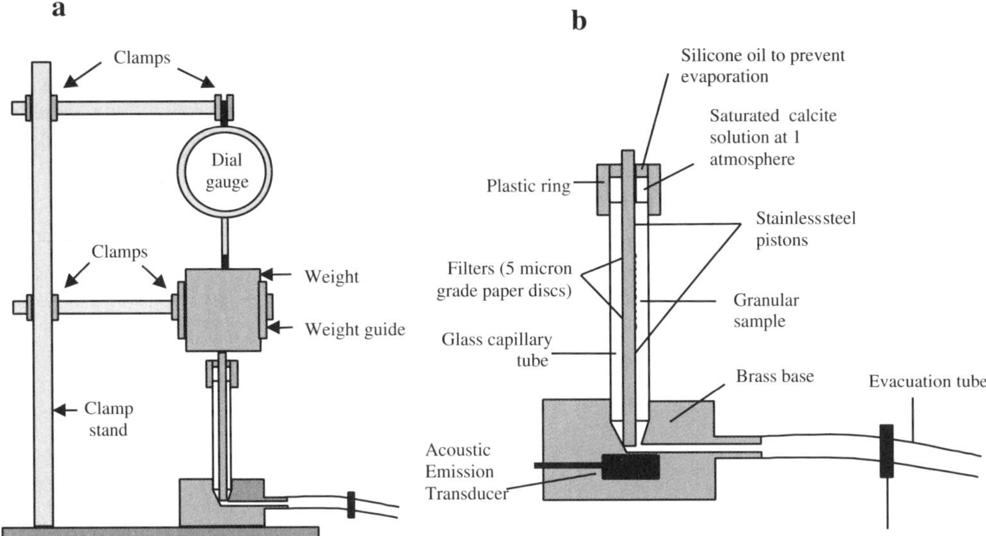

Fig. 1. Experimental apparatus. (**a**) Compaction apparatus. (**b**) Details of the sample assembly.

close to the bottom piston, to detect any grain cracking or brittle deformation. The acoustic emission (AE) transducer was connected via a preamplifier to a band-pass filter with further amplification before discrimination of events using an adjustable-threshold comparator. The total gain was 60 dB and the frequency band limited to between 100 kHz and 1 MHz, with a −36 dB per octave roll-off characteristic, above and below the band. The resonant frequency of the transducer, a PZT ceramic filter element, was 1–2 MHz. Individual events triggered a 200 ms monostable pulse stretcher (to eliminate ring-down counts and overlapping events), before accumulation by a digital computer-based counter and logging program. For maximum sensitivity, the trigger level, for event discrimination, was set just above the background electrical noise level. Signal-to-noise ratios for typical events were around 12 dB, as seen on a digital storage oscilloscope (Gould 840), and triggering by all significant events occurred, provided they were separated more than 200 ms, which was the time of the monostable pulse stretcher blanking circuit.

After loading into the apparatus, 60 mg samples were first pre-compacted for 30 minutes under dry conditions at an applied axial stress (σ_e) of 8 MPa. The aim was to obtain a well-controlled 'starting aggregate' with reproducible porosity, and to eliminate intergranular rearrangements during later compaction. The samples were then unloaded and the length (taken as the initial length) of the samples was measured using a travelling microscope. The porosity of the dry-loaded samples was calculated from the initial length of the sample, the weight of the sample and the density of calcite (2710 kg/m^3), giving a value of $50 \pm 1.5\%$. A plastic ring was then placed around the top of the capillary tube to form a reservoir for the fluid, and an evacuation tube was attached to the fluid outlet of the brass-base (see Fig. 1b). All joints were sealed with silicone grease to prevent evaporation.

After dry compaction, wet testing was carried out at applied axial stresses of 1–4 MPa. The required load was first set by adding different weights to the upper piston. The dial gauge was then brought into position (Fig. 1) and set to zero. Fluid was subsequently added via the fluid reservoir (see Fig. 1b), and drawn through the sample into the brass base, using the evacuation tube. From the moment when the sample was wetted (<30 seconds), creep of the sample was monitored as a function of time for up to 25 days, by reading the dial gauge. During creep, the system was sealed to reduce evaporation by clamping the evacuation pipe, and by putting a drop of silicone oil on top of the saturated solution in the upper reservoir ring. The wet experiments were all carried out in the 'drained' condition, so the applied stress was the effective stress. Note that in addition to the wet experiments performed using saturated $CaCO_3$ solution, a number of control experiments were also performed both dry and adding decane ($C_{10}H_{22}$) instead of saturated solution. Both the saturated solution phase and decane were observed to penetrate the samples rapidly, even without applying a vacuum, thus demonstrating that both fluids wet the calcite powder. In separate tests on cleaved calcite fragments, the wetting angle of decane on calcite was assessed relative to water and found to be significantly lower.

The apparatus (Fig. 1) and method described allowed strains to be measured with an accuracy of *c.* 10^{-4}. Runs performed without a sample

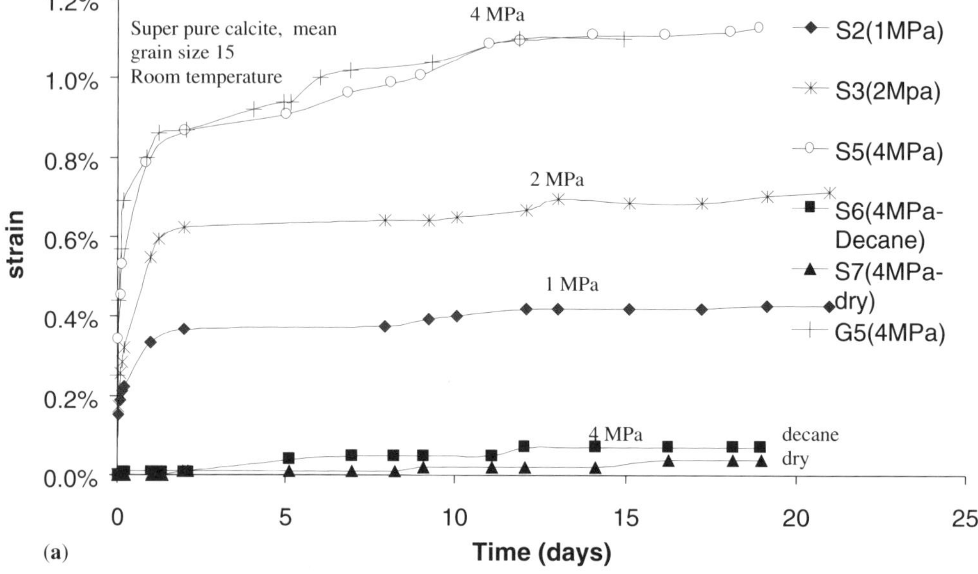

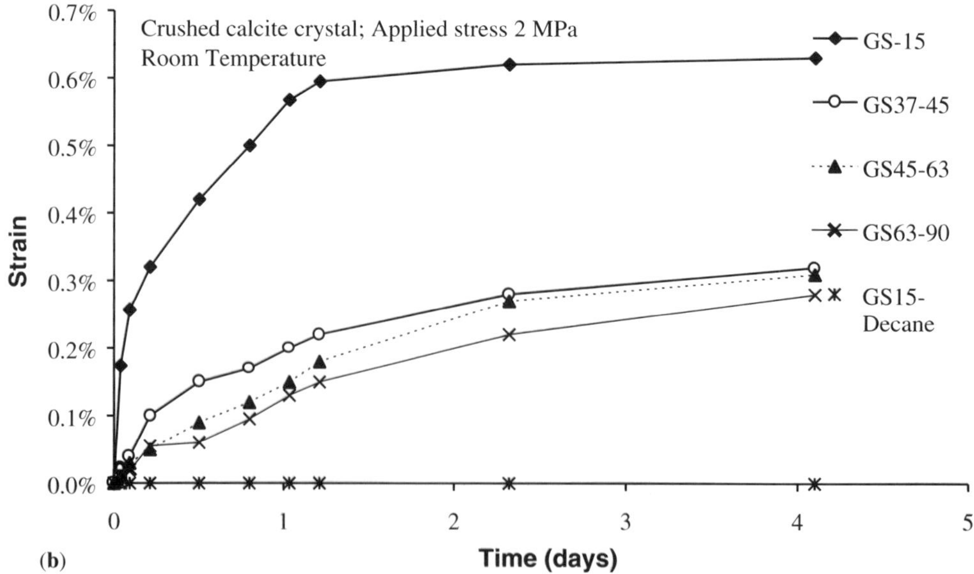

Fig. 2. Compaction creep curves for (**a**) different applied stresses and (**b**) different grain sizes.

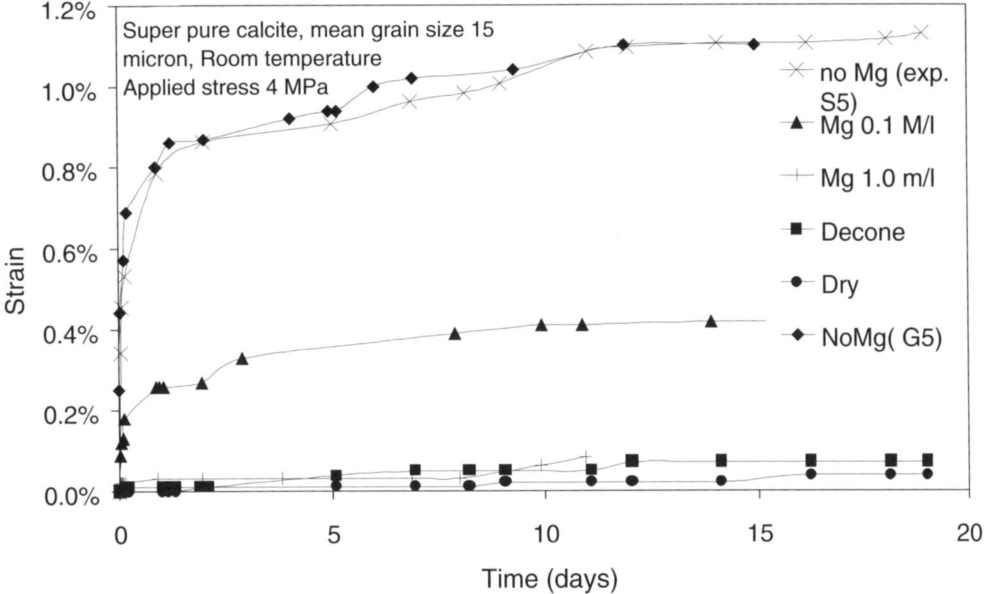

Fig. 3. Comparison of compaction behaviour of dry samples, samples flooded with inert fluid (decane), wet samples (containing saturated solution), and samples filled with solution containing Mg^{2+}.

showed that room temperature variation ($\pm 2\,^{\circ}C$) effects on the dial gauge reading were negligible. This is believed to be because the thermal expansion of the clamp stand counteracted the temperature effects on the dial gauge, dead weight, pistons and brass base (Fig. 1a). The apparatus enabled us to measure strains with a relative error of 1–5%. Strain rates were calculated as a function of strain and time, using the three-point central difference method. The relative errors in strain rate are estimated to be 1% to 10% at rate values of 10^{-7} to $10^{-9}\,s^{-1}$ respectively. Wall friction was measured in a series of experiments whereby the load transmitted through the sample was measured using a balance. This method showed that, in the wet and decane flooded samples, frictional forces were generally less than 10% of the applied load.

Results

Thirteen experiments were successfully performed. The pre-compaction stage at an axial stress of 8 MPa under dry conditions resulted in more or less instantaneous or time-independent compaction of the samples by a few percent (2–5%). Acoustic emission events were recorded only at the very beginning of the initial dry loading. No acoustic emissions were picked up in the subsequent wet stage of the experiments. In view of the difficulties of picking up acoustic emission signals in calcite (Wong, pers. comm.), we draw no further inference from this result. Control experiments in which samples were tested either dry or flooded with decane ($C_{10}H_{22}$, similar viscosity to water) showed hardly any creep (Fig. 2). Wet samples deformed surprisingly rapidly. Representative creep curves obtained for the wet samples are shown for different applied stresses in Fig. 2a and for different grain sizes in Fig. 2b. Note that compaction is enhanced by increasing the applied stress and by decreasing grain size. When Mg^{2+} was present in the pore fluid, compaction of the sample was strongly and systematically inhibited depending on Mg^{2+} concentration (Fig. 3). Since the compaction strains were small, microstructural evidence of IPS in the wet compacted samples was not expected, therefore no microstructural observations were attempted.

Discussion

Deformation mechanisms in dry and decane-flooded samples

Pre-compaction under dry conditions, at an axial stress of 8 MPa, caused a few percent of instantaneous strains. This compaction is presumably mainly due to grain reorientation, intergranular sliding and grain cracking. Small amounts of instantaneous strain are probably also caused by crystal plastic deformation as well as an elastic contribution. When samples were reloaded dry, at lower axial stress (1–4 MPa) (control experiments), almost no instantaneous compaction or compaction creep could be measured. The decane-flooded samples showed similar behaviour. We therefore infer that dry pre-compaction produced an aggregate which was 'locked' against deformation at lower stress (dry or flooded with decane) and that creep by conventional dislocation mechanisms was negligible in our experiments at stress of 1–4 MPa.

Deformation mechanisms in samples tested wet

Unlike the dry and decane-flooded samples, the samples tested wet under the same experimental conditions showed significant compaction. The behaviour of the decane-flooded samples suggests that the initially dry compacted or 'locked' aggregate cannot be 'unlocked' merely by reduction of friction or by physical surface (energy) effects at grain contacts through the addition of a fluid. Because our experimental conditions do not favour solid-state plastic deformation (Rutter 1976) and since no creep occurred in dry compaction, the creep mechanism operating in the wet samples must be a fluid-enhanced deformation mechanism. Given the observed effects of stress and grain size on compaction strain rates, IPS accompanied by intergranular sliding accommodation is the most obvious candidate. If IPS is indeed the rate-controlling deformation mechanism, the compaction rate should be influenced by manipulation of any of the serial process of dissolution, diffusion and precipitation, e.g. by adding chemical reaction inhibitors or enhancers to slow down or speed up dissolution or precipitation. From the literature, it is well known that Mg^{2+} ions in concentrations larger than 10^{-3} mol l^{-1} can significantly inhibit calcite precipitation (Reddy 1977; Mucci & Morse 1983), while concentrations less than 2×10^{-4} mol l^{-1} have no major effects. Our experimental results showed that Mg^{2+} with concentrations of 0.1 and 1 mol l^{-1} in the pore fluid drastically slowed down the compaction of our calcite samples. This finding suggests that deformation of the wet samples must involve processes of precipitation and dissolution, supporting our inference of IPS.

Another fluid-enhanced deformation mechanism which might have operated under our experimental conditions is subcritical cracking by a stress-corrosion mechanism (Atkinson 1982, 1984). Subcritical cracking can potentially cause grain and/or grain contact failure at low stresses if water is present in the sample, thus producing time-dependent compaction creep. The effect of Mg^{2+} in inhibiting compaction suggests that dissolution and precipitation of calcite are involved, e.g. through IPS. However, if subcritical cracking in calcite is through dissolution of calcite at the crack tips, as suggested by Atkinson (1984), we cannot rule out the possibility that creep deformation in our wet sample is caused by subcritical cracking. On the other hand, the dependence of the compaction strain rate on grain size would be expected to be different for subcritical cracking than for IPS. Finer grain size tends to promote IPS (Rutter 1976; De Meer & Spiers 1999) whereas it tends to inhibit grain cracking (Zhang *et al.* 1990; Brzesowsky 1995). In addition, the strain-rate dependence on applied stress for IPS is very different from the stress-corrosion crack growth velocity dependence on applied stress (Atkinson & Meredith 1989). In the following section, we compare the effects of strain, grain size and applied stress on our measured compaction strain rates with theoretical predictions of IPS models.

Comparison of experimental data with IPS theory

IPS theory (De Meer & Spiers 1999) predicts that the compaction strain rate of a porous aggregate increases as a function of decreasing grain size and increasing applied effective stress (Raj 1982; Spiers & Schutjens 1990). The strain rate is also predicted to decrease with increasing volumetric strain (Spiers and Brzesowsky 1993). Following De Meer & Spiers (1999), for low stress and small volumetric strains (up to 15–20%), the general forms of the theoretical IPS creep equations for hydrostatic or uniaxial compaction of a simple cubic grain pack of spherical grains are:

$$\dot{\beta}_s = AK_s \frac{\sigma_e^n}{d} f_s(e_v) \quad (1)$$

(dissolution control);

$$\dot{\beta}_d = BK_d \frac{\sigma_e}{d^3} f_d(e_v) \quad (2)$$

(diffusion control);

$$\dot{\beta}_p = CK_p \frac{\sigma_e^m}{d} f_p(e_v) \quad (3)$$

(precipitation control).

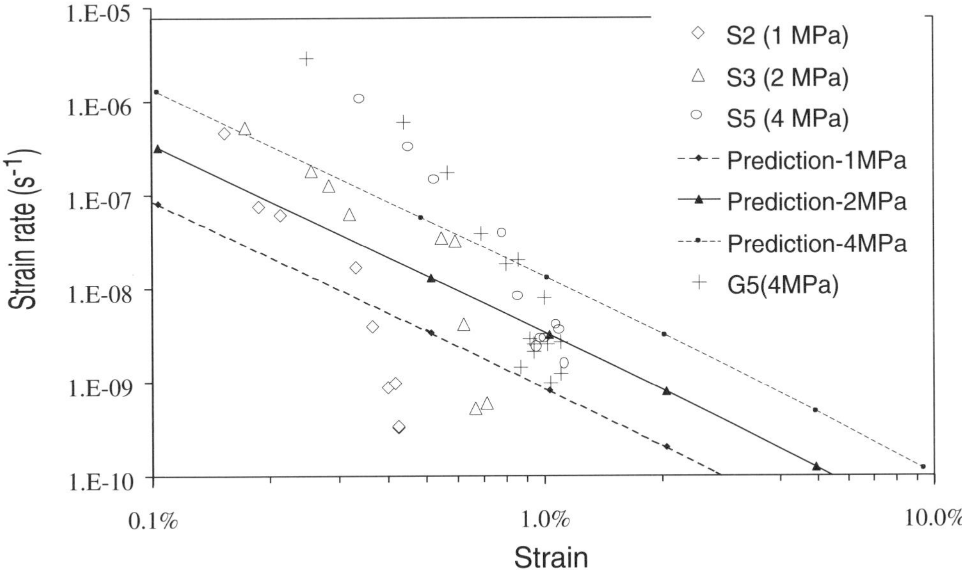

Fig. 4. Plots of strain rate versus strain for samples indicated in Fig. 2a. Dashed lines are our theoretical model predictions, using Equation (3) and parameters listed in Table 1.

Here $\dot{\beta}$ represents volumetric strain rate (defined as $\dot{\beta} = -\dot{v}/v$, with v being instantaneous volume); A, B, and C are constants related to the packing of the granular aggregate; K_s and K_p incorporate temperature-dependent dissolution and precipitation rate coefficients; K_d is a temperature-dependent term incorporating the product of grain boundary diffusivity, solid solubility and average grain boundary fluid thickness; σ_e is the applied effective stress; d is the grain size; and e_v is the volumetric compaction strain (Spiers & Schutjens 1990; De Meer & Spiers 1999). The exponents n and m respectively represent the order of the dissolution or precipitation velocity versus driving force relation.

Figure 4 shows a log–log plot of strain rate versus strain, constructed for the experimental data of Fig. 2a. The measured strain rates decrease first steadily and then rapidly with increasing strain. Theoretical predictions of the strain rate versus strain are also presented in Fig. 4, assuming that precipitation of calcite on the pore walls is the rate-limiting step, and inserting parameters values for K_p and m from the crystal growth literature on calcite into Equation 3. The parameter values and data sources used are given in Table 1. We consider the case of precipitation control only here, because calculations assuming dissolution or diffusion control indicate that these would be far faster than precipitation control, and therefore not rate controlling at steady state. Note that our experimental data show the same broad trend as predicted by our model, but that the experimental strain rates decrease with strain far more rapidly than predicted.

Table 1. *Parameter values used in the theoretical model of IPS for calcite*

Parameters/function	Expression/values	Sources
K_d	$0.22\,k\,K_{eq}\,(1/RT)^2\,\Omega^2$	Reddy & Wang (1980)
m	2	Reddy & Wang (1980)
K_{eq} ($mol^2\,l^{-2}$) solubility product	$a^{2+}_{Ca}a^{2-}_{CO_3}/a_{CaCO_3}$	
k ($l\,mol^{-1}\,s^{-1}$)/($m^2\,l^{-1}$)	1.03	Wiechers *et al.* (1975)
$Cf_p(e_v)$	$\approx[216(1-e_v)]/\{[\pi^2(0.25-(1-e_v)^{2/3}]^3\}$	
Ω ($m^3\,mol^{-1}$)	3.6934×10^{-5}	
R ($J\,mol^{-1}\,K^{-1}$)	8.314	
T (K)	294	

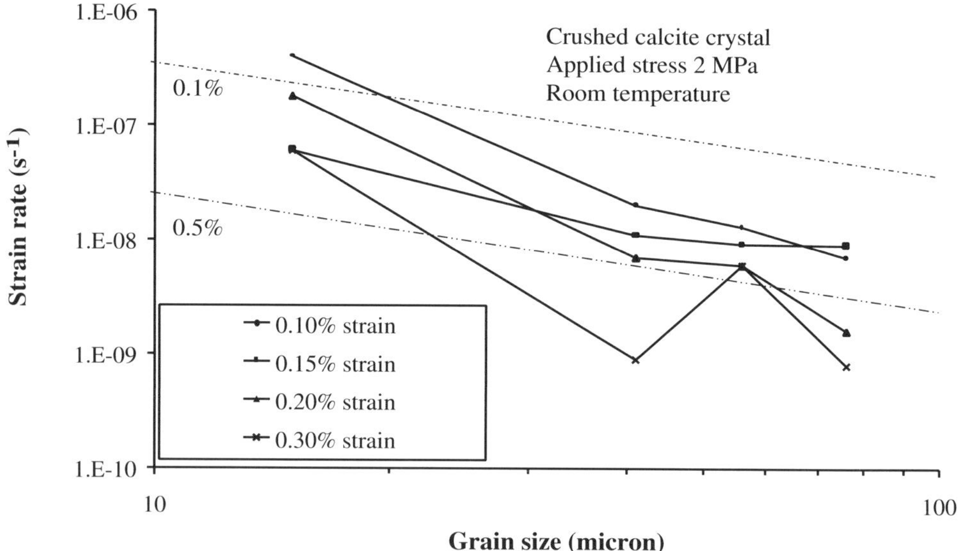

Fig. 5. Log–log plot of strain rate versus grain size for data in Fig. 2b. Dashed lines are the theoretical prediction (using Equation 3 and parameters listed in Table 1) at fixed strains of 0.1% and 0.5%.

The theoretical models of IPS (equations 1–3) predict that strain rate is inversely proportional to grain size for interfacial reaction controlled IPS, compared with an inverse cubic relation for diffusion controlled IPS. Turning to our experimental results, Fig. 5 shows a log–log plot of strain rate against grain size for fixed volumetric strains. This was constructed using the rate data obtained from the 2 MPa experiments on the crushed calcite crystal material with different grain sizes (Fig. 2b). The slopes of the data set fall in the range of 1–2. It is therefore not clear which intergranular pressure solution mechanism might control the compaction rate from the grain size dependence alone. However, it is clear that compaction is faster in the finer grained samples. This is important because it is consistent with IPS but makes it unlikely that creep is caused by grain cracking. In the latter case, strain rate is expected to increase with increasing grain size (Zhang *et al.* 1990; Brzesowsky 1995).

In a log–log plot of strain rate against stress, theory predicts that diffusion controlled intergranular pressure solution will show a slope of $n = 1$ and that interfacial reaction controlled IPS will show a n-value which depends on the kinetics of reaction. From the crystal growth literature, the precipitation rate of calcite is quadratically dependent on the degree of supersaturation or driving force (Reddy 1977; Reddy & Wang 1980, see Table 1). Figure 6 shows a comparison of our experimental data from Fig. 2a versus the theoretical dependence of strain rate on stress. The slope of our experimental data is about 2, which is roughly consistent with the theoretical prediction for precipitation controlled IPS. However, the stress-corrosion crack growth velocity depends on stress in a highly non-linear way, with a power law stress exponent larger than 10 (Atkinson 1984; Atkinson & Meredith 1989).

On the basis of the above, we infer that compaction in our wet experiments probably did occur by IPS (accompanied by inter-grain sliding) and that precipitation was most likely to have been rate controlling. The poor agreement between our experimental results and our precipitation-controlled IPS model, regarding the dependence of strain rate on strain (Fig. 4), could be due to the fact that the model assumes a spherical grain shape, whereas our calcite powders consisted of rough grains. An additional reason for discrepancies between theory and the experimental results could lie in the assumption of the model that IPS involves a series of strain-dependent steady states with no transient effects (e.g. transients associated with changes in pore fluid composition). For added confidence, our inference that precipitation-controlled IPS is probably the controlling mechanism in the present experiments should be tested in future by carrying out flow-through experiments following the approach of De Meer & Spiers (1997).

The fact that we seem to have obtained IPS in the laboratory suggests that the same process

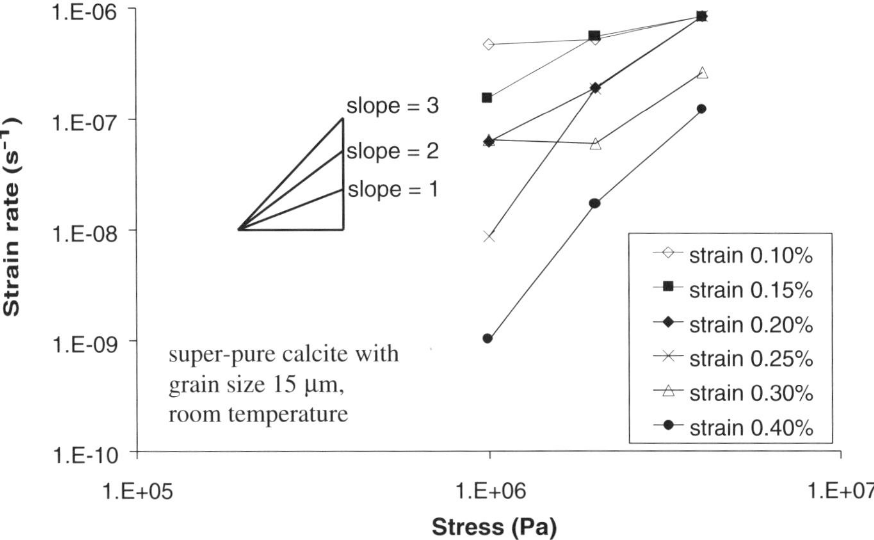

Fig. 6. Log–log plot of strain rate versus stress for data in Fig. 2a.

should operate under comparable conditions in nature, for example during carbonate diagenesis and fault sealing under shallow, low temperature, conditions. Since Mg^{2+} is always present in significant concentration in carbonate pore waters (Collins 1975), the effects of Mg^{2+} which we have observed may be very important in controlling the rate of IPS in natural situations. However, until we can confirm and better qualify the rate-controlling process (notably the effect of porosity), we make no attempt to estimate rates of IPS in calcite rocks in nature.

Conclusions

We carried out one-dimensional, dead-weight compaction creep experiments on calcite powder at room temperature. After pre-compaction at a stress of 8 MPa, the samples showed no further compaction when reloaded dry, or flooded with inert fluid (decane) at lower stresses (1-4 MPa). This suggests that grain reorientation, intergranular sliding and coupled plastic deformation do not proceed under dry conditions or in the presence of an inert fluid, after pre-compaction. However, samples tested with calcite-saturated solution showed significant time-dependent compaction creep. The compaction rate decreased rapidly with increasing volumetric strain. It increased with the applied stress squared and was inversely related to grain size to the power of 1 to 2. The behaviour of the wet versus the dry and decane-flooded samples demonstrated that wet deformation was certainly a fluid-enhanced process. Adding Mg^{2+} (a well-known inhibitor of calcite precipitation) into the saturated solution drastically slowed down the creep of compaction. This effect suggests that deformation of the wet samples involved some kind of dissolution/precipitation process. The two candidate deformation mechanisms are IPS and stress-corrosion cracking. The observed grain size and stress dependence contradicts expectations for creep via stress-corrosion cracking. However, the mechanical behaviour of the wet samples, that is the dependence of strain rate on applied stress and grain size, is broadly consistent with a precipitation-controlled IPS model. Furthermore, calculations based on kinetics data from the literature show that dissolution or diffusion controlled IPS in calcite should be much faster than a precipitation-controlled process. We infer that IPS was probably the main deformation in our wet experiments and it is likely to be precipitation controlled. Given the low temperature and stress conditions of our experiments, this suggests that precipitation-controlled IPS should play a role in carbonate diagenesis and fault sealing at very shallow depths of burial. However, further experimental work is needed to quantitatively characterize the process.

This research was funded by Shell Research (SIEP, Rijswijk). We thank F. Dula and J. Constanty of Shell for valuable discussions and for granting permission to publish this article, and F. Renard and B. Evans for their useful comments.

References

ATKINSON, B. K. 1982. Subcritical crack propagation in rocks: theory, experimental results and application. *Journal of Structural Geology*, **4**, 41–56.

ATKINSON, B. K. 1984. Subcritical crack growth in geological materials. *Journal of Geophysical Research*, **89**, 4077–4114.

ATKINSON, B. K. & MEREDITH, P. G. 1989. The theory of subcritical crack growth with applications to minerals and rocks. *In*: ATKINSON B. K. (ed) *Fracture Mechanics of Rock*. Academic Press, London, 111–166.

BAKER, P. A. KASTNER, M., BYERLEE, J. D. & LOCKNER, D. A. 1980. Intergranular pressure solution and hydrothermal recrystallization of carbonate sediments – an experimental study. *Marine Geology*, **38**, 185–203.

BATHURST, R. G. C. 1958. Diagenetic fabrics in some British Dinantian limestones. *Liverpool and Machester Geology*, **2**, 1–36.

BRZESOWSKY, R. H. 1995. *Micromechanics of Sand Grain Failure and Sand Compaction*. PhD dissertation, Utrecht University, Netherlands, *Geologica Ultraiectina*, **133**.

BURGER, H. R. 1975. "Pressure solution" or indentation?: Comment and reply. *Geology*, **3**, 292–293.

CAROZZI, A. V. & VON BERGEN, D. 1987. Stylolitic porosity in carbonate: a critical factor for deep hydrocarbon production. *Journal of Petroleum Geology*. **10**, 267–282.

COLLINS, A. G. 1975. *Geochemistry of Oilfield Water*. Developments in Petroleum Sciences I, Elsevier Science, 142–143.

COX, S. F. & PATERSON, M. S. 1991. Experimental dissolution–precipitation creep in quartz aggregate at high temperature. *Geophysical Research Letters*, **18**, 1401–1404.

COX, M. A. & WITHFORD, S. J. L. 1987. Stylolites in the Caballos Novaculite, west Texas. *Geology*, **15**, 439–442.

DAVIS, K. J., DOVE, P. M. & DE YOREO, J. J. 2000. The role of Mg^{2+} as an impurity in calcite growth. *Science*, **290**, 1134–1137.

DE BOER, R. B., NAGTEGAAL, P. J. C. & DUYVIS, E. M. 1977. Intergranular pressure solution experiments on quartz sand. *Geochimica et Cosmochimica Acta*, **41**, 257–264.

DE MEER, S. & SPIERS, C. J. 1997. Uniaxial compaction creep of wet gypsum aggregates. *Journal of Geophysical Research*, **102**, 875–891.

DE MEER, S. & SPIERS, C. J. 1999. On mechanisms and kinetics of creep by intergranular pressure solution. *In*: JAMTVEIT, B. & MEAKIN, P. (eds) *Growth Dissolution and Pattern Formation Geo-systems*. Kluwer Academic Publishers, Dordrecht, The Netherlands, 345–366.

DE MEER, S. & SPIERS, C. J. 2001. Diffusive properties of wetted grain contacts undergoing active intergranular pressure solution. *Abstract of Deformation Mechanisms, Rheology & Tectonics, Noordwijkerhout, The Netherlands*, April 2–4, 2001, p 113.

DEN BROK, S. W. J. & MOREL, J. 2001. The effects of elastic strain on the microstructure of free surface of stressed minerals in contact with an aqueous solution. *Geophysical Research Letters*, **28**, 603–606.

DEWERS, T. & HAJASH, A. 1995. Rate law for water-assisted compaction and stress-induced water–rock interaction in sandstones. *Journal of Geophysical Research*, **100**, 13093-13112.

DEWERS, T. & MUHARI, S. 2000. *Transition fom Plastic to Viscous Flow in Aggregate Granular Creep*. Abstract of EOS Transactions, AGU, **81**.

DURNEY, D. W. 1972. Solution transfer, an important geological deformation mechanism. *Nature*, **235**, 315–317.

ELLIOT, D. 1973. Diffusion flow laws in metamorphic rocks. *Bulletin of Geological Society of America*, **84**, 2645–2664.

GRATIER, J. P. 1993. Experimental intergranular pressure solution of halite by an indenter technique. *Geophysical Research Letters*, **20**, 1647–1650.

HEALD, M. T. 1956. Cementation of Simpton and St. Peter sandstones in parts of Oklahoma, Arkansas, and Missouri. *Journal of Geology*, **64**, 16–30.

HEYDARI, F. 2000. Porosity loss, fluid flow, and mass transfer in limestone reservoirs: Application to the upper Jurassic Smackover formation, Mississippi. *American Association of Petroleum Geologists Bulletin*, **84**, 100–118.

HICKMAN, S.H. & EVANS, B. 1991. Experimental pressure solution in halite: the effect of grain/interphase boundary structure. *Journal of the Geological Society, London*, **148**, 549–560.

HICKMAN, S. H. & EVANS, B. 1992. Growth of the grain contacts in halite by solution-transfer: implication for diagenesis, lithification and strength recovery. *In*: EVANS, B. & WONG, T. F. (eds) *Fault Mechanics and Transport Properties of Rocks*. Academic Press, San Diego, USA, 253–280.

HICKMAN, S. H. & EVANS, B. 1995. Kinetics of intergranular pressure solution at halite-silica interfaces and intergranular clay films. *Journal of Geophysical Research*, **100**, 13113–13132.

HOUSEKNECHT, D. W. 1987. Assessing the relative importance of compaction processes and cementation to reduction of porosity in sandstones. *American Association of Petroleum Geologists Bulletin*, **71**, 633–642.

HOUSEKNECHT, D. W. 1988. Intergranular pressure solution in four quartzose sandstones. *Journal of Sedimentary Petrology*, **58**, 228–246.

LEHNER, F. K. 1990. Thermodynamics of rock deformation by intergranular pressure solution. *In*: BARBER, D.J. & MEREDITH, P. G. (eds) *Deformation Processes in Minerals, Ceramics and Rocks*. Unwin Hyman, London, 296–333.

LEHNER, F. K. 1995. A model for intergranular pressure solution in open systems. *Tectonophysics*, **245**, 153–170.

MUCCI, F. & MORSE, J. W. 1983. The incorporation of Mg^{2+} and Sr^{2+} into calcite overgrowths: influence of growth rate and solution composition. *Geochimica et Cosmochimica Acta*, **47**, 217–233.

RAJ, R. 1982. Creep in polycrystalline aggregates by matter transport through a liquid phase. *Journal of Geophysical Research*, **87**, 4731–4739.

Reddy, M. M. 1977. Crystallization of calcium carbonate in the presence of trace concentrations of phosphorous-containing anions. *Journal of Crystal Growth*, **41**, 287–295.

Reddy, M.M. & Wang, K. K. 1980. Crystallization of calcium carbonate in the presence of metal ions. I. Inhibition of magnesium ion at pH 8.8 and 25 °C. *Journal of Crystal Growth*, **50**, 470–480.

Renton, J. J., Heald, M. T. & Cecil, C. B. 1969. Experimental investigation of pressure solution of quartz. *Journal of Sedimentary Petrology*, **39**, 1107–1117.

Robin, P. F. 1978. Intergranular pressure solution at grain-to-grain contacts. *Geochimica et Cosmochimica Acta*, **42**, 1383–1389.

Rutter, E. H. 1976. The kinetics of rock deformation by intergranular pressure solution. *Philosophical Transactions of the Royal Society, London*, **A282**, 257–291.

Rutter, E. H. 1983. Intergranular pressure solution in nature, theory and experiment. *Journal of the Geological Society, London*, **140**, 725–740.

Rutter, E. H. & Wanten, P. H. 2000. Experimental study of the compaction of phyllosilicate-bearing sand at elevated temperature and with controlled pore water pressure. *Journal of Sedimentary Research*, **70**, 107–116.

Schutjens, P. M. T. M. 1991. Experimental compaction of quartz sands at low effective stress and temperature conditions. *Journal of the Geological Society, London*, **148**, 527–539.

Schutjens, P. M. T. M. & Spiers, C. J. 1999. Intergranular pressure solution in NaCl: grain-to-grain contact experiments under the optical microscope. *Oil & Gas Science and Technology*, **54**, 729–750.

Sorby, H. C. 1863. On the direct correlation of mechanical and chemical forces. Proceedings of the Royal Society, London, **12**, 583–600.

Spiers, C. J. & Brzesowsky, R. H. 1993. Densification of wet granular salt: theory versus experiment. *In*: Kakihana, H., Hardy, H. R. Jr., Hoshi, T. & Toyokura, K. (eds) *Seventh Symposium on Salt*. Vol I. Elsevier, Amsterdam, 83–92.

Spiers, C. J. & Schutjens, P. M. T. M. 1990. Densification of crystalline aggregates by fluid-phase diffusional creep. *In*: Barber, D. J. & Meredith, P. G. (eds) *Deformation Processes in Minerals, Ceramics and Rocks*. Unwin Hyman, London, 334–353.

Sprunt, E. S. & Nur, A. 1977. Experimental study of the effects of stress on solution rate. *Journal of Geophysical Research*, **82**, 3013–3022.

Tada, T. & Siever, R. 1986. Experimental knife-edge intergranular pressure solution of halite. *Geochimica et Cosmochimica Acta*, **50**, 29–36.

Tada, R. & Siever, R. 1989. Intergranular pressure solution during diagenesis. *Annual Review of Earth and Planet Sciences*, **17**, 89–118.

Weyl, P. K. 1959. Intergranular pressure solution and force of crystallization – a phenomenological theory. *Journal of Geophysical Research*, **64**, 2001–2025.

Wiechers, H. N. S., Sturrock, P. & Marais, G. V. R. 1975. Calcium carbonate crystallization kinetics. *Water Research*, **9**, 835–845.

Zhang, J., Wong, T. F. & Davis, D. M. 1990. Micromechanics of pressure-induced grain crushing in porous rocks. *Journal of Geophysical Research*, **95**, 341–352.

Numerical modelling of pressure solution in sandstone, rate-limiting processes and the effect of clays

E. GUNDERSEN[1], D. K. DYSTHE[1], F. RENARD[2], K. BJØRLYKKE[1] & B. JAMTVEIT[1]

[1]*Fluid Rock Interaction group, Departments. of Geology & Physics, University of Oslo, P.O. Box 1047 Blindern 0316 Oslo (e-mail: elisabeth.gundersen@geologi.uio.no)*

[2]*LGIT, CNRS-Observatoire, Université J. Fourier, BP 53,38041 Grenoble, France (e-mail: francois.renard@lgit.obs.ujf-grenoble.fr)*

Abstract: Pressure solution is an efficient mechanism for ductile deformation and local mass transport in the upper crust. In this paper we model pressure solution as a mechanism involving four steps: (1) dissolution at the grain contacts; (2) diffusion of solutes through fluid films at the contact between two grains; (3) transport of solutes by diffusion through the pore fluid into other adjacent open pores; and (4) precipitation on the surface of grains at their contact with the pore fluid. In this study we constrain under which conditions pressure solution is limited by one of the four processes: dissolution; contact diffusion; precipitation; and global diffusion. From our model of pressure solution, based on thermodynamic relationships we derive three dimensionless numbers which represent the competetion between the four mentioned processes. With these numbers we can define the crossover from a situation where one process acts as the limiting process to a new situation controlled by another process. We also see how the different rate-limiting processes influence the amount of mass transported during the compaction process. In addition we study the effect of clays, as it has been suggested that these minerals speed up the rate of pressure solution. We propose two models, a chemical related and a mechanical model for how the clay particles may affect the dissolution process of quartz.

Pressure solution is a deformation mechanism, dependent on temperature and stress, which reduces the porosity of rocks in the upper crust by dissolution and precipitation of minerals. Pressure solution is usually described as a grain-scale deformation mechanism involving several successive steps: (1) dissolution at the grain-grain contact; (2) diffusion of solutes along a water film layer in the grain contact and into the pore volume; (3) solute transport by diffusion through the main porosity network, and finally (4) precipitation of solutes on grain surfaces, see e.g. Weyl (1959), Rutter (1976), Houseknecht (1987) and Gundersen *et al.* (2002). The four steps are schematically shown in Fig. 1. The third step is responsible for mass transfer at a local scale (centimeter to decimeter) (Giles *et al.* 2000). The result of pressure solution is deformation of the individual grains and cementation of the porespace, leading to a compaction of the rock as the porosity is reduced.

The main thermodynamical driving force of pressure solution is the difference in the chemical potential between a fluid–solid interface, subject to a normal stress, and a pore fluid–interface, subject to hydrostatic pressure (Kamb 1961; Paterson 1973). This driving force results in enhanced solubility at the stressed interfaces and the diffusional transfer of dissolved material to the pore space which are regions of lower stress (Weyl 1959; Rutter 1976). Here we assume that stress is the only driving force of dissolution through the effect of stress on the chemical potential of the solid. A more detailed discussion is given by Paterson (1973) and Lehner (1995).

Some authors have argued that dissolution of quartz in sandstones is driven by the enhanced solubility of quartz at quartz–mica interfaces: see Bjørkum (1996). This process may take place under conditions where temperature, rather than stress at grain contacts, is the main parameter controlling the pressure-solution rate. Geological observations indicate that clay or mica particles could greatly enhance the pressure solution compaction process. However, there is no clear understanding of the physics of the catalytic effect of clay particles.

A general discussion of how clay affects the process of pressure solution in sandstones and limestones is contained in Renard *et al.* (2001). They propose a mechanical effect where the transport properties of the contacts are enhanced, either due to a increased thickness of the water film at the contacts or equivalently an increase in the diffusion constant at the grain

From: DE MEER, S., DRURY, M. R., DE BRESSER, J. H. P. & PENNOCK, G. M. (eds) 2002. *Deformation Mechanisms, Rheology and Tectonics: Current Status and Future Perspectives.* Geological Society, London, Special Publications, **200**, 41–56. 0305-8719/02/$15

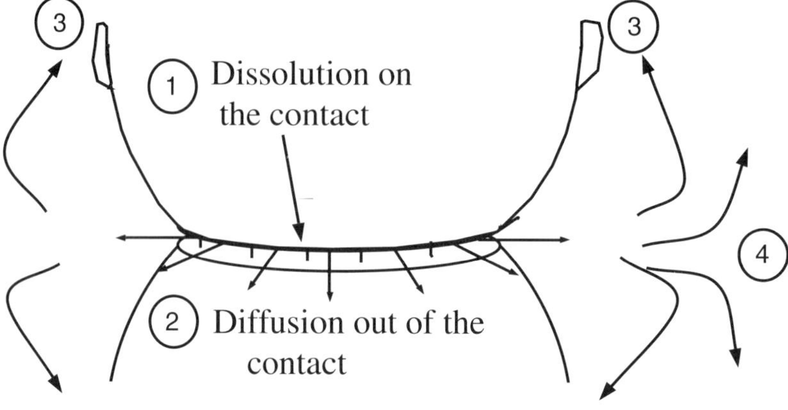

Fig. 1. At a grain scale, pressure solution is a three-step process: (1) stress-enhanced dissolution at grain contacts; (2) diffusion through the contact to the pore; and (3) precipitation on the grain surface. If one considers that the system is not closed at the grain scale, a fourth step adds a level of complexity to the process: (4) solute transport between pores, which can lead to mass transfer at a decimetre (dm) scale.

contact. A chemical effect where the presence of mica or clay increases the solubility at the quartz–mica and quartz–clay interface was quantified through petrographic observations by Oelkers *et al.* (2000). Here, they assumed that the dissolution is independent of differential stress.

In the present work we consider two minerals, quartz and an unspecified clay, and propose a phenomenological approach where we present a linear variation of the diffusion and kinetic constants of silica with the concentration of clay. These relations are then included as a part of a numerical model based upon several differential equations which describe the process of pressure solution on a local (grain size) and a global (rock size) level. The main equations are presented in this work. However a more detailed description of the analytical derivation of the mathematical framework is given in Gundersen *et al.* (2002).

In order to differentiate between the two proposed effects of clays we need to study situations where either dissolution or diffusion out of the contact acts as the the rate-limiting process of pressure solution. For example, an increase in the diffusion constant at the grain contact will only affect the process of pressure solution as long as contact diffusion is the rate-limiting process. However, precipitation-limited processes are also included in the simulations. By introducing dimensionless quantities we will define three dimensionless numbers which quantify under which conditions dissolution, precipitation or contact diffusion acts as the rate-limiting process. As we shall see, variations in the rate-limiting processes influence the process of pressure solution even in the absence of clay.

Modeling the effect of clay particles on pressure solution

The base of our model is the thermodynamic equations which relate normal stress and chemical potential. This relationship is then coupled to transport equations based on the fluid film at the grain contact model proposed by Rutter (1976). Finally, the equations are coupled to regular diffusion of solute in porous media. Advective fluid flow is not included in the model. The equations are derived by a global mass balance of the solute phase in the pore volume and a local mass balance at each grain contact (Gundersen *et al.* 2002) Those relationships are then coupled to equations which express the deformation of the grains and the evolution of the rock texture, where the grains have a truncated spherical geometry: see Fig. 2.

The grain framework is represented by a dense cubic packing of truncated spheres. The grain boundary is considered as a flat interface with a mean roughness averaged over the contact. This average takes into account indications that the interface can have a structure of channels and islands at a micrometer scale (de Boer 1977; Spiers & Brzesowsky 1993; Renard *et al.* 2001; Dysthe *et al.* in prep).

The whole cubic packing is subjected to a constant normal stress (σ). We then define a contact normal stress (σ_n) as the mean stress at each grain contact surface. This stress depends on the relation between the surface area and the diameter of the sphere: see equations (14–16). This relation is independent of the initial size of the individual grains (Renard *et al.* 2000).

The kinetics of the various steps in the model depends on the different assumptions made about the sediment mineralogy and grain size. By changing the various parameters involved we can investigate their individual influences upon the rate-limiting process for pressure solution. Moreover, we can also investigate the effect of clay particles in situations where either dissolution, precipitation or diffusion acts as the rate-limiting process.

For example, if at grain contacts the rate of dissolution is much slower than the rate of contact diffusion and precipitation (which may be the case for quartz at shallow burial depths), then contact stress might be the only important variable and there will be no significant transport over long distances. One should then obtain very small concentration gradients independent of particle size. On the other hand, if diffusion at the contact area is the rate-limiting process then variations in grain size as well as variations in the amount of clay particles will increase the concentration gradient of silica in solution and thereby promote transport over longer distances.

The resulting model is a set of highly coupled non-linear equations which can only be solved using numerical methods. All the parameters used in the following equations and their units are given in Table 1. The mass balance for the

Table 1. *Symbols used for parameters in Equations 1 to 17*

Parameter	Description	unit
$A_{cont,i}$, $i = x, y, z$	contact area in the *i*-direction	m^2
A_{pore}	pore surface area	m^2
$c_{cont,i}$, $i = x, y, z$	concentration of solute in a grain contact in the *i*-direction	mole/m^3
c_{pore}	concentration of solute in the pore fluid	mole/m^3
D_{cont}	diffusion constant in the grain contact	m^2/s
D_{cont0}	constant diffusion constant in the grain contact	m^2/s
D_{pore}	diffusion constant in the pore	m^2/s
m_{diff}	matter transported by diffusion out of the contact	mole
m_{diss}	matter dissolved from the grain contacts	mole
m_{prec}	precipitated matter on the pore surface	mole
k_{diss}	kinetic coefficient of dissolution	mole/(m^2s)
k_{diss0}	constant kinetic coefficient of dissolution	mole/(m^2s)
k_{prec}	kinetic coefficient of precipitation	mole/(m^2s)
K_{cont}	concentration of solute in a grain contact at equilibrium	mole/m^3
K_{pore}	concentration of solute in the pore at equilibrium	mole/m^3
CC	clay content in each element	%
L_i, $i = x, y, z$	length of the truncated sphere in the *i*-direction	m
L_f	radius of the spherical grain	m
P_f	pore fluid pressure	Pa
$R_{cont,i}$, $i = x, y, z$	radius of a grain contact in the *i*-direction	Pa
$\bar{V}_s$	molar volume of the dissolving mineral	m^3/mole
α	parameter in Equation 12 which regulates the effect of clay on the dissolution kinetics	
Δ_i	thickness of the water-film in the contacts, in the *i*-direction	m
ϕ	porosity	%
σ_i, $i = x, y, z$	normal stress component in the *i*-direction	Pa
σ_{ni}, $i = x, y, z$	normal stress to a grain-surface in the *i*-direction	Pa
U	dimensionless number, measuring the competition between dissolution and contact diffusion	
V	dimensionless number, measuring the competition between precipitation and global diffusion	
W	dimensionless number, measuring the competition between local and global transport	

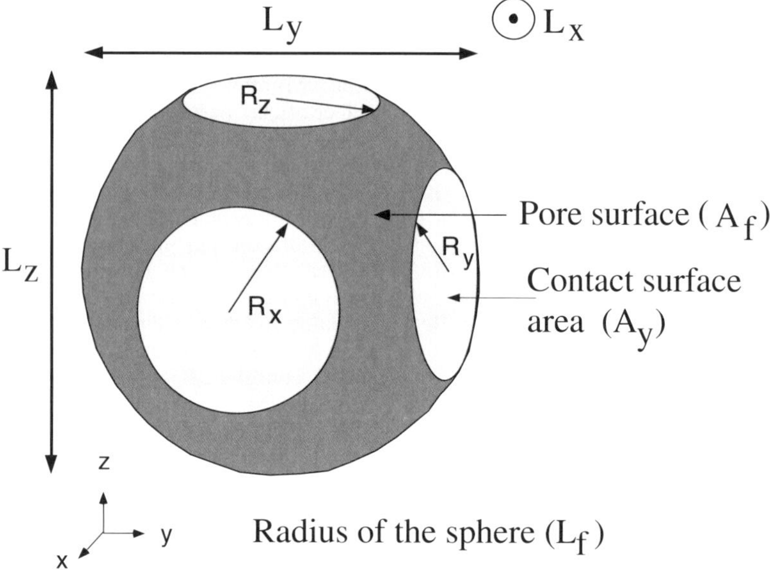

Fig. 2. Schematic representation of a single grain for which the surface is divided into contacts with other grains (light areas) and free face in contact with the pore fluid (dark area). Each grain has six neighbors, arranged in cubic packing.

solute phase in the pore volume is given as:

$$\phi\frac{\partial c_{pore}}{\partial t} = D_{pore}\nabla^2 c_{pore} + \frac{1}{L_x L_y L_z}\left[\frac{\partial m_{prec}}{\partial t} + 2\sum_{i=1}^{3}\frac{\partial m_{diff,i}}{\partial t}\right] \quad (1)$$

where c_{pore} is the concentration of solute in the pore fluid, ϕ is the porosity and the lengths L_x, L_y and L_z are defined in Fig. 2. The first term on the right side in Equation 1 describes the diffusive transport of solutes in the pore space by a typical Fick's law. The coefficient of diffusion D_{pore}, in the pore fluid is assumed to follow an Arrhenius law (Applin 1987; Nakashima 1995). The driving force for this transport is a concentration gradient related to the differences in stress concentration at grain contacts between domains with different grain sizes or different kinetics. The diffusion term represents how the solutes are transported through the pore space, assuming a constant coefficient of diffusion. A further refinement of this model would be to take into account some porosity-dependent tortuosity factor in the diffusion coefficient. However, numerical tests indicate that such tortuosity would not significantly modify the results of the simulations. In the model, the decrease in the length of the diffusion path is taken directly into account for porosity reduction due to compaction.

The second term in Equation 1 represents mass loss due to precipitation on the grain surfaces and is given as:

$$\frac{\partial m_{prec}}{\partial t} = k_{prec}A_{pore}\left(1 - \frac{c_{pore}}{K_{pore}}\right) \quad (2)$$

where k_{prec} is the kinetic coefficient of precipitation, K_{pore} is the equilibrium concentration of solute in the pore and A_{pore} is the grain surface area which is in contact with the pore fluid. Notice that as the spheres changes during simulation, $A_{pore} = g(L_f, L_x, L_y, L_z)$ varies correspondingly. Finally, the last term in Equation 1 describes the mass production due to diffusional mass transport out of the contact and reads:

$$\frac{\partial m_{diff,i}}{\partial t} = 2\pi\frac{\Delta_i}{2}D_{cont}(c_{cont,i} - c_{pore}) \quad (3)$$

where $c_{cont,i}$ is the concentration of solute in a grain contact in the *i*-direction. Note, that in Equation 1 this term is summarized over all the six contact areas of a truncated spherical grain: see Fig. 2. The thickness of the water film at the grain contacts, Δ_i is divided by two since it is shared between two contact surfaces. The coefficient of diffusion inside the water film D_{cont} follows an Arrhenius law: see Gratier & Guiguet (1986).

The global diffusion equation (Equation 1) is then coupled to six equations which represent the local mass balance for each contact on the spherical grain. The concentration in the trapped fluid phase at the contact is given as the difference between the mass produced by dissolution and the mass lost by diffusion, assuming that the solute concentration is constant over the grain boundary.

$$\frac{\partial c_{cont,i}}{\partial t} \frac{1}{\pi R_{cont,i}^2 \frac{\Delta_i}{2}} \left[\frac{\partial m_{diss,i}}{\partial t} - \frac{\partial m_{diff,i}}{\partial t} \right] \quad (4)$$

Here, the index i represents the space dimensions, $i = x, y, z$, and $R_{cont,i}$ is the radius of the contact surface in the i direction, which is a function of the spherical lengths: see Fig. 2.

The first term in Equation 4 represents local production by grain dissolution and is given as

$$\frac{\partial m_{diss,i}}{\partial t} = k_{diss} A_{cont,i} \left(1 - \frac{c_{cont,i}}{K_{cont,i}} \right) \quad (5)$$

where $A_{cont,i}$ is the surface area of the grain in contact with another grain in the i direction, k_{diss} is the kinetic coefficient of dissolution and $K_{cont,i}$ represents the equilibrium concentration of solute in a contact in the i direction and is a function of contact normal stress and temperature (Renard *et al.* 1999). The second term in Equation 4 is the local diffusional transport away from the contact and into the pore space along a trapped water film as given by Equation 3. The grain shape evolution is given by the four equations:

$$\frac{dL_i}{dt} = -2k_{diss} \bar{V}_s \left(1 - \frac{c_{cont,i}}{K_{cont,i}} \right) \quad (6)$$

$$\frac{dL_f}{dt} = -k_{prec} \bar{V}_s \left(1 - \frac{c_{pore}}{K_{pore}} \right) \quad (7)$$

where the index i is given as $i = x, y, z$ and $\bar{V}_s$ is the molar volume of the dissolving mineral. The kinetic constant for dissolution k_{diss} and precipitation k_{prec} are only functions of the temperature, while the equilibrium constants $K_{cont,i}$ and K_{pore}, and the width of the water film (Δ_i) are functions of the normal contact stress and pore pressure as well as the temperature (Renard *et al.* 1999).

The simulation domain is a 2D square rock sample located several kilometers below the surface and exposed to stress and a constant temperature field. The normal contact stress component σ_{ni} is related to the normal stress at the grain contacts σ_i by the three equations:

$$\sigma_{nx} = \sigma_x \frac{L_y L_z}{A_x} \quad (8)$$

$$\sigma_{ny} = \sigma_y \frac{L_x L_z}{A_y} \quad (9)$$

$$\sigma_{nz} = \sigma_z \frac{L_x L_y}{A_z}. \quad (10)$$

Assuming that the normal stress (σ_i) is kept constant, as the contact surface area A_i increases and the diameter of the truncated sphere (L_f) decreases, the normal contact stress (σ_{ni}) decreases as the grains become more and more truncated. Since the equilibrium constants and the width of the water film depend on the normal contact stress, the driving force for pressure solution diminishes with time.

In order to model the effect of clays, we have introduced the parameter CC which represents the amount of clay particles (in weight %) in each element. There are several possible ways in which clays might affect the dissolution of quartz (Dewers & Ortoleva 1991). Since there is no detailed microscopic model of how the various particles interact, we have based our model on experiments on salt aggregates (Renard *et al.* 2001), experiments on salt lenses (Hickman & Evans 1991) and observations of mica indenting quartz grains (Bjørkum 1996). All these data show that clay enhances pressure solution.

In our first model we assume that the clays enhance diffusion at the grain contact by increasing the diffusion constant D_{cont} such that if $CC > 0$ then:

$$D_{cont} = D_{cont0}(1 + 0.05CC). \quad (11)$$

The equation indicates that a 1% increase in CC corresponds to a 5% increase in the rate of contact diffusion. The form of Equation 11 is not theory based, but may be vewed as a local linearization that is consistent with the experimental results of Renard *et al.* (2001).

Another possible effect of the clay particles is an increase in the kinetic coefficient of dissolution, k_{diss}. The kinetic coefficient of dissolution can then be modelled such that if $CC > 0$ then:

$$k_{diss} = k_{diss0}(1 + \alpha CC) \quad (12)$$

where k_{diss0} is the dissolution constant given from the experiments done on quartz by Dove (1994). The constant α indicates the increase of quartz dissolution rate at the contact with clay particles. We have, so far no means for measuring α experimentally.

A large α, in an otherwise homogeneous environment, would indicate that most of the quartz is dissolved at the quartz–clay contact. The model can be used to investigate a dissolution-rate limited situation. As we will see in the following section, changes in the kinetic coefficient of precipitation influence both cases.

The final equation updates the porosity ϕ as the difference between the volume of a cubic box, with lengths L_x, L_y, L_z and the volume of the grain (Renard *et al.* 1999). Notice that the individual treatment of the grains supplies the flexibility necessary to investigate the process of pressure solution and diffusional transport under various rate-limiting processes.

Numerical models and boundary conditions

The equations presented in the previous section are highly coupled and need to be solved using different numerical approaches. We couple an individual treatment of the grains with a continuous description of the solute phase in the pore fluid. The space domain is taken to be a two-dimensional rock sample, modeled as a cubic packing of truncated spheres: see Fig. 3. Each element in the numeric grid consists of a chosen number of equal-sized grains. The heterogeneity of the rock sample is then given, as the variation in grain size and clay content can vary freely between the elements. In each element of the grid the number of grains is large, more than 100.

The geometrical evolution of the grains is modeled as a discrete process where the grains in each grid element are treated separately: see Equations 6–7. In order to visualize the deformation of the rock sample as the grains deform during pressure solution, an adaptive grid is developed. On this grid the nodes change position following each time step as the grains deform. Moreover, the contact surface area of each element is updated according to the lengths L_x and L_z of the truncated spheres and thereby represents the porosity loss in the xz-plane. The total change in volume of each element is then obtained by multiplying the volume variation of each grain with the number of grains in the element. The total deformation of the rock sample is acquired by a summation over all the elements. Figure 4 shows an example of a computational grid before and after deformation.

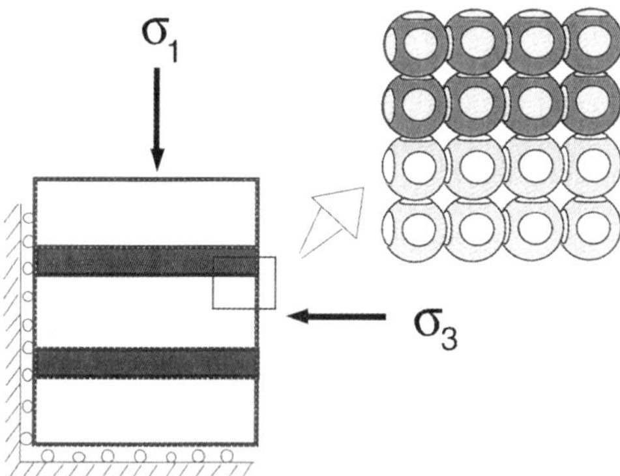

Fig. 3. A domain where coupled pressure solution creep and local solute transport are simulated. Each element in the numerical grid contains a given number of homogeneous grains. The domain is two-dimensional (x and z directions) and has a thickness of one layer of grains in the y direction. The grain size and clay content can be varied between the elements. In this simulation, the clay particles are located in two dark layers.

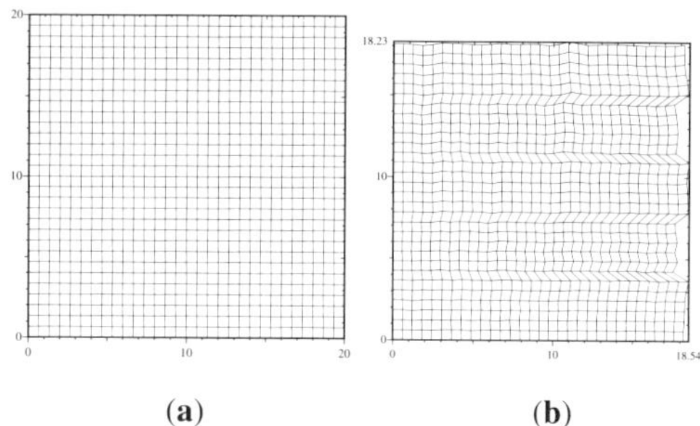

Fig. 4. A grid before and after a simulation. In this example each element consisted of grains whose sizes are randomly distributed. Furthermore, two different distributions alternate and the areas with the smallest grains have compacted more than the areas with larger ones.

The mass conservation equations, Equations 1 to 3 are derived from a continuum model, assuming a continuous solute phase. The equations are solved using a Galerkin finite element method, with linear basis functions on the spatial domain and a Crank-Nicolson scheme in the time domain. All the numerical schemes were programmed in C++ and make use of the numerical library Diffpack (Langtangen 1999).

Two of the boundaries of the grid are fixed as stiff walls on which the side of the simulated domain can only slip. The other two boundaries are free and deform during the simulation. No-flux conditions for the solute phase were imposed for the whole boundary.

For simplicity we have assumed constant temperature and no fluid pressure gradients due to gravity within the simulation domain. This assumption is valid because we are modelling processes at the meter scale. Furthermore, the stress at the boundaries is constant during the simulations. The program was extensively tested for numerical stability. In all the simulations we checked that the total mass of solid was conserved. Variations in the global mass were always less than 1%.

Rate-limiting processes

The progress of pressure solution, as it is described by Equations 1–5, depends on four highly coupled processes: precipitation; dissolution; local diffusion out of the contact; and global diffusion through the pore volume. The slowest process is the rate-limiting process.

In a steady-state situation local mass balance (Equation 4) reads:

$$\frac{\partial m_{diff,i}}{\partial t} = \frac{\partial m_{diss,i}}{\partial t} \qquad (13)$$

and applies to each of the six contacts of the truncated spherical grains. It is convenient to introduce dimensionless quantities. We introduce the following characteristic quantities: length l_c, diffusion constant $D_{cont,c}$, concentration K_c, dissolution constant $k_{diss,c}$ and thickness of the water film in the contacts Δ_c. Using an asterisk as superscript for representing dimensionless quantities, we have:

$$k_{diss} = k_{diss,c} k^*_{diss}$$
$$x = l_c x_*$$
$$D_{cont} = D_{cont,c} D^*_{cont}$$
$$c_{cont,i} = K_c c^*_{cont,i}$$
$$A_{cont} = l_c^2 A^*_{cont}$$
$$\Delta = \Delta_c \Delta^*$$

Equation 13 on a dimensionless form can be expressed as:

$$U k^*_{diss,i} A^*_{cont,i} \left(1 - \frac{c^*_{cont,i}}{K^*_{cont,i}}\right) = \Delta^*_i D^*_{cont,i} (c^*_{cont,i} - c^*_{pore}) \qquad (14)$$

where the quantities marked with an asterisk are normalized to the respective characteristic quantities. The dimensionless number U is given as:

$$U = \frac{k_{diss,c} l_c^2}{\Delta_c D_{cont,c} K_c}. \qquad (15)$$

U measures the relative rate of dissolution at the contact areas and diffusion out of the contact. If U is less than some reference value U_0 the process is limited by dissolution, while contact diffusion is the limiting process when $U \gg U_0$. Here U_0 represents the value of U where the rate-controlling process changes from diffusion limited to dissolution limited.

In the diffusion-controlled case, the rate of pressure solution is independent of the kinetics of dissolution. Moreover, an increase of the kinetic coefficient of dissolution (k_{diss}) does not affect the progress of pressure solution nor the distance over which cementation occurs. In the dissolution-controlled case an increase of the diffusion constant in the grain contact (D_{cont}) does not affect the pressure solution process.

Regardless of which of the two processes controls the rate of the pressure solution, the distance over which mass is transported does not change. This distance is given by the global mass balance and depends on the rate of precipitation and global diffusion. In order to define a regime where precipitation is the rate-limiting process we follow the procedure given below for defining the dimensionless numbers in the global mass balance (Eq. 1). We introduce the characteristic quantities: kinetic coefficient of precipitation $k_{prec,c}$; area of the grain in contact with the pore fluid $A_{pore,c}$; and the diffusion constant in the pore $D_{pore,c}$. Since the diffusion term in Equation 1 is on a global scale it may be convenient to introduce a global length scale with the characteristic value L_c.

The dimensionless quantities are then given by:

$$\nabla = \left(\frac{1}{L_c}\right)\nabla^*$$
$$k_{prec} = k_{prec,c} k^*_{prec}$$
$$A_{pore} = A_{pore,c} A^*_{pore}$$
$$D_{pore} = D_{pore,c} D^*_{pore}$$

On a dimensionless form, the global mass balance at steady state (Equation 1) reads:

$$L^*_x L^*_y L^*_z D^*_{pore} \nabla^2 c^*_{pore} = -V \frac{\partial m^*_{prec}}{\partial t} - W 2 \sum_{i=1}^{3} \frac{\partial m^*_{diff,i}}{\partial t} \qquad (16)$$

introducing the two dimensionless numbers:

$$V = \frac{k_{prec,c} A_{pore,c} L_c^2}{D_{pore,c} K_c l_c^3} \quad \text{and} \quad W = \frac{\Delta_c D_{cont,c}}{D_{pore,c} l_c}. \qquad (17)$$

A typical characteristic length l_c is the diameter of the spherical grains before compaction, and a characteristic L_c can be given as the length between two layers with clay particles.

The number V measures the ratio of the precipitation rate and the diffusion rate in the pore space. It is also a measure of the global transport of dissolved material in the whole domain. If V is less than some reference value V_0, pressure solution is precipitation limited on a global scale. This will be the case if either the rate of the precipitation kinetics is slow or the grain surface area in contact with the pore fluid available for precipitation is limited. In this situation, precipitation is much slower than the diffusion in the pore space and thus the concentration of dissolved material has sufficient time to homogenize on a global scale. As we shall see in the next section, changes in the in the dimensionless constant V which can be given as a reduction in the grain surface area influences the distance over which mass is transported.

The number W compares the two diffusion constants and links the global transport process to the local one through the product $D_{cont,c}\Delta_c$, which is a part of both the number U and W. The three dimensionless numbers are the only relevant parameters of the model. In the following, we demonstrate through several simulation cases how variations in the dimensionless numbers result in different rate controls and thereby affect the process of pressure solution for a sandstone with clay.

Simulation results

In this section we focus on the effects of clay particles in the three (extreme) regimes for pressure solution defined in the previous section. We also study how the scale of dissolved silica transport varies among the three regimes.

The clay particles are concentrated in parts of the simulation domain, and their influence on the compaction process will be described by Equations 11 or 12. In the following examples we modelled pressure solution in sandstone at pressure and temperature conditions typical at a burial depth of 3 km. The model system size is in the range of 20×40 cm. Two horizontal clay-enriched layers were introduced in the sandstone. These layers were typically 3.6 cm thick. The temperature in the domain was constant at 375°K. The pore fluid pressure was constant, $P_{pore} = 30$ MPa, the horizontal stresses, $\sigma_x = \sigma_y = 60$ MPa and the vertical stress, $\sigma_z = 66$ MPa. The initial radius of the grains, L_f was 0.01 cm, while L_x, L_y and L_z were given as $2 \times 0.90 \times L_f$. This corresponds to an initial porosity of 31%.

The simulations lasted until the porosity in the whole simulation domain was less than 3.5%. We assumed that when the porosity is below 3.5% in some element, a percolation threshold is reached (Feder 1988). In these elements, porosity is closed at grain scale and there is no more transport between the pores. The time step was adjusted such that the increment of the deformation was less than 0.1% between each time increment. Therefore the time step automatically adapted in the regions of the fastest deformation.

In the following we present three cases. First we study clay-enhanced contact diffusion as it is given by Equation 11. In order to isolate this effect, we look at situations where contact diffusion is the rate-limiting process. In the second case, we study clay-enhanced dissolution kinetics (Eq. 12) when dissolution is the rate-limiting process. In the final case we study the crossover from a situation where either contact diffusion or dissolution is the rate-limiting process to a situation where precipitation becomes the rate-limiting process.

Case I: clay-enhanced contact diffusion

The effect of clay content on the diffusion at the grain contact is given by Equation 11. A 1% increase in the clay content CC corresponds to a 5% increase in the rate of contact diffusion. In order to study only changes in the coefficient of diffusion we looked at situations where contact diffusion is the rate-limiting process. From the Equation 15 this is the case when $U \gg U_0$. When D_{cont} follows the Arrhenius law given by Gratier & Guiguet (1986) and the kinetics constant for dissolution k_{diss} is obtained from the experiments done on quartz by Dove (1994), a crossover from diffusion to dissolution control appears for $U_0 = 0.001$.

Starting with a sandstone with two horizontal layers containing 20% clay, we can measure the accumulated mass transport as the rock deforms during compaction. With an initial porosity of 31%, Fig. 5(a)–(d) show how the accumulated mass transport pattern develops during compaction. The plots indicate how mass is transported out of the clay-rich layers and spreads into the surrounding areas. For Figure (d) the porosity of the whole rock sample is less than 3.5%. If the local accumulated mass transport is positive the area has received mass and becomes cemented, whereas a negative value represents areas supplying this cement and hence lose mass during compaction. From Fig. 5(d) we see that the layers containing the clays have lost mass while the surrounding areas have gained mass. Figure 5(e)–(h) visualize how the porosity

Fig. 5. Figures (**a**)–(**d**) correspond to the variation of mass before deformation. Mass has been transported out of the layers containing the clay particles and precipitated at the areas around. Figures (**e**)–(**h**) shows how the porosity changes with time, while Figures (**i**)–(**l**) show how the concentration of silica changes during compaction. The reduction of the grid dimension indicates how the rock sample deforms during compaction. The scales of the two axes are in centimetres.

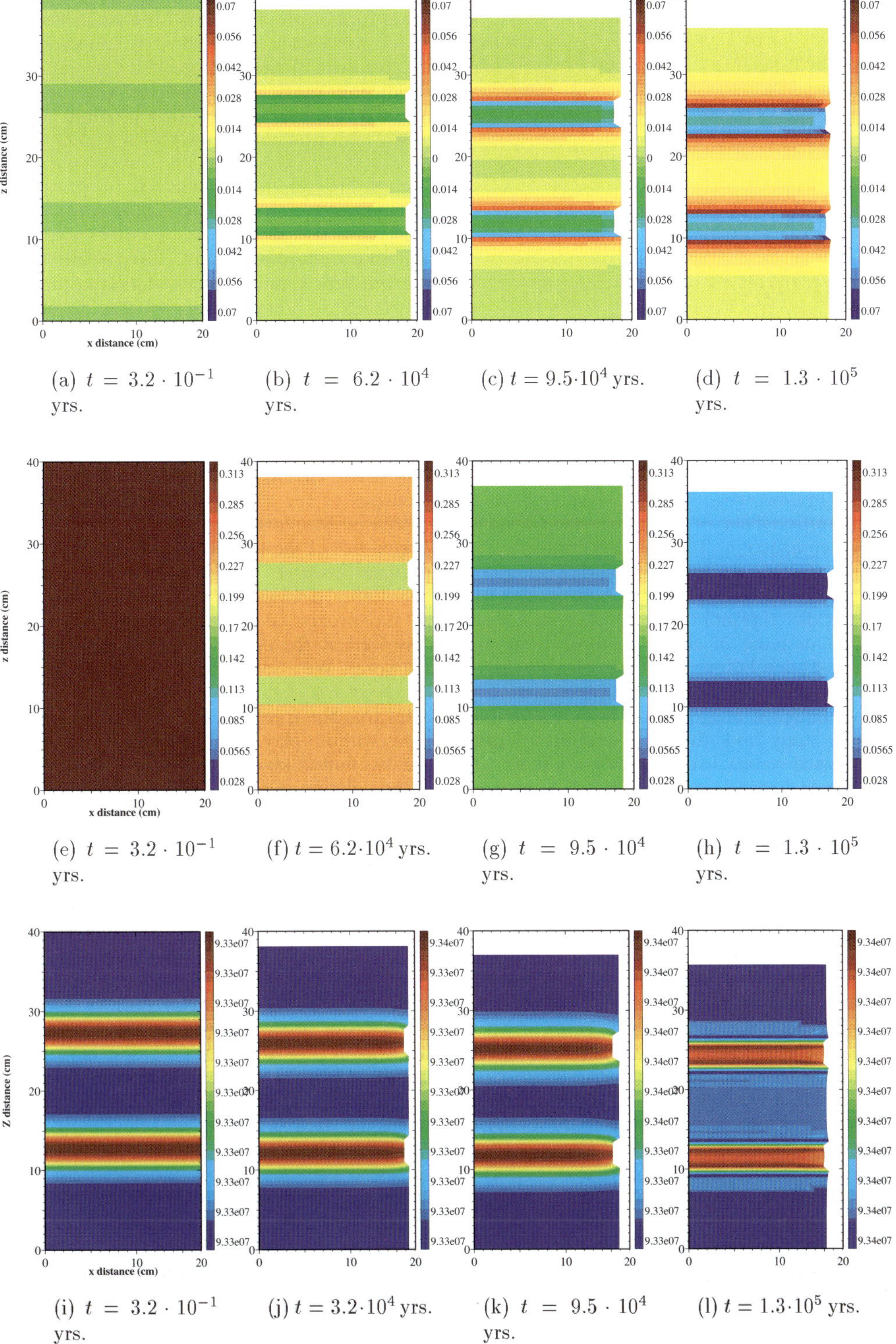

(a) $t = 3.2 \cdot 10^{-1}$ yrs.

(b) $t = 6.2 \cdot 10^{4}$ yrs.

(c) $t = 9.5 \cdot 10^{4}$ yrs.

(d) $t = 1.3 \cdot 10^{5}$ yrs.

(e) $t = 3.2 \cdot 10^{-1}$ yrs.

(f) $t = 6.2 \cdot 10^{4}$ yrs.

(g) $t = 9.5 \cdot 10^{4}$ yrs.

(h) $t = 1.3 \cdot 10^{5}$ yrs.

(i) $t = 3.2 \cdot 10^{-1}$ yrs.

(j) $t = 3.2 \cdot 10^{4}$ yrs.

(k) $t = 9.5 \cdot 10^{4}$ yrs.

(l) $t = 1.3 \cdot 10^{5}$ yrs.

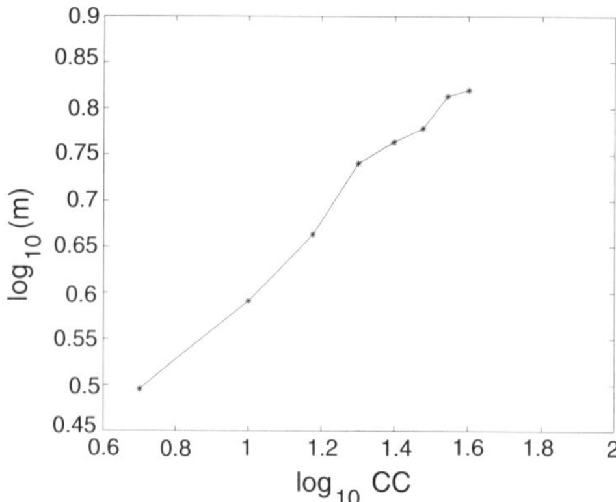

Fig. 6. An almost linear log–log relationship between mass transport out of the contact and clay content in the two chosen layers. In this situation diffusion out of the contact is the rate limiting process.

evolves in time, while Fig. 5(i)–(l) show how the concentration field of dissolved silica changes during compaction. The concentration of silica is largest in the two layers containing clay particles. At the end of the compaction process, the percolation threshold ($\phi = 3.5\%$) has been reached in most of the simulation areas and hence diffusion between the layers has ceased.

If the total mass transport is approximated by the accumulated mass next to the clay layer (m), we can study how the total amount of mass transported varies with the clay content, CC. An almost linear log–log relationship between the amount of mass transported (m) and clay content CC is seen in Fig. 6. This indicate that as long as contact diffusion is the rate-limiting process, mass transport will vary in a logarithmic manner with the fraction of clay particles in the two chosen layers. Notice that in this model a change in the clay content changes the diffusion constant D_{cont} which again affects the dimensionless numbers U and W.

In Fig. 7 we want to study the effect of variations in the distance between the two layers containing the clay. We study a 20×60 cm sandstone where the two clay layers are 5 cm thick. In the four different plots of Fig. 7, we represent cross sections along the vertical line $x = 10$ cm each curve is taken at different time steps during the compaction. The clay content in the two horizontal layers is 20% in all the plots. In Fig. 7(a) the distance between the two layers are 40 cm, which changes to 20 cm in Figure (b), 10 cm in Figure (c) and finally 5 cm in Figure (d).

Figure 7 shows that if the two layers which lose mass are located within the transport distance of each layer, intermediate areas gain mass from both layers and become cemented faster than areas further away. Figure 7 also indicates that the distance over which mass is transported is independent of the distance between the layers producing mass. This distance can only be affected by the rate of precipitation or the global diffusion constant.

Case II: clay-enhanced dissolution kinetics

We now study the effect of clay particles where the pressure solution is dissolution controlled, i.e. $U \ll U_0$. Enhancement of the dissolution rate by clay particles is described by Equation 12. Figures 8 and 9 show how variations in the constant of dissolution changes the result for $V \ll V_0$. In this example $\alpha = 0.05$, the clay content $CC = 20\%$ and k_{diss0} varies as seen in Fig. 8. From Fig. 9 we see that the time taken until the porosity in the model domain is less than 3.5% increases as the kinetic constant of dissolution decreases. Moreover, the amount of mass transported out of the two layers containing the clay particles increases when the kinetic constant of dissolution decreases. These results shows that as the process slows down more material is dissolved at the contacts and transported into the pores. We also notice that the shapes of the curves in Fig. 8 do not vary greatly. This indicates that the transport length of mass into the areas without clay is independent of changes in the dissolution kinetics.

By isolating this effect to only the two layers containing clays we keep k_{diss0} constant and instead vary α: see Fig. 10. In this way we concentrate on the effect of clay.

In Fig. 10, α varies over four orders of magnitude. A small α value indicates that the dissolution kinetics are only slightly faster in the areas containing clays than in the remaining areas (Figs 10(a) and (e)). On the other hand, with a large value of α the dissolution kinetics are much faster in the clay areas than in the pure silica layers. In this situation, the percolation threshold in the clay layers is reached while the porosity is still large in the surrounding areas. As all the silica in contact with the clay particles dissolves, these layers might collapse. Figure 10 shows a situation where the clay-rich layers which lose mass also lose porosity because of grain indentation. In the area along these layers mass is gained and the porosity is lost because of cementation. Figure 10(d) and (h) demonstrate a situation where the clay-rich layer starts to collapse while the porosity is still large in the areas outside the transport distance of the dissolved silica.

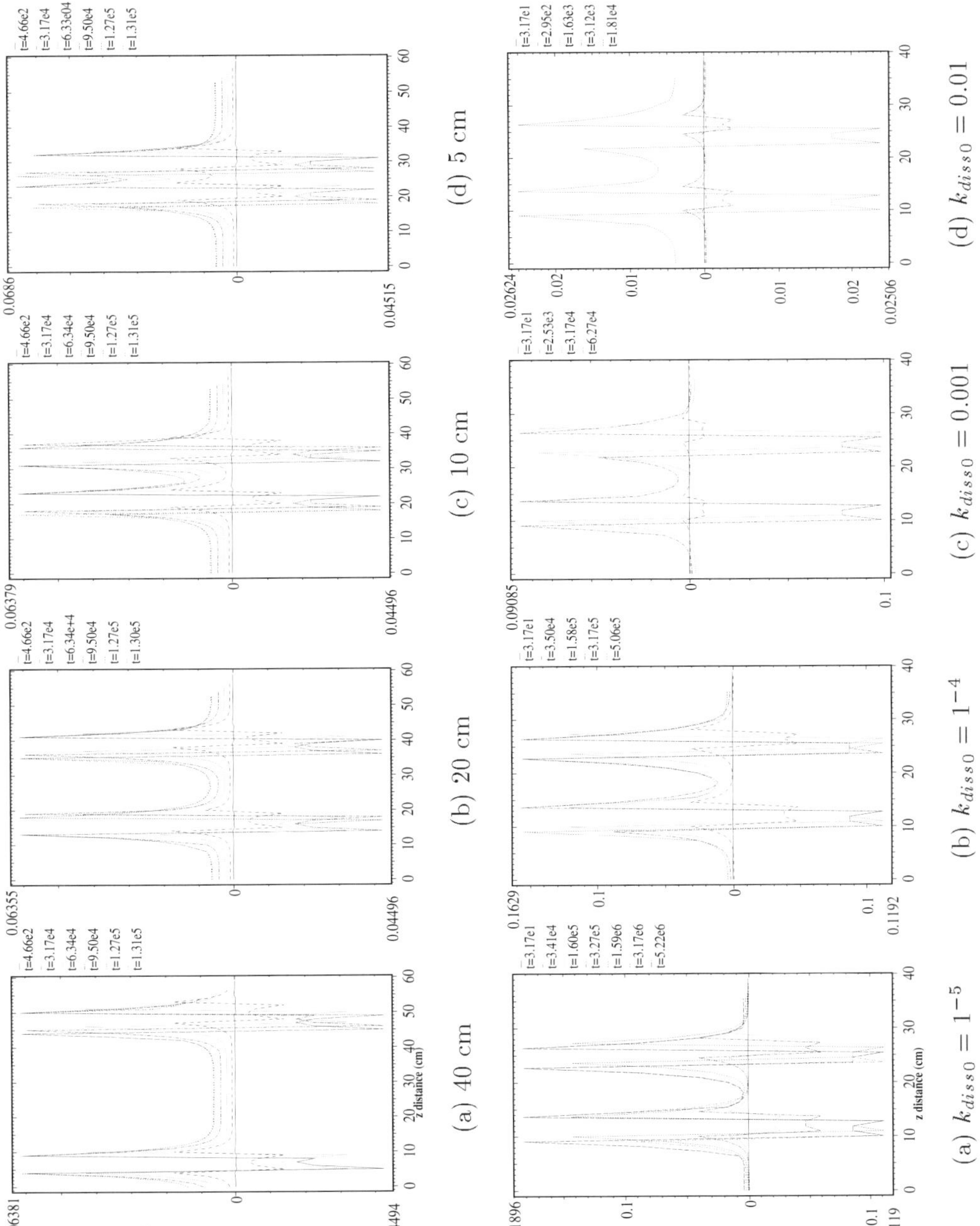

Fig. 7. Accumulated mass transport when the distance between the two layers containing the clay particles varies. It shows that if the two clay-rich layers which lose mass are located within the transport distance of each layer, intermediate areas gain mass from both layer. We also see that the distance between the two layers does not influence the amount of dissolved mass nor the transport distance.

Fig. 8. Accumulated mass transport for various values of the kinetic coefficient of dissolution. When the rate of dissolution decreases more mass dissolves and is transported out of the contact areas.

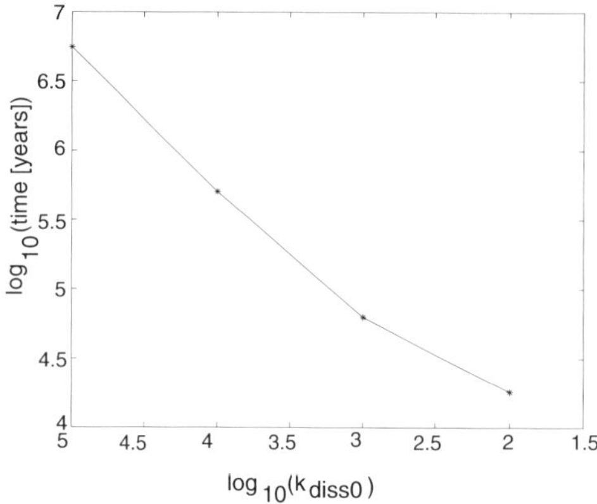

Fig. 9. The plot shows an almost linear log–log relationship between the time to reach the percolation threshold and the dissolution constant when dissolution is the rate-limiting process. The values are taken from Fig. 5.

Case III: Precipitation-limiting processes

In order to visualize the influence of precipitation we varied the dimensionless number V. Four different plots of the the accumulated mass transport for various values of V are shown in Fig. 11. The clay content in the two horizontal layers is 20% in all the plots.

When $V < V_0$ we have a situation where clay coatings or other factors reduce the grain surface such that the process of pressure solution becomes to a greater extent controlled by the precipitation. From Fig. 11(a)–(c) we see that when precipitation becomes the rate-limiting process the time required to reach the percolation threshold in the whole area increases. In Fig. 11(d) the precipitation rate is faster than the dissolution rate. There is almost no transport and the dissolved mass precipitates close to the clay layer. Such fast precipitation rate is not a probable situation under basin conditions: however it may be realistic during metamorphism when temperatures and stresses are much larger.

When the rate of precipitation becomes slower, the dissolved mass spreads out through the whole areas without clay. This is clearly shown in Fig. 11 as the gradient in accumulated mass away from the clay layers becomes flatter as V decreases.

From Fig. 10 it is clear that the distance over which material is transported is not affected by changes in the dissolution kinetics. Most of the mass has precipitated close to the areas where is was dissolved. In order to change this situation we need to slow down the precipitation rate (reduce V) as shown in Fig. 12.

As in Fig. 11, Fig. 12 demonstrates how the distance over which mass is transported increases as the rate of precipitation decreases, and the rate-limiting process changes from dissolution to precipitation ($V < V_0$).

In Fig. 13 we see how the distance between the two layers containing clays influences the transport pattern when precipitation is the rate-limiting process. Comparing this figure to Fig. 7 where contact diffusion is the rate-limiting process, Fig. 13 clearly shows how variations in the rate-controlling process influences the effect of clay particles.

Discussion

Both observations of natural rocks and laboratory experiments indicate that the presence of clays accelerate the rate of pressure-solution creep. Although neither the physical mechanism nor the chemistry involved is understood, we have modelled the possible consequences of a clay catalyst by introducing a simple phenomenological description of its possible effects on dissolution and diffusion rates at grain–grain contacts.

We have studied the effect of clays both in the case when the rate of pressure solution is controlled by inter-granular diffusion and where it is controlled by the precipitation rate. Figure 14 summarizes the various crossovers between a dissolution to a precipitation-controlled situation as the two dimensionless numbers are U and V varied. These two numbers control both the amount of exported cement and the distance over which this cement is transported.

Contact diffusion-limited compaction

Pressure-solution-controlled compaction of halite and carbonates or of quartz at depth larger than 3 km is known to be limited by grain boundary diffusion (Gratier and Guiguet 1986; Hickman & Evans 1991; Mullis 1993). In this case precipitation is rather fast and occurs in areas close to the areas of dissolution. If the clay content varies locally and enhances dissolution in particular layers, a diffusion-controlled profile in the mass gradient may develop between clay-rich and clay-free layers (see Fig. 5).

In this case the rate of pressure solution is independent of the rate of dissolution. Changes

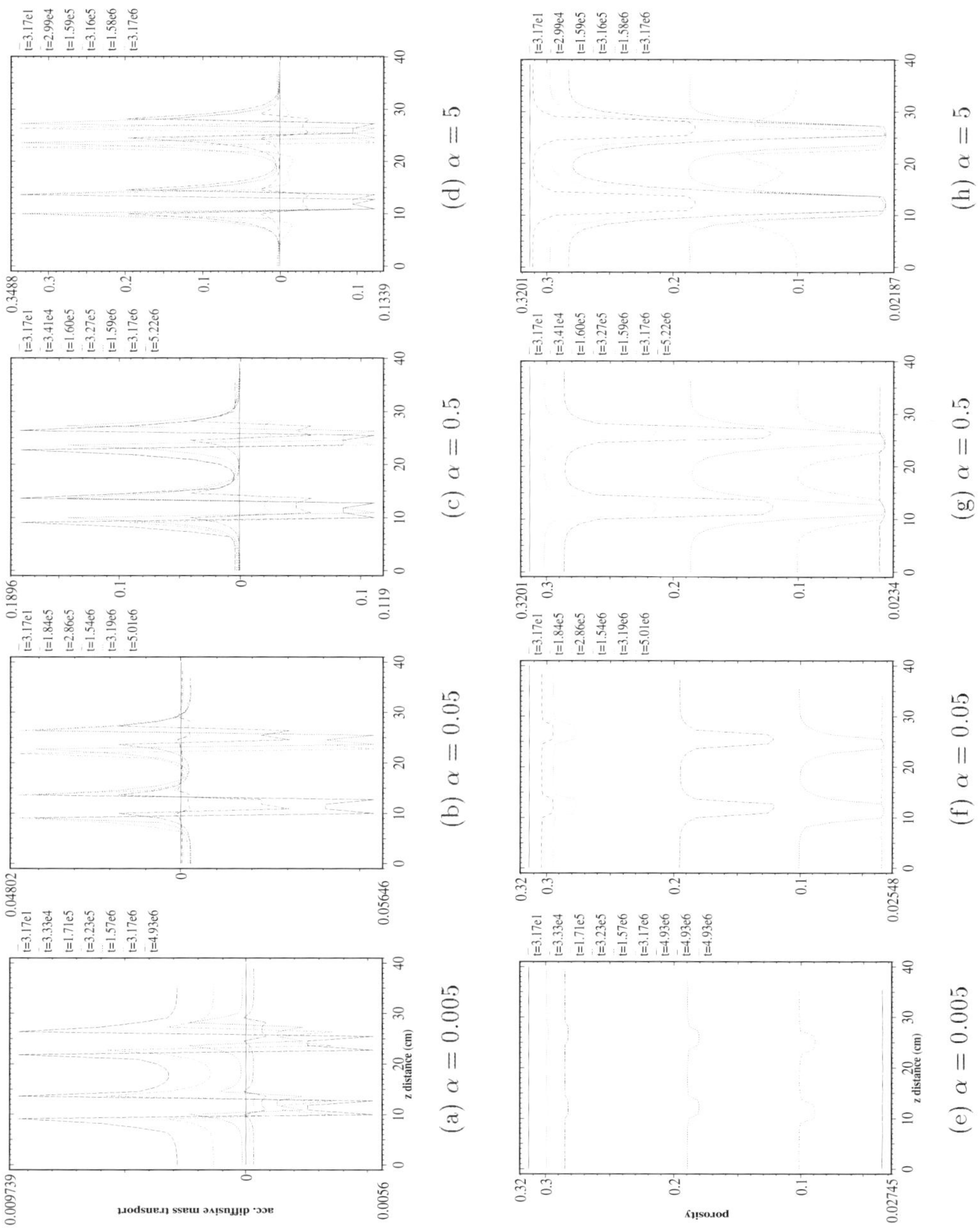

Fig. 10. Accumulated mass transport and porosity evolution for various values of α. A large α results in a fast dissolution rate in the clay-rich layers. Then, during compaction there will be large differences in porosity between the layers containing clay and the layers without clay. If the dissolution rate is fast enough the clay layers eventually collapse and become a thin clay layer with an irregular shape, see Figures (**d**) and (**h**).

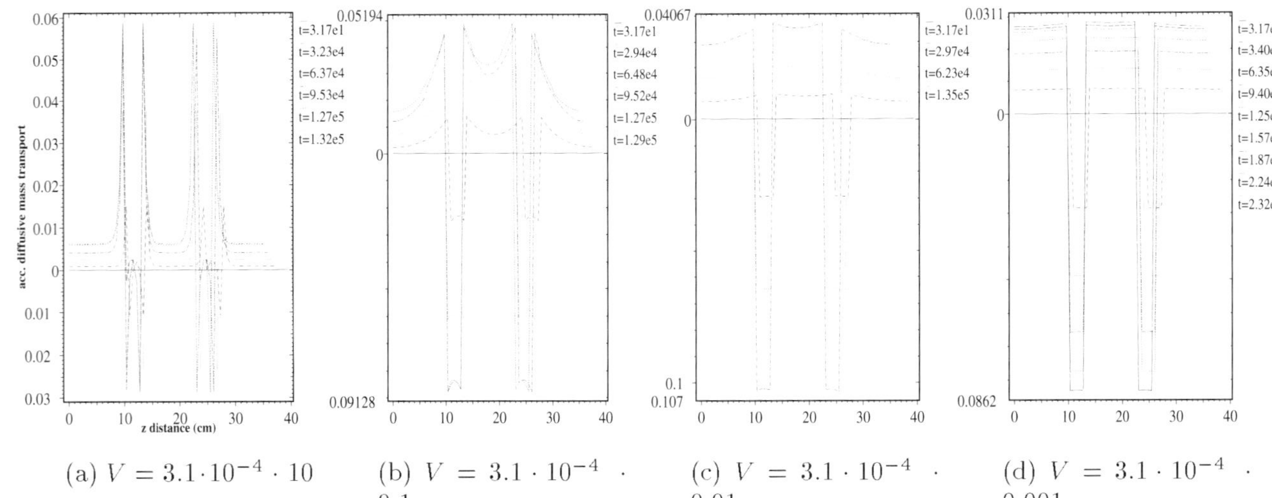

(a) $V = 3.1 \cdot 10^{-4} \cdot 10$ (b) $V = 3.1 \cdot 10^{-4} \cdot 0.1$ (c) $V = 3.1 \cdot 10^{-4} \cdot 0.01$ (d) $V = 3.1 \cdot 10^{-4} \cdot 0.001$

Fig. 11. Accumulated mass transport for various values of the dimensionless constant V. As the precipitation rate becomes slower, the distance over which mass is transported increases and the dissolved mass spreads out through the whole areas without clays. The figure demonstrates the crossover from diffusion to precipitation control.

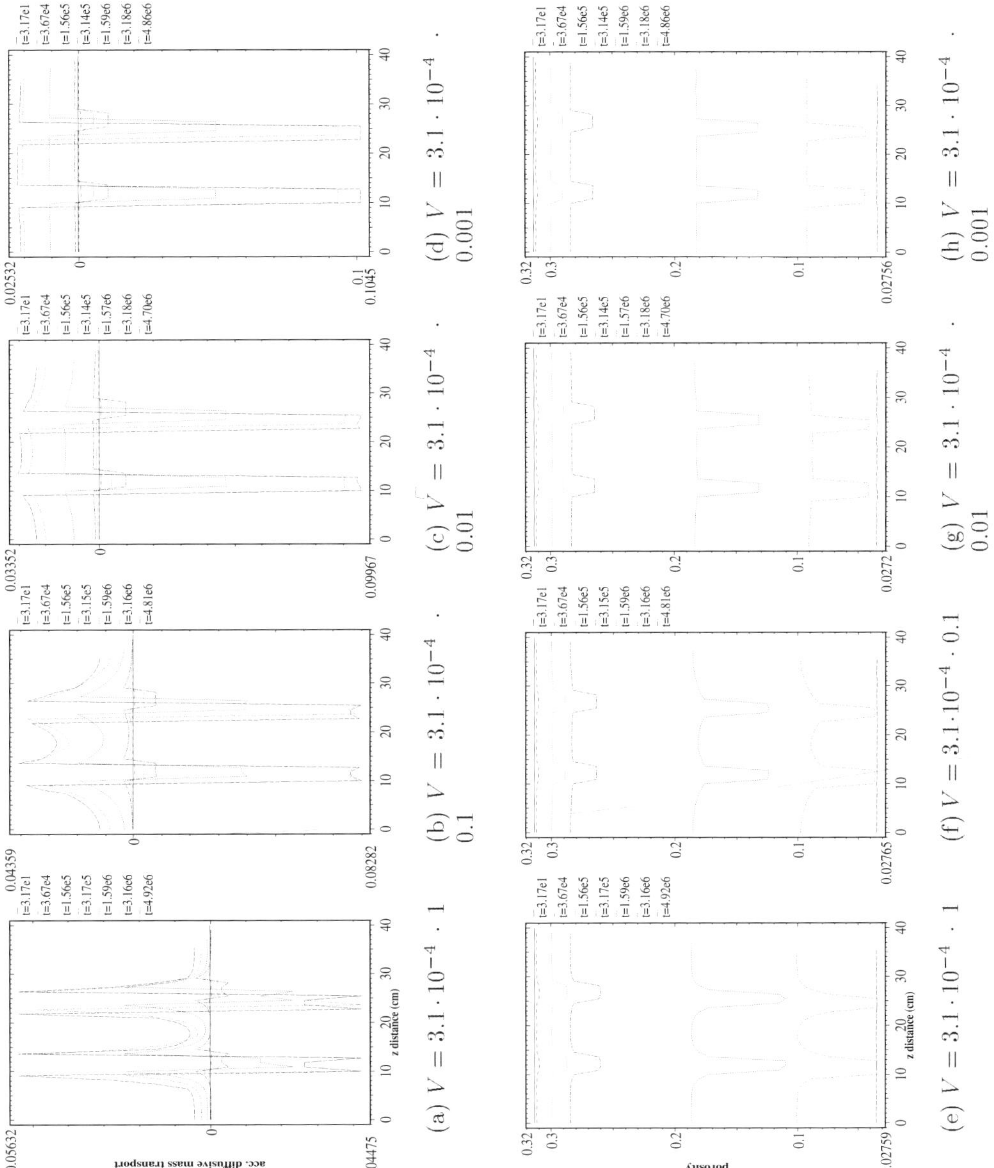

Fig. 12. Accumulated mass transport and porosity evolution for various values of V. The plots show how the distance over which mass is transported increases as the rate of precipitation kinetics decrease. The figure demonstrates the crossover from dissolution to precipitation control.

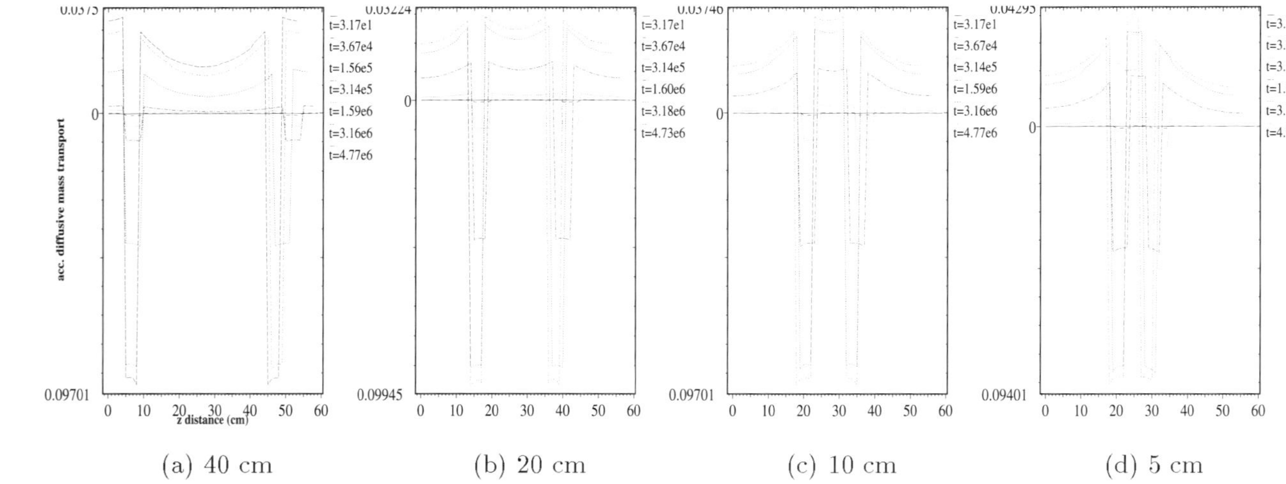

(a) 40 cm (b) 20 cm (c) 10 cm (d) 5 cm

Fig. 13. Accumulated mass transport when the distance between the two layers containing the clay particles vary and $k_{prec0} = 0.01$.

in the diffusion rate out of the contact controls the rate of the process. However the total amount of mass dissolved as well as the distance over which mass has been transported at the end of the process ($\phi = 3.5\%$) is constant.

Dissolution-limited compaction

If dissolution kinetics are much slower than diffusion and precipitation, contact stress might be the only important parameter and no significant transport over long distances can occur. This might be the case for quartz at shallow burial depths, where the temperature is low (80–90 °C).

In our model dissolution is the rate-limiting process when $U < U_0$ and when the surface area on the grain available for precipitation is larger than a given fraction of the contact surfaces (i.e. $V > V_0$). As seen in Fig. 8, changes in the kinetic coefficient of dissolution do not influence the distance over which dissolved mass is transported. However, if the dissolution kinetics are much faster in the clay areas than in the pure silica layers, the silica in contact with the clay particles dissolves quickly and the layers containing clays collapse and reduce to pure clay layers: see Fig. 10.

If dissolution is slower than precipitation, we have a situation where a dissolved molecule precipitates quickly and there is limited time available for transport: see Fig. 11(a).

Precipitation-limited compaction

When precipitation is the rate-limiting process we have the situation described by Bjørkum (1996), Walderhaug (1996), and Wangen (1999). In their models, quartz is dissolved only at regions with higher clay content (stylolites). The dissolved silica is then transported into the inter-stylolite region by diffusion where it precipitates homogeneously as cement. The precipitation rate is controlled by temperature, grain surface area and supersaturation. Geological environments corresponding to such a situation could be domains where precipitation is prevented by clay coating.

In our model the precipitation is balancing the global diffusion, which indicates that it will always globally affect the process of pressure solution. If precipitation is slower than the diffusion in the pore space, then precipitation is the rate-limiting process of pressure solution. Limitations on the grain surface areas available for precipitation could lead to this situation. Figures 11 and 12 show the crossover from either dissolution or local diffusion to a precipitation-controlled situation. It is seen that when precipitation is the rate-limiting process the whole process slows down and dissolved material is transported over larger distances. When the precipitation rate is slow, the concentration of dissolved material has sufficient time to homogenize on a global scale and there is almost no gradient of concentration between the clay layers and their surroundings. Moreover, the rock has enough time to adapt the changes given by the compaction process. Rapid and unstable changes which might result in patterns given by stylolites seem to be a unrealistic outcome under these circumstances.

Compaction of sediments

In sedimentary environments, deposition usually results in some variations in grain size or mineralogy. This leads to the horizontal layering of sedimentary rocks. During compaction, stress concentration at grain contacts drives dissolution and diffusion in the pore fluid. The result is that gradients of solutes can develop during diagenesis in sedimentary rocks or in any porous rocks that undergo compaction (as for example in a fault gouge). This process induces mass transfer at the scale of centimetres to decimetres. As a result the initial (small) heterogeneity of the rock is amplified through a self-organizing process (Ortoleva 1994). Moreover, this feedback is self-enhancing because the heterogeneities become more and more contrasted.

Experimental works on pressure solution (Dewers & Hajash 1995) indicate that there is a non-linear relationship between the stress and the strain rate, whereas classical theory proposes a linear relationship. This difference is of importance because it controls the pressure-solution-induced strain-hardening of the rocks and thereby the rheology properties of the crust. Such mechanical properties represent future areas for our numerical work.

Conclusions

In this article we have studied the possible role of clay particles during compaction of a quartz-rich sandstone. By using a pressure-solution model based on thermodynamic relationships, the fluid film model and ordinary diffusion in a porous media, we attempt to define under

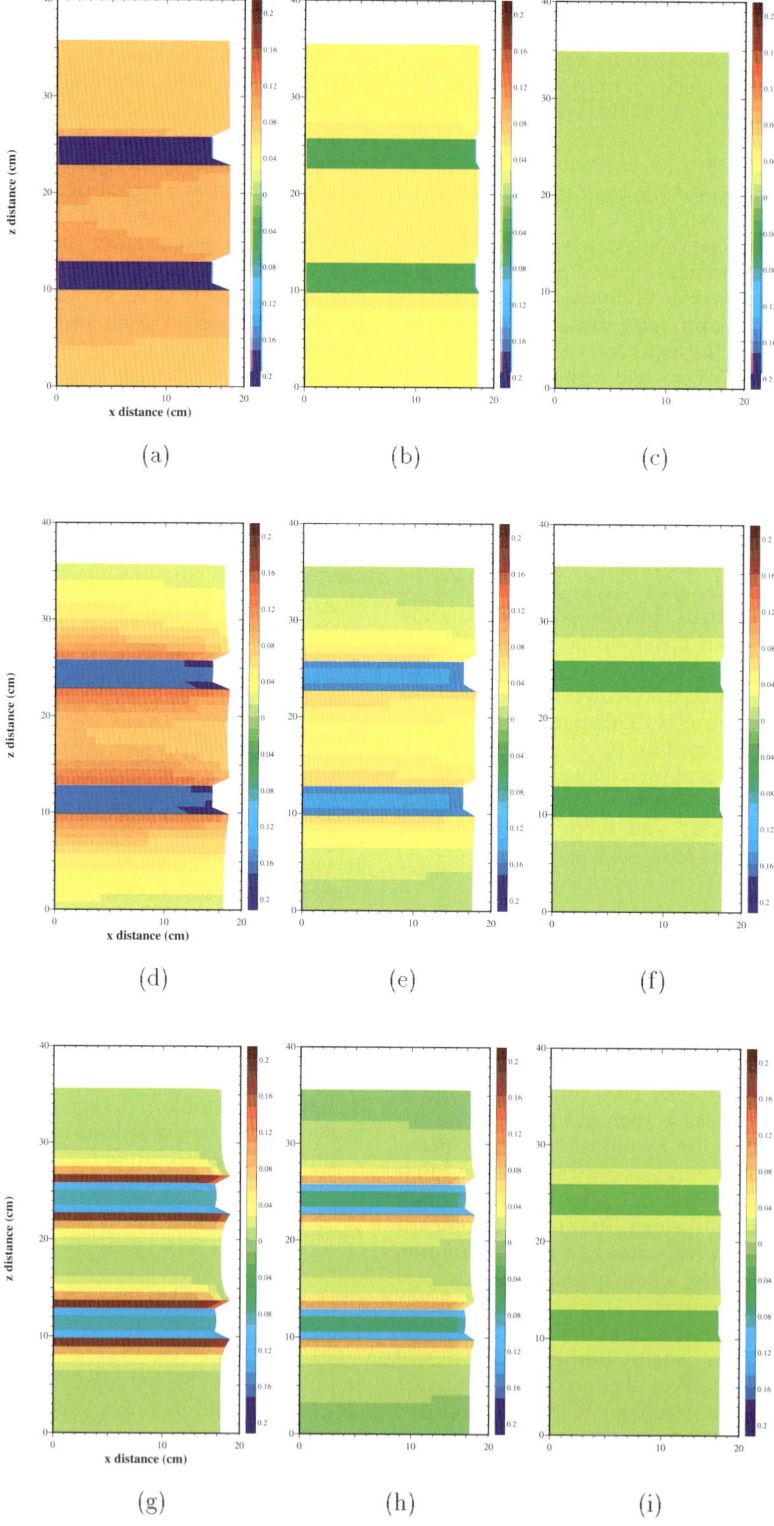
z distance (cm)
x distance (cm)
0
10
20
30
40
0.2
0.16
0.12
0.08
0.04
(a)
(b)
(c)
(d)
(e)
(f)
(g)
(h)
(i)

which conditions dissolution, contact diffusion or precipitation will act as the rate-limiting process.

Dissolution is a local process governing the amount of mass which dissolves at the grain–grain contact. Contact diffusion affects both local and global processes as it governs the rate of mass transport into the pore volume. Finally, precipitation controls global mass transport by limiting the fluid supersaturation. Using the dimensionless number U we have a clear definition of when dissolution or contact diffusion is the local rate-limiting process. However, the rate of precipitation always affects the distance over which mass is transported, and when the rate of precipitation becomes slower than the rate of global diffusion the process of pressure solution becomes precipitation controlled on a global scale. This situation is given by the dimensionless number V.

The aim of this study is to demonstrate how clay particles in contact with quartz might affect the process of pressure solution. During this work we first needed to analyse how variations in the rate-limiting processes influence the process of pressure solution even in the absence of clay.

The main results of these numerical simulations is to constrain how the various geological parameters affect pressure-solution creep in a true coupled system. We can then decipher under which conditions mass transfer may occur at the local scale in porous rocks. By using the result obtained from this study in combination with detailed petrographic studies on the extent of cementation in the various parts of a layered sandstone (for example by application of cathodoluminescence) we should be better equipped to identify the rate-limiting process during compaction of a rock sample and its consequences for the detailed dynamics of pressure solution.

We would like to thank B. Bos and an anonymous reviewer for several thoughtful comments which improved the quality of the paper. The project has been supported by the Norwegian Research Council through grant no. 113354/420 to the Fluid Rock Interaction group, the Centre National de la Recherche Scientifique (INSU Mi-Lourds), the GDR Corinthe and the PAI Aurora (France-Norway Collaboration).

References

APPLIN, K. R. 1987. The diffusion of dissolved silica in dilute aqueous solutions. *Geochimica et Cosmochimica Acta*, **51**, 2147–2151

BJØRKUM, P. A. 1996. How important is pressure solution in causing dissolution of quartz in sandstones? *Journal of Sedimentary Research*, **66**, 147–154.

DE BOER, R. B., NAGTEGAAL, P. & DUYVIS, E. 1977. Pressure solution experiments on quartz sand. *Geochimica et Cosmochimica Acta*, **41**, 257–264.

DEWERS, T. & HAJASH, A. 1995. Rate-laws for water-assisted compaction and stressed-induced water-rock interaction in sandstones. *Journal of Geophysical Research*, **100**, 13093–13112.

DEWERS, T. & ORTOLEVA, P. 1991. Influence of clay minerals on sandstone cementation and pressure solution. *Geology*, **19**, 1045–1048.

DOVE, P. M. 1994. The dissolution kinetics of quartz in sodium chloride solutions at 25 °C to 300 °C. *American Journal of Science*, **294**, 665–712.

FEDER, J. 1988. *Fractals*, Plenum.

GILES, M. R., INDRELID, S. L., BEYNON, G. V. & AMTHOR, J. 2000. The Origin of Large-Scale Quartz Cementation: Evidence From Large Data Sets and Coupled Heat-Fluid Mass Transport Modelling. *In*: WORDEN, R. & MORAD, S. (eds) *Quartz cementation in sandstones*. Blackwell Science, 21–38.

GRATIER, J. P. & GUIGUET, R. 1986. Experimental pressure solution-deposition on quartz grains: the crucial effect of the nature of the fluid. *Journal of Structural Geology*, **8**, 845–856.

GUNDERSEN, E., RENARD, F., DYSTHE, D., BJØRLYKKE, K. & JAMTVEIT, B. 2002. Coupling between pressure solution and mass transport in porous rocks. *Journal of Geophysical Research*, in press.

HICKMAN, S. H. & EVANS, B. 1991. Experimental pressure solution in halite: the effect of grain interphase boundary structure. *Journal of the Geological Society, London*, **148**, 549–560.

HOUSEKNECHT, D. W. 1987. Intergranular pressure solution in four quartzose sandstones. *Journal of Sedimentary Geology*, **58**, 228–246.

KAMB, W. B. 1961. The thermodynamic theory of non-hydrostatically stressed solids. *Journal of Geophysical Research*, **66**, 259–271.

LANGTANGEN, H. P. 1999. *Computational Partial Differential Equations*. Springer, Berlin.

LEHNER, F. K. 1995. A model for intergranular pressure solution in open systems. *Tectonophysics*, **245**, 153–170

MULLIS, A. M. 1993. Determination of the rate-limiting mechanism for quartz pressure dissolution, *Geochimica et Cosmochimica Acta*, **57**, 1499–1503.

Fig. 14. Accumulated mass transport for various combinations of the dimensionless numbers U and V. Starting with plot (a) the plots below represents increasing V values, while the two plots on the right side, represent increasing U values. The figure demonstrates the crossover between a dissolution-controlled to a precipitation-controlled situation where the effect of clay is constant. As V increases, the cemented areas become more and more localized close to the clay-rich layers where most of the dissolution occurred. As U increases the compaction of the whole area becomes more homogeneous.

NAKASHIMA, S. 1995. Diffusivity of ions in pore water as a quantitative basis for rock deformation rate estimates. *Tectonophysics*, **245**, 185–203.

OELKERS, E. H., BJØRKUM, P. A., NADEAU, P. H. & WALDERHAUG, O. 2000. Making diagenesis obey thermodynamics and kinetics: the case of quartz cementation in sandstones from offshore mid-Norway. *Applied Geochemistry*, **15**, 295–309.

ORTOLEVA, P. 1994. *Geochemical Self Organization*. Oxford University Press, New York.

PATERSON, M. S. 1973. Nonhydrostatic thermodynamics and its geologic applications. *Reviews of Geophysics and Space Physics*, **11**, 355–389.

RENARD, F., BROSSE, E. & GRATIER, J.-P. 2000. Quartz cementation in sandstones. *In*: *The Different Processes Involved in the Mechanism of Pressure Solution in Quartz-rich Rocks and Their Interactions*. Blackwell Science, pp. 67–78.

RENARD, F., DYSTHE, D. K., FEDER, J., BJØRLYKKE, K. & JAMTVEIT, B. 2001. Enhanced pressure solution creep rates induced by clay particles: experimental evidence from salt aggregate. *Geophysical Research Letters*, **28**, 1295–1298.

RENARD, F., ORTOLEVA, P. & GRATIER, J. P. 1999. An integrated model for transitional pressure solution in sandstones. *Tectonophysics*, **312**, 97–115.

RUTTER, E. H. 1976. The kinetics of rock deformation by pressure solution. *Philosophical Transactions of the Royal Society of London*, **283**, 203–219.

SCHUTJENS, P. M. 1991. Experimental compaction of quartz sand at low effective stress. *Journal of the Geological Society, London*, **148**, 527–539.

SPIERS, C. J. & BRZESOWSKY, R. H. 1993. Densification behaviour of wet granular salt: theory versus experiment. *Seventh Symposium of Salt*, **1**, 83–91.

WALDERHAUG, O. 1996. Kinetic model of quartz cementation and porosity loss in deeply buried sandstone reservoirs. *AAPG Bulletin*, **80**, 731–745.

WANGEN, M. 1999. Modelling quartz cementation of quartzose sandstones. *Basin Research*, **11**, 113–126.

WEYL, P. K. 1959. Pressure solution and the force of crystallization – a phenomenological theory. *Journal of Geophysical Research*, **69**, 2001–2025.

Compaction creep of quartz–muscovite mixtures at 500 °C: Preliminary results on the influence of muscovite on pressure solution

A. R. NIEMEIJER & C. J. SPIERS

HPT Laboratory, Faculty of Earth Sciences, Utrecht University, P.O. Box 80.021, 3508 TA, Utrecht, The Netherlands (e-mail: niemeyer@geo.uu.nl)

Abstract: It is widely claimed that the presence of phyllosilicates in sandstones increases intergranular pressure solution (IPS) rates in these rocks. However, this has not been experimentally confirmed. This study reports the results of isostatic hot-pressing compaction experiments at a temperature of 500 °C and an effective pressure of 100 MPa on mixtures of quartz and muscovite. Previous work has shown that under these conditions dissolution is rate controlling in pure quartz. No acceleration of compaction rates of quartz by the addition of muscovite was observed. Instead, a modest decrease in compaction rates was observed (factor 3–10), which we infer was due to a decrease in IPS rate. The effect of muscovite in slowing IPS may be due to the influence of dissolved aluminum (Al^{3+}) dominating over any accelerating effects of alkali-metal cations. From the geochemical literature, Al^{3+} in solution is expected to decrease the solubility, dissolution rates and precipitation rates of quartz. However, the effect of the addition of muscovite on IPS rates in quartz when controlled by diffusion or precipitation may be different. Experiments should be conducted on quartz sand under conditions where diffusion or precipitation is rate controlling to investigate these possible effects.

From microstructural evidence, it is widely accepted that intergranular pressure solution (IPS) is an important compaction, lithification and deformation mechanism in clastic sediments under diagenetic and low-grade metamorphic conditions (Rutter 1983; Tada & Siever 1989). IPS exerts a strong influence on porosity and permeability reduction in sandstones and mudstones, and hence on reservoir and top-seal evolution in sedimentary basins (Addis & Jones 1985; Houseknecht 1987; Rutter & Wanten 2000). It is also likely to play an important role in controlling strength recovery and seismic versus frictional–viscous slip on gouge-bearing crustal faults (Janssen *et al.* 1997; Bos *et al.* 2000*a*,*b*; Renard *et al.* 2000; Bos & Spiers 2000, 2001).

Despite the importance of IPS in siliciclastic sediments and quartz-phyllosilicate fault gouges, the rate-controlling mechanism and kinetics of IPS in quartz are still poorly constrained, principally because of the slowness of the process under lab conditions. Recent experimental work has suggested that when there is no long-range advective or diffusive transport, IPS in pure quartz sand is probably controlled by the kinetics of quartz dissolution, at least at porosities down to *c.* 15% (Niemeijer *et al.* 2002). Regarding the effects of phyllosilicates, numerous field and petrographic associations between phyllosilicate content and enhanced IPS or pressure solution seams (e.g. Heald 1959; Rutter 1983; Engelder & Marshak 1985; Tada & Siever 1989; Renard *et al.* 1997) have led to the view that phyllosilicates dramatically increase IPS rates in quartz, presumably via enhanced grain boundary diffusion or enhanced reaction kinetics. However, the influence of phyllosilicates on IPS rates in quartz has remained largely unexplored from both the experimental and the theoretical points of view.

Although few experiments have been carried out on the influence of phyllosilicates on IPS, two recent studies indicate strong enhancement effects. Hickman & Evans (1995) studied pressure solution rates at halite–silica contacts by pressing a convex halite lens against a fused silica plate under brine. One experiment was conducted in which the silica plate was coated with a 0.8 µm thick film of Na-montmorillonite. The presence of the clay film produced an approximately five-fold increase in pressure solution convergence rate over clay-free experiments conducted under otherwise identical conditions. Assuming diffusion to be rate limiting, as observed in their halite–silica experiments, Hickman & Evans (1995) calculated an effective grain boundary (clay layer) diffusivity about 30 times greater than for the halite–silica interfaces. Montmorillonite was used as the phyllosilicate phase because of its propensity for extreme swelling through incorporation of interlayer water. The observed increase in effective grain boundary diffusivity per unit clay layer thickness is therefore likely to represent the maximum effect of phyllosilicates on pressure solution rate enhancement, when diffusion is rate controlling. However, in other materials such as quartz, where interface

From: DE MEER, S., DRURY, M. R., DE BRESSER, J. H. P. & PENNOCK, G. M. (eds) 2002. *Deformation Mechanisms, Rheology and Tectonics: Current Status and Future Perspectives*. Geological Society, London, Special Publications, **200**, 61–71. 0305-8719/02/$15 © The Geological Society of London.

reactions may be rate controlling, where different surface forces may come into play at contacts with phyllosilicates, and where surface reactions may be influenced by the chemical effects of the phyllosilicates present, the influence of phyllosilicates on IPS may be quite different.

In the second recent study, Rutter & Wanten (2000) investigated IPS rates by hot isostatic pressing of samples consisting of granular quartz embedded in a matrix of fine-grained illite and muscovite, with a small adjacent volume of porous quartz sand to provide a sink for precipitation of silica dissolved from the phyllosilicate-quartz mixture. The presence of phyllosilicates in the matrix strongly enhanced compaction rates compared to pure quartz samples. Rutter & Wanten argue that this was caused by enhancement of pressure solution of the quartz at quartz/phyllosilicate contacts and by a strain contribution due to intracrystalline plasticity plus frictional flow in the phyllosilicates. However, they give no suggestions about the rate-limiting process controlling IPS. Their experiments are indeed difficult to interpret since, compared with the pure quartz sand samples, the introduction of the phyllosilicate matrix leads to changes in the stresses transmitted across quartz grain surfaces (and hence in the driving force for IPS), in the source-sink diffusion path, in the areas available for quartz dissolution and precipitation, in the chemistry and pH of the pore fluid and probably in the state of the intergranular fluid. As all of these factors can potentially influence the rate and/or rate-controlling mechanism of IPS, the exact effect of the phyllosilicates on IPS in quartz is hard to deduce from the experiments of Rutter & Wanten (2000).

It is thus evident that while there are ample indications in nature for an influence of phyllosilicates on IPS rates in quartz-rich rocks, the effect has yet to be confirmed, mechanistically identified, or systematically characterized in laboratory experiments. This preliminary study consists of compaction experiments on quartz–muscovite mixtures at a confining pressure of 300 MPa, a pore fluid pressure of 200 MPa and a temperature of 500 °C, and extends our previous study of IPS in pure quartz sand under similar conditions (Niemeijer *et al.* 2002). In that study, we found that IPS dominated over both cataclasis and dislocation creep, and we concluded that IPS rates were controlled by the quartz dissolution reaction at grain contacts. Here, we attempt to examine the effect of adding sufficient muscovite to coat grain contacts, without filling all the intergranular pore space, by four preliminary experiments.

Background considerations

In a chemically closed system (no removal or addition of solid mass by long-range diffusive or advective transport in the pore fluid) compaction of a fluid-saturated granular aggregate by IPS involves dissolution of solid material at stressed grain-to-grain contacts, diffusion of this material through the intergranular fluid into the open pores, followed by precipitation on the free pore walls. The process is driven by differences in effective normal stress, hence the normal component of the solid chemical potential, between grain contacts and free pore walls (e.g. Lehner 1990). Under steady-state conditions, any of the three mechanisms of dissolution, diffusion and precipitation may be rate controlling (de Meer & Spiers 1995). If we wish, as in the present study, to investigate the influence of phyllosilicates on IPS rates in quartz, it is therefore useful to first consider what, in theory, their influence may be on the three serial processes of IPS. These considerations form the basis for our choice of experimental method.

We begin by briefly considering the physical and chemical properties of phyllosilicates, especially their surface properties in the presence of water. Phyllosilicates are characterised by $[SiO_4]^{4-}$ tetrahedra linked together to form flat sheets with the composition $[Si_4O_{10}]_n$. This group of minerals includes muscovite, biotite, phlogopite, chlorite, clays, talc and serpentine, all of which are soft, plastically deformable minerals of variable but generally low density, and which always contain a specific amount of structurally bound water. Muscovite, the mineral used in this study, belongs to the mica sub-group and is characterized by a 2:1 layering, i.e. a periodic structure consisting of two tetrahedral sheets $[Si_4O_{10}]_n$ and one octahedral $[Al_2(OH)_6]_n$ sheet repeated throughout the mineral structure (Deer *et al.* 1977). Muscovite (and mica in general) always contains a variable amount of structurally bound water, in the form of hydroxyl ions in the octahedral sheets (Deer *et al.* 1977). Cation substitution in the tetrahedral sheets (i.e. Al^{3+} for Si^{4+}), results in an overall negative charge on the tetrahedral layers. The charge is compensated, and the layers bonded together by large, positively charged interlayer cations, most commonly K^+, Na^+ or Ca^{2+}. However, as a result of substitution of Si^{4+} by Al^{3+} in the tetrahedral sheets, freshly cleaved muscovite (and mica in general) has a negative surface charge under vacuum. In aqueous solution protons will sorb to the surface of the muscovite to balance the negative charge. The pH at the point of zero charge (pH_{pzc}, i.e. the pH of the

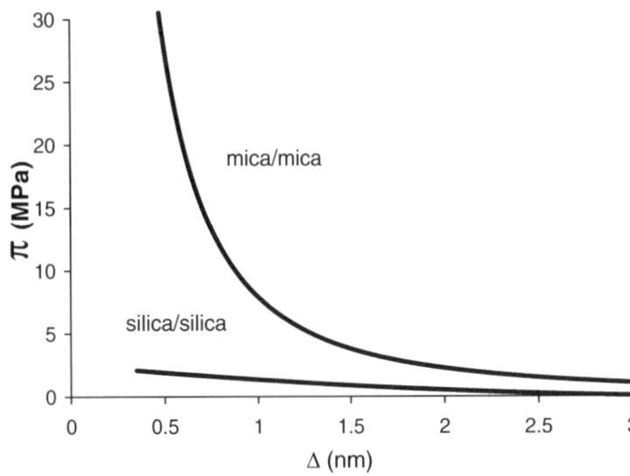

Fig. 1. Disjoining pressure π for a water film of thickness Δ confined between parallel silica and mica surfaces. The disjoining pressure is the difference between the pressure acting on the solid and the fluid pressure; it is similar to an effective stress (after Heidug 1995).

solution at which the surface is electrically neutral) of muscovite in an aqueous solution has been calculated to be 6.6 (Sverjensky 1994).

When diffusion is the rate-controlling process of IPS, the effect of adding phyllosilicates on IPS rates in quartz is expected to be one of accelerating pressure solution (Hickman & Evans 1995; Renard *et al.* 1997; Rutter & Wanten 2000). This expectation is mainly based on measurements of the disjoining pressure acting between silica–silica and mica–mica surfaces (Fig. 1). For a flat, interfacial fluid film, the disjoining pressure equals the amount by which the fluid pressure acting on the fluid–solid interface exceeds the hydrostatic pressure in the bulk fluid. In the case of films between similar materials, it is believed to be caused mainly by hydration forces (i.e. forces which result from the short range ordering of water molecules along the solid–liquid interface) rather than van der Waals interactions or electrical double-layer repulsion, since these cannot explain the magnitude of the measured dissoining pressure (Heidug 1995; Hickman & Evans 1995; Renard & Ortoleva 1997). In the case of stressed granular solids, the disjoining pressure in an interfacial fluid film balances the effective compressive stress to maintain an open, wetted grain boundary with a stress-dependent thickness. The various measurements on silica–silica and mica–mica interfaces imply that quartz–mica interfaces can support a thicker layer of water in the grain boundary than quartz–quartz interfaces at a given normal stress. At a film thickness of 1 nm a mica–mica interface can support ~10 times more stress than a silica–silica interface: see Fig. 1 (Pashley & Israelachvili 1984; Heidug 1995; Renard & Ortoleva 1997; Renard *et al.* 1997). This is because, at a quartz–quartz interface, van der Waals forces are attractive and the surface charge is low, resulting in only weak ordering of water molecules and thus in weak hydration forces. At a quartz–mica interface however, van der Waals forces are repulsive and the surface charge (of the mica) is high, resulting in strong ordering of water molecules and thus in strong hydration forces, ultimately leading to a higher disjoining pressure and fluid film thickness than at a quartz–quartz interface at similar stresses. Nonetheless, at sufficiently high intergranular stresses, water will be squeezed out, even from a quartz–mica interface (Israelachvili 1992; Heidug 1995).

When dissolution or precipitation is the rate-controlling process, the influence of muscovite on pressure solution rate can be chemical or physical. First, the presence of muscovite in quartzitic rocks changes the chemistry of the pore fluid by (in)congruent dissolution of the muscovite (a typical dissolution rate at 70 °C and a pH of 6.2 is $\approx 10^{-16}$ mol/cm^{2}s: Knauss & Wolery 1989). This releases cations (e.g. K^+, Al^{3+}) into the solution and/or modifies the pH of the pore fluid. When the pore fluid contains alkali-metal or alkali-earth ions, or has a non-neutral pH, the solubility and interfacial reaction rates of quartz are increased by typically one order of magnitude, depending on the concentrations of the cations (Dove & Rimstidt 1994; Renard *et al.* 1997; Dove & Nix 1997; Dove 1999). This will increase the IPS dissolution rate, the diffusive transport rate and the precipitation rate, so that IPS rates will increase. However, the presence of a minute amount of aluminum in the pore fluid (as low as 0.15 ppm equivalent of Al_2O_3) can strongly decrease the apparent solubility of quartz as well as dissolution and precipitation rates (Iler 1973, 1979; Mullis 1993; Dove & Rimstidt 1994), presumably by adsorption of Al^{3+} onto quartz surfaces. The magnitude of the decrease in dissolution–precipitation rates, however, remains unclear. Estimates vary from 3 to as much as 8 orders of magnitude, but the effect may be much smaller (*c.* 1 order of magnitude: Dove, pers. comm.). Moreover, the presence of muscovite in pore spaces, and on pore walls, decreases the area available for precipitation on the quartz surface (Tada & Siever 1989; Dewers & Ortoleva 1991; Mullis 1993), resulting in a decrease in overall precipitation rate. Thus, the presence of phyllosilicates in general, and muscovite in particular, can increase or decrease IPS rates in quartz. In addition, the reaction mechanism at a quartz–muscovite interface may be completely different from that at a quartz–quartz interface, due to a different surface charge, a different pH and the presence of cations such as Al^{3+} and K^+.

Finally, it is important to note that the presence of phyllosilicates may cause a change in the rate-controlling mechanism of IPS, compared with a pure quartz rock, through any of the above-mentioned effects on dissolution, diffusion and precipitation rates.

Experimental method

Our experiments consisted of isostatic compaction or hot-pressing (HIPing) experiments carried out on wet quartz sand with 0–20 wt% added muscovite. The temperature used was 500 °C, the confining pressure was 300 MPa, the pore fluid pressure 200 MPa and the effective pressure was thus 100 MPa. The experiments were carried out using the same starting material as used by Niemeijer *et al.* (2002) in their study of IPS in quartz, namely quartz sand from the Miocene 'Bolderiaan' formation, Belgium. The material had previously been refined in a ball mill, ultrasonically cleaned in distilled water and etched in HF solution to remove surface damage (Schutjens 1991). We sieved the product to obtain a starting grain size fraction of 45–75 μm. Muscovite additive was prepared using natural muscovite from Norway. This was crushed in a ball mill and the resulting powder sieved to obtain fractions of 45–90 and <28 micrometer. These were in turn mixed with the sieved quartz sand, in different ratios, to obtain 5, 10 and 20 wt% muscovite-quartz samples, in addition to several pure quartz samples.

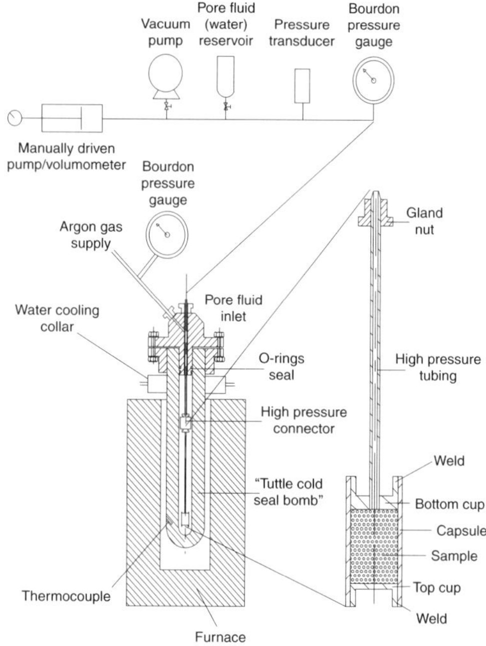

Fig. 2. Schematic diagram of the experimental apparatus with an enlargement showing the capsule/sample set-up.

Apparatus and procedure

The apparatus used is described by Niemeijer *et al.* (2002). It consists of a cold seal 'Tuttle bomb' pressurized with Argon, plus an internal sample/capsule assembly linked to an independent pore fluid system with volumetric pump (Fig. 2). The pore fluid and Argon pressures were measured with a Bourdon-type pressure gauge (resolution ~1 MPa) and/or a pressure transducer (resolution ~0.2 MPa). The temperature of the sample was measured using a pre-calibrated K-type thermocouple embedded in the bomb wall (error of ~2 °C).

In each experiment, ~2 grams of quartz–muscovite mixture were loaded into the annealed copper capsule assembly via the 'top cup' end (enlargement Fig. 2). The capsule was then sealed by pressing and welding in the copper 'top cup' (Fig. 2), producing a sample porosity of ~45-50%. The sample assembly was subsequently loaded into the Tuttle bomb, evacuated and flooded with water from the pore fluid reservoir. Each sample was then cold isostatically pressed (CIPed) at a confining pressure of 300 MPa for 30 mins, with the pore pressure maintained close to 0.1 MPa (1 atm). This was done to produce a reproducible starting porosity and microstructure for the hot pressing stage (cf. Spiers & Brzesowsky 1993; de Meer & Spiers 1995) and to minimize strain due to grain rearrangement and cataclasis during HIPing at 100 MPa. The confining pressure was subsequently reduced to ~140 MPa and the pore fluid pressure raised to ~135 MPa using the volumometer pump. The sample was then heated at ~15 °C/min. During heating, the effective pressure was maintained as low as possible and was always <10 MPa. After attaining thermal equilibrium (~1 hour) at the desired test temperature (500 °C) and confining pressure (300 MPa), the required effective pressure was applied by extracting fluid from the sample using the volumometer. The confining and pore fluid pressures were subsequently kept within ±5 MPa of the desired values. The minimum measurable fluid volume increment was ~2 μl, yielding an absolute resolution in volumetric strain at room temperature of ~0.02 %.

Experiments were terminated when the pore volume loss was no longer detectable via changes in pore fluid pressure. At this stage, the bomb was quenched, using compressed air, at an average cooling rate of ~30 °C/min. The indurated sample was then removed from the apparatus and capsule and dried in an oven at 60 °C for 24 hours. The final porosity (ϕ_f) of the sample was measured by wrapping it in ultra-thin plastic film and determining its volume (V_f) using the Archimedes method and density data on quartz and muscovite (Deer *et al.* 1977). The samples were finally impregnated with epoxy resin and sectioned for optical and SEM study.

Data acquisition and processing

As indicated above, pore volume loss during the experiments was determined incrementally by measuring the volume of fluid extracted from the sample to maintain constant pore pressure. The volume of muscovite was assumed to be constant and thus of no direct influence on porosity reduction. The measurements were accurately corrected for density changes associated with the cooling of the pore fluid during extraction, using appropriate P–V–T equations for water (Burnham *et al.* 1969). The 'starting porosity' (ϕ_0) of the sample, i.e. before HIPing at the test temperature, was calculated by adding the total amount of fluid expelled from the sample during HIPing to the final pore volume (ϕ_f, V_f) of the sample. The difference in the pore and sample volumes, measured at room temperature and pressure compared to the post-HIPing volumes at test conditions, can be estimated using the bulk modulus of sandstone (~2.5–7.5 × 10^{11} Pa: Birch 1966) and the volumetric thermal expansion coefficient of quartz (~4–5 × 10^{-5} K^{-1}: Skinner 1966) and was found to be negligible. The overall relative standard error in volume measurements was approximately 5%, implying an absolute standard error in the porosity measurements of less than 0.2%. The volumetric strain rate associated with individual data was calculated using the two-point central difference method.

Dry control runs

Control HIPing experiments on 'dry' sand fractions were performed in the same apparatus as the wet tests, but using fully sealed capsules with no added water. The 'dry' samples were first CIPed at 300 MPa for 30 min and their porosities determined at atmospheric pressure using the Archimedes method. The samples were then HIPed at 500 °C and pore volume loss was measured periodically by removing the capsule from the bomb and re-determining its total (current) volume.

Results

Mechanical data

The complete set of experiments reported here is listed in Table 1, along with data on the starting and final porosities of the samples and the total volumetric strains achieved.The CIPed-only quartz sand sample (D9) showed compaction

Table 1. *List of experiments*

Sample number	T (K)	wt% muscovite	P_c (MPa)	P_f (MPa)	P_e (MPa)	ϕ_0 (%)	ϕ_f (%)	Total volumetric strain (%)
D 9	298	0	300	*c.* 0.1	*c.* 300	30.31*	30.31	–
D 17	773	0	60	'dry'	60	27.17*	19.65	9.06
CPf 3	773	0	300	200	100	27.31†	12.71	16.70
CPf 10§	773	10	300	200	100	24.46†	10.52	15.56
CPf 11§	773	20	300	200	100	24.53†	10.30	15.84
CPf 12‡,§	773	10	300	200	100	22.17†	8.31	15.07
CPf 13§	773	5	300	200	100	24.15†	9.51	16.00

List of experiments performed showing initial (pre-HIPing) porosities (ϕ_0), final porosities (ϕ_0), wt% muscovite and isostatic pressing conditions. T denotes test temperature, P_c confining pressure, P_f pore fluid pressure and P_e the applied effective pressure. 'Dry' means no water added.

*Starting or pre-HIPing porosity measured under atmospheric conditions, i.e. after CIPing.

†Starting or pre-HIPing porosity calculated for loaded conditions from the final porosity and porosity change determined during HIPing.

‡Fraction of < 28 μm muscovite used.

§Fluid pressure measured with high pressure transducer.

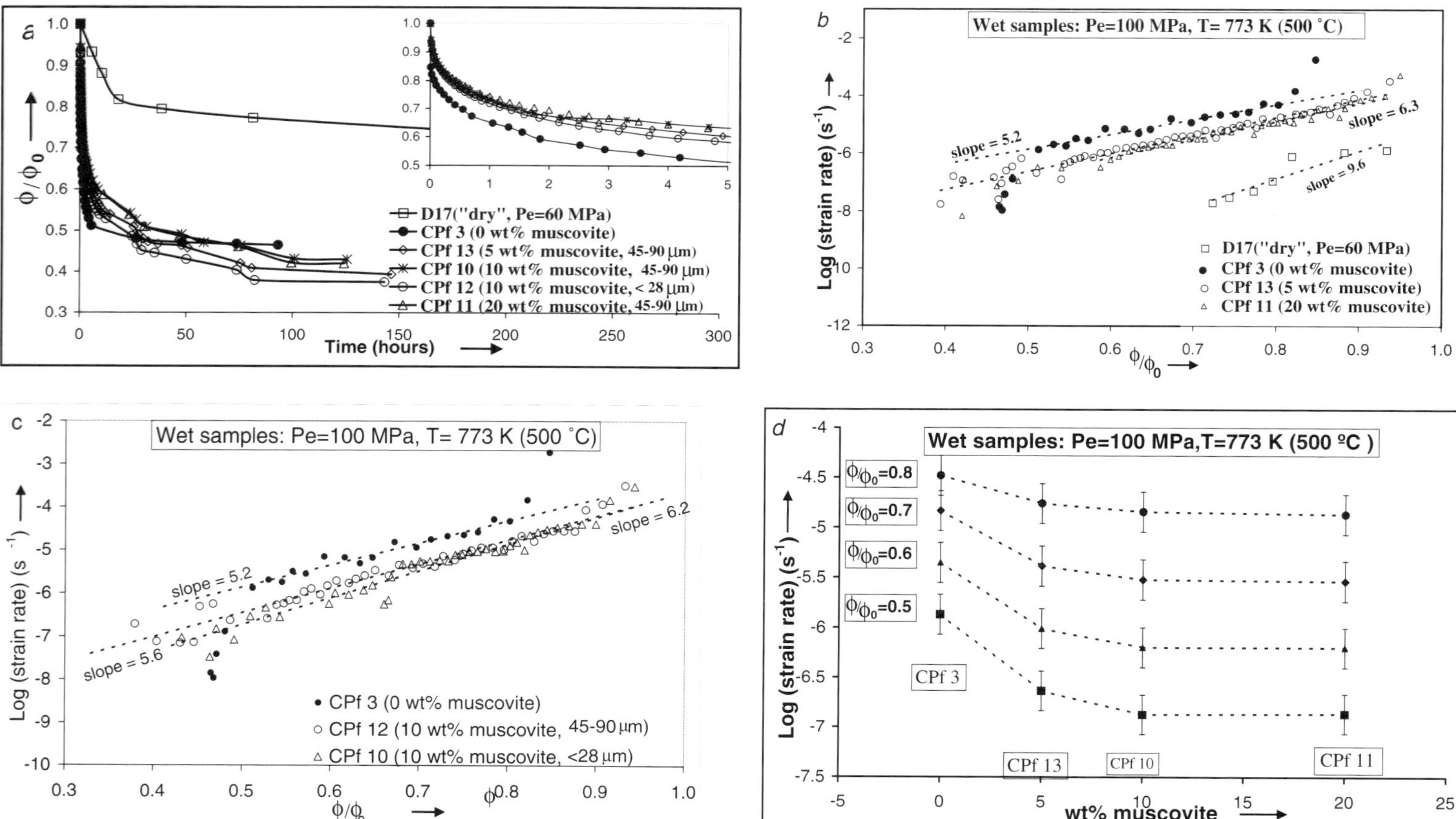

Fig. 3. Compaction creep data showing influence of muscovite. (**a**): Compaction creep curves in the form of ϕ/ϕ_0 versus time plots for the samples D17, CPf3, CPf10 and CPf11. Inset shows compaction creep curves in the first 5 hours. (**b**): Log (strain rate) versus ϕ/ϕ_0 plots with linear best fits for samples D17, CPf3, CPf11, and CPf13. (**c**): Log (strain rate) versus ϕ/ϕ_0 plots with linear best fits for samples CPf3, CPf10 and CPf12. (**d**): Log (strain rate) versus wt% muscovite curves showing the influence of muscovite on compaction rates in wet samples CPf3, CPf10, CPf11 and CPf13.

from ~45% porosity to ~30% (Table 1). Compaction creep data obtained during HIPing of the wet quartz–muscovite samples (CPf10, 11, 12 and 13), the wet quartz-only sample (CPf3) and the 'dry' sand sample (D17) are presented in Fig. 3a in the form of ϕ/ϕ_0 versus time curves. These demonstrate time-dependent compaction in all samples. The 'dry' control sample (D17) shows relatively slow compaction and modest final strain (Table 1). By contrast, the wet, quartz-only sample (CPf3) compacted very rapidly in the first ~12 hours of the experiment, halving its porosity in that time. The wet muscovite–quartz samples (CPf10, 11, 12, 13) compacted more slowly intially, but attained similar final porosities and porosity reduction rates after 3–4 days. Note that the curves of samples CPf10, 11 and 12 are closely similar, although sample CPf12 seems to maintain a higher strain rate to lower values of ϕ/ϕ_0. Sample CPf13 (5 wt% muscovite) compacted slightly faster than CPf10, 11 and 12, but still more slowly than the quartz-only sample (CPf3). Figures 3b and 3c show compaction rate data computed for the dry and wet samples (0 wt%, 5 wt%, 10 wt%, 45–75 µm and <28 µm and 20 wt% muscovite) versus normalized porosity ϕ/ϕ_0. Normalized porosity was used rather than volumetric strain or absolute porosity since this quantity reduces the influence of variations in starting porosity upon strain rate as discussed by Niemeijer *et al.* (2002). It demonstrates that the dry sample (D17) compacted at rates 1–3 orders of magnitude slower than the wet samples at given strain (ϕ/ϕ_0). The fits (of the linear part) shown in Figs 3b and c were used to construct plots of compaction strain rate versus muscovite content for the wet samples (CPf3, 10, 11, and 13) at constant ϕ/ϕ_0 values (Fig. 3d). Both Figs 3b, c and d show a decrease in strain rate of 0.5–1 order of magnitude, caused by addition of muscovite, with no further reduction in the range 10–20 wt% muscovite.

Microstructural observations and compositional data

The starting fraction of Bolderiaan Sand consisted of >99% quartz (Niemeijer *et al.* 2002). Our optical and SEM observations showed the quartz grains to be subangular to angular with surfaces characterized by triangular, pyramidal and sickel-shaped etch pits. The sieved fraction of the natural muscovite was analysed using x-ray diffraction (XRD), inductively coupled plasma mass spectrometry (ICP-MS) and thermogravimetric analysis (TGA). The material consists of >99% muscovite, with the chemical composition given in Table 2.

Table 2. *Chemical composition of natural muscovite*

Composition	wt%
SiO_2	43.35
Al_2O_3	34.53
K_2O	10.93
Fe_2O_3	4.40
MgO	1.21
Na_2O	0.53
CaO	0.10
BaO	0.14
TiO_2	0.15
H_2O	4.61
Total	99.85

Chemical composition of natural muscovite (fraction 45–90 µm) determined by ICP-MS analyses. Other elements are less than 0.1 wt%. H_2O content was determined by TGA.

Optical microscopy and SEM observations made on both the CIPed-only material (D9-equivalent to the 'starting material' for the HIPing samples) and the 'dry' HIPed sample (D17) show closely similar microstructures characterized by widespread intragranular and transgranular cracks, grain size reduction and sharp grain contact points (Fig. 4a). We found no microstructural evidence that dissolution–precipitation processes were active in these samples.

The final grain size of all wet-HIPed samples (CPf3, 10 and 11) was reduced compared to the starting sand fractions, but much less than in the cold compacted sample (compare Figs 4a–d, see also Niemeijer *et al.* 2002). In all wet-HIPed samples (Figs 4b–d) fewer fine grain fragments and fewer cracks were observed than in the CIPed-only material (Fig. 4a), suggesting that many of the finest cataclastic fragments seen in the cold-pressed material dissolved during heating and/or HIPing and that some cracks may have healed. All wet samples show tightly fitting, often micro-sutured grain boundaries and concavo-convex grain-to-grain contacts (indentations and truncations), indicative of IPS (Figs 4b–d). We observed only a few intragranular fractures associated with well-fitting or indented grain contact, suggesting that interaction between fracturing and pressure solution (Gratier *et al.* 1999) did not play an important role in our experiments. Minor undulatory extinction is visible optically in the quartz grains, but the total fraction of quartz grains showing this is less than 5%. Muscovites in the muscovite-bearing

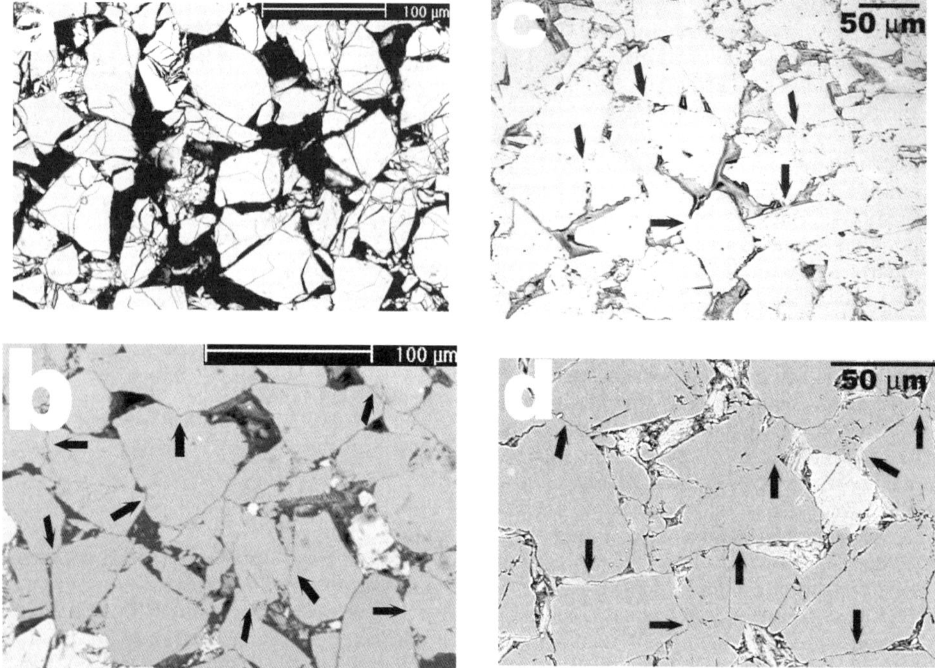

Fig. 4. Microstructures characterizing the various experiments. (**a**): SEM backscatter image of sample D9 (CIPed-only sample, $P_c = 300$ MPa, $P_f = 0.1$ MPa, $\phi_f = 30.31\%$) (**b**): SEM backscatter image of sample CPf3 (wet compacted sample, $P_c = 300$ MPa, $P_f = 200$ MPa, 0 wt% muscovite, $\phi_f = 12.71\%$), showing grain-to-grain indentations and truncations (black arrows) indicative of IPS. (**c**): Reflected light micrograph of sample CPf10 (wet compacted sample, $P_c = 300$ MPa, $P_f = 200$ MPa, 10 wt% muscovite, $\phi_f = 10.52\%$), showing grain-to-grain indentations and truncations (black arrows) indicative of IPS. (**d**): SEM backscatter image of sample CPf12 (wet compacted sample, $P_c = 300$ MPa, $P_f = 200$ MPa, 10 wt% muscovite <28 µm, $\phi_f = 8.31\%$), showing grain-to-grain indentations and truncations (white arrows) indicative of IPS.

wet-tested samples almost always show some undulatory extinction (Figs 4c–d). They are often sharply 'bent' around the quartz grains with no clear indication of fracturing. At least some muscovites are present in ~50% of quartz grain contacts that show indications of IPS. About ~25% of the 'pore space' between quartz grains is occupied with muscovite. In other pore spaces, muscovite is present but rarely coats the surface of the quartz grains (Figs 4c and d). No muscovites were observed that penetrated quartz grains 'end on' (cf. Bjørkum 1996). The muscovites have not developed a preferred orientation, nor do they form a continuous network in 2D section.

Discussion

Using identical methods and identical quartz sand, Niemeijer *et al.* (2002) provided evidence that IPS dominates the compaction of quartz sand at 400–600 °C and effective pressures of 50–150 MPa, and is probably controlled by the quartz dissolution reaction rate. Our present microstructural results show evidence for IPS in both quartz-only and quartz–muscovite samples under the same conditions. The mechanical data demonstrate that the presence of muscovite hinders rather than enhances compaction. The amount of muscovite in the samples was too low to form a continuous network or to completely fill the space between the quartz grains, suggesting that the muscovite could not have formed a load-supporting framework or matrix. Hence the quartz-grain framework must have supported almost as much of the applied effective pressure as in the pure quartz sand samples. Plastic deformation of muscovite in the pores could have modestly hindered compaction or enhanced it by promoting grain boundary sliding between quartz grains. However, since IPS was clearly important in all wet samples, we infer that the presence of muscovite did not enhance IPS rates dramatically, but probably decreased IPS rates.

Recalling the possible effects of phyllosilicates on IPS rates which we have already identified, a decrease in IPS rate due to the addition of muscovite could potentially be explained by dissolution of aluminum from the muscovite in the pore fluid, the dissolved Al^{3+} then lowering the solubility of quartz and slowing the dissolution and precipitation kinetics of quartz (Iler 1973, 1979; Dove & Rimstidt 1994). The amount of Al_2O_3 equivalent in solution after heating (~1 hour) was calculated using the only data we know of on the dissolution kinetics of muscovite (Knauss & Wolery 1996). We extrapolated Knaus & Wolery's dissolution rate data obtained at 70 °C to the temperature conditions of our experiments using their activation energy of 59 kJ/mol. The results obtained was of the order of *c.* 0.8 ppm Al_2O_3 equivalent for sample CPf 10, which is sufficient to lower dissolution rates of quartz by a measurable amount according to previous work on the effects of Al^{3+} (Iler 1973, 1979).

However, muscovite can also hinder the precipitation of quartz by decreasing the number of precipitation sites on pore walls (Tada & Siever 1989; Dewers & Ortoleva 1991; Mullis 1993). This could cause a shift in rate controlling mechanism from dissolution towards precipitation, decreasing IPS rate accordingly. If in a pure quartz aggregate the grain boundary fluid occupies an open island–channel structure, rather than an ultra-thin film maintained by surface forces, the presence of muscovite in the quartz grain boundaries could possibly retard diffusion of quartz out of the boundaries by supporting a thin fluid film which replaces the more open island–channel structure. This would cause a shift in rate controlling mechanism from dissolution to diffusion, again decreasing IPS rates. It is even possible that in our experiments contact stresses may have been high enough to squeeze water out of some mica–quartz contacts (Israelachvili 1992).

Our microstructural observations do not exclude any of the above explanations for a possible reduction in IPS rates due to the addition of muscovite, since it was present in quartz grain contacts actively undergoing IPS and at some pore walls. However, our mechanical data show that the decrease in compaction rates caused by adding muscovite does not seem to depend strongly on the amount of muscovite added (5 wt% versus 10 wt% and 20 wt%). This suggests that the inferred decrease in IPS rate may have been caused by a chemical effect, such as a decrease in dissolution rate due to saturation of the pore fluid with Al^{3+}, rather than an effect of muscovite coating grain boundaries or pore walls, or filling the pore space and thus supporting the load. Due to lack of experimental data on the combined influence of alkali-metal cations (notably K^+ in the case of muscovite) and Al^{3+} on dissolution rates in quartz, we cannot estimate the amount of decrease or increase in IPS rate that the presence of these cations would cause.

Our results do not confirm the inference of Rutter & Wanten (2000) regarding the accelerating influence of phyllosilicates on pressure solution rate in quartz, nor the work of Hickman & Evans (1995) on the accelerating influence of clay films on pressure solution at halite–silica contacts. However, the pressure solution process at clay-free halite–silica and halite–halite contacts is diffusion controlled (Spiers *et al.* 1990), rather than interface controlled so that clay in grain boundaries can potentially increase pressure solution rates if it supports a thicker fluid film than in a clay-free boundary. Rutter & Wanten (2000) used an experimental set-up which included a seperate reservoir of pure quartz sand as an (easy) sink for dissolved quartz. The large surface area available in this sink means that precipitation would be unlikely to become rate controlling in their experiments, and any retardation of precipitation kinetics caused by the presence of phyllosilicates would not be measured as a decrease in compaction rate. The increase of compaction rate that Rutter & Wanten (2000) reported could perhaps have been caused by an increase in dissolution rate by the presence of alkali-metal cations (e.g. K^+, Na^+), released from the clay minerals and/or the muscovite. This is purely speculative, however, since doubt remains whether the enhancement in dissolution kinetics by alkali-metal cations is larger than the retardation caused by cations such as Al^{3+} or Fe^{3+}. Alternatively, the enhancement of IPS reported by Rutter & Wanten (2000) could have been caused by a switch from precipitation or dissolution control, in pure quartz sand, to the parallel process of source-sink diffusion control in the sand-phyllosilicate samples with pure sand sink. Finally, Rutter & Wanten (2000) argue that the increase in IPS rates they observed may have been caused, in part, by the smaller total amount of quartz compared to the pure quartz samples, and by additional compaction of the clay matrix by frictional and intracrystalline plastic processes. These processes should also have played a role in our experiments, but we nonetheless observed a decrease in compaction rate in mica-bearing samples.

Clearly further systematic research is necessary to resolve the question of the effects of phyllosilicates on IPS in quartz. Nonetheless, our results have shown that muscovite causes no dramatic enhancement of IPS under conditions favouring dissolution-controlled IPS. IPS in nature may be diffusion or precipitation controlled, which may lead to a different effect of phyllosilicates on IPS rates in quartz. To assess if phyllosilicates have an accelerating effect in the diffusion or precipitation-controlled field, experiments need to be conducted on samples with sufficiently coarse grain size, sufficiently low porosity or using different pore fluids, such that these rate-controlling processes are favoured over dissolution. Experiments should also be conducted systematically to investigate the possible chemical effects of phyllosilicates independent of direct mechanical effects. This could be done by incorporation of phyllosilicates into the pore fluid system but not in the quartz sand sample.

Conclusions

(1) We have conducted a set of four isostatic compaction experiments on quartz–muscovite mixtures at a confining pressure of 300 MPa, a fluid pressure of 200 MPa, an effective pressure of 100 MPa and a temperature of 500 °C, in order to investigate the effect of muscovite on IPS rates in quartz sand. Under the chosen conditions, previous work indicates that dissolution controls the rate of IPS in pure quartz sand.

(2) The presence of muscovite does not accelerate IPS rates in quartz sand under the conditions used, but decreases compaction rates by a factor of 3–10, possibly due to a decrease in IPS rate.

(3) The effect of muscovite in slowing IPS may be due to the effect of dissolved aluminum (Al^{3+}) dominating over any accelerating effects of alkali-metal cations. This is based on the fact that the presence of Al^{3+} in concentrations of *c.* 0.15 ppm Al_2O_3 equivalent is known to decrease the solubility, dissolution rates and precipitation rates of quartz.

(4) The possibility that muscovite (and other phyllosilicates) strongly enhances grain boundary diffusion and hence IPS in quartz, under laboratory or natural conditions favouring grain boundary diffusion control in pure quartz, cannot be eliminated on the basis of our experiments. To test if this occurs, experiments need to be performed on coarse quartz sand, lower porosities or using different pore fluid(s) (e.g. acidic fluids).

(5) Enhanced IPS in nature and experiments could be due to chemical effects on the precipiation and/or dissolution reactions, depending on the composition of phyllosilicates present and competing effects of alkali-metals versus Al^{3+} or Fe^{3+}. Experiments should be conducted systematically to investigate the possible chemical effects of phyllosilicates independent of direct mechanical effects. This could be done by incorporation of phyllosilicates into the pore fluid system but not in the quartz sand sample.

We thank G. Kastelein for constructing the experimental apparatus and P. van Krieken and E. de Graaf for further technical support. We also thank the reviewers B. den Brok and J.-P. Gratier for their comments which improved this paper.

References

ADDIS, M. A. & JONES, M. E. 1985. Volume changes during diagenesis. *Marine and Petroleum Geology*, **2**, 241–246.

BIRCH, F. 1966. Compressibility; elastic constants. *In*: S. P. CLARK, Jr. (ed.), *Handbook of Physical Constants*. Geological Society of America, Memoir, **97**, 97–173.

BJØRKUM, P. A. 1996. How important is pressure in causing dissolution of quartz in sandstones? *Journal of Sedimentary Research*, **66**, 147–154.

BOS, B. & SPIERS, C. J. 2000. Effect of phyllosilicates on fluid-assisted healing of gouge-bearing faults. *Earth and Planetary Science Letters*, **184**, 199–210.

BOS, B. & SPIERS, C. J. 2001. Experimental investigation into the microstructural and mechanical evolution of phyllosilicate-bearing fault rock under conditions favouring pressure solution. *Journal of Structural Geology*, **23**, 1187–1202.

BOS, B., PEACH, C. J. & SPIERS, C. J. 2000*a*. Slip behavior of simulated gouge-bearing faults under conditions favoring pressure solution. *Journal of Geophysical Research*, **104**, 16699–16717.

BOS, B., PEACH, C. J. & SPIERS, C. J. 2000*b*. Frictional-viscous flow of simulated fault gouge caused by the combined effects of phyllosilicates and pressure solution. *Tectonophysics*, **327**, 173–194.

BURNHAM, C., HOLLOWAY, J. R. & DAVIS, N. F. 1969. *Thermodynamic Properties of Water to 1,000 °C* and 10,000 bars. Geological Society of America Special Papers, **132**..

DEER, W. A., HOWIE, R. A. & ZUSSMAN, J. 1977. *An Introduction to the Rock Forming Minerals, Part 3: Sheet Silicates*. Longman Group Limited, London.

DE MEER, S. & SPIERS, C. J. 1995. Creep of wet gypsum aggregates under hydrostatic loading conditions. *Tectonophysics*, **245**, 171–183.

DEWERS, T. & ORTOLEVA, P. 1991. Influences of clay minerals on sandstone cementation and pressure solution. *Geology*, **19**, 1045–1048.

DOVE, P. M. 1999. The dissolution kinetics of quartz in aqueous mixed cation solutions. *Geochimica et Cosmochimica Acta*, **63**, 3715–3727.

DOVE, P. M. & NIX, C. J. 1997. The influence of the alkaline earth cations, magnesium, calcium, and barium on the dissolution kinetics of quartz. *Geochimica et Cosmochimica Acta*, **61**, 3329–3340.

DOVE, P. M. & RIMSTIDT, J. D. 1994. Silica-water interactions. *In*: HEANEY, P. J., PREWITT, C. T. & GIBBS, G. V. (eds) *Silica: physical behaviour, geochemistry and materials applications*. Mineralogical Society of America, Washington, DC, 259–308.

ENGELDER, T. & MARSHAK, S. 1985. Disjunctive cleavage formed at shallow depths in sedimentary rocks. *Journal of Structural Geology*, **7**, 327–343.

GRATIER, J. P., RENARD, F. & LABAUME, P. 1999. How pressure solution creeep and fracturing process interact in the upper crust to make it behave in both a brittle and viscous manner. *Journal of Structural Geology*, **21**, 1189–1197.

HEALD, M. T. 1959. Significance of stylolites in permeable sandstones. *Journal of Sedimentary Petrology*, **30**, 251–253.

HEIDUG, W. K. 1995. Intergranular solid-fluid phase transformations under stress: the effect of surface forces. *Journal of Geophysical Research*, **100**, 5931–5940.

HICKMAN, S. H. & EVANS, B. 1995. Kinetics of pressure solution at halite-silica interfaces and intergranular clay films. *Journal of Geophysical Research*, **100**, 13113–13132.

HOUSEKNECHT, D. W. 1987. Assesing the relative importance of compaction processes and cementation to reduction of porosity in sandstones. *Bulletin of the American Association of Petroleum Geologists*, **71**, 633–642.

ILER, R. K. 1973. Effect of adsorbed alumina on the solubility of amorphous silica in water. *Journal of Colloid and Interface Science*, **43**, 399–408.

ILER, R. K. 1979. *The Chemistry of Silica*. John Wiley and Sons, New York.

ISRAELACHVILI, J. N. 1992. *Intermolecular and Surface Forces*. Academic, San Diego, California.

JANSSEN, C., MICHEL, W., BAU, M., LUDERS, V. & MUHLE, K. 1997. The North Anatolian Fault Zone and the role of fluids in seismogenic deformation. *Journal of Geology*, **105**, 387–404.

KNAUSS, K. G. & WOLERY, T. J. 1989. Muscovite dissolution kinetics as a function of pH and time at 70 °C. *Geochimica et Cosmochimica Acta*, **53**, 1493–1501.

LEHNER, F. K. 1990 Thermodynamics of rock deformation by pressure solution. *In*: BARBER D. J. & MEREDITH, P. G. (eds) *Deformation Processes in Minerals, Ceramics and Rocks*. Unwin Hyman, London.

MULLIS, A. M. 1993. Determination of the rate-limiting mechanism for quartz pressure solution. *Geochimica et Cosmochimica Acta*, **57**, 1499–1503.

NIEMEIJER, A. R., SPIERS, C. J. & BOS, B. 2002. Compaction creep of quartz sand at 400–600 °C: Experimental evidence for dissolution controlled pressure solution. *Earth and Planetary Science Letters*, **195**, 261–275.

PASHLEY, R. M. & ISRAELACHVILI, J. 1984. Molecular layering of water in thin films between mica surfaces and its relation to hydration forces. *Journal of Colloid and Interface Science*, **101**, 510–522.

RENARD, F. & ORTOLEVA, P. 1997. Water films at grain–grain contacts: Debye–Hückel, osmotic model of stress, salinity, and mineralogy dependence. *Geochimica et Cosmochimica Acta*, **61**, 1963–1970.

RENARD, F., GRATIER, J. P. & JAMTVEIT, B. 2000. Kinetics of crack-sealing, intergranular pressure solution and compaction around active faults. *Journal of Structural Geology*, **22**, 1395-1407.

RENARD, F., ORTOLEVA P. & GRATIER, J. P. 1997. Pressure solution in sandstones: influence of clays and dependence on temperature and stress. *Tectonophysics*, **280**, 257–266.

RUTTER, E. H. 1983. Pressure solution in nature, theory and experiment. *Journal of the Geological Society, London*, **140**, 725–740.

RUTTER, E. H. & WANTEN, P. H. 2000. Experimental study of the compaction of phyllosilicate-bearing sand at elevated temperature and with controlled pore water pressure. *Journal of Sedimetary Research*, **70**, 107–116.

SCHUTJENS, P. M. T. M. 1991. Experimental compaction of quartz sand at low effective stress and temperature conditions. *Journal of the Geological Society, London*, **148**, 527–539.

SKINNER, B. J. 1966. Thermal expansion. *In*: CLARK, S. P. Jr. (ed.), *Handbook of Physical Constants*. Geological Society of America, Memoir, **97**, 75–97.

SPIERS, C. J. & BRZESOWSKY, R. H. 1993. Densification behaviour of wet granular salt: theory vs experiment. *In*: KAKIHANA, H., HARDY, H. R. JR., HOSHI, T. & TOYOKURA, K. (eds), *Seventh Symposium on Salt, vol. 1*. Elsevier Science, Amsterdam, pp. 83–92.

SPIERS, C. J., SCHUTJENS, P. M. T. M., BRZESOWSKY, R. H., PEACH, C. J., LIEZENBERG, J. L. & ZWART, H. J. 1990. Experimental determination of constitutive parameters governing creep of rocksalt by pressure solution. *In*: KNIPE, R. J. & RUTTER, E. H. (eds.) *Deformation Mechanisms, Rheology and Tectonics*. Geological Society, London, Special Publications, **45**, 215–227.

SVERJENSKY, D. A. 1994. Zero-point-of-charge prediction from crystal chemistry and solvation theory. *Geochimica et Cosmochimica Acta*, **58**, 3123–3129.

TADA, R. & SIEVER, R. 1989. Pressure solution during diagenesis. *Annual Reviews Earth and Planetary Science Letters*, **17**, 89–118.

In situ experimental study of roughness development at a stressed solid/fluid interface

BAS DEN BROK[1], JACQUES MOREL[2,3] & MOHSINE ZAHID[2,4]

[1]*Geologisches Institut ETH Zentrum, Zürich, Switzerland (e-mail: denbrok@erdw.ethz.ch)*
[2]*Institut für Geowissenschaften, Johannes Gutenberg Universität, Mainz, Germany*
[3]*Present address: Institut de recherche pour le développement (IRD), Centre Nouméa, 98848 Nouméa, Nouvelle Calédonie*
[4]*Present address: Institute for Materials and Processes in Energy Systems (IWV), Forschungszentrum Jülich, Germany*

Abstract: Theory and experiments have demonstrated that the initially flat surface of an elastically strained solid is morphologically unstable. The elastic strain energy of a rough, corrugated surface is lower than that of a flat one. Hence, stress forces the surface into a rough structure, but the associated increase in surface energy counteracts this roughening. In this way an equilibrium surface roughness consisting of μm-scale grooves and ridges can develop if the solid is transported, e.g. by diffusion through an aqueous solution, from sites of high stress to sites of low stress. We report *in situ* experimental observations of the surface of elastically strained potassium (K-) alum single crystals held in K-alum solution. The observations confirm earlier reports of the development of stress-induced μm-scale grooves on the surface of this material. The *in situ* observations show, however, that the stress-induced surface morphology is not a static, but a dynamic structure. The grooves are mobile, and may for example propagate or increase or decrease in length. They may move upwards, downwards, or remain where they are. Others rotate and undulate. It is suggested that if stress is high enough, grain boundaries in (wet) rocks could posses a similar structure of channels, continuously changing position and orientation, in line with the so-called 'dynamically stable' island–channel grain boundary structure that is essential to several pressure solution models.

The mechanical behaviour of the Earth's crust is strongly affected by the presence of small amounts of interstitial water at grain boundaries and in micropores and fractures (e.g. Fyfe *et al.* 1976). Under the influence of stress gradients, the mineral grains may locally dissolve in the interstitial water and the dissolved material diffuse through – or flow with – the water to other locations, where it may precipitate again. This material transport may lead to deformation of the rocks, a mechanism commonly referred to as 'pressure solution' (PS) (e.g. Lehner 1990, Paterson 1995 and references therein). PS is widely regareded as a very important ductile deformation mechanism in the continental crust, at least up to low- to medium-grade metamorphic conditions. Yet, how precisely PS works, notably at the microscopical scale, is poorly understood. Reliable PS constitutive flow laws are therefore lacking for most rocks and rock-forming minerals (rocksalt forms a noticeable exception: Spiers & Schutjens 1990; Spiers & Carter 1998). For this reason PS is generally not taken into consideration in crustal strength profiles or deformation mechanism maps (e.g. Carter & Tsenn 1987; Ranalli 1987; Twiss & Moores 1992).

One of the major problems in defining a PS flow law is that the microstructure of grain boundaries during deformation by PS is not well understood. This is an important problem. It can be shown that different, but reasonable, assumptions made on the grain boundary structure may lead to up to 10 orders of magnitude difference in the predicted strain rate (den Brok 1998). A point of major uncertainty is whether or not grain boundaries can maintain a *rough* microstructure while PS is in progress. It is held by some (e.g. Raj 1982; Spiers & Schutjens 1990; Lehner 1990; Ghoussoub & Leroy 2001) that water at grain boundaries is present in some kind of 'island–channel' structure which provides easy access for water to enter the boundary and for dissolved material to diffuse out of it. Schutjens & Spiers (1999) recently reported *in situ* observations of such grain boundary roughening phenomena, and there are numerous 'post-mortem' observations of rough grain boundaries reported by various groups (e.g. Spiers & Schutjens 1990; Cox & Paterson 1991; Dewers & Hajash 1995). However, until recently, the idea of a rough grain boundary consisting of islands and channels could not be explained as a predicted feature of any model (Lehner 1995).

From: DE MEER, S., DRURY, M. R., DE BRESSER, J. H. P. & PENNOCK, G. M. (eds) 2002. *Deformation Mechanisms, Rheology and Tectonics: Current Status and Future Perspectives.* Geological Society, London, Special Publications, **200**, 73–83. 0305-8719/02/$15 © The Geological Society of London.

In the present paper we present experimental support for a model proposed by Leroy & Heidug (1994) and further developed by Ghoussoub & Leroy (2001) that, at least qualitatively, offers an explanation for the development of a rough boundary structure during deformation by PS. The model is based on recent theories and experimental observations on the effect of elastic strain on the stability of solid–fluid interfaces. The initially straight surface of an elastically strained solid is morphologically unstable. It turns into a rough structure if a local material transport is possible, e.g. by diffusional mass transfer through an aqueous solution. The elastic energy is responsible for the instability, while interfacial energy is the stabilising influence. Thus, an equilibrium structure is formed, consisting of μm-scale valleys and ridges. We have performed an *in situ* experimental study of such roughness development on free crystal faces held under stress in a slightly undersaturated aqueous solution.

Before presenting our results we will first discuss the different grain boundary models proposed in the literature to illustrate where the problems lie.

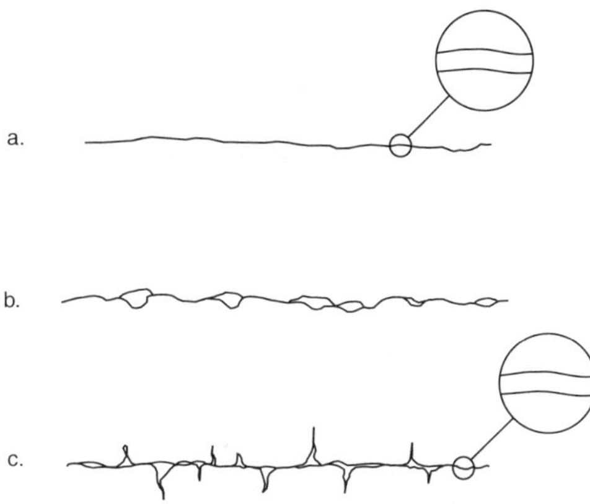

Fig. 1. Cartoon illustrating grain boundary models. The grain boundaries are depicted in cross section. (**a**) Thin film model (Weyl 1959; Rutter 1976). The thin aqueous film is assumed to be only a few nm wide. (**b**) Island–channel model (Raj 1982; Spiers *et al.* 1990; Lehner 1990). The boundary is commonly assumed to be effectively a few hundreds of nm wide. (**c**) Island-crack model (Gratz 1991; den Brok 1998). The boundary at the crack-bound islands is assumed to consist of a thin film (cf. Fig. 1a). Crack spacing several μm's.

Grain boundary models

Basically, three different grain boundary models models have been proposed in the PS literature (Fig. 1): (i) the thin film model; (ii) the island–channel model; and (iii) the island–crack model.

Thin film model

In the thin flim model (Fig. 1a) it is assumed that water is present at grain boundaries in the form of a strongly adsorbed, very thin, continuous fluid film capable of transmitting shear stresses (e.g. Weyl 1959; Rutter 1976). The molecular structure of such films remains obscure to date. Originally it was thought that they would typically be 1 or 2 nm thick (e.g. Rutter 1976) or possibly even thicker (e.g. 1–30 nm: Hickman & Evans 1991). Surface force measurements by Peschel *et al.* (1982), Israelachvili (1986, 1992, 1997) and Horn *et al.* (1988, 1989) have shown, however, that such films must be much thinner. At NPT-conditions, a 1–2 nm aqueous solution film squeezed between silica plates is able to resist an effective shear stress of ~0.1 MPa only. About 1 MPa was needed to squeeze out such a water film completely. According to Zhu & Granick (2001) water films more than about three monolayers thick (i.e. ≥6 Å) posses an intrinsic liquid structure. Hence, thin films capable of resisting geologically realistic stresses can only be expected to be at most one or two atomic layers thick. The diffusivity in such molecularly thin aqueous films is not well known. Originally, it was assumed that the diffusivity is several orders of magnitude lower than in the bulk fluid, because of electro-viscous effects (e.g. Rutter 1976). More recently it has appeared, however, that the viscosity increase due to the electro-viscous effect in thin solution films is very small (considerably less than a factor 10 according to, e.g. Rubie 1986). Moreover, molecular dynamics based computer simulations have predicted that the self-diffusivity in molecularly thin films is decreased only by a factor ≤2 (Magda *et al.* 1985; Vanderlick & Davis 1987; Schoen *et al.* 1988; Bitsanis *et al.* 1990).

Island–channel grain boundary model

It may be argued on the basis of the work of e.g. Israelachvili (1997) that at stresses higher than a few Mpa, thin films will be squeezed out from between the mineral grains leading to solid–solid contacts. In the island–channel model (Fig. 1b) it is assumed that the grain boundary has a microscopically *rough* structure, consisting of solid–solid contact 'islands' (or pillars) across which the stress is transmitted, and water-filled 'channels' containing free liquid at a uniform pressure equal to the macroscopic pore fluid pressure (e.g. Raj 1982; Lehner 1990; Spiers & Schutjens 1990, Ghoussoub & Leroy 2001). The channels form an interconnected network through which dissolved material can be transported away, out of the grain boundary. This island–channel structure is assumed to be dynamically stable, i.e. it is assumed to be a time-statistically constant, non-equilibrium structure. During continuous dissolution, islands and channels are assumed to change position continuously, while on average the structure remains the same. The island–channel structure may typically be up to a few hundred nm thick (Schutjens & Spiers 1999; 100–250 nm at the halite–glass contact following de Meer *et al.* 2001).

The island–channel model assumes that grain boundaries can maintain a rough structure during the PS process. Whether this is possible may be questioned. A rough structure may very well be unstable and smooth out with ongoing PS. The islands supporting the stress could be preferentially dissolved and the fluid mechanically expelled or trapped in inclusions by surface energy driven healing of the grain boundaries (e.g. Hickman & Evans 1991). In support of the island–channel model it was suggested that the roughness may smooth out for conditions approaching equilibrium, but not necessarily so for the non-equilibrium conditions corresponding to PS (Spiers & Schutjens 1990; Lehner 1990, 1995). A rough structure may possibly be treated as a quasi-stationary, non-equilibrium structure, because of: (1) local crystal plastic deformation; (2) the presence of defects in the solid; (3) heterogeneities in solid deformation; or (4) crystallographically controlled interface kinetics (Spiers & Schutjens 1990). All of this could continuously perturb the rate of dissolution and precipitation along the phase boundary. In this way, a rough surface with some average, dynamically stable non-equilibrium structure could, in principle, be maintained during deformation by PS. However, whether or not this is actually the case remains unclear (e.g. Lehner 1995). Moreover, different solids may be capable of supporting different rough structures maintained by different mechanisms.

Island–crack model (Gratz 1991)

Gratz (1991) studied the surface of quartz grains from sandstones compacted by PS in the nature. He observed that the surface consisted of numerous μm-scale islands, separated by crack-like features. On the basis of his observations he proposed a grain boundary model consisting of static islands supporting the stress, separated by microfracture-controlled fluid channels (Fig. 1c). He assumed that the grain-to-grain contacts at the fracture-bound islands contained a thin fluid film of the type assumed by Weyl (1959) and Rutter (1976), and that the diffusivity in the film on the islands was several orders of magnitude lower than the diffusivity in the channels. Thus, the rate of mass removal would be controlled by the diffusion in the fluid film at the island–island contacts. Compared to the thin film model, the overall diffusivity would be significantly increased due to the presence of the microfractures.

Compared to the dynamically stable island–channel grain boundary structure, the island-crack grain boundary structure proposed by Gratz is a static structure. Channels remain where they are at fixed positions. The roughness is maintained due to the growth of the fractures. With ongoing dissolution the roughness is not expected to be smoothed out.

Relationship between the models

It is important to note that the three different grain boundary models do not exclude each other. The dominance of either one of the grain boundary structures may depend on several factors, such as, e.g. the type of solid(s) involved, the P–T conditions prevailing, the amount of water present, or the intensity of the stress. At very low stresses, thin film diffusion could very well dominate the PS process, whereas at intermediate stresses, islands and channels could develop. At the highest stresses, microfractures could start to dominate the PS process.

Morphological (in-)stability of elastically strained surfaces

The question we wish to examine is whether or not a rough, dynamically stable island–channel type grain boundary structure can exist, while PS is going on, even in the absence of perturbing influences such as defects, plastic deformation and so on. Would the roughness not tend to be smoothed out? In this paper we will argue that this is not the case. It has been observed recently in experiments on various solids (e.g. Kim *et al.* 1999 and references therein) that a non-hydrostatically stressed, elastically strained solid, which is in contact with its own melt or vapor, can partially release its elastic strain energy by a morphological instability at the interface. This strain relief mechanism gives rise to what appears to be a roughening of the surface into grooves and ridges of a particular spacing. A corrugated surface has a lower elastic energy than a flat one. This phenomenon was first predicted by Asaro & Tiller (1972), but since the independent rediscovery of the instability by Grinfeld (1986) and Srolovitz (1989) it is often referred to as 'Grinfeld instability'.

Following Srolovitz (1989) the nominally flat surface of an elastically stressed solid is unstable with respect to the growth of perturbations with wavelengths (λ) greater than a critical wavelength (λ_c). For $\lambda > \lambda_c$ the energy of the system may be lowered by the formation of a rough surface of high-stress valleys and low-stress ridges, thus providing a driving force for material transport, e.g. by diffusion through an aqueous solution from the valley to the ridge (Fig. 2). The equilibrium geometry of the surface is then controlled by a balance between the elastic strain energy driving an increase in the amplitude of the valley-and-ridge structure, and the surface energy driving a decrease in the amplitude, i.e. a smoothing of the valley-and-ridge structure. According to Srolovitz (1989) the change in free energy (ΔF) in going from a flat surface to a surface consisting of valleys and ridges with an amplitude c and a wavelength λ (Fig. 2) may be approximated by:

$$\Delta F = \frac{-c\lambda\sigma^2}{4E} + 2c\gamma \tag{1}$$

where σ is the differential stress in the bulk, γ is the surface energy, and E is Young's modulus. The critical wavelength (i.e. the wave length at $\Delta F = 0$) is then equal to

$$\lambda_c = 8\gamma E/\sigma^2. \tag{2}$$

For a typical rock-forming mineral such as quartz, with $\gamma \approx 0.5\,\mathrm{J/m^2}$ and $E \approx 50\,\mathrm{GPa}$, λ_c would be of the order of $20\,\mu\mathrm{m}$ for $\sigma = 100\,\mathrm{MPa}$ and $5\,\mu\mathrm{m}$ for $\sigma = 200\,\mathrm{MPa}$. Hence, provided that dissolution and precipitation can take place, a microstructurally significant surface roughness could be developed by stressing the rocks within the elastic deformation regime.

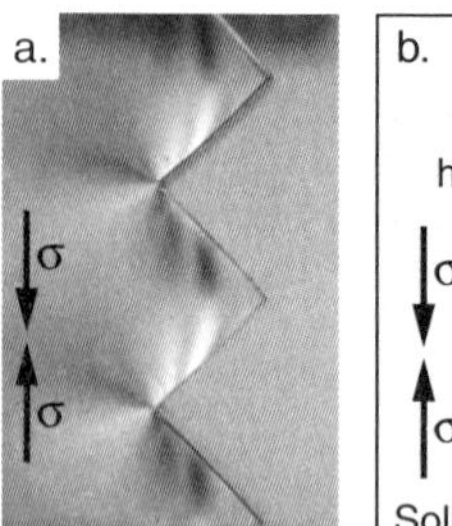

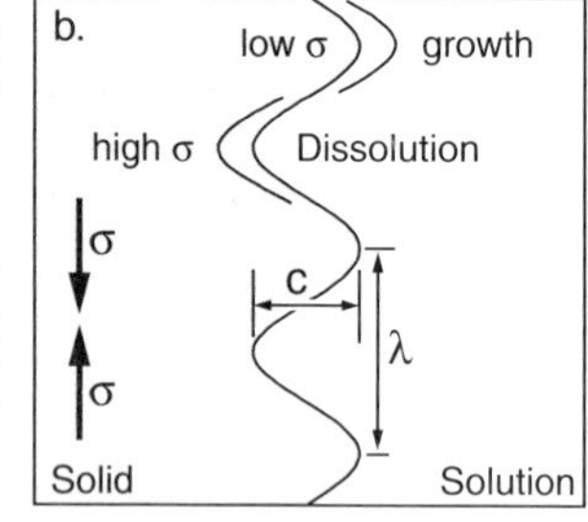

Fig. 2. (**a**) Plexiglass model under stress (plexiglass left, air right) seen through crossed polarizers and illustrating the concentration of stress at concavities. Crests between the grooves are almost free of stress. Grooves are about 0.5 mm apart. (**b**) Free face of a solid stressed with a bulk stress σ in contact with a solution and with a sinusoidal surface profile. The profile has wavelength l and amplitude c. The stress in the valleys is higher than the stress on the ridges, thus providing a driving force for further increase in amplitude of the valley-and-ridge structure, e.g. by dissolution at the high-stress valleys and growth at the low-stress ridges (after den Brok & Morel 2001).

Leroy & Heidug (1994) were the first to realise that Grinfeld-type instabilities could be relevant to PS phenomena. They used linear stability analysis to numerically study the interface evolution of a fluid-filled tubular pore embedded in a stressed solid matrix. They superposed small disturbances on a known equilibrium shape and then analysed their evolution in time, using kinetic laws to describe the relevant transport mechanisms. They investigated the alteration in pore geometry caused by a dissolution–precipitation reaction and showed that large fluid-filled pores in a stressed solid matrix are morphologically unstable and should break up into smaller pores. Ghoussoub (2000) and Ghoussoub & Leroy (2001) applied the theory to the evolution of grain boundary structure in aggregates deforming by PS. They numerically showed how the instability would lead to localized dissolution and penetration of water into the grain boundary by marginal dissolution. They suggested that the newly formed fluid layer would be highly instable, leading to a repeated reorganization or dynamic evolution of the grain boundary internal structure during the action of PS.

Experimental evidence for the development of Grinfeld-type instabilities at the surface of elastically strained, soluble solid in contact with an aqueous solution was first provided by Morel (2000) and den Brok & Morel (2001). These authors performed experiments on potassium (K-) alum (K Al $[SO_4]_2 \cdot 12H_2O$), which was used as a rock analogue material. Rectangular single crystals of this material were put in a vessel at room P-T conditions and elastically strained with a dead weight in a slightly undersaturated K-alum solution. Grooves developed in the initially flat surface of the crystals, oriented dominantly perpendicular to the maximum compressive stress. The grooves disappeared soon after the stress was taken off. The grooves were typically 20–40 μm wide and 10–20 μm deep, and their size was broadly in agreement with the above theory.

In this paper we report on *in situ* observations made on K-alum single crystals stressed in aqueous solution under conditions similar to those reported by den Brok & Morel (2001). From the experiments it appears that stress-induced surface roughness is not a static structure, but dynamic, with grooves constantly changing position and orientation.

In situ experiments: procedure

The experiments were carried out *in situ*, in a see-through vessel mounted under an optical microscope (Fig. 3; cf. den Brok *et al.* 1998). This vessel consists of two 50 × 50 mm glass slides of 1.6 mm thick, separated by a 1.1 mm thick U-shaped Ertalon spacer. The glass slides and the spacer were glued together with UV-activated loctite ('Loctite 350'). The slot-shaped vessel thus obtained measures about 10 × 20 × 1.2 mm in size. Samples used were rectangular platelets of about 6 × 8 × 1 mm, cut with a diamond blade (Buehler) saw from solution-grown potassium alum single crystals. Volitaile oil ('Shell S4919') was used as a lubricant. A 2 ± 0.05 mm diameter hole was drilled in the middle of each samples with a twist drill, again using volatile oil as a lubricant. This hole was drilled in order to obtain a geometrically well-defined stress concentration in the centre of the sample, as in experiments by Sprunt & Nur (1977) and Bosworth (1981). Without a hole, the surface phenomena of interest would develop dominantly at the stress concentrations at the corner of the contact between the sample and the piston, where it is difficult to make good observations.

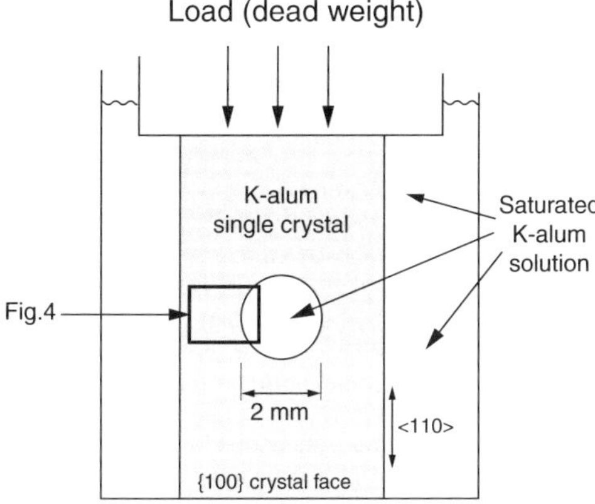

Fig. 3. Schematic diagram, approximately to scale, showing experimental set-up of *in situ* experiment 'Iskal-1' of Morel (2000). The K-alum single crystal platelet (*c.* 1 × 6 × 8 mm), with a central hole of *c.* 2 mm in diameter, was mounted in a slot-like see-through vessel consisting of two 5 × 5 mm wide and 1.6 mm thick glass plates, separated by a U-shaped 1.1 mm thick Ertalon spacer. Glass and spacers were glued together using Loctite UV-activated glue. The set-up is seen from the side. The sample was loaded using a dead weight, corresponding to a bulk stress of *c.* 2.8 MPa. Saturated K-alum solution was added from above with a syringe.

After the hole was drilled, the sides of the samples were polished using polishing paper (mesh size up to 1200), again using volatile oil as a lubricant. Finally, the samples were cleaned in destilled water and dried with tissue paper.

To perform an experiment, a sample was slid into the vessel and mechanically loaded with a 1 mm thick and 8 mm wide piston carrying a dead weight. The weight corresponded to a calculated bulk compressive stress in the sample of 2 to 3 MPa. Around the central hole, the maximum compressive stress would be 4 to 5 times higher (note that the compressive stress would be 3 times higher in case of a hole in an infinitely large crystal, but 4 to 5 times higher in case of our samples actually measuring 6 × 8 × 1 mm in size; Morel 2000).

The piston was made of Vanadium steel, because alum solutions are weakly acid (due to Al^{3+} in solution) and would corrode ordinary steel. After loading, the vessel was filled with a slightly undersaturated potassium alum solution, injected from above with the help of a syringe. A layer of silicon oil was added on top to prevent evaporization of the solution. Experiments were carried out at 'room' temperature, in a small room in which the temperature could be controlled sufficiently accurately to keep the temperature in the solution to within ±0.2 °C.

The optical microscope was coupled to a TK-1070E JVC digital colour video camera connected to a personal computer running 'NIH-image', a public domain image processing and analysis software available at http://rsb.info.-nih.gov/nih-image/. Thus, images of part of the sample, with a spatial resolution of 2–3 μm/pixel, were automatically captured at regular intervals during the entire experiment (Figs 3 and 4). Observations were made of the free crystal face next to the hole at the point where the compressive stress (vertical) was expected to be maximum. Between the crystal face and the glass (of the see-through vessel) a layer of 50–100 μm solution was present.

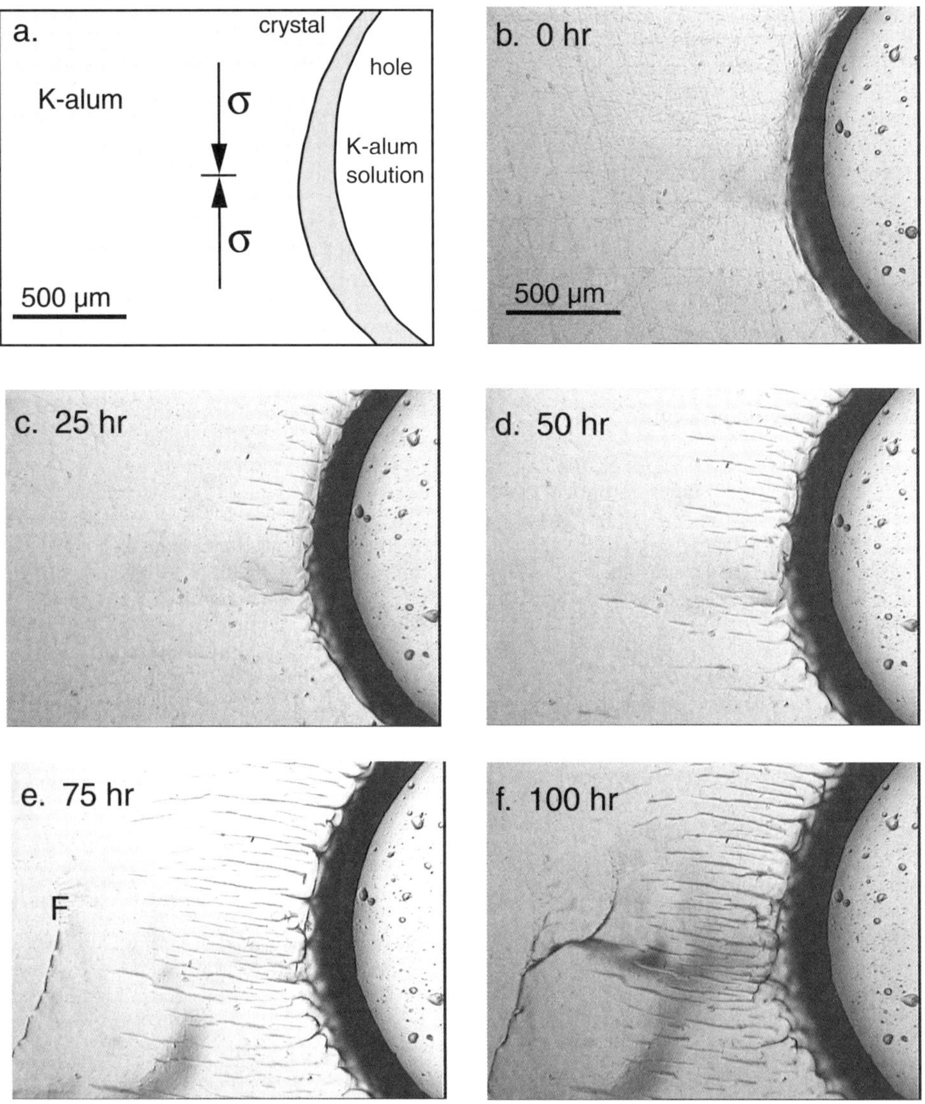

Fig. 4. Images obtained from experiment Iskal-1. (**a**) Schematic drawing of geometry. (**b**) At the beginning of the experiment. Note scratches at the K-alum surface. (**c**) After 25 hours, regularly spaced fine dissolution grooves have developed at left side of the hole. (**d**) After 50 hours, grooves have propagated further towards the left, away from the central hole. (**e**) After 75 hours, further propagation of grooves. Note that a fracture (F) has developed. Note also bending of grooves (above) towards the fracture. (**f**) Situation at the end after 100 hours. Renewed fracturing. Note that the fracture has migrated somewhat to the right. Note also increased bending of the grooves.

Six experiments were carried out in all (Iskal-1 to -6). In this paper only results from experiment Iskal-1 will be reported. This is the only experiment in which we did not vary the concentration of the solution and/or vary the stress during the experiment; see Morel (2000) for a description of all experiments. In experiment Iskal-1, the sample was cut parallel to the {100} face. It was loaded with the maximum compressive stress direction parallel to ⟨110⟩. The calculated maximum stress left and right of the hole was *c.* 10 MPa. Initially the added solution was undersaturated by 0.3°. The temperature during the experiment was 21.2±0.2 °C.

Potassium (K-) alum was used in these experiments as a rock analogue material, as in experiments by den Brok & Morel (2001), primarily because it is an elastic/brittle material at room temperature and atmospheric pressure. K-alum single crystals can be loaded up to 20 MPa without measurably deforming. At higher stresses crystals break cataclastically. In this way any possible effects due to crystal plastic deformation could be excluded. Furthermore, K-alum has a high solubility, and dissolution and growth takes place relatively fast (at rates comparable to NaCl) so that geologically relevant processes can be studied at room P-T conditions in experiments that do not require an excessively long period of time.

Observations

We started making observations immediately after the solution was added. Several stages of the experiment are dipicted in Figure 4 and some of the essential observations are depicted in Figure 5. Within an hour, some very fine grooves had already developed at the edge of the hole drilled in the crystal. These grooves seemed to have nucleated at tiny concave irregularities at the edge of the hole. As soon as they were discernible, the grooves were oriented with their long axis (g) approximately perpendicular to the maximum compressive stress (σ). The grooves propagated outwards jerkily (not steadily) at an average velocity of the order of 10 μm/hr, with peak velocities up to ~100 μm/hr. They propagated parallel to $\langle 110 \rangle$. Some grooves did not nucleate at the edge of the hole, but further to the left *on* the crystal face. In some cases, these grooves appeared to have nucleated at previously existing scratches in the surface. These grooves also had their long axis g oriented approximately perpendicular to σ and parallel to $\langle 110 \rangle$.

Grooves increased in length by propagating sideways (parallel to g and sub-perpendicular to σ). They did not nucleate at one moment, but grew during the entire experiment. At the end of the experiment (after 100 hr) the longest grooves had reached a length of almost 1 mm. During the entire experiment, groove width (15–25 μm) and depth remained approximately constant. Long grooves were not wider than short grooves. Grooves did not increase in width when they were propagating. Grooves also moved (translated) over the surface at velocities up to 10 μm/hr, in directions at various angles to g. Some grooves moved upwards, others moved downwards. Some grooves moved in a jerky, unsteady fashion, others moved more continuously. Some grooves moved, in the meanwhile increasing in length. Others moved while decreasing in length. The groove pattern was therefore not static, but dynamic. Grooves increased in length, then stopped increasing in length and sometimes decreased in length. Others moved upwards, then stopped and moved downwards again. Grooves linked up or split in two grooves. Some grooves started off straight, then the entire groove, or only part of it, rotated and attained an oblique orientation. Other grooves started in oblique orientation, then (partly) straightened out. Most of the grooves were slightly undulating during the experiment.

After 71.25 hours a fracture developed in the sample (see F in Fig. 4e). This fracture developed instantaneously. It must have lead to a sudden increase in compressive stress at the right side of the fracture (in Fig. 4e). Within an hour after the fracture had developed, the number of grooves per unit area had almost doubled there, the width and depth of the grooves remaining unaffected (cf. Fig. 4d with Fig. 4f). Also immediately after the fracture developed the grooves near the fracture tip rotated with their long axis towards the fracture tip, suggesting of a reorientation of the maximum compressive stress after the fracture developed. The fracture partly healed in the following hours. The crack tip turned into an array of elongated fluid inclusions. The lower part remained open, but translated slowly to the right, possibly driven by the difference in elastic strain energy between the crystal lattice to the left and the right of the fracture. It moved towards the right, i.e. in the direction of the highest strain, similar to the behaviour observed by den Brok *et al.* (1999) of grain boundaries in elastically strained sodium chlorate aggregates.

Finally a second fracture nucleated from the first one, and soon thereafter the entire sample broke.

In experiments Iskal-2-6 (Morel 2000) it was observed that the grooves dissappeared when either the stress was taken off (see also den Brok & Morel 2001) or when the solution was oversaturated.

Discussion and conclusion

The observations made essentially confirm those of den Brok & Morel (2001). Grooves developed on the elastically strained surface in orientations approximately perpendicular to the maximum compressive stress (σ) and parallel to $\langle 110 \rangle$. The grooves are also comparable in size. At the end, the surface structure was similar to that reported by den Brok & Morel (2001), who

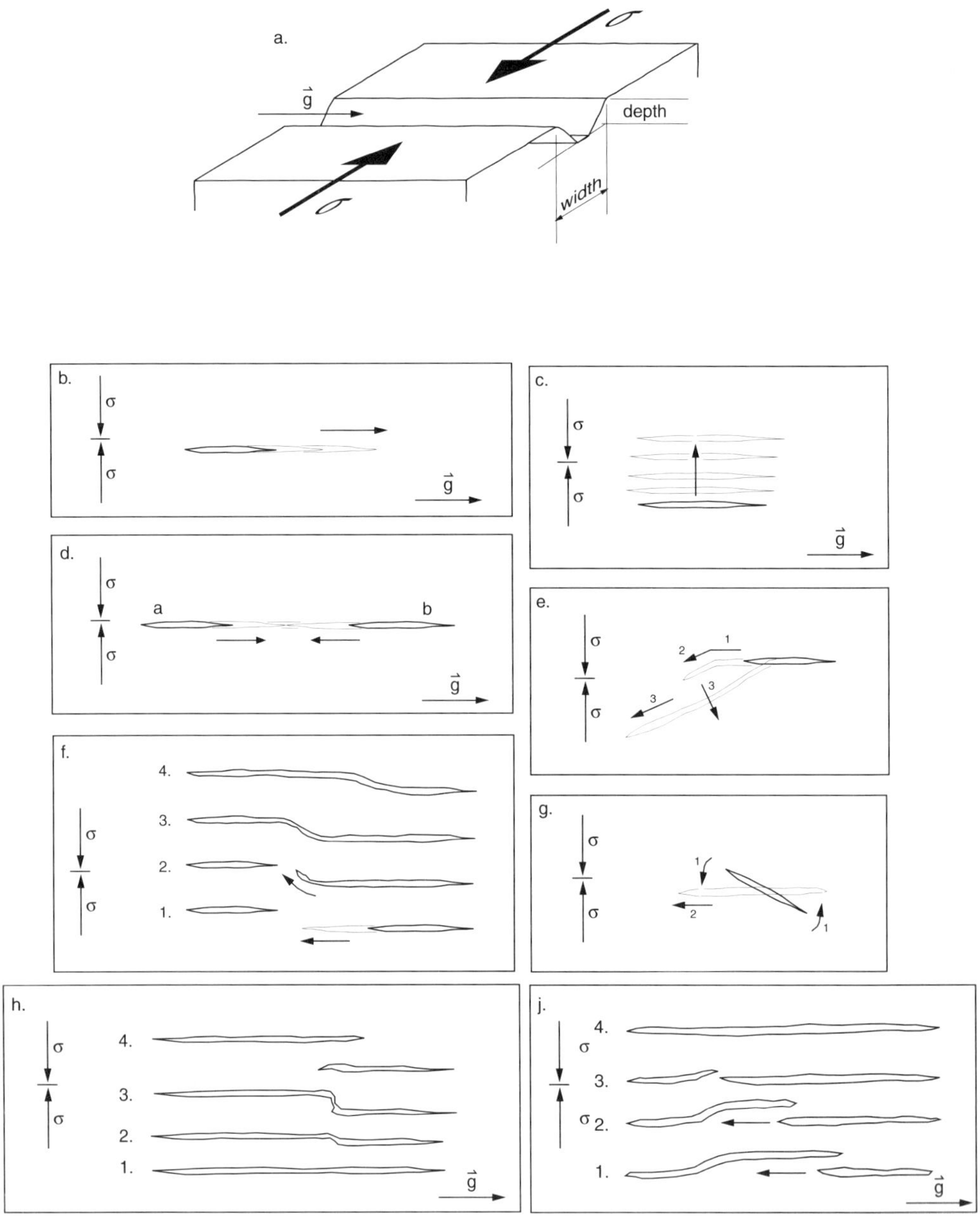

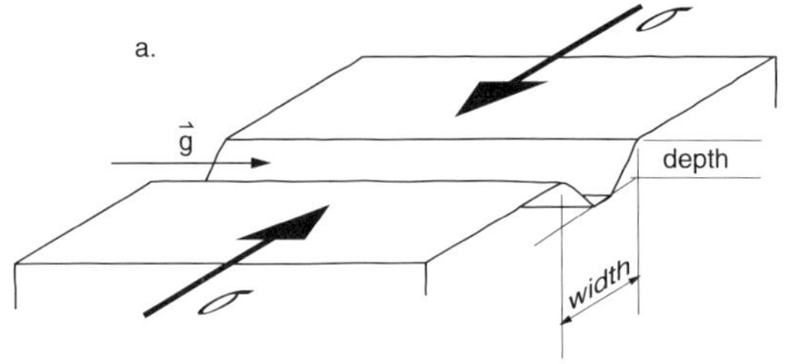

Fig. 5. (**a**) Schematic drawing of geometry (width, depth) of a single groove developed in a crystal face. Grooves developed mostly with their long-axis (g) sub-perpendicular to the maximum compressive stress (σ). (**b**) Propagation (growth) of groove parallel to g. This took place typically at velocities up to 100 μm/hr. (**c**) Translation of groove in a direction perpendicular to g. This took place typically at velocities up to 10 μm/hr. (**d**) Propagation of two grooves a and b linking up to form one single groove. (**e**) Propagation of groove, first parallel to g (1), then at an angle to the initial g (2). Finally the groove starts translating (3) while propagating further in oblique orientation (3). (**f**) Linking of two grooves that are slightly off-set. (**g**) Rotation (1) of initially oblique groove and further translation. (**h**) Splitting of groove. Only left part of the groove starts translating upwards while right part remains fixed. First the groove starts to bend (2 and 3) then it breaks apart into two grooves (4).

were only able to study the surface structure before and after the experiment. The ability to study to the grooves *in situ* and observe how the grooves developed and evolved in time provided essential and new information. The stress-induced roughness consisting of dissolution grooves appeared not to be a static, but a dynamic, structure. The instabilities observed were not static instabilities, fixed in space, but could move as the local stress field changed. The instabilities readily changed position and orientation. To our knowledge we are the first to have seen and documented such a dynamic system of stress-induced surface instabilities.

This process of stress-induced grooves or channels moving along a boundary is much like the type of process invisaged by e.g. Lehner (1990) and Spiers & Schutjens (1990) in their 'dynamically stable' island–channel grain boundary model. The difference is that in our case the instabilities are purely stress induced (i.e. due to elastic strain), whereas Lehner (1990) and Spiers & Schutjens (1990) proposed that differences in crystal plastic strain (crystals defects) would be responsible for generation and mobility of the channels. This is not unlikely in the case of NaCl, which they were discussing.

Leroy & Heidug (1994), Ghoussoub (2000) and Ghoussoub & Leroy (2001) proposed that Grinfeld instabilities (rather than gradients in crystal plastic strain) would play a role in the development of grain boundary roughness. Our observations support their idea. It is important to note, however, that in the models of Ghoussoub (2000) and Ghoussoub & Leroy (2001) the instabilities are not assumed to have an equilibrium shape, such as in our experiments and such as predicted by Grinfeld (1986) and Srolovitz (1989). In Ghoussoub's and Leroy's models, the grooves are not assumed to be in equilibrium. They develop as Grinfeld instabilities, but then deepen and turn into cracklike features, wetting the grain boundaries (cf. Yang & Srolovitz 1993). In our experiments equilibrium existed between, on the one hand, stress forcing the surface to roughen, and, on the other hand, surface forces forcing the surface to straighten (cf. Eq. 1). Stress (elastic strain) led to the development of equilibrium instabilities. Gradients in stress and changes in the magnitude and orientation of the stress field led to motion and reorientation of the instabilities. Thus, according to our findings, fluid channels in the island–channel grain boundaries would be stress-induced (equilibrium) Grinfeld-type instabilities that propagate and translate as the local stress is changing, e.g. due to mass removal by PS. This structure is not expected to smooth out as long as the stress is sufficiently high. Only if the stress is taken off it is to be expected that the roughness will disappear and fluid be trapped in individual inclusions. Den Brok & Morel (2001) observed that the grooves disappeared soon after the stress was taken off.

If stress is increased above a certain threshhold value, the grooves deepen and turn into fractures. This was predicted e.g. by Yang & Srolovitz (1993) and observed on stressed K-alum surfaces by Morel (2000). Thus, an island–crack grain boundary would develop (Gratz 1991; den Brok 1998).

It remains to be demonstrated (e.g. in experiments) that the instabilities do form in *tight* grain boundaries under normal stress. Our observations were made on *free* surfaces (50–100 μm water layer between the glass and the crystal) and caused by surface parallel stress. A preliminary *in situ* experimental study carried out by Zahid (2001) showed, however, that a mobile Grinfeld-type instability developed at a tight grain boundary under normal stress at the grain to grain contacts between two sodium chlorate ($NaClO_3$) crystals. (Like K-alum, $NaClO_3$ is an elastic/brittle salt with a relatively high solubility and fast interface kinetics.) It seems likely, that in addition to surface normal stress, surface parallel stress gradients are built up during PS, and that these may lead to the development of Grinfeld-type instabilities. If the boundary parallel stress changes in magnitude and orientation – a very likely case when PS at the boundary takes place – the instabilities should move, as in our experiments.

It also remains to be demonstrated that the instabilities do form in polycrystalline aggregates and in 'real' rocks under stress. Zahid (2001) observed that Grinfeld-type instabilities developed and moved at free grain boundaries in elastically strained polycrystalline sodium chlorate aggregates. Further experiments are planned to be carried out on quartz, NaCl and calcite.

We learned much from discussions with Y. Leroy, Y. Podladshikov and C. Spiers, and from reviews by J. Ghoussoub and C. Spiers. Our work was financially supported partially by the Deutsche Forschungsgemeinschaft (DFG project Br-1664-1) and the VW-Stiftung.

References

Asaro, R. J. & Tiller, W. A. 1972. Interface morphology development during stress corrosion cracking: Part I. Via surface diffusion. *Metallurgical Transactions*, **3**, 1789–1796.

Bitsanis, I., Somers, S. A., Davis, H. T. & Tirell, M. 1990. Microscopic dynamics of flow in molecularly narrow pores. *Journal of Chemical Physics*, **93**, 3427–3431.

Bosworth, W. 1981. Strain-induced preferential dissolution of halite. *Tectonophysics*, 78, 509–525.

Carter, N. L. & Tsenn, M. C. 1987. Flow properties of continental lithosphere. *Tectonophysics*, **136**, 27–63.

Cox, S. F. & Paterson, M. S. 1991. Experimental dissolution–precipitation creep in quartz aggregates at high temperatures. *Geophysical Research Letters*, **18**, 1401–1404.

de Meer, S., Spiers, C. J. & Nakashima, S. 2001. How thick are grain boundary thin films? An FTIR study on actively pressure dissolving halite-glass contacts. *Abstract Volume of Deformation Mechanisms, Rheology and Tectonics, Noordwijkerhout, The Netherlands*, April 2–4, 2001, p. 114.

den Brok, S. W. J. 1998. Effect of microcracking on pressure solution strain rate: the Gratz grain boundary model. *Geology*, **26**, 915–918.

den Brok, S. W. J. & Morel, J. 2001. The effect of elastic strain on the microstructure of free surfaces of stressed minerals in contact with an aqueous solution. *Geophysical Research Letters*, **28**, 603–606.

den Brok, S. W. J., Zahid, M. & Passchier, C. W. 1998. cataclastic creep of very soluble brittle salt as a rock analogue. *Earth and Planetary Science Letters*, **163**, 83–95.

den Brok, S. W. J., Zahid, M. & Passchier, C. W. 1999. Stress induced grain boundary migration in very soluble brittle salt. *Journal of Structural Geology*, **21**, 147–151.

Dewers, T. & Hajash, A. 1995. Rate laws for water-assisted compaction and stress-induced water-rock interaction in sandstones. *Journal of Geophysical Research, B, Solid Earth and Planets*, **100**, 13093–13112.

Fyfe, W. S., Price, N. J. & Thompson, A. B. 1978. *Fluids in the Earth's Crust : Their Significance in Metamorphic, Tectonic and Chemical Transport Processes*. Elsevier, Amsterdam.

Ghoussoub, J. 2000. *Solid-fluid Phase Transformation Within Grain Boundaries During Compaction by Pressure Solution*. Doctoral Thesis, École nationale des ponts et chausées – École polytechnique, Palaiseau, France.

Ghoussoub, J. & Leroy, Y. M. 2001. Solid-fluid phase transformation within grain boundaries during compaction by pressure solution. *Journal of the Mechanics and Physics of Solids*, **49**, 2385–2430.

Gratz, A. J. 1991. Solution-transfer compaction of quartzites – progress towards a rate law. *Geology*, **19**, 901–904.

Grinfeld, M. A. 1986. Instability of the interface between a non-hydrostatically stressed elastic body and a melt. *Doklady Akademii Nauk SSSR*, **290**, 1358–1363.

Hickman, S. H. & Evans, B. 1991. Experimental pressure solution in halite: the effect of grain/interphase boundary structure. *Journal of the Geological Society, London*, **148**, 549–560.

Horn, R. G., Clarke, D. R. & Clarkson, M. T. 1988. Direct measurement of surface forces between sapphire crystals in aqueous solutions. *Journal of Materials Research*, **3**, 413–416.

Horn, R. G., Smith, D. T. & Haller, W. 1989. Surface forces and viscosity of water measured between silica sheets. *Chemical Physics Letters*, **162**, 404–408.

Israelachvili, J. N. 1986. Measurement of the viscosity of liquids in very thin films. *Journal of Colloid and Interface Science*, **110**, 263–271.

Israelachvili, J. N. 1992. Adhesion forces between surfaces in liquids and condensable vapours. *Surface Science Reports*, **14**, 109–159.

Israelachvili, J. N. 1997. *Intermolecular and Surface Forces*. Academic Press, London.

Kim, K.-S., Hurtado, J. A. & Tan, H. 1999. Evolution of a surface-roughness spectrum caused by stress in nanometer-scale chemical etching. *Physical Review Letters*, 83, 3872–3875.

Lehner, F. K. 1990. Thermodynamics of rock deformation by pressure solution. *In*: Barber, D. J. & Meredith, P. G. (eds) *Deformation Processes in Minerals, Ceramics and Rocks*. Unwin Hyman, London, 296–333.

Lehner, F. K. 1995. A model for intergranular pressure solution in open systems. *Tectonophysics*, **245**, 153–170.

Leroy, Y. M. & Heidug, W. K. 1994. Geometrical evolution of stressed and curved solid-fluid phase boundaries, 2, Stability of cylindrical pores. *Journal of Geophysical Research*, **99**, 517–530.

Magda, J. J., Tirrell, M. & Davis, H. T. 1985. Molecular dynamics of narrow, liquid-filled pores. *J. Chem. Phys.*, **83**, 1888–1901.

Morel, J. 2000. *Experimental Investigation into the Effect of Stress on Dissolution and Growth of Very Soluble Brittle Salts in Aqueous Solution*. PhD Thesis, Johannes Gutenberg-Universität, Mainz, Germany.

Paterson, M. S. 1995. A theory for granular flow accommodated by material transfer via an intergranular fluid. *Tectonophysics*, **245**, 135–151.

Peschel, G., Belouschek, P., Müller, M. M., Müller, M. R. & König, R. 1982. The interaction of solid surfaces in aqueous systems. *Colloid and Polymer Science*, **260**, 444–451.

Raj, R. 1982. Creep in polycrystalline aggregates by matter transport through a liquid phase. *Journal of Geophysical Research*, **87**, 4731–4739.

Ranalli, G. 1987. Rheology of the Earth. Allen & Unwin, Boston.

Rubie, D. C. 1986. The catalysis of mineral reactions by water and restrictions on the presence of aqueous fluid during metamorphism. *Mineralogical Magazine*, **50**, 399–415.

Rutter, E. H. 1976. The kinetics of rock deformation by pressure solution. *Philosophical Transactions of the Royal Society of London*, **A283**, 203–219.

Schoen, M., Cushman, J. H., Diestler, D. J. & Rhykerd, C. L. 1988. Fluids in micropores. II. Self-diffusion in a simple classical fluid in a slit pore. *Journal of Chemical Physics*, **88**, 1394–1407.

Spiers, C. J. & Schutjens, P. M. T. M. 1990. Densification of crystalline aggregates by fluid-phase diffusional creep. *In*: Barber, D. J. & Meredith, P. G. (eds) *Deformation Processes in Minerals, Ceramics and Rocks*. Unwin Hyman, London, Unwin Hyman, Boston, 334–353.

Spiers, C. J. & Carter, N. L. 1998. Microphysics of rocksalt flow in nature. *In*: Aubertin, M. & Hardy, H. R. (eds) *The Mechanical Behavior of Salt, Proceedings of the 4th Conference*. Trans Tech Publ. Series on Rock and Soil Mechanics **22**, 115–128.

Schutjens, P. M. T. M. & Spiers, C. J. 1999. Intergranular pressure solution in NaCl: Grain-to-grain contact experiments under the optical microscope. *Oil & Gas Science and Technology – Rev. IFP*, **54**, 729–750.

Sprunt, E. S. & Nur, A. 1977. Experimental study of the effects of stress on solution rate. *Journal of Geophysical Research*, **82**, 3013–3022.

Srolovitz, D. J. 1989. On the stability of surfaces of stressed solids. *Acta Metallurgica*, **37**, 621–625.

Twiss, R. J. & Moores, E. M. 1992. *Structural Geology*. W. H. Freeman & Company, New York.

Vanderlick, T. K., & Davis, H. T. 1987. Self-diffusion in fluids in microporous solids. *Journal of Chemical Physics*, **87**, 1791–1795.

Weyl, P. 1959. Pressure solution and the force of crystallisation – a phenomenological theory. *Journal of Geophysical Research*, **64**, 2001–2025.

Yang, W. H. & Srolovitz, D. J. 1993. Cracklike surface instabilities in stressed solids. *Physical Review Letters*, **71**, 1593–1596.

Zahid, M. 2001. *Duktile Gesteinsdeformation durch zeitabhängige Kataklase – Eine experimentelle Untersuchung an einem elastisch/sprödem Gesteinsanalogmaterial*. Dissertation Johannes Gutenberg-Universität Mainz.

Zhu, Y. & Granick, S. 2001. Viscosity of interfacial water. *Physical Review Letters*, **87**, 96–104.

Fluid-assisted large strains in a crustal-scale décollement (Hercynian Belt of South Brittany, France)

F. LE HEBEL, D. GAPAIS, S. FOURCADE & R. CAPDEVILA

Géosciences Rennes, U.M.R. 6118 C.N.R.S., Université de Rennes 1, 35042, Rennes cedex, France (e-mail: Florence.Le-Hebel@univ-rennes1.fr)

Abstract: Crustal deformations often occur along gently-dipping shear zones, where the scale of fluid transfers and the coupling between deformations and fluid availability are important questions. We present a structural and geochemical study of felsic rocks deformed at about 350–400 °C within a crustal-scale décollement zone from the Hercynian Belt of South Brittany. At regional scale, deformations resulted in large distributed strains accumulated by dissolution–crystallization of quartz and feldspars. Mass transfers produced veins growing parallel to the foliation. Veins were major precipitation zones, while adjacent rocks were the main zones of dissolution. Deformation patterns, vein mineralogy, and oxygen isotope data show that both the amount of fluids and the scale of fluid flow were limited, and that the dominant transfer mechanism was diffusion. Results suggest that the deformation zone has acted as a trap for early fluids rather than as a syn-kinematic fluid channel.

The existence of close relationships between fluid movements and crustal deformation is well established (Beach 1976; Fyfe *et al.* 1978; Etheridge *et al.* 1984; Cox & Etheridge 1989, Newton 1990; Ferry 1994; Oliver 1996), especially where dissolution–crystallization processes are involved (Durney 1972; Rutter 1976; Gratier 1993). Furthermore, shear zones are often described as channels for fluids (e.g. Beach & Fyfe 1972; Beach 1976; Kerrich *et al.* 1977; Fourcade *et al.* 1989; Carter & Dworkin 1990; Marquer & Burkhard 1992; Ring 1999), allowing connections between superficial fluids and deep fluid reservoirs (Reynolds & Lister 1987; Pili *et al.* 1997). Our particular interest concerns crustal-scale décollement zones because of their gently-dipping attitude, at high angle to the dominant directions of fluid motions expected to occur throughout the metamorphic crust. Our purpose is to discuss how fluid-rock interactions within such zones can influence deformation processes and fluid paths within the crust. For this purpose, we discuss the example of a large-strain unit from the Hercynian belt of South Brittany, which was downthrust to about 25 km, exhumed, and then reworked during late-orogenic extension.

Geological setting

South Brittany (Western France) is part of the internal zones of the Hercynian belt of Western Europe. It is bounded to the north by a crustal-scale wrench zone (the South Armorican Shear Zone) and disappears to the south under the Aquitaine Basin (Fig. 1). From bottom to top, the main tectonic units are as follows (Fig. 1) (Ters 1972; Audren 1987; Gapais *et al.* 1993; Brown & Dallmeyer 1996).

(1) High temperature units, mainly made of migmatites, gneisses, and amphibolites. These rocks were exhumed during upper Carboniferous times (*c.* 320–290 Ma) (Jones & Brown 1989, 1990; Audren & Triboulet 1993; Gapais *et al.* 1993; Brown & Dallmeyer 1996).
(2) Metasediments affected by Barrovian-type metamorphism of upper Carboniferous age (325 Ma or less), and decreasing in grade from bottom to top (Ters 1972; Brillanceau 1978; Bossière 1988; Triboulet & Audren 1988; Goujou 1992; Brown and Dallmeyer 1996).
(3) Porphyritic felsic metavolcanics known as the Vendée and Belle-Ile-en-Mer 'Porphyroïdes', and affected by HP-LT metamorphism (7–9 kbar, 350–400 °C; Le Hébel *et al.* 2002). Sheets of upper-Carboniferous synkinematic two-mica granites (Bernard-Griffiths *et al.* 1985) mark out the transition zone between metasediments and metavolcanics (Gapais *et al.* 1993) (Fig. 1).
(4) Mafic rocks and metapelites affected by HP metamorphism (14–18 kbar, 500–550 °C) (blueschists of Ile-de-Groix) (Triboulet 1974; Guiraud *et al.* 1987; Bosse *et al.* 2000, 2002) during late Devonian times (380–370 Ma), and exhumed during the lower Carboniferous (Bosse *et al.* 2000, 2002).

At regional scale, the metasediments and the metavolcanics are affected by a subhorizontal to gently dipping foliation bearing a stretching

From: DE MEER, S., DRURY, M. R., DE BRESSER, J. H. P. & PENNOCK, G. M. (eds) 2002. *Deformation Mechanisms, Rheology and Tectonics: Current Status and Future Perspectives.* Geological Society, London, Special Publications, **200**, 85–101. 0305-8719/02/$15

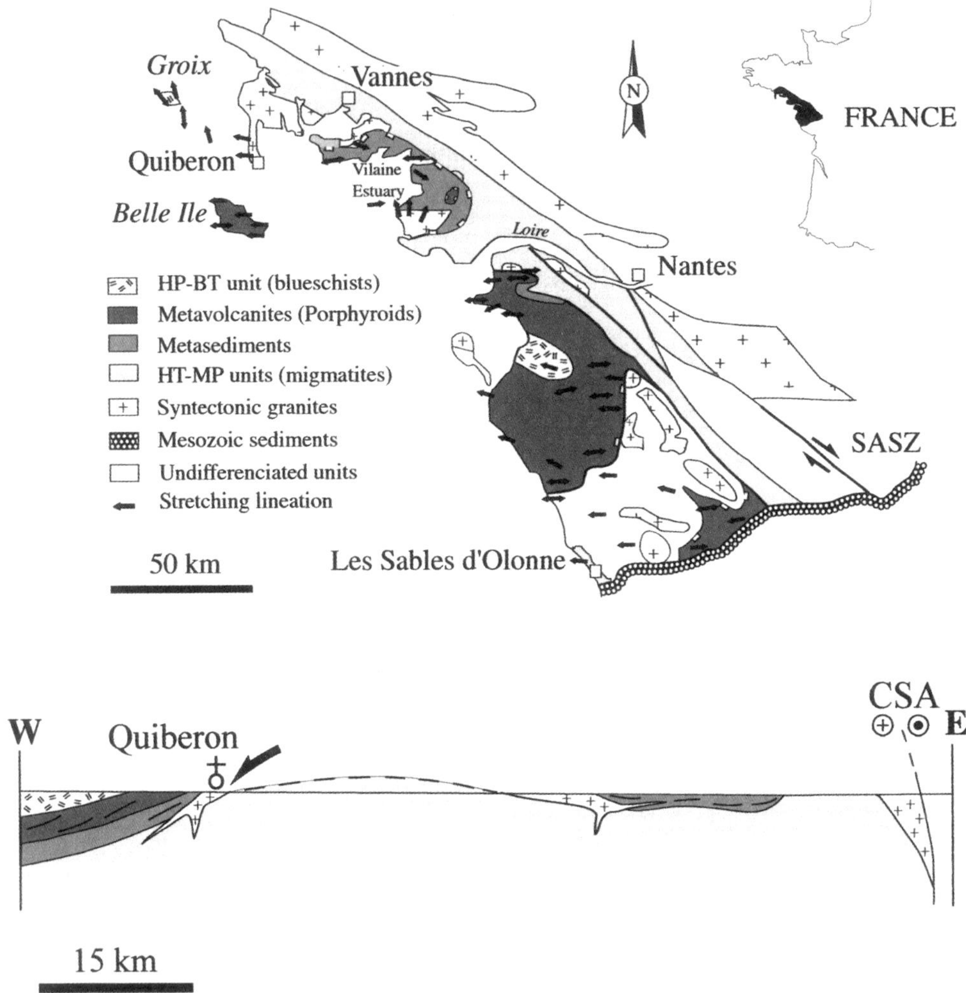

Fig. 1. Simplified geological map of South Brittany and schematic east–west cross-section in the Quiberon area. SASZ, South Armorican Shear Zone. Modified from Brun & Burg (1982) and Gapais *et al.* (1993).

lineation associated with overall westward-directed shear (Burg 1981; Brun & Burg 1982; Vauchez *et al.* 1987) (Fig. 1). Recent pressure estimates around 7–9 kbar in the metavolcanics (Le Hébel *et al.* 2002) indicate that their early tectonic history is comparable to that of the overlying blueschists, with late Devonian syn-collisional downthrusting, followed by syn-convergence exhumation during lower Carboniferous times (Bosse *et al.* 2000, 2002). The basal contact of the metavolcanics is thus part of an early thrust system allowing the exhumation of HP units. During the upper Carboniferous, the area was reworked by extensional tectonics, with crustal thinning, emplacement of leucogranites, and exhumation of underlying migmatites (Gapais *et al.* 1993). The present study focuses on the metavolcanic unit, which was located at the vicinity of the brittle–ductile transition during regional extension.

Petrography

The rocks studied are made of volcanogenic metasediments and metavolcanics (Boyer 1974; Chalet 1985) of calcalkaline affinity. Two main types of lithologies have been studied: coarse-grained rhyodacitic to rhyolitic augengneisses made of millimetre-scale to centimetre-scale phenocrysts of feldspar and quartz embedded in a fine-grained matrix, and fine-grained, phenocryst-free, dacitic lithologies. The present-day mineralogies of the two lithologies are essentially Qz–Kf–phg ± Ab and Qz–phg–Ab–chl, respectively. Widespread adularization of feldspars is observed, even in rare areas where the rocks are undeformed. This attests to

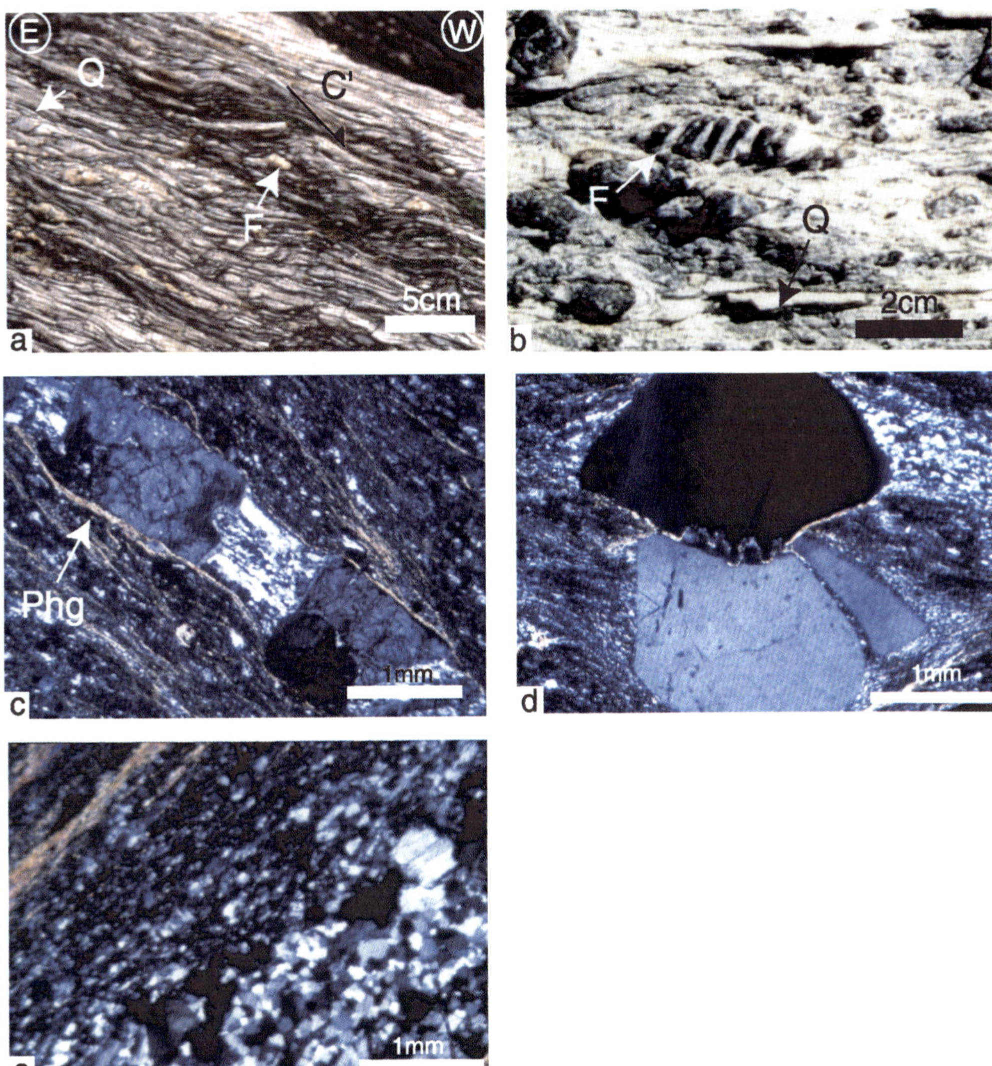

Fig. 2. Photographs showing structures in the metavolcanics. (**a**) Coarse-grained augengneiss viewed perpendicular to foliation and parallel to stretching lineation, showing subhorizontal foliation, associated extensional shear bands (C′) indicating top-to-the-west motion, and feldspar (F) and quartz (Q) porphyroclasts (Port Giraud, Loire Estuary). (**b**) Coarse-grained augengneiss showing quartz ribbons (Q) and truncated K-feldspars (F). The grey to black colour of the rock is due to the presence of graphitoid, subsequent to oil impregnation during early history (Pointe du Talus, Belle-Ile-en-Mer). (**c**) Thin section of augengneiss, showing truncated K-feldspar and quartz shadow-zones surrounded by thin phengite-rich rims (Phg). View parallel to stretching lineation and perpendicular to foliation. (**d**) Thin section of moderately deformed augengneiss showing two quartz porphyroclasts, with quartz crystallization in shadow zones, stylolites along grain contact at high angle to principal shortening (vertical), and thin white-mica rims around clasts. View perpendicular to foliation and parallel to stretching lineation. (**e**) Thin section of strongly deformed metavolcanics showing alternating quartz–feldspar bands and phengite-rich bands.

important early metasomatic processes, marked by very important chemical changes (K, Rb and Ba gains, Ca and Sr losses). Early impregnation with oil (now present as graphitoid) is also widespread within the unit.

Deformation

Metavolcanics

Metavolcanics are affected by a pervasive flat-lying foliation bearing a strong lineation marked by stretched quartz and feldspar phenocrysts (Fig. 2a, b). Associated shear criteria mainly indicate top-to-the-west shearing (Fig. 2a) (Burg 1981; Brun & Burg 1982; Vauchez *et al.* 1987). Finite strains are large, the stretching recorded by truncated feldspar phenocrysts being commonly of the order of 300–400% or more (Burg 1981). Remnants of low-strain domains are very rare (Vauchez *et al.* 1987).

At all scales, structures attest to extensive processes of dissolution–crystallisation (see also Vauchez *et al.* 1987). Feldspar porphyroclasts show crystallizations of quartz and (or) feldspar in shadow zones and in fractures at high angle to the principal stretch direction (Fig. 2b, c). Quartz pressure shadows also occur around quartz phenocrysts (Fig. 2d). In weakly deformed samples, quartz grains are globular. Stylolites are observed between adjacent grains, and discrete white mica rims occur around phenocryst boundaries at high angle to the direction of principal shortening (Fig. 2d). With increasing strain, quartz ribbons develop, stretching of feldspars increases, and phengite rims around phenocrysts become larger (Fig. 2c). Strongly deformed rocks consist of alternating quartz–feldspar bands and phengite-rich bands (Fig. 2e). The relationships between strain intensity and relative phengite content were further examined in thin sections. Strain intensity was estimated using the ratio between preserved primary magmatic quartz and total modal quartz. A plot of the relative proportion of phengite-rich bands against this strain indicator shows a positive correlation (Fig. 3). Error bars in this diagram are probably large. Nevertheless, strain estimates are qualitatively consistent with other strain indicators (degrees of elongation of quartz phenocrysts and of feldspar truncation).

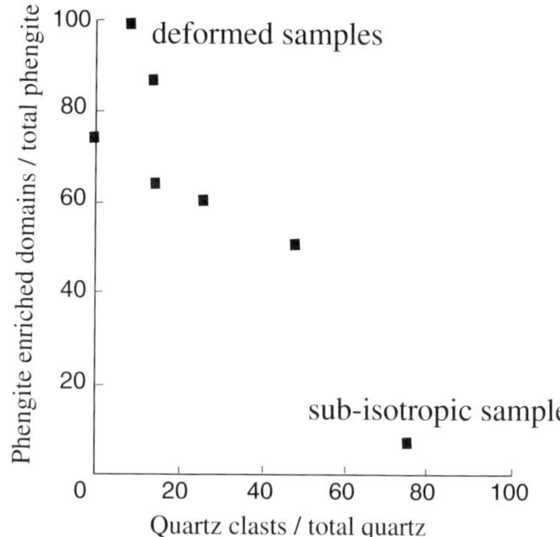

Fig. 3. Relative proportion of phengite versus strain intensity, estimated using ratio between relict magmatic quartz clasts and total modal quartz.

Syn-kinematic vein arrays

Outcrops show a spectacular quartz-rich vein system, generally pegmatoidic (quartz + feldspar) (Fig. 4a, b). Most veins are parallel to the foliation, but a few veins cut across the foliation as local swarms or tension gashes (Fig. 4b).

Foliation-parallel veins contain quartz and feldspar, and show phengite-rich selvages (Fig. 4c). Vein-selvage associations alternate with layers of fine-grained or coarse-grained metavolcanics. Veins appear to have formed parallel to the foliation. Indeed, veins parallel to the axial plane of folds associated with the regional foliation are observed (Fig. 4d). The range of maximum spacing of vein concentrations is metre-scale.

Pegmatoidic veins are affected by extensive cracking and sealing processes. Feldspar-rich domains and quartz-rich domains have rod shapes and ribbon shapes, respectively (Fig. 4e). Cracks are sealed by quartz and feldspar fibres elongate parallel to the stretching lineation (Fig. 4f). Evidence for substantial control of crack margin mineralogy on sealing minerals is common (Fig. 4e).

Various degrees of cracking, intracrystalline deformation (prismatic deformation bands, undulatory extinctions) and dynamic recrystallization (core and mantle structures) affect the fibres (Fig. 4f). The small size of recrystallized grains and the dominant prismatic attitude of deformation bands are consistent with a low temperature dislocation glide mechanism involving basal slip. Despite recrystallization, quartz fibres can show significant concentrations of ⟨C⟩ axes at low angle to the stretching lineation ($\lambda 1$, Fig. 5) (Vauchez *et al.* 1987). As intracrystalline ⟨C⟩ slip is not expected because of the low

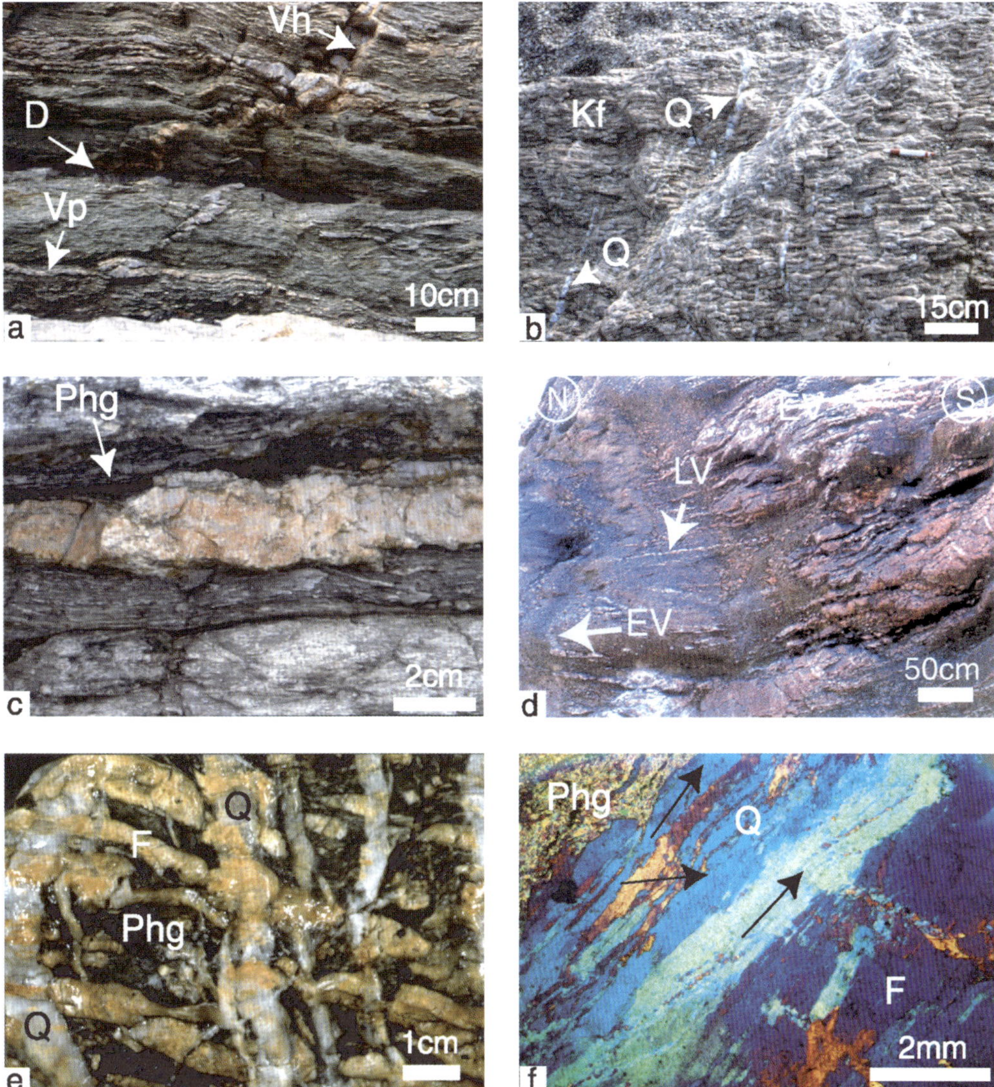

Fig. 4. Photographs showing structures associated with veins formed by dissolution-crystallization. (**a**) View perpendicular to foliation and parallel to stretching lineation, showing coarse-grained lithology (top) and finer-grained lithology (bottom), with veins parallel to foliation (Vp) and at high angle to foliation (Vh); oblique veins are cut or sheared along a discrete décollement zone (D) parallel to foliation (Piriac, Vilaine Estuary). (**b**) Top view of a quartz-feldspar vein parallel to foliation with late quartz-rich gashes (Q) at high angle to stretching lineation underlined by K-feldspar rods (Kf) (La Sauzaie, Vendée). (**c**) Quartz-feldspar vein with phengite-enriched selvages (Phg). View parallel to stretching lineation and perpendicular to foliation (Pointe du talus, Belle-Ile-en-Mer). (**d**) Fold with axial-planar flat-lying foliation. The fold deforms early quartz-feldspar veins (EV) and shows later axial-planar veins (LV), which indicates that successive generations of veins formed during progressive deformation and that veins grew parallel to the flattening plane. The view is at high angle to stretching lineation and fold axis (La Sauzaie, Vendée). (**e**) Vein structure showing rod-shaped feldspars (F), local patches of phengite-rich residues (Phg), and quartz-feldspar gashes at high angle to stretching lineation. The nature of crystallization in gashes is locally controlled by vein wall mineralogy. View parallel to foliation (La Sauzaie, Vendée). (**f**) Thin-section of a vein showing relict, partly recrystallized, quartz (Q) and feldspar (F) fibres, and some phengite-rich residue (Phg). Arrows underline traces of local quartz ⟨C⟩ axis projections, which lie at low angle to fibre long axis and stretching lineation (view parallel to foliation, cross-polar light, gypsum plate).

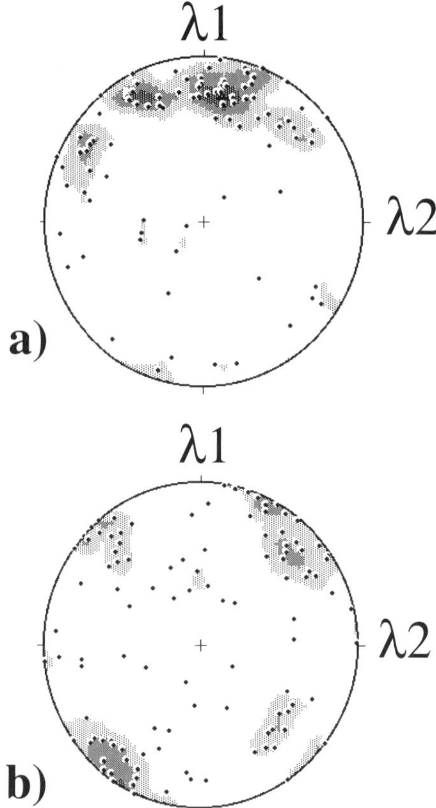

Fig. 5. Examples of ⟨C⟩ axis fabrics of quartz fibres in veins (lower hemisphere, equal area projection). Number of measured grains are 90 (**a**) and 110 (**b**). Contours of shaded zones are in area % (the limited number of measurements results from the large fibre sizes, see Fig. 4f). $\lambda 1$ and $\lambda 2$ are principal stretch axis and intermediate strain axis, respectively.

temperature, such fabrics are best explained as resulting from oriented growth during sealing (Cox & Etheridge 1983; Gapais & Barbarin 1986; Hippertt 1994; Stallard & Shelley 1995). Low angles between veins, fibre long axes, and primary quartz ⟨C⟩ axis preferred orientations confirm that veins grew parallel to the foliation.

The interface between veins and phengite selvages is generally marked by particularly well-developed quartz and feldspar fibres parallel to the stretching lineation. This can be attributed to local phyllosilicate-enhanced strain softening (Weyl 1959; Tada & Siver 1989; Dewers & Ortoleva 1991; Schwartz & Stöckert 1996), as a result of enhanced interphase boundary diffusivity (Rutter 1983; Hickman & Evans 1995; Renard *et al.* 1997; Farver & Jund 1999; Bos *et al.* 2000). Consistently, foliation, vein boundaries and lithological interfaces can act as zones of localized slip (Fig. 4a).

The latest deformation features are tension gashes at high angle to the principal stretch. They are mainly filled by quartz and never bounded by phengite-rich rims (Fig. 4a, b). They can cut across veins and host-rock, but are concentrated within vein-rich domains (Fig. 4b).

Geochemistry

Analytical techniques

Rock samples of about 3 to 15 kg (depending on grain size) were crushed and then powdered using an agate mortar. Major and trace elements were analysed by ICP AES and ICP MS, respectively (CRPG, Nancy). Uncertainties and their dependence on abundance levels are available on the web site: http://crpg.cnrs-nancy.fr.

Quartz from veins (weighing 10^{-3} to 10^{-2} kg) used for O isotopes was hand-picked and briefly treated with hydrofluoric acid to remove any trace of feldspar, and then powdered using a steel mortar. O was extracted in Ni tubes using BrF_5 (Clayton & Mayeda 1963), converted to CO_2, and analysed using a VG SIRA 10 instrument (Rennes University). Results (expressed with the δ notation versus the VSMOW scale) were normalized using internal standards (basaltic MORB glass Circé 93) and international standards (carbonate NBS 19, quartz NBS 28). During the course of this study, the NBS 28 reference was found in the range 9.3 to 9.4‰, and the results were normalized to the recommended value of 9.6‰. On the basis of duplicate extractions, the total uncertainties with respect to the VSMOW scale are estimated to be generally around 0.15‰.

Composition of veins versus lithology

In order to determine the relationships between the nature of feldspars in veins and in their host-rock, staining experiments were performed on rock slabs (Nold & Erickson 1967). This was done on veins from both coarse-grained and fine-grained rocks. K-feldspar was detected in nine out of ten vein samples from coarse-grained lithologies (Qz, Kfsp, Phg, +/− Ab mineralogy). In contrast, the fifteen vein slabs sampled in fine-grained lithologies (Qz-Phg-Ab-Chl mineralogy) contained quartz and albite only. These results are consistent with mineral electron microprobe analyses (Fig. 6).

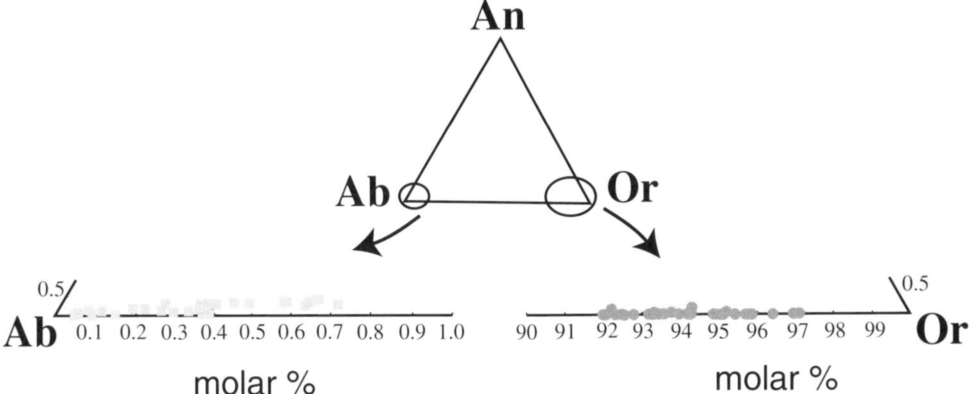

Fig. 6. Examples of feldspar compositions (microprobe analyses, Paris VI University) from veins embedded in Kf-bearing coarse-grained lithologies (dots), and Kf-free fine-grained lithologies (squares). Albite can locally occur in some veins from coarse-grained lithologies, but Kf never occurs in veins from fined-grained lithologies.

Mass transfers associated with vein development

At several sites, we applied Gresens (1967) and Grant (1986) methods on phengite selvage/host-rock pairs, in order to estimate mass transfers involved in vein formation. Chemical analyses are given in Table 1. For each sample pair, we compared the results of this approach with those of staining experiments and of mineral electron microprobe analyses on respective adjacent veins. According to Grant's method, we determined an isocon, which is a reference line corresponding to a zero concentration change obtained by plotting concentrations of immobile elements in the presumed phengite residue (Ca) against those in the initial protolith (Co) (Fig. 7). We found that Th, Ta, Nb, Zr, Hf, Y, REE and Ti were relatively well linearly correlated. These elements, which are reputed to be rather

Table 1. *Chemical analyses of phengitic selvage-protolith pairs used for Grant's analysis (electron microprobe analyses, Paris VI University)*

Sample	Protolith FL 159	Residue FL156	Protolith FL 167	Residue FL 166	Isotropic CG 13a	Deformed FL 23
(wt%)						
SiO_2	72.44	51.62	61.73	51.51	76.65	78.44
TiO_2	0.33	0.75	0.78	1.05	traces	0.08
Al_2O_3	14.75	25.17	18.53	23.65	11.89	11.32
Fe_2O_3	1.94	5.24	6.92	8.71	0.94	1.23
MnO	0	0.03	0.06	0.11	traces	traces
MgO	0.6	2.03	2.36	3.19	traces	0.08
CaO	0.21	0.21	0.16	0.20	traces	traces
Na_2O	2.93	0.57	1.69	1.20	1.68	1.09
K_2O	4.82	9.82	4.14	5.70	7.23	6.15
P_2O_5	0.12	0.11	0.10	0.14	traces	traces
Pf	1.75	4.42	3.37	4.45	1.16	0.93
Total	99.89	99.97	99.84	99.91	99.65	99.32
(ppm)						
Ta	1.21	2.28	1.37	1.91	2.78	2.64
Hf	5.31	11.65	4.24	6.41	4.50	4.20
Th	23.20	49.18	19.46	24.67	43.00	40.20
Nb	12.38	27.19	15.46	20.88	17.40	20.50
Zr	191.0	424.8	162.0	238.7	110.0	108.5
Y	36.74	72.20	27.71	33.87	41.30	40.10
Nd	39.58	89.1	41.39	53.28	45.80	48.70

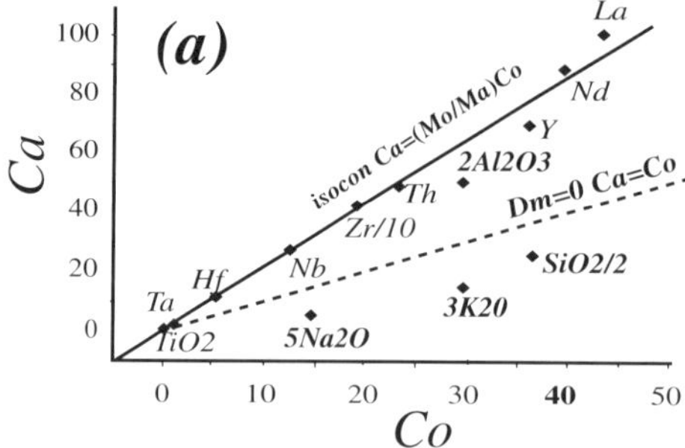

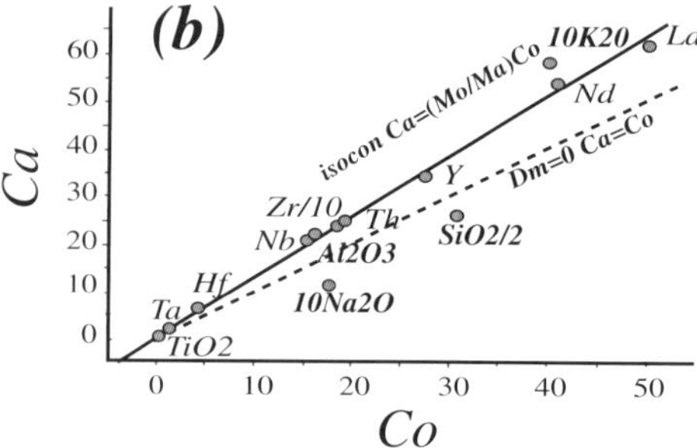

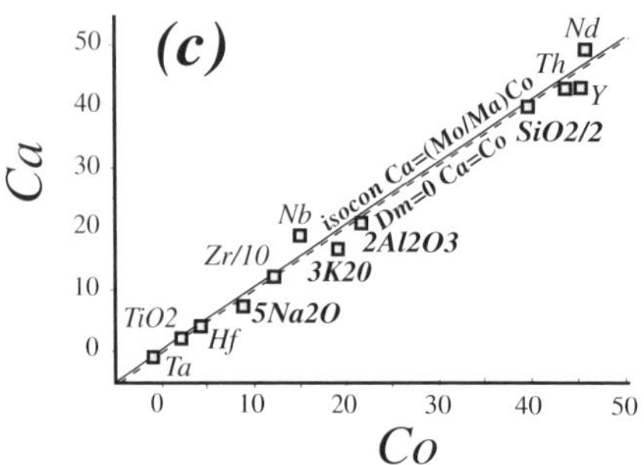

Fig. 7. Grant's diagrams showing element concentrations within phengite residue (Ca) against neighbouring host-rocks (Co) in coarse-grained layer (**a**) and fine-grained layer (**b**), and within deformed metavolcanics (Ca) against neighbouring isotropic metavolcanics (Co) (**c**). Major elements are expressed in oxide weight percentages; trace elements are in ppm and are arbitrarily scaled. Elements plotting above and below the isocon show relative gains and losses, respectively.

immobile in low-temperature aqueous hydrothermal systems, were thus used to define the isocon. The slope of the isocon defines the mass change, and deviation of a data point from the isocon shows the concentration change for the corresponding element. A first pair of phyllosicate selvage-protolith, sampled in a coarse-grained augengneiss (protolith mineralogy: Qz, Kf, phg, ±Ab), shows relative losses in SiO_2, Na_2O, K_2O, and Al_2O_3 (Fig. 7a). Consistently, the adjacent vein contains quartz, albite and K-feldspar. A second pair from a fine-grained lithology (protolith mineralogy: Qz, Phg, Ab, Chl) shows relative losses in Na_2O and SiO_2 (Fig. 7b), and adjacent veins are composed of quartz and albite.

Attempts to estimate volume changes between protoliths and phengitic residues were done using measurements of specific gravities (He pycnometer measurements in Orléans University, precision of the order of ±0.001). In the fine-grained lithology (Fig. 7b), a volume loss of *c.* 75% was obtained, which is compatible with entire removal of quartz and feldspar from the protolith. In the coarse-grained lithology (Fig. 7a), the estimated volume loss (*c.* 45%) is too low if one assumes a complete removal of quartz and feldspar from the protolith. This is in fact due to the occurrence of some quartz and feldspar remnants within phengite-rich selvages. In addition, the true protolith might have been more felsic than the one sampled.

Rare localities are moderately deformed, with low-strain domains affected by metre-scale shear zones (Vauchez *et al.* 1987) marked by the lack of veins. In one of these localities, we applied the Grant's method on an isotropic-deformed rock pair. Results indicate no relative loss or gain of elements (Fig. 7c). This emphasizes that where veins did not develop, no

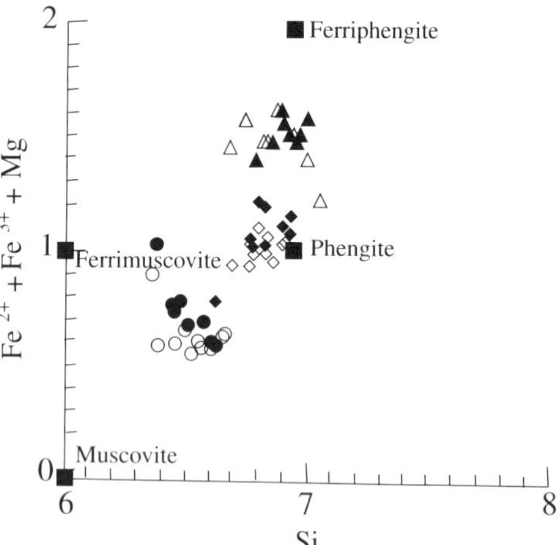

Fig. 8. Diagram showing compositional similarity of phengite grains in pairs of phengite-rich residue and adjacent protholith from three different areas (circles, triangles and diamonds). Full and open symbols are phengites in residue and protholith, respectively (electron microprobe analyses, Paris VI University).

significant element transfers occurred during deformation.

It is worth noting that the compositions of phengite grains in pairs of phengite-rich selvages and adjacent host-rocks vary from one sampling site to another, but are similar at a given sampling site (Fig. 8). This feature illustrates the 'chemical inertia' of phengite and probably chlorite.

Oxygen isotopes

Oxygen isotopes have been analysed on whole-rocks (WR) and on quartz veins sampled in different lithologies of the metavolcanics.

From regional scale to metre-scale, quartz shows variable isotopic compositions (Table 2). To document the relationships between isotopic compositions of veins and host lithology, we focused our study on a 1200 m long section in Belle-Ile-en-Mer (see Fig. 1), exhibiting representative lithological variations. From NE to SW, the $\delta^{18}O$ values of quartz from veins and of WR display concomitant variations, with jumps mostly located at lithological boundaries (Fig. 9a, b) (Le Hébel *et al.* 2000). Quartz $\delta^{18}O$ values are relatively uniform around 13 to 14‰ in the first coarse-grained layer, then decrease to 12 ± 0.5‰ in the fine-grained layer, and reach 16 ± 0.5‰ within the next 75 m thick coarse-grained layer. A local ^{18}O-enrichment ($\delta^{18}O = 18 \pm 0.5$‰) is observed on the profile, in a zone probably reworked by a late deformation associated with oxidization of local sulfides.

Table 2. *$\delta^{18}O$ values of whole-rock and quartz in the profile shown on Fig. 9*

Samples of quartz from veins parallel to foliation	$\delta^{18}O$ v. SMOW	Whole-rock samples	$\delta^{18}O$ v. SMOW
FL 3y Qz	16.36 (0.01)	FL 3b WR	12.1 (0.2)
FL 5d Qz	13.3	FL 3c WR	13.34 (0.01)
FL 40c Qz	13.46 (0.06)	FL 5b WR	11.9
FL 40d Qz	12.8	FL 5a WR	12.0
FL 40e Qz	13.39 (0.03)	FL 171 WR	10.3
FL40i Qz	12.9	FL 175 WR	10.3
FL 40j Qz	12.9	FL 176 WR	12.72 (0.02)
FL 41e Qz	15.9	FL 178 WR	14.5 (0.1)
FL 45a Qz	13.3	FL 180 WR	11.6
FL 45b Qz	13.6	FL 182 WR	9.0
FL 55a Qz	16.2	FL 183 WR	9.7
FL 55b Qz	17.3	FL 185 WR	11.9
FL 55c Qz	17.9	FL 237 WR	11.3
FL 57 Qz	11.6	FL 239 WR	10.6
FL 58 Qz	11.7	FL 241 WR	12.2
FL 59 Qz	13.5	FL 243 WR	11.4
FL 60 Qz	13.6	FL 246 WR	9.9
FL 81 Qz	11.4	FL 248 WR	15.0
FL 82a Qz	11.7	FL 250 WR	12.6
FL 82b Qz	11.9 (0.3)	FL 252 WR	12.5
FL 104 Qz	18.4		
FL 105 Qz	17.1		
FL 108 Qz	11.9		
FL 113 Qz	11.5		
FL 114 Qz	11.7		
FL 128 Qz	17.49 (0.09)		
FL 131 Qz	11.79 (0.02)		
FL 140	15.4		
FL 141	15.9		
FL 181 Qz	15.0		
FL 184 Qz	11.4		
FL 240 Qz	13.0		
FL 244 Qz	13.2		
FL 247 Qz	18.0		
FL 249 Qz	14.8		
FL 251 Qz	14.6		

Numbers in brackets indicate the deviation recorded for two separate O extractions.

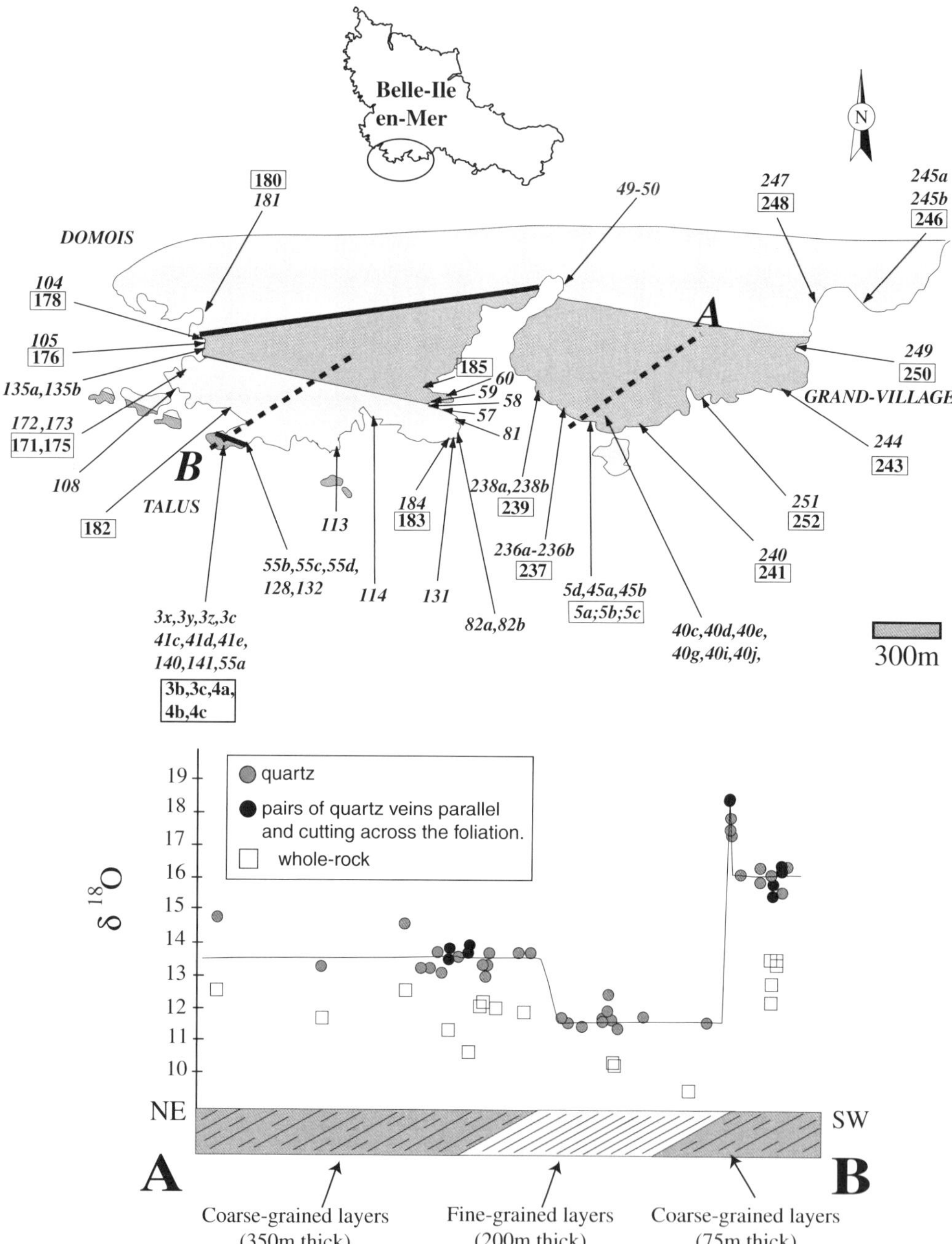

Fig. 9. Example of variations of $\delta^{18}O$ values of whole-rock and quartz veins in alternating lithologies (Belle-Ile-en-Mer area: see Fig. 1). Sampling localities are shown on map (**a**) (numbers in boxes are whole-rock data, others are quartz veins), and $\delta^{18}O$ values are shown on cross-section (**b**) (modified after Le Hébel *et al.* 2000).

At each sampling site, quartz from foliation-parallel veins and from late tension gashes show similar isotopic compositions ($\delta^{18}O$ variations are systematically less than 1 δ unit, a range similar to the isotopic variability within a given lithology: Fig. 9b, Table 3).

WR $\delta^{18}O$ values are clearly too high to reflect magmatic signatures since the highest $\delta^{18}O$

Table 3. *$\delta^{18}O$ values of quartz from early and late veins at various localities, and $\delta^{18}O$ values of quartz from early veins and from magmatic clasts in neighbouring host-rocks (bottom)*

Sample	Quartz $\delta^{18}O$ from vein parallel to foliation	Quartz $\delta^{18}O$ from vein cutting across foliation
Belle-Ile-en-Mer		
FL 3x-FL 3z	16.50 (0.02)	16.6
FL 41d-FL 41c	16.13 (0.07)	15.6(0.1)
FL 55d-FL 132	18.4	18.4
FL 236a-FL 236b	13.4	13.7
FL 238a-FL 238b	13.9	13.1
FL 172-FL173	12.7	13.4
FL 49-FL 50	12.2 (0.2)	13.5 (0.3)
FL123a -FL 123b	11.24 (0.02)	11.4 (0.2)
FL 135a-FL 135b	16.82 (0.05)	16.7
FL 138-FL 139	12.68 (0.04)	13.66 (0.04)
FL 245a-FL 245b	12.4	12.9
Piriac (Vilaine Estuary)		
FL 191a-191c	13.1	12.6
FL 190a-190c	13.0	12.6
FL193a-193b	13.1	14.3
La Sauzaie (Vendée)		
FL 221-FL 225	14.8	15.1
Sample	$\delta^{18}O$ clast	$\delta^{18}O$ vein
Belle-Ile-en-Mer		
FL 40g	11.8	11.3 (0.2)
FL 3c	15.4	15.9 (0.1)
Piriac (Vilaine Estuary)		
FL 196	12.19 (0.03)	12.95 (0.02)

Numbers in brackets indicate the deviation recorded for two separate O extractions.

values reported for crustal granitoids do not exceed 13‰ (e.g. O'Neil & Chappell 1977; Bernard-Griffiths *et al.* 1985; Fourcade *et al.* 2001). Such high $\delta^{18}O$ compositions are reminiscent of those recorded in some K-metasomatized ash flow tuffs elsewhere (e.g. Roddy *et al.* 1988) and are recorded even in undeformed rocks which also contain replacement adularia. They clearly reflect fluid-rock interactions under large fluid/rock ratios which are likely related to the early metasomatic event marked by adularization.

Discussion

Nature and scale of mass transfer

K-feldspar is only found within veins from coarse-grained, low Fe-Mg, augengneisses containing the Qz–Kf–Phg ± Alb paragenesis. In fine-grained dacitic lithologies, which have much higher Fe + Mg contents, K is totally combined as phengite and the only feldspar present in veins is albite (the excess Fe + Mg is mainly stored in chlorite). This feature demonstrates that vein mineralogy is controlled by the host-rock composition, and shows that mass transfers by a fluid were limited. Only alkalis contained in feldspars were redistributed into veins, and phengite behaved as an inert residual phase. This inference is consistent with structural observations and is further supported by phengite compositions (Fig. 8). An earlier study of the same rocks attributed the enrichment in phengite with increasing strain to fluid channelling along shear zones (Vauchez *et al.* 1987). Actually, the studied example differs from more common ones where water-assisted breakdown of feldspars produces white micas (Hemley 1959; Beach 1976), as argued in many LT or MT shear zones (e.g. White & Knipe 1978; Marquer *et al.* 1985; Wintsch *et al.* 1995; Wibberley 1999, van Staal *et al.* 2001). Mass transfer studies in such zones often suggest that Al is moderately mobile or immobile (Carmichael 1969; Ferry 1983; Ague 1991; Lentz 1999). In the present example, feldspar crystallization in veins attests to Al mobility. Geochemical data and the occurrence of residual phengite rims bounding quartz–feldspar veins are consistent with a chemically closed system in which redistribution of Si, Al, K and/or Na occurred by centimetre-scale to metre-scale mass transfers.

Scale of fluid transfers

Although crystallizations are concentrated in veins, dissolution and precipitation processes are distributed throughout the rock. Thus, the structures emphasize that the element sources and sinks are distributed, and that distances between sources and sinks are limited.

Quartz $\delta^{18}O$ values from veins are systematically higher than those from associated WR, which suggests that veins and minerals from the host-rock are close to isotopic equilibrium. This inference is consistent with the fact that quartz from veins and from magmatic clasts have comparable $\delta^{18}O$ values (Table 3). The prevalence of a near-equilibrium isotopic state was further checked on a few samples by mass-balance calculations using WR chemical compositions and the isotopic fractionation curves of Zheng (1993*a*, *b*) at temperatures of 350–400 °C. The concomitant isotopic variations of quartz from both veins and WR may be achieved in two extreme ways: (1) isotopic buffering of the vein-WR system under large fluid/rock ratios; or (2) vein segregation in a rock-dominated closed system (models e and a, respectively, in Oliver, 1996, Fig. 2). Achievement of relatively constant vein $\delta^{18}O$ values within each lithology (Fig. 9) in an advective context would require that fluid/rock ratios remain constant within each lithology, but change abruptly across lithological contacts. Although the different lithologies may have different permeabilities, this scenario is hardly tenable. Conversely, the second hypothesis is consistent with the chemical and geometrical complementarities of the vein/phengitic residue/protolith triads, and with the fact that quartz from late quartz-dominated tension gashes and from earlier quartz-feldspar veins show similar $\delta^{18}O$ values at outcrop scale. Indeed, this feature is best explained if the fluid was still rock-buffered when the late gashes formed in a mechanical and chemical context different from that of foliation-parallel veins. In addition, the persistence of organic matter (graphitoid dust scattered throughout dark to black-coloured rocks) emphasizes that the buffering capacity of the rock with respect to the fluid was not exceeded. More generally, mineralogical and isotopic data suggest that the syn-deformation fluids kept a substantial memory of early metasomatic fluids which percolated throughout the unit. Recently, van Staal *et al.* (2001) have documented the preservation of prograde metasomatic fluids within volcanic rocks mylonitized in a tectonic context comparable to that of the present example, with HP-LT downthrusting followed by exhumation.

Thus, our data indicate that both the amount of fluids and the scale of fluid and mass transfers remained limited during deformation (see the diagnostic features summarized by Oliver & Bons 2001). Therefore, the vein system developed essentially by diffusion of elements through the fluid rather than by fluid advection. This explains

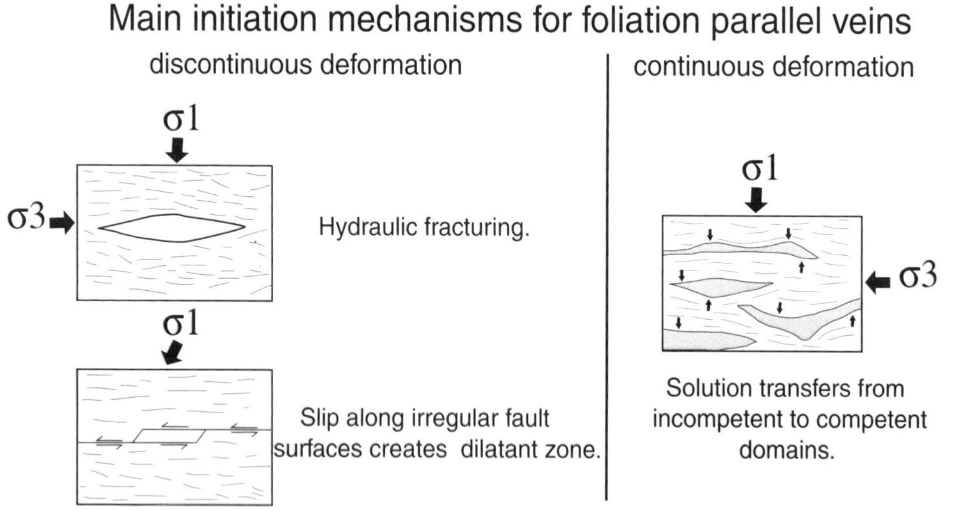

Main initiation mechanisms for foliation parallel veins

discontinuous deformation | continuous deformation

σ1, σ3 — Hydraulic fracturing.

σ1 — Slip along irregular fault surfaces creates dilatant zone.

σ1, σ3 — Solution transfers from incompetent to competent domains.

Fig. 10. Sketch of main types of mechanisms for the growth of foliation-parallel veins (see text for explanation).

the mobility of elements reputed for their low solubility in low temperature fluids, such as Al (see experiments of Vidal & Durin 1999).

Progressive deformation model

Axial-planar and foliation parallel veins have been described in numerous metamorphic belts (Yardley 1975; Sawyer & Robin 1986; Yardley & Bottrell 1992, Simpson 2000). It is often accepted that such veins result from hydrofracturing of a low tensile strength foliation under low differential stresses (Kerrich 1986; Gratier 1987) (Fig. 10). Particular relationships between values of the principal compressive stress (σ_1) and of the tensional strength (T_1 and T_3, perpendicular and parallel to the foliation, respectively) allowing the development of a vein perpendicular to σ_1 by hydraulic fracturing are $Pf > \sigma_1 + T_1$ and $\sigma_3 + T_3 > \sigma_1 + T_1$ (Kerrich 1986; Gratier 1987). The drop in fluid pressure associated with cracking, and subsequent increase in fluid pressure during sealing can account for a cyclic crack-seal mechanism (Ramsay 1980; Cox & Etheridge 1983; Etheridge *et al.* 1984; Gratier *et al.* 1999; Renard *et al.* 2000). Other models of vein development at high angle to the principal compressive stress invoke slip along fractures, with growth and sealing of pull-apart structures (Fig. 10) (Mackinnon *et al.* 1997). Continuous deformation can also produce foliation-parallel veins, provided that viscosity contrasts occur in the material (Robin 1979; Sawyer & Robin 1986). Models have emphasized that jumps in minimum principal stress from incompetent to competent layers can drive mass transfers and lead to metamorphic segregation. In Robin's (1979) model, where a constant σ_1 is applied perpendicular to layers of different competency, weak layers (e.g. mica-rich) require a smaller differential stress than the competent ones (e.g. quartz-rich) to accommodate the imposed strain rate. In response to the stress field, inter-layer diffusion can occur in order to equilibrate chemical potentials. In the more competent layers, the chemical potential of silica along interfaces perpendicular to σ_3 will be lower than those along interfaces in the less competent layers. Consequently, the matter migrates from the less competent layers to the low pressure interfaces in the more competent ones (Fig. 10).

In the rocks studied, fractures are concentrated in veins, which constitute competent domains acting as preferential sinks in the solution-transfer process; whereas weaker, mica-rich host rocks mostly acted as sources. Quartz from the veins shows evidence of dislocation-creep, whereas dissolution dominates in quartz from the host rocks. This, combined with the concentration of cracks in veins, is consistent with higher differential stress in the veins than in the matrix (Stöckhert *et al.* 1999), and with element transfers from matrix to veins (Sawyer & Robin 1986).

Interactions between pressure-solution and cracking are common in natural LT deformations (e.g. Stöckhert *et al.* 1999) and have been studied in experiments (Gratier *et al.* 1999). Such combinations of deformation mechanisms are best achieved under high fluid pressures, which lower the yield stress for failure to the range expected for pressure solution (e.g. Cox & Etheridge 1989). Hydraulic fracturing probably occurred in the rocks studied, but we did not find unequivoqual microstructural evidence. In particular, we have not observed fibres attesting to stretching at high angle to the vein plane.

In initial deformation stages, quartz and feldspar porphyroclasts can act as nuclei for the growth of mica-free competent domains, and initial variations of clast concentrations can provide sites for localization of brittle deformation (Fig. 11). Boudinage of feldspars, combined with fibre growth in shadow zones, can lead to coalescence of quartz–feldspar aggregates which can suffer further cracking and sealing. The whole process leads to vein growth along the stretching direction, while dissolution mainly occurs in the less competent matrix where micas concentrate. If dissolution produces more silica than that which can be accommodated by vein extension, the excess is expected to crystallize at a vein boundary, in the vicinity of sealed cracks, leading to vein thickening (Sawyer & Robin 1986).

As long as the applied differential stress is high enough for fracture, and fluids are available, this process can proceed and veins grow further. Moreover, removal of quartz and feldspar from the adjacent matrix serves to increase the competency contrasts between mica-free veins and mica-enriched selvages, so that the vein-forming process is self-propagating. However, an increase in thickness of the phengitic rims built up along vein walls can be a limiting factor for vein growth. Indeed, the coalescence of phyllosilicates preferentially oriented parallel to the vein boundary can produce a low diffusivity layer limiting transfers, by increasing the diffusion path for an element travelling from matrix to vein. On the other hand, this process could induce a lateral switch of precipitation sites, and thus coalescence of veins. This indeed appears to occur, as suggested by the common occurrence of phengite-rich remnants within veins (Fig. 11).

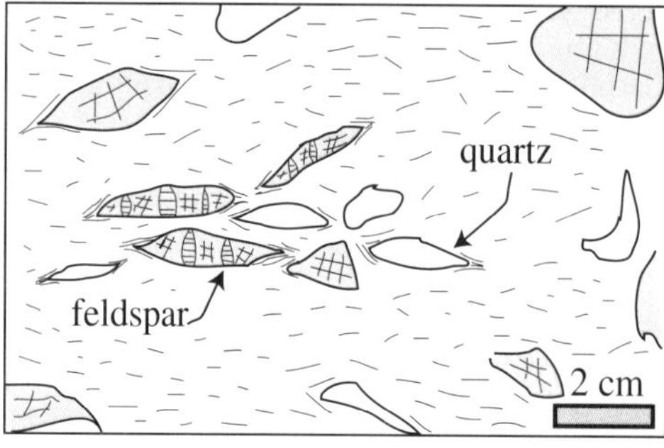

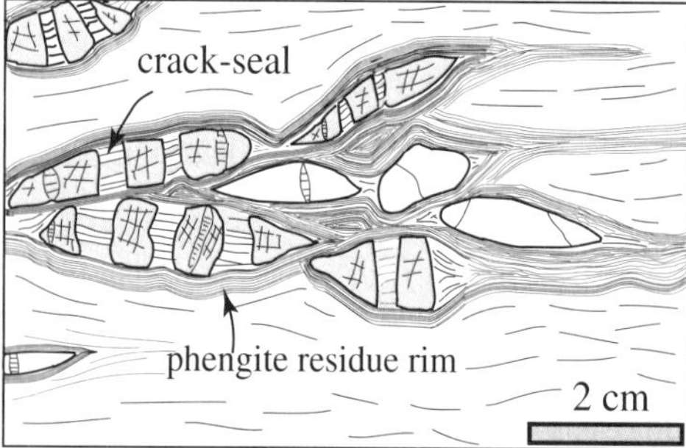

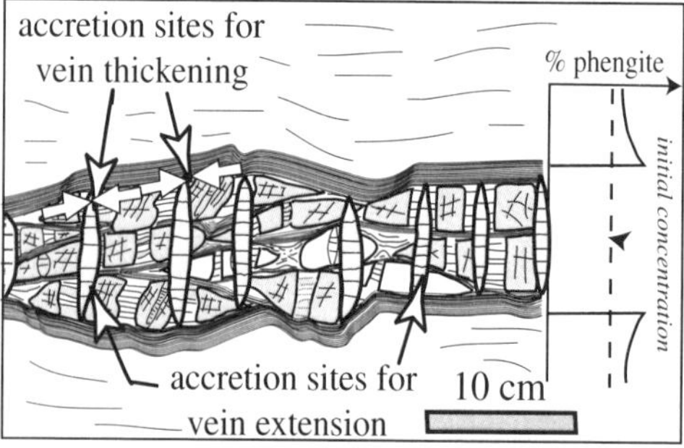

Fig. 11. Proposed scenario for the growth of foliation-parallel veins during progressive deformation (see text for explanation).

As discussed previously, LT deformation of felsic rocks marked by a water-assisted production of micas generally promotes strain localization. In the present case, the limited strain heterogeneity observed may be related to the prevalence of low fluid/rock ratios preventing extensive mica formation but allowing distributed dissolution–crystallization at aggregate scale.

Conclusions

The analysis of interactions between deformation and fluid transfers in the studied metavolcanics allows us to draw the following conclusions.

(1) Deformation was accommodated by extensive LT dissolution–crystallization processes involving both quartz and feldspars.
(2) Dissolution–crystallization involved combined ductile and brittle mechanisms, which suggests substantially high fluid pressures. On the other hand, compositional and geochemical data show that both the amount and the scale of fluid transfers were limited. From these, we infer that mass transfers resulted essentially from diffusion through the fluid in an overall closed system, rather than from fluid advection.
(3) Mass transfers driven by diffusion can account for the observed mobility of elements such as Al, which are reputed for their limited solubility in LT fluids. This peculiar feature differs from examples of shear zones with high fluid/rock ratios where Al is often rather immobile.
(4) The rocks studied emphasize that dissolution–crystallization can induce very large pervasive strains throughout wide areas, without involving large fluid/rock ratios due to extensive fluid percolation.
(5) Deformation resulted in the development of quartz-feldspar veins growing parallel to the regional foliation. During progressive deformation, veins were dominantly brittle, with extensive crack-seal processes. They constitute major crystallization zones, while the adjacent country rocks are the main zones of dissolution.
(6) During deformation, phengite behaved as a residual phase. The resulting relative enrichment in mica content with increasing strain differs from many LT or MT shear zones where water-assisted breakdown of feldspars induces an increase in the absolute amount of micas. This example underlines that interpretations of mica enrichment in terms of large fluid/rock ratios require caution.
(7) Mineralogical and isotopic data suggest that the crustal-scale deformation zone studied may have acted as a trap for early fluids. This interpretation differs from other examples where shear zones appear to be channels for migrating crustal fluids. It may be related to the low dip of the deformation zone, at high angle to dominant directions of fluid motions expected to occur throughout the metamorphic crust.
(8) Under LT to MT conditions and large fluid/rock ratios, deformation of felsic rocks is often heterogeneous, marked by strong strain localization within retrogressed faults or shear zones. The felsic rocks studied, which show distributed strains despite

thermal conditions of the order of 350–400 °C, appear therefore rather remarkable.

This work was part of the Géofrance 3D Program (ARMOR 2, BRGM and CNRS). We are grateful to M. Ballèvre, J. P. Brun, M. C. Boiron, M. Cathelineau and C. Marignac for helpful discussions. J. R. Kienast helped with microprobe analyses, and F. Martineau helped with oxygen isotope measurements. Comments and suggestions by G. Dresen, M. Drury and an anonymous reviewer helped to improve the final version of the manuscript.

References

AGUE, J. J. 1991. Evidence for major mass transfer and volume strain during regional metamorphism of pelites. *Geology*, **19**, 855–858.

AUDREN, C. 1987. Evolution structurale de la Bretagne méridionale au Paléozoïque. *Mémoires de la Société Géologique de Bretagne*, **31**.

AUDREN, C. & TRIBOULET, C. 1993. P-T-t-deformation paths recorded by kinzigites during diapirism in the western Variscan belt (Golfe du Morbihan, southern Brittany, France). *Journal of Metamorphic Geology*, **11**, 337–356.

BEACH, A. F. 1976. The interrelations of fluid transport, deformation, geochemistry and heat flow in early Proterozoic shear zones in the Lewisian Complex. *Philosophical Transactions of the Royal Society of London*, **A280**, 569–604.

BEACH, A. F. & FYFE, W. S. 1972. Fluid transport and shear zones at Scourie, Sutherland: evidence of overthrusting? *Contributions to Mineralogy and Petrology*, **36**, 175–180.

BERNARD-GRIFFITHS, J., PEUCAT, J. J., SHEPPARD, S. & VIDAL, P. 1985. Petrogenesis of Hercynian leucogranites from South Armorican massif. Contributions of REE and isotopic (Sr, Nd, Pb, O) geochemical data to the study of source rock characteristics and ages. *Earth and Planetary Science Letters*, **74**, 235–250.

BOS, B., PEACH, C. J. & SPIERS, C. J. 2000. Frictional-viscous flow of simulated fault gouge caused by the combined effects of phyllosilicates and pressure solution. *Tectonophysics*, **327**, 173–194.

BOSSE, V., BALLÈVRE, M., FÉRAUD, G. & PEUCAT, J. J. 2000. Petrological and geochronological constraints in the Ile de Groix blueschists (Armorican Massif, France). *In*: DIAZ GARCIA, F., GONZALES CUADRA, P., MARTINEZ CATALAN, J. R. & ARENAS, R. (eds) *Variscan-Applachian Dynamics: the Building of the Upper Palaeozoic Basement.* Basement Tectonics, A Coruña, Spain, Program and Abstracts, **15**, 63–66.

BOSSE, V., BALLÈVRE, M. & VIDAL, O. 2002. Ductile thrusting recorded by the garnet isograd from blueschist facies metapelites of the Ile de Groix, Armorican Massif, France. *Journal of Petrology*, **43**, 485–510.

BOSSIÈRE, G. 1988. Evolutions chimico-minéralogiques du grenat et de la muscovite au voisinage de l'isograde biotite-staurotide dans un métamorphisme prograde de type barrovien: un exemple en Vendée littorale (Massif Armoricain). *Comptes Rendus de l'Académie des Sciences, Paris, Série II*, **306**, 135–140.

BOYER, C. 1974. *Volcanismes Acides Paléozoïques dans le Massif Armoricain.* Thèse de l'Université d'Orsay-Paris Sud.

BRILLANCEAU, A. 1978. *Guide géologique régional Poitou-Vendée-Charente.* Masson, Paris.

BROWN, M. D. & DALLMEYER, R. D. 1996. Rapid Variscan exhumation and the role of magma in core complex formation: southern Brittany metamorphic belt, France. *Journal of Metamorphic Geology*, **14**, 361–379.

BRUN, J. P. & BURG, J. P. 1982. Combined thrusting and wrenching in the Ibero-Armorican arc: a corner effect during continental collision. *Earth and Planetary Science Letters*, **61**, 319–332.

BURG, J. P. 1981. Tectonique tangentielle hercynienne en Vendée littorale: signification des linéations d'étirement E-W dans les porphyroïdes à foliation horizontale. *Comptes Rendus de l'Académie des Sciences, Paris, Série II*, **293**, 849–854.

CARMICHAEL, D. M. 1969. On the mechanism of prograde metamorphic reactions in quartz-bearing pelitic rocks. *Contribution to Mineralogy and Petrology*, **20**, 244–267.

CARTER, E. K. & DWORKIN, S. I. 1990. Channelized fluid flow through shear zones during fluid-enhanced dynamic recrystallization, Northern Apennines, Italy. *Geology*, **18**, 720–723.

CHALET, M. 1985. *Contribution à l' Étude de la Chaine Hercynienne d'Europe Occidentale: Étude Lithostratigraphique et Structurale des 'Porphyroïdes' et Formations Paléozoïques Associées du Bas Bocage Vendéen (Région de Mareuil-sur-Lay).* Thèse de l'Université de Poitiers.

CLAYTON, R. N. & MAYEDA, T. K. 1963. The use of bromine pentafluoride in the extraction of oxygen from oxides and silicates for isotopic analysis. *Geochimica et Cosmochimica Acta*, **27**, 43–52.

COX, S. F. & ETHERIDGE, M. A. 1983. Crack-seal fibre growth mechanisms and their significance in the development of oriented layer silicate microstructures. *Tectonophysics*, **92**, 147–170.

COX, S. F. & ETHERIDGE, M. A. 1989. Coupled grain-scale dilatancy and mass transfer during deformation at high fluid pressures: examples from Mount Lyell, Tasmania. *Journal of Structural Geology*, **11**, 147–162.

DEWERS, T. & ORTOLEVA, P. 1990. A coupled reaction/transport/mechanical model for intergranular pressure solution, stylolites and differential compaction and cementation in clean sandstones. *Geochimica et Cosmochimica Acta*, **54**, 1609–1625.

DURNEY, D. W. 1972. Solution-transfer, an important geological deformation mechanism. *Nature*, **235**, 315–316.

ETHERIDGE, M. S., WALL, V. J., COX, S. F. & VERNON, R. H. 1984. High fluid pressures during metamorphism and deformation: implications for mass transport and deformation mechanisms. *Journal of Geophysical Research*, **89**, 4344–4358.

FARVER, J. R. & JUND, R. A. 1999. Oxygen bulk diffusion measurement and TEM characterization of a natural ultramylonite: implications for fluid transport in mica-rich bearing rocks. *Journal of Metamorphic Geology*, **17**, 669–683.

FERRY, J. M. 1983. On the control of temperature, fluid composition, and reaction progress during metamorphism. *American Journal of Science*, **238-A**, 201–232.

FERRY, J. M. 1994. A historical review of metamorphic fluid flow. *Journal of Geophysical Research*, **99**, 15487–15498.

FOURCADE, S., MARQUER, D. & JAVOY, M. 1989. $^{18}O/^{16}O$ variations and fluid circulation in a deep shear zone: the case of the ultramylonites from the Aar Massif (Central Alps, Switzerland). *Chemical Geology*, **77**, 119–131.

FOURCADE, S., CAPDEVILA, R., OUABADI, A. & MARTINEAU, F. 2001. The origin and geodynamic significance of the Alpine cordierite-bearing granitoids of northern Algeria. A combined petrological, mineralogical, geochemical and isotope (O, H, Sr, Nd) study. *Lithos*, **57**, 187–216.

FYFE, W. S., PRICE, N. J. & THOMPSON, A. B. 1978. *Fluids in the Earth Crust*. Elsevier, Amsterdam.

GAPAIS, D. & BARBARIN, B. 1986. Quartz fabric transition in a cooling syntectonic granite (Hermitage massif, France). *Tectonophysics*, **125**, 357–370.

GAPAIS, D., LAGARDE, J. L. *ET AL.* 1993. La zone de cisaillement de Quiberon: témoin d'extension de la chaîne varisque en Bretagne méridionale au Carbonifère. *Comptes Rendus de l'Académie des Sciences, Paris, Série II*, **316**, 1123–1129.

GOUJOU, J. C. 1992. Analyse pétro-structurale dans un avant-pays métamorphique: influence du plutonisme tardi-orogénique varisque sur l'encaissant épi à mézozonal de Vendée. *Document du BRGM*, **216**.

GRANT, J. A. 1986. The isocon diagram: a simple solution to Gresens'equation for metasomatic alteration. *Economic Geology*, **2**, 47–65.

GRATIER, J. P. 1987. Pressure solution-deposition creep and the associated tectonic differentiation in sedimentary rocks. *In*: JONES, M. E. & PRESTON, R. M. F. (eds) *Deformation of Sediments and Sedimentary Rocks*. Geological Society, London, Special Publications, **29**, 25–38.

GRATIER, J. P. 1993. Le fluage des roches par dissolution-cristallisation sous contrainte dans la croûte supérieure. *Bulletin de la Société Géologique de France*, **164**, 267–287.

GRATIER, J. P., RENARD, F. & LABAUME, P. 1999. How pressure solution creep and fracturing processes interact in the upper crust to make it behave in both a brittle and viscous manner. *Journal of Structural Geology*, **21**, 1189–1197.

GRESENS, R. L. 1967. Composition-volume relationships of metasomatism. *Chemical Geology*, **2**, 47–65.

GUIRAUD, M., BURG, J. P. & POWELL, R. 1987. Evidence for a Variscan suture zone in the Vendée, France: a petrological study of blueschist facies rocks from Bois de Cené. *Journal of Metamorphic Geology*, **5**, 225–237.

HEMLEY, J. J. 1959. Some mineralogical equilibria in the system K_2O-Al_2O_3-SiO_2-H_2O. *American Journal of Science*, **257**, 241–270.

HICKMAN, S. S. & EVANS, B. 1995. Kinetics of pressure-solution at halite–silica interfaces and intergranular clay films. *Journal of Geophysical Research*, **100**, 13113–13132.

HIPPERTT, J. F. 1994. Microstructures and c-axis fabrics indicative of quartz dissolution in sheared quartzites and phyllonites. *Tectonophysics*, **229**, 141–163.

JONES, K. A. & BROWN, M. 1989. The metamorphic evolution of the Southern Brittany Migmatite Belt, France. *In*: DALY, J. S., CLIFF, R. A. & YARDLEY, B. W. D. (eds) *Evolution of Metamorphic Belts*. Geological Society, London, Special Publications, **43**, 501–505.

JONES, K. A. & BROWN, M. 1990. High-temperature 'clockwise' P-T paths and melting in the development of regional migmatites: an example from Southern Brittany, France. *Journal of Metamorphic Geology*, **14**, 361–379.

KERRICH, R. 1986. Fluid infiltration into fault zones: chemical, isotopic and mechanical effects. *Pure and Applied Geophysics*, **124**, 225–268.

KERRICH, R., FYFE, W. S., GORMAN, B. E. & ALLISON, I. 1977. Local modification of rock chemistry by deformation. *Contribution to Mineralogy and Petrology*, **65**, 183–190.

LE HÉBEL, F., FOURCADE, S., GAPAIS, D., MARIGNAC, C., CAPDEVILA, R. & MARTINEAU, F. 2000. Fluid-assisted spreading of thickened continental crust: preliminary data from the Variscan belt of South Brittany (France). *Journal of Geochemical Exploration*, **69–70**, 561–564.

LE HÉBEL, F., VIDAL, O., KIENAST, J. R. & GAPAIS, D. 2002. Evidence for HP-LT Hercynian metamorphism within the ' porphyroïdes ' of South Brittany. *Comptes Rendus de l'Académie des Sciences, Paris*, 205–211.

LENTZ, D. R. 1999. Deformation-induced mass transfer in felsic volcanic rocks hosting the Brunswick no 6 massive-sulfide deposit, New Brunswick: Geochemical effects and petrogenetic implications. *The Canadian Mineralogist*, **37**, 489–512.

MACKINNON, P., FUETEN, F. & ROBIN, P. Y. F. 1997. A fracture model for quartz ribbons in straight gneisses. *Journal of Structural Geology*, **19**, 1–14.

MARQUER, D. & BURKHARD, M. 1992. Fluid circulation, progressive deformation and mass-transfer processes in the upper crust; the example of basement-cover relationships in the External Crystalline Massifs, Zwitzerland. *Journal of Structural Geology*, **14**, 1047–1057.

MARQUER, D., GAPAIS, D. & CAPDEVILA, R. 1985. Comportement chimique et orthogneissification d'une granodiorite en faciès Schistes verts (Massif de l'Aar, Alpes centrales). *Bulletin de Minéralogie*, **108**, 209–221.

NEWTON, R. C. 1990. Fluids and shear zones in the deep crust. *Tectonophysics*, **182**, 21–37.

NOLD, J. L. & ERICKSON, K. P. 1967. Changes in K-feldspar staining methods and adaptation for field use. *American Geologist*, **52**, 1575–1576.

OLIVER, N. H. S. 1996. Review and classification of structural controls on fluids flow during regional metamorphism. *Journal of Metamorphic Geology*, **14**, 477–492.

OLIVER, N. H. S. & BONS, P. D. 2001. Mechanisms of fluid flow and fluid-rock interaction in fossil metamorphic hydrothermal systems inferred from the vein-wallrock patterns, geometry and microstructure. *Geofluids*, **1**, 137–162.

O'NEIL, J. R & CHAPPELL, B. W. 1977. Oxygen and hydrogen isotope relations in the Berridale batholith. *Journal of the Geological Society, London*, **133**, 559–571.

PILI, E., RICARD, Y., LARDEAUX, J. M. & SHEPPARD, S. M. F. 1997. Lithospheric shear zones and mantle-crust connections. *Tectonophysics*, **208**, 15–29.

RAMSAY, J. G. 1980. The crack-seal mechanism of rock deformation. *Nature*, **284**, 135–139.

RENARD, F., GRATIER, J. P. & JAMTVEIT, B. 2000. Kinetics of crack-sealing, intergranular pressure solution, and compaction around active faults. *Journal of Structural Geology*, **22**, 1395–1407.

RENARD, F., ORTOLEVA, P. & GRATIER, J. P. 1997. Pressure solution in sandstones: influence of clays and dependence on temperature and stress. *Tectonophysics*, **280**, 257–266.

REYNOLDS, S. J. & LISTER, G. S. 1987. Structural aspects of fluid-rock interactions in detachment zones. *Geology*, **15**, 449–452.

RING, U. 1999. Volume loss, fluid flow, and coaxial versus noncoaxial deformation in retrograde, amphibolite facies shear zones, northern Malawi, east-central Africa. *Geological Society of America Bulletin*, **111**, 123–142.

ROBIN, P. Y. 1979. Theory of metamorphic segregation and related processes. *Geochimica et Cosmochimica Acta*, **43**, 1587–1600.

RODDY, M. S., REYNOLDS, S. J., SMITH, B. M. & RUIZ, J. 1988. K-metasomatism and detachment-related mineralization, Harcuvar mountains, Arizona. *Geological Society of America Bulletin*, **100**, 1627–1639.

RUTTER, E. H. 1976. The kinetics of rock deformation by pressure solution. *Philosophical Transactions of the Royal Society of London*, **A283**, 203–219.

RUTTER, E. H. 1983. Pressure solution in nature, theory and experiment. *Journal of the Geological Society, London*, **140**, 725–740.

SAWYER, E. W. & ROBIN, P. Y. 1986. The subsolidus segregation of layer-parallel quartz-feldspar veins in greenschist to upper amphibolite facies metasediments. *Journal of Metamorphic Geology*, **4**, 237–260.

SCHWARTZ, S. & STÖKHERT, B. 1996. Pressure solution in siliciclastic HP-LT metamorphic rocks – constraints on the state of stress in deep levels of accretionary complexes. *Tectonophysics*, **255**, 203–209.

SIMPSON, G. D. H. 2000. Synmetamorphic vein spacing distributions: characterization and origin of a distribution of veins from NW Sardinia, Italy. *Journal of Structural Geology*, **22**, 335–348.

STAAL, C. R. VAN, ROGERS, N. & TAYLOR, B. E. 2001. Formation of low-temperature mylonites and phyllonites by alkali-metasomatism weakening of felsic volcanic rocks during progressive, subduction-related deformation. *Journal of Structural Geology*, **23**, 903–921.

STALLARD, A. & SHELLEY, D. 1995. Quartz c-axes parallel to stretching directions in very low-grade metamorphic rocks. *Tectonophysics*, **249**, 31–40.

STÖCKHERT, B., WACHMANN, M., KÜSTER, M. & BIMMERMANN, S. 1999. Low effective viscosity during high pressure metamorphism due to dissolution precipitation creep: the record of HP-LT metamorphic carbonates and siliciclastic rocks from Crete. *Tectonophysics*, **303**, 299–319.

TADA, R. & SIEVER, R. 1989. Pressure solution during diagenesis. *Annual Review of Earth and Planetary Sciences*, **17**, 89–118.

TERS, M. 1972. Carte géologique à 1/80. 000 de 'Palluau-Ile d'Yeu'. no. 129, 2ème édition, B. R. G. M.

TRIBOULET, C. 1974. Les glaucophanites et roches associées de l'Ile de Groix (Morbihan, France): étude minéralogique et pétrologique. *Contributions to Mineralogy and Petrology*, **45**, 65–90.

TRIBOULET, C. & AUDREN, C. 1988. Controls on P-T-t deformation path from amphibole zonation during progressive metamorphism of basic rocks (estuary of the River Vilaine, South Brittany, France). *Journal of Metamorphic Geology*, **6**, 117–133.

VAUCHEZ, A., MAILLET, D. & SOUGY, J. 1987. Strain and deformation mechanisms in the Variscan nappes of Vendée, South Brittany, France. *Journal of Structural Geology*, **9**, 31–40.

VIDAL, O. & DURIN, L. 1999. Aluminium mass transfer and diffusion in water at 400-550 °C, 2 Kbar in the K_2O-Al_2O_3-SiO_2-H_2O system driven by a thermal gradient or by a variation of temperature with time. *Mineralogical Magazine*, **63**, 633–647.

WEYL, P. K. 1959. Pressure solution and the force of crystallization – a phenomenological theory. *Journal of the Geological Society, London*, **148**, 527–539.

WHITE, S. H. & KNIPE, R. J. 1978. Transformation and reaction enhanced ductility in rocks. *Journal of the Geological Society, London*, **135**, 513–516.

WIBBERLEY, C. 1999. Are feldspar-to-mica reactions necessarily reaction-softening processes in fault zones? *Journal of Structural Geology*, **21**, 1219–1227.

WINTSCH, R. P., CHRISTOFFERSEN, R. & KRONENBERG, A. K. 1995. Fluid-rock reaction weakening of fault zones. *Journal of Geophysical Research*, **100**, 13021–13032.

YARDLEY, B. W. D. 1975. On some quartz-plagioclase veins in the Connemara schists, Ireland. *Geological Magazine*, **112**, 183–190.

YARDLEY, B. W. D & BOTTRELL, S. H. 1992. Silica mobility and fluid movement during metamorphism of the Connemara schists, Ireland. *Journal of Metamorphic Geology*, **10**, 453–464.

ZHENG, Y. F. 1993*a*. Calculation of oxygen isotope fractionation in hydroxyl-bearing silicates. *Earth and Planetary Science Letters*, **120**, 247–263.

ZHENG, Y. F. 1993*b*. Calculation of oxygen isotope fractionation in anhydrous silicate minerals. *Geochimica et Cosmochimica Acta*, **57**, 1079–1091.

The origin of fibrous veins: constraints from geochemistry

MARLINA A. ELBURG[1,3], PAUL D. BONS[2,4], JOHN FODEN[1] & CEES W. PASSCHIER[2]

[1] *Department of Geology and Geophysics, Adelaide University, Adelaide SA5005, Australia*
[2] *Tektonophysik, Institut für Geowissenschaften, Johannes Gutenberg-Universität Mainz, D-55099 Mainz, Germany*
[3] *Present address: Department of Geochemistry, Max Planck Institute for Chemistry, PO Box 3060, 55020 Mainz, Germany (e-mail: elburg@mpch-mainz.mpg.de)*
[4] *Present address: Institut für Geowissenschaften, Eberhard Karls Universität, Sigwartstrasse 10, 72076 Tübingen, Germany*

Abstract: Several recent studies have suggested that antitaxial fibrous veins may form without fracturing, and not by the commonly invoked crack-seal mechanism. It has also been suggested that such veins would derive their nutrients locally by diffusional transport. This hypothesis was tested on carbonaceous shale-hosted antitaxial fibrous calcite veins from Oppaminda Creek in the northern Flinders Ranges, South Australia. Apart from their fibrous texture, these veins lack the classical features of crack-seal veins, such as wallrock-parallel inclusion bands.

Diffusional transport of locally derived calcite cannot explain all major and trace element data of the veins and their adjacent wallrock and indicate that part of the calcite was transported over distances of at least >decimetres, probably ≫100 m. Sr isotopic fingerprinting shows that an external fluid that carried radiogenic Sr must have percolated through the system. Fluid flow was pervasive as there is no evidence that this fluid preferentially percolated through the veins. Our data support the view that antitaxial fibrous veins of the type found at Oppaminda Creek grew in the absence of fractures, but show that such veins do not necessarily indicate local diffusional transport. Our data confirm a recently postulated basin-wide fluid flow event around 586 Ma that is probably related to copper mineralization in the area.

Veins are found in an abundant range of types, from huge quartz veins in Brazil and Australia (e.g. Hippertt & Massucatto 1998; Bons 2001*a*) to mineral-filled micro-cracks (e.g. Ramsay 1980; Mullenax & Gray 1984) and from planar veins to pressure fringes (Durney & Ramsay 1973). The large variety of vein types and their internal structures indicates that no single mechanism can explain the formation of all veins.

Excluding magmatic veins, proposed vein-growth mechanisms can be divided into two main classes: (1) vein formation dominated by advective fluid flow (e.g. Robert & Brown 1986; McCaig *et al.* 1995); and (2) vein formation dominated by diffusional transport (Durney & Ramsay 1973; Durney 1976; Ramsay & Huber 1983; Rutter 1983; Fisher & Brantley 1992; Jamtveit & Yardley 1997; Bons 2000; Oliver & Bons 2001). In the first class of vein-growth mechanisms, nutrients can be transported over long distances to a vein-growth site. Such a vein-growth system can be classified as 'open', whereby it should be noted that 'open' or 'closed' depends on the scale of the system under consideration. A number of processes can cause the actual precipitation of transported nutrients, such as temperature changes (Eisenlohr *et al.* 1989), pressure changes (Vrolijk 1987; Henderson & McCaig 1996), reactions with wall rock and fluid mixing (Boullier *et al.* 1994). The second class, diffusional transport, is typically associated with closed system vein growth, due to the limited distance of diffusional transport: typically restricted to centimetres–decimetres. The reasons for vein precipitation from diffusionally transported nutrients may be differences in porosity (Putnis *et al.* 1995) or pressure gradients, often regarded as being associated with deformation by dissolution–precipitation creep, as shown by the common close association of stylolites and veins (Beach 1974, 1977; Durney 1976; Mullenax & Gray 1984).

Veins are commonly equated with crack or fissure fillings, as can be seen in terminology like 'tension gash' (Ramsay & Huber 1983; Passchier & Trouw 1996). This view is strengthened by the paradigmatic 'crack-seal-theory' of Ramsay (1980), which has been applied extensively to explain vein textures (Cox 1987; Fisher & Byrne 1990; Urai *et al.* 1991; Boullier *et al.* 1994; Jessell *et al.* 1994; Foxford *et al.* 2000;

From: DE MEER, S., DRURY, M. R., DE BRESSER, J. H. P. & PENNOCK, G. M. (eds) 2002. *Deformation Mechanisms, Rheology and Tectonics: Current Status and Future Perspectives*. Geological Society, London, Special Publications, **200**, 103–118. 0305-8719/02/$15

Koehn & Passchier 2000; Renard *et al.* 2000). Crack-seal veins are veins that formed by repeated opening of a thin crack and subsequent sealing of the crack. The thickness of vein precipitate in one cycle is typically in the order of 10 μm, but may reach up to a centimetre (Ramsay 1980; Cox & Etheridge 1983; Robert & Brown 1986; Wiltschko & Morse 2001). Repeated crack-sealing fits well with theories of seismic pumping or fault-valve action that propose repeated cycles of: (1) fluid pressure increase resulting in hydraulic fracturing; and (2) subsequent fluid flow through the fractures, associated with precipitation of dissolved material (Sibson *et al.* 1975, 1988; Sibson 1987; Byerlee 1990; Cox 1995; Matthäi & Roberts 1997). Crack-seal veins typically have elongate or columnar vein crystals, as well as vein-parallel and opening trajectory-parallel bands of inclusions of wall rock fragments, neocrysts of secondary minerals and fluid inclusions (Ramsay 1980; Ramsay & Huber 1983; Cox 1987; Jessell *et al.* 1994; Passchier & Trouw 1996; Foxford *et al.* 2000; Koehn & Passchier 2000). However, based on early works of e.g. Mügge (1928), and Pabst (1931), several recent studies have challenged the idea that all fibrous veins are formed by crack-sealing, and suggest that some types may have formed without loss of cohesion between vein fibres and the wall rock; i.e. without fracturing (Bons & Jessell 1997; Bons 2000; Oliver & Bons 2001; Hilgers 2000; Means & Li 2001; Wiltschko & Morse 2001). Bons & Jessell (1997) and Bons (2000) suggested that veins that formed without fracturing can be recognized by truly fibrous vein crystals that show no indication of growth competition. Such fibres have very high length/width ratios (>10 to $\gg 10$) and all fibres have approximately the same shape and curvature, although the fibre-forming minerals (typically calcite or quartz) rarely grow fibrous crystals in other circumstances. This sets these fibrous veins apart from veins with crystallographically controlled fibrous crystals, such as satin-spar (gypsum) veins (Phillips 1974; Machel 1985). Truly fibrous veins commonly lack the typical inclusion bands that indicate cyclical growth by the crack-seal mechanism.

Bons & Jessell (1997) and Bons (2000) proposed that crack-seal veins have elongate vein crystals that show signs of growth competition and suggested the use of the term 'elongate-blocky' rather than 'fibrous'. Wiltschko & Morse (2001) and Hilgers (2000), however, suggested that even classical crack-seal veins according to Ramsay (1980), Ramsay & Huber (1983) and Cox (1987), with elongate-blocky crystals and wallrock-parallel inclusion bands, may form without crack-sealing. Wiltschko & Morse (2001), for example, proposed that crystal growth in a vein may push aside the vein walls by the force of crystallization.

Another argument against a crack-sealing origin is the fact that in antitaxial veins growth occurs on both outer surfaces of a vein, usually with a very high degree of symmetry. If this happened by the crack-seal mechanism, cracking would occur simultaneously, or alternating on both sides of the vein. Both are mechanically unlikely (Bons & Jessell 1997; Oliver & Bons 2001). Finally, Means & Li (2001) experimentally produced antitaxial fibrous veins without repeated crack-sealing.

The growth mechanism of antitaxial fibrous veins that form without fracturing, also called 'Taber veins' (after Taber 1918), remains unclear. Diffusional transport along concentration gradients is normally cited as driving the nutrient transport towards such a growing vein (e.g Bons & Jessell 1997; Wiltschko & Morse 2001; Means & Li 2001). Pressure gradients may cause the development of concentration gradients. This seems evident for pressure fringes that grow in the pressure shadows of rigid objects, such as pyrite crystals (Pabst 1931; Durney & Ramsay 1973; McKenzie & Holness 2001). Bons (2000) proposed that lenticular veins that are stiffer than their host rock may develop their own pressure shadow. Putnis *et al.* (1995) and Putnis & Mauthe (2001) observed that the equilibrium concentration of a dissolved mineral is inversely proportional to pore size. A dense rock with small pores can thus carry a fluid with a higher equilibrium concentration of a dissolved mineral than the fluid in larger pore spaces or fractures. This would provide a mechanism for transport of nutrients towards crack-seal veins, but not for Taber veins, although Means & Li (2001) report microscopic (m-scale) fractures between wall rock and growing fibres, which could perhaps play a role in causing a concentration gradient. A similar model explains the development of a concentration gradient by the lower supersaturation that is needed to add material to a fibre than that needed to nucleate or grow a small crystallite in the wall rock (Means & Li 2001). So far, all these models remain hypothetical.

In all the above models, nutrient transport is diffusional. Permeability is much lower in an unfractured rock than in a fractured rock and therefore diffusional transport may be more important in the restricted fluid flow environment where Taber veins are thought to grow. It would be useful if the microstructure within veins could be directly linked to their formation mechanism, such as whether the vein minerals grew in an open fracture or not. Such microstructures may possibly also be used to make inferences about the mode of fluid transport: advective transport through cracks or pores, or diffusional transport through a static intergranular fluid.

Inferences made about mode of transport are mostly based on the geometry of microstructures (e.g. Bons & Jessell 1997), but geochemical analysis may give additional insight into the genesis of veins. Bons & Jessell (1997) and Bons (2000) proposed that antitaxial fibrous calcite veins from Oppaminda Creek (northern Flinders Ranges, South Australia) are prime examples of veins that grew without the presence of cracks and indicated that therefore diffusion was the most likely transport mechanism of their nutrients. This hypothesis is tested in this paper, using major and trace element analyses and the Sr isotopic system. We show here that the growth of these veins cannot be simply explained by local diffusional mass transfer, but must have included various amounts of externally derived fluids and material.

Sr isotopes and their use

In order to determine whether fibrous veins have grown from locally or regionally derived material, we studied, among other things, the strontium (Sr) isotope composition of fibrous calcite veins. Sr is one of the alkaline earth elements and shows chemical behaviour that is very similar to Ca, and is therefore highly suited for the study of calcite veins. Sr has four naturally occurring stable isotopes, with masses 84, 86, 87 and 88. Three of these isotopes (84, 86 and 88) occur in fixed relative proportions in natural samples, but ^{87}Sr is formed by radioactive decay of ^{87}Rb. Therefore, the ^{87}Sr/^{86}Sr ratio of a sample will reflect both the age of the sample and its original Rb/Sr ratio. This has led to the application of the Rb–Sr isotopic system as a geochronological tool (e.g. Hahn *et al.* 1943). However, in chemical systems that are either very young (e.g. Whitford & Jezek 1979) or virtually free of Rb (e.g. Faure *et al.* 1963), the Sr isotope ratios of its components can be used as a fingerprinting tool. Rb is not incorporated into the calcite lattice, so radioactive decay over time will not affect its ^{87}Sr/^{86}Sr ratio. Since the differences in mass between the isotopes of Sr is much smaller than for oxygen or carbon isotopes, natural physical processes such as dissolution or precipitation will not fractionate the isotopes from each other (Faure 1986). Therefore, if calcite in a fibrous vein was derived from calcite in its host rock, the Sr isotopic ratio of the vein and the calcite in the host should be the same. Thus, comparing the Sr isotope composition of vein and host allows us to assess whether it is possible that fibrous calcite veins were formed from local material, or whether they must have formed by advective transport of material over longer distances (cf. McCaig *et al.* 1995).

Sample area and sample description

Antitaxial fibrous calcite veins were sampled from the Neoproterozoic Tapley Hill Formation in the northern Flinders Ranges (Fig. 1) (see Coats & Blissett 1971 and Drexel *et al.* 1993 for regional geology). The Tapley Hill Formation forms part of the Adelaidean (Neoproterozoic) sedimentary cover in South Australia, often referred to as the Adelaide Geosyncline (Preiss 1987) and has a poorly defined depositional age of 750 Ma (Preiss 1989). It consists of graphite-bearing carbonaceous shales, which change in character from pyritic in the lower section to more carbonate-rich in the upper part of the sequence. The Tapley Hill Formation overlies Sturtian tillites, which have been related to the 'snowball earth' theory (Hoffman *et al.* 1998). The study area was mildly deformed during the Delamerian orogeny (485–515 Ma), with subsequent reactivation at 400 and 325 Ma (McLaren *et al.* 2002) but the grade of metamorphism in the sample area did not exceed lower greenschist facies (Drexel *et al.* 1993; Paul *et al.* 1999). The Tapley Hill Formation has been the subject of several studies concerning its radiogenic (Veizer & Compston 1976; Turner *et al.* 1993; Barovich & Foden 2000; Foden *et al.* 2001) and stable isotope signature (McKirdy *et al.* 2001).

In the study area in Oppaminda Creek, veins are most frequent in the lower part of the succession. These veins are the typical fibrous veins (*sensu stricto*; Bons & Jessell 1997; Bons 2000) or Taber veins, as described in the introduction, with crystals of very high length to width ratios, and no evidence for growth competition. The width of the veins varies from a few millimetres to approximately 8 cm. All fibrous veins in the area are antitaxial (Durney & Ramsay 1973), with individual fibres that grew outwards from a median line or narrow zone. Fibres are often smoothly curved. Curvature is a primary growth feature, not the result of later deformation, as the crystals are not internally deformed (Williams & Urai 1989). Except for being fibrous, the veins lack all the features usually attributed to crack-sealing,

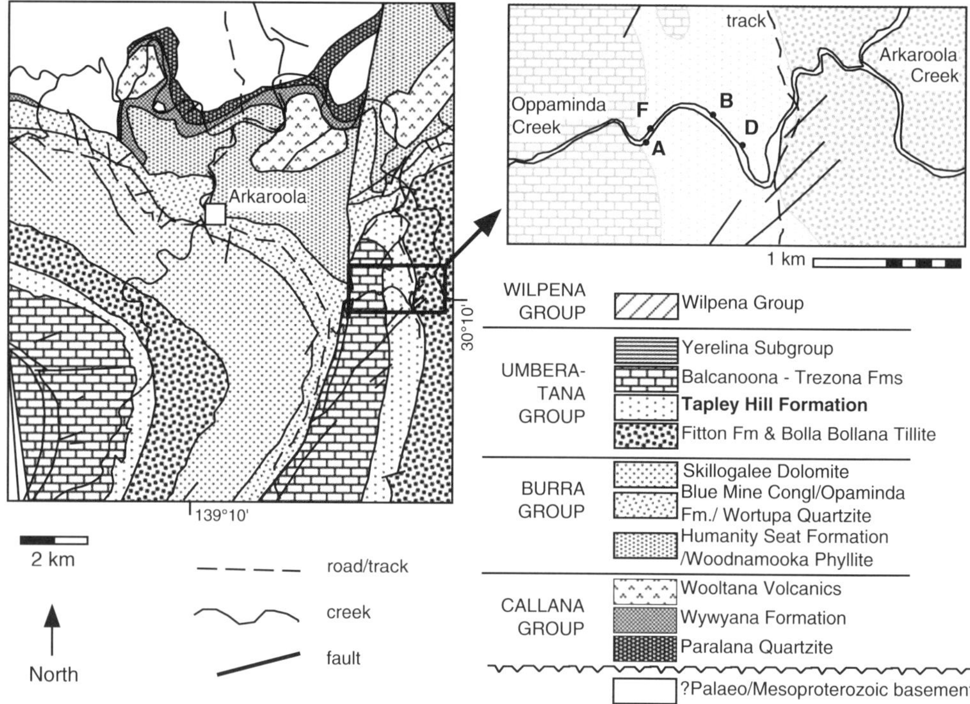

Fig. 1. Map of sample locations in Oppaminda Creek in the Arkaroola – Mt Painter Sanctuary in the northern Flinders Ranges, South Australia. Samples not shown on map were loose pieces found in the creek bed in the creek section shown on this map.

such as wallrock-parallel inclusion bands (Ramsay 1980; Wiltschko & Morse 2001). The veins are completely homogeneous in cathodoluminescence images. Fibrous veins sometimes laterally grade into stretching veins (Durney & Ramsay 1973), that consist of alternating slivers of wall rock and vein material, when veins intersect carbonaceous siltstone layers. Blocky veins are less common, and composite veins, with both blocky and fibrous parts, are rare.

Veins range in length from a few centimetres to several metres and are spaced from decimetres to more than 10 metres. Many veins appear as isolated flat lenticular bodies in outcrop, with no connection to other veins, although connectivity in the third dimension cannot be excluded. The veins lie in the limb of a km-scale syncline that is attributed to the Delamerian Orogeny. Host rock and veins show virtually no deformation, except for calcite twinning, but may have rotated passively during folding. The veins are sometimes, but not always, associated with dark alteration zones (Fig. 2a), where the beds can also be seen to decrease in thickness. This is likely to result from a depletion of certain components in the local wallrock, possibly during formation of the veins.

Sample A (original sample number ME99OPP10) was collected in the higher part of the Tapley Hill Formation (Fig. 2b). It contains an incomplete vein, of which we estimate that slightly less half has been preserved (no median line was observed in thin section). This half vein has a maximum width of 7 mm. The reason for sampling this incomplete vein was the field observation that the host rock beds towards the vein became thinner, and changed colour, giving the impression that local transport of calcite from the host rock was possibly responsible for vein formation. To estimate the changes in host rock composition towards the vein, two bulk samples were produced of host rock at 0–2 cm from the vein, and at 12–15 cm from the vein. The thickness in the bed sampled decreases from 41 mm at 12–15 cm, to 35 mm next to the vein.

Sample B (ME99OPP8) contains a complete fibrous vein of 1.75 cm width, from the lower part of the formation (Fig. 2c, d). This sample showed no field evidence for local calcite transport, since the appearance of the beds did not change towards the vein. Host rock samples were taken at 0–1.5 cm from the vein, and at 5–7.5 cm from the vein (for XRF and Sr isotope

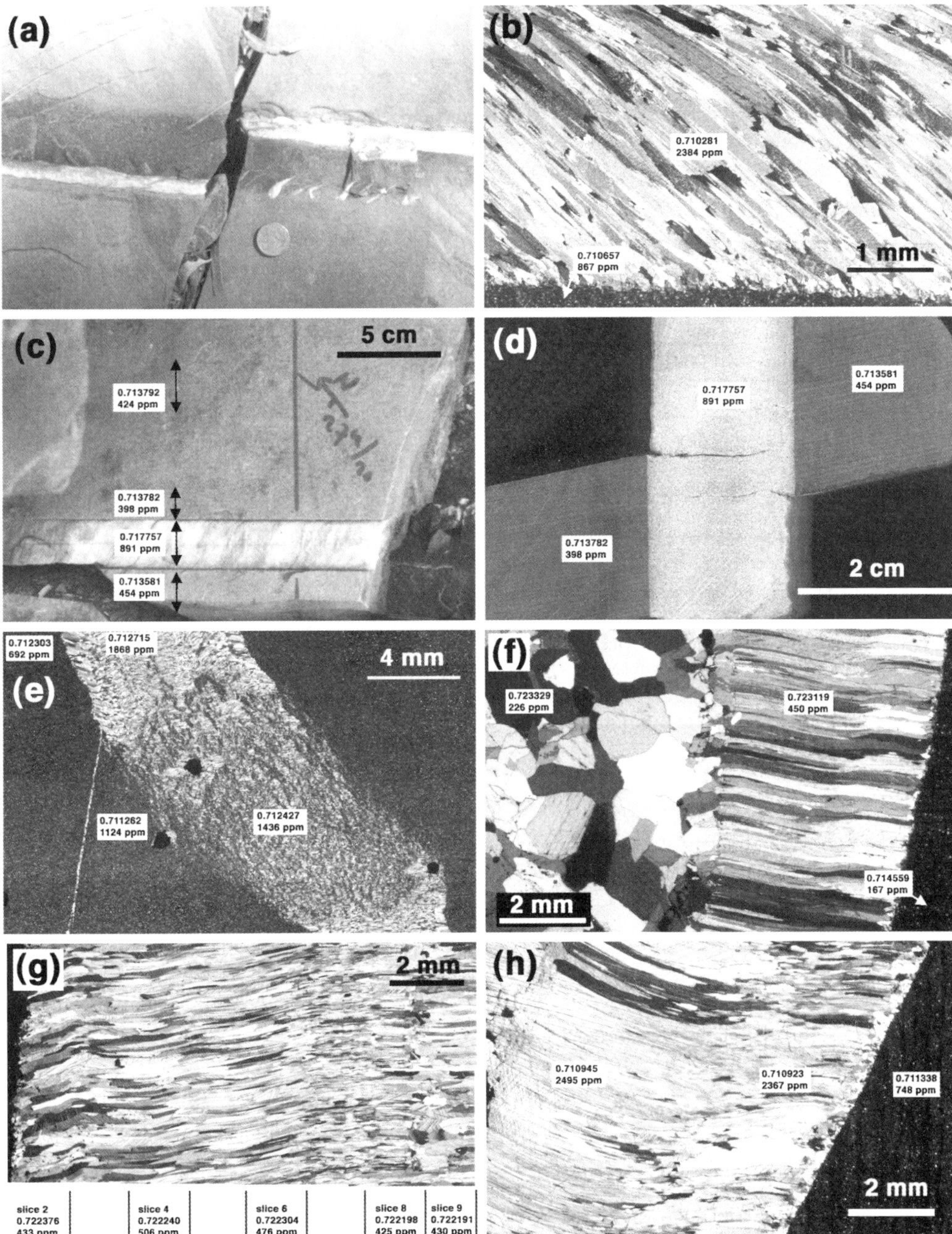

Fig. 2. Photographs of veins at Oppaminda Creek. Sr measurements within the field of view are given ($^{87}Sr/^{86}Sr$ ratio and Sr concentration). (**a**) Veins with typical dark discolouration in the adjacent shale. Diameter coin is 24 mm. (**b**) Thin section of the edge of vein A. (**c**) Vein B, looking down on west-dipping sedimentary layering. (**d**) Cut surface through vein B and adjacent shale, showing straight millimetre-scale layering. (**e**) Thin section of antitaxial fibrous vein C that becomes a diffuse stretched-type vein where it crosses a carbonaceous siltstone layer. The diffuse part contains three pyrite crystals with antitaxial pressure fringes. (**f**) Thin section of vein D that has an older blocky part (left) and a younger fibrous part (right). (**g**) Thin section of vein E that was sliced parallel to the vein wall (left). The 1 mm wide median zone has a slightly more blocky texture and overlaps with slices 8 and 9. (**h**) Vein C. All micrographs are with crossed polarizers.

Table 1. *Running conditions for microprobe analyses*

	Ca	Mg	Fe	Mn	Sr
Peak	Kα	Kα	Lα	Lα	Lα
Crystal	PET	TAP	PC1	PC0	TAP
Counting time (s)	30	30	100	100	60
Detection limits (ppm)	7522	2377	956	1981	538

analyses), and at 2 cm distance from the other side of the vein (Sr isotopes only).

Sample C (ME99OPP7) contains a vein that is 8 mm wide (Fig. 2e). Most of the vein is fibrous, but in the area where the vein crosses an apparently more carbonaceous layer the vein is bent and becomes diffuse. Pyrite inclusions are present both in the diffuse vein and in the host rock. This sample is also from the lower section of the Tapley Hill Formation.

Sample D (941115-23b) is a composite vein with a fibrous part of 0.5 cm, and a blocky part of at least 0.7 cm (Fig. 2f).

Sample E (Loose B) is a 2.3 cm thick fibrous vein without host rock, that was found loose in the sample area (Fig. 2g). This sample was divided into sixteen separate slabs, perpendicular to fibre growth, with a gem-cutters saw, so that individual vein-growth stages could be analysed to assess vein homogeneity on a millimetre scale.

Sample F (941111-7) is an asymmetric fibrous vein, with a total width of 2 cm (Fig 2h).

Analytical techniques

The four bulk host rock samples and the vein of sample B were crushed in a steel jaw crusher after removal of weathered rims and cleaning in an ultrasonic bath. A split was ground to <2 micron grainsize in a tungsten carbide ring mill, and this material was used for X-ray fluorescence (host rocks only) and Rb–Sr isotopic analyses. All other vein and host rock samples were of significantly smaller size, and were ground by hand in an agate mortar and pestle.

X-ray fluorescence (XRF) analyses were performed at the Department of Geology and Geophysics, Adelaide University. LOI was determined on approximately 4 g of pre-dried sample by heating to 960 °C overnight. Major elements were determined on fused discs of sample mixed with lithium meta/tetraborate flux (ratio sample:flux = 1:4) with a Philips PW1480 100 kV spectrometer. Trace elements were analysed on pressed powder pellets. Reproducibility is better than 5%, and generally better than 1%, for major elements and around 5% for trace elements. The accuracy of the measurements, as determined by analyses of international standards, is better than 5% for all elements, except Ba, Ni, Zn, Cu, Cr, for which accuracy is better than 10%.

Calcite was leached from 0.1 g of the ground host rocks with 8 ml of cold 1N acetic acid overnight. The supernatant liquid was pipetted off, and the residue washed in distilled water. This water was pipetted off and added to the leachate. The leachate was split in two fractions, one for Sr isotopic analyses, and one for Sr and Rb concentration analyses by isotope dilution. This leaching process ensures that only the calcite fraction of the vein and host rock is dissolved and analysed, as we assumed that only the calcite fraction of the host rock would possibly contribute to vein formation. The leached residue of the host rock was dissolved separately using a mixture of HF and HNO_3, and also analysed for Sr isotopic composition and Sr and Rb concentrations.

Sr isotopic compositions were measured on a Finnigan MAT 262 Thermal Ionisation Mass Spectrometer in static mode. The average $^{87}Sr/^{86}Sr$ ratio for SRM987 during the period when the samples were run was 0.710271 ± 20 (2 sigma, $n = 15$). Full procedure blanks are better than 1 ng for Sr. Full procedure blank for Rb is less than 50 pg. Sr and Rb concentrations were measured by isotope dilution on a Finnigan MAT 261 in dynamic mode. Reproducibility of Sr concentrations is 0.2% and of Rb 0.6%.

Microprobe analyses of the veins were performed with a Cameca SX-51 at CEMMSA at Adelaide University. Running conditions were 7 kV accelerating voltage and 20 nA beam intensity, with a focused (diameter <2 μ) beam. Peaks, crystals, counting times, and detection limits under these conditions are given in Table 1.

Results

Major and trace elements host rock samples by XRF

Four bulk samples of the host rocks to the veins were analysed: the area between 0–2 cm and between 12–15 cm away from vein A; and the

Table 2. *X-ray fluorescence analyses for host rocks near to and further away from the vein*

	Vein B 0–1.5 cm*	Vein B 5–7.5 cm	Vein A 0–2 cm	Vein A 12–15 cm
SiO_2 (%)	62.03	61.78	60.65	56.98
TiO_2 (%)	0.99	0.98	0.90	0.82
Al_2O_3 (%)	13.91	13.93	13.24	12.05
Fe_2O_3 (%)	6.98	7.02	6.98	6.81
MnO (%)	0.09	0.09	0.07	0.09
MgO (%)	5.41	5.53	5.22	5.44
CaO (%)	5.44	5.58	7.97	13.08
Na_2O (%)	1.64	1.65	1.62	1.60
K_2O (%)	3.05	3.03	2.93	2.66
P_2O_5 (%)	0.24	0.24	0.24	0.22
SO_3	0.08	0.09	0.16	0.34
LOI	7.75	7.95	10.95	14.27
total	99.85	99.93	99.98	100.07
V (ppm)	186	186	168	154
Cr (ppm)	98	99	92	84
Co (ppm)	30	30	38	36
Ni (ppm)	47	49	52	48
Cu (ppm)	43	42	15	22
Zn (ppm)	47	46	55	44
Ga (ppm)	19	18	18	14
Ba (ppm)	551	562	395	363
La (ppm)	29	30	26	26
Ce (ppm)	73	74	65	68
Nd (ppm)	28	32	25	26
Nb (ppm)	16	16	16	14
Zr (ppm)	231	228	185	164
Y (ppm)	16	16	16	14
Sr (ppm)	128	131	229	401
Rb (ppm)	123	122	137	125
U (ppm)	5	4	5	4
Pb (ppm)	9	10	24	24
Th (ppm)	14	16	15	14
Sr/CaO	23.55	23.48	28.70	30.69
Sc (ppm)	14	15	15	14

*Distance from vein.
All values have been normalized to 100% on a volatile-free basis, except LOI (loss on ignition: volatiles such as H_2O and CO_2) and total.

area between 0–1.5 cm and 5–7.5 cm away from vein B. The results are given in Table 2. All elements and oxides have been normalized to 100% on a volatile-free basis. The two analyses for the host rock samples from B are virtually undistinguishable, whereas those from A show important differences.

The host rock of A next to the vein has lower CaO contents and a lower 'loss on ignition' (LOI), which represents volatiles such as CO_2 and H_2O that escape from the sample during the heating step at 960 °C. The low CaO content and LOI suggests that calcite has been lost from the host rock near to the vein. This loss, combined with the normalisation to 100% gives the effect that all elements and oxides that have not been lost show higher values than in the host rock between 12–15 cm. This is called the closure effect. We can evaluate how much material was lost if we assume that certain elements should have remained immobile, and that their apparent increase towards the vein is solely a result of the closure effect. Typical elements with low solubilities in aqueous fluids, and therefore in general immobile behaviour are Al, Nb, Zr, Ti and P. All these elements appear to be 10% higher in the host rock next to the vein than in the host further away from the vein, indicating that there must have been a mass loss of approximately 10% (Fig. 3a). Elements that plot above the correlation line between the immobile elements have been lost from the host rock next to the vein. The most obvious of these are Ca, Sr, Cu and SO_3. Sr is an element that shows similar

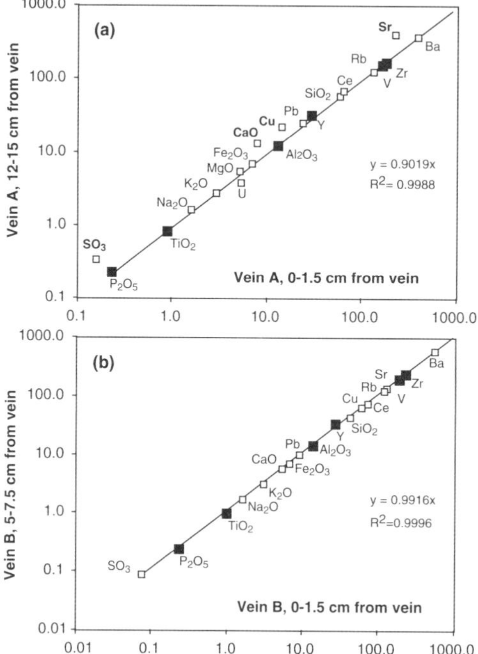

Fig. 3. Elemental diagrams of host rock samples adjacent and further away from vein. (**a**) Sample A shows a depletion of CaO, Sr, Cu and SO_3 close to the vein. (**b**) Sample B shows no difference between host rock adjacent and away from vein. Oxides in weight%, elements in ppm. The straight line represents the best fit through analyses of the least mobile elements and oxides (black squares).

chemical behaviour to calcium, and it is not surprising that the loss of calcium is coupled to that of Sr. However, CaO and Sr have not been lost to the same degree: the CaO/Sr ratio of the host rock near the vein is higher than that of the host rock further away, indicating that there has been preferential loss of Sr.

The two analyses of the host rock near to and further away from the vein in B do not show any significant differences, indicating that in this sample there has been no preferential loss of material from the host rock next to the vein (Fig. 3b).

Sr isotopic ratios, Sr and Rb concentrations

The mass spectrometry results for the veins, and the calcite and siliciclastic fractions of the host rock are given in Table 3. All values are as measured, so no age corrections have been made.

The two analyses of calcite from the host of sample A show very similar Sr isotope ratios to the Sr in the vein from this sample, which plots in between the two values for the host rocks. It is therefore permissible that the Sr in the vein is derived from the calcite fraction of the host, in agreement with our previous observations that calcite and Sr had been lost from the host rock near to the vein. The data also show that the calcite in the host near to the vein contains less Sr than the calcite in the host further away from the vein. This agrees with the fact that the vein has significantly higher Sr concentrations than the calcite in the host rock. We can perform a simple mass balance calculation, assuming that all leached material represents calcite (this is a simplification, since it will also represent adsorbed moisture on the silicate fraction, and potentially a small contribution from $FeCO_3$ and $MgCO_3$), and that the difference in Sr content is accounted for by the loss of approximately 9% of calcite with a higher Sr content. This calculation shows that the relative depletion in Sr in the calcite from the host rock near the vein cannot account for the complete increase in Sr in the vein, since the calculation yields 1680 ppm Sr, while the vein in reality contains 2380 ppm Sr. We note that there has also been a small loss of Sr in the silicate (leached) fraction of the host rock near to the vein. Although this is only 4 ppm, the fact that the silicate fraction forms more than 80% of the sample means that this could still contribute to the Sr content of the vein. The silicate fraction of the host rock, which has far higher Rb/Sr ratios and present-day Sr isotopic compositions than the calcite fractions, would have a Sr isotopic ratio very similar to that of the vein and calcite fractions at 592 Ma.

The Sr isotopic composition of the vein in sample B has a higher $^{87}Sr/^{86}Sr$ ratio (0.7177) than the Sr from the calcite in the host (0.7136–0.7138). This cannot be explained by any *in situ* decay of Rb, since the vein has a lower Rb/Sr ratio than the calcite in the host, the opposite from what would be expected if the elevated Sr isotopic ratio was the result of *in situ* decay. The amount of material that is leached from the host rock samples near to the vein and further away from the vein is very similar (13%), but less material was leached from the host rock 2 cm from the other side of the vein (10%). This could possibly reflect that the latter sample had a somewhat coarser grainsize when leached, but it could also be natural variation in the rock. Sr contents of the host rock calcite vary between 400 and 450 ppm, while the vein calcite contains 890 ppm. This is significantly lower than for the host rock calcite in A (870–1150 ppm). The leached fraction of the host

Table 3. *Mass spectrometry results for leachates and leached samples*

Sample	Description	Rb (ppm)	Sr (ppm)	% of sample	$^{87}Rb/^{86}Sr$	$^{87}Sr/^{86}Sr$	2 SE	Equilibrium age†
Vein B	Vein cc	0.02	891.14	100.00	0.0001	0.717757	0.000022	
Vein B	Host 0–1.5 cm cc	6.63	398.15	13.15	0.0482	0.713782	0.000012	
Vein B	Host 5–7.5 cm cc	6.74	423.77	13.04	0.0460	0.713792	0.000012	
Vein B	Host –2 cm cc	7.65	453.51	10.67	0.0488	0.713581	0.000014	
Vein B	Host 0–1.5 cm ld	127.87	76.62	86.85	4.85	0.756341	0.000012	557
Vein B	Host 5–7.5 cm ld	125.99	77.89	86.96	4.70	0.755496	0.000011	563
Vein B	Host –2 cm ld	123.85	74.99	89.33	4.80	0.755805	0.000019	556
Vein A	Vein cc	0.08	2383.75	95.81	0.0001	0.710281	0.000012	
Vein A	Host 0–2 cc	5.00	867.26	16.40	0.0167	0.710657	0.000014	
Vein A	Host 12–15 cc	2.49	1151.40	25.22	0.0063	0.710097	0.000012	
Vein A	Host 0–2 ld	144.16	77.36	83.6	5.417	0.755546	0.000012	586
Vein A	Host 12–15 ld	140.13	81.94	74.78	4.969	0.752618	0.000023	592
Vein F-a cc	Vein cc	0.71	1959.71	84.77	0.0010	0.710998	0.000015	
Vein F-b cc	Vein cc	0.20	2495.36	100.00	0.0002	0.710945	0.000011	
Vein F-c cc	Vein cc	0.15	2367.10	95.79	0.0002	0.710923	0.000011	
Vein F-d cc	Host cc	12.73	748.61	9.69	0.0492	0.711338	0.000011	
Vein F-e cc	Host cc	13.97	567.06	9.20	0.0713	0.711548	0.000011	
Vein F-d ld	Host ld	172.94	109.52	90.31	4.5855	0.745112	0.000011	523
Vein F-e ld	Host ld	162.65	96.85	90.80	4.8768	0.745439	0.000012	497
Vein E-2	Vein cc	0.05	433.32	100.00	0.0004	0.722376	0.000012	
Vein E-4	Vein cc	0.03	505.87	100.00	0.0001	0.722240	0.000011	
Vein E-6	Vein cc	0.03	476.49	100.00	0.0002	0.722304	0.000014	
Vein E-8	Vein cc	0.02	425.29	100.00	0.0002	0.722198	0.000012	
Vein E-9	Vein cc	0.11	429.91	99.90	0.0007	0.722191	0.000020	
Vein C-a cc	Vein cc	0.06	1868.90	99.23	0.00009	0.712715	0.000013	
Vein C-b cc	Host to vein a cc	9.47	692.36	8.67	0.0396	0.712303	0.000010	
Vein C-c cc	diffuse vein cc	0.29	1436.16	64.55	0.0006	0.712427	0.000014	
Vein C-d cc	Host to vein c cc	1.09	1123.70	32.33	0.0028	0.711262	0.000013	
Vein C-a ld	Vein ld	8.37	1658.12	0.77	0.0146	0.712893	0.000011	
Vein C-b ld	Host ld	118.38	90.59	91.33	3.79	0.745949	0.000015	614
Vein C-c ld	Vein ld	75.57	164.66	35.45	1.33	0.723453	0.000012	
Vein C-d ld	Host ld	80.37	102.23	67.67	2.28	0.732646	0.000011	622
Vein D-x cc	blocky vein cc	0.06	225.83	94.56	0.0008	0.723329	0.000015	
Vein D-y cc	fibrous vein cc	0.04	450.27	99.79	0.0003	0.723119	0.000013	
Vein D-z cc	Host cc	10.10	167.27	7.37	0.1748	0.714559	0.000020	
Vein D-x	Vein ld	3.75	106.56	5.44	0.1020	0.723106	0.000018	
Vein D-y	Vein ld	77.04	754.05	0.21	0.2961	0.724895	0.000022	
Vein D-z	Host ld	123.34	47.22	92.63	7.61	0.782398	0.000024	544

*No age corrections have been applied to the Sr isotopic ratios.
†The 'equilibrium age' is the age at which the Sr fraction from the leached host rocks (siliciclastics) would be in isotopic equilibrium with the vein Sr.

rock has Sr contents that are similar to the leached host of A.

Sample F was first analysed because in hand specimen it seemed that parts of the vein were blocky, and parts fibrous. However, the thin section shows that all three parts of the vein (samples A, B, C) are fibrous, although the vein shows a distinct asymmetry. The host rock samples were both from within 1 cm of the vein, but part D was closer to the vein than E. The Sr isotopic composition of the vein is similar to that of calcite from the host, but slightly lower. If this is the result of *in situ* decay of Rb in the host fraction, then the age of the vein would be approximately 578 Ma. Although the Sr isotopic data permit that the calcite in the vein is derived from that in the host rock, there is no evidence for Sr depletion in calcite near to the vein. This may reflect the smaller scale at which the host samples were taken compared to A. Again, Sr contents of vein calcite are much higher than that of host rock calcite (570–750 versus 1960–2500 ppm);

leached host rock fractions contain more Sr (100–110 ppm) than of A and B.

Sample D is a vein with a blocky and fibrous part (X and Y respectively). Sr isotopic compositions are fairly similar between the different crystal morphologies, and much higher than of the host calcite (0.723 versus 0.715). Sr contents are highest in the fibrous part, followed by the blocky part, and lowest in the host calcite.

Sample E was analysed to test isotopic heterogeneity on a smaller scale. For this purpose, the vein was cut into sixteen slices of approximately 1 mm, of which five were analysed. On this scale, the vein is quite homogeneous in terms of isotopic values (0.7222–0.7224) and Sr content (430–510 ppm). The small amount of heterogeneity cannot be explained by *in situ* decay of ^{87}Rb.

Finally, sample C was analysed because the vein in this sample shows a distinct transition from a clearly fibrous vein (a) in the more shale-rich part of the host rock (b), to a more diffuse stretched type of vein (c), mixed with parts of the host, in the more calcitic layer of host rock (d). This relationship is borne out by the leachate data, which shows the amount of leachable material is highest in the fibrous vein, followed by the diffuse vein, then in the host to the diffuse vein, and lowest in the host to the fibrous vein. Sr isotopic ratios are highest in the fibrous vein, then the diffuse vein, then the silicate-rich host, then the calcite-rich host rock. Again, correction for *in situ* decay of Rb can not bring the veins into equilibrium with the host rock calcite.

Microprobe data

Electron microprobe analyses were performed on the veins of sample A and B to further assess vein homogeneity. If the veins were formed by the crack-seal mechanism, we would expect to find cyclicity in elemental concentrations during each crack-seal episode. Most estimates for the width of each successive crack-seal episode are in the order of 10 µm (Ramsay 1980; Cox & Etheridge 1983; Wiltschko & Morse 2001). Wider crack-seal cycles are not expected, as the lack of growth competition between fibres indicates at the most a very narrow crack, if any crack at all (Urai *et al.* 1991; Bons 2001*b*). Initial small-scale chemical heterogeneities are likely to have been preserved, as there are no indications for any static or dynamic recrystallization of the fibres.

Our main interest was the element Sr, since this had also been the focus of our isotopic studies. Although great care was taken in microprobe calibrations, the only standard for Sr present was almost pure $SrSO_4$, and the difference in Sr concentration and crystal structure between this mineral and calcite may have introduced a systematic error in the data, by which Sr concentrations measured by the microprobe appeared to be 35% too high. This was found when the average Sr content of each sample was calculated from the microprobe measurements, and from our isotope dilution measurements. We have a check on the accuracy of the isotope dilution measurement, since the Rb and Sr contents of the host rock leachates and leached samples combined added up to the same concentrations as measured by X-ray fluorescence on the unleached host samples. Despite this systematic error, we are convinced that the relative concentration measurements by the electron microprobe are correct, since the ratio of the average Sr contents in the two vein samples measured by microprobe is the same as measured by isotope dilution. Most analyses were performed on traverses along single fibres within the vein. Analytical uncertainty is believed to be better than 5% for $CaCO_3$, and better than 10% for the other carbonates.

The greatest number of analyses were performed on the vein B, and these show little difference in Sr (or Mg, Mn, Fe) contents between different fibres (Fig. 4). Only very close (<70 µm) to the border with the host rock do Sr (and Mg, Fe) concentrations increase notably (Fig. 4 and 5a). Within a single fibre, Sr concentrations can vary by 100%, but there does not always appear to be a systematic pattern. We carried out traverses at different scales, varying

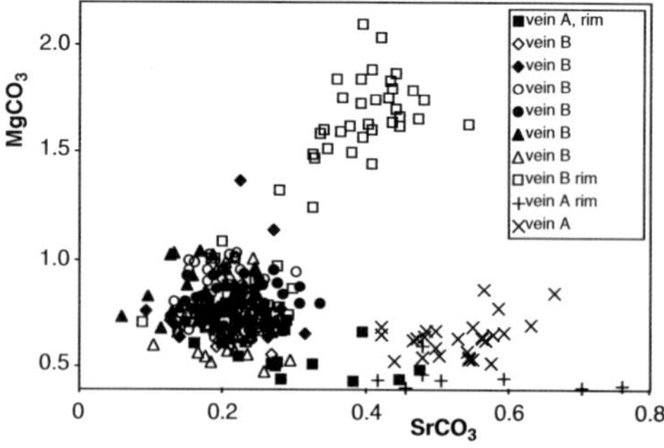

Fig. 4. $MgCO_3$ versus $SrCO_3$ of all microprobe analyses for the fibrous veins. Legend in figure. Different symbols are used for each along-fibre traverse in different parts of the veins. Average Sr contents are higher towards the rim of the veins for both samples, but overlap for all other parts of the vein in sample B.

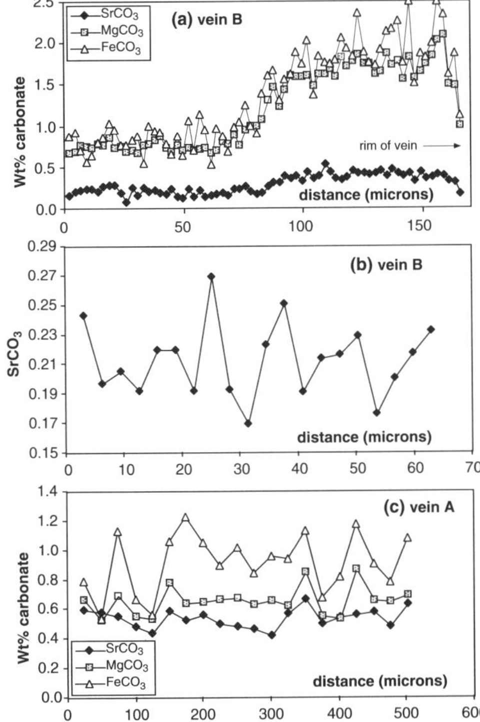

Fig. 5. (**a**) Traverse from rim of the vein (left) to further inside the vein along a single fibre in B. Sr, Mg and Fe are all higher near to the rim of the vein. (**b**) $SrCO_3$ analyses along a single fibre in vein B. No systematic behaviour is observed on the 10 μ scale, as would be expected from crack-seal behaviour. (**c**) Traverse along a single fibre within vein A. Although Sr shows two cycles of decreasing concentration on the 150 μ scale, this cyclicity appears to be reversed for the next 200 μ. Also, the changes in Sr content are not matched by similar cycles of changes in Mg or Fe content.

from tens of microns to hundreds of microns. Some traverses show apparent systematic behaviour for part of the fibre on a 10 μm scale (Fig. 5b), but for most of the traverses, no systematic variation was observed on either the 10 or the 100 μm scale.

Only two traverses were performed for vein A, one of which shows, again, enrichment in Sr near to the rim of the vein (Fig. 4). The other one, taken further away from the vein rim, shows apparent systematic behaviour for the first 300 μm (two cycles of decreasing Sr content; Fig. 5c), but this is followed by 200 μm in which the cyclicity appears reversed (increasing Sr content). Also, the apparent systematic behaviour in one element (Sr) is not followed by the other elements (Fe, Mg). Therefore, our data do not show regular elemental concentration patterns, as would be expected from veins formed by the crack-seal mechanism.

Discussion

The XRF and isotope data presented for vein A make a good case for derivation of the vein calcite from the calcite fraction of the host rock, and the isotope data of sample F are in agreement with this theory as well, if the age of vein formation was around 580 Ma. However, the other samples do not comply with this idea. We will now try to constrain the origin of the calcite in these veins, and the petrogenetic processes involved.

In Fig. 6a we plotted the present-day Sr isotopic composition versus the Sr content of fibrous veins and host calcite, and in Fig. 6b we also plotted the host siliciclastics (leached samples). In Fig. 6a we observe that the fibrous veins with the highest Sr content are hosted by calcite

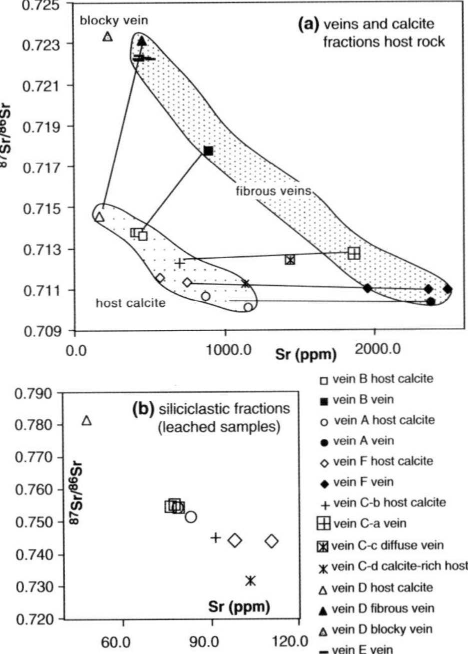

Fig. 6. (**a**) $^{87}Sr/^{86}Sr$ ratio versus Sr contents of veins and calcite fractions of host rocks. Although the range in isotopic ratios is larger in the veins, the host rock calcite and veins show matching behaviour of decreasing Sr content with increasing $^{87}Sr/^{86}Sr$ ratio, suggesting a relationship between veins and host calcite. (**b**) $^{87}Sr/^{86}Sr$ ratio versus Sr contents of siliciclastic fractions of the host rocks.

with the highest Sr contents; that an increase in Sr isotopic composition of the vein is mimicked by the host calcite (although to a more limited extent); and that there is a negative correlation between Sr content and Sr isotopic composition for both the veins and host calcites, i.e. tie-lines between vein and host calcite do not cross each other. The only two samples that have vein calcite close to equilibrium with host calcite are those with the highest Sr contents in the veins. For the siliciclastic fraction of the host rocks, we also observe a negative correlation between Sr content and Sr isotopic composition. If this were solely the result of radiogenic decay of ^{87}Rb, a better correlation would be shown by ^{87}Sr/^{86}Sr versus ^{87}Rb/^{86}Sr ratio (not shown), but this is not the case.

The fact that the vein calcite and the host calcite show similar trends with respect to Sr content and Sr isotopic composition suggests that the vein and the host have some 'knowledge' of each other's composition. This would argue against a completely external source for the vein Sr (and thereby the vein calcite), unless material from this external source with more radiogenic Sr also provided some of the Sr in the host rock calcite. If this were the case, then one could imagine that the vein and host calcite with lower Sr contents were susceptible to 'contamination' with Sr from a different, more radiogenic, source, while those with high Sr contents were less affected by interaction with this component. This would then explain the observed negative correlation between Sr content and the Sr isotopic composition of vein and host rock calcite. There are two basic choices for the origin of the radiogenic component: (1) a component that is foreign to the system and has been brought in by advection; or (2) a component that comes from within the system, i.e. from the siliciclastic fraction of the host rocks.

Radiogenic Sr from the siliciclastic fraction?

If the radiogenic Sr was sourced from the siliciclastic fraction of the host rock, we would expect to see a correlation between the Sr isotopic composition of the siliciclastic fraction of the host rock and that of the vein calcite. It is indeed true that the most radiogenic fibrous vein, sample D, also has the most radiogenic Sr in the host rock. However, we must remember that we are looking at present-day ratios, while the veins were formed at some time in the past. Since the Rb/Sr ratio of the siliciclastic fraction of this sample is also the highest measured, the Sr isotopic composition would have been appreciably lower in the past. At 544 Ma the Sr isotopic composition of the siliciclastic fraction of the host is the same as that for the fibrous vein, so if the Sr in the vein was solely formed from Sr in the siliciclastic fraction of the host rock, it would have happened at this time. If the Sr in the vein were derived from both the calcite in the host and the siliciclastic fraction, its formation age needs to be younger than that. This is permissible from a geological point of view, since the main deformation and metamorphism in the area occurred at <515 Ma, and probably as late as 420 Ma (McLaren *et al.* 2002).

The main argument against this idea is the very low measured Sr content of the siliciclastic fraction of the host in this sample: 47 ppm, versus 167 ppm in the host calcite. At 420 Ma, we would need 75% of host siliciclastic and 25% of host calcite to produce Sr with the isotopic composition of the vein. At any earlier age, we would need a greater relative proportion of siliciclastic host rock contributing its full Sr content to the vein. This argument is of course based on the assumption that the present-day Sr content of the siliciclastic fraction of the host is representative of that at the time of vein formation. If the siliciclastic fraction of the host did contribute Sr to the vein, it is likely that the present-day value is much lower than at the time of vein formation. Nevertheless, we think this scenario is not very likely, because of the high proportion of siliciclastic Sr. As it is far easier to mobilize Sr from the calcite than from the siliciclastics (as demonstrated by our laboratory leaching procedure), it is difficult to see how one could generate a vein with such a high proportion of siliclastic derived Sr compared to calcite-derived Sr.

Radiogenic Sr from an extraneous source?

If the radiogenic Sr component in the veins was derived externally, we do not have any control on its composition, except by assuming that a single component was contributing to all veins. In that case, the ^{87}Sr/^{86}Sr ratio of this component must have been at least as radiogenic as that of the most radiogenic vein, i.e. >0.723. The systematics of Sr content versus Sr concentration of the veins would also indicate that this component has a relatively low Sr concentration. The source of this radiogenic Sr component is unlikely to be any of the other carbonate units within in the Neoproterozoic sedimentary pile, since they all have compositions close to sea water (^{87}Sr/^{86}Sr $\leq$ 0.710; Veizer & Compston

1976). The most likely source of the Sr is either the siliciclastics from the sedimentary sequence, or the Mesoproterozoic basement underlying it. If this has influenced the isotopic composition of the veins, it is likely to also have affected the host calcite, since vein and host calcite show similar behaviour.

The data are reminiscent of those presented by Bickle (1992) and McCaig *et al.* (1995), who describe cases where extraneous fluids percolated along fracture networks and partly reset the isotopic signature of the wall rocks. The model of Bickle (1992) assumes a homogeneous fluid that enters the system and a homogeneous starting composition of the host rock. A fluid with $^{87}Sr/^{86}Sr \approx 0.723$ and in equilibrium with solid calcite with *c.* 450 ppm Sr would have been in equilibrium with the vein with the highest isotopic value. An appropriate choice of unaltered host would be one that has calcite with *c.* 2000 ppm Sr and $^{87}Sr/^{86}Sr \approx 0.710$ (highest Sr-concentration and lowest isotopic value found). At the upstream side, veins would be in equilibrium with the incoming fluid and the host would be a mixture of old (2000 ppm Sr) and new (450 ppm Sr) calcite, resulting in values between these two end-members. However, we find a value of 250 ppm Sr for the host of vein D, too low to be explained by the model. This suggests that the assumption of a homogeneous host cannot be valid.

Another problem with the fracture flow and partial wall rock alteration model is the actual absolute concentration of calcite in the wall rock. Where localized alteration is observed (vein A), the wall rock is actually depleted in calcite. This is difficult to reconcile with a fluid that is apparently supersaturated with respect to calcite inside a crack (as it formed a calcite vein), while simultaneously being undersaturated in the adjacent wall rock, as it removed calcite there. Finally, the arguments against a crack-sealing origin of antitaxial fibrous veins that are listed in the introduction and by Bons & Jessell (1997) and Bons (2000), and the apparent limited connectivity of the veins argue against percolation of fluids through the veins.

The alternative to fluid flow through fractures is pervasive flow through the host. The homogeneous isotopic composition of vein B supports this, as the host is not more strongly affected near the vein than further away. The pervasively percolating fluid may have added or removed calcite from the host. The veins would then not represent paths of fluid flow, but rather accumulations of calcite that precipitated from fluids carrying a mixture of extraneous (radiogenic) Sr and Sr from the calcite in the host rock. What caused this localized precipitation cannot be constrained from the geochemical data presented here.

Simple mixing calculations can be performed to estimate how much of the Sr in the host calcite would be derived from this percolating fluid and how much from the original calcite that was present. To avoid making assumptions about Sr contents in the fluid or in the host calcite, we only calculate the amount of 'foreign' Sr that must be added to a homogeneous host calcite to obtain the desired Sr isotopic composition, as measured in the host. To model the host calcite of D, with a Sr isotopic ratio of 0.7146, we assumed that its original isotopic composition 0.710, similar to that of the host rock calcite of A. The isotopic signature of the extraneous Sr was taken to be 0.723 (the maximum value measured in any vein). With these assumptions, the amount of 'original' Sr in the host calcite is 65%, and 35% is foreign. In this case, the Sr in the vein of this sample is 100% 'foreign'. It is therefore likely that the calculated amount of foreign Sr in the host is a maximum estimate. Using a similar fluid end member, the estimate for sample B would yield 29% foreign Sr in the host calcite, and 57% in the vein.

A basin-wide fluid flow event?

Although our work was only intended to shed light on the genesis of fibrous veins, we found that our interpretation of pervasive fluid flow through the Tapley Hill Formation agrees well with recent large-scale Sr isotope work on the Neoproterozoic sedimentary succession in the Adelaide Geosyncline (Foden *et al.* 2001). Sr isotope whole-rock analyses of pre-586 Ma sediments in the Adelaide Geosyncline define an errorchron with an age of 586 ± 30 Ma, and an initial $^{87}Sr/^{86}Sr$ ratio of 0.7183. This has been interpreted to reflect equilibration of the sedimentary succession around 586 Ma with a Sr-bearing fluid, with an average $^{87}Sr/^{86}Sr$ ratio of 0.7183.

On the basis of the Sr isotopic data for sample F, we suggested that the vein-forming event took place around 580 Ma, an age which falls well within the range of that suggested for the basin-wide fluid-flow event. Since we find veins with $^{87}Sr/^{86}Sr$ ratios in excess of 0.718, we interpret that the radiogenic Sr component in the fluid must have had an $^{87}Sr/^{86}Sr$ ratio of at least 0.723. This is not in disagreement with the study by Foden *et al.* (2001), since they only refer to the basin-wide *average* isotopic composition of the fluid, which is likely to be a mixture of radiogenic

Sr from basement lithologies, sea water, and carbonate rocks. Our much smaller-scale study identifies the minimum $^{87}Sr/^{86}Sr$ isotopic ratio of the radiogenic component. Its rather high initial $^{87}Sr/^{86}Sr$ value reflects a source with long-term enrichment of Rb over Sr, such as the Mesoproterozoic granites that form the basement to the Neoproterozoic cover rocks to which the Tapley Hill formation belongs (Coats & Blissett 1971).

The *c.* 580 Ma fluid-flow event precedes the only recognized major deformation phase that affected the vein host rocks: the Cambro-Ordovician Delamerian folding phase. These veins are thus pre-tectonic, which may explain the lack of any consistent pattern that can be related with the local large-scale folding. These veins would have passively rotated into their current orientation in the limb of a large syncline. Minor deformation by twinning may have occurred during this folding.

The host rock XRF-data for sample A showed that the two elements that were most mobile, apart from Ca and Sr, were copper and sulphur. The Adelaide Geosyncline contains a number of Cu-rich ore deposits, and their formation may have been related to the basin-wide fluid event at 586 Ma (Foden *et al.* 2001). Our data clearly support this conclusion.

Conclusions

Our data show clear evidence for derivation of vein calcite from the immediate host rock in one sample (A), but the isotopic data for most other veins shows that local host calcite cannot have been the only component contributing to the vein. The fibrous texture by itself is therefore no proof of local derivation of nutrients for the veins, contrary to the suggestions by Bons (2000) and Oliver & Bons (2001). The congruent elemental and isotopic behaviour of Sr in the host calcite and the vein argues against channelized advection of a foreign fluid through cracks in which the vein calcite grew. The lack of any cyclical patterns in microprobe traverses along a single fibre supports the previously reported microstructural arguments against crack-sealing in these veins (Bons & Jessell 1997; Bons 2000; Means & Li 2001). We therefore conclude that the fibrous veins grew without cracking in an environment of pervasive fluid flow, with the veins representing sites of local precipitation rather than fluid-flow conduits. The veins derived their nutrients partly from their local host and partly from the external fluid source. Our observations cannot shed more light on the detailed mechanism of vein growth. Diffusional transport along concentration gradients probably acted to extract nutrients from the percolating fluids.

Our conclusions agree with large-scale Sr isotope studies that identified a pervasive fluid-flow event in the Adelaide Geosyncline around 586 Ma (Foden *et al.* 2001). As this event precedes the only known major deformation event that could have affected the veins, the veins are pre-tectonic.

This work was done while M. Elburg was a recipient of an Australian Research Council postdoctoral fellowship. D. Bruce kept the isotope lab running while the analyses were carried out, and J. Stanley performed the XRF analyses. C. Noble was instrumental in setting up the microprobe to analyse Sr in carbonates. We thank the Sprigg family for access to their property. The paper benefited from thoughtful reviews by A. McCaig and P. Mason. This research project was supported by the Foundation 'Stichting Dr. Schümannfonds', grant no. 2000/02.

References

BAROVICH, K. M. & FODEN, J. 2000. A Neoproterozoic flood basalt province in southern-central Australia: geochemical and Nd isotope evidence from basin fill. *Precambrian Research*, **100**, 213–234.

BEACH, A. 1974. A geochemical investigation of pressure solution and the formation of veins in a deformed greywacke. *Contributions to Mineralogy and Petrology*, **46**, 61–68.

BEACH, A. 1977. Vein arrays, hydraulic fractures and pressure-solution structures in a deformed flysch sequence, S.W. England. *Tectonophysics*, **40**, 201–225.

BICKLE, M. J. 1992. Transport mechanisms by fluid-flow in metamorphic rocks: oxygen and strontium decoupling in the Trois Seigneurs Massif – a consequence of kinetic dispersion? *American Journal of Science*, **292**, 289–316.

BONS, P. D. 2000. The formation of veins and their microstructures. *In*: JESSELL, M. W. & URAI, J. L. (eds) *Stress, Strain and Structure, A Volume in Honour of W. D. Means. Journal of the Virtual Explorer*, **2**. World Wide Web Address: http://virtualexplorer.com.au/VEjournal/Volume2/www/contribs/bons.

BONS, P. D. 2001*a*. The formation of large quartz veins by rapid ascent of fluids in mobile hydrofractures. *Tectonophysics*, **336**, 1–17.

BONS, P. D. 2001*b*. Development of crystal morphology during unitaxial growth in a progressively widening vein: II. The numerical model. *Journal of Structural Geology*, **23**, 865–872.

BONS, P. D. & JESSELL, M. W. 1997. Experimental simulation of the formation of fibrous veins by localised dissolution–precipitation creep. *Mineralogical Magazine*, **61**, 53–63.

BOULLIER, A.-M., CHAROY, B. & POLLARD, P. J. 1994. Fluctuation in porosity and fluid pressure during hydrothermal events: textural evidence in the Emuford District, Australia. *Journal of Structural Geology*, **16**, 1417–1429.

BYERLEE, J. D. 1990. Friction, overpressure and fault normal compression. *Geophysical Research Letters*, **17**, 2109–2112.

COATS, R. P. & BLISSETT, A. H. 1971. Regional and economic geology of the Mount Painter Province. *Geological Survey of South Australia Bulletin*, **43**.

COX, S. F. 1987. Antitaxial crack-seal vein microstructures and their relationship to displacement paths. *Journal of Structural Geology*, **9**, 779–788.

COX, S. F. 1995. Faulting processes at high fluid pressures: an example of fault-valve behavior from the Wattle Gully Fault, Victoria, Australia. *Journal of Geophysical Research*, **100**, 841–859.

COX, S. F. & ETHERIDGE, M. A. 1983. Crack-seal fibre growth mechanisms and their significance in the development of oriented layer silicate microstructures. *Tectonophysics*, **92**, 147–170.

DREXEL, J. F., PREISS, W. V. & PARKER, A. J. (eds) 1993. The Geology of South Australia. Volume I: The Precambrian. *Geological Survey of South Australia Bulletin*, **54**, 242 pp.

DURNEY, D. W. 1976. Pressure-solution and crystallization deformation. *Philosophical Transactions of the Royal Society of London*, **A283**, 229–240.

DURNEY, D. W. & RAMSAY, J. G. 1973. Incremental strains measured by syntectonic crystal growths. *In*: DE JONG, K. A. & SCHOLTEN, R. (eds) *Gravity and Tectonics*. Wiley, New York, 67–95.

EISENLOHR, B. N., GROVES, D. & PARTINGTON, G. A. 1989. Crustal-scale shear zones and their significance to Archaean gold mineralization in Western Australia. *Mineralium Deposita*, **24**, 1–8.

FAURE, G. 1986. *Principles of isotope geology*. New York, John Wiley & Sons.

FAURE, G., HURLEY, P. M. & FAIRBAIRN, W. H. 1963. An estimate of the isotopic composition of strontium in rocks of the Precambrian Shield of North America. *Journal of Geophysical Research*, **68**, 2323–2329.

FISHER, D. & BRANTLEY, S. L. 1992. Models of quartz overgrowth and vein formation: deformation and fluid flow in an ancient subduction zone. *Journal of Geophysical Research*, **97**, 20043–20061.

FISHER, D. & BYRNE, T. 1990. The character and distribution of mineralized fractures in the Kodiak Formation, Alaska: implications for fluid flow in an underthrust sequence. *Journal of Geophysical Research*, **95**, 9069–9080.

FODEN, J., BAROVICH, K. M., JANE, M. & O'HALLORAN, G. 2001. Sr-isotopic evidence for late Neoproterozoic rifting in the Adelaide geosyncline at 586 Ma: implications for a Cu ore forming fluid. *Precambrian Research*, **106**, 291–308.

FOXFORD, K. A., NICHOLSON, R., POLYA, D. A. & HEBBLETHWAITE, R. P. B. 2000. Extensional failure and hydraulic valving at Minas da Panasqueira, Portugal: evidence from vein spatial distributions, displacements and geometries. *Journal of Structural Geology*, **22**, 1065–1086.

HAHN, O., STRASSMAN, F., MATTAUCH, J. & EWALD, H. 1943. Geologische Alterbestimmungen met der Strontiummethode. *Chemische Zeitung*, **67**, 55–56.

HENDERSON, I. H. C. & MCCAIG, A. M. 1996. Fluid pressure and salinity variations in shear zone-related veins, central Pyrenees, France: implications for the fault-valve model. *Tectonophysics*, **262**, 321–348.

HILGERS, C. 2000. *Vein Growth in Fractures – Experimental, Numerical and Real Rock Studies*. PhD thesis, Aachen University.

HIPPERTT, J. F. & MASSUCATTO, A. J. 1998. Phyllonitization and the development of kilometer-size extension gashes in a continental-scale strike-slip shear zone, north Goiás, central Brazil. *Journal of Structural Geology*, **20**, 433–446.

HOFFMAN, P. F., KAUFMAN, A. J., HALVERSON, G. P. & SCHRAG, D. P. 1998. A Neoproterozoic snowball earth. *Science*, **281**, 1342–1346.

JAMTVEIT, B. & YARDLEY, B. W. D. (eds) 1997. *Fluid Flow and Transport in Rocks – Mechanisms and Effects*. Chapman & Hall, London.

JESSELL, M. W., WILLMAN, C. E. & GRAY, D. R. 1994. Bedding parallel veins and their relationship to folding. *Journal of Structural Geology*, **16**, 753–767.

KOEHN, D. & PASSCHIER, C. W. 2000. Shear sense indicators in striped bedding-veins. *Journal of Structural Geology*, **22**, 1141–1151.

MACHEL, H. -G. 1985. Fibrous gypsum and fibrous anhydrite in veins. *Sedimentology*, **32**, 443–454.

MATTHÄI, S. K. & ROBERTS, S. G. 1997. Transient versus continuous fluid flow in seismically active faults: an investigation by electric analogue and numerical modelling. *In*: JAMTVEIT, B. & YARDLEY, B. W. D. (eds) *Fluid Flow and Transport in Rocks*. Chapman & Hall, London, 263–296.

MCCAIG, A. M., WAYNE, D. M., MARSHALL, J. D., BANKS, D. & HENDERSON, I. 1995. Isotopic and fluid inclusion studies of fluid movement along the Gavarnie Thrust, Central Pyrenees: reaction fronts in carbonate mylonites. *American Journal of Science*, **295**, 309–343.

MCKENZIE, D. & HOLNESS, M. 2001. Local deformation in compacting flows: development of pressure shadows. *Earth and Planetary Science Letters*, **180**, 169–184.

MCKIRDY, D. M., BURGESS, J. M. *ET AL*. 2001. A chemostratographic overview of the late Cryogenian interglacial sequence in the Adelaide Fold-Thrust Belt, South Australia. *Precambrian Research*, **106**, 149–186.

MCLAREN, S. N., DUNLAP, W. J., SANDIFORD, M. & MCDOUGALL, I. 2002. Thermochronology of high heat producing crust at Mount Painter, South Australia: implications for tectonic reactivation of continental interiors. *Tectonics*, in press.

MEANS, W. D. & LI, T. 2001. A laboratory simulation of fibrous veins: some first observations. *Journal of Structural Geology*, **23**, 857–863.

MÜGGE, O. 1928. Über die Entstehung faseriger Minerale und ihrer Aggregationsformen. *Neues Jahrbuch für Mineralogie, Geologie und Paläontologie*, **58A**, 303–348.

MULLENAX, A. G. & GRAY, D. R. 1984. Interaction of bed-parallel stylolites and extension veins in boudinage. *Journal of Structural Geology*, **6**, 63–71.

OLIVER, N. H. S. & BONS, P. D. 2001. Mechanisms of fluid flow and fluid-rock interaction in fossil metamorphic-hydrothermal systems inferred from vein-wallrock patterns, geometry, and microstructure. *Geofluids*, **1**, 137–163.

PABST, A. 1931. 'Pressure-shadows' and the measurement of the orientation of minerals in rocks. *The American Mineralogist*, **16**, 55–70.

PASSCHIER, C. W. & TROUW, R. A. J. 1996. *Microtectonics*. Springer Verlag, Berlin.

PAUL, E., FLÖTTMANN, T. & SANDIFORD, M. 1999. Structural geometry and controls on basement-involved deformation in the northern Flinders Ranges, Adelaide Fold Belt, South Australia. *Australian Journal of Earth Sciences*, **46**, 343–354.

PHILLIPS, W. J. 1974. The development of vein and rock textures by tensile strain crystallization. *Journal of the Geological Society, London*, **130**, 441–448.

PREISS, W. V. 1987. The Adelaide Geosyncline – late Proterozoic stratigraphy, palaeontology and tectonics. *Geological Survey of South Australia Bulletin*, **53**.

PREISS, W. V. 1989. A stratigraphic and tectonic overview of the Adelaide Geosyncline, South Australia. *In*: JAGO, J. B. & MOORE, P. S. (eds) *The Evolution of a Late Precambrian–Early Palaeozoic Rift Complex: the Adelaide Geosyncline*. Brisbane, Watson Furguson & Company. Geological Society of Australia, Special Publication, **16**, 1–33.

PUTNIS, A. & MAUTHE, G. 2001. The effect of pore size on cementation in porous rocks. *Geofluids*, **1**, 37–41.

PUTNIS, A., PRIETO, M. & FERNANDEZ-DIAZ, L. 1995. Fluid supersaturation and crystallization in porous media. *Geological Magazine*, **132**, 1–13.

RAMSAY, J. G. 1980. The crack-seal mechanism of rock deformation. *Nature*, **284**, 135–139.

RAMSAY, J. G. & HUBER, M. I. 1983. *The Techniques of Modern Structural Geology, Volume 1: Strain Analysis*. Academic Press, London.

RENARD, F., GRATIER, J. P. & JAMTVEIT, B. 2000. Kinetics of crack-sealing, intergranular pressure solution, and compaction around active faults. *Journal of Structural Geology*, **22**, 1395–1407.

ROBERT, F. & BROWN, A. C. 1986. Archaean gold bearing quartz veins at Sigma mine, Abitibi greenstone belt, Quebec: part II, vein paragenesis and hydrothermal alteration. *Economic Geology*, **81**, 593–616.

RUTTER, E. H. 1983. Pressure solution in nature, theory and experiment. *Journal of the Geological Society, London*, **140**, 725–740.

SIBSON, R. H. 1987. Earthquake rupturing as a mineralising agent in hydrothermal systems. *Geology*, **15**, 701–704.

SIBSON, R. H., MOORE, R. M. & RANKIN, A. H. 1975. Seismic pumping – a hydrothermal transport mechanism. *Journal of the Geological Society, London*, **131**, 653–659.

SIBSON, R. H., ROBERT, F. & POULSEN, K. H. 1988. High angle reverse faults, fluid pressure cycling, and mesothermal gold-quartz deposits. *Geology*, **16**, 551–555.

TABER, S. 1918. The origin of veinlets in the Silurian and Devonian strata of New York. *Journal of Geology*, **26**, 56–73.

TURNER, S., FODEN, J., SANDIFORD, M. & BRUCE, D. 1993. Sm-Nd isotopic evidence for the provenance of sediments from the Adelaide Fold Belt and southeastern Australia with implications for episodic crustal addition. *Geochimica et Cosmochimica Acta*, **57**, 1837–1856.

URAI, J. L., WILLIAMS, P. F. & VAN ROERMUND, H. L. M. 1991. Kinematics of crystal growth in syntectonic fibrous veins. *Journal of Structural Geology*, **13**, 823–836.

VEIZER, J. & COMPSTON, W. 1976. $^{87}Sr/^{86}Sr$ in Precambrian carbonates as an index of crustal evolution. *Geochimica et Cosmochimica Acta*, **40**, 905–914.

VROLIJK, P. 1987. Tectonically driven fluid flow in the Kodiak accretionary complex, Alaska. *Geology*, **15**, 466–469.

WHITFORD, D. J. & JEZEK, P. A. 1979. Origin of Late-Cenozoic lavas from the Banda Arc, Indonesia: Trace element and Sr isotope evidence. *Contributions to Mineralogy and Petrology*, **68**, 141–150.

WILLIAMS, P. F. & URAI, J. L. 1989. Curved vein fibres – an alternative explanation. *Tectonophysics*, **158**, 311–333.

WILTSCHKO, D. V. & MORSE, J. W. 2001. Crystallization pressure versus 'crack seal' as the mechanism for banded veins. *Geology*, **29**, 79–82.

Effects of stress on the anisotropic development of permeability during mechanical compaction of porous sandstones

WENLU ZHU[1], LAURENT G. J. MONTÉSI[1] & TENG-FONG WONG[2]

[1] *Department of Geology and Geophysics, Woods Hole Oceanographic Institution, Woods Hole, MA 02543, USA (e-mail: wzhu@whoi.edu)*
[2] *Department of Geosciences, State University of New York at Stony Brook, Stony Brook, NY 11794, USA*

Abstract: To investigate the influence of stress on permeability anisotropy during mechanical compaction, a series of triaxial compression experiments with a new loading configuration called hybrid compression were conducted on three porous sandstones. The effective mean and differential stresses in hybrid compression tests were identical to those in conventional triaxial extension tests. Permeability was measured along the axial direction in both hybrid compression and conventional extension tests, which corresponds to flow along the maximum principal stress direction in the former case and the minimum principal stress direction in the latter case. Since their loading paths coincide, the comparison of permeability values from the two types of tests provides quantitative estimates of the development of permeability anisotropy as a function of effective mean and differential stresses. Our data show that the permeability evolution is primarily controlled by stress. Before the onset of shear-enhanced compaction C^*, permeability and porosity reduction are solely controlled by the effective mean stress, with negligible stress-induced anisotropy. With the onset of shear-enhanced compaction and initiation of cataclastic flow, the deviatoric stress induces enhanced permeability and porosity reduction. The permeability tensor may show significant anisotropy. Our data indicate that the maximum principal component of permeability tensor k_1 is parallel to the maximum principal stress σ_1, and the minimum principal component k_3 is parallel to the minimum principal stress σ_3. During the initiation and development of shear-enhanced compaction, k_1 can exceed k_3 by as much as two orders of magnitude. With the progressive development of cataclastic flow, changes of permeability and porosity become gradual again, and the stress-induced permeability anisotropy diminishes as k_1 and k_3 gradually converge. Our data imply that permeability can be highly anisotropic in tectonic settings undergoing cataclastic flow, inducing the fluid to flow preferentially along conduits subparallel to the maximum compression direction. However, this development of permeability anisotropy is transient in nature, becoming negligible with an accumulation of strain of about 10%. The anisotropic development of permeability in a lithified rock is dominantly controlled by microcracking and pore collapse. This is fundamentally different from the mechanisms active in unconsolidated materials such as sediments and fault gouges, in which the permeability evolution is primarily controlled by the development of fabric and shear localization via the accumulation of shear strain.

Many efforts have been made to understand the effects of fluid on active tectonic environments. However, it is difficult to estimate permeability in many geological processes because permeability is very sensitive to pressure, temperature and stress. Laboratory measurements of permeability under crustal conditions of pressure, temperature and stress provide useful constraints on these problems (Brace 1980).

Experimental observations have shown that permeability can be significantly modified under hydrostatic or non-hydrostatic stresses (e.g. David *et al.* 1994; Zhu & Wong 1997; Main *et al.* 1996). Because permeability is a second-rank tensor, it depends sensitively on the anisotropic development of damage. Yet, in most theoretical models to date, permeability is commonly prescribed as either a constant or a function of the effective mean stress (e.g. Person *et al.* 1996; Ingebritsen & Sanford 1998). The lack of knowledge about permeability anisotropy, especially in relation to nonhydrostatic loading, is aggravated by the fact that it is very difficult simultaneously to measure fluid transport in several different directions at elevated pressures.

Review of laboratory data shows that the sensitivity of permeability to mean stress varies widely among geomaterials (e.g. David *et al.* 1994). Very high sensitivity has been observed in clay gouge and fractured crystalline rock, with the potential for development of significant pore pressure excess from the continuous influx of fluid into a fault system made up of such materials, by a mechanism such as that proposed by Rice (1992). In contrast, since relatively low

From: DE MEER, S., DRURY, M. R., DE BRESSER, J. H. P. & PENNOCK, G. M. (eds) 2002. *Deformation Mechanisms, Rheology and Tectonics: Current Status and Future Perspectives*. Geological Society, London, Special Publications, **200**, 119–136. 0305-8719/02/$15

sensitivity is observed in porous rocks and granular materials, it is unlikely that pore pressure excess may be generated by such a mechanism (Wong & Zhu 1999).

While an increase in mean stress generally results in porosity reduction and permeability decrease, the effects of deviatoric stresses on both mechanical deformation and permeability evolution are more complex. In low porosity rocks (with porosities <5% or so), stress-induced dilatancy and permeability enhancement are observed whether the rock fails by brittle faulting or cataclastic flow (e.g. Zoback & Byerlee 1975; Stormont & Daemen 1992; Zhang *et al.* 1994; Kiyama *et al.* 1996; Peach & Spiers 1996; Zhu & Wong 1999). In porous rocks (>5%, including sandstone, limestone, and shale), it has been shown that under overall compressive loading, a deviatoric stress can either induce dilation by brittle fracture or compaction by cataclastic flow (e.g. Wong *et al.* 1997; Baud *et al.* 1999), and the dependence of permeability on porosity and deformation mechanism is fundamentally different from that found in low porosity rock (e.g. Mordecai & Morris 1971; Holt 1989; Rhett & Teufel 1992; Zhu & Wong 1997). In the brittle faulting regime, an appreciable amount of permeability reduction is actually observed in a dilating rock (Zhu & Wong 1997, Main *et al.* 2000). In the cataclastic flow regime, significant porosity reduction with concomitant permeability decrease is generally observed (Zhu & Wong 1997).

In the compactive cataclastic flow regime, although the failure mode is macroscopically 'ductile', the microscopic deformation mechanism involves pervasive brittle cracking (Wong *et al.* 1992). Quantitative characterization of the damage has demonstrated that nonhydrostatic stresses can induce significant anisotropy in microcrack density, due to the preferential development of stress-induced cracking subparallel to the maximum principal stress σ_1 (Menéndez *et al.* 1996; Wu *et al.* 2000). As permeability is very sensitive to the geometric complexity of the pore space, to what extent does the stress-induced anisotropic cracking affect permeability? How does permeability anisotropy evolve as a function of the stress state? Furthermore, how does the permeability anisotropy correlate with the mechanical deformation and failure mode?

In a seminal study, Zoback & Byerlee (1976) conducted conventional triaxial compression and extension tests on Ottawa sand to characterize the permeability anisotropy. In these loading configurations, the minimum principal stress σ_3 is maintained constant. In a conventional triaxial compression test the maximum principal stress σ_1 is applied parallel to the axial direction, whereas in a conventional triaxial extension test, σ_1 is along the radial direction. While permeability is always measured in the axial direction, it is parallel to σ_1 in the former case and perpendicular to σ_1 in the latter case. Comparison of permeability data obtained in triaxial compression tests (k_1) to those measured in extension tests (k_3) provides useful information on stress-induced permeability anisotropy. A similar methodology was adopted by Bruno (1994) and Zhu *et al.* (1997) to study stress-induced anisotropy in permeability in porous sandstones deformed in the brittle faulting and cataclastic flow regimes, respectively.

However, the use of this methodology for characterizing permeability anisotropy in geomaterials is subject to at least two limitations. First, the strain that can be accumulated in a triaxial configuration is limited. In granular materials this poses a serious limitation since the anisotropic development of permeability in sediments (Dewhurst *et al.* 1996; Bolton *et al.* 2000) and fault gouge (Zhang *et al.* 1999, 2001) is intimately related to the fabric that evolves over large strain. Consequently, a rotary shear configuration is preferred for such measurements in unconsolidated materials.

Second, permeability anisotropy should be inferred from data obtained from tests conducted at identical stress states along similar loading paths. However, because the intermediate principal stress σ_2 equals σ_3 in a conventional compression test and equals σ_1 in a conventional extension test, the loading paths in these two cases do not coincide. To acquire data along a loading path identical to that of the conventional triaxial extension test, σ_3 should not be maintained constant in a triaxial compression test and instead it should be systematically adjusted to follow the desired loading path. We refer to such tests along the prescribed loading path as 'hybrid' triaxial compression tests. Data on the permeability component k_3 (perpendicular to σ_1) from conventional extension tests (Zhu *et al.* 1997) and k_1 (parallel to σ_1) from the hybrid compression tests then provide a common basis for inferring the stress-induced permeability anisotropy. In this study, we focus on the anisotropy that develops during mechanical compaction of porous sandstones.

Experimental procedure

Laboratory measurements were conducted on the Adamswiller, Berea, and Rothbach sandstones with nominal porosities of 23.0%, 21.0%

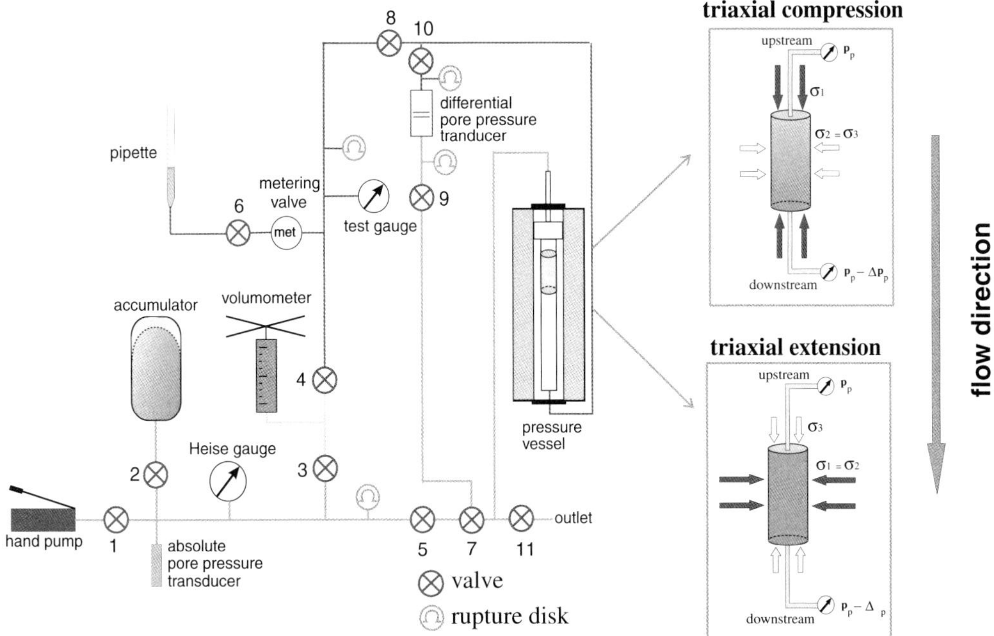

Fig. 1. Schematic diagram of the experimental set-up. Because the axial stress is applied separately by a servo-controlled hydraulic system, while the lateral stresses are applied by an air-driven pump, deformation tests can be conducted under both triaxial compression and extension conditions. A wide-range permeameter is included, which is designed to use either the steady-state flow or pulse transient technique for permeability measurements.

and 19.9% respectively. Petrophysical description of the sandstones (including porosity, grain size and modal analysis) was presented by Wong *et al.* (1997). Visual examination of the specimens reveal that the samples were free from macroscopic planar shear features. Tests on intact samples indicated that permeability anisotropy was insignificant in these porous sandstones. The Berea and Rothbach sandstone samples were cored perpendicular to their sedimentary bedding, whereas the Adamswiller sandstone samples were cored parallel to the bedding. All of the samples were ground to a cylindrical shape, with diameter of 18.4 mm and length 38.1 mm. Before each test, the sample was saturated with distilled water and jacketed with copper foil, and then it was positioned between two steel end-plugs, each of which had a concentric hole at the centre for fluid access to the upstream or downstream pore pressure system. Heat-shrink polyolefine tubings were used to isolate the sample assembly from confining pressure medium (kerosene).

The samples were then deformed at room temperature under hybrid triaxial compression with the starting hydrostatic pressures P_c at a value ranging from 90 to 260 MPa, while the pore pressure P_p was fixed at 10 MPa. A schematic diagram of our experimental set-up is shown in Fig. 1. At the beginning of each triaxial test, the initial permeability was measured at $P_c = 13$ MPa and $P_p = 10$ MPa (corresponding to an effective pressure of $P_c - P_p = 3$ MPa which is sufficient to load the jacket tightly around the sample, thus inhibiting leakage along the sample–jacket interface).

With this experimental set-up, the axial stress was applied by a servo-controlled hydraulic system, while the lateral stresses were applied by an air-driven pump. To investigate permeability evolution during cataclastic flow, mechanical loading was stopped at different stages of deformation, and the *in situ* permeability was measured as a function of the stress state. The average 'holding' period was about 3 hours. In a typical test, the permeability value of a sandstone sample would change by 3 to 5 orders of magnitude during deformation. We modified Bernabé's (1987) design of a wide-range permeameter (originally for hydrostatic loading) for operation in a triaxial system. Two different techniques can be implemented, depending on the permeability of interest. If the permeability at a certain stress state is relatively high

($k > 10^{-16}\,m^2$), it is measured by the steady-state flow technique. For $k < 10^{-16}\,m^2$, thermal fluctuation would render it difficult to achieve steady-state flow, and instead we used the pulse transient technique of Brace *et al.* (1968) for the relatively impermeable sample. In all of our tests, the uncertainty of permeability measurements is within 15%. A schematic diagram of the wide-range permeameter is shown in Fig. 1. More detailed description has been presented by Zhu & Wong (1997).

In addition to the hydromechanical tests for investigating the permeability anisotropy, a few mechanical tests without permeability measurements were conducted on Berea sandstone samples to understand the effects of loading configuration on mechanical deformation. In these mechanical tests, loading was continuous at a strain rate of 2.6×10^{-5}/s. To measure acoustic emission (AE) activity during the mechanical tests, we installed a piezoelectric transducer (PZT-7, 5.0 mm diameter, 1 MHz longitudinal resonant frequency) on the flat surface of one of the steel end-plugs attached to the sample. The AE signals were conditioned by a preamplifier (gain 40 dB, frequency response 1.5 kHz to 5 MHz) that distinguishes AE events from electric spikes. More technical details have been presented by Zhang *et al.* (1990) and Wong *et al.* (1997).

Loading configurations

Conventional triaxial extension versus conventional triaxial compression

Using the typical experimental set-up shown in Fig. 1, the three principal components of permeability will be one along the axial direction, and two in the transverse (radial) directions. To characterize permeability anisotropy during non-hydrostatic compaction, ideally flow would be measured along axial and transverse directions simultaneously during mechanical compaction. However, under high pressure this is very difficult to achieve. To circumvent this problem, we measured permeability in two separate sets of experiments following the methodology first proposed by Zoback & Byerlee (1976).

In a conventional triaxial extension test, the sample is first loaded hydrostatically (i.e. $\sigma_1 = \sigma_2 = \sigma_3$) to a desired hydrostatic pressure P_c, and then while maintaining the axial stress ($\sigma_3 = P_c$) constant, the radial stresses ($\sigma_1 = \sigma_2$) are increased either continuously (for mechanical tests) or intermittently (for permeability measurements). The loading path for the conventional triaxial extension test can be described by $Q = 3/2[P - (P_c - P_p)]$, where Q is the differential stress ($Q = \sigma_1 - \sigma_3$) and P is the effective mean stress ($P = (\sigma_1 + \sigma_2 + \sigma_3)/3 - P_p$). In a conventional triaxial compression test, once a desired hydrostatic pressure P_c is reached, the axial stress (σ_1) is increased while maintaining the radial stresses ($\sigma_3 = \sigma_2 = P_c$) constant. The loading path for the conventional triaxial compression can therefore be expressed as $Q = 3[P - (P_c - P_p)]$.

In both conventional extension and compression tests, the flow is always measured along the axial direction. Hence permeability values measured under extension correspond to the principal component (k_3) perpendicular to the maximum principal stress σ_1, whereas permeability values obtained under compression conditions correspond to the principal component (k_1) parallel to σ_1. Permeability measurements obtained for porous sandstones using the conventional extension and compression configurations have been presented by Zhu *et al.* (1997) and Zhu & Wong (1997), respectively.

Hybrid triaxial compression

The inelastic and failure behaviours of metals are solely controlled by the deviatoric stresses. In contrast, the mechanical deformation in porous rocks is also dependent on the mean stress. To describe the influence of stress on physical properties of rocks, it is necessary to introduce a function that depends on at least the first and second stress invariants. Under axisymmetric loading, these two invariants correspond to the quantities P and Q, respectively.

To infer permeability anisotropy from measurements of k_1 and k_3 in two different samples, the stress states of the samples should at least correspond to identical values of the first and second invariants. As indicated in Fig. 2a, the loading paths of a conventional extension and a conventional compression test have at most one coincident stress state in the (P, Q) space, and consequently data from two such tests allow one to infer permeability anisotropy at only one stress state. It is therefore desirable to pursue a different loading path in which both P and Q coincide during the entire test so that permeability k_1 measured under compression can be compared to k_3 obtained in extension tests at more than one coincident stress states.

To accomplish this, we designed the hybrid triaxial compression test in which σ_1 is along the axial direction (as in the conventional compression) with a P–Q trajectory that coincides

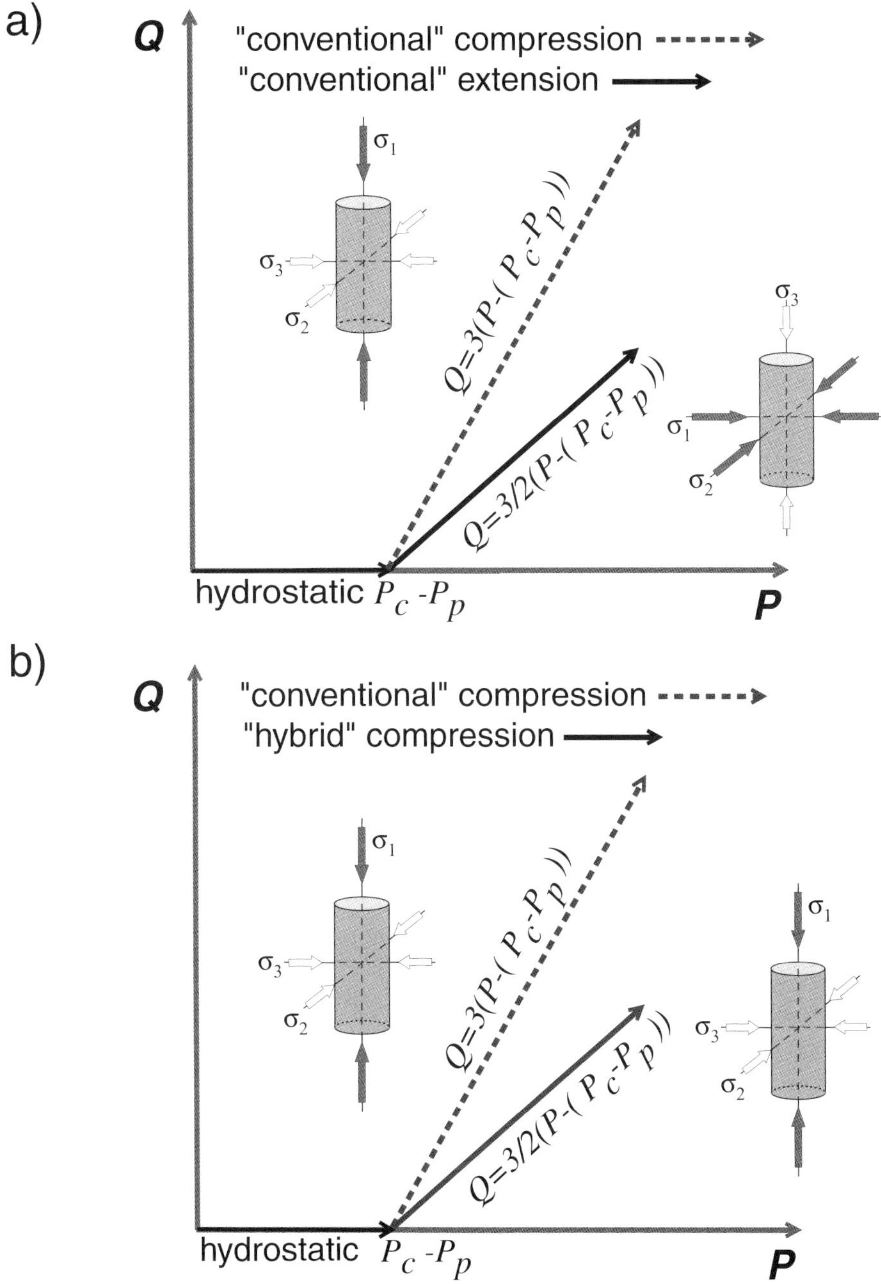

Fig. 2. Comparison of loading paths between: (**a**) conventional triaxial compression and conventional triaxial extension tests; (**b**) conventional and hybrid triaxal compression tests, in P–Q space. Note that the loading paths of conventional extension and compression in (**a**) have at most one coincident stress state, and the loading paths of conventional extension in (**a**) and hybrid compression in (**b**) coincide.

with that of the conventional extension test. In a hybrid test, after loading the sample hydrostatically to a desired confining pressure P_c, instead of maintaining the radial stresses constant, we increased proportionately the axial stress (σ_1) and radial stresses ($\sigma_3 = \sigma_2$). By controlling the increment of $\sigma_3 = \sigma_2$ to be ¼ of that of σ_1, the loading path of such a hybrid test (Fig. 2b) can be expressed as $Q = 3/2[P - (P_c - P_p)]$, which is the same as in a conventional extension test (Fig. 2a). For any conventional extension test, a hybrid compression test can be prescribed with an identical loading path in the (P, Q) space. The permeability k_1 measured under the hybrid triaxial compression test and k_3 obtained under conventional extension at identical stress states can be compared to infer permeability anisotropy.

Experimental results

Mechanical data

In this study, all of the experiments were conducted at sufficiently high effective pressure so that the deformation was distributed cataclastic flow throughout the sample. Because the deformation can be sensitive to stress space and stress path (e.g. Jamison 1992), and the hybrid loading path utilized in this study is not a common practice in deformation tests, we will first illustrate the mechanical behaviour using data for Berea sandstone at effective pressure $P_c - P_p$ of 200 MPa (Figs 3 and 4). For comparison, the data from conventional compression test at the same effective pressure and strain rate are also included. During nonhydrostatic loading, strain hardening (Fig. 3) and an enhanced porosity reduction (Fig. 4) were observed at stress levels beyond compactive yield stress C^*, which marked the onset of shear-enhanced compaction (Wong *et al.* 1992, 1997). Due to the difference in loading path (Fig. 2b) the onset of shear-enhanced compaction for hybrid compression occured at stresses levels that were lower than for conventional compression at the same effective pressure (Fig. 3). The onset of shear-enhanced compaction was manifested by a surge in AE activity, and the

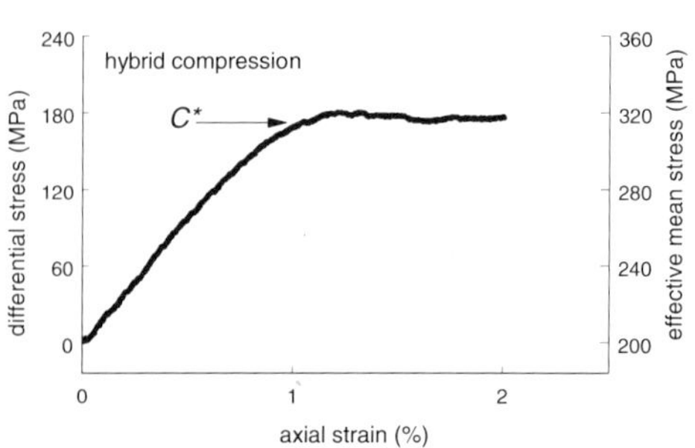

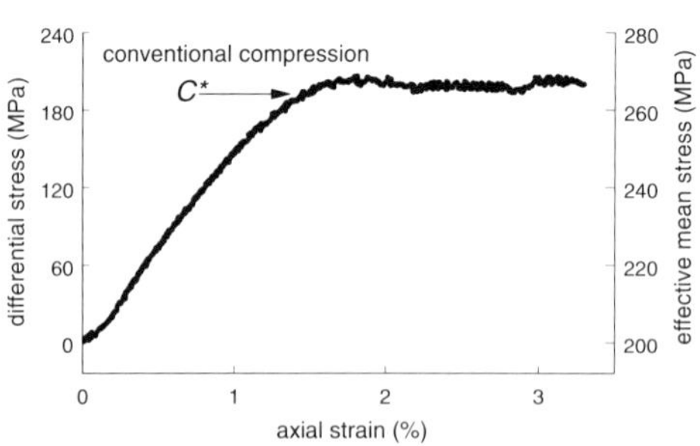

Fig. 3. Stress-strain curve. (**a**) Hybrid triaxial compression deformation data at effective confining pressure of $P_c - P_p = 200$ MPa. (**b**) Conventional triaxial compression deformation data at effective confining pressure of $P_c - P_p = 200$ MPa. The critical stress states C^* are indicated by arrows. Note that both of the samples underwent strain hardening when loaded beyond C^*.

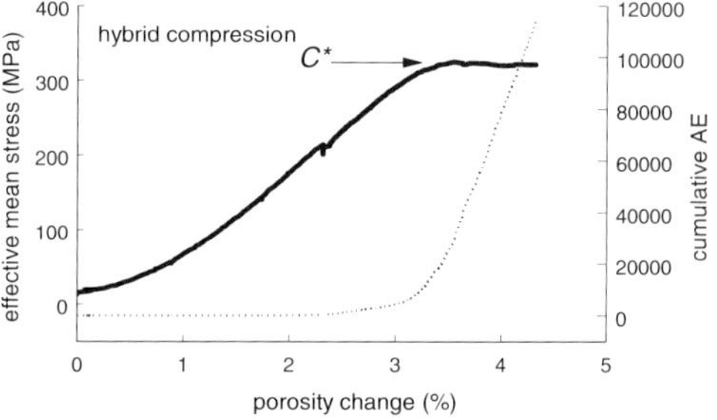

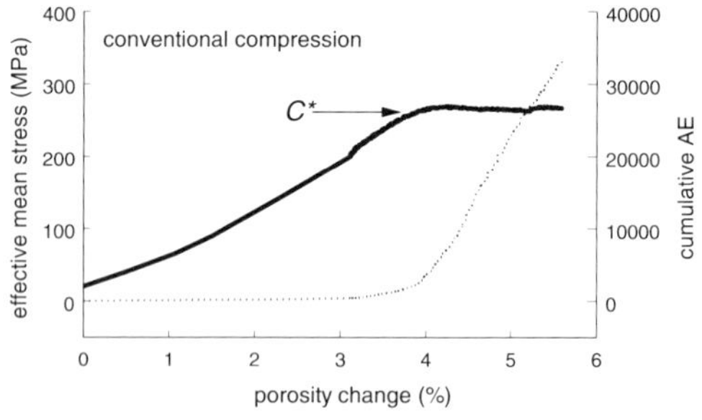

Fig. 4. Effective mean stress (solid curves) and cumulative AE (dotted curves) versus porosity change. (**a**) Hybrid triaxial compression deformation data at effective confining pressure of $P_c - P_p = 200$ MPa. (**b**) Conventional triaxial compression deformation data at effective confining pressure of $P_c - P_p = 200$ MPa. The critical stress states C^*, which are marked by surges in AE activity, are indicated by arrows.

compaction and cumulative AE count tracked one another (Fig. 4). While the mechanical data show that the compaction behaviour for hybrid and conventional compression tests are qualitatively similar, there are certain differences that should be noted. It is commonly assumed that before the onset of shear-enhanced compaction, the porosity change is solely controlled by the mean stress and therefore the data in Fig. 4 for the two tests should coincide at mean stresses lower than C^*. However, our measurements indicate that greater porosity reduction occurred in the conventional compression test. More extensive mechanical tests are required before we can confirm whether the apparent discrepancy arises primarily from variability among samples.

Permeability data

In this study, all of the hydromechanical tests with permeability measurements were terminated before 10% of axial strain, and samples retrieved after each test showed uniform deformation (with no bulging or shear localization in the deformed samples). Table 1 shows our data (from hybrid triaxial compression tests) as functions of stress, axial strain and porosity. In each experiment, the loading ram was locked for approximately 3 hours during several intervals while permeability was measured. Some stress relaxation was usually observed during the 'locked' periods. We report here the stress values recorded immediately after the loading ram was locked.

The permeability and porosity as functions of the effective mean stress for the three sandstones deformed by hybrid triaxial compression are shown in Fig. 5. The effective pressure $(P_c - P_p)$ used for each test are as indicated. For reference, the permeability evolution under hydrostatic loading is also included in Fig. 5 as dashed curves. During the hydrostatic loading, effective hydrostatic pressure by itself reduced the permeability by more than 3 orders of magnitude, with the most significant drop occurring after the onset of grain crushing at the critical effective pressure P^* (David *et al.* 1994; Zhu & Wong 1997), tracking closely the accelerated porosity reduction beyond P^*.

The nonhydrostatic and hydrostatic loadings are coupled in the hybrid compression experiments. Permeability evolution in our hybrid compression tests is qualitatively similar to that in conventional compression tests (Zhu & Wong 1997). The deformation curve coincides with the hydrostat up to C^*, beyond which there is an accelerated decrease in porosity in comparison to the hydrostat (Fig. 5). The critical stress state C^* also marks the onset of significant reduction of permeability induced by deviatoric stresses. Before a sample is loaded up to C^*, the permeability is primarily controlled by the effective mean stress, independent of the deviatoric stresses. However, as soon as the compactive yield stress level C^* is reached, the nonhydrostatic loading exerts dominant control over the permeability, reducing it by as much as 2 orders of magnitude during cataclastic flow. Finally, as pervasive pore collapse and grain crushing develop throughout the sample, permeability reduction becomes moderate again. As the sample deformed by the progressive cataclastic flow, permeability and porosity consistently decrease with increasing strain (Fig. 6).

Discussion

Influence of loading path on compactive yield stress and permeability

On the basis of data from conventional triaxial compression tests, Wong *et al.* (1997) suggested that the compactive yield stress C^* mapped out an elliptical yield envelope in the (P, Q) stress space, in accordance with the critical state (Schofield & Wroth 1968) and cap (DiMaggio & Sandler 1971) models. Yield stress data from conventional triaxial extension tests follow a similar trend (Zhu *et al.* 1997), implying that the stress for onset of shear-enhanced compaction is a function of only the first and second stress invariants. To test this further with our new hybrid compression data, we compiled in Fig. 7 the critical stress C^* obtained from permeability-deformation tests for all three sandstones under three different loading configurations. Due to the lack of space for the AE transducer in the sample column setup for fluid flow tests, we did not acquire AE activity as in the mechanical tests. Instead, we have to pick C^* from the porosity change data alone, which inevitably introduces larger uncertainty (Wong *et al.* 1997). Because of the stress relaxation during permeability measurements, yield stresses obtained during permeability-deformation tests are somewhat lower than those of deformation continuous loading (Fig. 8a). Notwithstanding the many uncertainties, there is reasonable agreement among yield stresses determined for the three loading paths, indicating that the yield envelope can be considered to depend on only the first and second invariants. However, as Zhu *et al.* (1997) pointed out, a more rigorous

Table 1. *Experimental results*

Maximum principal stress ($\sigma_1 - P_p$) (MPa)	Differential stress ($\sigma_1 - \sigma_3$) (MPa)	Porosity ϕ (%)	Axial strain ε_1 (%)	Permeability k_1 ($\times 10^{-15}$ m^2)
Adamswiller				
3	0	23.00		63.30
10	0	22.43		31.00
20	0	22.01		22.10
40	0	21.51		18.00
60	0	21.14		12.50
80	0	20.81	0	10.60
90	26.10	20.49	0.54	6.02
100	51.61	20.10	1.15	3.88
110	87.66	19.67	1.95	5.62
120	105.70	18.56	3.34	3.64
125	119.37	16.56	5.91	1.52
130	135.86	15.21	7.93	0.80
140	162.70	13.54	10.60	0.30
Berea				
3	0	21.00		232.60
10	0	20.40		180.20
20	0	20.09		145.00
40	0	19.71		116.00
70	0	19.23		107.90
100	0	18.85	0	102.50
115	35.96	18.53	0.56	92.44
130	77.51	18.14	1.33	71.29
140	110,06	17.95	1.73	69.22
150	140.17	17.72	2.31	34.81
160	169.04	17.48	2.83	19.38
170	190.98	16.96	3.44	0.66
175	204.82	16.08	4.50	0.64
177	205.04	14.04	7.16	0.51
185	227.64	12.45	9.51	0.40
3	0	21.00		187.10
10	0	20.53		157.60
20	0	20.15		133.10
40	0	19.67		124.30
60	0	19.36		118.40
90	0	18.96		115.70
120	0	18.20		105.20
160	0	17.84	0	100.80
175	55.63	17.54	0.75	76.24
190	95.61	17.27	1.38	63.74
200	130.72	17.09	1.93	80.55
210	165.72	16.90	2.48	57.99
220	186.48	16.36	2.93	40.44
230	207.92	15.29	3.57	29.67
235	220.84	13.46	4.64	1.25
240	238.79	12.28	6.78	0.66
250	258.22	2.80	8.27	0.65
3	0	21.00		233.00
10	0	20.67		213.00
20	0	20.36		193.00
40	0	19.93		171.00
60	0	19.60		163.00
90	0	19.19		161.00
120	0	18.83		153.00
160	0	18.40		134.00
200	0	17.92	0	115.00
215	46.05	17.54	0.63	106.00
230	97.71	17.03	1.43	87.10

Table 1. *Continued*

Maximum principal stress ($\sigma_1 - P_p$) (MPa)	Differential stress ($\sigma_1 - \sigma_3$) (MPa)	Porosity ϕ (%)	Axial strain ε_1 (%)	Permeability k_1 ($\times 10^{-15}$ m^2)
Berea				
245	130.64	16.58	2.09	57.70
260	178.96	15.98	2.98	14.00
268	195.28	14.28	4.86	1.60
275	219.07	12.70	6.71	0.68
285	252.14	11.49	8.19	0.42
300	289.79	10.29	9.93	0.22
3	0	21.00		361.00
10	0	20.24		224.00
20	0	19.61		196.00
40	0	19.13		190.00
60	0	18.75		171.00
90	0	18.38		153.00
130	0	17.83		144.00
170	0	17.42		127.00
210	0	17.02		103.00
250	0	16.59	0	87.30
265	41.96	16.36	0.55	79.20
280	90.68	16.09	1.27	44.80
290	115.09	15.41	1.69	16.60
300	140.94	14.99	2.20	0.67
310	158.01	13.82	3.21	0.44
316	183.97	12.11	4.75	0.37
330	233.42	9.95	6.87	0.29
Rothbach				
3	0	19.90		105.00
10	0	19.21		56.60
20	0	18.70		42.40
40	0	18.04		21.40
60	0	17.55		12.20
90	0	17.05		6.60
130	0	16.42		2.82
165	0	15.86	0	1.83
175	26.21	15.63	0.35	1.54
185	62.23	15.39	0.93	1.43
195	79.84	14.45	1.89	0.82
205	111.44	13.04	3.35	0.38
215	131.85	11.68	4.83	0.14
225	171.47	10.49	6.40	0.05
240	219.59	9.11	8.54	0.03

test of this assumption requires additional tests using non-axisymmetric loading.

The permeability component k_1 along the maximum principal stress direction has been measured along two different loading paths (Fig. 8b) in conventional and hybrid triaxial compression tests. Berea sandstone data from conventional (Zhu & Wong 1997) and hybrid (Table 1) triaxial compression tests are plotted together in Fig. 8b. Taking into account variability among samples, the permeability k_1 at C^* from the hybrid compression test with effective pressure 160 MPa is similar to that of the conventional compression test with effective pressure 250 MPa. Because of the different loading configurations, the stress state (P and Q) at C^* of these two tests is identical in spite of their different effective pressures. This implies that the permeability at C^* is a function only of the current P and Q, not of the loading history.

Development of permeability anisotropy

Data for hybrid compression (this study) and conventional extension tests (Zhu *et al.* 1997) are for identical loading paths but with flow

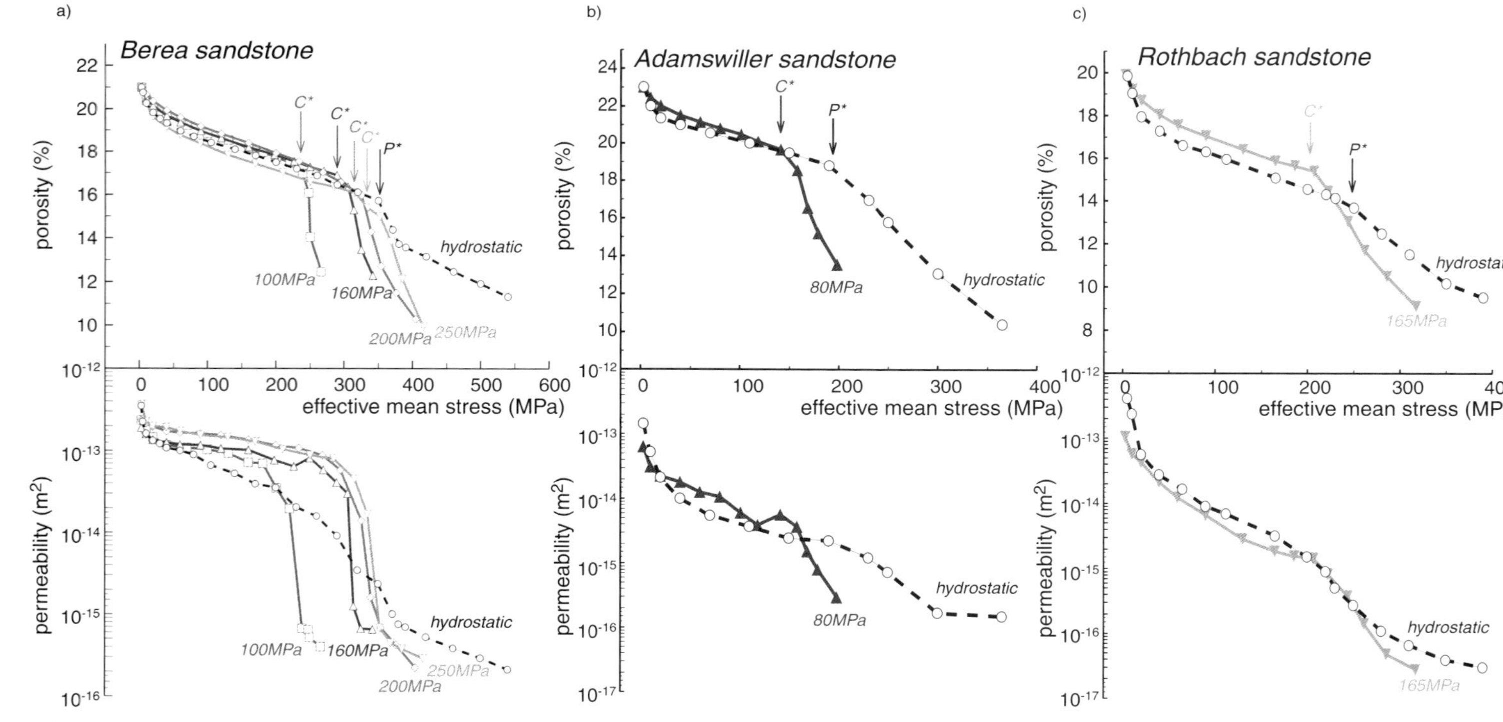

Fig. 5. Porosity and permeability as functions of effective mean stress for: (**a**) Berea; (**b**) Adamswiller; and (**c**) Rothbach sandstones under hybrid triaxial compression. The critical stress states C^* for the onset of shear-enhanced compaction are indicated by the arrows, and $(P_c - P_p)$ for each experiment is as indicated. The critical stress states C^* are marked by arrows. For reference the hydrostatic compression data are shown as dashed curves, and P^* indicates the onset of grain crushing and pore collapse.

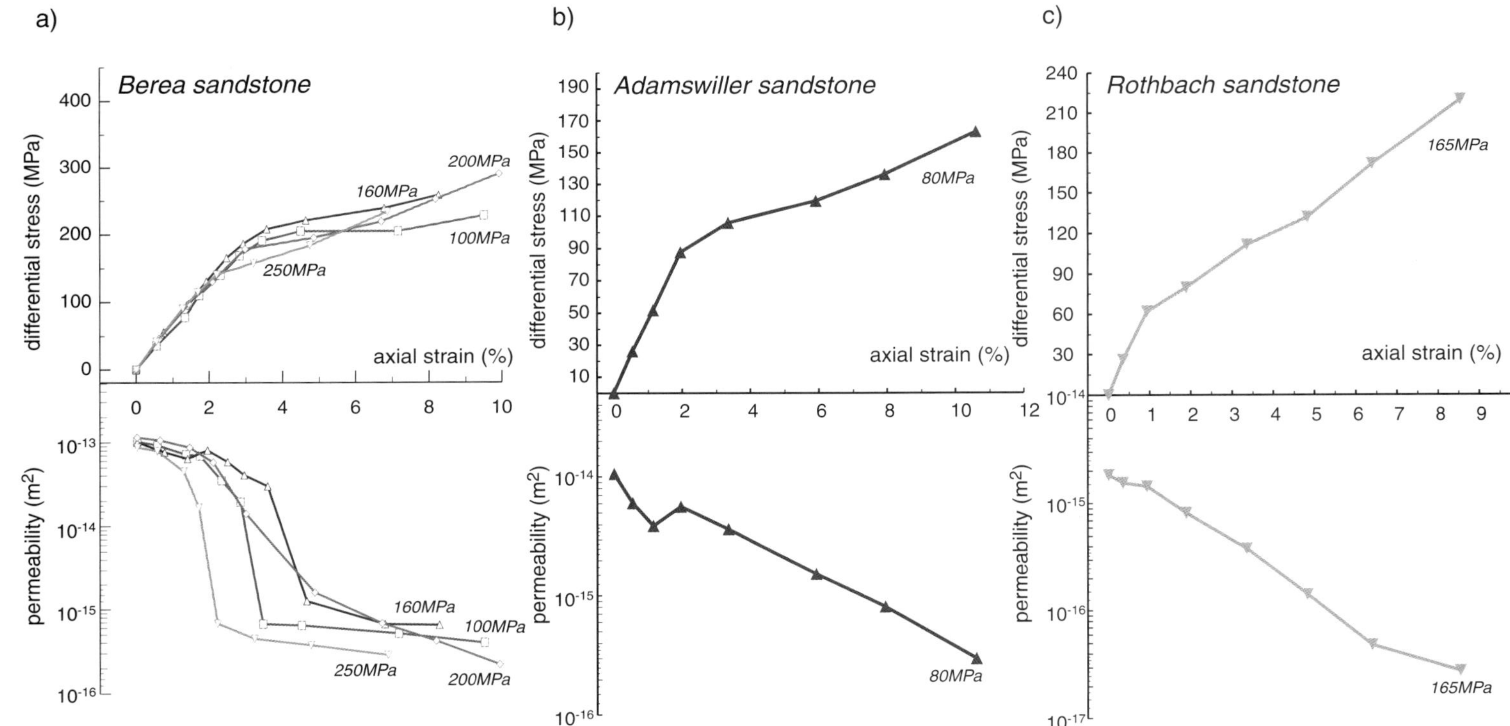

Fig. 6. Experimental data for: (**a**) Berea; (**b**) Adamswiller; and (**c**) Rothbach sandstones under hybrid triaxial compression with permeability measurements. The differential stress $Q = (\sigma_1 - \sigma_3)$ and permeability k_1 are plotted versus axial strain. The effective hydrostatic pressure $(P_c - P_p)$ for each experiment is as indicated. The critical stress states C^* are marked by arrows.

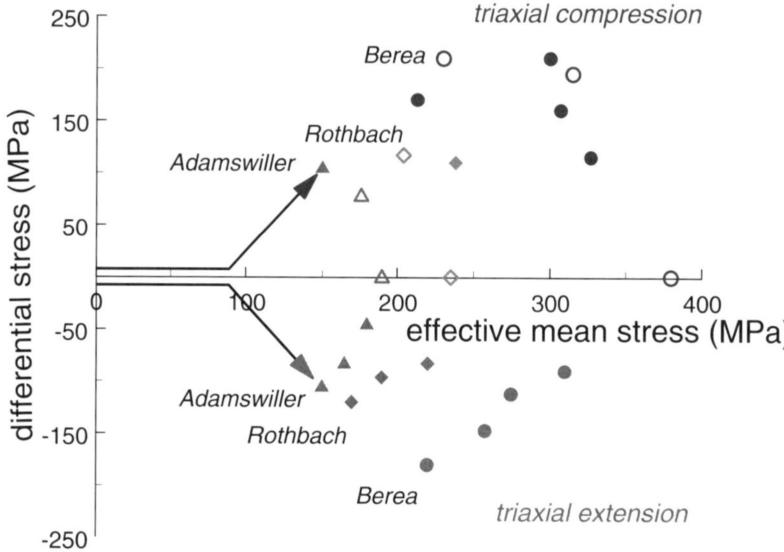

Fig. 7. Stress state C^* at the onset of shear enhanced compaction are shown in the (P, Q) stress space. Data from compression tests are plotted in the upper half of the graph. For comparison, data from extension tests (Zhu *et al.* 1997) are plotted in the lower half of the graph. Different sandstones are represented by different symbols. Hybrid compression data and extension data are represented by solid symbols, whereas conventional compression data from Zhu & Wong (1997) are represented by corresponding empty symbols.

along two orthogonal directions. When these two sets of data are compared, it is clear that before the onset of shear-enhanced compaction C^*, the changes of porosity and permeability are solely controlled by the effective mean stress P. As a result, k_1 (from hybrid compression) and k_3 (from conventional extension) are comparable to each other and stress-induced anisotropy is negligible.

Beyond the yield stress C^*, an enhanced decrease can be seen in both k_1 (Figs 5 and 6) and k_3 (Zhu *et al.* 1997, Figs 2 and 3). Our data for k_1 and k_3 as functions of the differential and effective mean stresses are compiled in Fig. 9. In the vicinity of C^*, experimental data for Berea sandstone indicate that k_1 may exceed k_3 by as much as 2 orders of magnitude. However, because the estimate of C^* for all of the permeability tests is rather inaccurate due to the lack of AE and the stress relaxation, the difference between k_1 and k_3 should not be considered as a robust feature. Interestingly, a transient behaviour of the permeability anisotropy is observed at C^*. The reduction of k_1 at C^* is consistently more rapidly than that of k_3, thus the anisotropy in permeability becomes less pronounced with the further development of strain hardening and shear-enhanced compaction. A qualitatively similar trend was observed for both Adamswiller and Rothbach sandstones, but the permeability anisotropy is not as pronounced.

The porosity of an intact porous rock is primarily equant and tubular pores. Recent studies on 3D pore structure (e.g. Fredrich *et al.* 1995; Lindquist *et al.* 2000) also suggest that these pre-existing pores are well connected. Quantitative characterization of damage associated with shear-enhanced compaction in Berea and Darley Dale sandstones has been reported by Menéndez *et al.* (1996) and Wu *et al.* (2000), respectively. On the one hand, it was observed that the average pore size decreases with increasing effective mean stress; there is no evident anisotropy in pore size distribution during cataclastic flow. On the other hand, significant anisotropy in microcrack density was observed in samples loaded beyond C^*. There is a preferential development of stress-induced microcracking sub-parallel to the maximum principal stress σ_1. Whereas the overall decrease in pore size is probably responsible for the drastic decrease of permeability with the onset of shear-enhanced compaction, the microcrack anisotropy induces the permeability components to evolve with stress at somewhat different rates, which is manifested by the anisotropic development of permeability in the vicinity of the compactive yield stress C^*.

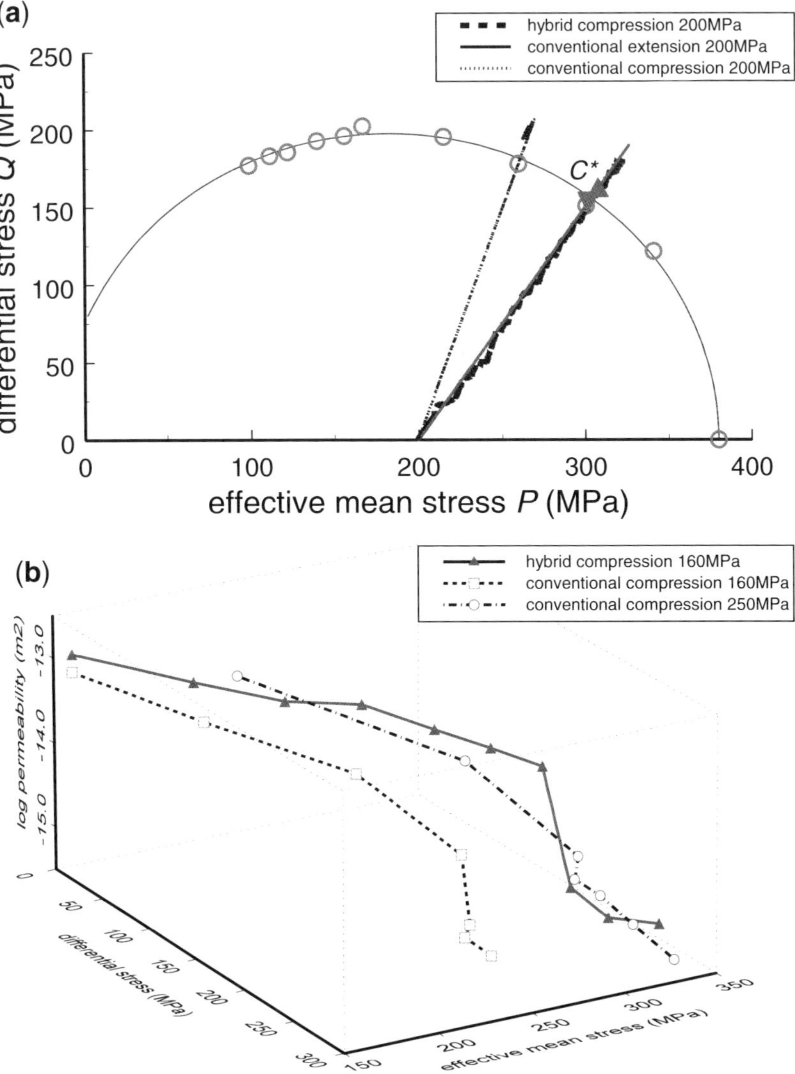

Fig. 8. (**a**) Comparison of actual loading paths of mechanical tests on Berea sandstone at effective pressure of 200 MPa under conventional compression (thin dotted line), conventional extension (solid line) and hybrid compression (thick dashed line). Solid upward triangle indicated the C^* in the hybrid compression test. Solid downward triangle indicats the C^* in the extension test. For reference, C^* in conventional compression tests at various effective pressures are plotted as empty circles. Note that the critical stress states in both hybrid compression and conventional extension tests coincide and show a very good agreement with the previously published yielding envelope (e.g. Wong *et al.* 1997; Baud *et al.* 2000). (**b**) Comparison of the evolution of k_1 as a function of effective mean and differential stresses in hybrid and conventional compression tests. Data of a hybrid compression test at $P_c - P_p = 160$ MPa are represented by a solid curve with solid triangles. Data from a conventional compression test at $P_c - P_p = 250$ MPa are represented by a dashed curve with open circles. Because of the different loading paths, the stress states at C^* (P and Q) are identical in these two tests. For reference, data from a conventional compression test at $P_c - P_p = 160$ MPa are also plotted by a dotted curve with open squares.

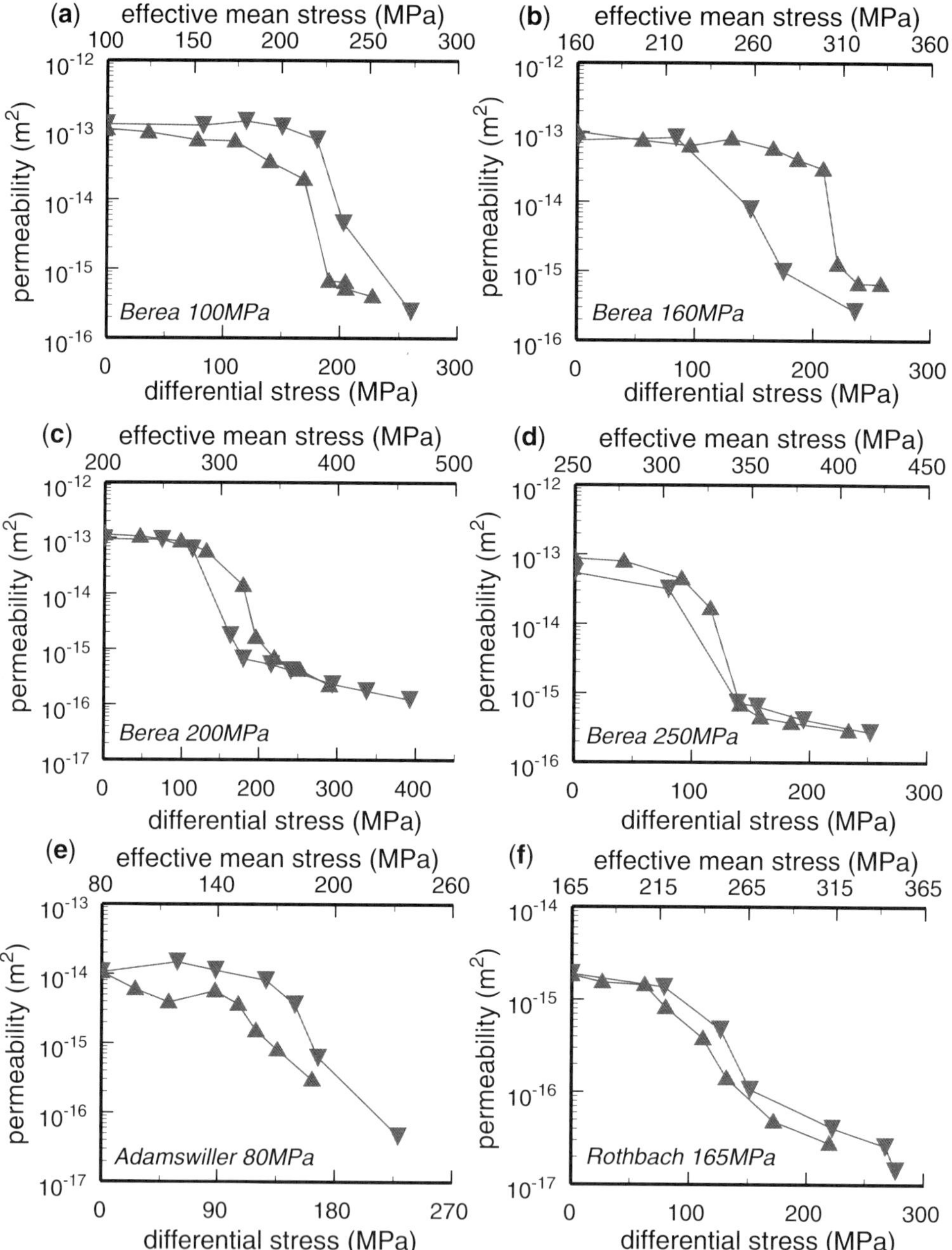

Fig. 9. Evolution of permeability anisotropy during the shear-enhanced compaction. Experimental results include: Berea sandstone at effective pressure of: (**a**) 100 MPa; (**b**) 160 MPa; (**c**) 200 MPa; (**d**) 250 MPa; (**e**) Adamswiller sandstone at 80 MPa; and (**f**) Rothbach sandstone at 165 MPa. Upward triangles represent results from hybrid tests, and downward triangles represent results from extension tests (Zhu *et al.* 1997).

Is anisotropic development of permeability controlled by stress or strain?

By manipulation of the loading paths to achieve identical mean and differential stress states, in both extensional and compressional tests, we determined the principal components of the permeability tensor assuming that a simple axisymmetric geometrical relation exists between the permeability tensor and stress tensor. Because the intact samples are free of macroscopic planar shear features and the deformation is pervasive cataclasis, the axisymmetric and coaxial assumptions are validated. The new data allow us to draw more specific conclusions on the anisotropic development of permeability during mechanical compaction of porous sandstones. The loading path for conventional triaxial extension and hybrid compression tests is particularly relevant to an extensional tectonic environment in sedimentary basins. Our data (Fig. 9) show that the anisotropic development of permeability in such lithified rocks is primarily controlled by stress. The compactive yield stress C^* defines the stress state beyond which permeability anisotropy undergoes a drastic increase. Generally the permeability for flow parallel to the maximum compression direction is greater than perpendicular ($k_1 > k_3$). Therefore, an appropriate model should capture the evolution of the permeability tensor as a function of the effective mean and differential stresses. Such a stress-dependent permeability model was developed by Zhu & Wong (2000), the details of which will be presented in a future publication.

Our data also indicate that permeability anisotropy becomes negligible with increasing strain, with both permeability components k_1 and k_3 converging to almost constant values. This suggests that in many compaction problems related to extensional tectonics and reservoir mechanics in sedimentary basins, it may not be as important to consider the anisotropic development of permeability with the accumulation of large strain.

However, it should also be noted that there are other situations in which strain clearly is the dominant factor in controlling the anisotropic development of permeability. In the brittle faulting regime, the first inception of shear localization occurs at relatively small strain, but further deformation will result in development of complex arrays of shear bands or the widening of the shear bands. The overall permeability will then depend on the geometric complexity of the localized structures that evolve with cumulative strain.

In unconsolidated materials such as sediments, strain seems to play a key role in the evolution of permeability anisotropy (Dewhurst *et al.* 1996; Bolton *et al.* 2000). In a mature fault zone, a relatively wide zone of gouge may exist. Recent

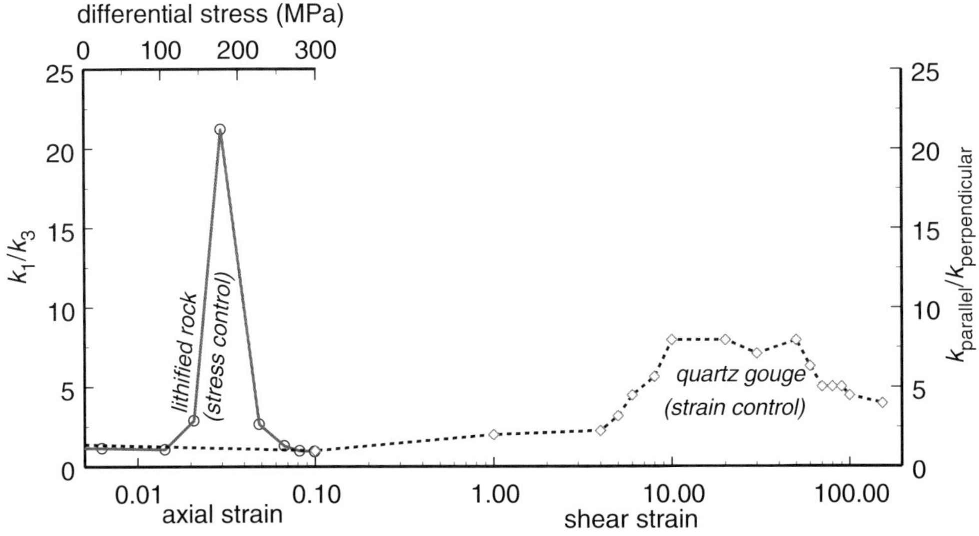

Fig. 10. Evolution of permeability anisotropy as a function of strain. The solid curve shows the ratio of k_1/k_3 as a function of axial strain (data obtained from Berea sandstone samples deformed at effective pressure of 200 MPa). For reference, we also plotted the differential stress as the second x-axis. The permeability measurements of quartz gouge underwent large shear strain were conducted by Zhang *et al.* (1999). The dashed curve shows the ratio of permeability measured parallel versus perpendicular to the gouge layer.

experimental data from rotary shear tests on simulated gouge layers show that permeability for flow parallel to the shear directions can exceed the perpendicular component by one order of magnitude (Zhang *et al.* 1999, 2001). Such permeability anisotropy (Fig. 10) seems to persist over large shear strains of the order of 50–100, possibly due to the permanent alignment of clay minerals and Riedel shears along the shear direction. Indeed measurements on cores (Faulkner & Rutter 1998) parallel and perpendicular to fault zones that have accumulated large strains corroborate this trend. It seems therefore that in unconsolidated materials such as sediments and fault gouge, the anisotropic development of permeability should be described by a model for the permeability tensor as a function of strain or slip. Such fundamental differences in evolution of permeability anisotropy should be accounted for in a crustal model that couples mechanical deformation and fluid transport.

Conclusions

A laboratory novel loading strategy, the hybrid compression test, is designed to characterize the stress controlled anisotropic development of permeability. In hybrid compression, the radial stresses and the axial stress are increased in a ratio of 1:4. The resulting trajectory in terms of first and second stress invariants is identical to that of conventional extension tests, in which the axial stress remains constant and the radial stresses are progressively increased. While permeability is always measured along the axial direction, hybrid compression experiments give the evolution of k_1, the permeability along the direction of maximum compression, and conventional extension experiments give k_3, the permeability along the direction of minimum compression. Given that the deformation is homogeneous and axisymmetric, it is reasonable to assume that k_1 and k_3 are the principal components of the permeability tensor.

The general behaviour of the hybrid compression tests is qualitatively similar to that of the conventional compression and extension tests. The porosity and permeability decrease with increasing stress at a moderate rate until a critical point, C^*, marking the onset of shear-enhanced compaction. Immediately after C^*, significant porosity reduction and permeability decrease occur. However, the rate of compaction and permeability decrease become gradual again when loaded beyond ~10% of axial strain. Within experimental uncertainty, C^* is similar for all types of loading.

Before C^*, permeability and porosity reduction depend only on the effective mean stress. No significant anisotropy was detected in our experiments. However, our data suggest drastic anisotropic development of permeability with the onset of shear-enhanced compaction and initiation of cataclastic flow. The maximum principal component of permeability tensor k_1 inferred from the hybrid compression tests can exceed the minimum principal component of permeability tensor k_3 inferred from the extension tests by up to 2 orders of magnitude. The preferred orientation of microcracks along the maximum principal stress is likely to be the origin of this anisotropy. However, the difference of k_1 and k_3 diminishes at a cumulative strain of ~10%. Permeability anisotropy appears as a transient phenomenon in our experiments. However, in contrast to a natural system, there is no healing process in our deformation experiments. It is conceivable that permeability anisotropy may appear cyclically in the Earth with effective diagenesis during the deformation. This may be particularly important at low pressure, when deformation is localized and faults can heal.

We are grateful to P. Baud and V. Vajdova for their help in conducting the mechanical tests. I. Main and C. Peach provided insightful reviews that clarified the manuscript. The first author is grateful for support from the J. Lamar Worzel Assistant Scientist Fund and The Penzance Endowed Fund in Support of Assistant Scientists at WHOI. The research was partially supported by the National Science Foundation under grants NSF-EAR9814796 (WHOI) and EAR98005072 (SUNYSB).

References

Baud, P., Schubnel, A. & Wong, T.-F. 1999. Dilatancy, compaction and failure mode in Solnhofen limestone. *Journal of Geophysical Research*, **105**, 19289–19303.

Baud, P., Zhu, W. & Wong, T.-F. 2000. Failure mode and weakening effect of water on sandstone. *Journal of Geophysical Research*, **105**, 16371–16389.

Bernabé, Y. 1987. A wide range permeameter for use in rock physics. *International Journal of Rock Mechanics and Mining Sciences*, **24**, 309–315.

Bolton, A. J., Maltman, A. J. & Fisher, Q. 2000. Anisotropic permeability and bimodal pore-size distributions of fine-grained marine sediments. *Marine and Petroleum Geology*, **17**, 657–672.

Brace, W. F. 1980. Permeability of crystalline and argillaceous rocks. *International Journal of Rock Mechanics and Mining Sciences*, **17**, 241–251.

Brace, W. F., Walsh, J. B. & Frangos, W. T. 1968. Permeability of granite under high pressure. *Journal of Geophysical Research*, **73**, 2225–2236.

BRUNO, M. S. 1994. Micromechanics of stress-induced permeability anisotropy and damage in sedimentary rock. *Mechanics of Materials*, **18**, 31–48.

DAVID, C., WONG, T.-F., ZHU, W. & ZHANG, J. 1994. Laboratory measurement of compaction-induced permeability change in porous rock: implications for the generation and maintenance of pore pressure excess in the crust. *Pure and Applied Geophysics*, **143**, 425–456.

DEWHURST, D. N., CLENNELL, M. B., BROWN, K. M. & WESTBROOK, G. K. 1996. Fabric and hydraulic conductivity of shear clays. *Geotechnique*, **46**, 761–768.

DIMAGGIO, F. L. & SANDLER, I. S. 1971. Material model for granular soils. *Journal of Engineering Mechanics*, American Society of Civil Engineers, **97**, 935–950.

FAULKNER, D. R. & RUTTER, E. H. 1998. The gas permeability of clay-bearing fault gouge at 20 °C. *In*: JONES, G., FISHER, Q. & KNIPE, R. J. (eds) *Faults, Fault Sealing and Fluid Flow in Hydrocarbon Reservoirs*. Geological Society, London, Special Publications, **147**, 147–156.

FREDRICH, J. T., MENÉNDEZ, B. & WONG, T.-F. 1995. Imaging the pore structure of geomaterials. *Science*, **268**, 276–279.

HOLT, R. M. 1989. Permeability reduction induced by a nonhydrostatic stress field. *Proceedings of Annual Technical Conference and Exhibition of Society of Petroleum Engineers of AIME*, **64**, SPE19595.

INGEBRITSEN, S. E. & SANFORD, W. E. 1998. *Groundwater in Geologic Processes*. Cambridge University Press, New York.

JAMISON, W. R. 1992. Stress spaces and stress paths. *Journal of Structural Geology*, **14**, 1111–1120.

KIYAMA, T., KITA, H., ISHIJIMA, Y., YANAGIDANI, T., AOKI, K. & SATO, T. 1996. Permeability in anisotropic granite under hydrostatic compression and triaxial compression including post-failure region. *Proceedings of North American Rock Mechanics Symposium*, **2**, 1161–1168.

LINDQUIST, W. B., VENKATARANGAN, A., DUNSMUIR, J. & WONG, T.-F. 2000. Pore and throat size distributions measured from synchrotron X-ray tomographic images of Fontainebleau sandstones. *Journal of Geophysical Research*, **105**, 21509–21527.

MAIN, I. G., KWON, O., NGWENYA, B. T. & ELPHICK, S. C. 2000. Fault sealing during deformation band growth in porous sandstone. *Geology*, **28**, 1131–1134.

MAIN, I. G., NGWENYA, B. T., ELPHICK, S. C., SMART, B., CRAWFORD, B. & POUX, C. 1996. Scale limits on fluid pressure diffusion during rapid self-sealing deformation and fluid flow. *In*: AUBERTIN, M., HASSANI, F. & MITRI, H. (eds) *Rock Mechanics*, **2**. Balkema, Rotterdam, 1161–1168.

MENÉNDEZ, B., ZHU, W. & WONG, T.-F. 1996. Micromechanics of brittle faulting and cataclastic flow in Berea sandstone. *Journal of Structural Geology*, **18**, 1–16.

MORDECAI, M. & MORRIS, L. H. 1971. An investigation into the changes of permeability occurring in a sandstone when failed under triaxial stress conditions. *Proceedings of the 12th U.S. Symposium on Rock Mechanics*, **12**, 221–239.

PEACH, C. J. & SPIERS, C. J. 1996. Influence of crystal plastic deformation on dilatancy and permeability development in synthetic salt rock. *Tectonophysics*, **256**, 101–128.

PERSON, M., RAFFENSPERGER, J. P., GE, S. & GARVEN, G. 1996. Basin-scale hydrogeologic modeling. *Review of Geophysics*, **34**, 61–87.

RHETT, D. W. & TEUFEL, L. W. 1992. Stress path dependence of matrix permeability of North Sea sandstone reservoir rock. *Proceedings of the 33rd U.S. Symposium on Rock Mechanics*, **33**, 345–354.

RICE, J. R. 1992. Fault stress states, pore pressure distributions, and the weakness of the San Andreas Fault. *In*: EVANS, B. & WONG, T.-F. (eds) *Fault Mechanics and Transport Properties of Rocks*, Academic Press, San Diego, 475–504.

SCHOFIELD, A. N. & WROTH, C. P. 1968. *Critical State Soil Mechanics*. McGraw Hill, London.

STORMONT, J. C. & DAEMEN, J. J. K. 1992. Laboratory study of gas permeability changes in rock salt during deformation. *International Journal of Rock Mechanics and Mining Sciences*, **29**, 323–342.

WONG, T.-F. & ZHU, W. 1999. Brittle faulting and permeability evolution: hydromechanical measurement, microstructural observation, and network modeling. *In*: HANBERG, W., MOZLEY, P., MOORE, J. & GOODWIN, L. (eds) *Faults & Subsurface Fluid Flow in the Shallow Crust*. AGU Geophysical Monograph, **113**, 83–99.

WONG, T.-F., DAVID, C. & ZHU, W. 1997. The transition from brittle faulting to cataclastic flow in porous sandstones: Mechanical deformation. *Journal of Geophysical Research*, **102**, 3009–3025.

WONG, T.-F., SZETO, H. & ZHANG, J. 1992. Effect of loading path and porosity on the failure mode of porous rocks. *Applied Mechanics Reviews*, **45**, 281–293.

WU, X. Y., BAUD, P. & WONG, T.-F. 2000. Micromechanics of compressive failure and spatial evolution of anisotropic damage in Darley Dale sandstone. *International Journal of Rock Mechanics and Mining Sciences*, **37**, 143–160.

ZHANG, S., COX, S. F. & PATERSON, M. S. 1994. The influence of room temperature deformation on porosity and permeability in calcite aggregates. *Journal of Geophysical Research*, **99**, 15761–15775.

ZHANG, S., TULLIS, T. E. & SCRUGGS, V. J. 1999. Permeability anisotropy and pressure dependency of permeability in experimentally sheared gouge materials. *Journal of Structural Geology*, **21**, 795–806.

ZHANG, S., TULLIS, T. E. & SCRUGGS, V. J. 2001. Implications of permeability and its anisotropy in a mica gouge for pore pressures in fault zones. *Tectonophysics*, **335**, 37–50.

ZHANG, J., WONG, T.-F., YANAGIDANI, T. & DAVIS, D. M. 1990. Pressure-induced microcracking and grain crushing in Berea and Boise sandstones: acoustic emission and quantitative microscopy measurements. *Mechanics of Materials*, **9**, 1–15.

ZHU, W., MONTÉSI, L. G. J. & WONG, T.-F. 1997. Shear-enhanced compaction and permeability reduction: Triaxial extension tests on porous sandstone. *Mechanics of Materials*, **25**, 199–214.

ZHU, W. & WONG, T.-F. 1997. The transition from brittle faulting to cataclastic flow: permeability evolution. *Journal of Geophysical Research*, **102**, 3027–3041.

ZHU, W. & WONG, T.-F. 1999. Network modeling of the evolution of permeability and dilatancy in compact rocks. *Journal of Geophysical Research*, **104**, 1963–2971.

ZHU, W. & WONG, T.-F. 2000. *Effects of Stress on Permeability (Abstract)*. EOS, Transactions, American Geological Union, 2000 Fall Meeting, **81**, F1085.

ZOBACK, M. D. & BYERLEE, J. D. 1975. The effect of microcrack dilatancy on the permeability of Westerly granite. *Journal of Geophysical Research*, **80**, 752–755.

ZOBACK, M. D. & BYERLEE, J. D. 1976. Effect of high-pressure deformation on permeability of Ottawa Sand. *The American Association of Petroleum Geology Bulletin*, **60**, 1531–1542.

The numerical simulation of microstructure

MARK W. JESSELL[1,2] & PAUL D. BONS[3,4]

[1] *Department of Earth Sciences, Monash University, Box 28E, Melbourne, Victoria, 3800, Australia*

[2] *Present address: Laboratoire des Mécanismes de Transfert en Géologie , UMR 5563 – CNRS-Université Paul Sabatier-OMP, 38, rue des 36-Ponts, 31400 TOULOUSE CEDEX, France (e-mail: mjessell@lmtg.ups-tlse.fr)*

[3] *Institut für Geowissenschaften – Tektonophysik, Johannes Gutenberg-Universität Mainz, Becherweg 21, 55099 Mainz, Germany*

[4] *Present address: Institut für Geowissenschaften, Eberhard Karls Universität, Sigwartstrasse 10, 72076 Tübingen, Germany*

Abstract: This review discusses the attempts that have been made by geologists to numerically simulate the evolution of microstructures in rocks. The strengths and weaknesses of the differing techniques are compared and equivalent materials science results are included. In particular we focus on the application of techniques that have been used to predict texture development, grain boundary geometries, deformation in one and two-phase systems and crystal growth.

The role of deformation in modifying structures can be seen at all scales, from the formation and movement of dislocations in single grains to changes in the structure of the whole Earth (Hobbs *et al.* 1976; Poirier 1985; Passchier & Trouw 1996). Attempts to simulate the evolution of these structures at each scale have been undertaken, from molecular dynamics at the atomic scale (Schiøtz *et al.* 1998; Yamakov *et al.* 2001), through to simulations of mantle convection (Houseman 1988; Moresi & Solomatov 1995). At the grain scale, an understanding of the coupling of deformation processes, mechanical properties and microstructural evolution underpins our ability correctly to interpret natural microstructures in terms of rock history and to develop relevant geodynamic models. The numerical simulation of microstructures in geological materials has progressed in parallel with techniques developed in the wider materials science community (see Raabe 1998 for a comprehensive bibliography). In recent years there has been an upsurge in interest in this field due to increased access to powerful computers. Equally important has been the introduction of new measurement techniques such as orientation imaging microscopy (OIM) and electron backscattered diffraction (EBSD) (Lloyd & Freeman 1991; Panozzo-Heilbronner & Pauli 1993; Fueten 1997; Leiss *et al.* 2000). These techniques allow us systematically to characterize the grain boundary topologies and textures of rocks with relative ease and in much greater detail (Trimby *et al.* 1998), thus provoking us to reconsider the microstructural parameters that should be used as indicators of specific processes.

It is beyond the scope of this review to describe in detail the numerous numerical techniques that have been developed to simulate microstructure evolution. Instead, we will focus on their application to the simulation and prediction of some geologically important microstructures, namely: texture (lattice preferred orientation) development; grain boundary geometries; deformation in two-phase systems; and crystal growth.

Numerical simulation of microstructures

Texture development

The earliest microstructural numerical simulation studies in geology were focussed on predicting the texture patterns or lattice preferred orientations found in naturally deformed rocks. This work was based on metallurgical models first developed by Taylor (1938) which, with modifications, are still widely applied today (Tóth *et al.* 1997). Lister and co-workers used a standard Taylor-Bishop-Hill formulation to study the development of textures in quartz and calcite rocks (Lister 1978, 1982; Lister *et al.* 1978; Lister & Paterson 1979; Lister & Hobbs 1980): see Fig. 1. Other groups have since extended this work to a broader range of minerals and to allow refinements of the Taylor-Bishop-Hill scheme such as the relaxed-constraints (Ord 1988) and self-consistent

From: DE MEER, S., DRURY, M. R., DE BRESSER, J. H. P. & PENNOCK, G. M. (eds) 2002. *Deformation Mechanisms, Rheology and Tectonics: Current Status and Future Perspectives.* Geological Society, London, Special Publications, **200**, 137–147. 0305-8719/02/$15

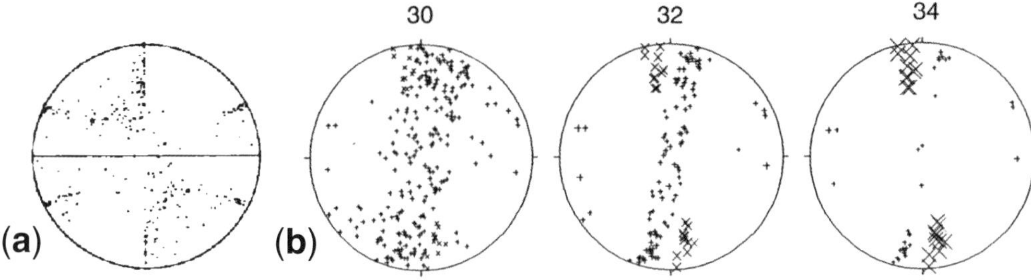

Fig. 1. Simulation of *c*-axis texture development in model quartzites. (**a**) Classical Taylor calculation for simple shear up to a shear strain of 3 (Lister 1981). (**b**) Texture development caused by both intracrystalline slip and dynamic recrystallization (i.e. nucleation and growth). Steps 30, 32 and 34 are shown (at shear strains of 0.75, 0.8 and 0.85 respectively) , recrystallization starts after step 30, the texture illustrated in step 34 does not change appreciably with further straining. Equal-area projection. The symbol + indicates old grains and the symbol × grains that have nucleated at least once; the size of the symbols is proportional to the grain volume (Takeshita *et al.* 1999).

approaches (Wenk *et al.* 1989*a*, *b*; Canova 1994; Takeshita *et al.* 1999). Their work has demonstrated that, even though many minerals do not fulfil the requirement for the strict application of the Taylor model (due to the limited number of slip systems available), many of the key features of natural textures can be accounted for.

An alternative approach recognizes that in some minerals, such as trigonal symmetry quartz at low temperatures, only one slip system may be activated readily. This necessitates incorporation of intra- and intergranular strain heterogeneity in models (Etchecopar 1977; Etchecopar & Vasseur 1987; Zhang *et al.* 1993, 1994*a*, *b*, 1996; Wilson & Zhang 1996; Wilson *et al.* 1996; Zhang & Wilson 1997). These models can reproduce some aspects of the behaviour of, for example, ice and low temperature quartz. The transition from a low temperature (single slip) behaviour to a high temperature (multi-slip) response is beyond the scope of both the Taylor (uniform strain) and Etchecopar (uniform stress) models.

Grain boundary sliding was shown to enhance texture development in model materials which allowed sliding interfaces between grains deforming with one slip system (Zhang *et al.* 1994*a*, *b*). The model of Ribe (1989) has been extended to simulate texture development in quartz when grain boundary sliding is the dominant deformation mechanism (Casey & McGrew 1999).

Under a wide range of crustal conditions, rocks deforming by crystalline plasticity are also modifying their microstructure as a result of dynamic recrystallization processes (Urai *et al.* 1986; Hirth & Tullis 1992). Jessell drew upon experiments on cold-worked copper (Kallend & Huang 1984) to support the assumption that the level of stored work is orientation dependent, and that the low crystal symmetry of the rock-forming minerals would result in a larger anisotropy of stored work than is present in metals (Jessell 1988*a*, *b*; Jessell & Lister 1990). He developed a hybrid scheme that combined the Taylor code of Lister with a Monte Carlo simulation that simulated the evolution of textures in quartz polycrystals by iterating between small increments of lattice rotations and grain boundary migration and subgrain formation (Fig. 2). The recrystallization processes in these models were driven by a simple stored work term for each element in the Monte Carlo simulation, as well as the more normal boundary energy derived neighbour relations (Anderson *et al.* 1984; Weaire & Rivier 1984). These simulations showed a better correspondence with natural textures than that predicted by models with lattice rotations only, even though there was still no mechanical interaction between grains.

Comparable hybrid lattice rotation/recrystallization schemes replace the Taylor-Bishop-Hill scheme entirely with finite element codes capable of heterogeneous intra-crystalline deformation (Radhakrishnan, *et al.* 1998; Raabe & Becker 2000; Bate 2001). Another approach to modelling the effects on texture of dynamic recrystallization in rocks has been to build in a grain-size weighting to simulate the changes in grain size that may result from dynamic recrystallization driven by orientation dependent stored work terms (Takeshita *et al.* 1999) (Fig. 1b).

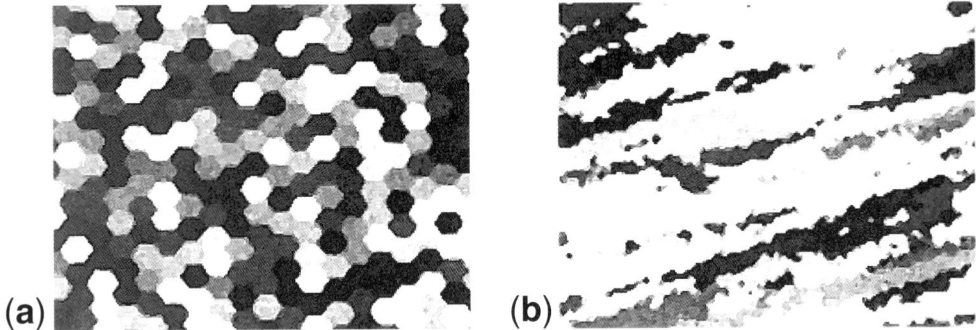

Fig. 2. The simulation of microstructure development in quartz deformed in simple shear. The technique combines a Taylor analysis that predicts new lattice orientations and stored work, with a Potts model which uses the stored work term to drive grain boundary migration and sub-grain formation. (**a**) Initial grain geometry, grey colours reflect *c*-axis orientations. (**b**) Microstructure after a shear strain of 3.0. (Jessell & Lister 1990)

Texture development is classically attributed to result from movement of dislocations (e.g. Wenk 1985). However, other deformation processes, such as dissolution–precipitation creep (Rutter 1976) may, hypothetically, also affect textures (Hippertt 1994; Stallard & Shelley 1995; den Brok 1996). Recently, Bons & den Brok (2000) explored this possibility with a numerical model. They showed that reaction-controlled dissolution–precipitation creep, coupled with grain rotation, can also lead to strong textures, similar to those described above. Their model used a constant stress approach, the opposite to the constant strain rate or Taylor approach, which may overestimate the strain rate heterogeneity and rotation rates of grains.

Many of the above mentioned models predict a texture in terms of a distribution of lattice orientations, without any predictions of the shape of grains or the locations of certain lattice orientations within a grain aggregate. This is a severe limitation if textures are also affected by processes where the relation with neighbouring grains is important, such as in case of dynamic recrystallization, grain boundary sliding and dissolution–precipitation creep. The recently enhanced ability to map out lattice orientations (as in Sander's (1950) Achsenverteilungsanalyse or AVA) with techniques such as the aforementioned EBSD and OIM (Leiss *et al.* 2000) also necessitates the development of models that include actual grain boundary topology and geometry.

Grain boundary topology and geometry

There have been a plethora of numerical simulation studies of grain boundary geometry in engineering materials, mostly focusing on grain growth, and recrystallization subsequent to deformation. The investigation of the grain growth phenomenon has in particular seen the application of a wide range of simulation techniques, in both 2D and 3D. These schemes can broadly be grouped into three classes:

(a) Discrete schemes mapped onto regular grids, in which local changes in grain boundary position occur as a result of individual grid points switching their orientations, as a function of the local distribution of orientations. These include Potts Models (Anderson *et al.* 1984; Kunaver & Kolar 1998) and Cellular Automata (Davies & Hong 1998).

(b) Continuous schemes mapped onto regular grids, where the local change in orientation property is a function of the properties of the whole system, such as Phase Field techniques (Fan & Chen 1997; Le Bouar *et al.* 1998).

(c) Discontinuous schemes mapped onto arbitrary networks. These schemes include Vertex Models (Soares *et al.* 1985; Cleri 2000); Front Tracking models (Wakai *et al.* 2000) and Finite Element techniques (Cocks & Gill 1996).

In comparison, there have been relatively few geological studies. Some of the studies mentioned in the previous section did make some predictions about grain geometries (Etchecopar 1977; Etchecopar & Vasseur 1987; Jessell & Lister 1990); however, the main focus of these studies was texture development. Bons & Urai applied a modified version of the Vertex model of Soares *et al.* (1985) to dynamic grain growth (Bons & Urai 1992), an approach also taken by

Bate & Mackenzie (2001). This allowed them to study grain size and grain geometries for the case of normal grain growth in a system where homogeneous strain is combined with grain growth and to demonstrate that the application of a bulk strain does not significantly alter the measured grain growth rates, but can slow down the development of a grain shape foliation (cf. Ree 1991) and play a role in the establishment of an oblique grain shape foliation (cf. Berthé *et al.* 1979).

As yet, there has been very little work on the development of microstructures during chemically driven nucleation and growth of new phases in rocks, but this area has enormous potential for further study (Foster 1999).

Deformation of one-phase systems

Even with only a single phase present, the simulation of polycrystalline deformation is far from trivial. Hybrid techniques are often used since markedly different microprocesses need to be simulated, either concurrently or consecutively. These techniques combine two or more of the techniques outlined in the previous sections, so in general there needs to be a re-mapping of the underlying description of the microstructure geometry from one simulation scheme to the other. Examples of hybrid schemes are combined Taylor lattice rotation and Potts dynamic recrystallization models (Jessell & Lister 1990), a combined Taylor model and a dimensionless recrystallization algorithm (Takeshita *et al.* 1999), combined Finite Element deformation/lattice rotation and Potts primary recrystallization (Radhakrishnan *et al.* 1998; Raabe & Becker 2000; Bate 2001) and a generalized hybrid scheme that attempts to combine a whole range of simulation techniques (Jessell *et al.* 2001).

The hybrid simulation of Jessell *et al.* (2001), named 'ELLE', consists of an open system of individual modules, each of which represents a specific grain-scale deformation process, including a Finite Element description of deformation, and a Front Tracking model for grain boundary migration driven by surface and defect energy terms. Microstructure evolution is achieved by passing a data file (which describes the topology and geometry of the polycrystal) through each module in turn (Fig. 3). This system has been applied to the problem of grain growth to investigate the role of surface energy anisotropy in modifying textures and grain boundary geometries (Bons *et al.* 2000). Piazolo *et al.* have applied this same simulation system to a systematic study of grain boundary geometries in deformed coarse and fine grain rocks (Piazolo *et al.* 2002). At present the mechanical modelling in this scheme cannot represent full elastic–plastic behaviour, but this is not an intrinsic limitation of the system.

Deformation of two-phase systems

While much of the work in geology has sensibly concentrated on single-phase systems, in reality most rocks consist of a mixture of two or more minerals. Several studies have looked at the specific case of a single rigid grain surrounded by a matrix of a weaker mineral. This case is important because it has been extensively used as an indicator of the kinematics and mechanics of deformation in natural settings. Apart from a number of kinematic models based on analytical solutions for viscous flow around objects (Bjørnerud 1989; Passchier *et al.* 1993; Beam 1996), finite-element formulations have been used to study the flow field around the rigid object (Bjørnerud & Zhang 1994, 1995; Bons *et al.* 1997; Kenkmann & Dresen 1998; Pennacchioni *et al.* 2000; Treagus & Lan 2000; Biermeier *et al.* 2001), and the pressure state around the rigid grain (Tenczer *et al.* 2001).

In terms of more general polyphase deformation, Finite Element codes have been used to develop models predicting the flow properties of materials in terms of the end-member behaviour of its constituent components (Tullis *et al.* 1991; Tóth *et al.* 1993; Bons & Cox 1994; Treagus & Treagus 2001; Treagus 2002). Compared to disciplines such as composite materials engineering, the geological modelling of polyphase rocks is still in its infancy.

Crystal growth

In a geological context, crystal growth at free surfaces plays a role in microstructure development in two main environments: during crystallization from igneous melts; and during precipitation from fluids. An originally 2D model has been developed into a 3D simulation of crystallization from a melt where an arbitrary number of seeds with different crystallographies may be specified (Amenta 2001). Schemes based on Potts or Ising models have been used by Mizuseki *et al.* (2000) to simulate crystal growth on planar surfaces and by Baker & Freda (1999) to simulate texture development in cooling silicate melts.

Microstructures of veins and pressure fringes potentially store much information on tectonic

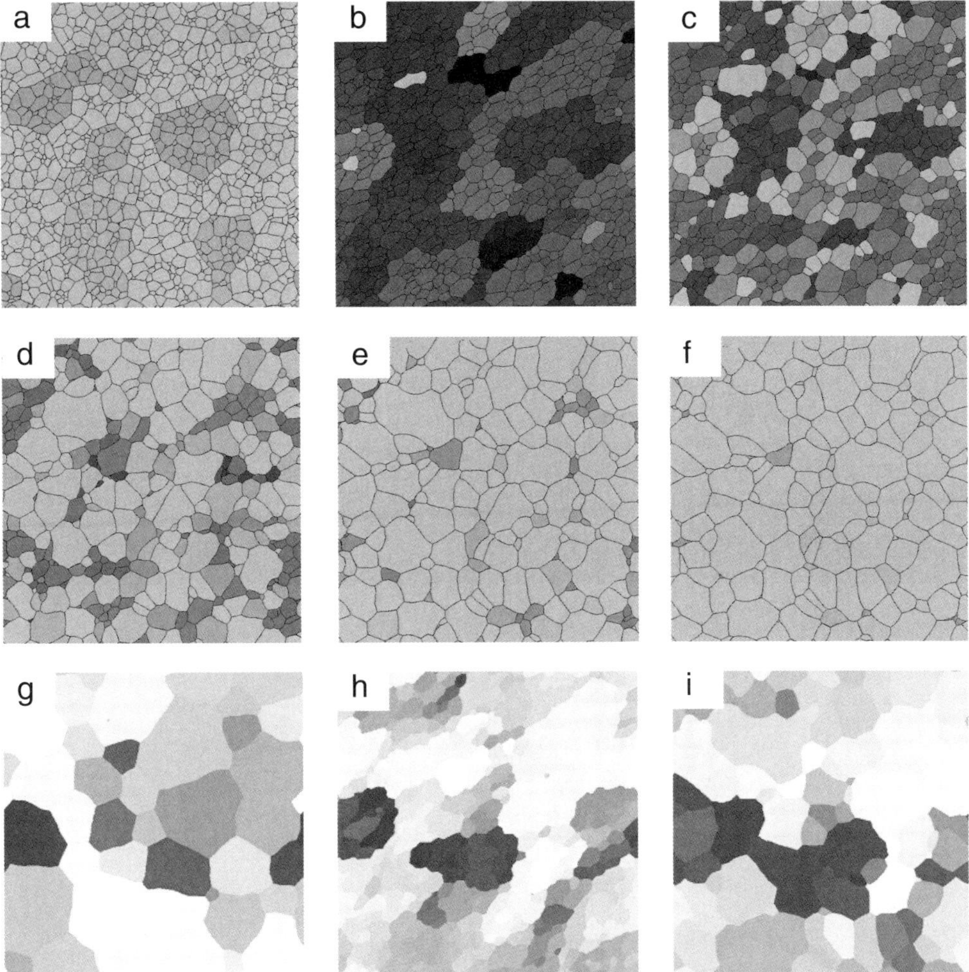

Fig. 3. Simulation of dynamic and meta-dynamic recrystallization in quartz, modelled with ELLE. (**a**) Starting grain boundary geometry. Shading reflects internal energy of grains and subgrains (darker equals higher). (**b**) After dextral shear strain of 0.65, modelled in thirteen steps. (**c–f**) Post deformation recovery and grain growth in intervals of twenty-six steps. (**g**, **h**, **i**) Same microstructure as in (**a**), (**b**) and (**f**) respectively, but with shading reflecting the lattice orientations. Notice that although the grain shapes are restored to a foam texture, the lattice orientations are preserved.

conditions during their formation (Durney & Ramsay 1973; Ramsay & Huber 1983; Bons 2000) and several numerical models have been developed to simulate their development (Etchecopar & Malavieille 1987; Kanagawa 1996; McKenzie & Holness 2000; Bons 2001). Bons (2001) uses a front-tracking approach in which linked nodes that describe the surfaces of vein crystals can move into the space that is produced by a gradually opening crack, essentially simulating the crack–seal mechanism of Ramsay (1980). This model formed the base for several studies in which the numerical simulations beautifully reproduced the vein and fringe structures under investigation (Koehn *et al.* 2000, 2001*a*, *b*; Hilgers *et al.* 2001) (Fig. 4). A similar approach has been taken by Paritosh *et al.* (1999) to investigate thin film diamond crystal growth. The investigation of growth kinetics when coupled with fluid flow has recently been tackled by Miller & Schröder (2001) using a Lattice Boltzmann approach. As with many other areas of microstructure simulation, the simulations listed here ignore the implications of the mechanical contrasts between different elements.

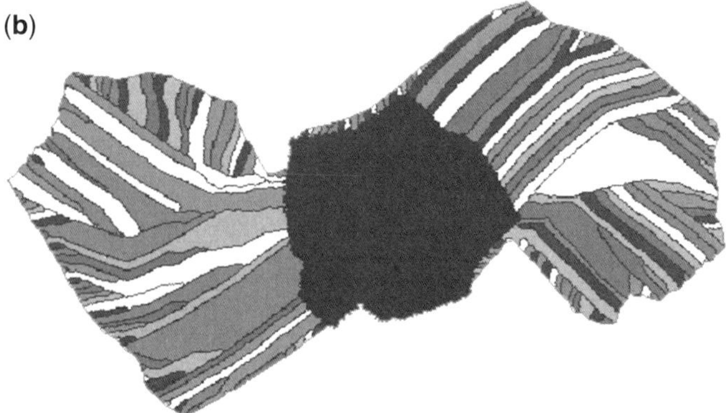

Fig. 4. Fibrous crystal growth around a rigid mineral as it rotates clockwise during progressive dextral simple shear. (**a**) Example from Lourdes (France) of a quartz + calcite + chlorite fringe system that developed around a pyrite grain. Width of image is 8 mm. (**b**) Numerical simulation of fringe growth around an object of the same shape, shading is used to distinguish different fibres. (Koehn *et al.* 2001*a*).

Discussion and conclusions

The distribution of the publication dates in the references section of this review shows that the numerical simulation of microstructures is a blossoming field in the Earth sciences. Many of the techniques that have been applied have direct parallels in the materials science literature; however, the complexity of geological systems has meant that in many geological studies the problems involve poly-phase materials and multi-process behaviour, areas that are relatively unexplored in the wider materials science literature. The modelling scheme ELLE (Jessell *et al.* 2001) is currently the only attempt in geology to fundamentally address this problem, as its open structure puts no limits on the number and types of processes that are modelled. The boundless opportunities this provides, in principle, have one serious draw-back: the more processes that are involved, the harder it becomes to check and test the validity of a model.

The solution to these problems will, ultimately, involve not simply numerical simulation

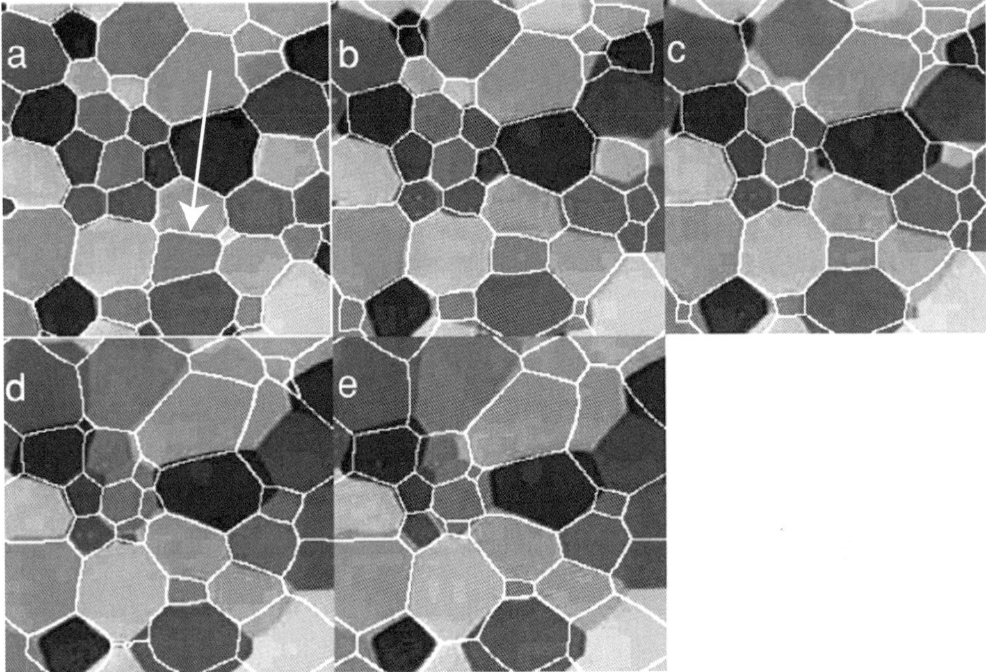

Fig. 5. Comparison between a transmitted light microscopy thin-film grain growth experiment in the polycrystalline organic compound octachloropropane (grey scale images) (Park, Ree & Means 1995, unpublished), and a numerical experiment with ELLE that used the same starting grain boundary geometry (white lines) (Jessell *et al.* 2001). Although the simulated isotropic growth successfully follows the development of some grains (arrow), it fails at other places, probably because growth in octachloropropane is not isotropic (Bons *et al.* 2000).

work, but instead an integrated approach that draws upon theory, field observations, classical mechanical experiments, and transmitted light experiments. The latter class of experiments are particularly useful, because they can be used as constraints, not just on the final microstructure, but also on the evolutionary path that that sample has undergone (Fig. 5).

Some geological processes are fundamentally difficult, if not impossible, to simulate experimentally in a laboratory. These are processes or combinations thereof that span large differences in scale, in both time and/or space, or those processes that involve different and conflicting scaling between experiment and nature. An example of the first could be deformation by occasional fracturing (strongly localized, occurring in seconds) punctuating slow ongoing ductile deformation (pervasive, time scales up to millions of years). An example of the second could be deformation during lower greenschist facies metamorphism of a pelite: to make an experiment possible in a reasonable time frame, deformation experiments need to be carried out at enhanced strain rate and temperature well above that for greenschist facies metamorphism. Here lie both a great challenge and an opportunity for numerical modelling.

The application of new numerical models should always include a discussion about the verification and testing of these models. Models based on single processes can often draw upon analytical solutions as tests; however, when several processes interact, these solutions are not often available. Finally the inputs to these models are the fundamental properties of geological materials at geologically relevant conditions. While some parameters, such as the elastic tensor, are quite well known, others, such as the surface energy properties of minerals under a range of conditions, are difficult to determine. Perhaps the greatest hope for the determination of these fundamental properties is the field of molecular dynamics, where they may be calculated via numerical experimentation. Alternatively, numerical modelling may actually help to constrain possible values for such parameters.

Although many of the simulation techniques are specifically applied to deforming rocks, most simulations discussed here are kinetic

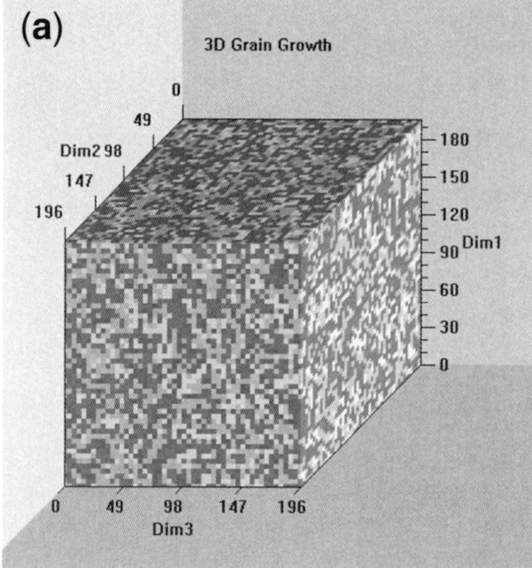

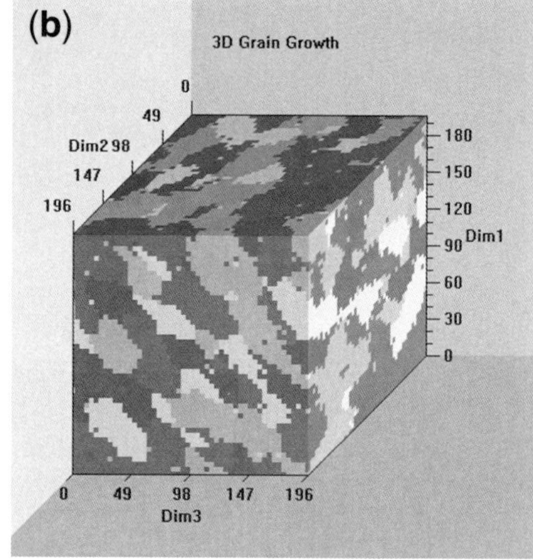

Fig. 6. Simple 3D Potts simulation of normal grain growth. (**a**) Starting configuration, with six possible lattice orientations randomly distributed through 50 × 50 × 50 site sample. (**b**) After 100 time steps, with 125,000 trials per time step (Jessell 2001, unpublished).

models (e.g. Bons 2001) and fail to consider the full dynamic or mechanical aspects of the deformation. Finally, we have to recognize that most of the simulation schemes that have been developed so far are two-dimensional, and there can be real limitations to two-dimensional schemes, especially when it comes to fluid flow and diffusional transport of materials. While some schemes are easily extended into three dimensions, such as particle codes and Potts Models (Fig. 6), there are real computational challenges in similarly extending others, including Front Tracking codes.

There is little doubt that the field of numerical simulation of microstructures will expand rapidly over the next few years, drawing upon techniques developed in the realm of materials science, and that the goal of a 'virtual rock' may eventually become a reality.

We are indebted to the Australian Research Council for support of the ELLE simulation code over the last five years. T. D. Barr, L. Evans, D. Koehn, K. Stüwe, R. Heilbronner and J. L. Urai are thanked for discussions and comments that improved this manuscript. Finally I would like to thank T. Takeshita, H.-R. Wenk, G. L. Lister and D. Koehn for granting permission to reproduce their figures.

References

AMENTA, R. V. 2001. Three-dimensional computer modeling of fabric evolution in igneous rocks. *Computers and Geosciences*, **27**, 477–483.

ANDERSON, M. P., SROLOVITZ, D. J., GREST, G. S. & SAHNI, P. S. 1984. Computer simulation of grain growth – I. Kinetics. *Acta Metallurgica*, **32**, 783–791.

BAKER, D. R. & FREDA, C. 1999. Ising models of undercooled binary system crystallization; comparison with experimental and pegmatite textures. *American Mineralogist*, **84**, 725–732.

BATE, P. 2001. Modelling microstructure development in annealing and deformation. *In*: GOTTSTEIN, G. & MOLODOV, D. A. (eds) *Recrystallization and Grain Growth. Proceedings of the First Joint International Conference*. Springer Verlag, Berlin, 39–48.

BATE, P. & MACKENZIE, L. 2001. Dynamic growth in superplastic AA5083. *In*: GOTTSTEIN, G. & MOLODOV, D. A. (eds) *Recrystallization and Grain Growth. Proceedings of the First Joint International Conference*. Springer Verlag, Berlin, 269–274.

BEAM, E. C. 1996. Modeling growth and rotation of porphyroblasts and inclusion trails. *In*: DE PAOR, D. G. (ed) *Structural Geology and Personal Computers*. Computer Methods in the Geoschiences, **15**, Pergamon, Elsevier, Oxford, 247–258.

BERTHÉ, D., CHOUKROUNE, P. & JEGOUZO, P. 1979. Orthogneiss, mylonite and non-coaxial deformation of granites: the example of the South Armoricain shear zone. *Journal of Structural Geology*, **1**, 31–42.

BIERMEIER, C., STÜWE, K. & BARR, T. D. 2001. The rotation rate of cylindrical objects during simple shear. *Journal of Structural Geology*, **23**, 765–776.

BJØRNERUD, M. G. 1989. Mathematical model for folding of layering near rigid objects in shear deformation. *Journal of Structural Geology*, **11**, 245–254.

BJØRNERUD, M. G. & ZHANG, H. 1994. Rotation of porphyroblasts in non-coaxial deformation; insights from computer simulations. *Journal of Metamorphic Geology*, **12**, 135–139.

BJØRNERUD, M. G. & ZHANG, H. 1995. Flow mixing, object-matrix coherence, mantle growth and the development of porphyroclast tails. *Journal of Structural Geology*, **17**, 1347–1350.

BONS, P. D. 2000. The formation of veins and their microstructures. *In*: JESSELL, M. W. & URAI, J. L. (eds) *Stress, Strain and Structure. A Volume in Honour of W D Means*. Journal of the Virtual Explorer, **2**. World Wide Web Address: http://www.virtualexplorer.com.au/VEjournal/Volume2

BONS, P. D. 2001. Development of crystal morphology during antitaxial growth in a progressively opening fracture: I. The numerical model. *Journal of Structural Geology*, **23**, 865–872.

BONS, P. D. & COX, S. J. D. 1994. Analogue experiments and numerical modelling on the relation between microgeometry and flow properties of polyphase materials. *Materials Science and Engineering*, **A175**, 237–246.

BONS, P. D. & DEN BROK, S. W. J. 2000. Crystallographic preferred orientation development by dissolution–precipitation creep. *Journal of Structural Geology*, **22**, 1713–1722.

BONS, P. D. & URAI, J. L. 1992. Syndeformational grain growth: microstructures and kinetics. *Journal of Structural Geology*, **14**, 1101–1109.

BONS, P. D., BARR, T. D. & TEN BRINK, C. E. 1997. The development of delta-clasts in non-linear viscous materials; a numerical approach. *Tectonophysics*, **270**, 29–41.

BONS, P. D., JESSELL, M. W., EVANS, L., BARR, T. D. & STÜWE, K. 2000. Modelling of anisotropic grain growth in minerals. *In*: KOYI, H. A. & MANCKTELOW, N. S. (eds) *Tectonic Modeling: A Volume in Honor of Hans Ramberg*. Geological Society of America Memoir, **193**, 45–49.

CANOVA, G. R. 1994. Self-consistent methods: application to the prediction of the deformation texture of polyphase materials. *Materials Science and Engineering*, **A175**, 37–42.

CASEY, M. & MCGREW, A. J. 1999. One-dimensional kinematic model of preferred orientation development. *Tectonophysics*, **303**, 131–140.

CLERI, F. 2000. A stochastic grain growth model based on a variational principle for dissipative systems. *Physica A*, **282**, 339–354.

COCKS, A. C. F. & GILL, S. P. A. 1996. A variational approach to two-dimensional grain growth-I. Theory. *Acta Materialia*, **44**, 4765–4775.

DAVIES, C. H. J. & HONG, L. 1999. The cellular automaton simulation of static recrystallization in cold-rolled AA1050. *Scripta Materialia*, **40**, 1145–1150.

DEN BROK, S. W. J. 1996. The effect of crystallographic orientation on pressure solution in quartzite. *Journal of Structural Geology*, **18**, 859–860.

DURNEY, D. W. & RAMSAY, J. G. 1973. Incremental strains measured by syntectonic crystal growths. *In*: DE JONG, K. A. & SCHOLTEN, R. (eds) *Gravity and Tectonics*. Wiley & Sons, New York, 67–95.

ETCHECOPAR, A. 1977. A plane kinematic model of progressive deformation in a polycrystalline aggregate. *Tectonophysics*, **39**, 121–139.

ETCHECOPAR, A. & MALAVIEILLE, J. 1987. Computer models of pressure shadows: a method for strain measurement and shear-sense determination. *Journal of Structural Geology*, **9**, 667–677.

ETCHECOPAR, A. & VASSEUR, G. 1987. A 3-D kinematic model of fabric development in polycrystalline aggregates. Comparisons with experimental and natural examples. *Journal of Structural Geology*, **9**, 705–717.

FAN, D. & CHEN, L.-Q. 1997. Topological evolution during coupled grain growth and ostwald ripening in volume-conserved 2-D two-phase polycrystals. *Acta Materialia*, **45**, 4145–4154.

FOSTER, C. T. 1999. Forward modeling of metamorphic textures. *Canadian Mineralogist*, **37**, 415–429.

FUETEN, F. 1997. A computer-controlled rotating polarizer stage for the petrographic microscope. *Computers and Geosciences*, **23**, 203–208.

HILGERS, C., KÖHN, D., BONS, P. D. & URAI, J. L. 2001. Development of crystal morphology during unitaxial growth in a progressively opening fracture: II. Numerical simulations of the evolution of antitaxial fibrous veins. *Journal of Structural Geology*, **23**, 873–885.

HIPPERTT, J. F. 1994. Microstructures and c-axis fabrics indicative of quartz dissolution in sheared quartzites and phyllonites. *Tectonophysics*, **229**, 141–163.

HIRTH, G. & TULLIS, J. 1992. Dislocation creep regimes in quartz aggregates: *Journal of Structural Geology*, **14**, 145–160.

HOBBS, B. E., MEANS, W. D. & WILLIAMS, P. F. 1976. *An Outline of Structural Geology*. John Wiley & Sons, New York.

HOUSEMAN, G. 1988. The dependence of convection planform on mode of heating, *Nature*, **332**, 346–349.

JESSELL, M. W. 1988*a*. A simulation of fabric development in recrystallizing aggregates – I: Description of the model. *Journal of Structural Geology*, **10**, 771–778.

JESSELL, M. W. 1988*b*. A simulation of fabric development in recrystallizing aggregates – II: Example model runs. *Journal of Structural Geology*, **10**, 779–793.

JESSELL, M. W. & LISTER, G. S. 1990. A simulation of the temperature dependence of quartz fabrics. *In*: KNIPE, R. J. & RUTTER, E. H. (eds) *Deformation Mechanisms, Rheology and Tectonics*. Geological Society, London, Special Publications **54**, 353–362.

JESSELL, M. W., BONS, P. D., EVANS, L., BARR, T. & STUWE, K. 2001. Elle: a micro-process approach to the simulation of microstructures. *Computers and Geosciences*, **27**, 17–30.

KALLEND, J. S. & HUANG, Y. C. 1984. Orientation dependence of stored energy of cold work in 50% cold rolled copper. *Metal Science*, **18**, 381–385.

KANAGAWA, K. 1996. Simulated pressure fringes, vorticity, and progressive deformation. In: Structural geology and personal computers. *In*: PAOR, D. G. (ed) *Structural Geology and Personal Computers*. Computer Methods in the Geosciences, **15**, Pergamon, Elsevier, Oxford, 259–283.

KENKMANN, T. & DRESEN, G. 1998. Stress gradients around porphyroclasts: palaeopiezometric estimates and numerical modelling. *Journal of Structural Geology*, **20**, 163–173.

KOEHN, D., AERDEN, D. G. A. M., BONS, P. D. & PASSCHIER, C. W. 2001*a*. Computer experiments to investigate complex fibre patterns in natural antitaxial strain fringes. *Journal of Metamorphic Geology*, **19**, 217–232.

KOEHN, D., BONS, P. D., HILGERS, C. & PASSCHIER, C. W. 2001*b*. Animations of progressive fibrous vein and fringe formation. *Journal of the Virtual Explorer*, **3**, World Wide Web Address: http://www.virtualexplorer.com.au/VEjournal/Volume4

KOEHN, D., HILGERS, C., BONS, P. D. & PASSCHIER, C. W. 2000. Numerical simulation of fibre growth in antitaxial strain fringes. *Journal of Structural Geology*, **22**, 1311–1324.

KUNAVER, U. & KOLAR, D. 1988. Three-dimensional computer simulation of anisotropic grain growth in ceramics. *Acta Materialia*, **46**, 4629–4640.

LE BOUAR, Y., LOISEAU, A. & KHACHATURYAN, A. G. 1998. Origin of chessboard-like structures in decomposing alloys. Theoretical model and computer simulation. *Acta Materiala*, **46**, 2777–2788.

LEISS, B., ULLEMEYER, K., *ET AL*. 2000. Recent developments and goals in texture research of geological materials. *Journal of Structural Geology*, **22**, 1531–1540.

LISTER, G. S. 1978. Texture transitions in plastically deformed calcite rocks. *Proceedings of the Fifth International Conference on Texture of Materials, Aachen, Germany*, Volume 2 Springer-Verlag, Berlin, 199–210.

LISTER, G. S. 1981. The effect of the basal-prism mechanism switch on fabric development during plastic deformation of quartzite. *Journal of Structural Geology*, **3**, 67–75.

LISTER, G. S. 1982. A vorticity equation for lattice reorientation during plastic deformation. *Tectonophysics*, **82**, 351–366.

LISTER, G. S. & HOBBS, B. E. 1980. The simulation of fabric development during plastic deformation and its application to quartzite; the influence of deformation history. *Journal of Structural Geology*, **2**, 355–370.

LISTER, G. S. & PATERSON, M. S. 1979. The simulation of fabric development during plastic deformation and its application to quartzite; fabric transitions. *Journal of Structural Geology*, **1**, 99–115.

LISTER, G. S., PATERSON, M. S. & HOBBS, B. E. 1978. The simulation of fabric development in plastic deformation and its application to quartzite: the model. *Tectonophysics*, **45**, 107–158.

LLOYD, G. E. & FREEMAN, B. 1991. SEM electron channelling analysis of dynamic recrystallization in a quartz grain. *Journal of Structural Geology*, **13**, 945–953.

MCKENZIE, D. & HOLNESS, M. 2000. Local deformation in compacting flows: development of pressure shadows. *Earth and Planetary Science Letters*, **180**, 169–184.

MILLER, W. & SCHRÖDER, W. 2001. Numerical modeling at the IKZ: an overview and outlook. *Journal of Crystal Growth*, **230**, 1–9.

MIZUSEKI, H., TANAKA, K., OHNO, K. & KAWAZOE, Y. 2000. A new crystal growth model based on a stochastic method under an external field. *Modelling and Simulation in Materials Science and Engineering*, **8**, 1–11.

MORESI, L. & SOLOMATOV, V. S. 1995. Numerical investigations of 2D convection with extremely large viscosity variations. *Physics of Fluids*, **7**, 2154–2162.

ORD, A. 1988. Deformation texture development in geological materials. *In*: KALLEND, J. S. & GOTTSTEIN, G. (eds) *Proceedings of the 8th International Conference on Textures of Materials*. The Metallurgical Society, Warrendale, Penn., 765–776.

PANOZZO-HEILBRONNER, R. & PAULI, C. 1993. Integrated spatial and orientation analysis of quartz c-axes by computer-aided microscopy. *Journal of Structural Geology*, **15**, 369–382.

PARITOSH, D., SROLOVITZ, J., BATTAILE, C. C., LI, X. & BUTLER, J. E. 1999. Simulation of faceted film growth in two-dimensions: microstructure, morphology and texture. *Acta Materialia*, **47**, 2269–2281.

PASSCHIER, C. W. & TROUW, R. A. J. 1996. *Microtectonics*. Springer Verlag, Berlin.

PASSCHIER, C. W., TEN BRINK, C. E., BONS, P. D. & SOKOUTIS, D. 1993. δ-objects as a gauge for stress sensitivity of strain rate in mylonites. *Earth and Planetary Science Letters*, **120**, 239–245.

PENNACCHIONI, G., FASOLO, L., CECCHI, M. M. & SALASNICH, L. 2000. Finite-element modelling of simple shear flow in Newtonian and non-Newtonian Fluids around a circular rigid particle. *Journal of Structural Geology*, **22**, 683–692.

PIAZOLO, S., BONS, P. D., JESSELL, M. W., EVANS, L. & PASSCHIER, C. W. 2002. Dominance of Microstructural Processes and their effect on microstructural development: Insights from Numerical Modelling of dynamic recrystallization. *In*: DE MEER, S., DURY, M. R., DE BRESSER, J. H. P. & PENNOCK, G. M. (eds) *Deformation Mechanisms, Rheology and Tectonics: Current Status and Future Perspectives*. Geological Society, London, Special Publications, **200**, 149–170.

POIRIER, J.-P. 1985. *Creep of Crystals: High-temperature Deformation Processes in Metals, Ceramics, and Minerals*. Cambridge University Press, Cambridge.

RAABE, D. 1998. *Computational 'Materials Science: The Simulation of Materials Microstructures and Properties'*. Wiley-VCH, Weinheim.

RAABE, R. & BECKER, R. C. 2000. Coupling of a crystal plasticity finite-element model with a probabilistic cellular automaton for simulating primary static recrystallization in aluminium. *Modelling and Simulation in Materials Science and Engineering*, **8**, 445–462.

RADHAKRISHNAN, B., SARMA, G. B. & ZACHARIA, T. 1998. Modeling the kinetics and microstructural evolution during static recrystallization – Monte Carlo simulation of recrystallization. *Acta Materialia*, **46**, 4415–4433.

RAMSAY, J. G. 1980. The crack-seal mechanism of rock deformation. *Nature*, **284**, 135–139.

RAMSAY, J. G. & HUBER, M. I. 1983. *The techniques of modern structural geology, 1: Strain analysis*. Academic Press, London.

REE, J. H. 1991. An experimental steady-state foliation. *Journal of Structural Geology*, **13**, 1001–1011.

RIBE, N. M. 1989. A continuum theory for lattice preferred orientation. *Geophysical Journal*, **97**, 199–207.

RUTTER, E. H. 1976. The kinetics of rock deformation by pressure solution. *Philosophical Transactions of the Royal Society, London*, **283**, 203–219

SANDER, B. 1950. *Einführung in die Gefügekunde der geologischen Körper*. Springer Verlag, Vienna.

SCHIØTZ, J., DI TOLLA, F. D. & JACOBSEN, K. W. 1998. Softening of nanocrystalline metals at very small grain sizes. *Nature*, **391**, 561–563.

SOARES, A., FERRO, A. C. & FORTES, M. A. 1985. Computer simulation of grain growth in a bimodal polycrystal. *Scripta Metallurgica*, **19**, 1491–1496.

STALLARD, A. & SHELLEY, D. 1995. Quartz c-axes parallel to stretching directions in very low-grade metamorphic rocks. *Tectonophysics*, **249**, 31–40.

TAKESHITA, T., WENK, H.-R. & LEBENSOHN, R. 1999. Development of preferred orientation and microstructure in sheared quartzite: comparison of natural data and simulated results. *Tectonophysics*, **312**, 133–155

TAYLOR, G. I. 1938. Plastic strain in metals. *Journal of the Institute of Metals*, **62**, 307.

TENCZER, V., STÜWE, K. & BARR, T. D. 2001. Pressure anomalies around rigid objects in simple shear. *Journal of Structural Geology*, **23**, 777–788.

TÓTH, L. S., MOLINARI, A. & BONS, P. D. 1993. Self-consistent modelling of the creep behavior of mixtures of camphor and octachloropropane. *Materials Science and Engineering*, **A175**, 231–236.

TÓTH, L. S., MOLINARI, A. & RAABE, D. 1997. Modelling of rolling texture development in a ferritic chromium steel. *Metalurgical Transactions*, **A28**, 2343.

TREAGUS, S. H. 2002. Modelling the bulk viscosity of two-phase mixtures in terms of clast shape. *Journal of Structural Geology*, **24**, 57–76.

TREAGUS, S. H. & LAN, L. 2000. Pure shear deformation of square objects, and applications to geological strain analysis. *Journal of Structural Geology*, **22**, 105–122.

TREAGUS, S. H. & TREAGUS, J. E. 2001. Effects of object ellipticity on strain, and implications for clast-matrix rocks. *Journal of Structural Geology*, **23**, 601–608.

TRIMBY, P. W., PRIOR, D. J. & WHEELER, J. 1998. Grain boundary hierarchy development in a quartz mylonite. *Journal of Structural Geology*, **20**, 917–935.

TULLIS, T. E., HOROWITZ, F. G. & TULLIS, J. 1991. Flow laws of polyphase aggregates from end-member flow laws. *Journal of Geophysical Research*, **96**, 8081–8096.

URAI, J. L., MEANS, W. D. & LISTER, G. L. 1986. Dynamic recrystallization of minerals. *In*: HOBBS, B. E. & HEARD, H. C. (eds) *Mineral and Rock Deformation: Laboratory Studies (The Paterson Volume)*. Geophysical Monograph, **36**, 161–199.

WAKAI, F., ENOMOTO, N. & OGAWA, H. 2000. Three-dimensional microstructural evolution in ideal grain growth: general statistics. *Acta Materialia*, **48**, 1297–1311.

WEAIRE, D. & RIVIER, N. 1984. Soap, cells and statistics – random patterns in two dimensions. *Contemporary Physics*, **25**, 59–99.

WENK, H.-R. 1985. *Preferred Orientation in Deformed Metals and Rocks: An Introduction to Modern Texture Analysis*. Academic Press, Orlando.

WENK, H.-R., CANOVA, G., MOLINARI, A. & KOCKS, U. F. 1989*a*. Viscoplastic modelling of texture development in quartzite. *Journal of Geophysical Research*, **94**, 17895–17906.

WENK, H.-R., CANOVA, G., MOLINARI, A. & MECKING, H. 1989*b*. Texture development in halite: comparison of Taylor model and self-consistent theory. *Acta Metallurgica*, **37**, 2017–2029.

WILSON, C. J. L. & ZHANG, Y. 1996. Development of microstructure in the high temperature deformation of ice. *Annals of Glaciology*, **23**, 293–302.

WILSON, C. J. L., ZHANG, Y. & STÜWE, K. 1996. The effects of localized deformation on melting processes in ice. *Cold Regions Science and Technology*, **24**, 177–189.

YAMAKOV, V., WOLF, D., SALAZAR, M., PHILLPOT, S. R. & GLEITER, H. 2001. Length-scale effects in the nucleation of extended dislocations in nanocrystalline Al by molecular-dynamics simulation. *Acta Materialia*, **49**, 2713–2722.

ZHANG, Z. & WILSON, C. J. L. 1997. Lattice rotation in polycrystalline aggregates and single crystals with one slip system: a numerical and experimental approach. *Journal of Structural Geology*, **19**, 875–885.

ZHANG, Y., HOBBS, B. E. & JESSELL, M. W. 1993. Crystallographic preferred orientation development in a buckled single layer: a computer simulation. *Journal of Structural Geology*, **15**, 265–276.

ZHANG, Y., HOBBS, B. E. & JESSELL, M. W. 1994*a*. The effect of grain boundary sliding on fabric development in polycrystalline aggregates. *Journal of Structural Geology*, **16**, 1315–1325.

ZHANG, Y., HOBBS, B. E. & ORD, A. 1994*b*. Numerical simulation of fabric development in polycrystalline aggregates with one slip system. *Journal of Structural Geology*, **16**, 1297–1313.

ZHANG, Y., JESSELL, M. W. & HOBBS, B. E. 1996. Experimental and numerical studies of the accommodation of strain incompatibility on the grain scale. *Journal of Structural Geology*, **18**, 451–460.

Dominance of microstructural processes and their effect on microstructural development: insights from numerical modelling of dynamic recrystallization

S. PIAZOLO[1,2], P. D. BONS[1,3,4], M. W. JESSELL[3], L. EVANS[3] & C. W. PASSCHIER[1]

[1]*Johannes Gutenberg-Universität, Institut für Geowissenschaften, FB 22, Becherweg 21, 55099 Mainz, Germany (e-mail: spi@geus.dk)*
[2]*Present address: GEUS, Thoravej 8, 2400 Copenhagen NV, Denmark*
[3]*Epsilon Laboratory, School of Earth Sciences, P.O. Box 28E, Monash University, Victoria, 3800, Australia*
[4]*Present address: Institut für Geowissenschaften, Eberhard Karls Universität, Sigwartstrasse 10, 72076 Tübingen, Germany*

Abstract: The influence of the dominance of different processes on the microstructural development of a quartzite was investigated using the numerical model 'ELLE'. Dynamic recrystallization of a polycrystalline aggregate was simulated by the concurrent operation of viscous deformation, lattice rotation, subgrain formation, rotational recrystallization, nucleation of new grains from strongly strained grains and recovery. The different observed microstructural characteristics depend on the relative rates at which grain boundary migration, subgrain formation, recrystallization by rotation and nucleation affect the microstructure. Observed sizes of recrystallized grains are significantly influenced by these different relative rates of processes. These rates are determined by parameters that mainly depend on temperature, fluid absence or presence, shear stress and strain rate. Therefore, the specific conditions at which deformation took place have to be taken into account if recrystallized grain sizes are used for palaeopiezometry. Comparison and combination of our results with experimental data and observations in natural examples provide the possibility of interpreting microstructures quantitatively in terms of temperature and shear strain rate.

Microstructural features that result from deformation-induced dynamic recrystallization are commonly used to evaluate the dominance of different processes active at the grain scale. Such dominance sheds light on the conditions prevailing during rock deformation (e.g. Guillopé & Poirier 1979; Urai *et al.* 1986; Drury & Urai 1990; Hirth & Tullis 1992; Takeshita *et al.* 1999; Shigematsu 1999). Two main microstructural processes can be distinguished: (a) subgrain rotation; and (b) grain boundary migration (Urai *et al.* 1986; Drury & Urai 1990). Microstructures that develop from subgrain rotation recrystallization are in general characterized by clusters of grains with similar orientations, although local deviations from this smooth pattern due to heterogeneous strain seem to be rather common. Core-and-mantle structures commonly develop (White 1976; Gifkins 1976). Grain boundary migration microstructures often show irregular grain shapes. Grains bulge into neighbouring grains, and 'old' grains are consumed by new grains with low dislocation density (e.g. Urai *et al.* 1986; Jessell 1987; Drury & Urai 1990).

Hirth and Tullis (1992) experimentally determined three different dislocation creep regimes in quartz aggregates according to dominance of processes, strain rate and temperature. Each one of these regimes shows characteristic microstructural features such as curvature of grain boundaries, shape and size of porphyroclasts and recrystallized grains, and spatial distribution of recrystallized grains. However, in order to apply experimental data of rock deformation to natural samples, there is a need to extrapolate over several orders of magnitude, which poses certain difficulties (Hirth & Tullis 1992; Stöckhert *et al.* 1999).

In this paper we establish a method which we hope will help to bridge the gap between experiments and nature. To do this we explore the role of the dominance of grain boundary migration, subgrain formation, rotational recrystallization, and nucleation of new grains from strained grains (Doherty *et al.* 1997; Humphreys & Hatherly 1995) on the microstructural development of a polycrystalline aggregate, using the numerical modelling system ELLE (Jessell *et al.* 2001). Different combinations of the relative

From: DE MEER, S., DRURY, M. R., DE BRESSER, J. H. P. & PENNOCK, G. M. (eds) 2002. *Deformation Mechanisms, Rheology and Tectonics: Current Status and Future Perspectives*. Geological Society, London, Special Publications, **200**, 149–170. 0305-8719/02/$15

rates at which processes affect a microstructure are examined. The results are briefly compared to results of experimental deformation of rock materials (Hirth & Tullis 1992; Jung & Karato 2001) and observations in naturally deformed quarzite veins by Stipp *et al.* (2002).

Numerical Model ELLE

General description of model

The modelling system ELLE (Evans *et al.* 1999, 2000; Jessell *et al.* 2001) is designed to model the 2D evolution of a microstructure that is a result of several concurrent (sub-) grain scale processes. It is based on a data structure that describes a polycrystalline material using a 2D network of nodes and connecting boundaries at the grain scale (Fig. 1, Jessell *et al.* 2001). Data on the geometry of the structure and attributes or properties such as dislocation density, viscosity and stress-state of nodes and polygons are stored in a data file. To simulate the progress of a process, individual algorithms can interact with this data structure: (a) by using it to determine the local values of driving forces; (b) by repositioning, creating and removing nodes; (c) by reconnecting boundary segments; and (d) by altering attributes. A central shell program controls the evolution of extrinsic variables, such as temperature, and determines which microprocesses will be involved by controlling the order and rate of execution of individual process algorithms.

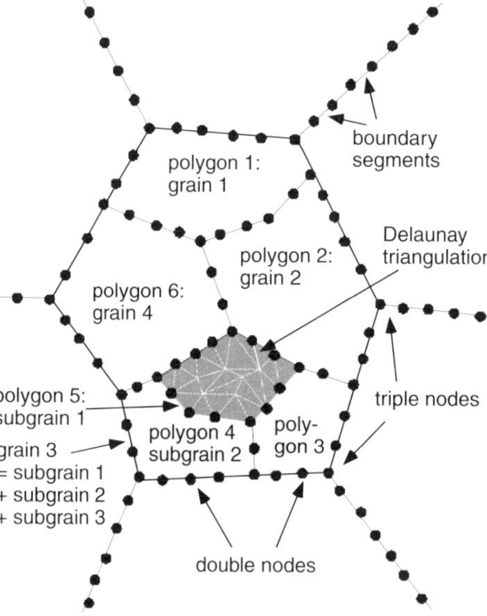

Fig. 1. Representation of the 2D finite element map of a grain aggregate in the ELLE model. Each polygonal domain can be subdivided into a number of polygonal sub-domains (subgrains; polygon 3, 4, 5), each of which in turn has a number of properties assigned to them, such as crystal lattice orientation. All polygons can be triangulated to smaller elements using a Delaunay triangulation routine (Shewchuk 1999). These elements may also have individual properties, including the stress and strain state during deformation, and/or trace element concentration.

In the model presented here, we simulate dynamic recrystallization during plane strain simple shear deformation of a polycrystal, involving crystal lattice rotation, formation of subgrains, recrystallization by nucleation, recrystallization by subgrain rotation, grain boundary migration and recovery.

In the following, we describe the routines and values as used in the simulations presented. Formulas and values are not specific to the ELLE modelling system. In contrast, formulas can be changed and the results of different formulas on the microstructural development can be investigated. This is, however, not the subject of this contribution.

Description of individual routines and parameters used

In the following a short description of the routines used in the model is given. A more extensive description is provided by Piazolo (2001). Tables 1 and 2 list the values and main formulas used. The formulas are used to calculate specific values e.g. driving force per one time step which represents 1000 years in our model.

Viscous deformation is modelled by a 2D finite element code which calculates stress and strain in linear and non-linear incompressible viscous, non-elastic materials undergoing irreversible large-strain deformation (Barr & Houseman 1992, 1996, 1999). During deformation, the ELLE microstructure is distorted into a parallelogram. Since the unit cell has periodic boundaries, nodes that move outside the unit cell can be repositioned to re-enter the square unit cell. Therefore, deformation up to high strains can be modelled. In the model the unit cell represents an area of 10 mm × 10 mm. The flow behaviour is Newtonian viscous. The commonly cited power-law behaviour observed in rocks (e.g. Carter & Tsenn 1987; Kirby & Kronenberg 1987) is to some extent modelled separately within the lattice rotation routine and the associated accumulation of dislocations in which the increase

Table 1 *Values of parameters and variables*

Parameters/settings	Value	Unit	Comments/examples of relevant literature
Minimum distance between nodes	0.00625×10^{-3}	m^2	—
Maximum distance between nodes	0.03175×10^{-3}	m^2	—
Temperature T	723	°K	—
Pressure P	300	MPa	—
Type of deformation W_n	1; simple shear		—
Flow behavior of material	Newtonian viscous		—
Time scaling factor t_{scale}	$3.1536e^{10}$ (1000a)	s	—
Length scaling factor l_{scale}	10^{-3}	m	Observed area: usual thin section scale area
Stress scaling factor τ_{scale}	10^8	Pa	Resulting in simulated viscosity values in accordance with Kirby (1983), Carter & Tsenn (1987), Clark & Royden (2000)
Dislocation density scaling factor ρ_{scale}	10^{13}	m^{-2}	Resulting in a range of simulated dislocation densities in accordance to Gottstein & Shvindlermann (1999), Hacker & Kirby (1993)
Strain per time step $\Delta\gamma$	0.05	—	—
Real world strain rate	1.6×10^{-12}	s^{-1}	Prior *et al.* 1990; Pfiffner & Ramsay 1982
Finite strain γ	2	—	—
Critical resolved shear stress (*CRSS*) for different slip systems	greenschist to lower amphibolite facies conditions	—	See for details Hobbs (1985); Jessell & Lister (1990)
Energy per dislocation E_ρ	7×10^{-9}	Jm^{-1}	Hacker & Kirby 1993; Gottstein & Shvindlermann 1999
Preferred orientation of subgrain walls	parallel & perpendicular to *c*-axis		Lloyd *et al.* 1992; Stöckert *et al.* 1999
Energy threshold value for subgrain formation τ_{split}	2.4×10^5	J/m^2	Gottstein & Shvindlermann 1999
Allowed area range for newly formed subgrain	3.12×10^{-4}–1.25×10^{-3}	mm^2	Twiss 1977; White 1979; Christie & Ord 1980; Kronenberg & Tullis 1984
Critical misorientation angle for recrystallization by rotation	10	°	White 1979; Lloyd *et al.* 1992; Stöckhert *et al.* 1999
Dislocation density of newly recrystallized grain ρ_{nucl} (used in recrystallization by nucleation routine)	10^{10}	m^{-2}	Hirth & Tullis 1992; Hacker & Kirby 1993; Stöckhert *et al.* 1999
Surface energy E_Γ	7×10^{-2}	Jm^{-2}	Gottstein & Shvindlermann 1999; Urai *et al.* 1986; Stöckhert & Dyster 2000
Recovery factor $R_{recover}$	0.94	—	
Base viscosity η_{base}	3.15×10^{18}	Pa s	Carter & Tsenn 1987; Clark & Royden 2000
Minimum driving force threshold value for recrystallization by nucleation τ_{nucl}	a) 4.8×10^5 b) 3.6×10^5	Jm^{-2}	Gottstein & Shvindlermann 1999
Grain boundary mobility M_{gb}	a) 1×10^{-12} b) 20×10^{-12} c) 40×10^{-12}	$m^2 s^{-1} J^{-1}$	Resulting migration rates of grain boundaries are in accordance to Prior *et al.* (1990)

of the dislocation density and therefore viscosity with stress is taken into account (see below).

The lattice rotation of polygons that represent either grains or subgrains and accumulation of dislocations within these is the result of two phenomena arising from dislocation glide during progressive deformation. Rotation of the crystal lattice is required to accommodate an arbitrary deformation because deformation by slip is possible only along a small number of

Table 2. *Main formulas and/or pseudo-codes used*

Process	Formula/pseudocode	Explanation/symbols (see Table 1 for symbols and units)
Accumulation of dislocations	$\rho_{new} = 0.4 \times \rho_{initial} + 0.7 \times W$	ρ_{new}: new dislocation density; $\rho_{initial}$: initial dislocation density; W: work term from Taylor-Bishop-Hill code
Subgrain formation & recrystallization by nucleation	$E_\gamma = \rho \times E_\rho \times \rho_{scale}$	ρ: dislocation density of grain/subgrain E_γ: strain energy of grain E_ρ: energy of dislocation ρ_{scale}: dislocation scaling
Subgrain formation/ recrystallization by nucleation	if $E_\gamma > \tau_{split}/\tau_{nucl}$ then *DoSubgrain*/ *DoNucleation*	If the internal energy is higher than the threshold value, a grain will – at a certain probability – be split into subgrains (subgrain formation routine) or a new grain will nucleate (recrystallization by nucleation routine)
Grain boundary migration	$\Delta E = \Delta E_\Gamma + \Delta E_\rho$	ΔE: total energy change by movement of node ΔE_Γ: energy change due to surface energy ΔE_ρ: energy change due to differences in strain energy
Grain boundary migration	$v_{node} = M_{gb} \times \Delta E$	v_{node}: movement velocity of node M_{gb}: grain boundary mobility
Grain boundary migration	$D = v_{node} \times t_{scale}$	D: distance of node movement t_{scale}: time scaling factor; here: 1000 years
Recovery	$\rho_{new} = \rho_{initial} \times R_{recover}$	ρ_{new}: new dislocation density $\rho_{initial}$: initial dislocation density $R_{recover}$: recovery factor (temperature dependent); here: 0.94
Calculation of kinematic viscosity	$\eta = (\eta_{base} + \mathrm{SQRT}(\rho))$	η: viscosity η_{base}: base viscosity of grain; here: 3.15×10^{18} Pa s ρ: dislocation density of grain/subgrain

slip systems. The resistance to dislocation movement increases as dislocation pile-ups or tangles form, and additional stress is needed further to move dislocations. The crystal is said to 'work-harden', i.e. the dislocation density increases due to the build up of tangles within the crystal (Kocks 1976, Kocks 1985; Mecking & Kocks 1981; Barber 1985).

The routine that models the lattice rotations and the accumulation of dislocations consists of two parts. The first part is the calculation of the crystal lattice rotation and the associated work using the Taylor-Bishop-Hill calculation method (Taylor 1938; Bishop & Hill 1951*a*, 1951*b*; Lister & Paterson 1979). The second part is the calculation of the accumulation of dislocations. Within the Taylor-Bishop-Hill calculation it is assumed that no deformation takes place until the entire strain increment can be achieved for a grain in the direction of the slip vector. The stress gradually increases until the critical resolved shear stress (CRSS) threshold is surpassed on just enough slip systems to allow the specific strain to take place (for details see Lister *et al.* 1978; Jessell & Lister 1990). Once this state is achieved, the work term can be calculated as the product of the small strains achieved by each slip system and the imposed stress. No lattice reorientation due to rigid body rotation is taken into account and strain is assumed to be plane strain. The calculation of the lattice rotation is done for each polygon, i.e. grain or subgrain. This means that for each subgrain the Taylor-Bishop-Hill calculation may lead to misorientations (for further details see below) between subgrains. The amount of accumulated dislocations is then determined using the work term from the first calculation. Theory (Argon 1970; Kohlstedt & Weathers 1980) and experimental observations (e.g. Durham *et al.* 1977; Beemann & Kohlstedt 1988; de Bresser 1996) suggest a general positive correlation of dislocation density and differential stress. In the routine, the change in dislocation density is assumed to be linearly proportional to the amount of work per unit area. The new dislocation density ρ_{new} of a polygon undergoing work is calculated according to the following equation:

$$\rho_{new} = a\rho_{initial} + bW \quad (1)$$

where $\rho_{initial}$ corresponds to the 'initial' dislocation density of a polygon given in the ELLE data file before the onset of the routine and the dimensionless W is the calculated work for deformation of the grain according to the Taylor-Bishop-Hill code. The values of a and b are chosen to be 0.4 and 0.7, respectively. The sum of the constants a and b is chosen to be 1.1 to ensure that there is an increase of dislocations with progressive deformation provided that W is larger than CRSS. If one simply uses a direct function of W, without any reference to the previous dislocation density, a strong oscillation of dislocation density occurs from one time step to the next. In a physical sense, this means that the number of dislocations within a crystal is a function of the number of dislocations that were present within the crystal before deformation and the number of dislocations that were generated due to the deformation of the crystal by crystal slip.

The formation of subgrains which is, in other words, the discontinuous development of subgrain boundaries is modelled by repeated 'splitting' of a 'parent' polygon (i.e. grain) into two smaller 'daughter' polygons (i.e. subgrains). The driving force for subgrain formation is the total strain energy per unit area, which depends on the dislocation density of the polygon. The driving force E_γ for subgrain formation is calculated according to:

$$E_\gamma = \rho E_\rho \rho_{scale} \qquad (2)$$

where ρ is the dislocation density, E_ρ the energy of dislocations and ρ_{scale} the dislocation density scaling factor. If $E_\gamma > \tau_{split}$ (energy threshold value for splitting of a polygon), there is a probability that the grain will 'split' into subgrains. The probability of 'splitting' increases with increasing E_γ. The new subgrain boundaries are chosen to be preferrentially parallel or perpendicular to the crystallographic c-axis of the 'parent' grain according to observations of subgrain boundary orientations by Lloyd *et al.* (1992) and Stöckhert *et al.* (1999). τ_{split} is chosen according to values derived for subgrain formation in metals (Table 1, Gottstein & Shvindlermann 1999) since few data are available for silicates. The newly formed subgrains have material properties identical to those of the 'parent' grain (e.g. crystallographic orientation, dislocation density and viscosity). In the coarse-grained microstructures, subgrain boundaries are initially put into the microstructure, for computational efficiency. In the initially fine-grained aggregate no subgrains are present. In experiments with both initially fine and coarse-grained microstructure, subgrains are formed. The 'allowed' subgrain size is 3.12×10^{-4} to 1.25×10^{-3} mm^2, which is in the range of subgrains observed in silicates (Twiss 1977; White 1979; Christie & Ord 1980; Kronenberg & Tullis 1984).

The process which we call in this paper 'recrystallization by nucleation' models the discontinuous nucleation and growth of relatively strain- and dislocation-free grains which develop from highly strained subgrains. This process is based on two apparently distinct steps: (1) the initial formation of the new grain; and (2) its growth. During this process the dislocation structure in an 'old' subgrain is cleared out and the new grain exhibits a low dislocation density. Such recrystallization by nucleation has been observed in albite, magnesium and quartz (Knipe & White 1979; Fitz Gerald *et al.* 1983; Urai *et al.* 1986; Drury & Urai 1990). In metallurgy (e.g. Doherty *et al.* 1997; Solas *et al.* 2001) it is now accepted that in this process the new grain grows from small regions of recovered subgrain or cells, that are already present in the deformed microstructure. A subgrain that has a high dislocation density is characterized by a high number of even smaller subgrains (cells) to very small scales (μm) which themselves may have low internal dislocation density values (Kocks 1985; Humphreys & Hatherly 1995; Doherty *et al.* 1997). One of these minute cells can act as a nucleus, that will grow rapidly at the cost of its neighboring cells (Fig. 2, Gottstein & Mecking 1985; Gottstein & Svindlerman 1999; and references therein). To allow such a nucleation the total free energy must decrease during expansion of the nucleus. For this: (a) a critical nucleus size r_c has to be exceeded; (b) there must be an instability of the microstructure, i.e. differences in dislocation density; and (c) the boundary must be a mobile grain boundary, i.e. a high-angle boundary (Gottstein & Mecking 1985) on at least one side of the nucleus. The possibility that a strain-free nucleus satisfies the prerequisites as stated above increases with increasing dislocation density (Kocks 1985). The new grain exhibits a low dislocation density because during nucleation and growth of the nucleus dislocations are swept into the moving tilt wall. The mobility of dislocations, which is directly related to the growth rate of a nucleus is material and temperature dependent: the boundaries of a new nucleus sweep over its neighbouring grains at greater speed at high temperature than at low temperature (Frost & Ashby 1983).

The driving force for recystallization by nucleation is the strain energy of the grain or subgrain E_γ which is calculated according to Equation 2. The critical threshold value τ'_{nucl} of a specific mineral species at a specific temperature is calculated from a base nucleation

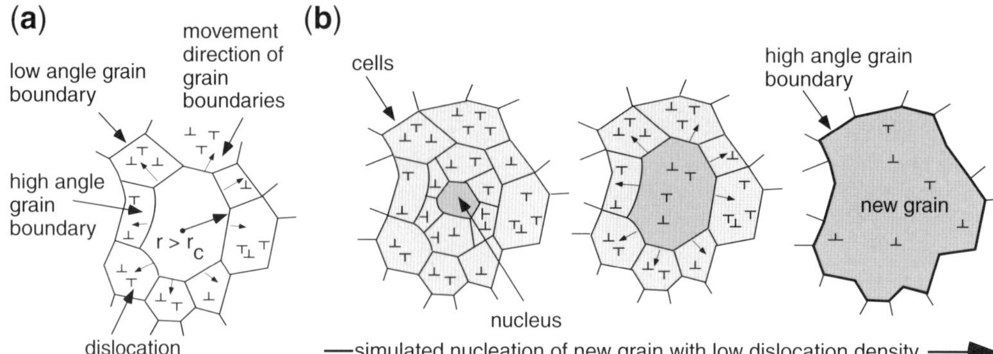

Fig. 2. Schematic illustration of the mechanism of recrystallization by nucleation (Gottstein & Svindlermann 1999; Humphreys & Hatherly 1995; Doherty *et al.* 1997). (**a**) Recrystallization nucleus with growth potential in a deformed aggregate (modified after Gottstein & Svindlermann 1999). A nucleus can only grow if the total free energy decreases during expansion of the nucleus of size r. For the rapid growth of a nucleus, a critical nucleus size r_c has to be exceeded, the internal energy of the nucleus must be significantly lower than that of its neigbouring cells and the angle of misorientation between the nucleus and its neighbouring cells must be high (Gottstein & Svindlermann 1999). (**b**) Schematic illustration of how the process recrystallization by nucleation is modelled. A grain with a high strain energy i.e. high dislocation density is envisaged as a group of small cells. The higher the strain energy the higher the possibility that one of these cells has a significant lower strain energy than its neighbours. Such a cell then acts as a nucleus, the boundaries of which may rapidly sweep its neighbours. Eventually a new grain develops with low internal energy and high-angle grain boundaries.

threshold value τ_{nucl} specific to a mineral species:

$$\tau'_{nucl} = (\tau_{nucl})1000/(T + 1000) \qquad (3)$$

where T is the temperature. If $E\gamma > \tau'_{nucl}$ there is a probability that a nucleus which fulfils the three requirements for recrystallization by nucleation is present in a polygon (see above). In simulations two different τ_{nucl} values are used. The dislocation density of newly nucleated grain is set to a specified low dislocation density value (10^{10} m^{-2}) according to: (1) the range of values observed in calcite (Hacker & Kirby 1993) and quartz (Stöckhert *et al.* 1999); and (2) a new randomly picked crystallographic orientation. The randomly picked orientation must satisfy the condition that the boundaries at all sides of the grain of the recrystallized grain are mobile high-angle boundaries, to take account of the fact that a new nucleus will only be able to grow if at least some of its boundaries are high-angle boundaries (Fig. 2). In simulations all subgrain boundaries must be high-angle boundaries to allow the 'successful' nucleation of the grain as, due to computational limitations, the position of the subgrain boundaries does not change.

A subgrain undergoes recrystallization by rotation if all of its subgrain boundaries develop from low-angle boundaries to high-angle boundaries. For each time step, the routine calculates the misorientation angle between adjacent polygons following the procedure described by Randle (1992) and Lloyd *et al.* (1992). Then, the routine checks if a polygon is completely surrounded by high-angle boundaries. The value of the critical angle between a high and low-angle boundary was chosen to be 10° according to values given for quartz (e.g. White 1979; Lloyd *et al.* 1992; Trimby *et al.* 1998). Once the program 'decides' that a subgrain should become a true grain, the subgrain is promoted to grain status, without changes to any other attributes such as dislocation density or crystallographic orientation.

In crystalline materials grain boundary migration occurs due to the difference in stored internal strain energy and chemical potential due to curvature between adjacent grains (Table 2). Here, grain boundary migration is modelled by incremental movement of nodes. The direction of the movement is chosen to achieve a maximum reduction of the total energy of the system (Table 2). The distance of movement is a function of the product of grain boundary mobility, which is strongly temperature and fluid presence dependent, and the driving force (Table 2). The area swept by the grain boundary by movement of the nodes is devoid of dislocations, as the dislocations move into the moving grain boundary (Barber 1985; Kocks 1985). The calculation of the change in energy is divided in two main components. These are: (1) the change in strain energy stored in adjacent polygons; and (2) the change in surface energy. The change in dislocation density due to the movement of a node takes the initial dislocation density of the two or three adjacent grains and the new areas and new dislocation densities into account. Therefore, the dislocation density of a grain that sweeps over another high dislocation density grain decreases. In the model the calculation of the surface energy has two components. These are: (1) surface energy due to length of the boundary segments; and (2) the anisotropy of this surface energy due to the crystallographic *c*-axis misorientation angle between two adjacent grains or subgrains. The influence of the *c*-axis misorientation angle is chosen so that the calculated surface energy of a specific boundary is zero at a *c*-axis misorientation angle of 0°, 0.1 × surface energy E_Γ (Table 1) at 10° and 0.99 × E_Γ at 20° and above. These relationships were chosen according to the general relationships observed in metals (Humphreys & Hatherly 1995; Gottstein & Svindlerman 1999). This means that a boundary between subgrains, hence a low-angle boundary, is virtually immobile due to the very low energy change accompanying movement of the grain boundary. Therefore, the low-angle boundaries remain largely stable during deformation. E_Γ is chosen to be 7×10^{-2} Jm^{-2} according to measurements of surface energies provided by Gottstein & Shvindlermann (1999), Urai *et al.* (1986) and Stöckhert & Dyster (2000).

Thermodynamically unfavourable lattice defects are removed during recovery. This process is simulated by a reduction of the dislocation density of each grain or subgrain per time step according to:

$$\rho_{new} = \rho_{initial} R_{recover} \quad (4)$$

where $R_{recover}$ is a temperature dependent recovery factor. In our model (at $T = 450\,^\circ C$) $R_{recover}$ is 0.94 (Table 1).

The rheology of the polycrystal is assumed to be viscous, rather than plastic. In the model, it is assumed that the activation energy required to overcome an obstacle in the crystal lattice is the rate-limiting factor for the deformation of the material and therefore the viscosity is directly related to the square root of the dislocation density (Argon 1970; Frost & Ashby 1983):

$$\eta = (\eta_{base} + SQRT(\rho)) \quad (5)$$

where η is the viscosity, η_{base} a base viscosity specific to the mineral species and ρ the dislocation density. η_{base} is chosen to be 3.15×10^{18} Pa s so that the viscosities in our simulations correspond to the viscosity which is thought to be typical for greenschist to amphibolite facies conditions (e.g. Carter & Tsenn 1987; Clark & Royden 2000).

Numerical simulations

A set of twelve simulations was performed. Values were generally chosen in accordance with values provided in the literature for ductile deformation of wet quartz at greenschist to amphibolite facies conditions (Table 1). If no such values existed, values were chosen to fit the general proposed relationship between different values. The value of the strain increments ($\Delta\gamma = 0.05$) and the finite strain ($\gamma = 2$) were chosen to keep the calculation time within a limit of about 5 days. For calculations a Beowulf cluster was used which consists of eleven DEC/Compaq Alpha nodes. Each was configured with 128M RAM and 5G of IDE disk and had a 533 Mhz 21164A (EV56) CPU with 2M tertiary cache on an LX motherboard.

To investigate the effect of the relative rates at which the specific processes of grain boundary migration, subgrain formation, recrystallization by subgrain rotation and nucleation act upon a microstructure we decided to vary two parameters which strongly influence the rate of: (1) grain boundary migration; and (2) recrystallization by nucleation. Accordingly: (1) the value for grain boundary mobility (M_{gb}); and (2) the energy threshold value for recrystallization by nucleation (τ_{nucl}) (Table 3) were varied. All other values were kept constant. Parameters such as temperature (450 °C) and pressure

Table 3. *Settings for simulations*

Grouping	Abbreviation for experimental conditions	Minimum threshold for τ_{nucl} [Jm^{-2}]	Grain boundary mobility M_{gb} [$m^2 s^{-1} J^{-1}$]
Group A	1-1	4.8×10^5	1×10^{-12}
	2-1	3.8×10^5	1×10^{-12}
Transition A/B	2-2	3.8×10^5	20×10^{-12}
Group B	1-2	4.8×10^5	20×10^{-12}
Ttransition B/C	2-3	3.8×10^5	40×10^{-12}
Group C	1-3	4.8×10^5	40×10^{-12}

Note all simulations were performed starting with both a fine and a coarse-grained aggregate.

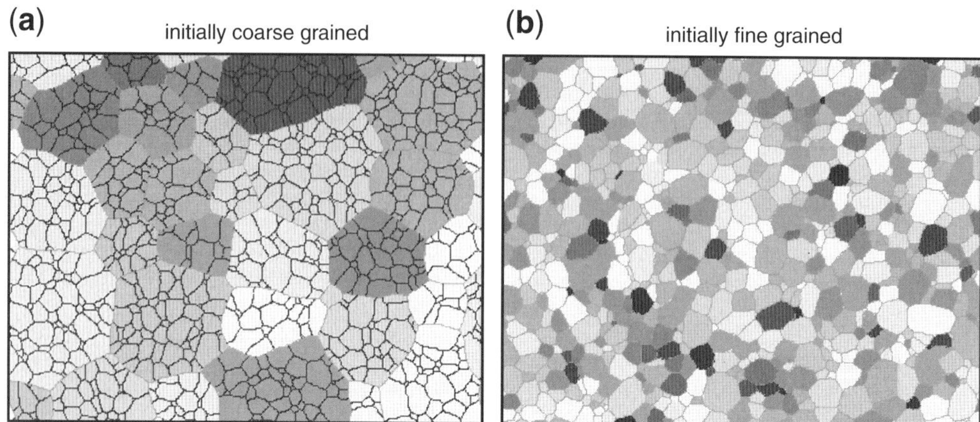

Fig. 3. Pictures of the two starting grain aggregates. Each grain is represented by an arbitrary grey-level. (**a**) Coarse grained and (**b**) fine-grained aggregate. A grain is defined as a polygon that has high-angle boundaries (angle of misorientation between neighbouring grains >10°), which are represented by light grey lines. Dark lines are low-angle (<10°) subgrain boundaries.

(3 kbar) corresponding to greenschist to lower amphibolite facies conditions were kept constant, so that the effect of a change in the value of the two parameters mentioned above could be investigated exclusively. In addition, fine and coarse-grained starting aggregates were used (Fig. 3). The fine-grained starting aggregate was chosen to illustrate fabric development that may occur at strains higher than 2. For simplicity, in the following sections we will denote the different τ_{nucl} values as high and low, and M_{gb} values as low, medium and high.

A variety of sequential images of the developing microstructures during progressive deformation is provided on http://www.uni-mainz.de/FB/Geo/Geologie/tecto/elle_movies. These sequences give a visual idea of the microstructural development during progressive deformation and the features described and discussed in the sections below.

Results

Simulations of either of the two initial microstructures show that similar characteristic features develop when deformed under the same conditions (Fig. 4). For different combinations of M_{gb} and τ_{nucl}, characteristic features are observed. For the sake of illustration and to provide a basis for comparison with other data, three main groups of microstructural development are distinguished. There are also transitions between these groups which will not be further considered in detail in this paper. The groups and associated parameter values used are provided in Table 3.

Microstructural observations and interpretation

In this section a summary of the microstructural changes with respect to their deformation conditions is given. With increasing grain boundary mobility, the grain size of recrystallized grains increases (Table 4, Fig. 4), mean grain size (Fig. 5) increases, and aspect ratios decrease (Fig. 6). The overall dislocation density decreases and the distribution of dislocations becomes more homogeneous with increasing grain boundary mobility (Figs 7 & 8). The decrease in dislocation density is further enhanced by lowering the energy threshold value for recrystallization by

Fig. 4. Selected simulated microfabrics where each grain is represented by an arbitrary grey level. High-angle (>10°) boundaries are light grey and low-angle (<10°) boundaries are dark grey lines. Width of view is 10 mm. (**a**) Microstructures of group A; symbols shown in the graph of the microstructure cg 2-1, $\gamma = 1$ (upper left corner of figure): thick dark line represents the outline of an original grain, thick white line represents the remaining porpyroclasts. In the area between the black and white line, recrystallized grains are seen. This resembles a mantled porphyroclast. Symbols shown in the graph of the microstructure: cg 1-1, $\gamma = 1$ (left side of figure): thick black line represents outline of an original grain, crosses (x) mark grains which develop within the porphyroclast.

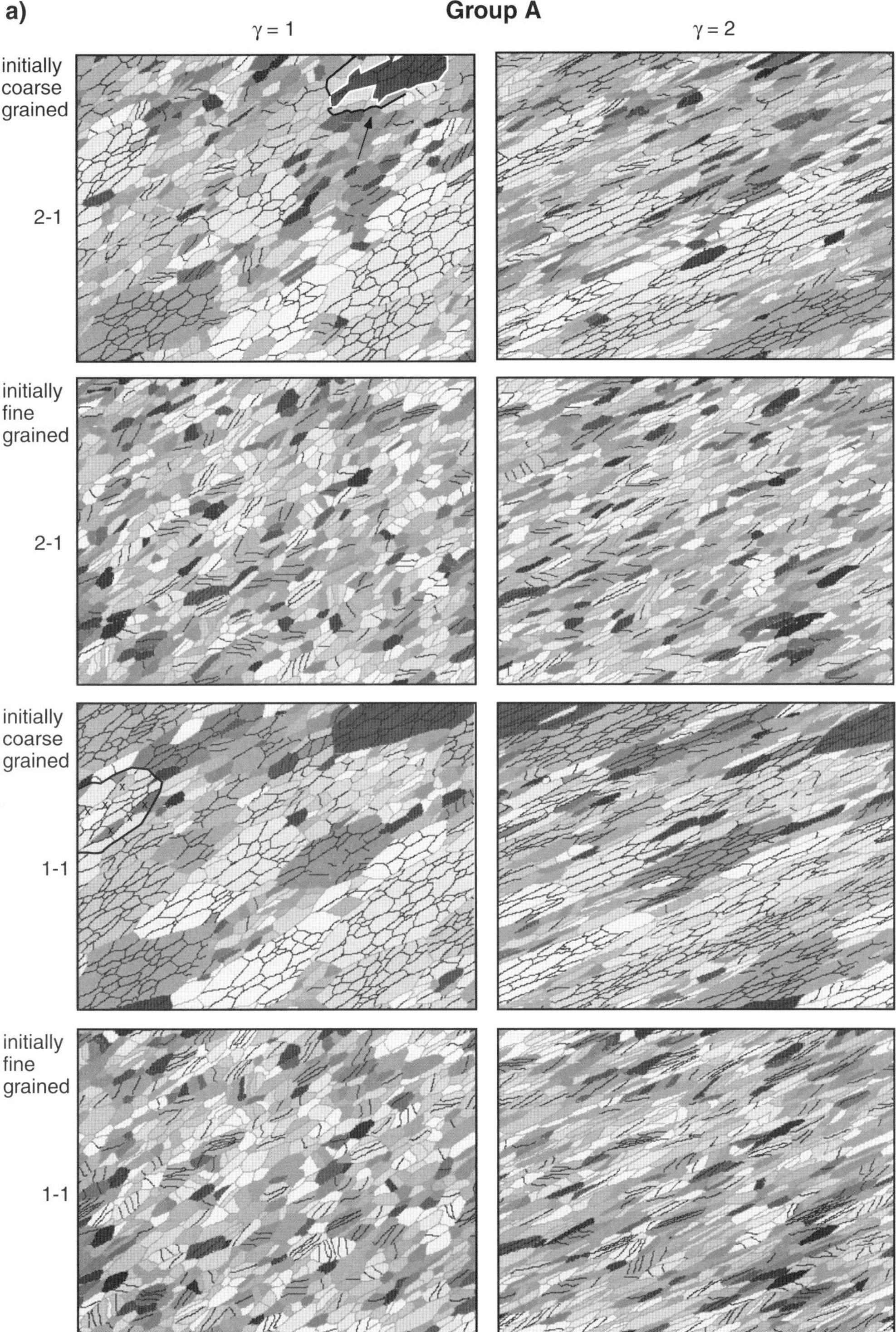
a)
Group A
γ = 1
γ = 2
initially coarse grained
2-1
initially fine grained
2-1
initially coarse grained
1-1
initially fine grained
1-1

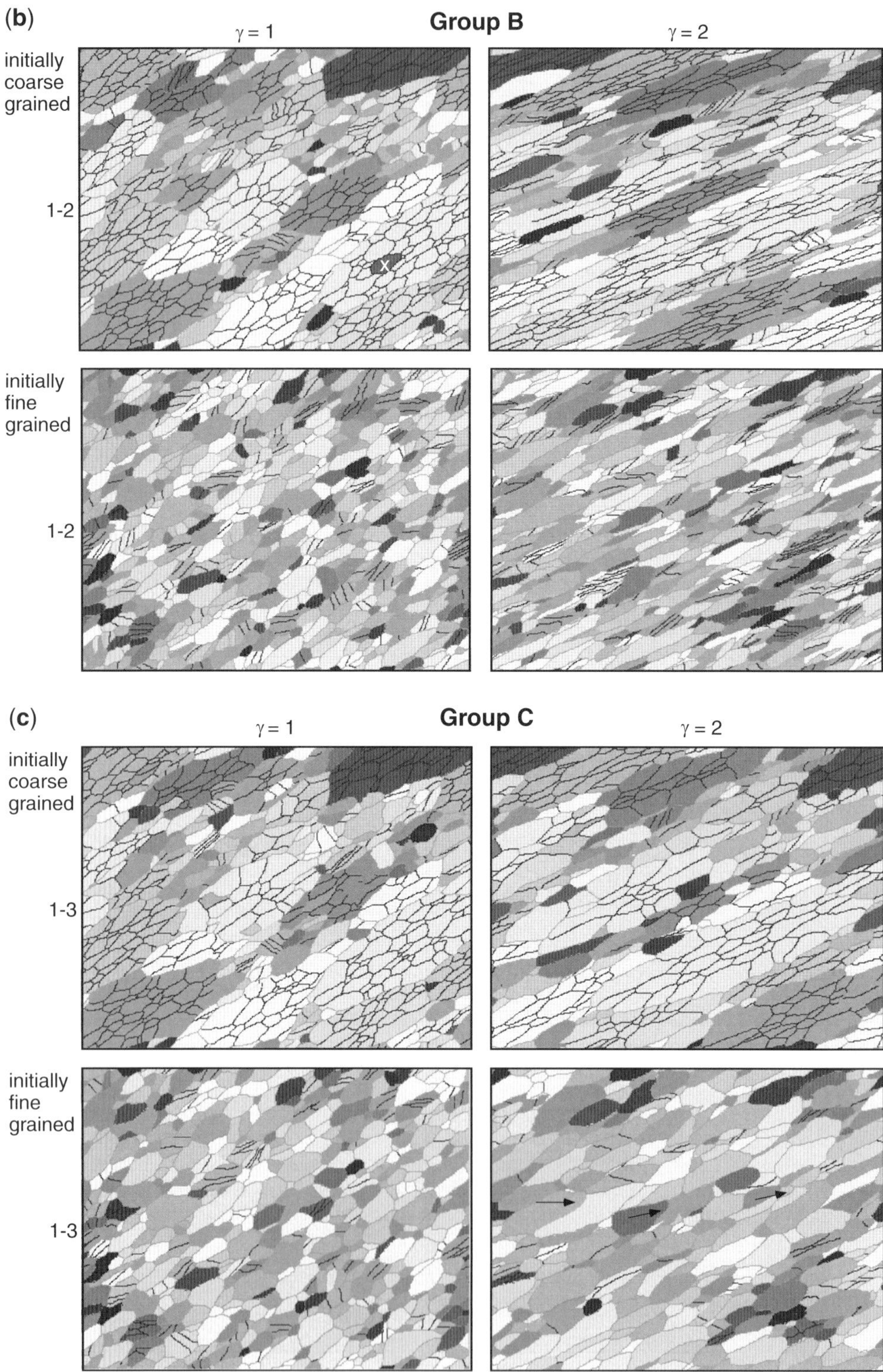
(b)
Group B
γ = 1
γ = 2
initially
coarse
grained
1-2
initially
fine
grained
1-2
(c)
Group C
γ = 1
γ = 2
initially
coarse
grained
1-3
initially
fine
grained
1-3

nucleation (τ_{nucl}). Another effect of the lowering of τ_{nucl} is a decrease in the ratio of the length of low and high-angle grain boundaries R_{gbl} (Fig. 9). A detailed description of the resultant microstructural groups is given below.

Group A: variable τ_{nucl}, *low* $\mathrm{M_{gb}}$

The microstructural group A develops at low M_{gb}, independently of the value of τ_{nucl}. Developing microstructures are characterized by straight grain boundaries, the development of core-and-mantle structures (cf. White 1976; Gifkins 1979; Passchier & Simpson 1986) and a fine recrystallized grain size (Table 4, Fig. 4a). Recrystallized grains are in the same size range as subgrains ($0.8 - 1 \times 10^{-2}\ \mathrm{mm}^2$) and predominantly develop at the rim of larger porphyroclasts, regardless of whether they formed by rotation or nucleation. In some cases, an array of recrystallized grains cross-cuts a porphyroclast (Fig. 4a). The preferred recrystallization at the rims of large grains can be attributed to a more pronounced build up of dislocations at the rim than in the centre of a grain. At a rim of a grain, stress perturbations occur due to different crystallographic orientation and viscosity of adjacent grains. This results in differences in the work necessary to deform a grain as simulated by the Taylor-Bishop-Hill code (see above) and therefore differences in the build up of dislocations. The total area of recrystallized grains is a function of τ_{nucl}; it is low (32% of total area) at high τ_{nucl} and high at low τ_{nucl} (60%). The mean aspect ratio of grains is highest in initially coarse-grained microstructures (Fig. 6).

During progressive deformation the grain size decreases significantly in the initially coarse-grained aggregates (Fig. 5) as recrystallization of subgrains to grains occurs during deformation. The mean grain size remains constant in the initially fine-grained aggregates as the size range of subgrains is similar to the original grain size and therefore no grain size reduction can occur during recrystallization by nucleation and rotation. Processes that would result in grain size increase, such as grain boundary migration, do not play a significant role, because M_{gb} is low. The ratio of the measured grain aspect ratio and the aspect ratio of the theoretical strain ellipse R_m/R_e show that the aspect ratios in experiments only slightly deviate from the theoretical value (Fig. 6). This can be attributed to the fact that most grains deform by viscous deformation and the effect of processes which tend to produce more equant grains such as grain boundary migration and recrystallization by rotation and nucleation are minor. R_{gbl} decreases with increasing strain (Fig. 7), due to the progressive rotation of subgrains and associated change of low-angle boundaries to high-angle boundaries. The average dislocation density is high and local densities are strongly variable (Fig. 8, Fig. 9). In simulations with low τ_{nucl} the total dislocation density is lower than in the ones with high τ_{nucl}. This shows that under the considered conditions, recrystallization by nucleation that results in sudden appearance of grains with low dislocation density strongly influences the distribution and total amount of dislocations. Dislocation densities of adjacent grains or subgrains vary significantly because the grain boundary migration rates are low. At low migration rates the migration of grain boundaries into high dislocation density regions is limited and thus differences in dislocation density of adjacent grains are not levelled out. The rates of recrystallization by rotation and nucleation are significantly higher than grain boundary migration rates.

Group B: high τ_{nucl}, *medium* $\mathrm{M_{gb}}$

Microstructures of group B develop at high τ_{nucl} and medium M_{gb}. Microstructures are characterized by slightly curved grain boundaries, some core-and-mantle structures and recrystallized grains which are notably larger than the initial subgrain size (Table 4, Fig. 4b). Recrystallized grains appear at the rim of large porphyroclasts but may also be seen nucleating within a grain (Fig. 4b). The remaining porphyroclasts are smaller than those seen in group A. The mean grain size in originally coarse-grained fabrics decreases significantly at low strain, but remains constant or even increases slightly at higher strain (Fig. 5). At first, recrystallization by nucleation and rotation significantly alter the microstructure and cause the development of a crystallographic preferred orientation. With an

Fig. 4. (**b**) Microstructures of group B; symbol shown in the graph of the microstructure: cg 1-2, $\gamma = 1$, (upper left corner of figure): white cross (x) marks a recrystallized grain within a large grain. (**c**) Microstructures of group C; symbol shown in the graph of the microstructure: fg 1-3, $\gamma = 2$ (lower right corner of figure): arrows point to bulging grain boundaries. See Tables 1 and 3 for abbreviations, settings and conditions of simulations for all subsequent figures.

Table 4. *Summary of conditions, features and interpretation of different microstructural groups*

Authors	Hirth & Tullis (1992)	Stipp *et al.* (2002)	This study
Grouping Conditions	*Regime 1* Low T/high strain rate	*I* 280–400 °C	*Group A* τ_{nucl}: 4.8×10^5 Jm^{-2}; M_{gb}: 1×10^{-12} m^2 s^{-1}J^{-1}; τ_{nucl}: 3.8×10^5 Jm^{-2}; M_{gb}: 1×10^{-12} m^2 s^{-1}J^{-1}
Microstructural features	Small recrystallized grains, some flattening of grains, patchy undulose extinction, generally high dislocation density, large variability of dislocation density, no subgrains	Recrystallized grain size: 5 µm (300 °C)–25 µm (400 °C), some flattening of porphyroclasts, increase of aspect ratio with T, patchy undulose extinction, less pronounced with increasing T, volume proportion of recrystallized grains increases with T, at high T: core-and-mantle structures	Small recrystallized grain size (same grain size range as subgrains), some flattening of porphyroclasts, recrystallized grains mainly at grain boundaries, well developed core- and mantle structures, generally high dislocation density, large variability of dislocation density
Main process	Grain boundary migration recrystallization	Bulging rerystallization (BLG)	Rotation recrystallization and recrystallization by nucleation strongly affect microstructure; no significant influence of grain boundary migration
Groupings Conditions	*Regime 2* Medium T/medium strain rate	*II* 400–500 °C	*Group B* τ_{nucl}: 4.8×10^5 Jm^{-2}; M_{gb}: 20×10^{-12} m^2 s^{-1}J^{-1}
Microstructural features	Strongly flattened grains, core-and-mantle structure, recrystallized grains at grain boundaries, generally medium dislocation density	Recrystallized grain size: 60 (400 °C) – 85 µm (500 °C), strongly flattened grains, core-and-mantle structure, volume proportion of recrystallized grains is up to 90%, subgrains progressively affect porphyroclasts, recrystallized grains show almost no intracrystalline deformation, average aspect ratios: 1.79–2.63 at 440–510 °C, at high temperatures broad range of grain sizes (85 + 60/ – 30 µm)	Medium recrystallized grain size (larger than subgrain grain size range), flattened grains/porphyroclasts, recrystallized grains often at grain boundaries, core-and-mantle structure, subgrains progressively affect porphyroclasts, medium mean aspect ratios, medium total dislocation density, medium variability in dislocation density between different grains
Main processes	Rotation recrystallization important	Subgrain rotation (SGR)	Rotation recrystallization and especially recrystallization by nucleation mainly effect the microstructure development; some effect of grain boundary migration
Grouping Conditions	*Regime 3* High T/low strain rate	*III* >500–550 °C	*Group C* τ_{nucl}: 4.8×10^5 Jm^{-2}; M_{gb}: 40×10^{-12} m^2 s^{-1}J^{-1}
Microstructural features	Relatively large recrystallized grains, larger than subgrains, no strongly flattened grains, bulging grain boundaries, low dislocation density and small variability of dislocation density	Grain size: 220 µm (550 °C) – few mm (700 °C), broad range of grain sizes, lobate grain boundaries, interfingering grains, typical grain boundary migration microstructures (cf. Jessell 1987), no strongly flattened grains	Large mean grain size, increasing with progressive deformation at high strains, large grain size of recrystallized grains, larger than subgrains, rare core-and-mantle structure, lobate grain boundaries, only few subgrains present, no strongly flattened grains, low aspect ratios, both for recrystallized and old grains, low total dislocation density and small variability in dislocation density
Main processes	Grain boundary migration important	Grain boundary migration (GBM)	Strong influence of grain boundary migration, some effect of recrystallization by nucleation and least influence of rotation recrystallization

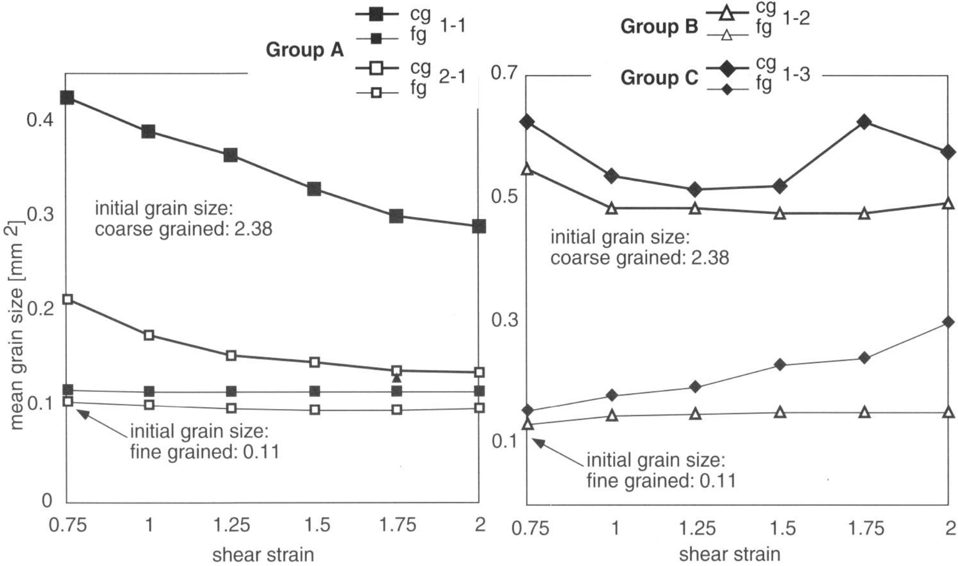

Fig. 5. Mean grain size as a function of shear strain, from $\gamma = 0.75$ to 2. The mean grain size of coarse-grained microstructures (cg) decreases rapidly with the onset of recrystallization by rotation and nucleation (finite strain: 0.20–0.5). In fine-grained microstructures (fg) mean grain sizes significantly increase in the strain interval 0–0.75 Group B only.

increasing number of grains in easy-slip orientations, the average dislocation density decreases and hence the driving force for further nucleation and rotational recrystallization decreases. Migrational recrystallization may eventually cause an increase in grain size. Microstructures that develop from fine grained fabrics show a very weak increase in grain size at high strains. In these fabrics, recrystallization by rotation and nucleation do not play a significant role during the first stage of the experiment. Mean aspect ratios are still high, but lower than those of group A (Fig. 6a), especially for the remaining porphyroclasts. This is more pronounced in group B than in group A (Fig. 6b). The grain shapes are modified not only by subgrain formation and subsequent rotation and nucleation by recrystallization but also to some extent by grain boundary migration. In initially coarse-grained textures, R_{gbl} decreases at low strains ($\gamma > 1.25$) but increases slightly at higher strain ($\gamma > 1.5$). This points to an initial stage with a high rate of recrystallization by rotation and nucleation that produces high-angle boundaries at the expense of low-angle boundaries, and at high strains, the formation of new subgrains which increases the relative number of low-angle boundaries. In the initially fine-grained aggregate, R_{gbl} remains low throughout the simulations. Only a few new subgrains form per time step and many grains are affected by nucleation recrystallization that results in the production of high-angle boundaries.

The average dislocation density is lower and less variable in group B than in group A (Figs 8 & 9). The lowering and smoothing out of dislocation density differences between grains can be attributed to two processes. First, recrystallization by nucleation significantly lowers the dislocation density locally within a grain or subgrain. Second, a medium M_{gb} results in a higher rate of grain boundary migration, causing a relative increase in low dislocation density grains at the expense of high dislocation density grains. The influence of subgrain formation and rotation recrystallization is less pronounced in group B than in group A. The low rate of subgrain formation is attributed to two main factors. (1) The grain size range of subgrains is in a similar range as the initial grain size. Accordingly, only a few grains develop to a size which allows them to split into subgrains larger than the set minimum subgrain size. (2) The overall low dislocation density which develops due to the factors mentioned above results in a low driving force for subgrain formation. The resultant low rate of subgrain formation has a direct effect on the rate of rotation recrystallization, as rotational recrystallization can only occur if subgrains are present.

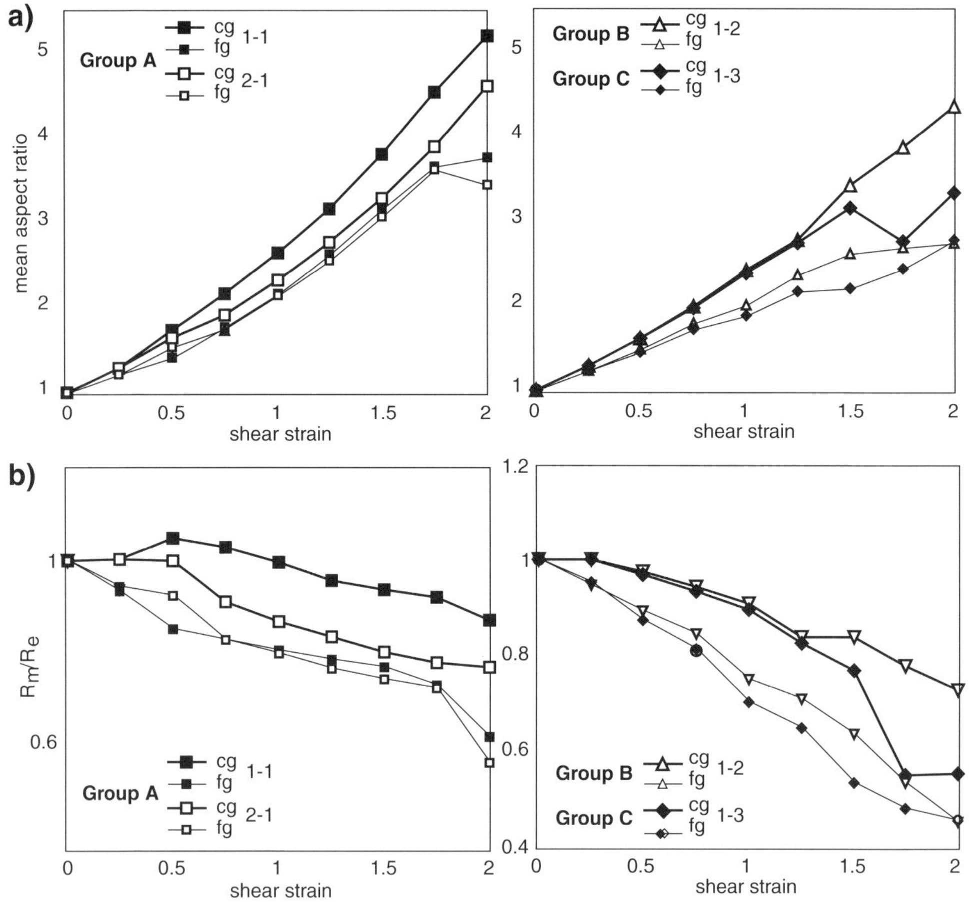

Fig. 6. (**a**) Graph showing mean grain aspect ratio (R_m) with increasing finite strain. R_m is determined using the method of Panozzo (1983, 1984). (**b**) Ratio of measured R_m and axial ratio of the finite strain ellipse R_e. If $R_m/R_e < 1$, processes which tend to modify the grain shapes to more equant shapes, such as grain boundary migration, must have affected the microstructure.

Group C: high τ_{nucl}*, high* M_{gb}

Group C microstructures develop at high τ_{nucl} and high M_{gb}. The microstructures are characterised by bulging, irregular grain boundaries, scarcity of subgrains and only rare core-and-mantle structures (Fig. 4c). The mean grain size increases with progressive deformation (Fig. 5) which reflects the high rates of grain boundary migration, as grain boundary migration results in the growth of large grains at the expense of small grains. In contrast to the microstructures of groups A and B, the change in grain size during progressive deformation is not always smooth but can rapidly change between different finite strain values. During progressive deformation a broad range of grain sizes is observed in the coarse-grained aggregate. There is a marked increase in grain size at $\gamma = 1.5$ (Fig. 5) as very little recrystallization by nucleation and rotation has occurred and the effect of grain boundary migration is, in comparison, more dominant than in other stages of the experiment. Recrystallized grains are significantly larger (1.4×10^{-2}–1.5×10^{-2} mm^2) than the subgrain size (Table 1). The high grain boundary migration rates significantly affects the grain shape. Grains become more equant because grain boundaries move in such a way that the total surface energy is reduced, hence grain boundaries are reduced in length and curvature. Therefore, aspect ratios are generally significantly lower than the strain ellipse (Fig. 6) and the remaining porphyroclasts are smaller and less flattened than those seen in group A and B.

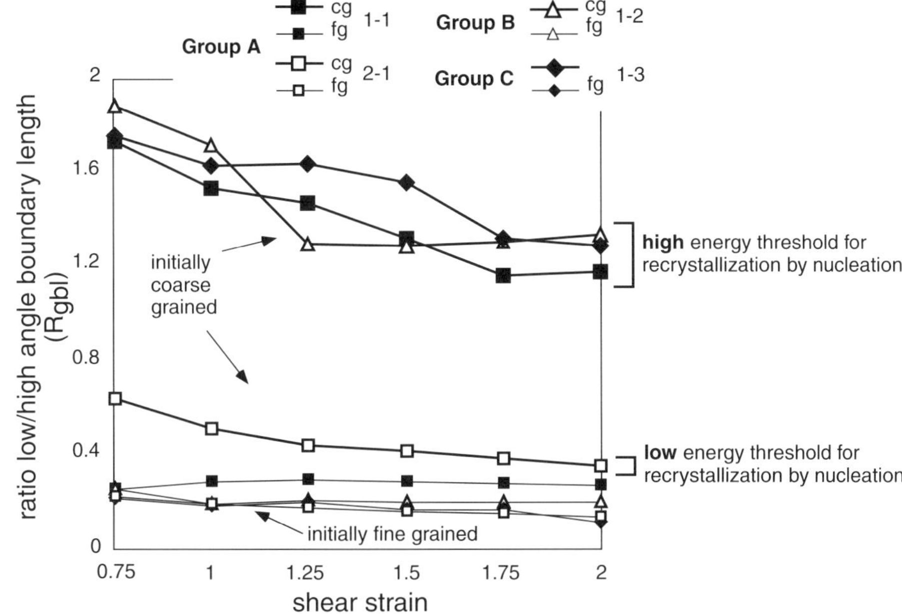

Fig. 7. Graph of the ratio of low to high misorientation angle boundary length R_{gbl} against shear strain, from $\gamma = 0.75$ to 2. In the strain interval 0.2–0.5, R_{gbl} in the coarse-grained microstructures decreases rapidly with the onset of recrystallization by rotation and nucleation. In the fine-grained microstructures simulated with high τ_{nucl}, R_{gbl} increases, but otherwise remains approximately constant.

R_{gbl} decreases with progressive deformation (Fig. 7). It is generally high in the initially coarse-grained aggregate and low in the fine-grained aggregate. These trends suggest a slow rate of subgrain formation, which is related to the observed low dislocation density values (Figs. 8 & 9). High grain boundary migration rates result in a noticable reduction of the dislocation density (Fig. 8) per time step, and therefore the effective rate of accumulation of internal strain energy is also reduced. This, in turn, reduces the driving force and hence rate of recrystallization by rotation and nucleation and subgrain formation. In addition, a high rate of grain boundary migration results in an effective levelling out of the variability in dislocation density between different grains (Fig. 9).

Discussion

Comparison with experimental data

Hirth & Tullis (1992) delineated three dislocation creep regimes from experimentally deformed quartz by optical and transmission electron microscopy and mechanical data. Each of these regimes is characterized by certain microstructures that are interpreted to originate from three distinct mechanisms which result in an overall decrease in dislocations (Table 4). Experiments at different strain rates, temperatures and with different initial fabrics (coarse and fine-grained) enabled them to construct a temperature, strain rate, and dislocation creep regime map. Our experiments are only to a certain extent comparable with these experiments as we did not model the low temperature grain boundary migration recrystallization regime (regime I) of Hirth & Tullis (1992). Nevertheless, comparison of their data and our results (Table 4) show that the transition from microstructural groups A and B to C is to a certain degree comparable to transition from regime II and III of Hirth & Tullis (1992) (Table 4). Hirth & Tullis (1992) used two different starting materials to overcome the problem of the experimental limitation on the achievable finite strain. Just as in our simulations, the fabric development is in some respects different in different starting materials. In simulations of initially coarse-grained microstructures, strong grain shape fabrics develop, whereas in initially fine-grained microstructures the shape fabric is weaker (Fig. 4). In our

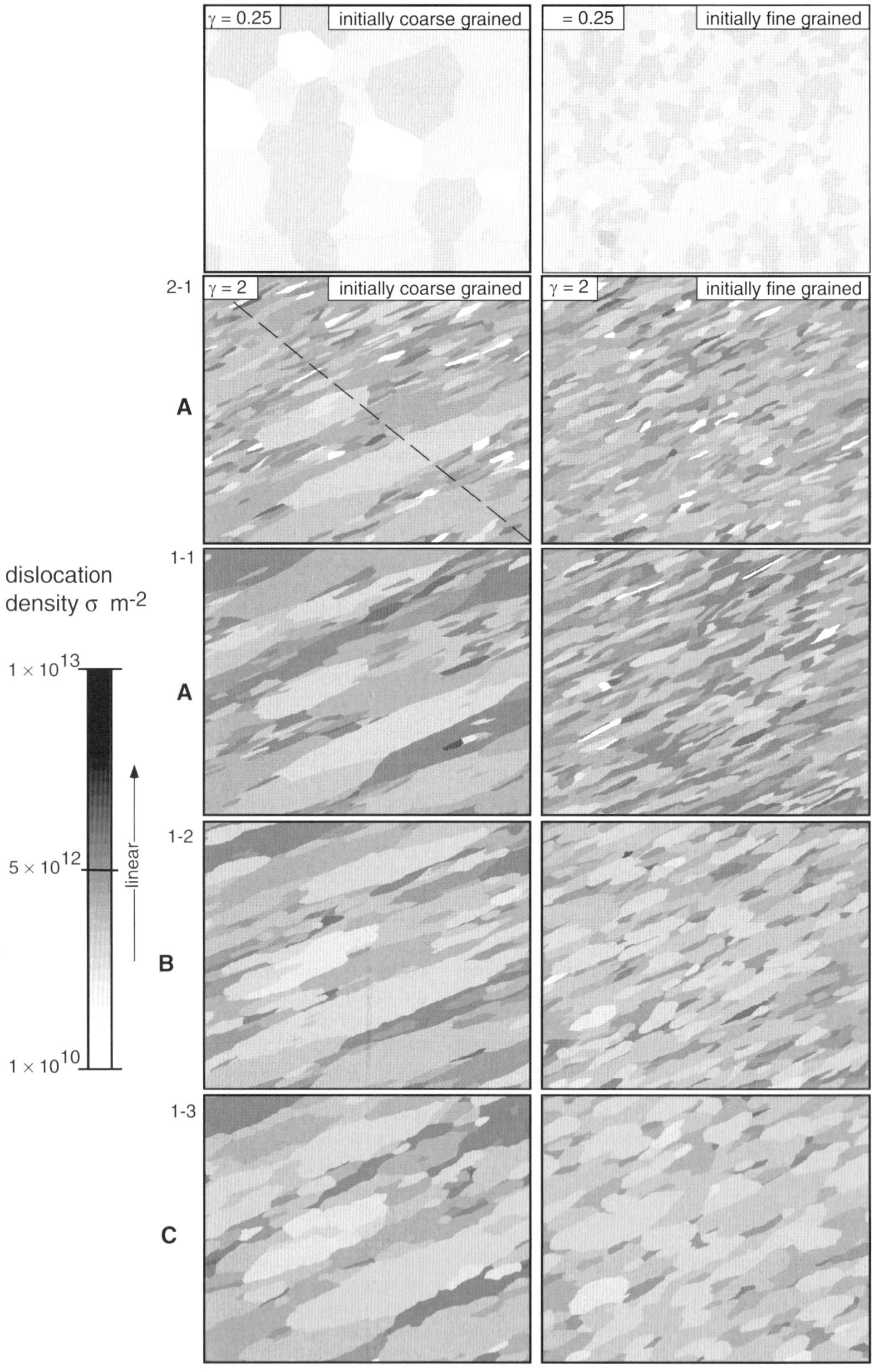
γ = 0.25
initially coarse grained
= 0.25
initially fine grained
2-1
γ = 2
initially coarse grained
γ = 2
initially fine grained
A
1-1
A
1-2
B
1-3
C
dislocation
density σ m-2
1 × 10^13
5 × 10^12
1 × 10^10
linear

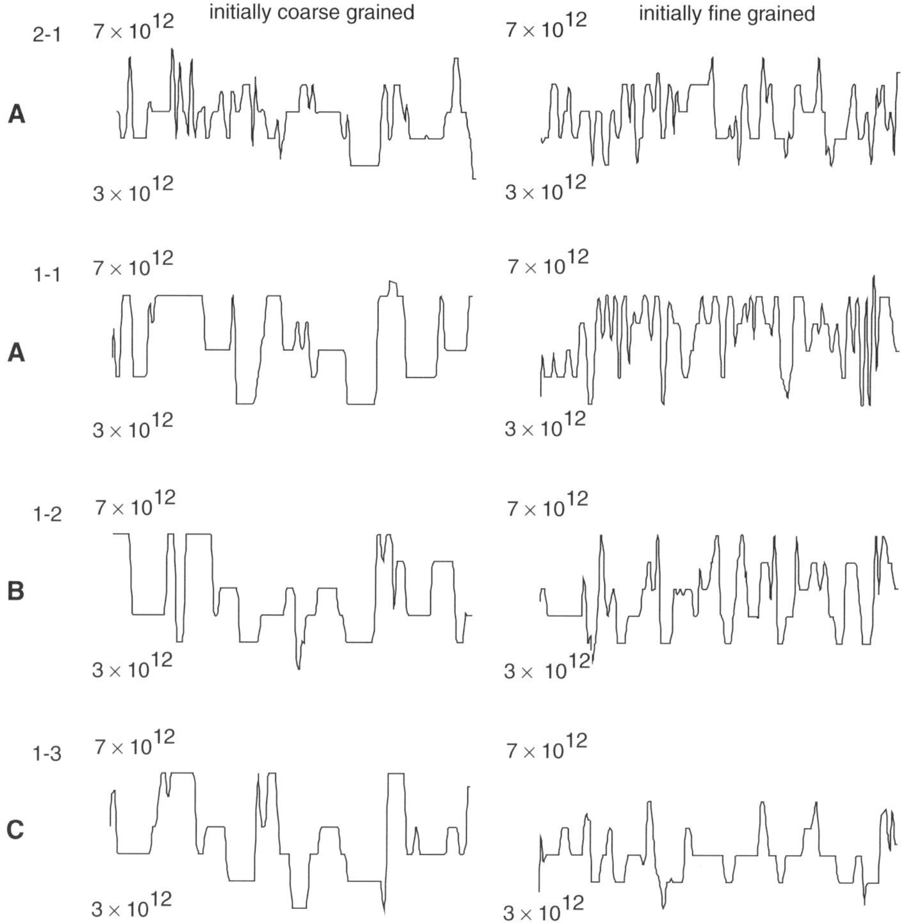

Fig. 9. Profiles showing the variability of dislocation density across a microstructure. The orientation of the line extends from the left hand upper corner to the right hand lower corner of the unit cell (see Fig. 8).

simulations, which go to significantly higher strains than are possible in experiments, a steady-state microstructure is also not achieved. One of the reasons may be the absence of grain boundary sliding in the simulation model, which might play a significant role in the development of steady-state microstructures (Means 1981; Herwegh, pers. comm.). An additional possibility is that higher finite strains are necessary.

Comparison with observations in natural examples

Stipp *et al.* (2002) reported different groups of microstructures that developed from initially coarse-grained, mylonitic quartz veins that were deformed in a temperature interval from about 280 °C to 700 °C. They distinguished three different main zones characterized by different microstructures, which they interpreted to have

---

Fig. 8. Distribution of dislocation density in simulations during progressive deformation. The grey-scale bar on the side depicts absolute values of dislocation density. Initially, all grains have a low dislocation density of $1 \times 10^{10}\ m^{-2}$. During progressive deformation, dislocations accumulate within grains. In most cases, the highest total dislocation density values are seen at finite strains of 0.3–0.7. The dashed line shown in the graph of the microstructure cg 2-1, $\gamma = 1$ represents the profile along which the dislocation density was measured (cf. Fig. 9).

developed due to different dynamic recrystallization mechanisms (Table 4). Comparison of their observations and our results shows some general correspondence of changes of simulated microstructures developed at different grain boundary mobilities with microstructural changes observed in rocks (Table 4).

Influence of grain boundary mobility values and their relationship to the presence of fluid and temperature

The effective mobility of a boundary is largely a function of temperature and the presence of fluid, where the mobility is high at high temperatures and in water/fluid-rich conditions (Frost & Ashby 1983). Accordingly, microstructures developing in simulations with a set high grain boundary mobility are expected to correspond to microstructures seen in rocks undergoing deformation at higher temperature and/or water-rich conditions. Comparison of microstructures developing at high temperatures is limited, as in simulations the rate of recovery and relationship of viscosity and dislocation density are kept constant, although these are known to be to some extent temperature dependent (e.g. Kocks 1985). Despite this drawback, simulations show the same general trends as are seen at high temperature deformation of quartz (Table 4, e.g. Mancktelow 1990; Hirth & Tullis 1992; Stipp *et al.* 2002).

Simulations can be directly compared to rock deformation experiments which study the effect

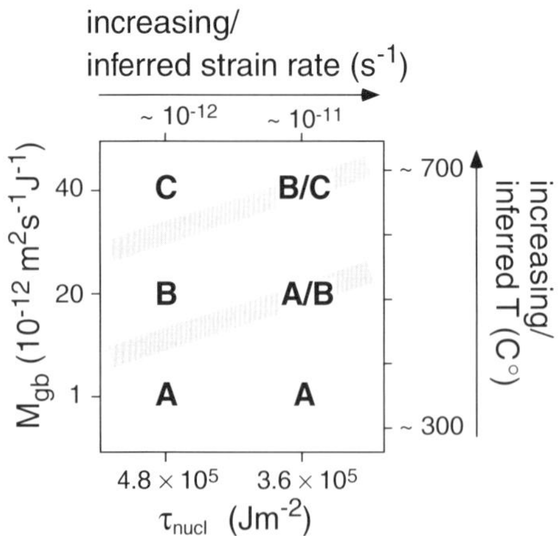

Fig. 10. Deformation microstructure maps. (**a**) Microstructural grouping derived from simulations as a function of τ_{nucl} and M_{gb}. These groups can be linked to strain rate and temperature, respectively.

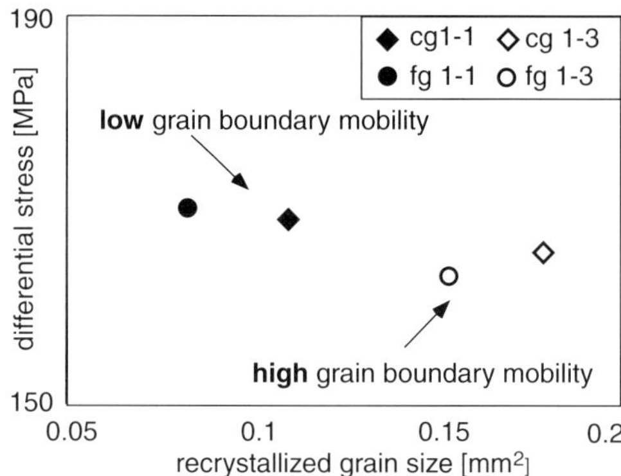

Fig. 11. Graph showing differential stress versus recrystallized grain size for simulations with different grain boundary mobilities. In experiments with low grain boundary mobility, grain sizes are significantly lower than in those with high grain boundary mobility.

of increased mobility due to water-rich conditions in a deforming rock. Jung & Karato (2001) experimentally deformed olivine aggregates at water-poor and water-rich conditions and showed that the size of recrystallized grains is strongly dependent on the water fugacity. This dependence is also closely linked to the dependence of recrystallized grain size on differential stress. Results from simulations with different grain boundary mobility values show very similar features (Fig. 11). In addition, the recrystallized grain size in simulations with initially coarse-grained microstructures is larger than the recrystallized grain size in the fine-grained counterpart. This relationship may also need to be considered when using recrystallized grain size for differential stress estimates.

Implications

Extrapolation of simulated conditions and simulated microstructures to natural conditions

As a summary of our results, the relationship between the different grain boundary mobility values and energy threshold for nucleation values τ_{nucl} and the microstructural groups is shown in Fig. 10. As stated above, with some limitations, an increase in temperature can be correlated to an increase in grain boundary mobility M_{gb}. Therefore, we have included in the graph a tentative relationship to increasing temperature. In addition, at a first approximation a decrease in the energy threshold value for nucleation τ_{nucl} can be correlated with an increase in strain rate. This assumption is based on the following reasoning. With increasing strain rate the rate of accumulation of strain energies within grains increases (Argon 1970). Such an increase in the built up of internal strain energy results in a higher rate of recrystallization by nucleation since an increased number of grains exhibit high dislocation densities and can therefore nucleate. The variation of τ_{nucl} has, to a limited extent, a similar effect on the rate of recrystallization by nucleation. At low τ_{nucl} the nucleation rate is significantly higher than at high τ_{nucl}; hence the general effect on microstructural development of low τ_{nucl} may correspond to that of high strain rates and the effect of high τ_{nucl} to that of low strain rates. The extrapolation of the $M_{gb} - \tau_{nucl}-$ microstructure graph to a tentative temperature–strain rate–microstructure graph (Fig. 10) gives a general idea of the link between results from numerical modelling and natural deformation. Further refinement of the current model is necessary to develop a more reliable temperature–strain rate–microstructure graph.

Dynamic recrystallization microstructures: result of relative rates of several concurrent processes

Our results show that a change in the parameter values, such as grain boundary mobility and/or the energy threshold for recrystallization by nucleation, is sufficient to produce microstructures that are commonly interpreted to be the result of the strong dominance of one single process. In our experiments, the same processes operate in each of the simulations. The important difference between the different simulations is that the relative rates of the simulated processes are varied. Although all processes are active and have some effect on the microstructure at all times, the rate at which one process affects the microstructure may be significantly higher than that of the other processes. Consequently, the developing microstructure predominantly shows the effect of the process with the highest rate. Accordingly, changes in parameters that are related to different conditions, with respect to e.g. temperature, strain rate and fluid activity, result in differences in developing microstructures caused by the different relative rates of concurrent processes.

Use of dynamic recrystallization grain size as an indicator for stress

Recrystallized grain size has been used as a measure of stress (e.g. Twiss 1977; Christie & Ord 1980; Kronenberg & Tullis 1984). In these studies, it is assumed that the recrystallized grain size is exclusively related to the imposed stress. However, White (1979) and Urai *et al.* (1986) pointed out that the use of recrystallized grain size as a measure of flow stress can be complicated. The size of recrystallized grains produced by different mechanisms, e.g. rotational recrystallization and grain boundary migration, may be different at the same flow stress (e.g. Poirier & Nicolas 1975; Guillopé & Poirier 1979). A temperature and water presence dependence of recrystallizated grain sizes has been experimentally shown by de Bresser *et al.* (1998) and Jung & Karato (2001), respectively. Furthermore, Hirth & Tullis (1992) pointed out that it is necessary to know in which dislocation creep system the sample was deformed if the dynamically recrystallized grain size is to be used for piezometry. Their interpretation is supported by our simulations, in which the relative rates of active processes directly affect the recrystallized grain size. These relative rates are directly linked to grain boundary mobility values, which are temperature and fluid presence dependent. Therefore, to derive flow stress values from recrystallized grain size, the conditions of deformation in terms of temperature and fluid presence have to be taken into account.

Comparison of the grain size development of initially coarse-grained and fine-grained aggregates shows that in experiments the grain size stabilizes at higher grain sizes in initially coarse-grained aggregates than in initially fine-grained aggregates. This suggests that the recrystallized grain size also depends on the initial grain size. This would put a further limitation to the use of grain size as a palaeopiezometer. Further investigations are needed to verify this feature.

Conclusions

Numerical simulations of microstructures that undergo progressive dynamic recrystallization show that different values of grain boundary mobility and energy threshold for recrystallization by nucleation result in microstructures that resemble those observed in natural examples and in experimentally deformed samples. A microstructure develops due to the effect of a number of concurrent processes. There is no switch between different processes at different conditions, but there is a change in relative rates at which the different processes affect the resulting microstructure.

Results show a significant dependence of the recrystallized grain size on different relative rates of the effect of different processes i.e. grain boundary migration versus subgrain formation and recrystallization by nucleation and rotation. These relative rates are closely related to temperature and water presence or absence. Therefore, this dependence has to be taken into account if the recrystallized grain size is used as a palaeopiezometer.

Microstructural groups defined by typical microstructural features are characteristic for a specific combination of parameters that control the rates of the dominant processes. These parameters are themselves dependent on conditions such as temperature, water presence or absence and strain rate. Combining our simulations, experimental data and observations in natural examples, the relationship of microstructures and conditions during their formation can be summarised in a grain boundary mobility–energy threshold for recrystallization by nucleation and a tentative temperature–strain rate–characteristic microstructure map. Such maps may help the field geologist to interpret observed microstructures in terms of the interaction and dominance of different processes and, with this understanding, help to derive conditions of deformation.

S. Piazolo wishes to thank the members of the Earth Science Department at Monash University, Australia for making facilities available used for this study. This work was funded by the Deutsche Forschungsgemeinschaft (GRK 392/1) and the Deutsches Akademisches Auslandsamt Deutschland. Constructive reviews by M. Drury and J. L. Raphanel, helpful editorial comments by G. Pennock and discussion and comments on an earlier version of the manuscript by J. Tullis and M. Stipp helped to improve the manuscript considerably and are gratefully acknowledged. M. Jessell would like to thank the LGIT, Grenoble for support during part of this work.

References

Argon, A. S. 1970. Internal stresses arising from the interaction of mobile dislocations. *Scripta Metallurgica*, **4**, 1001–1004.

Barber, D. J. 1985. Dislocations and Microstructures. *In:* Wenk, H.-R. (ed.) *Preferred Orientation in Deformed Metals and Rocks: An Introduction to Modern Texture Analysis.* Academic Press, London, 149–182.

Barr, T. D. & Houseman, G. A. 1992. Distribution of deformation around a fault in a non-linear dutile medium. *Geophysical Research Let*ters, **19**, 1145–1148.

Barr, T. D. & Houseman, G. A. 1996. Deformation fields around a fault embedded in a non-linear ductile medium. *Geophysical Journal International*, **125**, 473–490.

Barr, T. D. & Houseman, G. A. 1999. Basil – Finite element program. World Wide Web Address: http://www. earth. monash. edu. au/Research/Basil

Beeman, M. I. & Kohlstedt, D. I. 1988. Dislocation density: stress relationships in natural and synthetic sodium chloride. *Tectonophysics*, **149**, 147–161

Bishop, J. F. W. & Hill, R. 1951*a*. A theory of plastic distortion of a polycrystalline aggregated under combined stresses. *Philosophical Magazine*, **42**, 414–427.

Bishop, J. F. W. & Hill, R. 1951*b*. A theoretical derivation of the plastic properties of a polycrystalline face-centered metal. *Philosophical Magazine*, **42**, 1298–1307.

Carter, N. L. & Tsenn, M. C. 1987. Flow properties of continental lithosphere. *Tectonophysics*, **136**, 27–63.

Christie, J. M. & Ord, A. 1980. Flow stress from microstructures of mylonites: example and current assessment. *Journal of Geophysical Research*, **85**, 6253–6262.

Clark, M. K. & Royden, L. H. 2000. Topographic ooze: building the eastern margin of Tibet by lower crustal flow. *Geology*, **28**, 703–706.

de Bresser, J. H. P., Peach, C. J., Reijs, J. P. J. & Spiers, C. J. 1998. On dynamic recrystallization during solid state flow: effects of stress and temperature. *Geophysical Research Letters*, **25**, 3457–3460.

Doherty, R. D., Hughes, D. A. *et al.* 1997. Current issues in recrystallization: a review. *Materials Science and Engineering*, **A238**, 219–274.

Drury, M. R. & Urai, J. L. 1990. Deformation-related recrystallization processes. *Tectonophysics*, **172**, 235–253.

Durham, W. B., Goetze, C. & Blanke B. 1977. Plastic flow of oriented single crystals of olivine: 2. Observations and interpretations of the dislocation structures. *Journal of Geophysical Research*, **82**, 5755–5770.

Evans, L., Jessell, M. W., Bons, P. D., Barr, T. D., Stüwe, K. & Piazolo, S. 1999. *Elle – Microprocess based simulation of metamorphic and deformation texture development.* World Wide Web Address: http://www.earth.monash.edu.au/Research/Elle

Evans, L., Jessell, M. W. & Piazolo, S. 2000. *Elle Manual.* World Wide Web Address: http://orion.earth.monash.edu.au/Research/Elle/Monash/index.html

FitzGerald, J. D., Etheridge, M. A. & Vernon, R. H. 1993. Dynamic recrystallization in a naturally recrystallized albite. *Textures and Microstructures*, **5**, 219–237.

FROST, H. J. & ASHBY, M. F. 1983. *Deformation – Mechanism Maps: The Plasticity and Creep of Metals and Ceramics*. Pergamon Press, Oxford.

GIFKINS, R. C. 1976. Grain boundary sliding and its accommodation during creep and superplasticity. *Metallurgical Transactions*, **7A**, 1225–1232.

GOTTSTEIN, G. & MECKING, H. 1985. Recrystallization. *In*: Wenk, H.-R. (ed) *Preferred Orientation in Deformed Metals and Rocks: An Introduction to Modern Texture Analysis*. Academic Press, London, 183–214.

GOTTSTEIN, G. & SHVINDLERMAN, L. S. 1999. *Grain Boundary Migration in Metals: Thermodynamics, Kinetics, Applications*. CRC Press, LLC, Boca Roca.

GUILLOPÉ, M. & POIRIER, J. P. 1979. Dynamic recrystallization during creep of single crystalline halite: an experimental study. *Journal of Geophysical Research*, **4**, 5557–5567.

HACKER, B. R. & KIRBY, S. H. 1993. High-pressure deformation of calcite marble and its transformation to aragonite under non-hydrostatic conditions. *Journal of Structural Geology*, **15**, 1207–1222.

HIRTH, G. & TULLIS J. 1992. Dislocation creep regimes in quartz aggregates. *Journal of Structural Geology*, **14**, 145–159.

HOBBS, B. E. 1985. The geological significance of microfabric analyses. *In*: H.-R. Wenk (ed) *Preferred Orientation in Deformed Metals and Rocks: An Introduction to Modern Texture Analysis*. Academic Press, Orlando, 463–484.

HUMPHREYS, F. J. & HATHERLY, M. 1995. *Recrystallization and related annealing phenomena*. Pergamon Press, Oxford.

JESSELL, M. W. 1987. Grain-boundary migration microstructures in a naturally deformed quartzite. *Journal of Structural Geology*, **9**, 1007–1014.

JESSELL, M. W. & LISTER, G. S. 1990. A simulation of the temperature dependence of quartz fabrics. *In*: Knipe, R. J. & Rutter, E. H. (eds) *Deformation Mechanisms, Rheology and Tectonics*. Geological Society, London, Special Publications, **54**, 353–362.

JESSELL, M. W., BONS P. D., EVANS L., BARR T. D. & STÜWE K. 2001. Elle: the numerical simulation of metamorphic and deformation microstructures. *Computers and Geosciences*, **27**, 17–30.

JUNG, H. & KARATO, S.-I. 2001. Effects of water on dynamically recrystallized grain-size of olivine. *Journal of Structural Geology*, **23**, 1337–1344.

KIRBY, S. H. 1983. Rheology of the lithosphere. *Reviews in Geophysics and Space Physics*, **21**, 1458–1487.

KIRBY, S. H. & KRONENBERG, A. K. 1987. Rheology of the lithosphere: selected topics. *Reviews in Geophysics*, **25**, 1219–1244.

KNIPE, R. J. & WHITE, S. H. 1979. Deformation in low grade shear zones in the old red sandstone, S.W. Wales. *Journal of Structural Geology*, **1**, 53–66.

KOCKS, U. F. 1985. Dislocation interactions: flow stress and strain hardening. *In*: *Proceedings of the Conference to Celebrate the Fiftieth Anniversary of the Concept of Dislocation in Crystals: Dislocations and Properties of Real Materials*. The Institute of Metals, London, 125–143.

KOCKS, U. F. 1976. Laws for work-hardening and low-temperature creep. *Journal Engineering, Materials and Technology*, **98**, 76–85.

KOHLSTEDT, D. L. & WEATHERS, M. S. 1980. Deformation induced microstructures, paleo-piezometers and differential stresses in deeply eroded fault zones. *Journal of Geophysical Research*, **85**, 6269–6285.

KRONENBERG, A. K. & TULLIS, J. 1984. Flow strengths of quartz aggregates: grain size and pressure effects due to hydrolic weakening. *Journal of Geophysical Research*, **89**, 4281–4297.

LISTER, G. S. & PATERSON, M. S. 1979. The simulation of fabric development during plastic deformation and its application to quartzite: fabric transitions. *Journal of Structural Geology*, **1**, 99–115.

LISTER, G. S., PATERSON, M. S. & HOBBS, B. E. 1978. The simulation of fabric development in plastic deformation and its application to quartzite, the model. *Tectonophysics*, **45**, 107–158.

LLOYD, G. E., LAW, R. D., MAINPRICE, D. & WHEELER, J. 1992. Microstructural and crystal fabric evolution during shear zone formation. *Journal of Structural Geology*, **14**, 1079–1100.

MANCKTELOW, N. S. 1990. The Simplon Fault Zone. *Beiträge zur Geologischen Karte der Schweiz*, **163**.

MEANS, W. D. 1981. The concept of steady-state foliation. *Tectonophysics*, **78**, 179–199.

MECKING, H. & KOCKS, U. F. 1981. Kinetics of flow and strain-hardening. *Acta Metallurgica*, **29**, 1865–1875.

PANOZZO, R. 1983. Two dimensional analysis of shape-fabric using projections of digitized lines in a plane. *Tectonophysics*, **95**, 279–294.

PANOZZO, R. 1984. Two-dimensional strain from the orientation of lines in a plane. *Journal of Structural Geology*, **6**, 215–221.

PASSCHIER, C. W. & SIMPSON, C. 1986. Porphyroclast systems as kinematic indicators. *Journal of Structural Geology*, **8**, 831–844.

PFIFFNER, O. A. & RAMSAY, J. G. 1982. Constraints on geological strain rates: arguments from finite strain states of naturally deformed rocks. *Journal of Geophysical Research*, **87**, 311–321.

PIAZOLO, S. 2001. Shape Fabric Development During Progressive Deformation. Unpublished PhD thesis, University of Mainz, Germany. World Wide Web Address: http://ArchiMeD.uni-mainz.de/pub/2001/0032/diss. pdf

POIRIER, J.-P. & NICOLAS A. 1975. Deformation-induced recrystallization due to progressive misorientation of subgrains with special reference to mantle peridotites. Journal of Geology, **83**, 707–720.

PRIOR, D. J., KNIPE, R. J. & HANDY, M. R. 1990. Estimates of the rates of microstructural changes in mylonites. *In*: Knipe, R. J. & Rutter, E. H. (eds) *Deformation Mechanisms, Rheology and Tectonics*. Geological Society, London, Special Publications, **54**, 309–319.

RANDLE, V. 1992. *Microtexture Determination and its Applications*. The Institute of Materials, London.

SHEWCHUCK, J. R. 1999. Triangle: A Two-Dimensional Quality Mesh Generator and Delaunancy Triangulator. World Wide Web Address: http://www.cs.cmu.edu/%7Equake/triangle.html

SHIGEMATSU, N. 1999. Dynamic recrystallization in deformed plagioclase during progressive shear deformation. *Tectonophysics*, **305**, 437–452.

SOLAS, D. E., TOMÉ, C. N., ENGLER, O. & WENK, H. R. 2001. Deformation and recrystallization of hexagonal metals: modeling and experimental results for zinc. *Acta Materialia*, **49**, 3791–3801.

STIPP, M., STUENITZ, H., HEILBRONNER, R. & SCHMID, S. M. (this volume). Quartz: Correlation between natural and experimental conditions. *In*: DE MEER, S., DRURY, M. R., DE BRESSER, J. H. P. & PENNOCK, G. M. (eds) *Deformation Mechanisms, Rheology and Tectonics: Current Status and Future Perspectives*, Geological Society, London, Special Publications, **200**, 171–189.

STÖCKHERT, B. & DUYSTER, J. 2000. Discontinuous grain growth in recrystallised vein quartz – implications for grain boundary structure, grain boundary mobility, crystallographic preferred orientation, and stress history. *Journal of Structural Geology*, **21**, 1477–1490.

STÖCKHERT, B., BRIX, M. R., KLEINSCHRODT, R., HURFORD, A, J. & WIRTH, R. 1999. Thermochronometry and microstructures of quartz – a comparison with experimental flow laws and prediction on the temperature of the brittle–plastic transition. *Journal of Structural Geology*, **21**, 351–369.

TAKESHITA, T, WENK, H.-R. & LEBENSOHN, R. 1999. Development of preferred orientation and microstructure in sheared quartzite: comparison of natural data and simulated results. *Tectonophysics*, **312**, 133–155.

TAYLOR, G. I. 1938. Plastic strain in metals. *Journal of the Institute of Metals*, **62**, 307–324.

TRIMBY, P. W., PRIOR, D. J. & WHEELER, J. 1998. Grain boundary hierarchy development in a quartz mylonite. *Journal of Structural Geology*, **20**, 913–935.

TWISS, R. J. 1977. Theory and applicability of recrystallized grain size paleopiezometer. *Pure and Applied Geophysics*, **115**, 199–226.

URAI, J. L., MEANS, W. D. & LISTER, G. S. 1986. Dynamic recrystallization of minerals. *American Geophysical Union Geophysical Monograph*, **36**, 161–199.

WHITE, S. H. 1976. The effects of strain on the microstructures, fabrics and deformation mechanisms in quartz. *Philosophical Transactions Royal. Society of London*, **A283**, 69–86.

WHITE, S. H. 1979. Grain and sub-grain size variations across a mylonite zone. *Contributions to Mineralogy and Petrology*, **70**, 193–202.

Dynamic recrystallization of quartz: correlation between natural and experimental conditions

MICHAEL STIPP, HOLGER STÜNITZ, RENÉE HEILBRONNER & STEFAN M. SCHMID

Department of Earth Sciences, Basel University, Bernoullistrasse 32, 4056 Basel, Switzerland (e-mail: Michael.Stipp@unibas.ch)

Abstract: Quartz veins in the Eastern Tonale mylonite zone (Italian Alps) were deformed in strike-slip shear. Due to the synkinematic emplacement of the Adamello Pluton, a temperature gradient between 280 °C and 700 °C was effected across this fault zone. The resulting dynamic recrystallization microstructures are characteristic of bulging recrystallization, subgrain rotation recrystallization and grain boundary migration recrystallization. The transitions in recrystallization mechanisms are marked by discrete changes of grain size dependence on temperature. Differential stresses are calculated from the recrystallized grain size data using paleopiezometric relationships. Deformation temperatures are obtained from metamorphic reactions in the deformed host rock. Flow stresses and deformation temperatures are used to determine the strain rate of the Tonale mylonites through integration with several published flow laws yielding an average rate of approximately $10^{-14}\,s^{-1}$ to $10^{-12}\,s^{-1}$. The deformation conditions of the natural fault rocks are compared and correlated with three experimental dislocation creep regimes of quartz of Hirth & Tullis. Linking the microstructures of the naturally and experimentally deformed quartz rocks, a recrystallization mechanism map is presented. This map permits the derivation of temperature and strain rate for mylonitic fault rocks once the recrystallization mechanism is known.

Introduction

Dynamic recrystallization is in many materials primarily the result of two processes: (1) the formation and progressive rotation of subgrains; and (2) grain boundary migration (e.g. Guillopé & Poirier 1979; Urai *et al.* 1986). The interaction of these two processes is responsible for the occurrence of three different mechanisms of dynamic recrystallization forming characteristic microstructures. These are termed bulging recrystallization (BLG; e.g. Bailey & Hirsch 1962; Drury *et al.* 1985), subgrain rotation recrystallization (SGR; e.g. Hobbs 1968; White 1973; Guillopé & Poirier 1979) and grain boundary migration recrystallization (GBM; e.g. Guillopé & Poirier 1979; Means 1983; Urai *et al.* 1986; Jessell 1987). We observe similar recrystallization microstructures in quartz as have been described for other materials; thus we infer that the recrystallization mechanisms vary with temperature in the same way as they do in those materials (cf. Stipp *et al.* 2002). During BLG at low temperature conditions, local grain boundary migration is the dominant process. The contribution of subgrain rotation increases with temperature, until subgrain rotation dominates the recrystallization, and the resulting microstructures are those of SGR. At still higher temperatures above the range of SGR, the dominant recrystallization mechanism is GBM (cf. Guillopé & Poirier 1979; Urai *et al.* 1986). The dominant recrystallization process is again grain boundary migration, but boundaries sweep entire grains, so that in many cases there is no reduction in grain size.

In an experimental study, Hirth & Tullis (1992) have identified three dislocation creep regimes for quartz. The definition of the dislocation creep regimes is based on mechanical data and on recrystallization mechanisms identified by TEM and light microscope observations. Hirth & Tullis (1992) have demonstrated that the dominant recrystallization mechanism is controlled by temperature, stress and strain rate. The delineation of the different microstructural regimes reflects the relative rates of grain boundary migration, dislocation climb and dislocation production. In regime 1, recovery and recrystallization are mainly accommodated by strain-induced grain boundary migration, in regime 2 by climb-controlled dislocation creep (subgrain rotation) and in regime 3 by both grain boundary migration and subgrain rotation (Hirth & Tullis 1992). In a number of studies, the experimentally produced microstructures have been correlated with dynamic recrystallization microstructures of naturally deformed rocks (Dunlap *et al.* 1997; Snoke *et al.* 1998; Stöckhert *et al.* 1999; Hirth *et al.* 2001; Zulauf 2001). However, due to the given geological settings and outcrop conditions, none of these studies included

From: DE MEER, S., DRURY, M. R., DE BRESSER, J. H. P. & PENNOCK, G. M. (eds) 2002. *Deformation Mechanisms, Rheology and Tectonics: Current Status and Future Perspectives*. Geological Society, London, Special Publications, **200**, 171–190. 0305-8719/02/$15

the full range of dynamic recrystallization microstructures of quartz occurring in natural fault rocks. Some microstructural transitions were missing and the field samples were derived from different shear zones, so that the deformation conditions may have varied significantly.

The Eastern Tonale fault zone exposes a set of outcrops where the whole range of natural dynamic recrystallization microstructures of quartz can be observed within a single shear zone (Stipp *et al.* 2002). Deformation temperatures ranging from approximately 280 °C to 700 °C were derived from synkinematic mineral assemblages. The recrystallized grain sizes were measured in samples of different deformation temperatures. In this contribution, differential stresses have been inferred from paleopiezometry, and strain rates have been inferred from published flow laws of quartz. Combining the data of these naturally deformed rocks with the experimental data of Hirth & Tullis (1992), a recrystallization mechanism map will be constructed. Such a map, in conjunction with careful microstructural observations, will help geologists to determine or constrain the deformation conditions in quartz-rich mylonites from natural shear zones.

Geological setting

The Eastern Tonale Line in the Italian Alps is a dextral strike-slip segment of the Periadriatic fault system. In the area of interest, the fault zone was heated by the synkinematic Oligocene emplacement of the Presanella intrusion of the Adamello pluton (Fig. 1). The contact aureole of the pluton extends across the fault zone, which is 800 m wide and exhibits a vertical mylonitic foliation and a subhorizontal stretching lineation. Farther west, the Tonale Line is deformed entirely by brittle deformation (Werling 1992; Stipp & Schmid 1998). At the eastern border of the Adamello pluton, the Tonale Line is cut by the sinistral Giudicarie strike-slip zone. Near the pluton, the strike-slip fault includes the mylonitic rim of the Presanella intrusion in the south, followed by the Tonale mylonites of the Southern Mylonite Zone and the cataclasites towards the north (Fig. 1). The Stavel mylonites of the Northern Mylonite Zone (Fig. 1) situated north of the cataclasites are not part of this study as they belong to an older deformation stage (Werling 1992).

From north to south there is a temperature gradient ranging from 250 °C to about 700 °C in the Tonale mylonites (Fig. 1) at a constant confining pressure of 250 to 300 MPa. These P–T-estimates are based on critical mineral assemblages and related reaction isograds in the metasediments of the Tonale mylonites (Werling 1992; Stipp *et al.* 2002). Quartz microstructures show that the transition from dominantly brittle to dominantly crystal plastic deformation occurs at about 280 °C. Three microstructural zones have been recognized corresponding to the three different dynamic recrystallization mechanisms of quartz (Fig. 1). Bulging recrystallization (zone of BLG) dominates from approximately 280–400 °C, subgrain rotation recrystallization (zone of SGR) from approximately 400–500 °C and grain boundary migration recrystallization (zone of GBM) from approximately 500 °C to about 700 °C at the magmatic contact (Stipp *et al.* 2002).

Quartz veins taken from the Southern Mylonite Zone were used for the microstructural analysis (Fig. 1) because they contain fewer inclusions than quartzites or quartz layers in polymineralic mylonites. Only foliation-parallel veins were sampled and it was ensured that the dynamic recrystallization microstructures in the veins were the same as those in quartz within the metasedimentary host rocks.

Determination of recrystallized grain size

The mylonitic quartz veins have been analysed in X–Z sections, i.e. normal to the foliation and parallel to the stretching lineation. The microstructural characterization and the grain size analyses have been carried out using image analysis methods. In the zone of BLG, grain sizes have been determined using the autocorrelation function (ACF; Panozzo Heilbronner 1992). In the zones of SGR and GBM, the line-intercept method (Smith & Guttman 1953, cited in Ord & Christie 1984) was used. This method was chosen because it has been used for most of the experimentally calibrated piezometers as well as for previous studies on natural quartz

Fig. 1. Map of investigated area of the Tonale fault zone; temperatures and reaction isograds are inferred from mineral assemblages of metasediments (data from Werling 1992 and Stipp *et al.* 2002). Three zones of dynamic recrystallization microstructures are distinguished: BLG (bulging recrystallization); SGR (subgrain rotation recrystallization); and GBM (grain boundary migration recrystallization). Geographical coordinates refer to sheet 'Passo del Tonale' of the autonomous province of Trento (Italy; Carta Topografica Generale No. 041120, 1987). Inset: Regional setting of the investigated area (arrow).

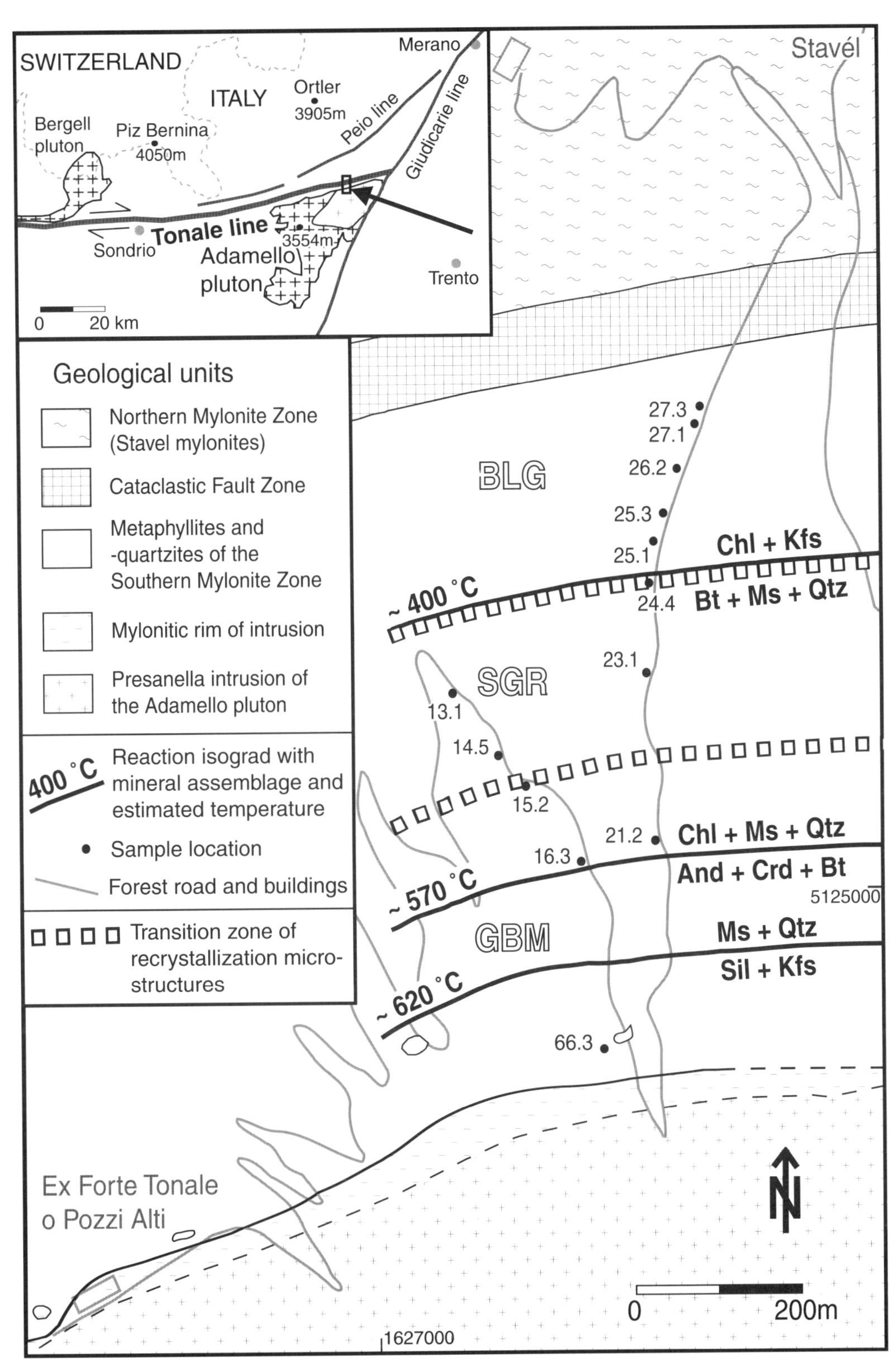
SWITZERLAND
ITALY
Merano
Ortler
3905m
Peio line
Giudicarie line
Bergell
pluton
Piz Bernina
4050m
Tonale line
Sondrio
3554m
Adamello
pluton
Trento
0
20 km
Stavél
Geological units
Northern Mylonite Zone
(Stavel mylonites)
Cataclastic Fault Zone
Metaphyllites and
-quartzites of the
Southern Mylonite Zone
Mylonitic rim of intrusion
Presanella intrusion of
the Adamello pluton
400 °C
Reaction isograd with
mineral assemblage and
estimated temperature
Sample location
Forest road and buildings
Transition zone of
recrystallization micro-
structures
BLG
27.3
27.1
26.2
25.3
25.1
Chl + Kfs
~ 400 °C
24.4
Bt + Ms + Qtz
SGR
23.1
13.1
14.5
15.2
21.2
Chl + Ms + Qtz
16.3
~ 570 °C
And + Crd + Bt
5125000
GBM
Ms + Qtz
Sil + Kfs
~ 620 °C
66.3
Ex Forte Tonale
o Pozzi Alti
N
0
200m
1627000

mylonites, thus ensuring the consistency of our data with previous studies. The analyses have been carried out on standard polished thin sections of a thickness of approximately 20 μm. For high numerical aperture of the objective, the optically resolved depth of focus becomes small. For example, for a numerical aperture of 0.75 (as used for sample 27-3 with the smallest grain size) the resolved depth of focus is below 1 μm. Thus, the grain size is significantly larger than the resolved depth of field and stereological corrections (cf. Panozzo Heilbronner 1992) are therefore not necessary.

Autocorrelation function (ACF)

Photomicrographs of six samples have been recorded under crossed polarizers in three different orientations (0°, 30°, 60°) with respect to the polarizers. Using NIH Image and the FFT macro (Rasband 1996) the ACFs were calculated for selected regions of interest (ROI) of 128 × 128 pixels and 256 × 256 pixels. A large number of measurements could be made and processed using the ACF. A minimum of 16 ROIs were selected in each sample, and for each ROI all three differently oriented images were used. Using the three different orientations for each ROI helped to avoid orientational bias. The average ACFs for each set of three and for the whole sample were determined. The average ACF for each sample was thresholded at 52% of the maximum ACF value. This thresholding value was determined by calibrating the ACF method against the line-intercept method by analysing the BLG sample with the largest grain size (sample 24-4). The long and short axes of the best fit ellipses were measured and scaled using the 'Analyze-tools' of NIH Image.

Line-intercept method

The sizes of the recrystallized grains in the SGR zone were determined by the line-intercept method (Ord & Christie 1984) carried out on *c*-axis orientation images (for optical orientation imaging, see Panozzo Heilbronner & Pauli 1993, 1994). Because the grains are anisotropic the line-intercepts were counted in directions parallel and normal to the preferred long axis orientation. Measurement grids of approximately 2 mm per side (real size) with a mesh width of approximately 100 μm in both directions were used. Areas of porphyroclastic ribbon grains have been avoided for the measurements or – if this was not possible they have been subtracted from the grid length.

The size of the recrystallized grains in the GBM zone is hard to determine due to the extremely sutured and lobate grain boundaries. The amplitude of the lobate boundaries can be so large that sections of neighbouring grains occur within other grains ('island grains'), an effect which was called a 'dissection microstructure' by Urai *et al.* (1986). Thus, using the line-intercept method in this zone involves counting different segments of the same grain as individual grains. Coherent grains can only be recognized on the basis of identical crystal lattice orientation. Since lobateness and the occurrence of dissection microstructures and island grains increase with increasing temperature in the GBM zone, the grain size based on line-intercepts becomes more and more inadequate with increasing temperature. Furthermore, line-intercepts, which are based on the assumption of equant grain shapes, represent an increasingly less reliable method to determine the size of irregularly shaped grains.

In order to get an approximate estimation of the grain size at least in the lower temperature part of the GBM zone, two samples (21-2, 16-3) have been measured by the same method as described for the zone of SGR including the intercepts with the island grains. A measurement grid (real size) of approximately 4 mm per side with a mesh width of 200 μm was used for sample 21-2 and a grid of approximately 10 mm square length with a mesh width of 500 μm was used for sample 16-3. Next to the magmatic contact the grain size is a few millimetres in diameter and grain shapes are very complex. Island grains and dissection microstructures render a grain size determination by the line-intercept method meaningless and are therefore not included here.

From the average long and short axes the geometric mean grain size has been calculated (Ranalli 1984). In the following we refer to this mean grain size if not stated otherwise. The average long and short diameter and the average grain sizes of the recrystallized grains of the BLG, SGR and GBM zones are summarized in Table 1. The recrystallized grain sizes of the various samples are plotted versus the deformation temperature in Fig. 2.

Microstructural correlation between nature and experiment

Temperature and strain rate conditions in experiments differ from those of natural deformation. The correlation of experimental and natural

Table 1. *Recrystallized grain sizes*

Sample	Longitudinal coordinate	Latitudinal coordinate	Long axis [μm]	Short axis [μm]	Mean grain size [μm]
BLG					
27-3	1627425	5125605	6.2	4.9	5.5
27-1	1627415	5125595	9.3	6.6	7.8
26-2	1627395	5125555	9.9	7.6	8.7
25-3	1627375	5125500	17.1	11.9	14.3
25-1	1627365	5125475	25.1	9.5	15.5
24-4	1627355	5125440	28.6	20.6	24.3
SGR					
13-1	1627090	5125220	78	43	58
23-1	1627330	5125345	89	42	61
14-5	1627140	5125155	113	49	74
15-2	1627185	5125125	134	53	84
GBM					
21-2	1627320	5125190	289	166	219
16-3	1627250	5125040	440	285	354

Recrystallized grain sizes of the investigated samples taken from different zones of dynamic recrystallization. Mean grain size is the geometric mean of grain long and short axes. For method of grain size measurement see text. Geographical coordinates refer to sheet 'Passo del Tonale' of the autonomous province of Trento (Italy; Carta Topografica Generale No. 041120, 1987).

deformation depends upon the operation of the same deformation mechanisms, which should produce identical microstructures at strain-independent steady state flow stress conditions. Transitions from one recrystallization mechanism to another cause microstructural changes which can be observed in both nature and experiments. We therefore focus on microstructures

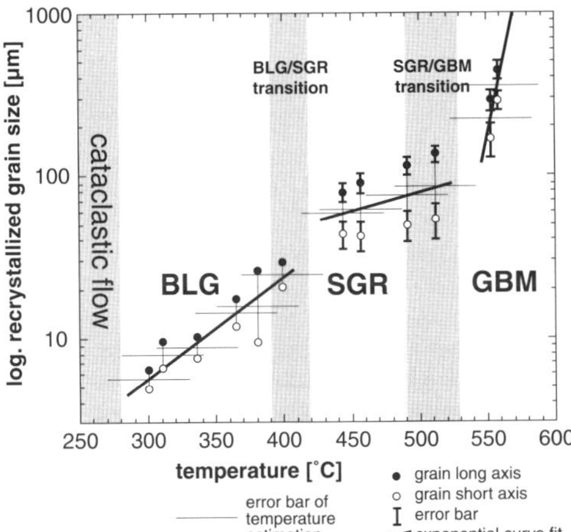

Fig. 2. Plot of recrystallized grain size versus deformation temperature; long and short axes of recrystallized grains and best fits of geometric mean grain size are shown. Error bars are one standard deviation; for technical reasons there are no error bars for the BLG data.

when relating nature to experiments. A reliable correlation of microstructures rests on clearly identifiable microstructural criteria. Of particular interest are distinctly different microstructures because they indicate switches in the dominant dynamic recrystallization mechanisms. Mercier (1980), Hirth & Tullis (1992) and Zulauf (2001) reported a step in the grain size–temperature relationship as the most obvious microstructural discontinuity. Our data indicate that this discontinuity is not a step, but rather a discrete change in the slope of the grain size–temperature relationship.

In the Tonale mylonites, there are two slope changes (Fig. 2) which coincide with transitions in the dominant dynamic recrystallization mechanism. Other microstructural criteria, e.g. shape and lobateness, change more slowly or gradually so that they can only be used in combination to define transitions in the dominant recrystallization mechanism. As a first approximation the three dislocation creep regimes defined by Hirth & Tullis (1992) can be correlated with the three dominant dynamic recrystallization mechanisms inferred from the deformed Tonale quartz veins (Stipp *et al.* 2002). However, the transitions from BLG to SGR and from SGR to GBM do not coincide with the transitions from regime 1 to regime 2 and from regime 2 to regime 3, respectively. In the following we will describe the microstructural transitions which are the basis for our correlation of the dynamic recrystallization mechanisms in the Tonale

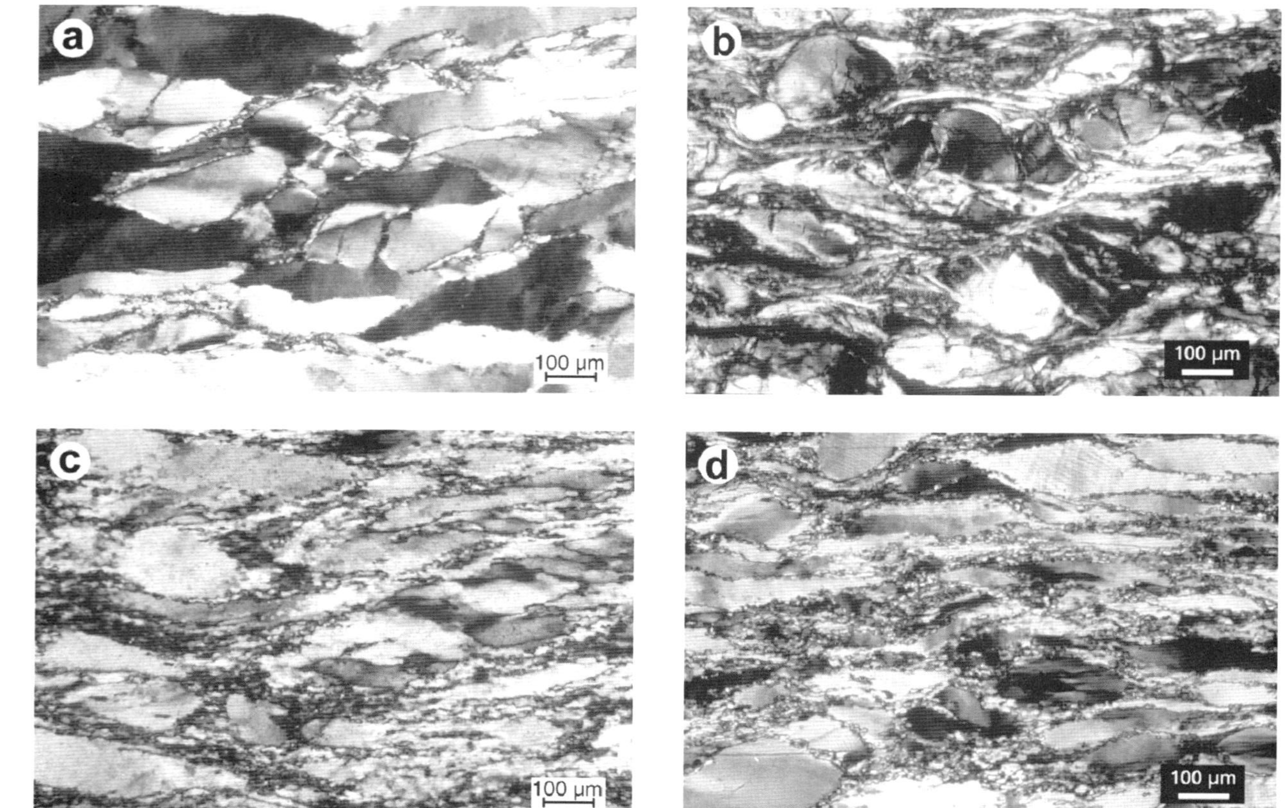

Fig. 3. Light micrographs of natural and experimental samples (crossed polarizers). Mylonitic quartz veins from the Eastern Tonale line (BLG, SGR, GBM) and experimental samples (regime 1, 2, 3) from Hirth & Tullis (1992). Samples are oriented normal to the main foliation. In natural samples the stretching lineation is horizontal. In experimental samples the shortening direction is vertical (micrographs taken from Tullis *et al.* 2000). (**a**) BLG I (*c.* 310 °C): Quartz porphyroclasts displaying undulose extinction and serrated grain boundaries, recrystallization along grain boundaries and fractures (sample 27-1). (**b**) Regime 1: Heavitree quartzite shortened 65% at 850 °C, $10^{-5}\,s^{-1}$ and 1200 MPa confining pressure (equivalent to BLG I). (**c**) BLG II (*c.* 370 °C): Recrystallization preferentially along serrated grain boundaries of porphyroclasts (core and mantle structures, sample 25-3). (**d**) Regime 2: Heavitree quartzite shortened 64% at 800 °C, $10^{-6}\,s^{-1}$ and 1200 MPa confining pressure (equivalent to BLG II).

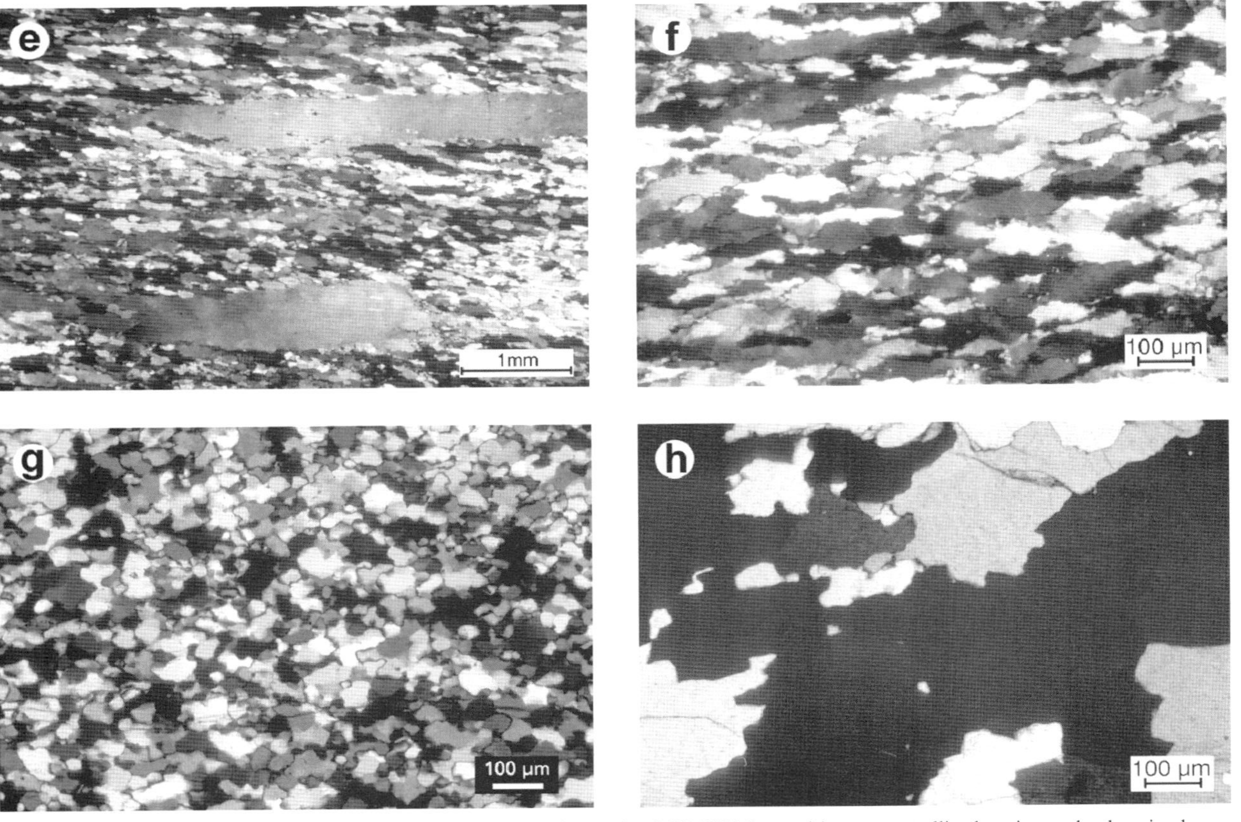

Fig. 3. (**e**) SGR (*c.* 510 °C): Polygonization and recrystallization of ribbon grains at the SGR/GBM-transition; recrystallized grains and subgrains have nearly the same grain sizes and shapes (sample 15-2; note the different scale). (**f**) SGR (*c.* 510 °C): Detail of (**e**) showing recrystallized grains only; recrystallized grains display sutured and weakly lobate grain boundaries indicating a contribution of grain boundary migration close to the SGR/GBM-transition (sample 15-2). (**g**) Regime 3: Black Hills quartzite shortened 50% at 1200 °C, $10^{-6}\,s^{-1}$ and 1200 MPa confining pressure. (**h**) GBM II (*c.* 650 °C): Amoeboid grain shapes, islands grains, very large recrystallized grain sizes (sample 66-3).

mylonites with the experimentally derived dislocation creep regimes of Hirth & Tullis (1992).

Variation within the zone of BLG

BLG corresponds to regime 1 (Fig. 3a, b) and the lower temperature part of regime 2 (Fig. 3c, d) of Hirth & Tullis (1992). We will call the part that corresponds to regime 1, BLG I, and the part that corresponds to the lower temperature part of regime 2, BLG II. The transition from BLG I to BLG II is not marked by a discontinuity in the recrystallized grain size–temperature relationship (Fig. 2). Instead, there are differences in dynamic recrystallization microstructures, which allow us to define a change between BLG I and BLG II. Hirth & Tullis (1992, 1994) attribute the recrystallization in regime 1 only to grain boundary migration processes without subgrain rotation. In the natural samples from the Eastern Tonale fault zone, BLG I porphyroclasts show (in the light microscope) some very small subgrains along the serrated grain boundaries which may correspond to the cell misorientation mechanism suggested for regime 1 (Hirth & Tullis 1992). Furthermore, BLG I microstructures commonly occur in conjunction with fractures. Fractures have been excluded for regime 1 of the experimental dislocation creep regimes and attributed to the brittle–plastic transition (Hirth & Tullis 1994). In nature, however, the confining pressure can be low at the brittle–plastic or frictional–viscous (Schmid & Handy 1991) transition. When local differential stresses reach the magnitude of the effective confining pressure, fracturing can occur in alternation with dynamic recrystallization in natural mylonites (cf. coseismic creep of Küster & Stöckhert 1999).

Microfracturing and related solution-deposition (cf. Hippert & Egydio-Silva 1996) can be neglected for the formation of new grains in the Tonale mylonites. We think that in nature, apart from subgrain rotation, microfracturing is another important potential process for separating bulges from porphyroclasts (van Daalen *et al.* 1999; Stipp *et al.* 2002). Fracture zones are bulged, and bulges and new grains along these fractures have the same size as bulges, subgrains and recrystallized grains along grain boundaries. The increasing amount and size of recrystallized grains with increasing temperature within the zone of BLG occurs together with a change in the preferred sites of dynamic recrystallization between BLG I and BLG II. In BLG I, recrystallization occurs preferentially at triple junctions of porphyroclasts (cf. Stipp *et al.* 2002), whereas the porphyroclasts in BLG II are completely surrounded by recrystallized grains (Fig. 3c, d). Subgrains of the same size as the bulges are detectable in the light microscope near the grain boundary region of the porphyroclasts, defining core and mantle structures which are characteristic of BLG II and SGR (White 1976; Fitz Gerald & Stünitz 1993). These microstructures indicate that progressive subgrain rotation is more important in the zone of BLG II compared to BLG I. The porphyroclasts within BLG I corresponding to regime 1 show finely serrated grain boundaries and undulatory extinction and some of the porphyroclasts are fractured (Fig. 3a, b). In the zone of BLG II microfracturing is rare and disappears completely with increasing temperature. The porphyroclasts of BLG II show a more pronounced elongation, and undulatory extinction and deformation lamellae are frequent, features, which correspond to those observed in regime 2 by Hirth & Tullis (1992; Fig. 3c, d).

BLG/SGR-transition

The BLG/SGR transition in the Tonale mylonites occurs within the dislocation creep regime 2 as defined by Hirth & Tullis (1992). The transition in nature is marked by a discontinuity in the recrystallized grain size-temperature relationship (Fig. 2). Porphyroclasts within the zone of SGR consist of extremely elongated ribbon grains (Fig. 3e) contrasting with the angular or only moderately elongated porphyroclasts found in the zone of BLG (Fig. 3a–d). Internal deformation features such as undulatory extinction and deformation lamellae occur occasionally within the ribbon grains (Fig. 3e); they may partly be due to overprinting during cooling. At the BLG/SGR transition, both the recrystallized grain size and the volume fraction of recrystallized grains increase significantly. In the zone of BLG, typically less than 20% of the quartz veins are recrystallized. In the zone of SGR about 30–80% are recrystallized (Stipp *et al.* 2002). The ribbon grains are progressively consumed by polygonization (progressive subgrain rotation, Fig. 3e).

Geometric dynamic recrystallization (GRX; McQueen *et al.* 1985; Doherty *et al.* 1997) which has been described as a type of dynamic recrystallization which occurs in metals at high deformation temperature and large strain can produce microstructures similar to those referred to as BLG II and SGR (Fig. 3c–g). If GRX occurs at low deformation temperature, as pointed out by Humphreys & Hatherly (1996), it cannot be distinguished from continuous recrystallization phenomena, i.e. subgrain rotation recrystallization. Hence, the application of the term GRX to the investigated quartz samples might be possible, but would not change the microstructural correlation, because microstructures similar to GRX also occur in the experimental sample set of Hirth & Tullis (1992; cf. Fig. 3).

SGR/GBM-transition

The SGR/GBM transition in nature (Fig. 3f) falls within regime 3 of the experimental dislocation creep regimes (Fig. 3g). Microstructures typical for GBM in nature are not attainable experimentally due to the onset of melting. Within the natural samples, this transition is again marked by a discontinuity in the recrystallized grain size–temperature relationship (Fig. 2). Recrystallized grains of the GBM zone show irregular grain shapes and a broad grain size distribution within a single sample. In contrast, the recrystallized grains of the SGR zone have rather constant grain shapes and sizes, which are almost identical to those of the subgrains observed in the light microscope. At the transition to GBM, the recrystallized grain size is somewhat larger than the subgrain size as visible in the light microscope. The recrystallized grain size distribution in the zone of GBM is broader and the grains display weakly sutured grain boundaries (Fig. 3e, f). GBM-microstructures (Fig. 3h) differ from regime 3 (Fig. 3g) in that they show larger grain sizes and more irregular grain shapes with more sutured grain boundaries and amoeboid grain shapes. Dissection microstructures (Urai *et al.* 1986) and 2D-island grains, which are typical for GBM (Fig. 3h), have not been found in regime 3 of Hirth & Tullis (1992). Instead, in regime 3 (Fig. 3g) grains are of similar size and shape, and grain sizes are comparable or even smaller than those of the uppermost SGR (compare grain sizes in Fig. 3f, g). Natural samples of the GBM zone are completely recrystallized. The transition to completely recrystallized microstructures occurs in the experiments of Hirth & Tullis (1992) and in the samples of the Tonale mylonites at the SGR/GBM transition.

The zone of GBM of the Tonale mylonites can be divided into a lower and a higher temperature part, GBM I and GBM II, respectively (Stipp *et al.* 2002). However, GBM II samples were not considered in this study because: (1) these microstructures are not attainable experimentally; and (2) due to the lobateness of grain boundaries, it is very difficult to determine grain sizes in such microstructures, as pointed out above.

Stress and strain rate calculations

The correlation of the experimental dislocation creep regimes and the natural zones of dynamic recrystallization is summarized in a schematic diagram (Fig. 4). In experiments, the same microstructural transitions take place at higher temperatures and at strain rates that are 5 to 10 orders of magnitude higher than in nature (e.g. Hobbs *et al.* 1976; Mercier *et al.* 1977; Suppe 1985; Twiss & Moores 1992). Assuming that the rheological properties are conformable in nature and experiment, a strain rate/temperature extrapolation can be made (cf. Paterson 1987). Through a stress determination (paleopiezometry) and using flow law equations (deformation temperature is known), the strain rate in the natural mylonites can be calculated.

Paleopiezometry

The paleostress, which is inferred to be the steady-state flow stress at the time of deformation, can be determined by different microstructural paleopiezometers. These are either based on the density of unbound dislocations (e.g. Weertman 1970; Goetze & Kohlstedt 1973; Takeuchi & Argon 1976), on the subgrain size (e.g. Raleigh & Kirby 1970; Twiss 1986), or on the recrystallized grain size (e.g. Luton & Sellars 1969; Mercier *et al.* 1977; Twiss 1977). Although only the first two types of piezometers are based on physical models, we have used a recrystallized grain size piezometer which is largely empirical. Both the subgrain size and the dislocation density are much more sensitive to annealing and retrograde overprinting and thus the recrystallized grain size has been characterized as the feature with the highest inherent stability in naturally deformed rocks (Mercier *et al.* 1977; White 1979; Kohlstedt & Weathers 1980). Recrystallized grain size piezometers have been calibrated in the form

$$\Delta\sigma = BD^{-x}, \quad (1)$$

where $\Delta\sigma$ is the steady state differential stress, D is the recrystallized grain size and B and x are empirical constants. More recently, several microphysical models have been proposed to explain the piezometric relationship (Edward *et al.* 1982 for subgrain size; Derby & Ashby 1987; Derby 1990; Shimizu 1998; de Bresser

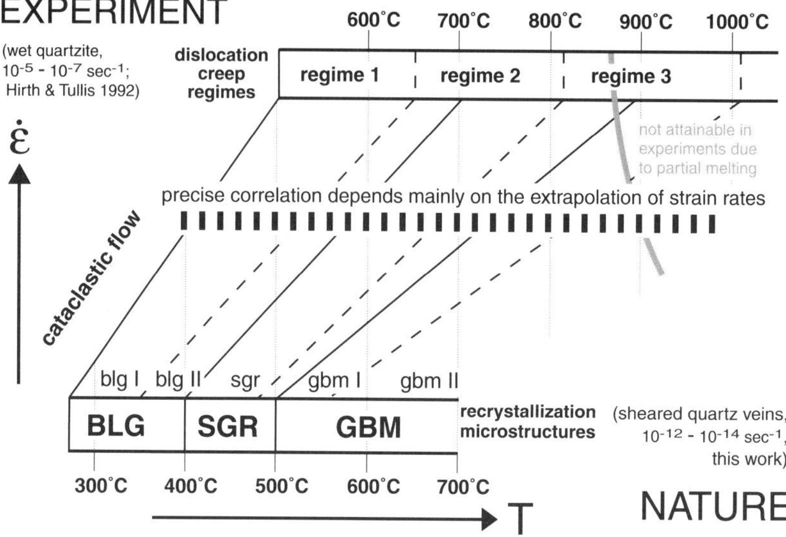

Fig. 4. Schematic diagram illustrating the ranges of temperature and strain rate of the natural zones of dynamic recrystallization and of the experimental dislocation creep regimes of Hirth & Tullis (1992). At slow strain rates, three main types of microstructures can be distinguished: bulging recrystallization (BLG); subgrain rotation recrystallization (SGR); and grain boundary migration recrystallization (GBM). Subdivisions of these main recrystallization microstructures can be correlated with the experimental dislocation creep regimes. Solid lines mark the transitions between BLG, SGR and GBM and their corresponding experimental microstructures; dashed lines mark the transitions in experimental dislocation creep regimes and their corresponding microstructures in nature.

et al. 1998 for recrystallized grain size). Since all these theoretical models have not yet been calibrated and used for dynamic recrystallization of quartz in natural fault rocks, we chose not to use piezometric relationships corresponding to these models. However, a recent review on piezometric relationships (de Bresser *et al.* 2001) emphasizes the need to test and calibrate theoretical models with new experimental studies. When such studies are available they can be used to critically evaluate and expand the considerations made below.

For any given grain size the piezometers for quartz (Mercier *et al.* 1977; Twiss 1977, 1980; Christie *et al.* 1980; Koch 1983 and Mainprice 1981 in Koch 1983) yield a stress range for the dynamically recrystallized grain sizes of more than one order of magnitude (Fig. 5). Wet quartzite conditions are much more realistic for the deformation in natural mylonitic shear zones and we therefore consider only piezometers calibrated for 'wet' conditions and the one which is based on the theoretical assumptions of Twiss (1977, 1980) for Fig. 6. The data of Mercier *et al.* (1977) and Mainprice (1981) are not reliable with respect to the water content of the samples. Furthermore, either the experimental constraints on the mechanical data are poor (solid confining medium for the experiments used in Mercier *et al.* 1977), or the choice of the starting material is not very satisfying (Dover flint in Mainprice 1981), because the finite deformation microstructures differ significantly from those found in natural quartz mylonites. Therefore, these calibrations have not been included in Fig. 6.

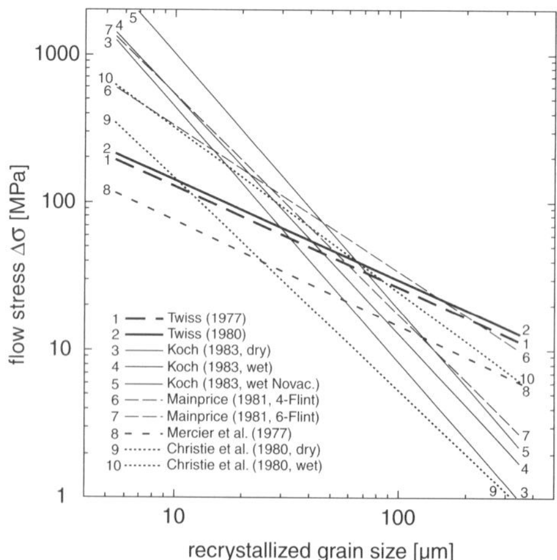

Fig. 5. Diagram of recrystallized grain size paleopiezometers of quartz; the piezometers of Mainprice (1981) were calculated by Koch (1983).

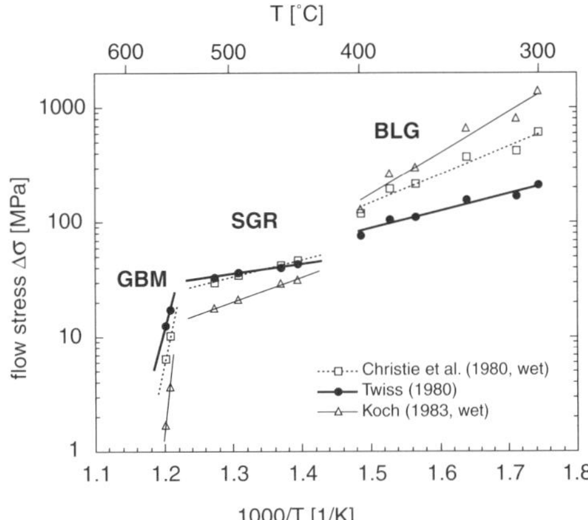

Fig. 6. Flow stresses calculated from recrystallized grain sizes using different piezometers. Empirical constants B, x are 4090, 1.11 for the piezometer of Christie *et al.* (1980), 676, 0.68 for Twiss (1980) and 21829, 1.61 for Koch (1983). The slopes of the best fits vary for each of the recrystallization mechanisms.

The wet piezometers of Christie *et al.* (1980) and Koch (1983) yield unrealistically high values for stress at the small grain sizes of BLG. This can probably be related to the poor experimental constraints on the mechanical data (see Gleason & Tullis 1993, 1995 for discussion). For sample 27-3, which has been deformed under the lowest temperature conditions, the flow stresses predicted by the different paleopiezometers in Fig. 6 range from 212 MPa (Twiss 1980) to 1396 MPa (Koch 1983, wet). The piezometers of Koch (1983) and Christie *et al.* (1980) predict differential stresses for BLG which are much higher than the confining pressure (250 to 300 MPa) derived from coexisting mineral assemblages (Stipp *et al.* 2002). Such high stresses would produce dominant brittle deformation (Hirth & Tullis 1994; Kohlstedt *et al.* 1995) instead of low temperature crystal plasticity. Dominantly brittle deformation microstructures have not been observed. Hence, the piezometers of Christie (1980, wet) and Koch (1983, wet) can be excluded.

For the entire sample set, including BLG I through GBM I, the piezometer of Twiss (1980) yields stresses from 212 MPa to 12.5 MPa (Fig. 6). The extrapolation of this flow stress range to the brittle–plastic transition, corresponding to the transition from the Cataclastic Fault Zone to the Southern Mylonite Zone of the Tonale Line at 280 °C (Fig. 1), yields a differential stress of about 250 MPa (Fig. 6). The confining pressure of 250 to 300 MPa derived from critical mineral assemblages is approximately equal to this differential stress magnitude at the brittle–plastic transition. Empirically, the high confining pressure limit of semibrittle deformation is reached when confining pressure and differential stress are of approximately the same magnitude (Kirby 1980; Evans *et al.* 1990; Kohlstedt *et al.* 1995). Thus the paleopiezometer of Twiss (1980) and our microstructural observations yield consistent results for the brittle–plastic transition. Paleopiezometers appropriate for the individual different recrystallization mechanisms are not calibrated for quartz as has been found experimentally for other minerals (see evaluation of paleopiezometers). The paleopiezometer by Twiss (1977, 1980) appears to be the most suitable calibration for our samples in the light of the geological constraints for the data and therefore it has been used in this study (cf. Gleason & Tullis 1993, 1995).

Strain rate estimation

The strain rate is related to flow stress and temperature by experimentally derived flow laws (e.g. Heard & Carter 1968; Parrish *et al.* 1976; Kronenberg & Tullis 1984; Koch *et al.* 1989). We have used the most recent experimental calibrations carried out by Luan & Paterson (1992) in the gas-apparatus and by Gleason & Tullis (1995) in the molten salt cell-technique in the Griggs-type piston-cylinder apparatus since the mechanical data of their studies are quite well constrained. Additionally, the geologically constrained, 'theoretical' flow laws of Paterson & Luan (1990) and Hirth *et al.* (2001) are considered in this study. The necessary coefficients for the flow laws are listed in Table 2.

In Fig. 7a the lines of constant strain rates corresponding to the flow laws of Luan & Paterson (1992) and Gleason & Tullis (1995; Table 2) as well as to the geologically constrained flow law of Paterson & Luan (1990) are compared with the data from this study. The comparison yields an estimate of the strain rate in the range of 10^{-14} to $10^{-12}\,s^{-1}$ (Fig. 7a). This estimate is within the traditionally inferred range for natural fault zone conditions (e.g. Pfiffner & Ramsay 1982; Suppe 1985; Twiss & Moores 1992). The two orders of magnitude difference in the estimated strain rate (Fig. 7a) is due to the different flow law coefficients, which are partly caused by experimental differences in water fugacity. For H_2O-saturated quartz aggregates the confining pressure is equal to the H_2O-pressure (for both the Tonale fault zone samples and the experiments of Luan & Paterson 1992 it is about 300 MPa). The experiments of Gleason & Tullis (1995)

Table 2. *Quartz flow law coefficients of flow laws used in this study*

	Q [kJ mol^{-1}]	A [MPa^{-n} s^{-1}]	n
Luan & Paterson (1992)	152	4×10^{-10}	4
Gleason & Tullis (1995)	223	1.1×10^{-4}	4
Paterson & Luan (1990)	135	6.5×10^{-8}	3.1
Hirth *et al.* (2001)	135	6.30957×10^{-12}	4

Experimental calibrations from Luan & Paterson (1992) and Gleason & Tullis (1995); coefficients presented by Paterson & Luan (1990) and Hirth *et al.* (2001) are constrained by quartzite rheology and microstructural comparison.

were carried out at 1.5 GPa confining pressure with 0.15 weight% of added water. Kohlstedt *et al.* (1995) introduced a water fugacity term into the flow law equation:

$$\dot{\varepsilon} = A\Delta\sigma^{-n}(f_{H_2O})^m \exp(-Q/RT), \qquad (2)$$

where $\dot{\varepsilon}$ is the strain rate, $\Delta\sigma$ is the differential stress, T is the temperature, R is the Boltzmann constant per mole, Q is the creep activation energy per mole, f_{H_2O} is the water fugacity, A is a material constant, n is the stress exponent and m the water fugacity exponent. Saturated water conditions are a reasonable assumption for the samples of the eastern Tonale Fault Zone because of the H_2O-releasing metamorphic reactions in the metasediments of the shear zone, the quite homogeneous microstructures of dynamic recrystallization (local changes in water content may cause a switch in the dominant deformation mechanism and may thus produce a different microstructure) and the abundance of fluid inclusions in the porphyroclasts.

In order to reduce differences in water fugacity the flow laws of Gleason & Tullis (1995) and Hirth *et al.* (2001) were normalized to the water fugacity at a pressure of 300 MPa, i.e. the confining pressure of the Tonale mylonites and the experiments of Luan & Paterson (1992), using a fugacity exponent of $m = 1$ (cf. Kohlstedt *et al.* 1995; Hirth *et al.* 2001). For the flow law of Gleason & Tullis (1995) the water fugacity coefficients after Tödheide (1972) were extrapolated to a confining pressure of 1.5 GPa (Fig. 7b). All flow laws indicate that the strain rate increases somewhat towards higher temperatures for the Tonale mylonites (using the stresses indicated by the Twiss-paleopiezometer relation). The strain rates, however, show considerable difference between the flow laws (Fig. 7b). Kohlstedt *et al.* (1995) come to a similar conclusion.

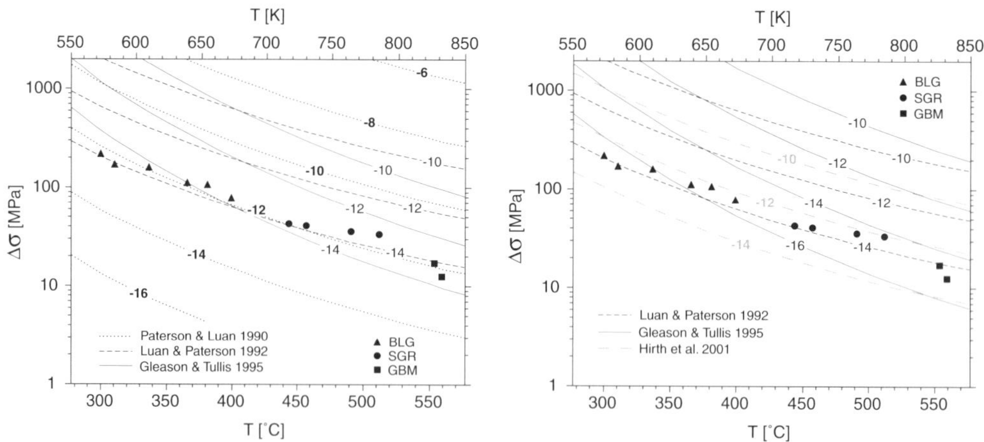

Fig. 7. (**a**) Flow stress data of the Tonale line samples (using the piezometer of Twiss 1980) plotted versus deformation temperature. Curves of constant strain rate are calculated from the flow laws of Luan & Paterson (1992), Gleason & Tullis (1995) and Paterson & Luan (1990). No account is taken of water fugacity differences. (**b**) Flow stress data of the Tonale line samples (using the piezometer of Twiss 1980) plotted versus deformation temperature. The curves of constant strain rate from the flow laws of Gleason & Tullis (1995) and Hirth *et al.* (2001) are normalized to a water fugacity at a confining pressure of 300 MPa and a fugacity exponent $m = 1$ so that all data sets have equivalent water fugacity conditions.

Their extrapolation of the laboratory data of Gleason & Tullis (1995) and Luan and Paterson (1992) to crustal deformation conditions at a constant strain rate leads to substantial differences in the predicted rock strength between the two data sets.

Discussion

Natural constraints on deformation along the Tonale Fault Zone

In the larger scale geological framework, the Tonale Line is a predominantly cataclastic fault zone (Schmid *et al.* 1989), which was active over at least 15 Ma (Müller 1998) with a dextral displacement of approximately 100 km (Laubscher 1991; Steck & Hunziker 1994; Frisch *et al.* 1998; Schmid & Kissling 2000) and a mean width of the fault zone of approximately 200 m. Hence, a reasonably fast shear strain rate estimate of approximately $10^{-12}\,s^{-1}$ results for this fault zone. Within that fault system, the investigated local Southern Mylonite Zone has formed as a consequence of a small local heat anomaly caused by the Adamello intrusion (Werling 1992). Thus, the deformation of the investigated mylonites is controlled by the larger kinematic framework of the entire Tonale fault zone and it can be inferred that the local mylonite strain rate results from the external kinematic boundary conditions related to plate movements. The actual local shear strain rate of the mylonites may vary from the overall estimate because of strain partitioning, variable shear zone width etc. The externally imposed shear strain rate of approximately $10^{-12}\,s^{-1}$ agrees best with the strain rate derived from the geologically constrained flow law of Hirth *et al.* (2001; Fig. 7b). Strain rates of $10^{-15}\,s^{-1}$ and slower are derived from the flow law of Gleason and Tullis (1995; Fig. 7b). Such slow strain rates are not in the range of typical geological strain rates (e.g. Pfiffner & Ramsay 1982) and are probably too low for the Tonale Line (see above).

Recrystallized grain size

Grain size reduction during dynamic recrystallization is observed in the case of BLG and SGR microstructures. As pointed out by Schmid and Handy (1991) syntectonic grain boundary migration recrystallization may lead to a decrease or an increase in size of the recrystallized grains. The very large grains in the GBM II zone, close to the magmatic contact, may indicate such a grain size increase. The recrystallized grain sizes may have been modified after deformation ceased, either by annealing or by deformation at decreasing temperatures. Lower temperature overprints have been excluded, since only samples in which the microstructure could clearly be related to the synkinematic mineral assemblages used for the temperature estimates (cf. Stipp *et al.* 2002) have been selected. Microstructural features indicative of annealing, e.g. straight grain boundaries and increased grain sizes, are rare and only occur in the GBM II zone. The fact that subgrains and recrystallized grains in the zone of SGR (observed in the light microscope) have identical sizes is further evidence that postdeformational grain growth is not important. Therefore, corrections for grain growth (Hacker *et al.* 1990, 1992) are not necessary for the investigated sample set. Microstructural evidence for posttectonic grain growth in the Tonale mylonites is absent and indicates that posttectonic grain growth had only a minor effect because of the relatively rapid cooling after contact metamorphism. Fast cooling was caused by: (1) the low depth (approximately 8 km) of mylonitization of the investigated section; and (2) because colder rocks were continuously passing along the intrusion during the strike-slip movement of the fault zone (advective cooling). The rapid cooling is confirmed by the small difference between U/Pb-ages on zircon (32.0 ± 2.3 Ma; Stipp 2001) Rb/Sr-ages on muscovite (31.3 ± 1.5 Ma; Del Moro *et al.* 1983) and K/Ar and Rb/Sr ages on muscovite ($\sim 29.5 \pm 2$ Ma; Del Moro *et al.* 1983) of the Presanella granitoids in the Adamello pluton, indicating fast cooling of the magmatic body.

Evaluation of paleopiezometers

The different slopes of the flow-stress versus temperature curves calculated for different recrystallization mechanisms (Fig. 6) indicate that, ideally, piezometers should be adjusted for different recrystallization mechanisms. The need to define different piezometers for different recrystallization mechanisms was first stated by Poirier and Guillopé (1979) and experimentally calibrated for halite (SGR, GBM) by Guillopé and Poirier (1979). A number of piezometers, which depend on the dynamic recrystallization mechanism, have been proposed for other rock-forming minerals, e.g. for calcite (SGR, GBM; Schmid *et al.* 1980; Rutter 1995), olivine (GBM; van der Wal *et al.* 1993). Post and Tullis (1999) showed that the slope of the stress–grain size relationship for the regime 1 piezometer of feldspar varies from that for regime 2 for other minerals. In the case of quartz, however, none of the proposed piezometers takes the recrystallization mechanism into account (cf. White 1982). Only the piezometer of Twiss (1977, 1980) is in quite good agreement with the most recent experimental studies on crystal plastic deformation of quartz aggregates (Gleason & Tullis 1993). Gleason and Tullis (1995) and Hirth *et al.* (2001) find the piezometer of Twiss (1977) the most accurate for the transition between regime 2 and 3 and for regime 3, and they recommend the use of this relationship for SGR and at the SGR/GBM transition until new experimental calibrations are available. This study also finds that the relation of Twiss (1977, 1980) is the most consistent paleopiezometer for geological deformation conditions within the zone of BLG. We are aware of the fact that the piezometer of Twiss (1977) is based on questionable assumptions of equilibrium thermodynamics, as pointed out in Poirier (1985), Derby (1990) and de Bresser *et al.* (2001). Yet we recommend its application because of its best fit to currently published experimental and natural data on the dynamic recrystallization of quartz.

The comparison between subgrains and recrystallized grains in the Tonale mylonites (observed in the light microscope) does not show a significant difference in size. However, subgrains can only be measured grain by grain and such measurements have not been included in this study because a profound statistical base is lacking. Furthermore, the difference between the larger subgrains determined in the light microscope compared to those determined in the TEM is not yet understood. Nevertheless, if we assume that the light optically determined subgrains represent the precursors towards progressive rotation and recrystallization without a significant variation in size, the same stress dependency for subgrains and recrystallized grains could apply (Schmid *et al.* 1980). In that case, the temperature seems not to have a major influence on the recrystallized grain size during ongoing recrystallization. Equal diameters of bulges and recrystallized grains in the samples also suggest the absence of thermally induced grain growth during and after dynamic recrystallization for the zone of BLG. The absence of thermally induced grain growth is in agreement with the considerations of Post and Tullis (1999) who point out that migration rate and strain rate have the same temperature dependence and, thus, the resulting recrystallized grain size is not temperature dependent. In addition, in experiments on Carrara marble, Rutter (1995) does not find a significant influence of temperature and strain rate on the recrystallized grain size. Instead, the same author finds a clear stress dependence and a difference between the piezometer for subgrain rotation recrystallization and that for grain boundary migration recrystallization. Hence, there is, so far, no experimental evidence for an important temperature effect on the dynamically recrystallized grain size of major rheologically relevant minerals. This is despite the theoretical models of Derby & Ashby (1987), Derby (1990), Shimizu (1998) and de Bresser *et al.* (1998), all of which include a (weak) temperature dependency via the activation energy term. Our results indicate that well constrained experimental data on quartz in terms of physical parameters (T, $\Delta\sigma$, ε, $\dot{\varepsilon}$, f_{H_2O}), recrystallized grain and subgrain size, and steady state recrystallization microstructures, will be required, before we can evaluate microphysical models of piezometers for quartz, as presented and discussed in de Bresser *et al.* (2001).

A deformation mechanism map of dynamic recrystallization

Hirth and Tullis (1992) plotted the results of their experiments on a strain rate versus temperature diagram. This type of diagram has also been proposed by Frost and Ashby (1982) as one possible presentation of deformation mechanism maps. The presentation of dynamic recrystallization mechanisms in a strain rate versus temperature plot has several advantages: (1) the extrapolation of deformation experiments to natural conditions is an extrapolation in strain rate; and (2) as pointed out by Handy (1989), time and strain can sometimes be derived directly from geological observations and isotopic studies so that in certain cases the natural strain rate can be determined independent of experimental flow laws.

The data set from the Tonale fault zone is plotted in such a strain rate versus temperature diagram (Fig. 8) together with the experimental data of Hirth and Tullis (1992) and another natural sample set from the Ruby Gap Duplex in Central Australia (Dunlap *et al.* 1997). The strain rates of the natural samples were calculated using flow laws, as discussed above (Fig. 8a). The two natural sample sets (Dunlap *et al.* 1997 and the Tonale mylonites) cover a range of strain rates from 10^{-20} to $10^{-11}\,s^{-1}$. Strain rates of 10^{-20} to $10^{-15}\,s^{-1}$ appear unrealistically slow, because strain rates of $10^{-15}\,s^{-1}$ and

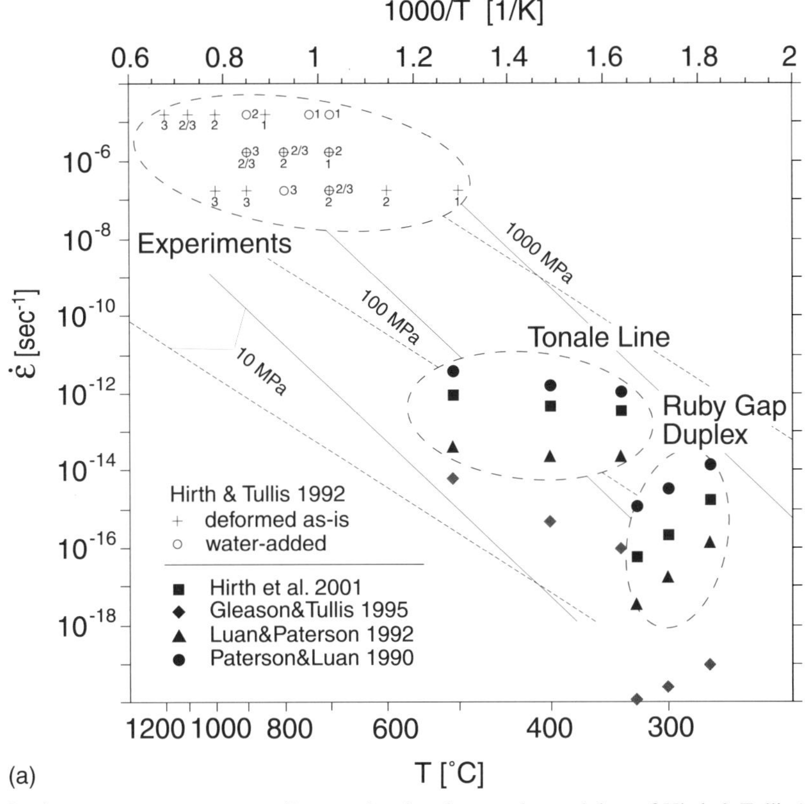

Fig. 8. (**a**) Strain rate versus temperature diagram showing the experimental data of Hirth & Tullis (1992), natural samples from the Tonale Line (this study) and those of the Ruby Gap Duplex (Dunlap *et al.* 1997). Only microstructures at transitions BLG I – BLG II (sample 26-2), BLG II – SGR (sample 24-4) and SGR – GBM I (sample 15-2) are used; data from Dunlap *et al.* (1997) have been reinterpreted using microstructural criteria discussed here. Deformation temperatures are taken from Dunlap *et al.* (1997) and Stipp *et al.* (subm.), and the flow stress data required for flow law calculations were derived from the piezometer of Twiss (1977; 1980). The natural strain rates have been calculated from four different flow laws. Experiments of Hirth & Tullis (1992) were carried out 'as-is' (crosses) or with added water (open circles); the dislocation creep regimes are indicated by the numbers. Lines of constant flow stress from the flow laws of Luan & Paterson (1992, dashed lines) and Gleason & Tullis (1995, solid lines) are displayed.

slower are not believed to leave microstructural traces in the rock (Pfiffner & Ramsay 1982). Thus, strain rates derived from the experimental flow laws of Gleason & Tullis (1995) and Luan & Paterson (1992) produce unrealistic results for the two natural sample sets. The geologically constrained empirical flow laws of Paterson & Luan (1990) and Hirth *et al.* (2001) show a strain rate of approximately $10^{-12}\,s^{-1}$ which is consistent with other geological constraints for the Tonale Line. The faster strain rates for the Ruby Gap samples of approximately $10^{-15}\,s^{-1}$ to $10^{-14}\,s^{-1}$ predicted by Paterson & Luan (1992) and Hirth *et al.* (2001) are also in the range of natural rock deformation. From the study of Hirth and Tullis (1992), both 'as-is' and water-added experiments are plotted (Fig. 8a). Together with the naturally deformed samples of the Ruby Gap Duplex (Dunlap *et al.* 1997) and the Tonale samples, there are now three data sets available to define the transitions of recrystallization mechanisms in a strain rate versus temperature diagram (Fig. 8a, b). Straight lines in a log strain rate versus $1/T$ plot imply constant stress conditions (Hirth *et al.* 2001). As a consequence of constant stress conditions, the grain sizes along the transitions should be constant. However, the recrystallized grain size in the experimental samples is much smaller than in the Tonale samples for corresponding recrystallization mechanisms (compare the experimentally and naturally deformed samples in Fig. 3). Thus, it remains questionable that the transitions between different recrystallization

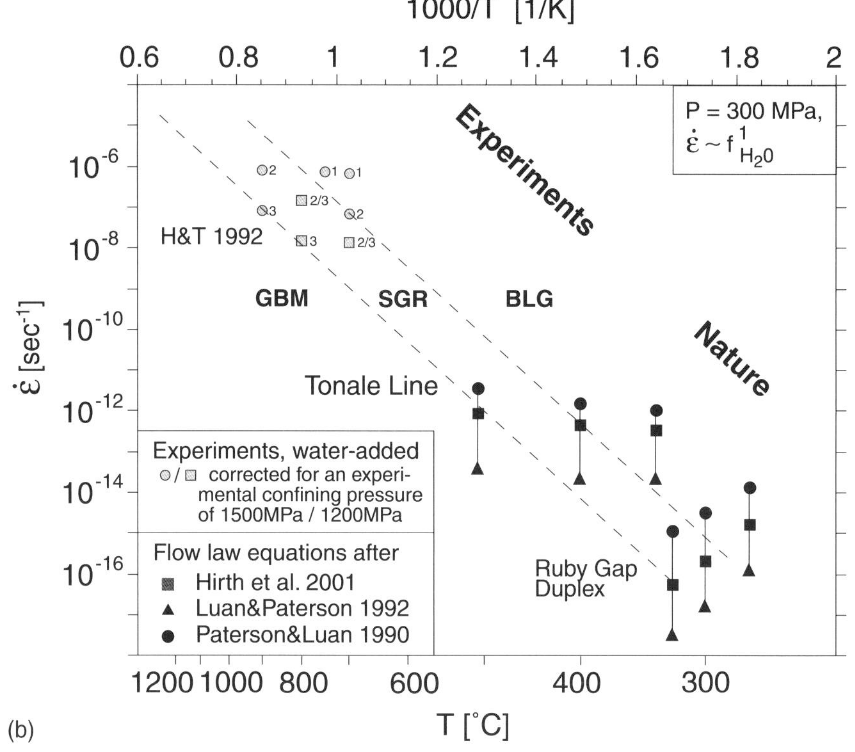

Fig. 8. (**b**) Strain rate versus temperature diagram (as in **a**) with the microstructural correlations made in this study indicated by dashed lines. From the experimental data set of Hirth & Tullis (1992, dislocation creep regimes are indicated by the numbers) only the water-added samples are used. Water fugacity is normalized to a confining pressure of 300 MPa.

mechanisms correspond to constant stress curves for a given flow law. The great temperature difference between corresponding natural and experimental recrystallization mechanisms might also effect a difference in the recrystallized grain size as discussed by de Bresser *et al.* (1998, 2001). At present, however, there is no experimental proof for this difference in quartz, as pointed out above.

Several adjustments must be made for correlating experimental and natural sample sets. The experiments of Hirth and Tullis (1992) were carried out in a solid medium apparatus, and the derived flow stresses tend to be overestimated. Water has pronounced weakening effects at high confining pressures and a correction needs to be applied to extrapolate flow stresses to lower pressures. One procedure for such a correction is to introduce the water fugacity with an exponent m as a coefficient into the flow law, as has been described before (cf. Kohlstedt *et al.* 1995; Post *et al.* 1996; Hirth *et al.* 2001). Thus, the experimental data points in Fig. 8a are shifted to slower strain rates (or higher temperatures) if normalized to the confining pressures of the naturally deformed samples at 300 MPa. The iso-stress lines of the Luan and Paterson (1992) flow law (dashed lines) and of the Gleason and Tullis (1995) flow law (solid lines) for uncorrected experimental data points are added for comparison.

The microstructural similarities between all sample sets are striking, so that it appears safe to correlate the natural and experimental conditions of temperature and strain rate for dynamic recrystallization. It is suggested that natural and experimental transitions of dynamic recrystallization mechanisms may follow a straight curve fit in a logarithmic strain rate versus $1/T$ plot. Such a fit of the boundaries between the three recrystallization mechanisms is indicated by the two dashed lines in Fig. 8b. Only the water-added samples have been used from the data set. These experiments have been carried out in NaCl- and pyrophyllite-cells (Jan Tullis, pers. comm.). Johannes (1978) found that the effective confining pressure in NaCl-cell assemblies corresponds to the externally measured pressure and that the effective confining pressure in pyrophyllite assemblies is about 20% lower than in NaCl-assemblies. Based on his findings we recalculated strain rates for the Hirth & Tullis (1992) data set with a water fugacity for an effective confining pressure of 1.5 GPa and 1.2 GPa for the salt and the pyrophyllite assemblies, respectively (Fig. 8b).

The diagram shows that the recrystallization mechanism transitions primarily depend on temperature and strain rate. Changing one of these two variables will change the dominant dynamic recrystallization mechanism. According to the reported predominance of relative constant natural strain rate conditions on the order of 10^{-14} to $10^{-12}\,s^{-1}$ (e.g. Mercier *et al.* 1977; Pfiffner & Ramsay 1982; Suppe 1985; Twiss & Moores 1992) microstructural changes have usually been related to temperature changes as is obvious for e.g. the Tonale mylonites or the Ruby Gap Duplex (Dunlap *et al.* 1997). However, major changes in the strain rate and related flow stress conditions at relatively small temperature variations have also been observed in conjunction with changes in the dominant recrystallization microstructure (e.g. White 1982; Stipp 2001). Because of the importance of dynamic recrystallization of quartz in mylonitic shear zones this recrystallization mechanism map is applicable to most natural deformation conditions. In contrast to the deformation mechanism maps previously constructed for quartz (e.g. Rutter 1976; White 1976; Etheridge and Wilkie 1979), which are largely based on theoretical assumptions and calculations, the recrystallization mechanism map in Fig. 8b is derived from experimental as well as natural data. The map includes the complete field of dislocation creep and covers the dynamic recrystallization microstructures of quartz occuring in natural mylonitic fault rocks. Additional experimental and geological field studies encompassing a broad spectrum of metamorpic conditions (e.g. Stöckhert *et al.* 1999; Zulauf 2001) are necessary to better constrain this dynamic recrystallization mechanism map.

We consider this study as an attempt to demonstrate that estimations of temperature and strain rate may be made by characterizing dynamic recrystallization microstructures. The geological application is fairly straightforward. The dynamic recrystallization mechanism (BLG, SGR, GBM) has to be determined from the microstructure of quartz (preferably in pure quartz aggregates). If the deformation temperature can be determined (e.g. from syndeformational mineral assemblages), then the range of strain rates can be determined from the recrystallization mechanism map. Alternatively, if age data are available which allow for an estimate of the strain rate within a shear zone, the recrystallization mechanism map provides an independent tool for the determination of deformation temperatures. The recognition of the recrystallization mechanism is valuable because it may help to constrain deformation temperatures and strain rates within a usually narrow range of natural fault zone conditions in the upper crust.

Summary and conclusions

Dynamic recrystallization microstructures of quartz have been used as a link between nature and experiment. The recrystallization microstructures of the eastern Tonale mylonites have been correlated with the experimental dislocation creep regimes of Hirth & Tullis (1992). From the Tonale mylonites, temperature, stress and strain rate data can be determined, using the Twiss theoretical recrystallized grain size piezometer for all recrystallization mechanisms and quartzite flow laws adjusted for water fugacity. The data from naturally deformed quartz can help to constrain piezometers and flow laws. There is a need for quartz paleopiezometers to be calibrated for different recrystallization mechanisms. Natural and experimental data sets of dynamic recrystallization microstructures and dislocation creep regimes can be plotted on a strain rate versus temperature diagram. The fields of the different recrystallization microstructures are separated by transition zones, and their trend is consistent in nature and experiment. Using microstructural correlations, a recrystallization mechanism map can be derived from this diagram. This map may be used for either estimating strain rates (if the temperature is known) or temperature (if the strain rate is known) from studying the recrystallization microstructure.

We are greatly indebted to J. Tullis and G. Hirth for their continued and enthusiastic support and their critical scrutiny toward this study. J. Tullis critically reviewed an early version of this paper and suggested a number of important improvements. The constructive reviews of J. L. Urai and J. C. White and the editorial comments of J. H. P. de Bresser and G. Pennock are gratefully acknowledged. This work is funded by the Swiss National Science Foundation (grants 20-49562.96 and 2000-055420.98).

References

Bailey, J. E. & Hirsch, P. B. 1962. The recrystallization process in some polycrystalline metals. *Proceedings of the Royal Society of London*, **A267**, 11–30.

CHRISTIE, J. M., ORD, A. & KOCH, P. S. 1980. Relationship between recrystallized grain size and flow stress in experimentally deformed quartzite. *EOS Transactions*, **61/17**, 377.

DE BRESSER, J. H. P., PEACH, C. J., REIJS, J. P. J. & SPIERS, C. J. 1998. On dynamic recrystallization during solid state flow: effects of stress and temperatures. *Geophysical Research Letters*, **25**, 3457–3460.

DE BRESSER, J. H. P., TER HEEGE, J. H. & SPIERS, C. H. 2001. Grain size reduction by dynamic recrystallization: can it result in major rheological weakening? *International Journal of Earth Sciences*, **90**, 28–45.

DEL MORO, A., PARDINI, G., QUERCIOLI, C., VILLA, I. M. & CALLEGARI, E. 1983. Rb/Sr and K/Ar chronology of Adamello granitoids, Southern Alps. *Memorie della Società Geologica Italiana*, **26**, 285–299.

DERBY, B. 1990. Dynamic recrystallization and grain size. *In*: BARBER, D. J. & MEREDITH, P. G. (eds) *Deformation Processes in Minerals, Ceramics and Rocks*. Unwin Hyman, London, 354–364.

DERBY, B. & ASHBY, M. F. 1987. On dynamic recrystallization. *Scripta Metallurgica*, **21**, 879–884.

DOHERTY, R. D., HUGHES, D. A. *ET AL.* 1997. Current issues in recrystallization: a review. *Materials Science and Engineering*, **A238**, 219–274.

DRURY, M. R., HUMPHREYS, F. J. & WHITE, S. H. 1985. Large strain deformation studies using polycrystalline magnesium as a rock analogue. Part II: dynamic recrystallisation mechanisms at high temperatures. *Physics of the Earth and Planetary Interiors*, **40**, 208–222.

DUNLAP, W. J., HIRTH, G. & TEYSSIER, C. 1997. Thermomechanical evolution of a ductile duplex. *Tectonics*, **16**, 983–1000.

EDWARD, G. H., ETHERIDGE, M. A. & HOBBS, B. E. 1982. On the stress dependence of subgrain size. *Textures and Microstructures*, **5**, 127–152.

ETHERIDGE, M. A. & WILKIE, J. C. 1979. Grainsize reduction, grain boundary sliding and the flow strength of mylonites. *Tectonophysics*, **58**, 159–178.

EVANS, B., FREDRICH, J. T. & WONG, T.-F. 1990. The brittle-ductile transition in rocks: recent experimental and theoretical progress. *In*: DUBA A. G., DURHAM, W. B., HANDIN, J. W. & WANG, H. F. (eds) *The Brittle-Ductile Transition in Rocks. Geophysical Monograph*, **56**, 1–20.

FITZ GERALD, J. D. & STÜNITZ, H. 1993. Deformation of granitoids at low metamorphic grade. I: reactions and grain size reduction. *Tectonophysics*, **221**, 269–297.

FRISCH, W., KUHLEMANN, J., DUNKL, I. & BRUGEL, A. 1998. Palinspastic reconstruction and topographic evolution of the eastern Alps during late Tertiary tectonic extrusion. *Tectonophysics*, **297**, 1–15.

FROST, H. J. & ASHBY, A. F. 1982. *Deformation Mechanism Maps*. Pergamon Press, Oxford.

GLEASON, G. C. & TULLIS, J. 1993. Improving flow laws and piezometers for quartz and feldspar aggregates. *Geophysical Research Letters*, **20**, 2111–2114.

GLEASON, G. C. & TULLIS, J. 1995. A flow law for dislocation creep of quartz aggregates determined with the molten salt cell. *Tectonophysics*, **247**, 1–23.

GOETZE, C. & KOHLSTEDT, D. L. 1973. Laboratory study of dislocation climb and diffusion in olivine. *Journal of Geophysical Research*, **78**, 5961–5971.

GUILLOPÉ, M. & POIRIER, J. P. 1979. Dynamic recrystallization during creep of single-crystalline halite: an experimental study. *Journal of Geophysical Research*, **84**, 5557–5567.

HACKER, B. R., YIN, A., CHRISTIE, J. M. & DAVIS, G. A. 1992. Stress magnitude, strain rate, and rheology of extended middle continental crust inferred from quartz grain sizes in the Whipple Mountains, California. *Tectonics*, **11**, 36–46.

HACKER, B. R., YIN, A., CHRISTIE, J. M. & SNOKE, A. W. 1990. Differential stress, strain rate, and temperatures of mylonitization in the Ruby Mountains, Nevada: implications for the rate and duration of uplift. *Journal of Geophysical Research*, **95**, 8569–8580.

HANDY, M. R. 1989. Deformation regimes and the rheological evolution of fault zones in the lithosphere: the effects of pressure, temperature, grain size and time. *Tectonophysics*, **163**, 119–152.

HEARD, H. C. & CARTER, N. L. 1968. Experimentally induced 'natural' intragranular flow in quartz and quartzite. *American Journal of Science*, **266**, 1–42.

HIPPERT, J. & EGYDIO-SILVA, M. 1996. New polygonal grains formed by dissolution-redeposition in quartz mylonite. *Journal of Structural Geology*, **18**, 1345–1352.

HIRTH, G. & TULLIS, J. 1992. Dislocation creep regimes in quartz aggregates. *Journal of Structural Geology*, **14**, 145–159.

HIRTH, G. & TULLIS, J. 1994. The brittle-plastic transition in experimentally deformed quartz aggregates. *Journal of Geophysical Research*, **99**, 11731–11747.

HIRTH, G., TEYSSIER, C. & DUNLAP, W. J. 2001. An evaluation of quartzite flow laws based on comparisons between experimentally and naturally deformed rocks. *International Journal of Earth Sciences*, **90**, 77–87.

HOBBS, B. E. 1968. Recrystallization of single crystals of quartz. *Tectonophysics*, **6**, 353–401.

HOBBS, B. E., MEANS, W. D. & WILLIAMS, P. F. 1976. *An Outline of Structural Geology*. John Wiley & Sons, New York.

HUMPHREYS, F. J. & HATHERLY, M. 1996. *Recrystallization and Related Annealing Phenomena*. Pergamon Press, Oxford.

JESSELL, M. W. 1987. Grain-boundary migration microstructures in a naturally deformed quartzite. *Journal of Structural Geology*, **9**, 1007–1014.

JOHANNES, W. 1978. Pressure comparing experiments with NaCl, AgCl, talc, and pyrophyllite assemblies in a piston cylinder apparatus. *Neues Jahrbuch Mineralogische Monatshefte*, **2**, 84–92.

KIRBY, S. H. 1980. Tectonic stresses in the lithosphere: Constraints provided by the experimental deformation of rocks. *Journal of Geophysical Research*, **85**, 6353–6363.

KOCH, P. S. 1983. *Rheology and Microstructures of Experimentally Deformed Quartz Aggregates*. PhD thesis, University of California.

KOCH, P. S., CHRISTIE, J. M., ORD, A. & GEORGE, R. P. 1989. Effect of water on the rheology of experimentally deformed quartzite. *Journal of Geophysical Research*, **94**, 13975–13996.

KOHLSTEDT, D. L. & WEATHERS, M. S. 1980. Deformation induced microstructures, paleopiezometers and differential stresses in deeply eroded fault zones. *Journal of Geophysical Research*, **85**, 6269–6285.

KOHLSTEDT, D. L., EVANS, B. & MACKWELL, S. J. 1995. Strength of the lithosphere: constraints imposed by laboratory experiments. *Journal of Geophysical Research*, **100**, 17587–17602.

KRONENBERG, A. & TULLIS, J. 1984. Flow strengths of quartz aggregates: grain size and pressure effects due to hydrolytic weakening. *Journal of Geophysical Research*, **89**, 4281–4297.

KÜSTER, M. & STÖCKHERT, B. 1999. High differential stress and sublithostatic pore fluid pressure in the ductile regime – microstructural evidence for short-term post-seismic creep in the Sesia Zone, Western Alps. *Tectonophysics*, **303**, 263–277.

LAUBSCHER, H. 1991. The arc of the Western Alps today. *Eclogae geologicae Helvetiae*, **84**, 631–659.

LUAN, F. C. & PATERSON, M. S. 1992. Preparation and deformation of synthetic aggregates of quartz. *Journal of Geophysical Research*, **97**, 301–320.

LUTON, M. J. & SELLARS, C. M. 1969. Dynamic recrystallization in nickel and nickel-iron alloys during high temperature deformation. *Acta Metallurgica*, **17**, 1033–1043.

MAINPRICE, D. H. 1981. *The Experimental Deformation of Quartz Polycrystals*. PhD thesis, Australian National University.

MCQUEEN, H. J., KNUSTAD, O., RYUM, N. & SOLBERG, J. K. 1985. Microstructural evolution in Al deformed to strains of 60 at 400 °C. *Scripta Metallurgica*, **19**, 73–78.

MEANS, W. D. 1983. Microstructure and micromotion in recrystallization flow of Octachloropropane: a first look. *Geologische Rundschau*, **72**, 511–528.

MERCIER, J.-C. C. 1980. Magnitude of continental lithospheric stress inferred from rheomorphic petrology. *Journal of Geophysical Research*, **85**, 6293–6303.

MERCIER, J.-C. C., ANDERSON, D. A. & CARTER, N. L. 1977. Stress in the lithosphere: inferences from steady state flow of rocks. *Pure and Applied Geophysics*, **115**, 199–226.

MÜLLER, W. 1998. *Isotopic Dating of Deformation Using Microsampling Techniques: The Evolution of the Periadriatic Fault System (Alps)*. PhD thesis, ETH Zürich.

ORD, A. & CHRISTIE, J. M. 1984. Flow stresses from microstructures in mylonitic quartzites of the Moine Thrust zone, Assynt area, Scotland. *Journal of Structural Geology*, **6**, 639–654.

PANOZZO HEILBRONNER, R. 1992. The autocorrelation function: an image processing tool for fabric analysis. *Tectonophysics*, **212**, 351–370.

PANOZZO HEILBRONNER, R. & PAULI, C. 1993. Integrated spatial and orientation analysis of quartz c-axes by computer-aided microscopy. *Journal of Structural Geology*, **15**, 369–382.

PANOZZO HEILBRONNER, R. & PAULI, C. 1994. Orientation and misorientation imaging: integration of microstructural and textural analysis. *In*: BUNGE, H. J., SIEGESMUND, S., SKROTZKI, W. & WEBER, K. (eds) *Textures of Geological Materials*. DGM Informationsgesellschaft Verlag, Oberursel, 147–164.

PARRISH, D. K., KRIVZ, A. L. & CARTER, N. L. 1976. Finite-element folds of similar geometry. *Tectonophysics*, **32**, 183–207.

PATERSON, M. S. 1987. Problems in the extrapolation of laboratory rheological data. *Tectonophysics*, **133**, 33–43.

PATERSON, M. S. & LUAN, F. C. 1990. Quartzite rheology under geological conditions. *In*: KNIPE, R. J. & RUTTER, E. H. (eds) *Deformation mechanisms, rheology and tectonics*. Geological Society, London, Special Publication, **54**, 299–307.

PFIFFNER, O. A. & RAMSAY, J. G. 1982. Constraints on geological strain rates: arguments from finite-strain states of naturally deformed rocks. *Journal of Geophysical Research*, **87**, 311–321.

POIRIER, J.-P. 1985. *Creep of Crystals*. Cambridge University Press, Cambridge.

POIRIER, J. P. & GUILLOPÉ, M. 1979. Deformation-induced recrystallization of minerals. *Bulletin de la Société française de Minéralogie et de Cristallographie*, **102**, 67–74.

POST, A. & TULLIS, J. 1999. A recrystallized grain size paleopiezometer for experimentally deformed feldspar aggregates. *Tectonophysics*, **303**, 159–173.

POST, A. D., TULLIS, J. & YUND, A. 1996. Effects of chemical environment on dislocation creep of quartzite. *Journal of Geophysical Research*, **101**, 22143–22155.

RALEIGH, C. B. & KIRBY, S. H. 1970. Creep in the upper mantle. *Mineralogical Society of America Special Paper*, **3**, 113–121.

RANALLI, G. 1984. Grain size distribution and flow stress in tectonites. *Journal of Structural Geology*, **6**, 443–447.

RASBAND, W. 1996. *NIH Image*. National Institute of Health, Research Services Branch NIMH.

RUTTER, E. H. 1976. The kinetics of rock deformation by pressure solution. *Philosophical Transactions of the Royal Society of London*, **A283**, 203–219.

RUTTER, E. H. 1995. Experimental study of the influence of stress, temperature, and strain on the dynamic recrystallization of Carrara marble. *Journal of Geophysical Research*, **100**, 24651–24663.

SCHMID, S. M. & HANDY, M. R. 1991. Towards a genetic classification of fault rocks: geological usage and tectonophysical implications. *In*: MÜLLER, D. W., MCKENZIE, J. A. & WEISSERT, H. (eds) *Controversies in Modern Geology*. Academic Press, London, 339–361.

SCHMID, S. M. & KISSLING, E. 2000. The arc of the Western Alps in the light of new data on deep crustal structure. *Tectonics*, **19**, 62–85.

SCHMID, S., AEBLI, H. R., HELLER, F. & ZINGG, A. 1989. The role of the Periadriatic Line in the tectonic evolution of the Alps. *In*: COWARD, M. P., DIETRICH, D. & PARK, R. G. (eds) *Alpine Tectonics*. Geological Society, London, Special Publications, **45**, 153–171.

SCHMID, S. M., PATERSON, M. S. & BOLAND, J. N. 1980. High temperature flow and dynamic recrystallization in Carrara marble. *Tectonophysics*, **65**, 245–280.

SHIMIZU, I. 1998. Stress and temperature dependence of recrystallized grain size: a subgrain misorientation model. *Geophysical Research Letters*, **25**, 4237–4240.

SMITH, C. S. & GUTTMAN, L. 1953. Measurement of internal boundaries in the three-dimensional structures by random sectioning. *Transactions of the American Institute of Mining and Metallurgical Engineers, Journal of Metals*, **197**, 81–87.

SNOKE, A. W., TULLIS, J. & TODD, V. R. (eds.) 1998. *Fault-Related Rocks*. Princeton University Press, Princeton.

STECK, A. & HUNZIKER, J. 1994. The Tertiary structural and thermal evolution of the Central Alps – compressional and extensional structures in an orogenic belt. *Tectonophysics*, **238**, 229–254.

STIPP, M. 2001. *Dynamic Recrystallization of Quartz in Fault Rocks from the Eastern Tonale Line (Italian Alps)*. PhD thesis, University of Basel.

STIPP, M. & SCHMID, S. M. 1998. Evidence for the contemporaneity of movements along the Tonale line and the intrusion of parts of the Adamello batholith. *Memorie di Scienze Geologiche*, **50**, 89–90.

STIPP, M., STÜNITZ, H., HEILBRONNER, R. AND SCHMID, S. M. 2002. The eastern Tonale fault zone: a 'natural laboratory' for crystal plastic deformation of quartz over a temperature range from 250 °C to 700 °C. *Journal of Structural Geology*, **24**, 1861–1884.

STÖCKHERT, B., BRIX, M. R., KLEINSCHRODT, R., HURFORD, A. J. & WIRTH, R. 1999. Thermochronometry and microstructures of quartz – a comparison with experimental flow laws and predictions on the temperature of the brittle-plastic transition. *Journal of Structural Geology*, **21**, 351–369.

SUPPE, J. 1985. *Priniciples of Structural Geology*. Prentice-Hall, Englewood Cliffs, New Jersey.

TAKEUCHI, S. & ARGON, A. S. 1976. Steady state creep of single-phase-crystalline matter at high temperatures. *Journal of Materials Science*, **11**, 1542–1566.

TÖDHEIDE, K. 1972. Water at high temperatures and pressures. *In*: FRANKS, F. (ed) *Water – A Comprehensive Treatise*, **1**, Plenum Press, New York, 463–514.

TULLIS, J., STÜNITZ, H., TEYSSIER, C. & HEILBRONNER, R. 2000. Deformation microstructures in quartzofeldspathic rocks. *In*: JESSEL, M. W. & URAI, J. L. (eds) *Stress, Strain and Structure, a Volume in Honor of W D Means. Journal of the Virtual Explorer* **2**. World Wide Web Address: http://www.virtualexplorer.com.au/VEjournal/2000Volumes/Volume2/www/contribs/tullis/SlideSet/deformation_micro. html

TWISS, R. J. 1977. Theory and applicability of a recrystallized grain size paleopiezometer. *Pure and Applied Geophysics*, **115**, 227–244.

TWISS, R. J. 1980. Static theory of size variation with stress for subgrains and dynamically recrystallized grains. *In*: USGS (ed) *Proceedings of the IX. Conference, Magnitude of Deviatoric Stresses in the Earth's Crust and Upper Mantle. Open File Report*, **80–625**, Menlo Park, California, 665–683.

TWISS, R. J. 1986. Variable sensitivity piezometric equations for dislocation density and subgrain diameter and their relevance to olivine and quartz. *In*: HOBBS, B. E. & HEARD, H. C. (eds) *Mineral and Rock Deformation: Laboratory Studies*. Geophysical Monograph, **36**, 247–261.

TWISS, R. J. & MOORES, E. M. 1992. *Structural Geology*. W. H. Freeman and Company, New York.

URAI, J. L., MEANS, W. D. & LISTER, G. S. 1986. Dynamic recrystallization of minerals. *In*: HOBBS, B. E. & HEARD, H. C. (eds) *Mineral and Rock Deformation: Laboratory Studies*. Geophysical Monograph, **36**, 161–199.

VAN DAALEN, M., HEILBRONNER, R. & KUNZE, K. 1999. Orientation analysis of localized shear deformation in quartz fibres at the brittle-ductile transition. *Tectonophysics*, **303**, 83–107.

VAN DER WAL, D., CHOPRA, P., DRURY, M. & FITZ GERALD, J. 1993. Relationship between dynamically recrystallized grain size and deformation conditions in experimentally deformed olivine rocks. *Geophysical Research Letters*, **20**, 1479–1482.

WEERTMAN, J. 1970. The creep strength of the Earth's mantle. *Reviews of Geophysics and Space Physic*, **8**, 145–168.

WERLING, E. 1992. *Tonale-, Pejo- und Judicarien-Linie: Kinematik, Mikrostrukturen und Metamorphose von Tektoniten aus räumlich interferierenden aber verschiedenaltrigen Verwerfungszonen*. PhD thesis, ETH Zürich.

WHITE, J. C. 1982. Quartz deformation and the recognition of recrystallization regimes in the Flinton Group conglomerates, Ontario. *Canadian Journal of Earth Sciences*, **19**, 81–93.

WHITE, S. 1979. Grain and sub-grain size variations across a mylonite zone. *Contributions to Mineralogy and Petrology*, **70**, 193–202.

WHITE, S. H. 1973. Syntectonic recrystallisation and texture development in quartz. *Nature*, **244**, 276–277.

WHITE, S. H. 1976. The effects of strain on the microstructures, fabrics and deformation mechanisms in quartz. *Philosophical Transactions of the Royal Society of London*, **A283**, 69–86.

ZULAUF, G. 2001. Structural style, deformation mechanisms and paleodifferential stress along an exposed crustal section; constraints on the rheology of quartzofeldspathic rocks at supra- and infrastructural levels (Bohemian Massif). *Tectonophysics*, **332**, 211–237.

The effect of static annealing on microstructures and crystallographic preferred orientations of quartzites experimentally deformed in axial compression and shear

RENÉE HEILBRONNER[1] & JAN TULLIS[2]

[1]*Department of Earth Sciences, Basel University, Bernoullistrasse 32, CH-4056 Basel, Switzerland*
(e-mail: Renee.Heilbronner@unibas.ch)
[2]*Department of Geological Sciences, Brown University, Providence RI 02912, USA*

Abstract: Quartzite samples were experimentally deformed with partial to complete dynamic recrystallization by axial compression (strain magnitude of 0.8 to 1.4) and by general shear (strain magnitude of 1.3 to 2.8) in each of the three dislocation creep regimes, and subsequently annealed with complete static recrystallization at the deformation temperature for 120 hours. The *c*-axis crystallographic preferred orientation (CPO), 3D grain size distribution, grain boundary surface shape, and misorientation density were measured before and after annealing. The effect of annealing on the CPO was minor, but the microstructure was greatly changed. All of the annealed samples were completely recrystallized. The recrystallized grain size increased by a factor of 2 to 5, and was greatest for samples deformed at lowest temperature. The grain boundary lobateness (PARIS factor) and misorientation density were reduced significantly. The CPOs for all the deformed samples were relatively unchanged by annealing, although the strengths are somewhat decreased; for sheared samples the asymmetry was preserved. The results suggest microstructural criteria for recognizing the occurrence of static annealing and for estimating the dynamically recrystallized grain size relevant for paleopiezometry from annealed samples.

Experimental deformation studies have documented that the microstructures produced by dislocation creep, including original and recrystallized grain size and shape as well as crystallographic preferred orientations (CPOs), contain a great deal of useful information concerning deformation conditions. One of the difficulties in obtaining similar information from the microstructures preserved in naturally deformed rocks is the uncertainty as to whether they have been modified by post-deformation static annealing. In particular it may be difficult to distinguish the dynamic recrystallization microstructures produced by high temperature, slow strain rate deformation from post-deformation static recrystallization. Such distinctions are important if one contemplates using the recrystallized grain size piezometer to infer the flow stress (e.g. White 1979). In addition, the patterns of CPOs contain useful information concerning the deformation temperature and symmetry, but there has been some indication in the geological literature that static annealing might randomize or change the CPO (Griggs *et al.* 1960; Green 1967), although more recent experimental studies on several materials indicate that annealing has relatively minor effects on the CPO (e.g. ice, Wilson 1982; octochloropropane, Ree & Park 1997; calcite, Covey-Crump 1997).

Annealing includes several processes: static recovery and recrystallization, involving reduction in stored strain energy by intragranular annihilation and rearrangement of dislocations as well as grain boundary migration recrystallization driven by local differences in strain energy; and grain growth, involving reduction in interfacial energy by coarsening of the recrystallized grain size. Annealing is important in materials processing and there is a large amount of literature on the process, including its effects on CPOs (for a recent review see Humphreys & Hatherly 1995). For 'cold-worked' materials that have been deformed by crystal plasticity at relatively low temperature with no dynamic recrystallization, annealing involves recovery as well as the 'nucleation' of strain-free recrystallized grains and their growth into the cold-worked material. All 'nuclei' are pre-existing parts of the deformed structure. For 'hot-worked' materials which have been deformed at relatively high temperatures and which have undergone partial or complete dynamic recrystallization, annealing involves recovery within the remaining porphyroclasts as well as growth of the recrystallized grains ('metadynamic recrystallization'). In lightly hot-worked materials with no dynamic recrystallization, annealing may not produce any static recrystallization.

From: DE MEER, S., DRURY, M. R., DE BRESSER, J. H. P. & PENNOCK, G. M. (eds) 2002. *Deformation Mechanisms, Rheology and Tectonics: Current Status and Future Perspectives*. Geological Society, London, Special Publications, **200**, 191–218. 0305-8719/02/$15

Quartz is abundant in the continental crust, generally forming one of the weaker phases in the ductile regime, and considerable experimental work has been done on its properties and deformation behavior. The distinctive microstructures produced by different mechanisms of dynamic recrystallization in the three regimes of dislocation creep identified in experimentally deformed samples by Hirth & Tullis (1992) have been found in quartzites naturally deformed at lower temperatures and slower strain rates (Dunlap *et al.* 1997; Stoeckhert *et al.* 1999; Stipp *et al.* 1999, 2002). The recognition of these distinctive microstructures, in combination with experimental flow laws (Luan & Paterson 1992; Gleason & Tullis 1995; Hirth *et al.* 2001), dynamically recrystallized grain size piezometer relations (Twiss 1977) and petrological and geochronological constraints, can provide significant information on the thermomechanical history of an area (e.g. Dunlap *et al.* 1997). However, accurate interpretation of the microstructures requires information on the possible effects of post-deformational modifications. We have conducted a study to determine the changes in grain size and shape and CPOs produced by annealing for quartzites experimentally deformed in axial compression and in general shear in each of the three dislocation creep regimes. We have not evaluated the kinetics of the microstructural changes, as that was outside the scope of the present study.

Methods

Experiments

Experiments were conducted on Black Hills quartzite, a fine-grained (d ≈ 100 m) quartzite (Fig. 1a, b) with ~1% impurities and a negligible shape and crystallographic preferred orientation (Heilbronner, 2002). This material has been used in several previous experimental deformation studies (e.g. Kronenberg & Tullis 1984; Tullis & Yund 1989; Hirth & Tullis 1992).

Both axial compression and general shear experiments were performed in a modified Griggs-type deformation apparatus, using all-NaCl assemblies (Fig. 1c). Samples were mechanically sealed using an inner Pt jacket and an outer thin Ni sleeve. Brazil quartz pistons cut at 45° were used as the forcing blocks for the shear experiments; platinum foil between these pistons and the upper and lower zirconia pistons allowed for sideways slip of the quartz pistons to compensate for the shear offset. We call these experiments general shear to emphasize that during the experiments, there is siginificant compaction across the shear zone, i.e., it is not true simple shear.

All experiments were performed at a confining pressure of 1.5 GPa. The temperature, strain rate and water content were chosen so as to produce one axially shortened and one general shear sample in each of the dislocation creep regimes of Hirth & Tullis (1992). Experimental conditions are listed in Table 1; the strain magnitudes of the sheared samples (γ of 2.1 to 7.2) are significantly higher than those of the axially compressed samples (shortened 60 to 80%).

The samples were deformed in all-NaCl assemblies at temperatures close to the melting temperature for NaCl. Previous comparisons of sample strengths in such assemblies with those in molten salt assemblies indicate that the confining medium contributes very little to the externally measured sample strength (Gleason & Tullis 1993). In order to eliminate all frictional contributions to the measured sample strengths, we performed experiments using NaCl as the sample in assemblies otherwise identical to those used for our deformed quartzites, for the two different strain rates used in each geometry. We subtracted the weak slope of the appropriate force record for these NaCl sample experiments from the force records of the quartzite experiments.

We plotted stress–strain curves in terms of strain magnitude so that the axial compression and general shear samples can be more directly compared; the amounts of shortening strain and shear strain for the samples are listed in Table 1. Following Schmid *et al.* (1987), the

Fig. 1. Starting material and sample assembly. (**a**) Optical micrographs of undeformed Black Hills quartzite (BHQ). (**b**) Grain size distribution of undeformed Black Hills quartzite, plotted as volume % versus radius of equivalent sphere (magnification ×2.5, 530 grains evaluated). Misorientation image magnification used was ×5, average grain diameter ~100 μm. (**c**) Geometry for experiments performed in axial compression (top; sample diameter is 6.3 mm) and shear (bottom; sample thickness is 1.27 mm). (**d**) Optical micrograph of entire sample (w858), after axial deformation and before annealing. Note heterogeneous deformation; analysed area (box) is centered in the zone of highest strain and greatest recrystallization. (**e**) Optical micrograph of Brazil quartz forcing blocks and sample (w920) after shearing deformation and before annealing; analysed area indicated by black box; cracks in quartz pistons developed during unloading, same scale as (**d**). Micrographs (a), (d) and (e) are taken with circular polarization (see caption, Fig. 11).

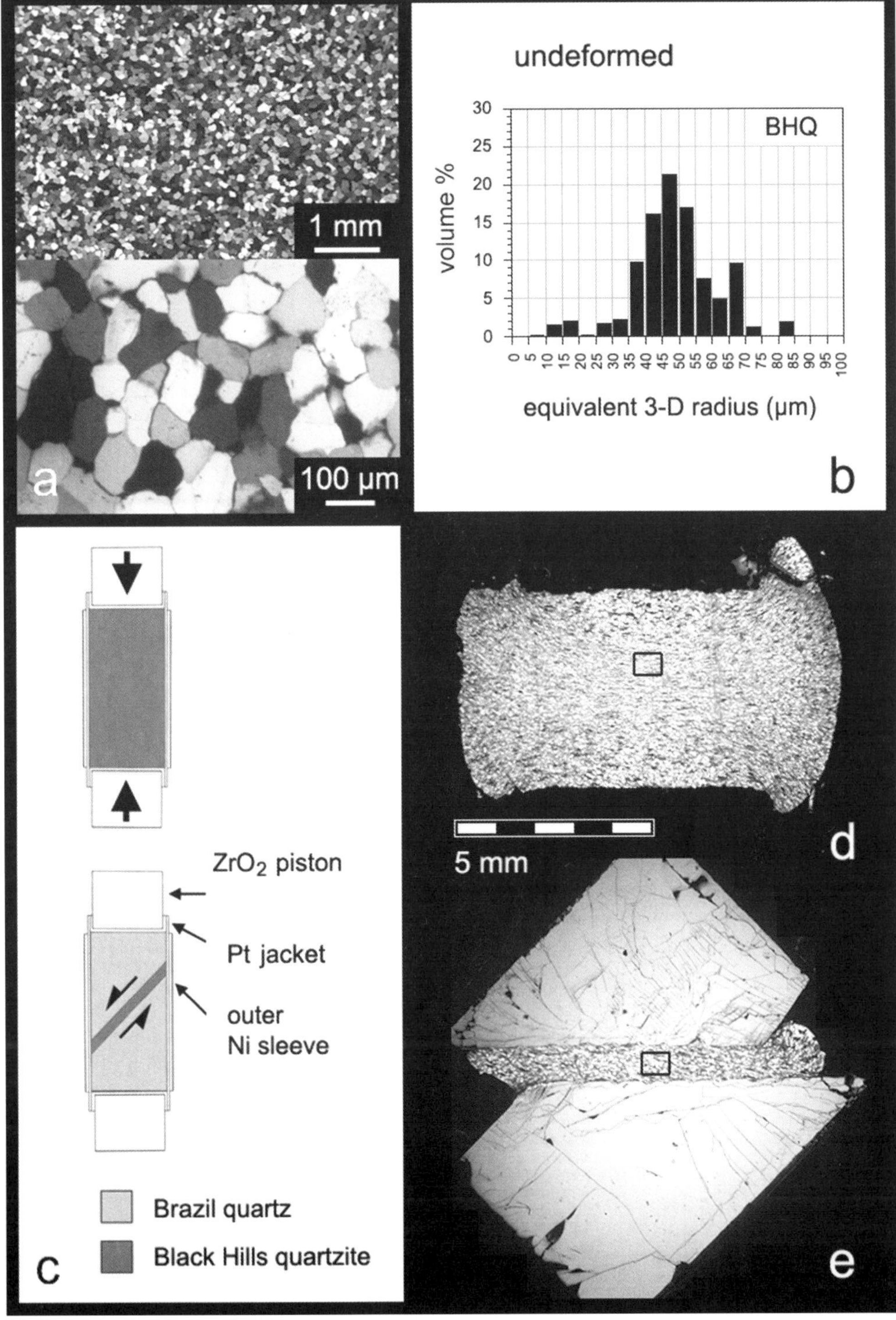
1 mm
100 μm
a
undeformed
BHQ
volume %
0
5
10
15
20
25
30
0 5 10 15 20 25 30 35 40 45 50 55 60 65 70 75 80 85 90 95 100
equivalent 3-D radius (μm)
b
ZrO2 piston
Pt jacket
outer
Ni sleeve
Brazil quartz
Black Hills quartzite
c
5 mm
d
e

Table 1. *Experimental conditions*

Regime	Deformed sample #	Confining pressure (GPa)	T (°C)	H_2O (wt%)	Axial strain rate (s^{-1})	Max. short. strain ε (%)	Max. strain magn.		Flow stress $\Delta\sigma$ (MPa)[‡]	Annealing T (°C)§	Annealed sample #
1	w871	1.5	850	0	1.5×10^{-5}	77	1.82		650	850	w875
2	w872	1.5	900	0.17	1.5×10^{-5}	58	1.07		310	900	w874
3	w858	1.5	900	0.17	1.5×10^{-6}	78	1.84		180	900	w860
					Shear strain rate (s^{-1})	Max. shear strain γ[*]	Max. strain magn.	Flattening strain (%) (magn.)[†]	Flow stress $\Delta\sigma = 2\tau$ (MPa)		
1	w940	1.5	850	0	3×10^{-5}	4.32	2.13	31.6 (0.46)	510	850	w943
2	w946	1.5	875	0.17	3×10^{-5}	7.18	2.81	48.8 (0.82)	420	875	w948
3	w920	1.5	900	0.17	1.5×10^{-5}	2.13	1.31	20.0 (0.27)	190	900	w921
3	w935	1.5	915	0.17	3×10^{-5}	5.65	2.49	44.0 (0.71)	210	915	w938

[*] Sample w920 has lower strain rate due to thicker shear zone.
[†] Strain magnitude equivalent to flattening strain is given in brackets.
[‡] Flow stress of axial samples is given as differential stress $\Delta\sigma$. For sample w871 final stress is indicated. Flow stress of sheared samples is given in terms of differential stress $\Delta\sigma = 2\tau$.
§Annealing time is 4 days for all samples.

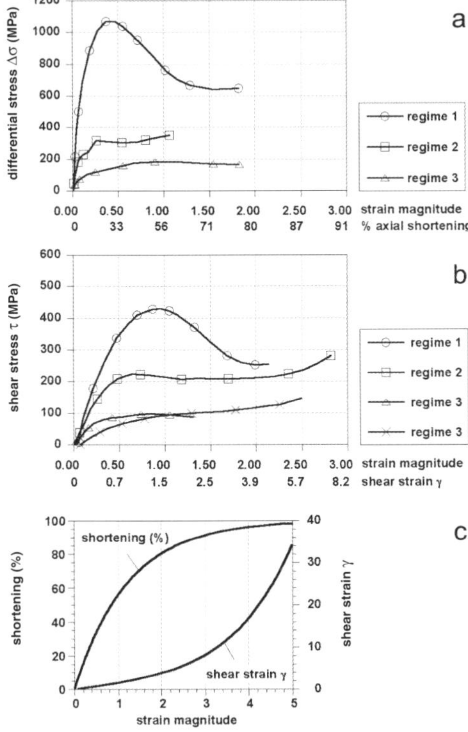

Fig. 2. Curves of stress versus strain magnitude for axial and shearing experiments. See text for methods of calculation. (**a**) Differential stress versus strain magnitude and % axial shortening. (**b**) Shear stress versus strain magnitude and shear strain, γ. (**c**) Relationship between strain magnitude and axial shortening (%) and shear strain (γ). For calculation of strain magnitude, see text.

strain magnitude is calculated as:

$$3^{-1/2}[(e_1 - e_2)^2 + (e_2 - e_3)^2 + (e_3 - e_1)^2]^{1/2}$$

where e_1, e_2 and e_3 are the logarithmic strains (after Nadai 1963). The relationships between shortening and shear strains and strain magnitude are shown in Fig. 2c. For the axially compressed samples, stress–strain curves were prepared using the standard assumption that samples remained right circular cylinders of constant volume. However, two modifications were introduced, as follows. (a) The length of the undeformed part of the sample near the piston-sample interfaces was subtracted from the original length. (b) At high strains the fraction of the sample cross sectional area that remains between the pistons (i.e. the physically relevant cross section that is subjected to the compressive stresses between the piston) is less than what would be calculated using the standard constant volume assumption; therefore only half of the standard area correction was applied (Fig. 2a). For the sheared samples, we corrected the stress–strain curves for the measured thinning that accompanied progressive shear as well as for the decreasing area of overlap of the pistons (Fig. 2b). The work hardening shown by the two samples with highest shear strain reflects the relatively small remaining area of piston overlap.

Following deformation, samples were rapidly quenched under load to 300 °C, and then the load and pressure were gradually reduced. Each sample was cut in half in a vertical plane (parallel to the shortening direction or the shear direction). One half was impregnated with epoxy and used for making an optical thin section, and the other half was trimmed and weld-sealed in Pt, placed into a modified all-NaCl assembly, and annealed at 1.5 GPa and the deformation temperature for 120 hours. A thin section was prepared for each annealed sample that corresponded exactly to the facing side of the original cut through the deformed sample, to enable the closest possible correlation of the deformed and annealed microstructures. The analysed areas in the deformed and annealed samples were chosen to be in the central homogeneously strained region to avoid mechanical and thermal end effects (Figs 1d, e).

Image analysis

On doubly polished thin sections (20 µm thickness), areas displaying typical microstructures were selected for analysis, as shown in Figs 1d and e. A Zeiss microscope was used with magnifications of ×5, ×10, and ×20 to acquire images of 1.5, 0.4, and 0.1 mm^2 size, respectively. Using a Canon EOS DCS3ir digital camera for capture, the matrix size of the digital version of the images was typically of the order of 1200 × 900 pixels. For detailed observations, in particular for grain shape analysis, it was necessary to use a high magnification (×20) and a relatively small area that includes a relatively small number of grains. For statistical evaluations, such as 3D grain size distributions and *c*-axis pole figures, we used lower magnifications (×5, ×10) which included larger numbers of grains.

For grain size analysis and CPO determination of very small grains, it is best to use ultra-thin sections in order to avoid situations where many small grains are superposed. It is not always possible to prepare ultra-thin sections, but by using high aperture objectives (high magnifications) and matching condensor settings, the depth of field can be decreased such that only

Fig. 3. *C*-axis orientation images of axial deformed samples, before annealing (left column) and after annealing (right column). Compression direction vertical. Scale bar and colour look-up table also apply to Figure 4. Black parts are holes, cracks, etc. which have been masked and excluded from analysis.

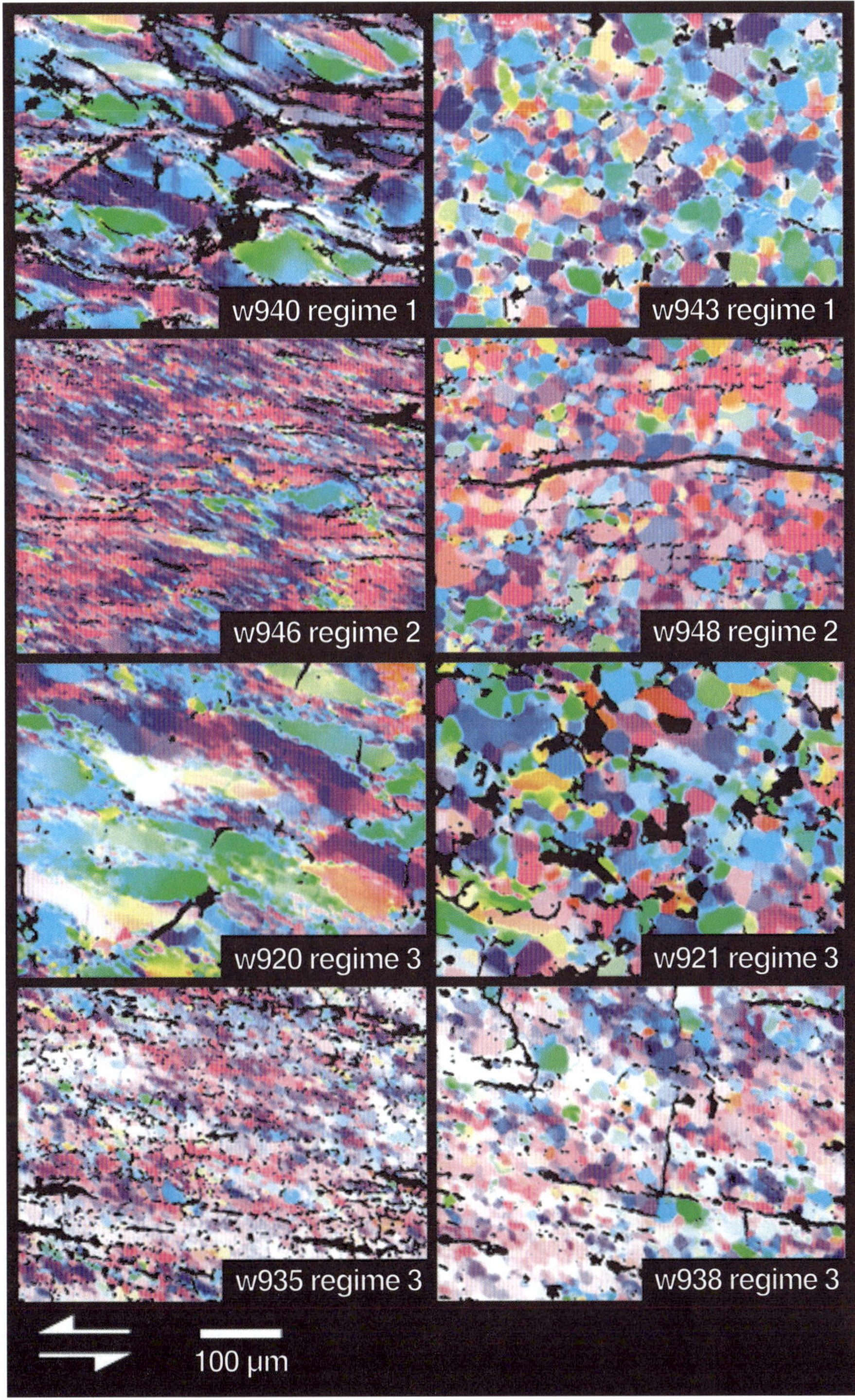

Fig. 4. *C*-axis orientation images of samples deformed by shearing, before annealing (left column) and after annealing (right column). Same scale bar and colour look-up table as in Figure 3. Black parts are holes, cracks, etc. which have been masked and excluded from analysis.

the top layer of the thin section (thickness less than 1 μm) contributes any significant information to the image. The situation is one of 'optical thinning' (similar to the laser-scanning microscope where the concept of 'optical sectioning' is used), and enables both the definition of small-scale curvatures of grain boundaries as well as the recognition of the interference colours of small grains. We made use of high aperture optics and maximum aperture settings, especially for grain size determinations and CPO calculations of samples with very fine recrystallized grains.

The azimuth and inclination planes of the *c*-axis orientation images (COIs) were calculated using the CIP method (computer integrated polarization microscopy) introduced by Panozzo-Heilbronner & Pauli (1993) and described by Heilbronner (2000*a*). A standard colour look-up table (CLUT) was used to colour code the *c*-axis directions, depicting vertical trending axes in blue, horizontal axes in yellow, and axes normal to the plane of the thin section in white (see Figs 3 and 4). Misorientation images, representing the angle between the *c*-axis orientation of any given pixel and an external reference direction (east, north, up), and orientation gradient images, representing the angular difference between any given pixel and its neighbouring pixels, were also calculated.

From the misorientation images, grain boundary maps were derived using the NIH Image

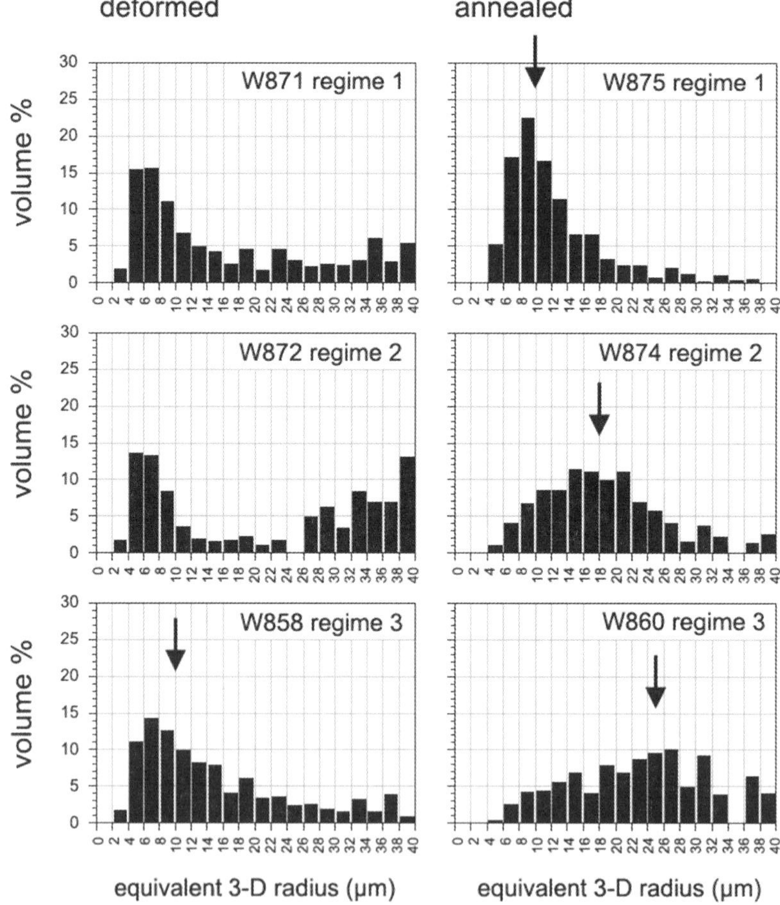

Fig. 5. Grain size distributions of axially deformed samples, before annealing (left column) and after annealing (right column), plotted as volume % versus radius of equivalent sphere. 2D grain boundary maps were prepared from misorientation images (magnification ×5 for deformed samples, ×5 for annealed samples); from these the distributions of cross sectional areas were determined; from these the 3D grain size distributions were calculated. Note: the maximum radius included is 40 μm, corresponding to the largest remaining porphyroclast. Arrows indicate mode of recrystallized grain size.

public domain software and the Lazy Grain Boundary macro (Heilbronner 2000*b*). On any given map, up to 8000 grain cross sections were measured and compiled as histograms of equivalent radii (i.e., radii of circles having the same areas as the measured shapes). The StripStar program (Heilbronner & Bruhn 1998; Heilbronner 2000*b*) was used to calculate the distributions of parent spheres that give rise to the measured distributions of 2D radii. The results are presented as volume-weighted histograms of 3D radii (Figs 5, 6 and 7).

The axially deformed samples were evaluated using magnifications of ×5 and ×20, resulting

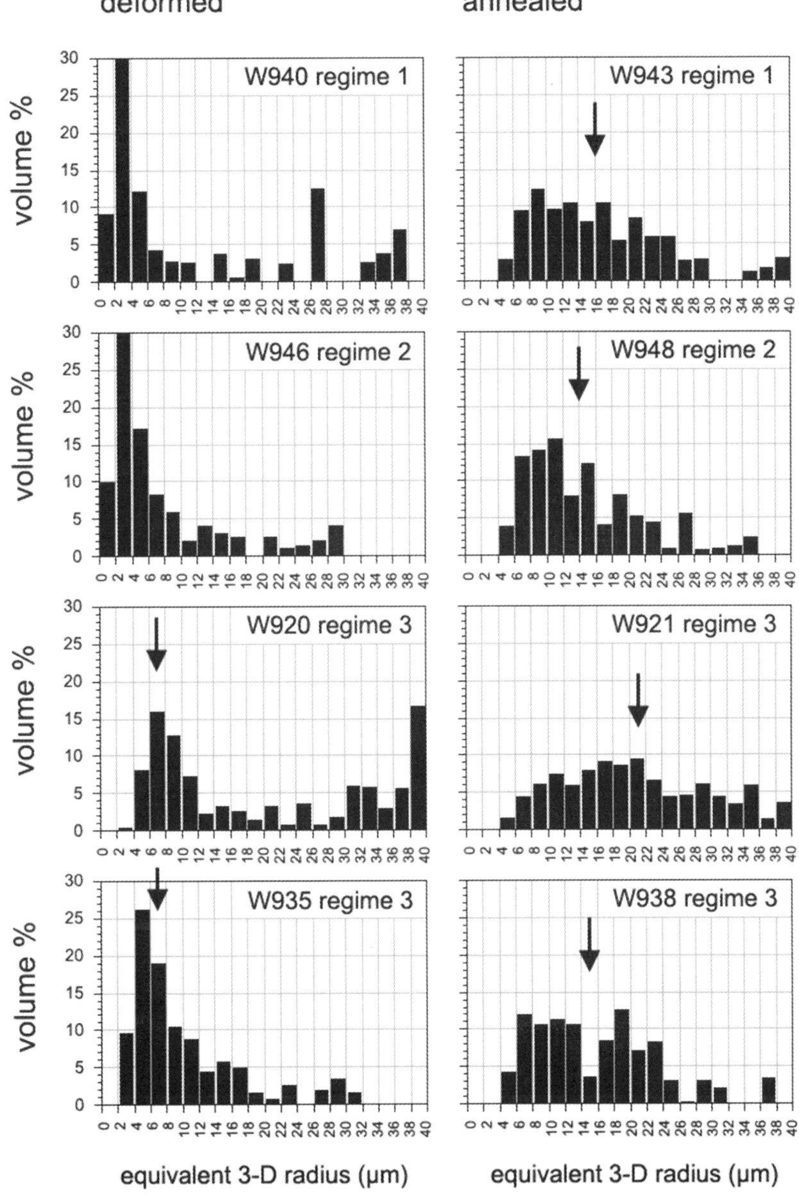

Fig. 6. Grain size distributions of samples deformed by shearing, before annealing (left column) and after annealing (right column), plotted as volume % versus radius of equivalent sphere. 2D grain boundary maps were prepared from misorientation images (magnification ×10 for deformed samples, ×5 for annealed samples). For explanation see Figure 5. Arrows indicate mode of recrystallized grain size.

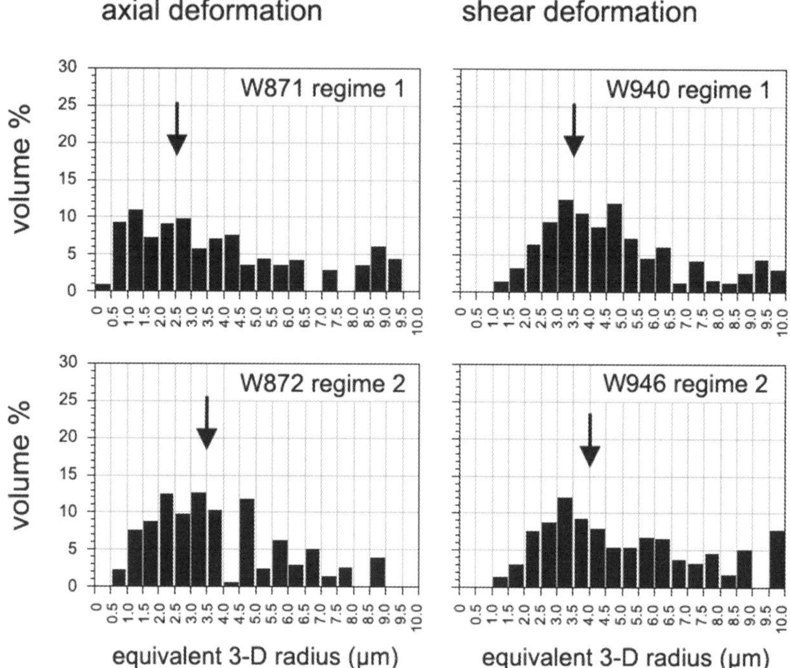

Fig. 7. Details of grain size distributions of axially deformed samples (left) and samples deformed by shearing (right), plotted as volume % versus radius of equivalent sphere. Misorientation image magnifications used were ×20 for axial samples, ×10 for sheared samples. The maximum radius is 10 μm, corresponding to the largest recrystallized grain. Arrows indicate mode of recrystallized grain size.

in pixel sizes of 1.27 μm and 0.32 μm, respectively, whereas the coarser-grained annealed samples were evaluated using ×5 only. The sheared samples (deformed and annealed) were all evaluated using a magnification of ×10, with a pixel size of 0.64 μm. Low magnifications of ×5 or ×10 are well suited for analysis of the annealed samples, which have relatively large average grain sizes, ranging between 20 and 50 μm. For the small grain sizes of samples deformed in regimes 1 and 2, however, a minimum magnification of ×10 to ×20 or even larger is necessary (Fig. 7).

The minimum magnification for grain size analysis (not for grain size detection) can be estimated by the following consideration. We have estimated that a statistically significant analysis requires the mode of the 3D distribution to be about 3 times larger than the smallest detected grain (in 2D cross section), and that there should be 10 to 20 pixels (3–4 pixel diameter) to define the minimum grain cross section. In other words, the smallest grain size (smallest mode of distributed grain size) that can be detected reliably by magnifications of ×5, ×10 and ×20 is 12 μm, 6 μm and 3 μm, respectively.

To represent the full range of the grain size distributions, including recrystallized grains and porphyroclasts (Figs 5 and 6), magnifications of ×5 and ×10 were used. For the exact determination of very small grain sizes of samples deformed in regimes 1 and 2, magnifications of ×10 and ×20 were used, and only the fraction below 20 μm diameter was evaluated (Fig. 7).

C-axis pole figures were obtained by evaluating the histograms of the azimuth and inclination values of the orientation images (see Heilbronner 2000*a*). The resulting pole figures (crystallographic preferred orientations, or CPOs) are area-weighted, as is done with CPOs determined using texture goniometry (Figs 8 and 9). By masking out the areas (and orientations) of the porphyroclasts, partial CPOs of the recrystallized fraction of the deformed samples were obtained. The maximum of the pole density is listed on the figures. The significance of the texture index does not depend on the number of pixels per image, but on the number of independent measurements, i.e. on the number of grains that are analysed in a given area. Assuming an original grain diameter of ~100 μm, the number of original grains per sampling area is

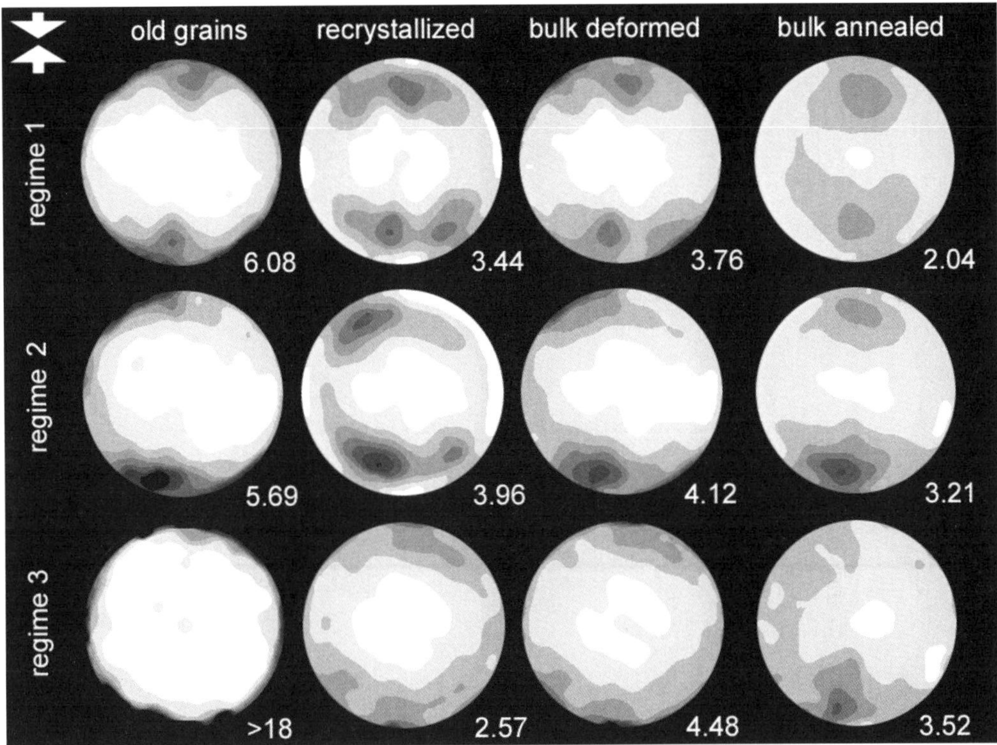

Fig. 8. Bulk and partial *c*-axis pole figures (CPOs) of axially deformed samples before annealing (left 3 columns) and after annealing (right column). CPOs were calculated from *c*-axis orientation images (CIP method); measured area = 1.5 mm^2 (*c*. 300 original grains). CPO maximum density is indicated in lower right of each figure. Contours at 0.5 times uniform.

300 for a magnification of ×5, 80 for ×10, and 20 for ×20. The actual number, of course, depends on the proportion of old grains still present after recrystallization. The number of recrystallized grains depends on the grain size and the percentage of recrystallized volume. For a grain size of 10 μm and 50% recrystallized volume, the number of recrystallized grains per image is approximately 15000, 4000, and 1000 for magnifications of ×5, ×10 and ×20, respectively.

Orientation gradients depict the angular difference in the *c*-axis orientation between any given pixel and its four neighbours (Heilbronner 2000*a*). The maximum difference is 90° (*c*-axes are assumed to be non-polar). If the orientation gradient is zero, the neighbouring pixels are parallel; if the gradient is small, minor lattice distortions are present; if the gradient is intermediate, subgrain boundaries may be present; if the gradient is large, grain boundaries are indicated. Since the number of pixels corresponding to grain boundaries or subgrain boundaries is small compared to the number of pixels representing grain interiors, the relatively high values of orientation gradients that typically occur along grain boundaries do not significantly influence the histogram of orientation gradients. Thus histograms of orientation gradient images, representing the 'density of misorientation' (Fig. 10), are a good measure of the strain energy that is stored in the microstructure.

Results

General description of deformed samples and correlation with creep regimes

In both axial compression and general shear, we have deformed a sample in each of the three dislocation creep regimes identified by Hirth & Tullis (1992), because these regimes are characterized by different processes of dynamic recrystallization which impart distinctive deformation microstructures and might have different responses to annealing. Optical micrographs of the analysed areas of the deformed and annealed samples (taken with circular polarization) are

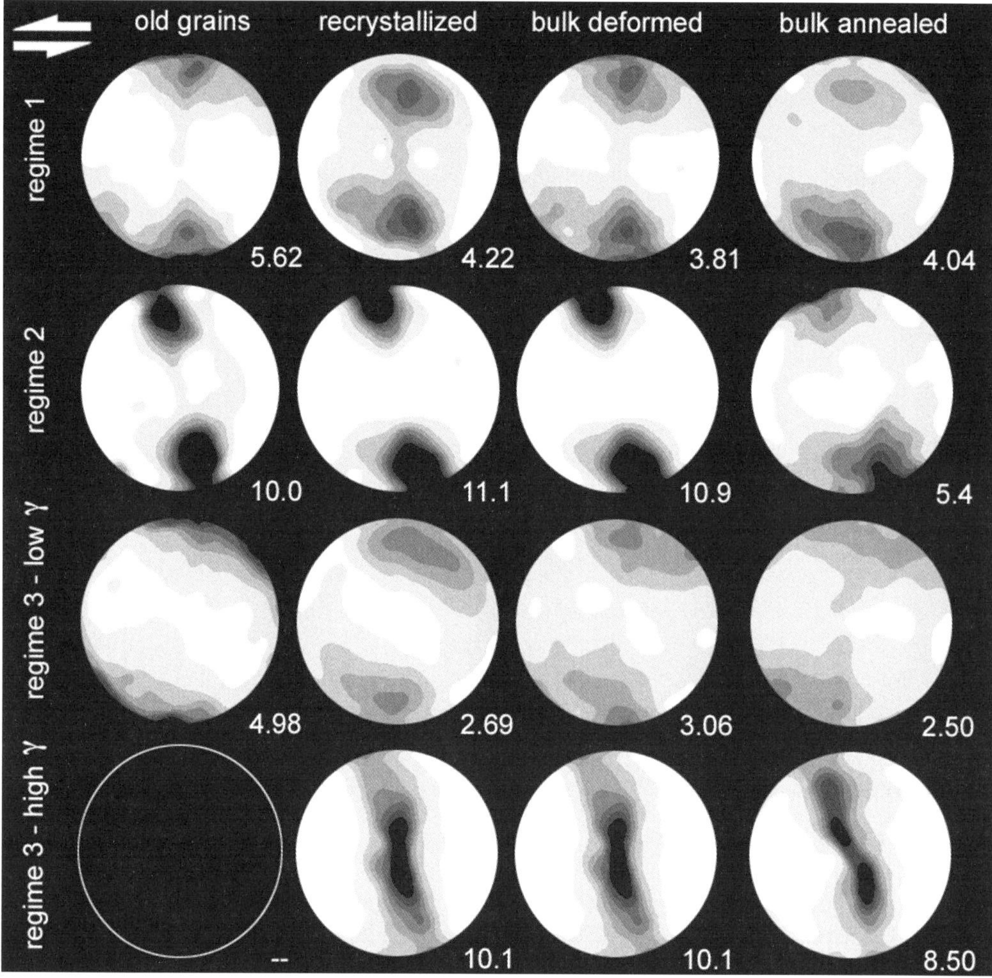

Fig. 9. Bulk and partial c-axis pole figures (CPOs) of sheared samples before annealing (left 3 columns) and after annealing (right column). Measured area = 0.4 mm^2 (*c.* 80 original grains) except for w935, where a combination of 3 areas (1.2 mm^2, *c.* 240 original grains) was used. Contours at 0.5 times uniform.

shown in Figure 11 (axial compression) and Figure 12 (general shear).

Regime 1 dislocation creep occurs at the lowest temperatures and/or fastest strain rates and is characterized by the highest flow stress, regime 2 is intermediate, and regime 3 occurs at the highest experimental temperatures and/or slowest strain rates and is characterized by the lowest flow stress. It is important to realize that the regime boundaries are transitional rather than sharp, and that additional and distinctive higher temperature microstructures have been described from naturally deformed quartz aggregates (Stipp *et al.* 2002).

Curves of differential stress and shear stress versus strain magnitude are shown in Figs 2a and b for the axially compressed and sheared samples, respectively. To better compare the flow stresses, however, we will use the equivalence of $2\tau = \Delta\sigma$ (Schmid *et al.* 1987), and in the following we will discuss the flow stresses in terms of differential stress for both axial and shear experiments (see also Table 1). For regime 1, the axial compression and sheared samples were deformed at almost identical conditions except for geometry; the overall behavior of both is the same, in that they both show initial work hardening, a peak stress and subsequent weakening to about 60% of the peak stress. The levels of the peak and flow stress, however, are higher in the axial experiments. The difference between the axial and sheared samples may be within experimental reproducibility, but they may also be due in part to the different

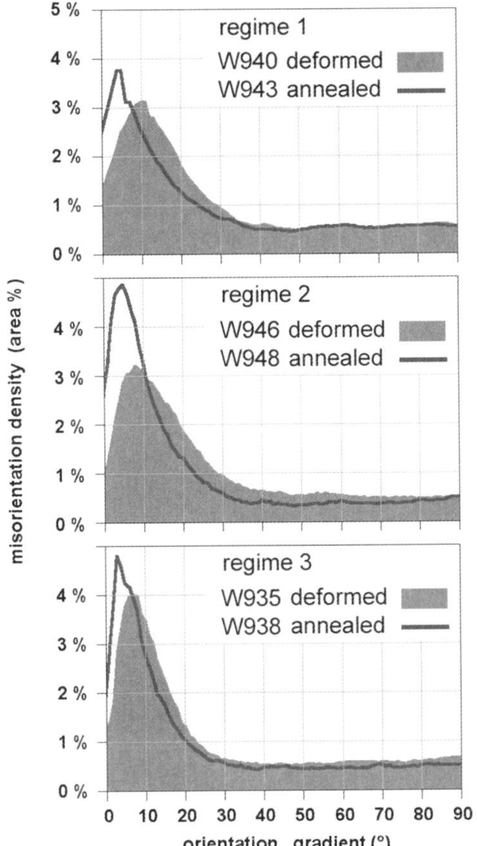

Fig. 10. Distribution of orientation gradients (= density of local misorientation) for sheared (grey fill) and annealed (heavy line) samples, obtained from histograms (bin size = 1°) of orientation gradient images (at ×10 magnification, 100 μm = 156 pixels, true size of pixel = $(0.64\,mm)^2$). Orientation gradient images show the angular difference of *c*-axis orientation between all pixels and their four neighbours. (The maximum possible difference is 90°.) High gradients indicate grain boundaries, low gradients (*c.* 10°) are typical for subgrain boundaries, very low gradients indicate strain-free interiors of grains.

corrections that were applied when calculating the stress–strain curves. The strain magnitude attained at peak stress is also different for the axial (0.4) and shear experiments (0.9), corresponding to 30% shortening and a shear strain γ of 1.5, respectively. The reason for this difference is not clear.

For regime 2, the axial and sheared samples were deformed at somewhat different conditions; the flow stress of the sheared sample was higher (i.e. it was closer to the transition from regime 1 to 2), and at the end of the experiment ($\gamma \approx 6$) its strength actually rose above that of the sheared regime 1 sample. For regime 3, the two sheared samples were deformed at a faster strain rate and had a higher flow stress than the axial sample (i.e. they were closer to the transition from regime 2 to 3), and at the end of the higher shear strain experiment ($\gamma \approx 5$) the strength was actually above that of the axial compression regime 2 sample. These complexities are pertinent to understanding the microstructures and especially the measured sizes of the dynamically recrystallized grains in these samples.

The microstructures characteristic of the three dislocation creep regimes are briefly reviewed below, with reference to the optical microstructures of our deformed samples.

Regime 1: In regime 1 dislocation creep, dislocation climb is very difficult and dynamic recovery occurs by grain boundary bulging recrystallization; dynamic recrystallization produces strain weakening and mechanical steady state requires substantial recrystallization. In the lower strain axial compression sample (Figs 3, 11) 50% of the volume is recrystallized, and the original grains are heavily work hardened, with strong and patchy undulatory extinction. Grain boundaries are sutured on a very fine scale, producing tiny (1–2 μm) recrystallized grains (at the limit of optical resolution), which are initially strain-free and thus accommodate most of the sample strain. The higher strain general shear sample (Figs 4, 12) has approximately the same amount of recrystallization (50%); continued deformation within the recrystallized regions has allowed the old grains to deform at a slower strain rate and lower stress than initially, so that some of them contain subgrains and are slightly elongate.

Regime 2: In regime 2 dislocation creep, the dominant process of dynamic recovery is dislocation climb, resulting in subgrain formation. Dynamic recrystallization occurs by subgrain rotation and does not produce weakening; thus mechanical steady state is achieved after low strain, and well before complete recrystallization. In the lower strain axial compression sample (Figs 3, 11) 40% of the volume is recrystallized, and the original grains are fairly uniformly flattened, with some deformation lamellae (mostly sub-basal in orientation) and sweeping undulatory extinction. The subgrains within deformed original grains are roughly the same size as the recrystallized grains that occur dominantly along grain boundaries. The higher strain (and higher flow stress) general shear sample (Figs 4, 12) is 90% recrystallized; it has some remnant original grains which are elongate oblique to

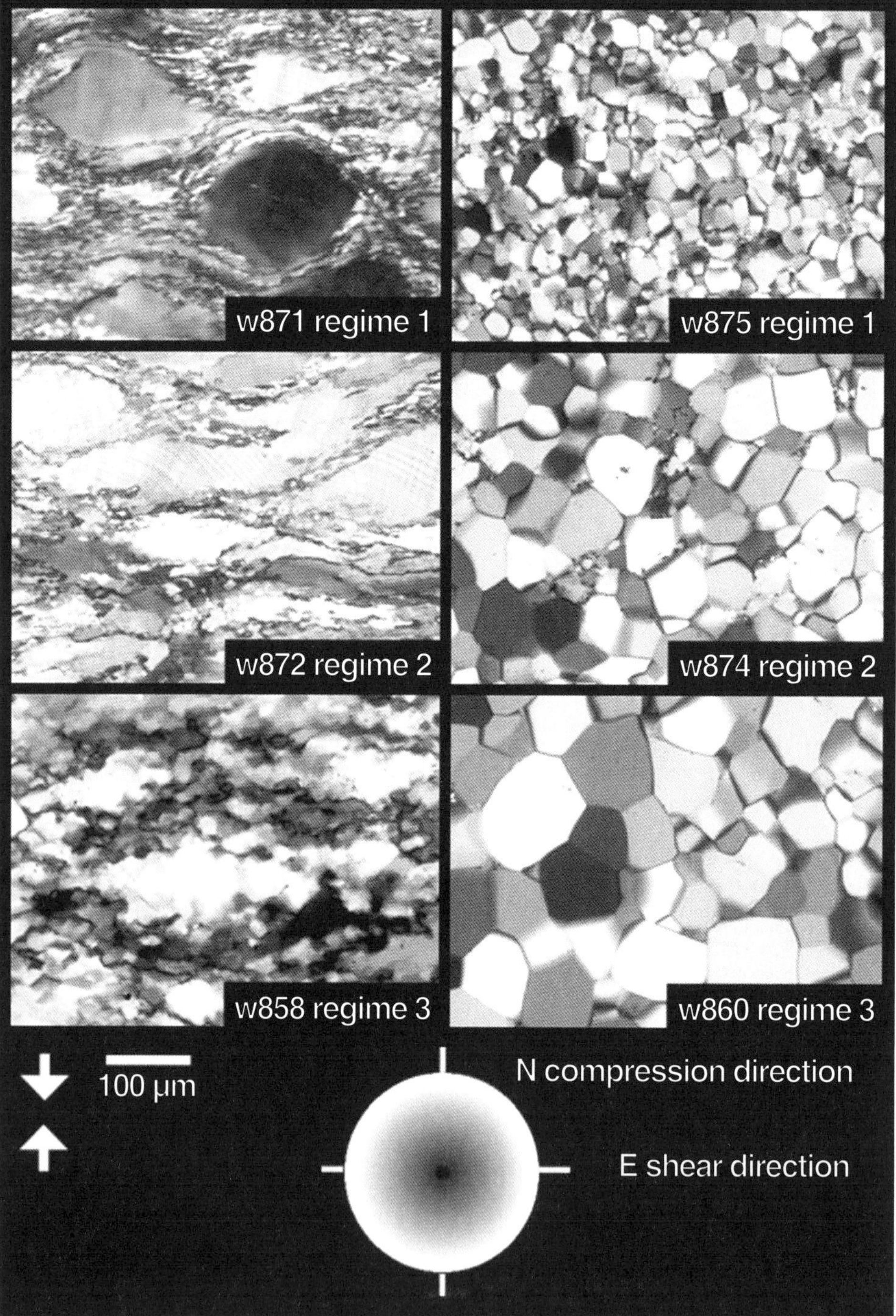

Fig. 11. Detailed optical micrographs of axially deformed samples, before annealing (left column) and after annealing (right column) using circular polarization (light grains = *c*-axes parallel to plane of thin section, dark grains = *c*-axes parallel to viewing direction; see look-up table). Compression vertical; flattening plane horizontal. Scale bar and colour look-up table also apply to Figure 12. Note: areas shown in this figure are not the same as those in Figure 3.

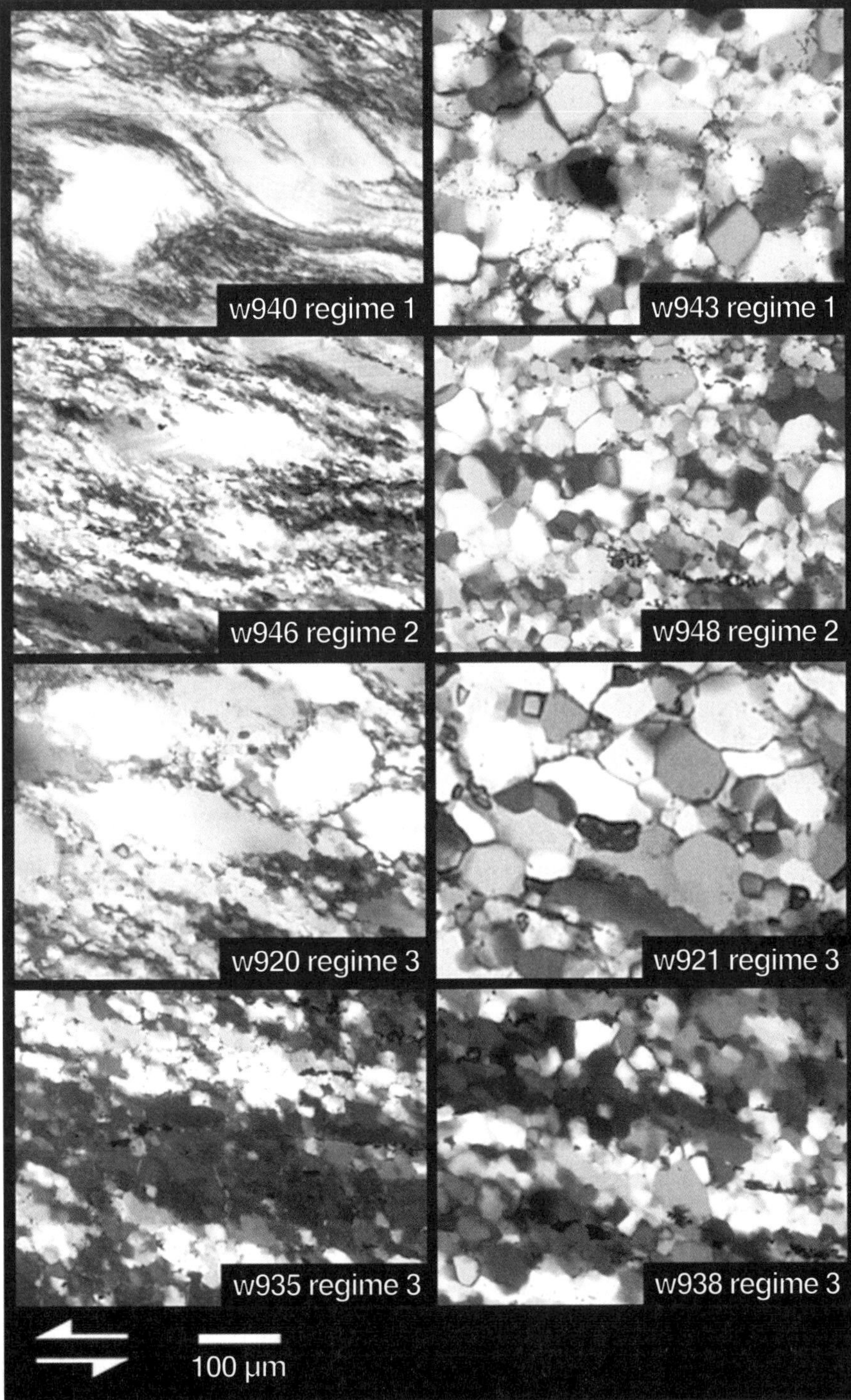

Fig. 12. Detailed optical micrographs of samples deformed by shearing, before annealing (left column) and after annealing (right column), using circular polarization. Shear plane horizontal; shearing sense sinistral. Same scale bar and colour look-up table as in Figure 11. Note: areas shown in this figure are not the same as those in Figure 4.

the foliation, defining an SC fabric (e.g. Lister & Snoke 1984). Some fine-scale grain boundary bulging occurred toward the end of the experiment as the stress increased.

Regime 3: In regime 3 dislocation creep, dislocation climb remains the dominant recovery process, producing larger subgrains than in regime 2. However, grain boundary mobility is very high; most recrystallized grains form by grain boundary migration and dynamic recrystallization is almost complete after low sample strain. In the low strain axial compression sample (Figs 3, 11) 85% of the volume is recrystallized; only a few remnants of original grains remain unrecrystallized, and the recrystallized grain size is much larger than the corresponding subgrain size. The two general shear samples both have higher strengths than the axial compression sample (Fig. 2), indicating that they were deformed at conditions closer to the boundary with regime 2. In the lower strain sample (Figs 4, 12) 45% of the volume is recrystallized; somewhat deformed remnants of original grains reside in a matrix of moderately large recrystallized grains. The higher strain sample (Figs 4, 12) is completely recrystallized, and the recrystallized grains are slightly elongate oblique to the foliation, again defining an SC fabric but with a higher angle

General description of annealed samples

Three factors influence the annealed microstructures: the deformed microstructure (grain size and shape, CPO, and defect structure), the annealing temperature and the annealing time. In this study, the annealing temperature was always the same as the deformation temperature, and the annealing time was always 120 hours. The only factor that was varied was the deformation microstructure. The micrographs of the annealed samples (Figs 11, 12) show complete recrystallization as well as grain growth. For the axial compression samples, the grain size of the annealed samples increases from regime 1 to 2 to 3, even though the annealing temperature was the same for the regime 2 and 3 samples. The annealing microstructures for the sheared samples are somewhat more complex, reflecting the stress increase at the end of the two high-strain experiments. The grain size distributions of the statically recrystallized (annealed) samples are broader than those of the dynamically recrystallized grains, and there is no a clear increase in average grain size with increasing temperature. These points are discussed in more detail below.

Grain sizes and grain size distribution

The 3D grain size distributions for the compressed and sheared samples are shown in Figs 5 and 6, respectively. The histograms of the deformed samples are unimodal (if recrystallization is complete) or bimodal (if recrystallization is incomplete). The relative proportions of recrystallized and unrecrystallized volume can be estimated from these histograms. In the axial regime 1 and 2 samples (w871 and w872, Fig. 5), for example, the histogram bars below 12 µm (diameter of 24 µm) amount to *c.* 50% and *c.* 40%, respectively, corresponding to 50% and 40% recrystallized volume (Table 2).

For bimodal grain size distributions, the mode of the smaller peak is the average recrystallized grain size, and the mode of the larger peak is the average size of the remaining porphyroclasts. The latter has a tendency to be smaller than the original grain size because the original grains get smaller as they are consumed by recrystallization. Moreover, if the proportion of recrystallized grains is relatively high, the number of porphyroclasts becomes very small, such that the mode of the larger grain sizes may not have much meaning.

The peak for the recrystallized grains is statistically more stable than the peak for the large grains. From our previous experience with dynamic recrystallization, we expect the peak to be symmetric, but in reality it is commonly skewed on account of several factors. (1) If the mode of the recrystallized grain size is very small, the peak is skewed because of insufficient resolution towards the small end of the distribution. (2) If the flow stress is not constant, the equilibrium grain size may increase or decrease during the deformation. Because of the insufficient resolution of ×5 magnification for analyzing recrystallized grain sizes in samples deformed in regimes 1 and 2, additional evaluations at higher magnification (×20) were carried out, allowing better resolution in the histograms (Fig. 7). The results of the evaluations at low and at high magnification were compared to visual inspection, and the best estimates for recrystallized grain size are listed in Table 2.

The histograms quantify the trend of increasing recrystallized grain size progressing from regime 1 (highest flow stress) to regime 3 (lowest flow stress) that is visible in the optical micrographs of Figs 11 and 12. For the axial compression samples, the average recrystallized grain diameter is 5 µm in regime 1, 7 µm in regime 2, and 20 µm in regime 3 (see histograms, Figs 5, 7). For the general shear samples, the recrystallized grain sizes are 7 µm for regime 1, 8 µm for regime 2, and 14 µm for regime 3 (see

Table 2. *Results of analyses*

Sample # (regime)	Vol % recryst.	Vol % annealed	Mode grain diameter (μm)	CPO max. density (bulk texture)	CPO max. density of recryst. fraction*	Mode of orient. gradient distrib. (°)	Measured perimeter/ perimeter of equivalent circle	PARIS factor (%)	Grain boundary surface per volume (μm^{-1})
w871 (1)	50		5	3.76	3.44		1.80	33.3	1.08
w872 (2)	40		7	4.12	3.96		1.79	35.6	0.99
w858 (3)	85		20	4.48	2.57		1.87	14.4	0.53
w875 (1 ann.)		100	20	2.04	=		1.53	0.2	0.23
w874 (2 ann.)		100	36	3.21	=		1.37	0.8	0.15
w860 (3 ann.)		100	50	3.52	=		1.32	0.9	0.12
w940 (1)	50		7	3.81	4.22	10			0.60
w946 (2)	90		8	10.9	11.1	9			0.56
w920 (3)	45		14	3.06	2.69				
w935 (3)	100		14	10.1	=	7			0.34
w943 (1 ann.)		100	32	4.04	=	4			0.25
w948 (2 ann.)		100	28	5.40	=	4			0.27
w921 (3 ann.)		100	42	2.50	=				
w938 (3 ann.)		100	30	8.50	=	4			0.25

*The annealed samples and sample w935 are 100 % recrystallized, therefore those CPOs are the same as in column 5.

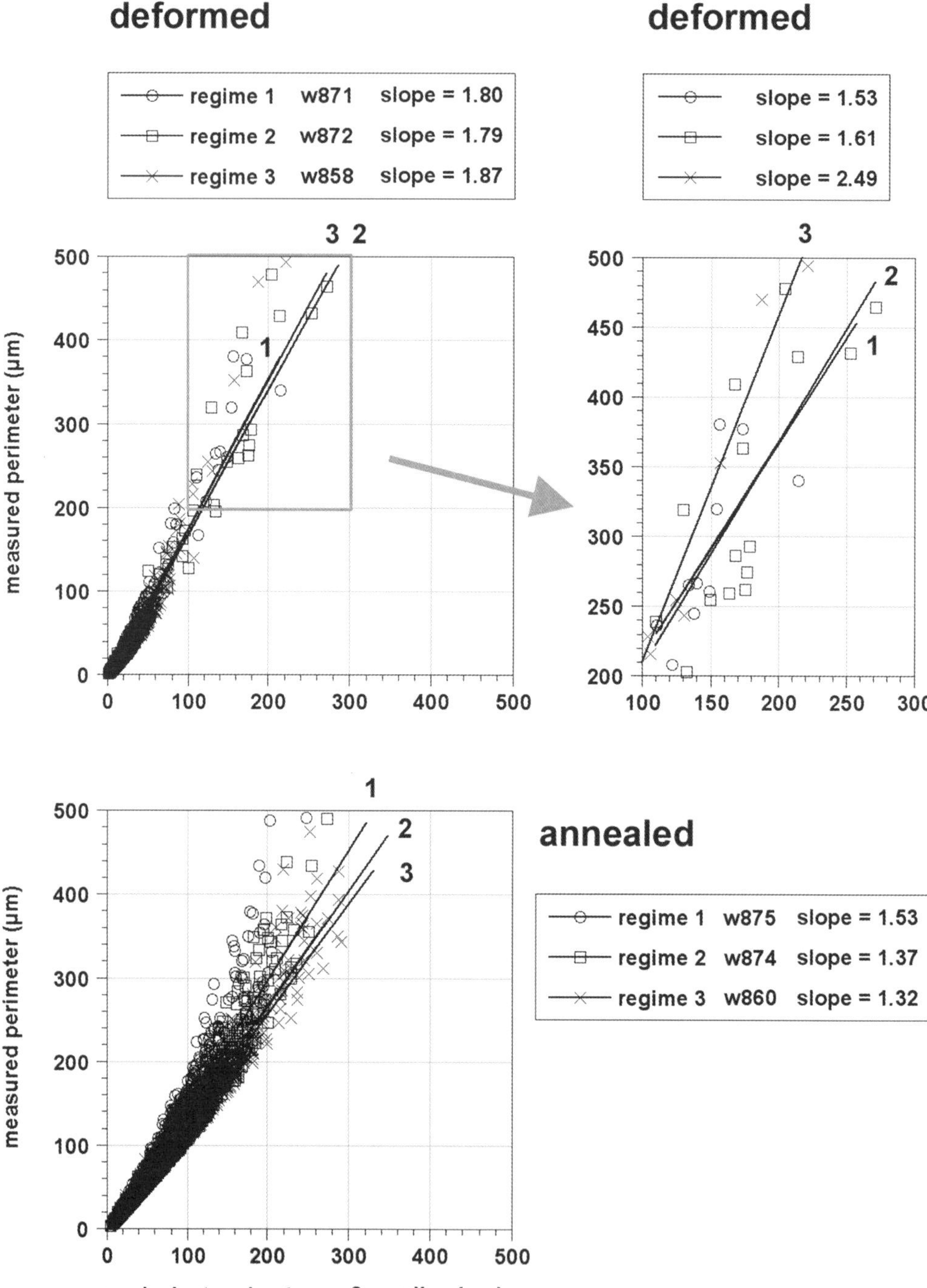

Fig. 13. Grain shape analysis for axially deformed samples. Grain boundary maps were prepared from misorientation images (×20 magnification) using NIH Image and the Lazy Grain Boundary macro (Heilbronner, 2000*b*). The measured grain perimeters, in μm, are plotted against the perimeters of circles having the same area; thus steeper slopes indicate a more lobate grain shape. Upper left: All grains of deformed samples (recrystallized and porphyroclasts) are evaluated. Upper right: Subset of grains with perimeters >200 μm (porphyroclasts) are evaluated. Bottom: All grains of annealed samples are evaluated.

histograms, Figs 6, 7). Several of the sheared samples were taken to such high shear strains that the strength increased significantly at the end of the experiment, possibly resulting in an 'overprinting' of lower regime microstructures onto those that had been developed earlier. The regime 2 sample had the highest final strength; the fact that it does not have the smallest recrystallized grain size means re-equilibration is not instantaneous. There has been some late overprinting with a smaller recrystallized grain size, but the overprinting is not complete.

A question remains concerning the difference between the TEM grain sizes and the grain sizes determined optically. This problem has been noted repeatedly but to our knowledge no explanation has been given in the literature. For example, Schmid *et al.* (1980) noted that the recrystallized grain size of 'equi-axed new grains' of dynamically recrystallized Carrara marble is one order of magnitude smaller when measured in the TEM than when measured optically. The regime 1 samples, while having optical grain sizes of 5–7 μm, have TEM grain sizes of ~1 μm (e.g. Hirth & Tullis 1992). It should be emphasized that the optical grain size of 5–7 μm is not an artifact resulting from insufficient optical resolution.

The annealed samples are all 100% recrystallized, and their average grain size is significantly larger than that of recrystallized grains in the equivalent deformed samples. The grain sizes of the annealed axial samples were evaluated at ×5 magnification, and those of the annealed sheared samples at ×10 (Figs 5, 6); a separate evaluation using higher magnification was not necessary. The grain size histograms are all unimodal, reflecting the fact that the samples are 100% recrystallized. The modal grain size of the axial samples increases from 20 μm, to 36 μm, to 50 μm for the annealed regime 1, 2 and 3 samples, respectively. For the sheared samples, the final grain size in the annealed samples varies from 32 μm (regime 1) to 28 μm (regime 2) to 42 μm and 30 μm (low and high shear strain in regime 3).

The skewed nature of some of the annealed grain size histograms may indicate that the microstructure had not reached equilibrium. For example, the grain size distributions of the annealed regime 1 and 2 samples tend to be positively skewed. In general, positive skewness indicates that a few grains have grown very large. During deformation, these samples were not completely dynamically recrystallized; perhaps during the anneal a few grains were able to rapidly consume the remaining porphyroclasts which had high internal energy.

Grain boundary character

On the grain boundary maps used for grain size determination we also determined a grain shape factor by measuring the perimeter of each grain and comparing it to the perimeter of a circle with equivalent area. Figure 13 shows the results for the axial compression samples. If the entire grain size population is considered (i.e., recrystallized grains and porphyroclasts), the slopes of the curves are between 1.8 and 1.9, showing no significant difference between the regimes. If, however, only the larger grains are considered (e.g. those with diameters of >30 μm), the slope is clearly larger for the regime 3 sample, indicating that the porphyroclasts are more lobate. This makes sense because grain boundary mobility is highest in regime 3, and most recrystallized grains have formed by fast grain boundary migration.

In the annealed samples all of the grains are polygonal; i.e., they are not lobate, and the grain shape slopes for the annealed samples are significantly lower than those for the deformed samples. A slope of 1.3 (Fig. 13) indicates zero lobateness, because for angular grains the actual perimeter is always larger than the perimeter of the equivalent circle (for squares the ratio is $4/\pi = 1.27$). A more meaningful analysis is provided by the PARIS factor (Panozzo & Hürlimann 1983) which compares the grain boundary outline length with the length of the smallest envelope to the grain. For any fully convex shape, this measure returns a value of 0%, corresponding to a slope of 1.00. Using the iSHAPES program (Heilbronner 2001) a number of large grains from the deformed and the annealed microstructures were analysed; the PARIS factors are listed in Table 2. The values for the deformed samples are between 15 and 35%; those of the annealed samples are all below 1%.

A further aspect of grain shape, surface density, was quantified by measuring the total grain boundary length within a given area (Fig. 14). Using the stereological relationship between density of intersection points and line density (SV = 2PL, where SV = surface area per volume and PL = intersection points per line length), the surface density was determined for the deformed and the annealed samples. The sheared samples were analysed using a magnification of ×10 for both the deformed and the annealed samples; the axial samples were analysed using ×20 for the deformed and ×5 for the annealed samples.

For both axial and sheared samples, there is a significant decrease of surface density from the

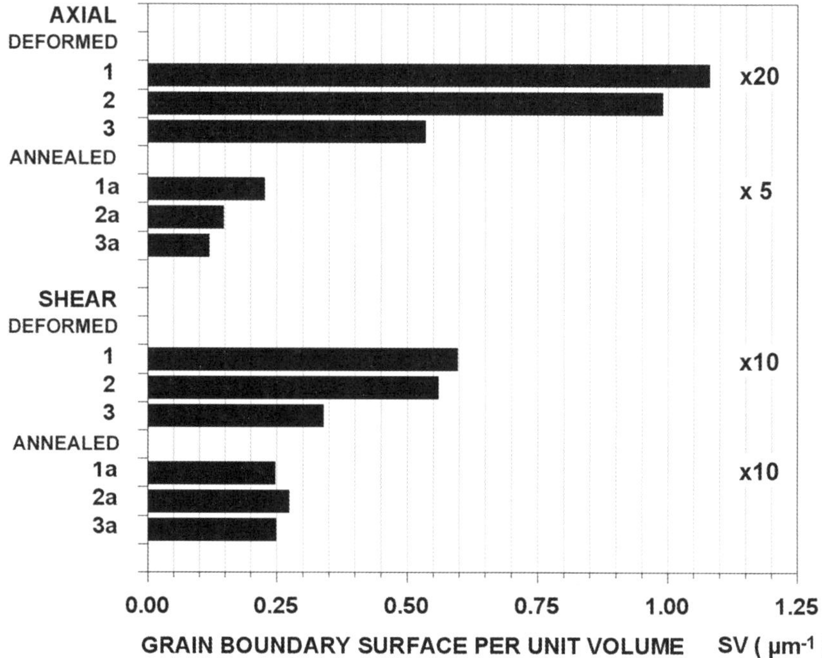

Fig. 14. Grain shape expressed in terms of boundary surface area per unit volume for samples deformed in axial compression and shear and their annealed equivalents (only the high strain regime 3 samples, w935 and w938, were used). The magnifications of the grain boundary maps from which the measurements were taken are indicated at the right. Annealed samples have more equant polygonal grain shapes.

deformed to the annealed samples (Fig. 14), as expected from that fact that annealing is in part driven by a decrease in interfacial energy. The decrease in surface density appears to be stronger for the axial samples, but this is due to the different magnifications used for the measurements. At higher magnification, grain boundaries appear to be longer because of their fractal nature. For the sheared samples, both the deformed and the annealed samples have been evaluated at the same magnification of $\times 10$. The surface density of the deformed samples decreases from regime 1 to regime 3, indicating larger grain size for higher temperature regimes. The surface density of the annealed samples does not vary as a function of deformation regime, indicating that minimization of interfacial energy is achieved regardless of the initial grain size distribution.

Crystallographic preferred orientations

The three samples deformed in axial compression all show a bulk pattern of a broad weak maximum or paired maxima of *c*-axes within a rough small circle girdle about the compression direction (Fig. 8). Samples deformed to higher strain in general shear (Fig. 9) show distinctively different patterns. In regime 1 there is an asymmetric broad single maximum, rotated slightly against the sense of shear. In regime 2 there is an asymmetric strong single maximum, rotated with the sense of shear. In regime 3 the lower strain sample has a broad peripheral maximum rotated slightly against the sense of shear, but the high strain sample has a single girdle with a strong Y maximum, rotated slightly with the sense of shear. (The development of these shear CPOs will be described in more detail in a separate paper.) The corresponding optical orientation image for the high shear strain regime 3 sample shows that there has been considerable coalescence of similarly oriented recrystallized grains; regions of recrystallized grains having the same orientation are much larger than original grains.

It is of interest to measure the CPO of dynamically recrystallized grains separately from that of the porphyroclasts in the same sample, in terms of distinguishing which component gives rise to the CPO of annealed samples (see Figs 8, 9). The bulk CPO of the axial regime 1 sample is dominated by the CPO of the porphyroclasts, whereas that of the regime 2 sample has equal contributions from porphyroclasts and recrystallized grains. In both cases, the CPOs of the porphyroclasts are not axially symmetric. They show two maxima in the compression direction, but off the periphery, while the CPOs of the recrystallized grains have four symmetrically disposed maxima. The significance of these deformation CPOs will not be discussed here, as it is the topic of a forthcoming paper. However, it should be noted that the orthorhombic bulk CPO is preserved, although somewhat weakened, during annealing. The bulk CPO of the axially deformed regime 3 sample is dominated by the recrystallized CPO, because this sample is 85% recrystallized. Again the annealed CPO is a slightly diffused version of the deformation CPO.

For the sheared regime 1 and 2 samples (Fig. 9), analogous observations can be made, although the symmetry of the bulk CPOs is monoclinic as opposed to orthorhombic, reflecting the rotational component of the deformation. Since the sheared regime 2 sample is nearly completely recrystallized, the bulk CPO is approximately equal to the partial CPO of the recrystallized grains. The CPO of the porphyroclasts in this sample is very similar. Again, all annealed CPOs are very similar to the bulk CPOs for the deformed samples, although typically displaying lower texture maxima.

The pattern and strength of the CPO for the low shear strain regime 3 sample remain very similar after annealing. However, whereas the pattern for the high shear strain regime 3 sample is slightly changed after annealing (compare corresponding CPOs in Fig. 9); the strength of the maximum at Y is decreased and the strengths of the intermediate maxima along the single girdle are increased. This apparent change, however, is due to a sampling artifact. Sheared regime 3 samples have a strong tendency for domain formation (Y domains and symmetrical domains of inclined *c*-axis orientation). Therefore, the bulk CPO may vary from one site to another, depending on whether both of the domain types are included in the measured area. The CIP image of the annealed high shear strain sample contains less than the typical amount of the Y maximum domain type and therefore the strength of the Y maximum appears to be reduced. In any case, the asymmetry of the CPO is again preserved during annealing.

Misorientation density

Orientation gradient images, representing the angular difference between the c-axis orientation of a given pixel and those of its four neighbours, provide information concerning the amount of strain energy stored in the grains. Histograms of orientation gradient images for the deformed and annealed shear samples were calculated (Fig. 10).

The maximum density of the annealed samples always occurs at 4° which is probably the 'noise' or background level of the measurement and actually means that within grains, i.e. throughout the bulk of the microstructure, no significant misorientations occur. Any misorientations that still exist after annealing have 'coalesced' to form subgrain and grain boundaries, and as such contribute only a very minor percentage to the misorientation density function. It is important to note that the misorientation distributions presented in this study depend mostly on misorientations within grains and not on those between grains (as would be the case in EBSD).

For the deformed samples, in regime 1 the peak occurs at 10°, in regime 2 at 9° and in regime 3 at 7°. Annealing obviously reduces the average of the orientation gradients by a mechanism which seems to preferentially eliminate misorientations in the range of 10° to 30° (compare histograms of deformed and annealed samples in Fig. 10). However, a minimum misorientation appears to be stable, as distributions peaking at 4° characterize all the annealed microstructures. Note also that these misorientations refer to misorientations of *c*-axes only, and thus are different from misorientation measurements obtained by EBSD analysis (e.g. Prior 1999; Wheeler *et al.* 2001).

Discussion

Grain size

The final grain size of annealed samples depends on the deformation microstructure. It could be the same as the original grain size, if the deformation occurred at high temperatures where dynamic recovery was rapid and the total strain was small enough that there was no dynamic recrystallization. An example of this situation has been described by Ree & Park (1997) for one set of their annealed OCP samples, and a similar result was found for annealing of experimentally deformed Carrara marble (Schmid, pers. comm.). However, for annealed samples that undergo static recrystallization, the final grain size reflects both nucleation and growth; for the same annealing time and temperature, higher rates of nucleation will produce smaller average grain sizes. Thus the grain size will be smaller for deformed samples with higher stored strain energy, because the nucleation rate is more affected by stored energy than is the growth rate (Humphreys & Hatherly 1995). For a given amount of stored strain energy, the grain size will be smaller for smaller starting grain sizes (or greater degrees of dynamic recrystallization), because most nucleation occurs at grain boundaries.

All of our samples were at least partially dynamically recrystallized prior to the anneal, and are fully recrystallized afterward. For all the samples, annealing produced an increase in the average grain size (see arrows indicating averages of grain size distributions of recrystallized grains in Figs 5 and 6). The trends are best seen in the axially compressed samples. The grain size increase for regimes 1, 2 and 3 is ×4, ×5 and ×2.5, respectively, leading to annealed grain sizes between 20 μm and 50 μm.

The grain size increase is least for the regime 3 sample, yet the resulting annealed grain size is largest. This sample had the largest starting recrystallized grain size and the lowest stored strain energy, and thus contained fewer 'nuclei'. The dynamically recrystallized grains were able to grow and the few remaining porphyroclasts had only a small amount of stored strain energy, probably insufficient to cause new recrystallized grains to form. Thus the final microstructure probably results entirely from growth of the pre-existing recrystallized grains.

The grain size increase is greatest for the lowest temperature regime 1 sample, yet the resulting annealed grain size is smallest. This sample had a low fraction and the smallest size of dynamically recrystallized grains, and the greatest stored strain energy in the porphyroclasts. The stored energy was probably sufficient to cause nucleation of new recrystallized grains early in the anneal. These statically recrystallized grains plus the pre-existing dynamically recrystallized grains created a large number of mostly strain-free 'nuclei', which could not grow very much before impingement occurred. The subsequent process of coarsening is slower than static recrystallization, especially as the annealing temperature was slightly lower for the regime 1 sample than the regime 3 sample. Thus for the same annealing time the final grain size is smaller.

The intermediate temperature regime 2 sample also had a low fraction of dynamically recrystallized grains, but an intermediate level of stored strain energy in the porphyroclasts. Thus there were probably few new recrystallized grains nucleated within these porphyroclasts, and the dynamically recrystallized grains first grew into the strained material before impinging and undergoing grain growth, producing an intermediate final grain size.

A similar difference in final grain size was observed by Wilson (1982) for ice aggregates experimentally deformed at −10 and −1 °C and then annealed at −1 °C. The cold-worked samples had a higher nucleation rate during annealing, causing earlier impingement, and a lower rate of grain size increase. Annealing of the hot-worked sample produced a faster rate of grain size increase.

Annealing of the sheared samples produced a similar behavior to that described above for the axial samples. The grain size increase for regimes 1, 2 and 3 (low and high strain sample) is ×4.5, ×3.5, ×3 and ×2, respectively, leading to annealed grain sizes between 30 μm and 40 μm. The complexities in grain size of the annealed sheared samples and their differences from the grain sizes of the annealed axial samples result from two factors. First, for regime 1, the sheared sample was deformed at a higher temperature (lower stress) portion of that regime, whereas for regime 3, the sheared samples were deformed at a slightly lower temperature (higher stress) portion of that regime, compared to the axial samples. Second, the high shear strain regime 2 and 3 samples underwent late strain hardening which reduced the size of dynamically recrystallized grains and increased the stored strain energy in the porphyroclasts. At the end of the experiments, the regime 2 sample actually had a slightly higher flow stress than the regime 1 sample (Fig. 2). Thus the grain size distribution of the annealed regime 1 sample (w943) is very similar to that of the annealed regime 2 samples (w948 and w874), with modes ~30 μm, and the increase of grain size is ~×4. The late hardening of the high shear strain regime 3 sample produced a defect structure similar to that of the axial regime 2 sample. As a consequence the grain size distribution after annealing is intermediate between that of the high shear strain regime 2 sample and that of the low shear strain regime 3 sample, and it is similar to that of the annealed axial regime 2 sample (compare right columns of Figs 5 and 6).

There is no evidence in any of our annealed samples for secondary or abnormal grain growth, in the form of abnormally large grains, perhaps with an irregular shape, resulting from high energy and very mobile grain boundaries of a special orientation. Possibly such grains might have grown if longer annealing times had been utilized. Abnormal grain growth has been observed in calcite aggregates experimentally deformed with no dynamic recrystallization and annealed at temperatures higher than the deformation temperature (Covey-Crump 1997), and it has also been observed in flints deformed with dynamic recrystallization and annealed at temperatures higher than the deformation temperature (Green 1967). Several studies have noted the growth of rhomb-bounded square grains in flint samples heated hydrostatically, or deformed, or annealed (e.g. Green 1967; Mainprice 1981; Gleason *et al.* 1993). It appears that {10-11} grain boundaries in quartz are highly mobile, especially in the presence of water and at high temperature, and they grow preferentially into grains of all other orientations. During axial compression only those rhomb-bounded grains grow which have their *c*-axes parallel to compression, because they have a 'hard' orientation and thus no stored strain energy (Gleason *et al.* 1993). During a subsequent anneal those same grains continue to grow to impingement, resulting in an extremely strong CPO (Green 1967).

We are not aware of any reports of rhomb-bounded porphyroblasts in naturally deformed and/or annealed quartz aggregates. However Stoeckhert & Duyster (1999) have described abnormal (or discontinuous) grain growth in recrystallized vein quartz. They find that the larger and irregularly shaped grains have *c*-axes $>30^{\circ}$ to the *c*-axes of the surrounding grains. They postulate that after the development of a CPO during dynamic recrystallization, a few 'hard' orientation grains with a lower dislocation density were favored for growth during the anneal, and remained favored for growth because they had higher energy, higher misorientation boundaries.

Grain shape

After the impingement of recrystallized grains during an anneal, including pre-existing dynamically recrystallized grains as well as any newly nucleated recrystallized grains, normal grain growth tends to produce a 'foam texture' of polygonal grains. During this period of relatively slow coarsening, larger grains with concave outward boundaries and >6 neighbours tend to grow and smaller grains to disappear (see review by Evans *et al.* 2001). In some circumstances and materials there may be 'abnormal grain growth' of some grains that grow larger than others and may have highly irregular boundaries.

The statically recrystallized grains produced by annealing are equant, straight-sided polyhedra, whereas dynamically recrystallized grains have more irregular and lobate grain boundaries and can be somewhat elongate, especially in sheared samples (see Figs 11 and 12). In other words, the effect of annealing is to reduce the amount of grain boundary surface area per unit volume (Fig. 14). A reduction of grain boundary surface may be achieved by increasing the grain size (Figs 5 and 6), or by reducing the lobateness of grains (Fig. 13) or by a combination of both processes.

In this study, we have used three measures to characterize the geometry of grain boundaries: the ratio of measured perimeter to equivalent perimeter (Fig. 13), the relative surface density (Fig. 14), and the PARIS factor (Table 2); all these measures are sensitive to the magnification at which the analysis is carried out. This dependence is largely due to the fractal nature of boundaries, and also to the depth of focus, i.e., the sharpness of the grain boundary outline in the plane of the image.

For the samples deformed in axial compression, the lobateness of porphyroclast boundaries in regime 1 and 2 samples is indistinguishable, both in terms of the slope shown in Fig. 13 ($\sim$1.8) and in terms of the PARIS factor (Table 2), even though the latter was evaluated at a magnification of $\times$50. At this relatively high magnification it is clear that in regime 1 the average wavelength of the serrations is much smaller than that in regime 2, yet the total 'excess' boundary length is the same for both (PARIS $\approx$35%). For regime 1, it should be noted that these optical measurements do not correlate with the fine scale of grain boundary bulging (*c.* 1 μm) seen in TEM (Hirth & Tullis 1992).

For regime 3, the measurements at high magnification seem to be in conflict with the results shown in Fig. 13: the PARIS factor (*c.* 15%) is smaller than that for regimes 1 and 2, while the slope is higher (1.87). However, this result indicates that the porphyroclast are elongate (a feature that contributes to the perimeter, but not to the PARIS factor). The smooth yet lobate shape of the porphyroclasts in regime 3 reflects their greater grain boundary mobility (no pinning).

The relatively high surface density measured for regimes 1 and 2 (Fig. 14) indicates that the deformed samples have small grain sizes and intense serration, while the somewhat lower density of regime 3 reflects a larger grain size and smoother lobes. Because of the fractal nature of the grain boundaries of the dynamically recrystallized samples, the surface density of the axial samples determined at $\times$20 (*c.* $1.0\,\mu m^{-1}$) is approximately twice the density determined for the sheared samples at $\times$10 (*c.* $0.5\,\mu m^{-1}$). The surface density of the annealed axial samples is between 0.125 and 0.25 μm^{-1}, and that of the sheared samples is 0.25 μm^{-1}. The surface densities of the annealed samples do not depend on the magnification because they have smooth grain boundaries; the different values of surface density for the axial samples thus reflect only the different grain sizes.

In a study of dynamically recrystallized quartz in experimentally deformed agates, Takahashi *et al.* (1998) found that the fractal dimension of the grain boundary surface decreases from $D = 1.22$ for an equivalent of regime 1 (10^{-5} s^{-1} and 800 °C) to $D = 1.05$ for an equivalent of regime 3 (10^{-6} s^{-1} and 900 °C). If our grain boundary surface measurements are converted to fractal dimensions (using logarithmic plots, and including all grains as in Fig. 13), the values for regimes 1, 2 and 3 are $D = 1.145$, 1.125 and 1.115, respectively. If we use only grains with a diameter >10 μm, the values are 1.23, 1.17, 1.15. The trend for our samples is the same as that of Takahashi *et al.*, but the decrease of fractal dimension from regime 1 to 3 is only 0.03 or 0.08 for our quartzites as opposed to 0.17 for the agates. In our study, the values are derived for complete populations ($N \approx 2000$–5000) evaluated at ×20 magnification (31 pixels/10 μm), while Takahashi *et al.* (1998) used between 70 and 150 carefully selected grains at a pixel resolution of 160 pixels/10 μm. Thus, the difference in absolute values may be explained by: (a) the fact that Takahashi *et al.* used only porphyroclasts which are more lobate than the recrystallized grains and (b) the effect of higher magnification on fractal boundaries.

Crystallographic preferred orientations

Despite large numbers of studies in materials science, a recent review stated that 'our present understanding of the origin of (static) recrystallization textures (CPOs) is little more than qualitative' (Humphreys & Hatherly 1995). In some cases the deformation CPO is retained or sharpened and in some cases it is changed; a change may occur during static recrystallization or during subsequent coarsening, especially if there is abnormal grain growth. There is still debate concerning the relative importance of 'oriented nucleation' versus 'oriented growth' for influencing the CPOs of annealed samples, but there is agreement that the nuclei from which static recrystallized grains originate are always small regions which pre-exist in the deformed state (Humphreys & Hatherly 1995).

Crystallographic preferred orientations: this study

For our quartzites, the pattern and strength of the CPOs in the annealed samples are very similar to those in the equivalent deformed samples, for deformed samples that had partial or complete dynamic recrystallization, and for samples with three different processes of dynamic recrystallization (in the three dislocation creep regimes). It is difficult to infer the exact processes that have contributed to the annealing CPO of our experimentally deformed quartzites, because for each deformed sample there is only one annealing time and temperature represented.

First consider the lower strain axial compression samples. In as much as the bulk CPOs of the axial regime 1 and 2 samples are dominated by the CPO of the porphyroclasts and the annealed CPOs are the same as the bulk CPOs (Fig. 8), it is the CPO of the porphyroclasts which is being preserved, although weakened, throughout annealing, and the partial CPO of the dynamically recrystallized grains is lost. For the regime 3 sample, the bulk CPO is dominated by the partial CPO of the recrystallized grains, and this pattern is preserved after annealing (Fig. 8). The CPOs of the recrystallized grains of regime 1 and 2 appear to form small circles about Y. This, however is an artifact introduced in the course of separating prophyroclasts and recrystallized grains on the CIP image. Recrystallized grains which formed along the boundaries of the porphyroclasts and which therefore have orientations very similar to them are often masked together with the prophyroclasts and are therefore missing in the partial CPO.

The true weakening of the CPOs is difficult to evaluate and probably subject to statistical artifacts. For example, in the case of regime 3, the grain size increases from 20 μm to 50 μm and therefore the number of grains in the evaluated area is reduced from 7500 to 1200. Although our texture measurements are based on *c*-axes, and not on the complete orientation distribution functions (ODFs), one may assume that the 1/n law applies, where n is the number of independent measurements (Burlini & Kunze 2000; Mathies & Wagner 1996). In other words, one may expect that for a decreasing number of independent measurements, i.e. for a decreasing number of grains represented on a given image, the measured maximum increasingly overestimates the true maximum. Therefore, if the maximum decreases even if the grain size increases (as it does in all regimes) the result may be accepted as significant. However, it is not correct to conclude that the weakening effect is stronger for regime 1 than for regime 2 and 3, even though the apparent decrease is strongest in regime 1 (Fig. 8, Table 2).

In other words, we do not observe a difference in how annealing affects the CPOs of the lowest (regime 1) and highest (regime 3) temperature axial compression samples, even though the former began the anneal with only *c*. 40% dynamic recrystallization and *c*. 60% porphyroclasts having relatively high stored strain energy and significant internal misorientation, whereas the latter began the anneal with *c*. 90% dynamic recrystallization and porphyroclasts as well as recrystallized grains characterized by low stored strain energy and little internal misorientation. For the relatively low strain axial regime 2 sample, it is not clear how the remaining porphyroclasts were replaced during static recrystallization; we assume it was growth of pre-existing dynamically recrystallized grains rather than by nucleation of any new recrystallized grains, because the existence of easy climb in this regime should mean that there were no regions of high dislocation density contrast. In the relatively low strain axial regime 3 sample, dynamic recrystallization is almost complete and the dynamically recrystallized grains continued to grow during the anneal.

Now consider the general shear samples. The bulk CPOs of all the sheared samples are dominated by the partial CPO of the recrystallized grains. In the regime 2 and high strain regime 3 samples this is because recrystallization is complete or nearly complete. In the case of the regime 1 and low strain regime 3 samples, where rather large fractions of porphyroclasts remained (50% in w940, 55% in w920) there is no significant difference between the partial CPO of the porphyroclasts and that of the recrystallized grains (Fig. 9). In all cases, the patterns of the annealed CPOs are the same as the deformed ones, although the strength is somewhat decreased.

For the high shear strain regime 2 and 3 samples, there seems to be a significant although slight change in pattern of the CPO and decrease in its strength upon annealing. Both samples were almost completely dynamically recrystallized at the start of the anneal, and thus one would expect that all of the grains in the annealed samples were already present at the start of the anneal. For the regime 2 sample, the strength of the peripheral maximum is significantly reduced and slightly elevated densities appear in the center of the pole figure. This pattern is mainly due to a broadening of the range of *c*-axis azimuths (which centered about the preferred orientation) from 45° to more than 120°. This effect can also be seen in Fig. 4 as a transition from a predominantly purplish CIP image (w946) to a more colourful one (w948). The physical process by which the c-axes are 'smeared out' is not yet clear. For the regime 3 sample, annealing appears to have decreased the strength of the Y maximum and increased the strength of the two maxima at intermediate positions within the single girdle. However, as mentioned previously, this apparent change is likely an artifact resulting from the domainal nature of the CPO in the deformed sample. If the analysed area in the annealed samples included a somewhat different proportion of domains, the bulk pattern could be quite different.

The sheared samples all have asymmetric CPOs. Although annealing results in small changes in the strength and pattern of the CPO, the asymmetry is preserved. Thus, even in samples with completely polygonal recrystallized grains, a prior shearing deformation can be inferred from the CPO pattern. Together with the shape analysis of aggregates of similarly oriented grains preserved after annealing, the approximate shear plane orientation may be inferred, in addition to the sense of shear across it.

Crystallographic preferred orientations: relation to previous studies

In the general tendency for the CPO produced during a relatively high temperature deformation to be preserved during annealing, our experimental results are similar to those of other recent studies. For example, Covey-Crump (1997) found that calcite aggregates deformed with no dynamic recrystallization at 426 °C and statically recrystallized during annealing at 500–700 °C retained their deformation CPO. There was an initial 'incubation' stage of annealing, involving recovery, and then a stage of rapid recrystallization and grain growth. Obviously the 'nuclei' for the statically recrystallized grains were already present in the deformed material. Similarly, ice samples deformed and then annealed at −1 °C retained their deformation CPO (Wilson 1982); OCP samples deformed with partial dynamic recrystallization and then annealed at the deformation temperature also retained their CPO (Ree & Park 1997); and naturally deformed and dynamically recrystallized galena aggregates underwent grain growth with retention of the pre-existing CPO upon experimental annealing (Siemes 1977).

One of the earliest experimental studies of annealing in the geological literature involved cold-worked calcite aggregates and single crystals and concluded that 'annealing recrystallization, in contrast to syntectonic recrystallization, tended to produce a random orientation' (Griggs *et al.* 1960). In that study, most samples were deformed at room temperature and annealed at considerably higher temperatures (400–800 °C). The cold-worked samples were mechanically twinned, creating a strong CPO, but they had no dynamic recrystallization and judging from what is now known about calcite deformation, they must have had extremely high dislocation densities as well as arrays of microcracks and even crush zones (e.g. Fredrich *et al.* 1990). It seems likely that small crushed grains as well as highly misoriented regions in the deformed samples provided abundant and close to randomly oriented 'nuclei' for the annealing recrystallization. Similarly, Siemes (1977) found that for galena samples deformed at room temperature, annealing at 700 °C caused complete recrystallization and randomization of the CPO. Again it seems possible that semi-brittle deformation may have created randomly oriented nuclei.

Situations where aggregates are deformed cold and then annealed at much higher temperatures are probably rare in nature, and it is likely that deformation CPOs are commonly preserved during subsequent annealing. However, it is possible for CPOs to be changed over time during grain growth, as shown in experimental studies of ice deformed at −10 °C and annealed at −1 °C (Wilson 1982). The deformed samples had a high nucleation rate upon annealing, so that recrystallized grains impinged relatively soon, and then underwent grain growth. Although the deformation CPO was initially preserved, it became more diffuse as grain growth continued.

In certain cases annealing can result in an extreme strengthening of the deformation CPO, for example the 'cube texture' in some metals (e.g. Humphreys & Hatherly 1995). An example of such strengthening was reported by Green (1967) for flints that had been deformed with dynamic recrystallization. As mentioned above, a moderate to strong maximum of *c*-axes parallel to compression develops during deformation, due to the growth of 'hard' orientation porphyroblasts bounded by rhomb {10-11} planes (Gleason & Tullis 1993). Annealing of low to moderate strain samples causes these porphyroblasts to grow to impingement and then to coarsen, thus greatly strengthening the *c* maximum CPO. Interestingly, we saw no tendency for rhomb-bounded porphyroblasts to grow in our deformed or annealed quartzites, although we observed the development of rhomb-bounded porphyroblasts in Black Hills quartzite deformed at higher temperatures (1200 °C and $10^{-6}\,s^{-1}$). Perhaps the development and mobility of these boundaries requires water on the grain boundaries and/or very high temperatures.

In some cases annealing can result in a change in the pattern of CPO. For example Green *et al.* (1970) found that in flints experimentally deformed in axial compression with little or no dynamic recrystallization, the CPO produced during deformation was a broad c maximum, whereas after annealing it was a broad great circle girdle. A similar change in the pattern of CPO was observed by Gleason & Tullis (1990) for novaculites deformed in regime 1 and annealed at 100 °C higher than the deformation temperature. At present we do not understand the difference in annealing behavior of the fine-grained novaculites and our coarser-grained quartzites deformed at almost identical conditions. However, for deformation in regime 1 dislocation creep, novaculites approach steady state flow stress much faster than do quartzites, because their initial grain size is much closer to the final steady state recrystallized grain size.

Implications for naturally deformed quartzites

The results of our experiments indicate that static annealing following deformation in any of the three dislocation creep regimes for quartz does not greatly affect the pattern or strength of the CPO, even though it completely changes the grain size and shape. Admittedly, longer annealing times would be required to give more confidence in this result. However, the rates of recovery and recrystallization processes depend on temperature, and the absolute temperature of any anneal following deformation in nature will be significantly lower than our annealing temperatures of 800–900 °C. Thus we believe that even for 'foam texture' quartz aggregates, a shear geometry and sense of shear can reliably be inferred from an asymmetric CPO.

The results of our study also have implications for using recrystallized grain size as paleopiezometer. If recrystallized grains in naturally deformed rocks have been annealed, then they will be larger than the dynamically recrystallized grains, and the inferred stresses will be too small. Therefore it would be useful to have grain shape criteria for determining whether recrystallized grains are the products of static annealing as opposed to being the products of high temperature deformation.

At this point we cannot provide exact criteria for such a distinction but we can suggest a procedure that may constrain the possible errors more narrowly in the case of quartzite. In all cases of dynamic recrystallization (regimes 1, 2 and 3) the microstructure of the deformed sample can be distinguished from that of the annealed sample. The most reliable distinction is the grain boundary shape (PARIS factor), but misorientation density is also useful. If a given natural microstructure appears to be annealed (PARIS factor <1%, mode of the misorientation density $\approx 5°$), the CPO pattern should indicate which of the dislocation creep regimes was active during deformation. After that, a first order estimate of the dynamic recrystallized grain size may be obtained by reducing the measured annealed grain size by a factor of 4 to 5 for regime 1 and 2, and a factor of 2 to 3 for regime 3. These suggestions are based on a small set of experiments, and may have to be modified if investigations over increased annealing times provide kinetic data.

Conclusions

For quartzites experimentally deformed with partial to complete dynamic recrystallization by axial compression (strain magnitude of 0.8 to 1.4) and by general shear (strain magnitude of 1.3 to 2.8) in each of the three dislocation creep regimes of Hirth & Tullis (1992) and subsequently annealed with complete static recrystallization at the deformation temperature for 120 hours, the following is concluded:

(1) The grain size increase on annealing was greatest for the samples deformed at lowest temperature and least for samples deformed at highest temperatures.
(2) For axial samples, the largest annealed grain size was attained by samples deformed at the highest temperature, the smallest by samples deformed at lowest temperature.
(3) There was no evidence of abnormal grain growth in our samples.
(4) The annealed samples all have equant, straight-sided, polygonal grains; annealing appears to reduce grain boundary lobateness to a common value, identical for all regimes.
(5) The change of grain boundary shape (smoothing of grain boundary surface) during annealing was greatest for the samples deformed at highest temperature (regime 3), since these deformed samples had the greatest grain boundary mobility.
(6) Annealing reduces the misorientation density, producing histograms of orientation gradients with significantly lowered modal values.
(7) The patterns of the CPOs are relatively unchanged and the strengths are somewhat decreased by annealing. For sheared samples, the asymmetry of the CPO is retained.

We thank Gayle Gleason for performing preliminary annealing experiments, Bill Collins for excellent thin sections, Holger Stunitz for helpful discussions, and Florian Heidelbach, Chris Wilson and Hans De Bresser for their constructive reviews. We are grateful for support from NSF EAR 9628348, EAR 9725622 and NF2000-055420.98.

References

BURLINI, L. & KUNZE, K. 2000. Fabric and seismic properties of Carrara marble mylonite. *Physics and Chemistry of the Earth (A)*, **25**, 133–139.

COVEY-CRUMP, S. J. 1997. The high temperature static recovery and recrystallization behavior of cold-worked Carrara marble. *Journal of Structural Geology*, **19**, 225–241.

DUNLAP, W. J., HIRTH, G. & TEYSSIER, C. 1997. Thermomechanical evolution of a ductile duplex. *Tectonics*, **16**, 983–1000.

EVANS, B., RENNER, J. & HIRTH, G. 2001. A few remarks on the kinetics of static grain growth in rocks. *International Journal of Earth Science*, **90**, 88–103.

FREDRICH, J., EVANS, B. & WONG, T.-F. 1990. Micromechanics of the brittle to plastic transition in Carrara marble. *Journal of Geophysical Research*, **94**, 4129–4145.

GLEASON, G. C. & TULLIS, J. 1990. The effect of annealing on the lattice preferred orientations of deformed quartz aggregates (abstract). *EOS Transactions of the American Geophysical Union*, **71**, 1657.

GLEASON, G. C. & TULLIS, J. 1993. Improving flow laws and piezometers for quartz and feldspar aggregates. *Geophysical Research Letters*, **20**, 2111–2114.

GLEASON, G. C. & TULLIS, J. 1995. A flow law for dislocation creep of quartz aggregates determined with the molten salt cell. *Tectonophysics*, **247**, 1–23.

GREEN, H. W. 1967. Quartz: extreme preferred orientation produced by annealing. *Science*, **157**, 1444–1447.

GREEN, H. W., GRIGGS, D. T. & CHRISTIE, J. M. 1970, Syntectonic and annealing recrystallization of fine-grained quartz aggregates. *In*: PAULITSCH, P. (ed.), *Experimental and Natural Rock Deformation*. Springer, Berlin, 272–335.

GRIGGS, D. T., PATERSON, M. S., HEARD, H. C. & TURNER, F. J. 1960. Annealing recrystallization in calcite crystals and aggregates. *In*: GRIGGS, D. T. & HANDIN, J. (eds) *Geological Society of America Memoir*, **79**, 21–37.

HEILBRONNER, R. 2000*a*. Optical orientation imaging. *In*: JESSEL, M. W. & URAI, J. L. (eds) *Stress, Strain and Structure, A volume in honor of W. D. Means. Journal of the Virtual Explorer*, **2**. World Wide Web Address: http://virtualexplorer.com.au/VEjournal/Volume 2/

HEILBRONNER, R. 2000*b*. Automatic grain boundary detection and grain size analysis using polarization micrographs or orientation images. *Journal of Structural Geology*, **22**, 969–981.

HEILBRONNER, R. 2001. iSHAPES, software. World Wide Web Address: http://www.unibas.ch/earth/micro/downloads/

HEILBRONNER, R. 2002. Local and global shape analysis of halftone images using the autocorrelation function. *Computers and Geosciences*, **28**, 447–455.

HEILBRONNER, R. & BRUHN, D. 1998. The influence of three-dimensional grain size distributions on the rheology of polyphase rocks. *Journal of Structural Geology*, **20**, 695–707.

HIRTH, G. & TULLIS, J. 1992. Dislocation creep regimes in quartz aggregates. *Journal of Structural Geology*, **14**, 145–159.

HIRTH, G., TEYSSIER, C. & DUNLAP, W. J. 2001. An evaluation of quartzite flow laws based on comparisons between experimentally and naturally deformed rocks. *International Journal of Earth Science*, **90**, 77–87.

HOBBS, B. E. 1968. Recrystallization of single crystals of quartz. *Tectonophysics*, **6**, 353–401.

HUMPHREYS, F. J. & HATHERLY, M. 1995. *Recrystallization and Related Annealing Phenomena*. Elsevier, Oxford.

KRONENBERG, A. K. & TULLIS, J. 1984. Flow strengths of quartz aggregates: grain size and pressure effects due to hydrolytic weakening. *Journal of Geophysical Research*, **89**, 4281–4297.

LISTER, G. S. & SNOKE, A. W. 1984. S-C mylonites. *Journal of Structural Geology*, **6**, 617–638.

LUAN, F. & PATERSON, M. 1992. Preparation and deformation of synthetic aggregates of quartz. *Journal of Geophysical Research*, **97**, 301–320.

MAINPRICE, D. H. 1981. *The Experimental Deformation of Quartz Polycrystals*. PhD thesis, Australian National University.

MATHIES, S. & WAGNER, F. 1996. On the 1/n law in texture related single orientation analysis. *Physica Status Solidi*, **196**, K11.

NADAI, A. 1963. *Theory of Flow and Fracture of Solids*. McGraw-Hill, New York.

PANOZZO, R. & HÜRLIMANN, H. 1983. A simple method for the quantitative discrimination of convex and convex-concave lines. *Microscopica Acta*, **87**, 169–176.

PANOZZO-HEILBRONNER, R. & PAULI, C. 1993. Integrated spatial and orientation analysis of quartz c-axes by computer-aided microscopy. *Journal of Structural Geology*, **15**, 369–382.

PRIOR, D. J. 1999. Problems in determining the misorientation axes, for small angular misorientations, using electron backscatter diffraction in the SEM. *Journal of Microscopy*, **195**, 217–225.

REE, J.-H. & PARK, Y. 1997. Static recovery and recrystallization microstructures in sheared octochloropropane. *Journal of Structural Geology*, **19**, 1521–1526.

SCHMID, S. M., PANOZZO, R. & BAUER, S. 1987. Simple shear experiments on calcite rocks: rheology and microstructure. *Journal of Structural Geology*, **9**, 747–778.

SCHMID, S. M., PATERSON, M. S. & BOLAND, J. N. 1980. High temperature flow and dynamic recrystallization in Carrara marble. *Tectonophysics*, **65**, 245–280.

SIEMES, H. 1977, Recovery and recrystallization of deformed galena. *Tectonophysics*, **39**, 171–174.

STIPP, M., STÜNITZ, H. & HEILBRONNER, R. 1999. Dynamic recrystallization microstructures of quartz in nature and experiment (abstract). *EOS Transactions of the American Geophysical Union*, **80**, F1053.

STIPP, M., STÜNITZ, H., HEILBRONNER, R. & SCHMID, S. 2002. The Eastern Tonale fault line: a 'natural laboratory' for crystal plastic deformation of quartz over a temperature range from 250 °C to 700°. *Journal of Structural Geology*, **24**, 1861–1884.

STOECKHERT, B. & DUYSTER, B. 1999. Discontinuous grain growth in recrystallized vein quartz – implications for grain boundary structure, grain boundary mobility, crystallographic preferred orientation, and stress history. *Journal of Structural Geology*, **21**, 1477–1490.

STOECKHERT, B, BRIX, M. R., KLEINSCHRODT, R., HURFORD, A. J. & WIRTH, R. 1999. Thermochronometry and microstructures of quartz – a comparison with experimental flow laws and predictions on the temperature of the brittle–plastic transition. *Journal of Structural Geology*, **21**, 351–369.

TAKAHASHI, M., NAGAHAMA, H., MASUDA, T. & FUJIMURA, A. 1998. Fractal analysis of experimentally, dynamically recrystallized quartz grains and its possible applications as a strain rate meter. *Journal of Structural Geology*, **20**, 269–275.

TULLIS, J. & YUND, R. A. 1989. Hydrolytic weakening of quartz aggregates: the effects of water and pressure on recovery. *Geophysical Research Letters*, **16**, 1343–1346.

TWISS, R. J. 1977. Theory and applicability of recrystallized grain size paleopiezometer. *Pure and Applied Geophysics*, **115**, 227–244.

WHEELER, J., PRIOR, D. J., JIANG, Z., SPIESS, R. & TRIMBY, P. W. 2001. The petrological significance of misorientations between grains. *Contributions to Mineralogy and Petrology*, **141**, 109–124.

WHITE, S. H. 1979. Difficulties associated with paleostress estimates. *Bulletin de Mineralogie*, **102**, 210–215.

WILSON, C. 1982. Texture and grain growth during the annealing of ice. *Textures and Microstructures*, **5**, 19–31.

Textures and microstructures of naturally deformed amphibolites from the northern Cascades, NW USA: methodology and regional aspects

BERND LEISS[1], HEIKE R. GRÖGER[1,2], KLAUS ULLEMEYER[3] & HERMANN LEBIT[4]

[1]*Geowissenschaftliches Zentrum der Universität Göttingen (GZG), Goldschmidtstraße 3-5, 37077 Göttingen, Germany (e-mail: bleiss1@gwdg.de)*

[2]*Present address: Geologisch-Paläontologisches Institut, Bernoullistrasse 32, 4056 Basel, Switzerland*

[3]*Geologisches Institut, Universität Freiburg, Albertstraße 23-B, 79104 Freiburg, Germany*

[4]*Department of Geology, 340 Kell Hall, Georgia State University, Atlanta, USA*

Abstract: Neutron texture analyses of quartz-bearing and quartz-free amphibolite mylonites from the Windy Pass thrust, Cascades Crystalline Core (Washington/USA) reveal pronounced textures of plagioclase and clino-amphiboles (hornblende, cummingtonite) but no preferred orientation of quartz. A reliable strategy for amphibolite fabric analysis is presented by a systematic analytical approach to the experimental diffraction data processing. Clino-amphiboles show transitional textures between ideal single crystal orientations and axial symmetric great circle distributions. Plagioclase reveals *a*-axes distributions scattering along a great circle approximating the foliation plane as well as *a*-axes maxima close to the macroscopic lineation. Correlation of the textures with grain shape anisotropies of hornblende and plagioclase and comparison with data from the literature suggest that the texture variations are due to different strain regimes rather than due to different crystallographic reorientation mechanisms. The kinematic directions deduced from the microfabric correlate well with the regional tectonic interpretations. In contrast, individual deformation paths are not yet established for the different tectonic units, as the significance of the separating Windy Pass thrust requires further structural analysis and fabric studies.

Quantitative texture analyses (QTA, texture = lattice preferred orientation) of polyphase rocks are still not routinely performed, although most rocks are composed of several minerals. When applying diffraction techniques, texture analyses of polyphase rocks are hampered by complex diffraction patterns with high peak densities. Therefore, sophisticated techniques (high resolution, recording of *d*-patterns etc.) and adequate analytical software are required. New developments in neutron diffraction methods and easier access to appropriate facilities have significantly improved the possibilities for systematic texture analyses of polymineralic rocks in the past years (Wenk & Pannetier 1990; Siegesmund *et al.* 1994; Chateigner *et al.* 1999; Leiss *et al.* 1999; Ullemeyer & Weber 1999; Wenk *et al.* 2001).

Fabric and texture analyses of amphibolites are expected to extend our knowledge on the deformational history of the Windy Pass area in the northern Cascades (Gröger 2001; Tischler 2001), which is part of the Cascades Crystalline Core exposed in Washington State (USA) and British Columbia (Canada; Fig. 1). The Cascades Crystalline Core belongs to the Cretaceous to lower Tertiary magmatic belt formed along the western active margin of the North American continent (Misch 1966; Brown & Talbot 1989). Kinematics along this major plate boundary are still the subject of debate concerning on the so-called 'Baja-British Columbia Problem' (Cowan *et al.* 1997; Butler *et al.* 2001).

The main focus of this study will be the correlation of microstructures and textures, especially in view of the different degrees of mylonitisation, different compositions and different tectonic situations in the field. The mechanisms of microfabric and texture development will be discussed. Due to the complexity of texture analyses of polyphase rocks, the reliability of quantitative texture analyses of polyphase rocks will also be addressed.

Geological setting and structural analysis

The present study focuses on the Windy Pass area (Fig. 1), which is located approximately 15 km west of Leavenworth, where the similarly named thrust placed the hanging-wall Ingalls Complex (Northwest Cascade System, Fig. 1a)

From: DE MEER, S., DRURY, M. R., DE BRESSER, J. H. P. & PENNOCK, G. M. (eds) 2002. *Deformation Mechanisms, Rheology and Tectonics: Current Status and Future Perspectives*. Geological Society, London, Special Publications, **200**, 219–238. 0305-8719/02/$15

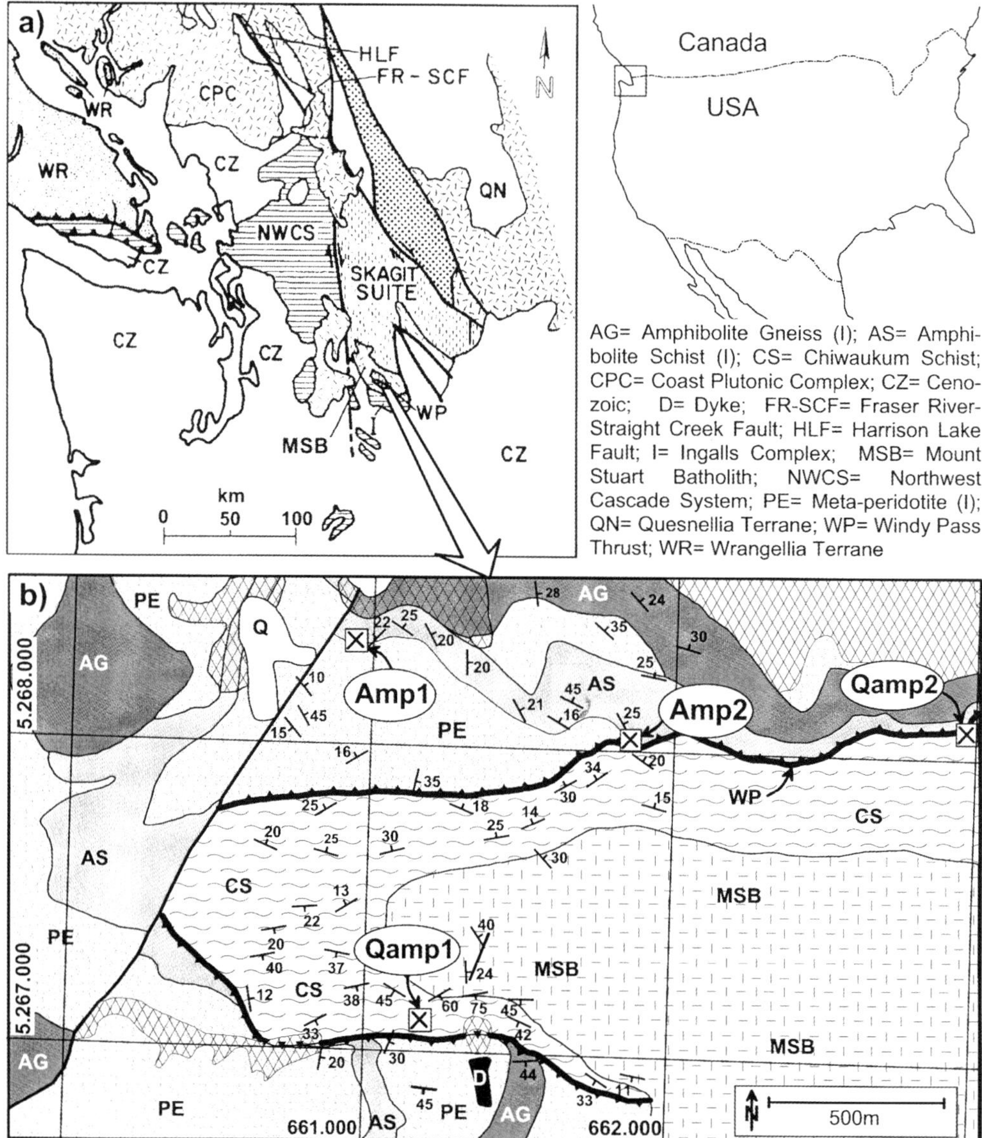

Fig. 1. Geological overview. (**a**) Structural subdivision of the Northern Cascades (simplified after Brown 1987). (**b**) Geological map of the Windy Pass area (Gröger 2001). Sample locations are indicated.

onto Chiwaukum Schists (Skagit Suite, Fig. 1a). The Ingalls Complex represents a metamorphic ophiolite nappe comprising Late Jurassic to Early Cretaceous meta-peridotites, interlayered with amphibolites and biotite schists (Miller 1980; 1985). The succession displays a duplex-like internal architecture formed by major imbrications of meta-peridotite layers. Ductile deformation is mainly concentrated in layers of amphibolite and biotite schists, which sigmoidally surround individual meta-peridotite bodies with varying sizes of 1–100 m. At the tips of these boudin-like bodies local instabilities have produced asymmetric folds with sub-horizontal to gently plunging NW–SE trending fold axes. Some of these folds were subjected to high shear strain, and the fold axes were successively rotated towards NE–SW directions. The average orientation and vergence of these folds coincides with the orientation of boudin gaps and the sigmoidal asymmetry of the meta-peridotite bodies. From these observations, the geometry of the large-scale imbrications and the orientation of stretching lineations, we infer a top-to-the-NE/ENE directed displacement for the formation of the internal architecture of the Ingalls Complex (Tischler 2001; Gröger 2001).

The Chiwaukum Schists are exposed in a half-window, formed by a WNW–ESE trending open culmination of the Windy Pass thrust (Fig. 1b). The Chiwaukum Schists consist predominantly of garnet-mica schists in the uppermost part of its litho-section. The whole sequence comprises amphibolite facies meta-sediments, which are mainly of detritic origin with a few layers of marble occurring as well as rare layers of mafic rocks. These rocks are considered to represent an originally siliciclastic sequence with intercalations of carbonates and mafic igneous rocks, presumably deposited in deep-marine fans close to a magmatic arc (Tabor *et al.* 1993).

Schistosity is very well developed in the Chiwaukum Schists and coincides with a dominant compositional layering. Sub-parallelism of schistosity and layering corresponds to at least one generation of isoclinal folds, whose NW–SE trending hinges are occasionally found in well-exposed outcrops. On schistosity planes (or compositional layer surfaces) a W–E to NW–SE trending sub-horizontal mineral lineation is present throughout the entire region (Plummer 1980; Paterson *et al.* 1994; Lebit *et al.* 1999). The schistosity forms locally open to tight upright folds with axial trends similar to those of the mineral lineation. This geometrical relationship is frequently used as the major argument for a transpressional regional deformation regime (e.g. Brown 1987; Brown & Talbot 1989). Lebit *et al.* (1999) relate the described structural morphologies to the co-axial superposition of horizontal isoclinal folds and upright folds. Numerical modelling of such a situation provides hinge line parallel finite X-axes, while the principal XY-planes of finite strain mimic the geometry of the imposed folds. Texture analysis of the involved rocks might be an appropriate tool to elucidate the competing models. In this respect the relationship between mineral lineation and fabric development is of particular interest.

Major displacements along the Windy Pass thrust took place prior to the emplacement of the Mount Stuart Batholith whose diorites to tonalites in fact truncate the thrust. However, shear zones are preserved in the continuation of the Windy Pass thrust in the igneous rocks of the batholith (Paterson *et al.* 1994). Igneous rocks were formed during late Cretaceous to Early Tertiary at mid-crustal level that corresponds with amphibolite facies regional metamorphism in the country rocks with temperatures up to 700 °C and pressures between 3–4 kb. Similar P–T conditions were predominant for internal imbrication of the Ingalls Complex and its displacement along the Windy Pass thrust (Miller 1980; Paterson *et al.* 1994 and references therein).

Microstructural analysis

Quartz-free amphibolite mylonites

Two samples (Amp1 and Amp2) came from the Ingalls Complex (Fig. 1b). In the vicinity of a competent meta-peridotite body a fine-grained, mylonitic amphibolite was collected about 200 m above the Windy Pass thrust (sample Amp1). The mylonite fabric was presumably formed by localized shear strain in the amphibolite due to flow around the competent meta-peridotite bodies. The second sample (Amp2) was taken at the base of the Ingalls Complex, approximately 6 m above the Windy Pass thrust. From the literature (Miller 1980; 1985; Paterson *et al.* 1994) and from our own field observations, an effect of thrusting is estimated up to 30 m away from the main thrust plane. Under the given metamorphic conditions, a discrete shear zone is not expected. Thus, fabric development in sample Amp 2 is likely to be dominated by shear along the Windy Pass thrust. Both samples exhibit a homogeneous composition, a well-developed foliation, and a weak, E–W trending mineral lineation. Their mineral composition is given in Table 1. Individual quartz layers of about 1–2 mm thickness occur only rarely. Retrograde alterations like sericitization and saussuritization of plagioclase and chloritization of hornblende are occasionally observed at grain boundaries.

The fabric of both samples is characterized by a homogeneous, fine-grained granoblastic microstructure of hornblende and plagioclase (An40-50) with an average grain size of ~0.05 mm (Fig. 2a). The hornblende partly exhibits a hypidiomorphic grain shape, whereas the plagioclase occupies the interstitial grain space, displaying xenomorphic grain shape. The plagioclase grains are not twinned but sometimes concentric undulose extinction can be observed in polarized light. The hornblende and plagioclase grains are disc-shaped with aspect ratios of ~3:1 in the principal XZ and YZ section (Fig. 2). The preferentially oriented disc-shaped grains define the foliation plane.

Table 1. *Mineralogical composition of the samples*

Sample	Structural unit	Composition (Volume %)			
Amp1	Ingalls C.	54% plg (An40)	44% hbl		2% acs (ore, biotite)
Amp2	Ingalls C.	37% plg (An50)	33% hbl	20% cpx	10% acs (zoisite/epidote, sphene, ore)
Qamp1	Chiwaukum S.	30% plg (An55)	50% cum	18% qtz	2% rut
Qamp2	Chiwaukum S.	27% plg (An35)	45% hbl	25% qtz	3% acs

acs, accessories; cpx, clino-pyroxene; cum, cummingtonite; hbl, hornblende; plg, plagioclase (anorthite content); qtz, quartz; rut, rutile.
The An-content was optically determined in the thin section and estimated from the peak positions in the neutron diffraction patterns.

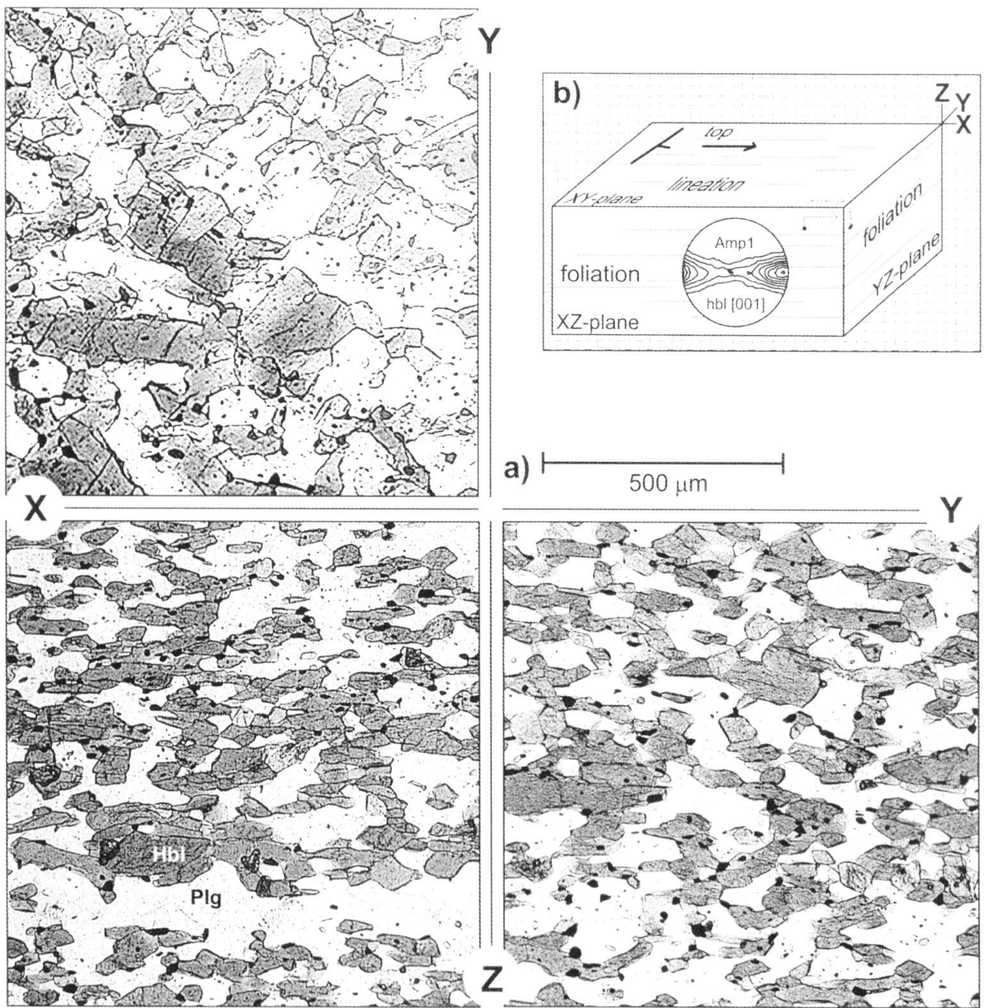

Fig. 2. (**a**) Three mutally perpendicular thin section microphotographs of sample Amp1 (plain polarized light). The orientations of the sections are indicated by the axes of the applied sample reference system. Hbl: hornblende, Plg: plagioclase. (**b**) Definition of the sample reference system *X*, *Y*, *Z* and its relation to foliation and lineation.

Quartz-bearing amphibolite mylonites

The quartz-bearing amphibolites (Qamp1 and Qamp2) come from the Chiwaukum Schists (Fig. 1b). Sample Qamp1 was collected at the southern limb of the WNW–ESE striking culmination that forms the half-window exposing the Chiwaukum Schists underneath the Windy Pass thrust. The second sample (Qamp2) comes from the northern limb (Fig. 1b). Sample Qamp1 was found approximately 30 m below the Windy Pass Thrust, while sample Qamp2 was located almost directly at the thrust zone. Thus, it is expected that sample Qamp2 suffered a similar deformation as that of sample Amp2, which was located in the Ingalls Complex just above the thrust. Both quartz-bearing amphibolites display well-developed foliation and a weak stretching lineation.

Sample Qamp1 consists of approximately 30% plagioclase, 50% cummingtonite instead of hornblende, 18 % quartz and 2 % rutile (Table 1). The colourless cummingtonite (Ca-free clino-amphibole) is responsible for the light colour of the specimen in contrast to the other three dark-coloured samples. The microfabric is fine-grained (grain size ~0.1 mm) and homogeneous for all mineral phases (Fig. 3a). A grain shape anisotropy of all mineral phases defines the stretching lineation. Among those, the

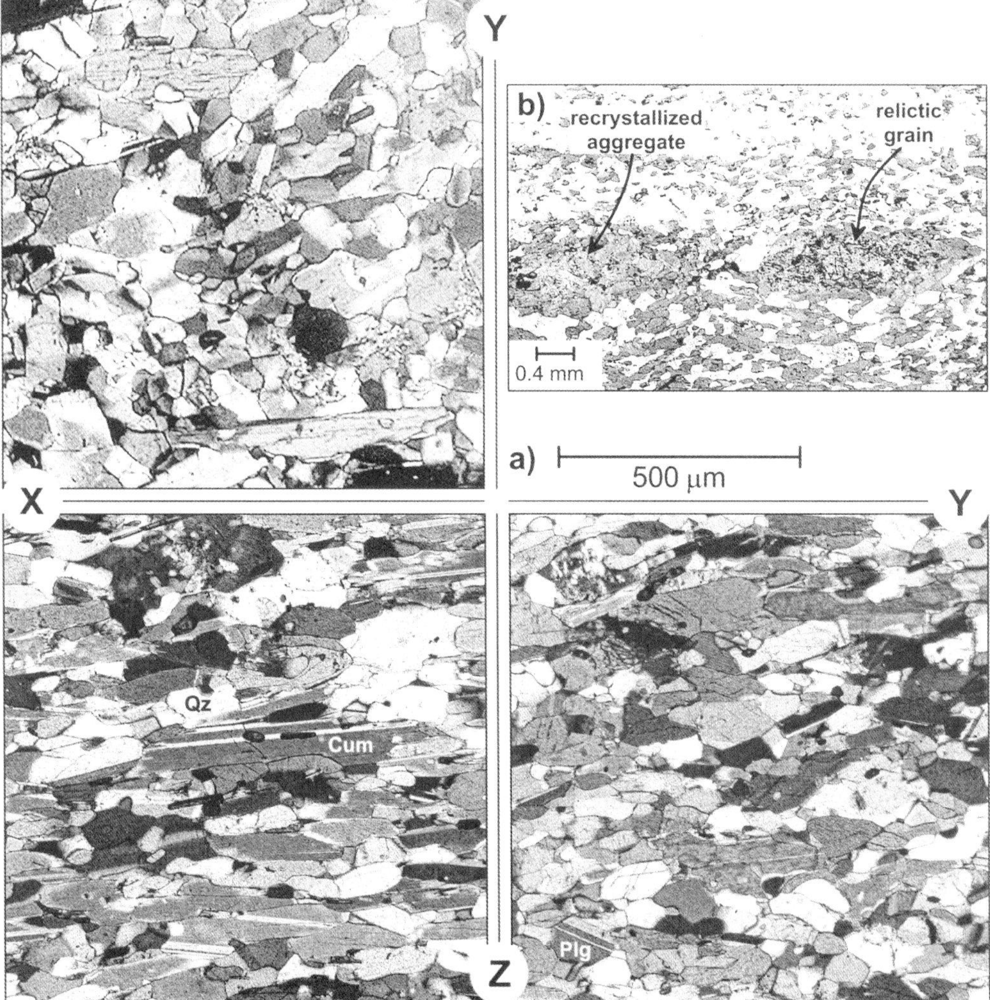

Fig. 3. (**a**) Three mutally perpendicular thin section microphotographs of sample Qamp1 (crossed polarizers). For the definition of the sample reference system *X*, *Y*, *Z* see Fig. 2b. Cum–cummingtonite, Plg–plagioclase, Qz–quartz. (**b**) *XY* section of sample Qamp2 (plain polarized light).

needle-shaped cummingtonite grains form the most obvious grain shape anisotropy with aspect ratios of up to 8:1 in the XZ section. Cummingtonite frequently forms twins on the (100) plane. Plagioclase (An55) also shows a hypidiomorphic grain shape. Only few grains reveal twinning according to the Albite law. Quartz is concentrated mainly in disc-shaped aggregates parallel to the foliation. The quartz grain boundaries are sutured.

Sample Qamp2 consists of approximately 27% plagioclase, 45% hornblende, 25% quartz and 3% of accessory minerals such as ore minerals, biotite and rutile (Table 1). The sample shows a heterogeneous microstructure formed by elongated aggregates of quartz/plagioclase and hornblende parallel to the compositional layering and foliation, respectively (Fig. 3b). The grain size of the component minerals varies with the compositional layers with an average grain size of ~0.3 mm. Hornblende shows a hypidiomorphic grain shape with grain sizes between 0.1 and 2 mm (Fig. 3b). Besides poikiloblastic hornblende, large hornblende grains and fine-grained, recrystallized aggregates can be found (Fig. 3b). Occasionally, chloritization can be observed along the hornblende grain boundaries. Plagioclase (An35) also shows hypidiomorphic grain shape and rare twinning according to the Albite-law. Sericitization of plagioclase is a common retrograde alteration feature, particularly in sample Qamp2. Quartz fills the interstitial space. In monophase aggregates, the quartz grains show sutured grain boundaries. In addition, layered accumulations of hornblende and plagioclase grains with homogeneous microstructures can be observed. Within these layers, the grains are strongly elongated (axial ratios: about 4:1 in the XZ section, about 3:2 in the YZ section, about 3:1 in the XY section) and define a particular strong macroscopic lineation.

Texture analysis

Data acquisition and quantitative texture analysis (QTA)

Texture measurements were performed at the time-of-flight (TOF) neutron texture diffractometer SKAT at Dubna, Russia (Ullemeyer *et al.* 1998). Neutron diffraction provides methodological advantages such as the high penetration depth of neutron radiation. The particular design of the SKAT diffractometer offers further advantageous properties such as the high *d*-resolution of the instrument ($\Delta d/d \approx 0.5\%$ at $d = 2$Å; *d*: lattice spacing), which is required to identify individual Bragg peaks in the large number of peaks produced in polyphase diffraction patterns. For the measurements, cylindrical samples were prepared ($h = l = 3$ cm). To ensure sufficiently high counting statistics, an exposure time of 30 min was selected for each of the 72 sample positions in a corresponding 5×5-degree grid. The bulk measuring time was 36 hours. After background subtraction, peak intensities were determined from the TOF-patterns by simple integration over all intensities at a pre-defined *d*-interval.

Two different methods were applied for the QTA. The WIMV algorithm (Matthies & Vinel 1982; included in the BEARTEX software package; Wenk *et al.* 1998) is a discrete method, as it directly leads to ODF (orientation distribution function) values in the 3D orientation space. The component method (Helming & Eschner 1990; MULTEX software) is a continuous method, where an ODF is transformed into a sequence of 3D Gauss distributions. In the case of high-quality experimental pole figures, both approaches should lead to similar results. Quantitative differences must be expected, however, in the case of weak textures or experimental pole figures of poor quality, e.g., a high level of statistical noise (see discussion in Kallend 1998; Ullemeyer *et al.* 2000). Consequently, an influence on the results arising from these fundamentally different methods of QTA cannot be excluded and will be qualitatively examined on the basis of the experimental and recalculated pole figures.

Implications from neutron diffraction patterns

A summary of the diffraction patterns of all 1368 sample directions of sample Amp1 is displayed in Fig. 4a. Hornblende and plagioclase (An40) can be identified as the main rock constituents. Since

Fig. 4. Experimental and synthetic TOF neutron diffraction patterns. The most significant reflections of the mineral phases are labelled, and peaks used for the quantitative texture analyses are indicated in bold (see also Table 2). (**a**) Experimental pattern of sample Amp1 (summary of all 1368 sample directions, normalized with respect to the energy distribution). (**b**) Theoretical patterns for hornblende (solid line) and plagioclase (dotted line). (**c**) Theoretical pattern for the mineral assemblage of sample Amp1 (44% Hbl + 54% Plg). (**d**) Experimental pattern for sample Qamp1 (summary of all 1368 sample directions, normalized). For explanation of numbers refer to the text.

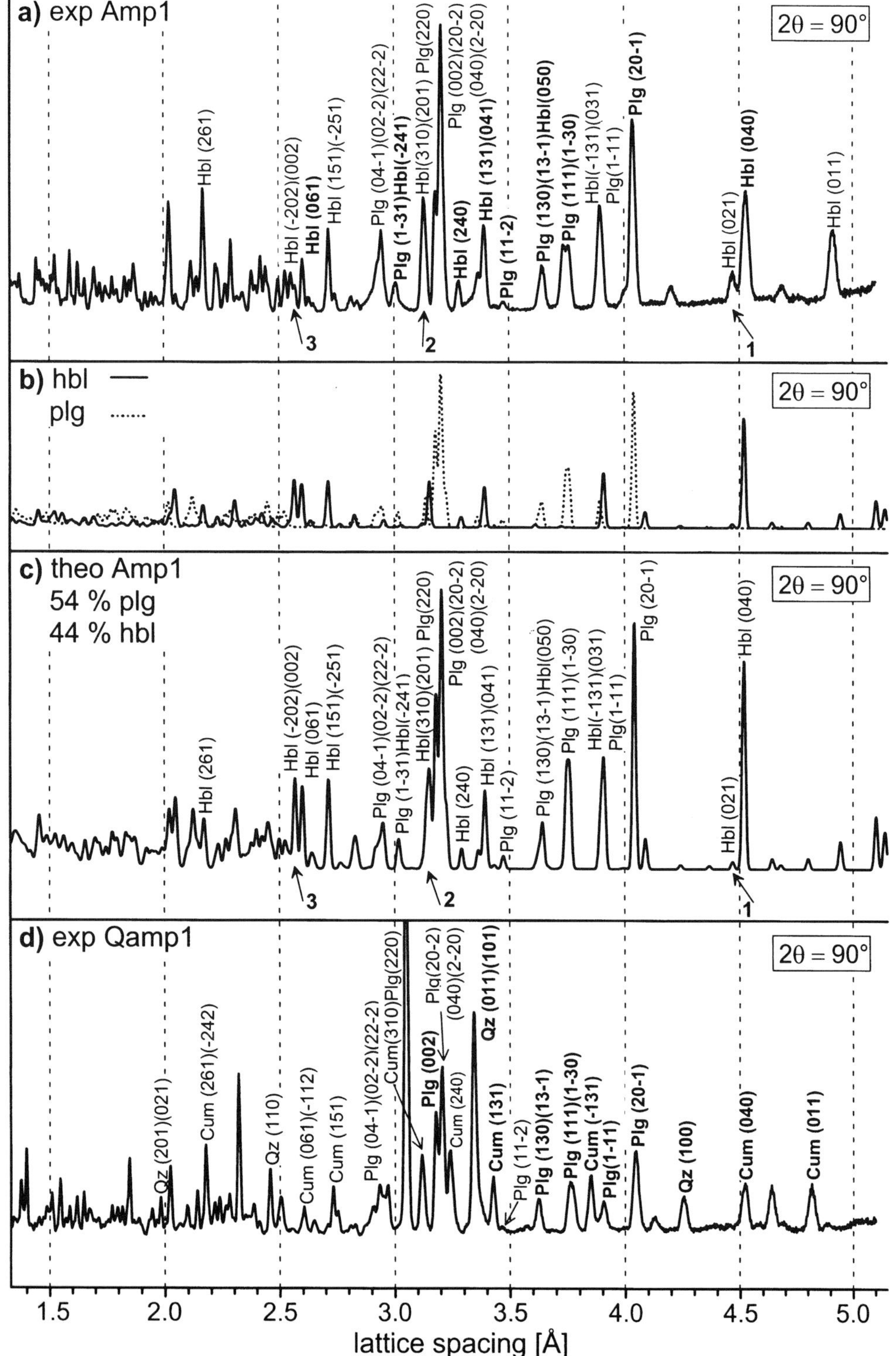
a) exp Amp1
2θ = 90°
Hbl (261)
Hbl (-202)(002)
Hbl (061)
Hbl (151)(-251)
Plg (04-1)(02-2)(22-2)
Plg (1-31)Hbl(-241)
Hbl(310)(201) Plg(220)
Plg (002)(20-2)
(040)(2-20)
Hbl (240)
Hbl (131)(041)
Plg (11-2)
Plg (130)(13-1)Hbl(050)
Plg (111)(1-30)
Hbl(-131)(031)
Plg(1-11)
Plg (20-1)
Hbl (021)
Hbl (040)
Hbl (011)
3
2
1
b) hbl
plg
2θ = 90°
c) theo Amp1
54 % plg
44 % hbl
2θ = 90°
Hbl (261)
Hbl (-202)(002)
Hbl (061)
Hbl (151)(-251)
Plg (04-1)(02-2)(22-2)
Plg (1-31)Hbl(-241)
Hbl(310)(201) Plg(220)
Plg (002)(20-2)
(040)(2-20)
Hbl (240)
Hbl (131)(041)
Plg (11-2)
Plg (130)(13-1)Hbl(050)
Plg (111)(1-30)
Hbl(-131)(031)
Plg(1-11)
Plg (20-1)
Hbl (021)
Hbl (040)
3
2
1
d) exp Qamp1
2θ = 90°
Qz (201)(021)
Cum (261)(-242)
Qz (110)
Cum (061)(-112)
Cum (151)
Plg (04-1)(02-2)(22-2)
Cum(310)Plg(220)
Plg (002)
Plg(20-2)
(040)(2-20)
Cum (240)
Qz (011)(101)
Cum (131)
Plg (11-2)
Plg (130)(13-1)
Plg (111)(1-30)
Cum (-131)
Plg(1-11)
Plg (20-1)
Qz (100)
Cum (040)
Cum (011)
1.5
2.0
2.5
3.0
3.5
4.0
4.5
5.0
lattice spacing [Å]

the summarized peak intensities approximate the peak intensities of a powder sample with random distribution, the volume fractions of the main rock constituents can be estimated from the comparison with calculated scattered intensities. The estimate is in good agreement with the compositions determined from the thin sections (see Table 1). For a better evaluation of the experimental diffraction patterns, a theoretical diffraction pattern was calculated from the crystal structure (hornblende: Walitzi & Trojer 1965; plagioclase (An38): Wenk *et al.* 1986) considering the estimated volume fractions of hornblende and plagioclase (Fig. 4b–c). The experimental (Fig. 4a) and the theoretical patterns (Fig. 4b) correlate very well. However, some hornblende reflections show a slight peak shift (arrow 2 in Figs 4a, c) or differ in intensity (arrows 1 and 3 in Figs 4a, c). Such variations are attributed to small differences in the chemical composition of the hornblende in the sample and the hornblende used for structure determinations. Such differences make the indexing of peaks and the proper assignment of scattered intensities to overlapped lattice planes more difficult. The peak identification usually can easily be achieved for low-index lattice planes, whereas intensity assignment is more difficult and can result in a systematic error during further data processing.

Figure 4d displays the summarized diffraction patterns of sample Qamp1. In contrast to sample Amp1, quartz occurs as an additional mineral phase and increases the number of diffraction lines. Furthermore, cummingtonite can be indexed instead of amphibole.

For the QTA of polyphase materials it is usually difficult to find a sufficient number of suitable peaks from each mineral phase. This holds especially in the case of low lattice symmetries. Single peaks are preferable, because the reliability of the QTA decreases with increasing number of superposed peaks. Lattice planes with less than 5% of the whole peak intensity are not considered, since intensity variations are close to the statistical error level.

If symmetry of the diffraction peaks is assumed, one half of a partly overlapped peak may be used as representative of the whole peak intensity (Braun 1994). By applying this principle to the right half of the peak 'plg $(20\bar{1})$' in sample Amp1, this pole figure becomes also accessible for the QTA. Considering these prerequisites, the reflections $(20\bar{1})$, $(111)(1\bar{3}0)$ and $(11\bar{2})$ were used for the QTA of plagioclase. The reflections $(130)(13\bar{1})$ and $(1\bar{3}1)$ are superposed by hornblende with an intensity contribution of 8–10%. Hence, the reliability of these pole figures for the QTA must be discussed in more detail (see discussion below). For hornblende, the QTA was based on reflections (040), (240), (061) and the right half of (131)(041).

Moreover, further restrictions are due to the applied algorithm for the QTA. WIMV processes only single phases and allows a maximum of three overlaps for each single phase. The component method allows overlaps between different phases to be handled. From the theoretical point of view this would mean that more pole figures can be used for the QTA. But in practice, restrictions apply for the number of overlaps. Too many overlapping peaks lead to a more or less random intensity distribution, which does not contribute usable information for the QTA.

Effect of experimental pole figures on ODF reproduction

The results of the plagioclase QTA of sample Ampl (Fig. 5) are independent of the applied QTA method (compare Figs 5b and c, d and e), but change when using three ($(20\bar{1})$, $(111)(1\bar{3}0)$, $(11\bar{2})$, Figs 5b–c) or five pole figures (additionally: $(130)(13\bar{1})$, $(1\bar{3}1)$, Figs 5d–e) as input. The differences are especially pronounced for the recalculated pole figures $(130)(13\bar{1})$. When calculating the kinematically significant pole figures [a], [b], [c] and (001), extreme differences are found for the b-axis intensity distributions (see

Fig. 5. Plagioclase texture of sample Amp1. The pole figures are equal area projections onto the **XZ** plane (lower hemisphere). The line marks the foliation and the square the lineation of the sample. Contours are multiples of a random distribution (m.r.d.), and the intensity maximum is given at the bottom right. Contouring starts at one time random. (**a**) Experimental pole figures. (**b**)–(**c**) Pole figures recalculated from components (**b**) and from the WIMV orientation distribution function (**c**) on the basis of three experimental pole figures (except pole figures *). (**d**)–(**e**) Pole figures recalculated from components (**d**) and from the WIMV orientation distribution function (**e**) on the basis of five experimental pole figures (including pole figures *). (**f**) Orientation distributions of the crystallographic axes *a*, *b*, *c* and (001)-poles, recalculated from the WIMV orientation distribution function obtained from three input pole figures (except pole figures *). (**g**) Orientation distributions of the crystallographic axes *a*, *b*, *c* and (001)-poles, recalculated from the orientation distribution function obtained from five input pole figures (including pole figures *).

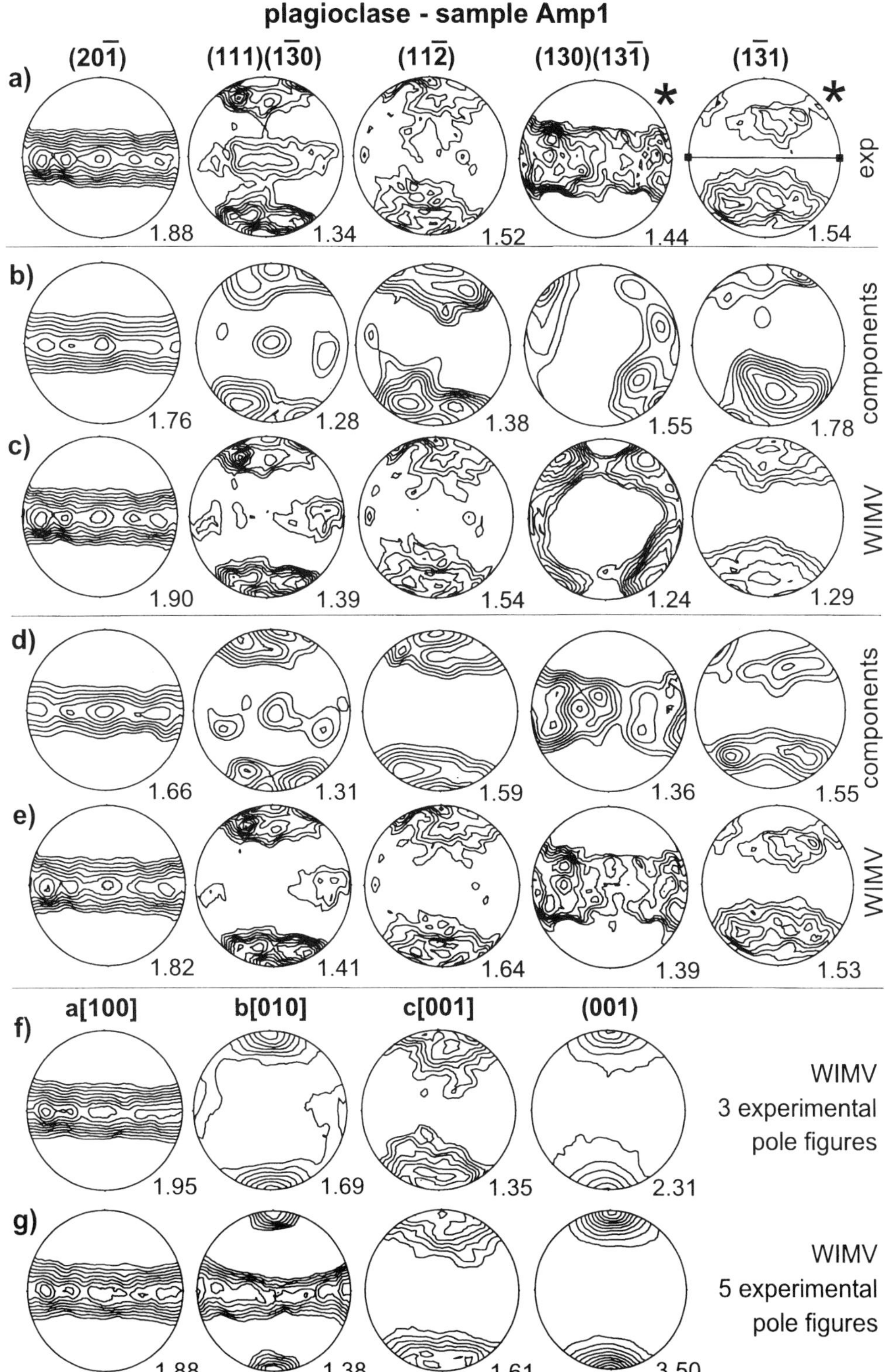
plagioclase - sample Amp1
a)
(20$\bar{1}$)
(111)(1$\bar{3}$0)
(11$\bar{2}$)
(130)(13$\bar{1}$)
(1$\bar{3}$1)
*
*
exp
1.88
1.34
1.52
1.44
1.54
b)
components
1.76
1.28
1.38
1.55
1.78
c)
WIMV
1.90
1.39
1.54
1.24
1.29
d)
components
1.66
1.31
1.59
1.36
1.55
e)
WIMV
1.82
1.41
1.64
1.39
1.53
f)
a[100]
b[010]
c[001]
(001)
WIMV
3 experimental
pole figures
1.95
1.69
1.35
2.31
g)
WIMV
5 experimental
pole figures
1.88
1.38
1.61
3.50

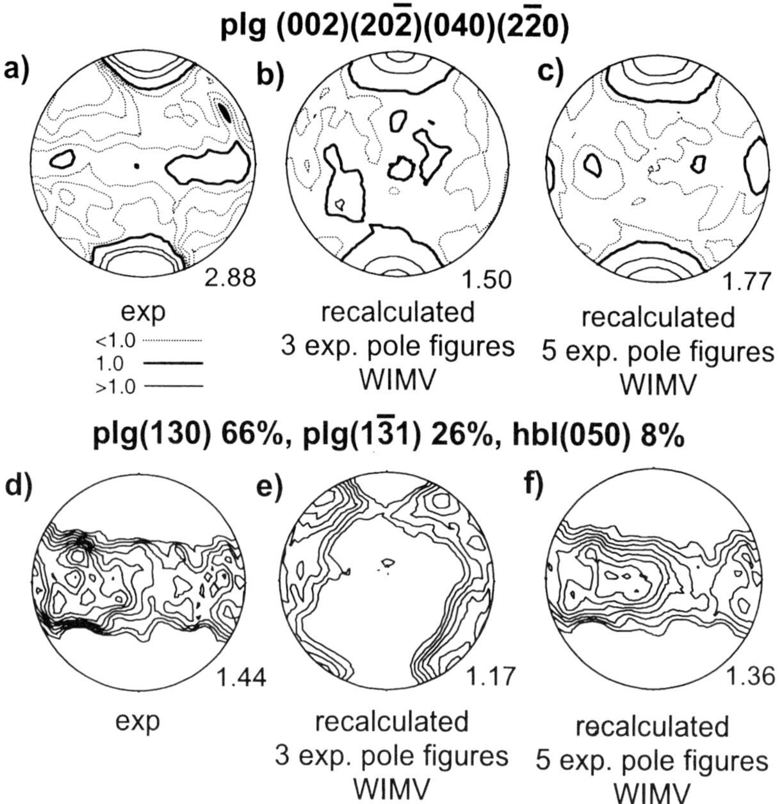

Fig. 6. Detailed inspection of the pole figures Plg (002)(20$\bar{2}$)(040)(2$\bar{2}$0) and Plg (130)(13$\bar{1}$) + Hbl (050) to explore the reliability of texture analyses (sample Amp1). For the conventions of representation refer to Figure 5. (**a**)–(**c**) Various pole figures Plg (002)(20$\bar{2}$)(040)(2$\bar{2}$0). (**a**) Experimental pole figure. Pole figure recalculated with (**b**) three and (**c**) five input pole figures from the WIMV orientation distribution function. In contrast to the other pole figures contours below random distribution are shown. (**d**)–(**f**) Various pole figures Plg (130)(13$\bar{2}$) + Hbl (050). (**d**) Experimental pole figure. Pole figure recalculated with (**e**) three and (**f**) five input pole figures from the WIMV orientation distribution function.

the WIMV-based pole figures in Figs 5f and g). This could result in different kinematic interpretations of the textures.

Criteria for the reliability of the results can be obtained from pole figures which are not used for QTA (see Table 2). Since Plg (040) is approximately parallel to the *b*-axis, it may be an independent indicator for reliability. Its intensity contribution to the superposed pole figure (002)(20$\bar{2}$)(040)(2$\bar{2}$0) is about 27% and therefore should have significant influence (Figs 6a–c). The experimental pole figure (002)(20$\bar{2}$)(040)(2$\bar{2}$0) shows a slight tendency to form a girdle in the foliation plane, if the whole intensity range – even intensities <1 – is considered (Fig. 6a). In the recalculated pole figures, this girdle becomes more clear when using five (Fig. 6c) instead of three (Fig. 6b) input pole figures for the QTA. Also for 'plg (130)(13$\bar{1}$)/hbl (050)', the input of only three pole figures leads to large differences while the result on the basis of five input pole figures fits the experimental pole figure very well (Fig. 6d–f). These observations support the conclusion that only the QTA on the basis of five experimental pole figures leads to reliable results.

The ODF of hornblende was calculated from four experimental pole figures (Fig. 7a). Both the component and WIMV method lead to comparable results for the experimental and recalculated input pole figures (Fig. 7b–c) and the recalculated pole figures [a], [b], [c] and (100) (Fig. 7d–e). In contrast to plagioclase, variations of the number of input pole figures for the QTA did not lead to significant modifications of recalculated pole figures.

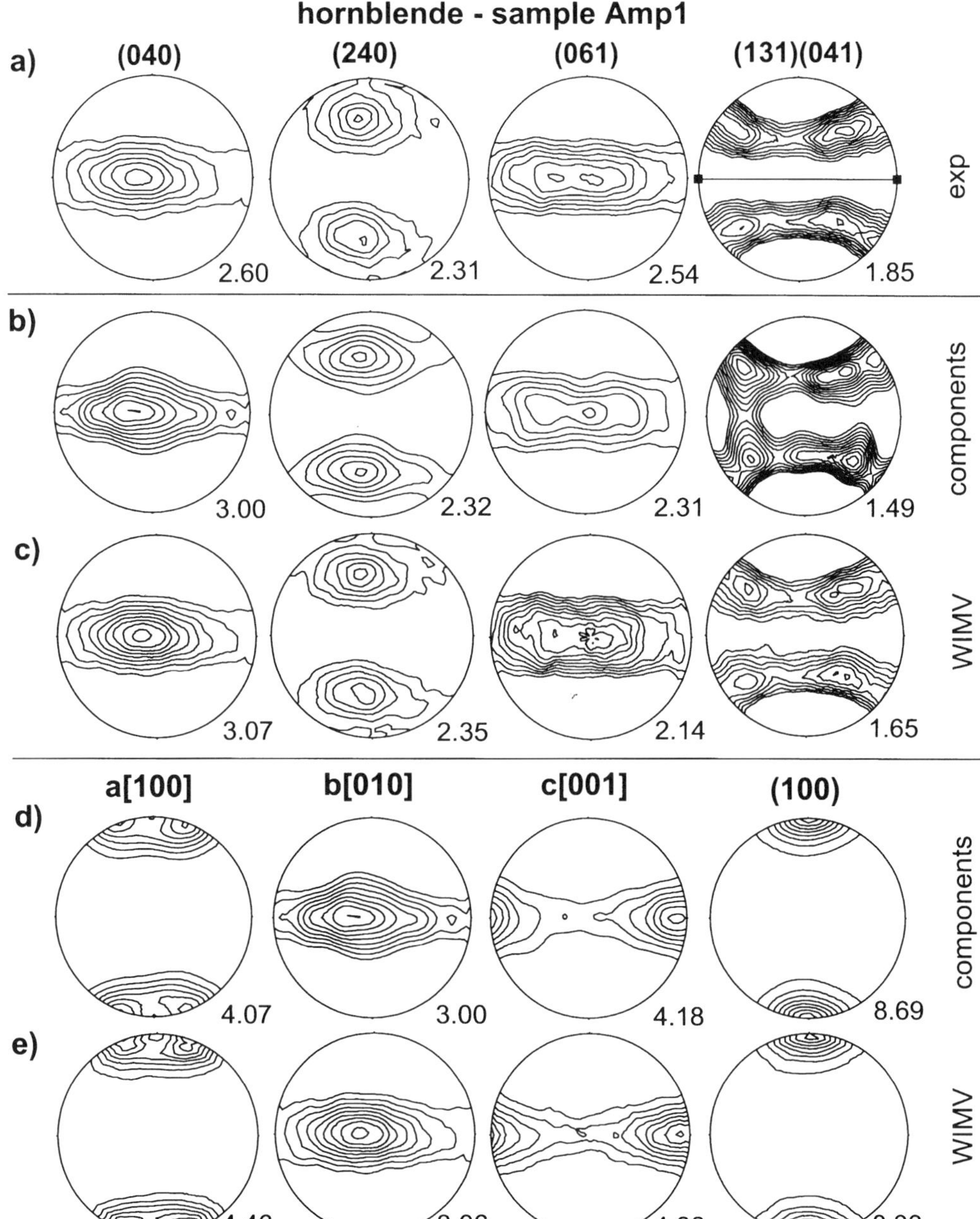

Fig. 7. Hornblende texture of sample Amp1. For the conventions of representation refer to Figure 5. (**a**) Experimental pole figures used for QTA. (**b**) Pole figures recalculated from components. (**c**) Pole figures recalculated from the WIMV orientation distribution function. (**d**) Orientation distributions of the crystallographic axes *a*, *b*, *c* and (001)-poles recalculated from components. (**e**) Orientation distributions of the crystallographic axes *a*, *b*, *c* and (001)-poles recalculated from the WIMV orientation distribution function.

Textures of the samples investigated

Except quartz, all minerals in the four samples display a well-developed texture. The clino-amphiboles show transitional texture types between two idealized end-members.

Type 1. Intensity distributions approximating a single crystal orientation (Fig. 8a) with *c*-axes

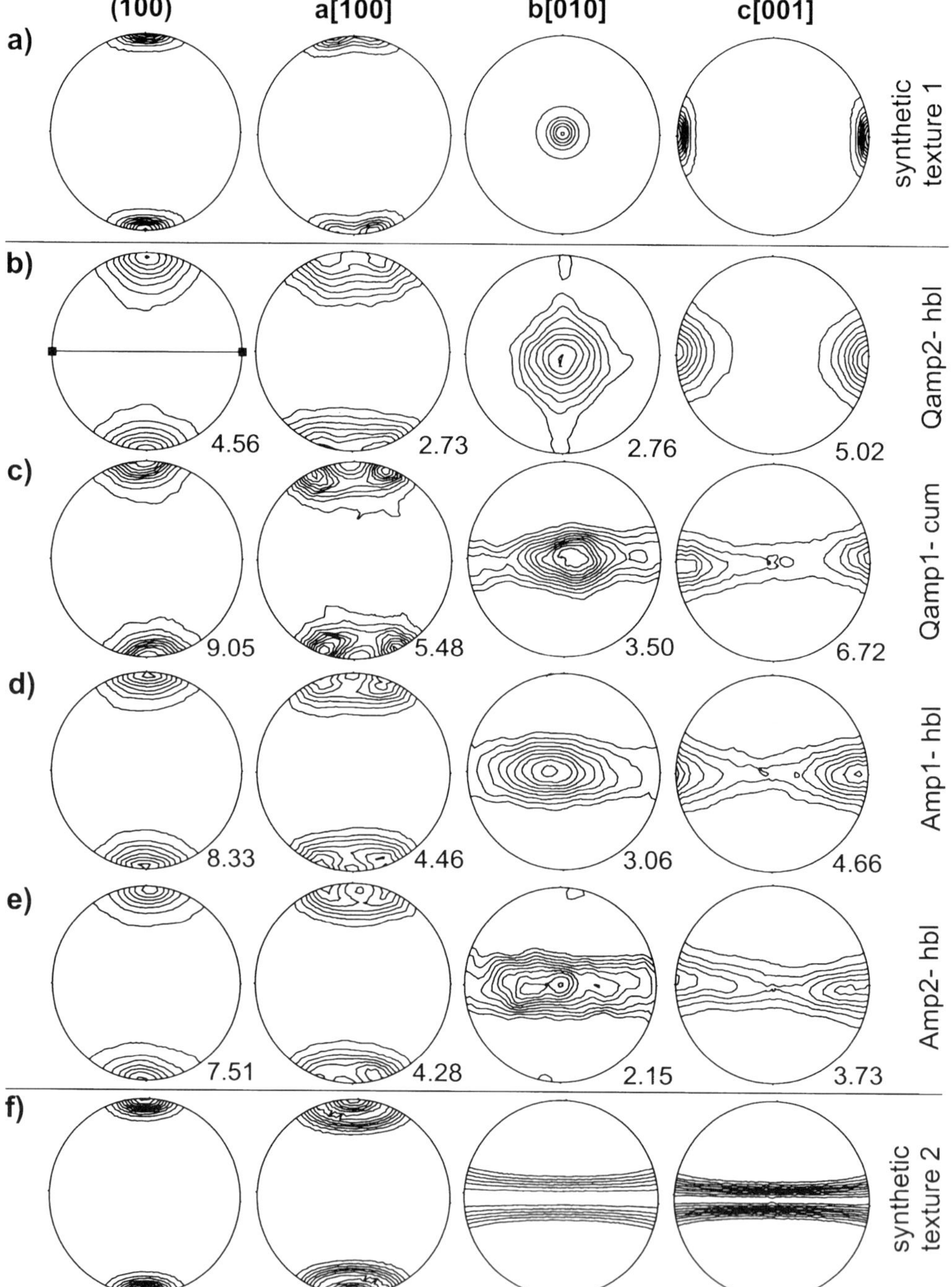

Fig. 8. Compilation of the clino-amphibole textures. For the conventions of representation refer to Figure 5. (**a**) Synthetic texture approximating an ideal single crystal preferred orientation. (**b–e**) Recalculated pole figures of the samples. (**f**) Synthetic texture approximating an ideal axial symmetric intensity distribution.

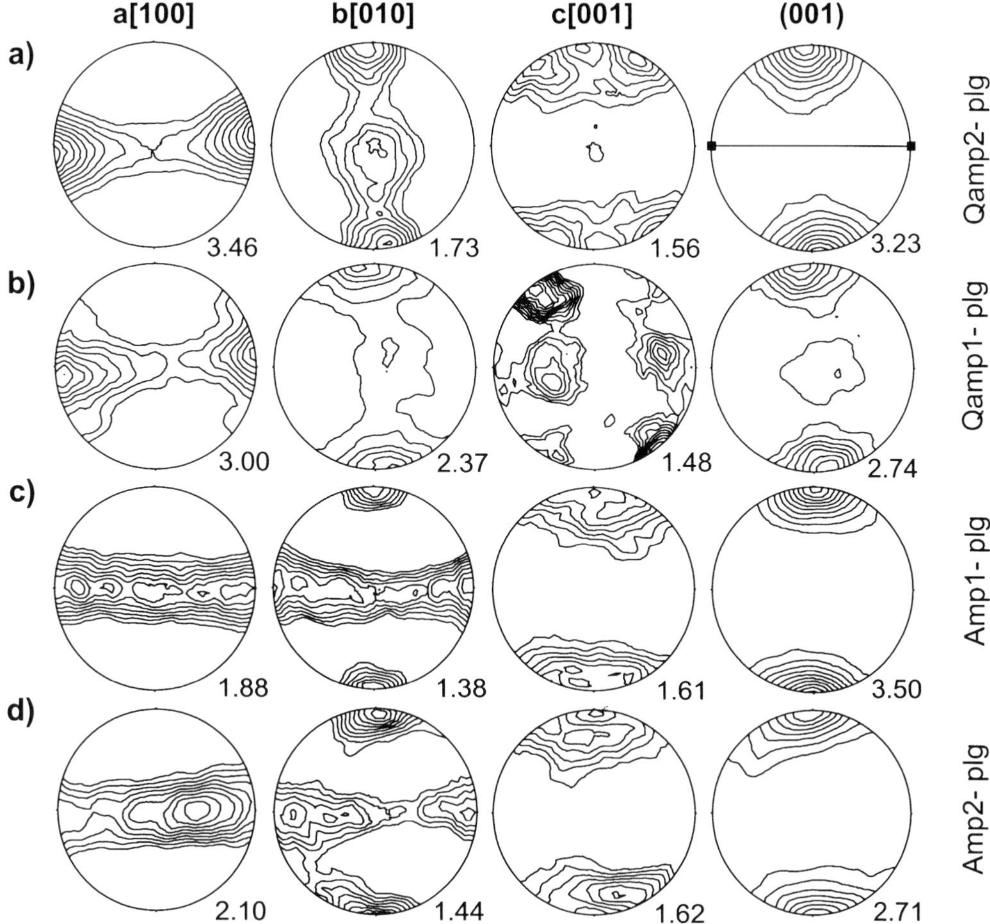

Fig. 9. Compilation of the plagioclase textures. For the conventions of representation refer to Figure 5.

parallel to the lineation and *b*-axes lying in the foliation plane normal to the lineation. The (100) planes are predominantly parallel the foliation. The *a*-axes distribution shows a double maximum on both sides of the foliation pole due to the monoclinic crystal symmetry.

Type 2. Axial symmetric intensity distributions (Fig. 8f). The (100)-planes are also mainly parallel to the foliation, but the crystallographic axes show circle distributions around the foliation pole.

Type 1 clino-amphibole texture is only found in the quartz-bearing hornblende–amphibolite sample Qamp2 (Fig. 8b). Samples Qamp1 and Amp1 show elements of both texture types (Fig. 8c–d). In sample Amp2 the circle distributions are nearly regular (Fig. 8e) and coincide with the Type 2 texture (Fig. 8f).

Compared to clino-amphibole, the plagioclase texture is more complex. In the two quartz-bearing amphibolites (Fig. 9a–b) the (001) planes are parallel to the foliation. The *b*-axes distribution shows also a single maximum perpendicular to the foliation and a submaximum perpendicular to the lineation. The *a*-axes display a single maximum parallel to the lineation with a week tendency of a great circle distribution parallel to the foliation. In sample Qamp2, the *c*-axes distribution shows a double maximum on both sides of the foliation pole due to the triclinic crystal symmetry (Fig. 9a), whereas in sample Qamp1 different maxima are developed. In the quartz-free amphibolite mylonites circle distributions of all crystallographic axes are developed around the single maximum of the (001)-poles normal to the foliation pole (Fig.

9c–d). The *b*-axes distributions show a submaximum perpendicular to the foliation.

It should be noted that a tendency to form more pronounced girdle distributions is found for both hornblende and plagioclase in the quartz-free amphibolite mylonite. In contrast, a tendency to form single crystal distributions is found for both phases in the quartz-bearing amphibolites (Fig. 10).

Discussion

Methodical aspects of QTA

The plagioclase texture in sample Amp1 (Figs 5 and 6) demonstrates well the difficulty of QTA of low symmetry mineral phases. Since the texture analyses with different mathematical methods (WIMV and component deconvolution) led to comparable results even in quantities, observed discrepancies should be related mainly to the experimental data. Errors may be introduced due to instrument misalignment, counting statistics (noise), grain size statistics, erroneous correction procedures during pole figure construction, and so on (Mücklich & Klimanek 1994). Quantitative error estimates are difficult to determine without repeated measurements of a sample or knowledge of a reference function (Luzin & Nikolayev 1996). Hence, error estimates are restricted to empirical methods. Smoothing procedures may be applied (and actually have been performed on the experimental pole figures as a routine procedure) to reduce the level of noise. Use of a Gaussian filter with a half width in the order of the grid distance is considered to cause no data falsification (see discussion in Kallend 1998; Nikolayev & Ullemeyer 1996). However, noise reduction did not prevent the large discrepancies between the plagioclase ODFs based on different numbers of input pole figures, so other sources of error must be assumed. In contrast to the elimination of noise, subsequent correction of errors like improper instrument adjustment is difficult to achieve. In the light of such restrictions, performing multiple texture analyses with various input pole figures and subsequent comparison of the results seems to be a practicable approach. This method provides reliable results on the basis of experimental data only.

Mechanisms of microstructural and textural development

The rock types analysed possess particular mylonitic fabrics, characterized by the macroscopic foliation, small grain size and frequently elongated grain shape, as well as pronounced textures of the major rock-forming minerals hornblende and plagioclase. In the following, we focus on the interpretation of the deformation mechanisms and/or deformation geometries that might have caused the observed textures. Discussion of the active mechanisms is hampered by the lack of experimental data or numerical models for texture development of the considered mineral assemblages. Most of the texture types observed have already been described. Wenk (1936), Schwerdtner (1964), Mainprice & Nicolas (1989), Siegesmund *et al.* (1994), Shelley (1994) and Dornbusch (1995) observed clino-amphibole textures of type 1 (Fig. 8a) characterized by pole figures approximating a single crystal orientation with *c*-axes and (100)-planes parallel to lineation and foliation, respectively. Textures of type 2 (Fig. 8f) – an axial symmetric intensity distribution with the *b*- and *c*-axes around the foliation normal (100) – were reported by Gapais & Brun (1981). Tendencies towards such great circle distributions of clino-amphibole textures were found by Kruhl & Huntemann (1991) and Dornbusch (1995) in granulite facies rocks, but they were not discussed in terms of potential texture forming mechanisms or their kinematic significance. An additional texture (type 3) may be derived from textures published by Schwerdtner (1964), Schwerdtner *et al.* (1971), Kern & Fakhimi (1975), Gapais & Brun (1981) and Shelley (1994). Type 3 represents an axial symmetric ideal texture with the clino-amphibole *c*-axes aligned with the lineation, whereas (100), (110) and [b] form great circle distributions around the lineation.

Several mechansims have been suggested to explain the development of texture types 1 and 3.

(1) Crystal plasticity with a general slip system of {hk0}[001] and a dominant slip system of (100)[001] (Biermann & van Roermund 1983; Cumbest *et al.* 1989; Skrotzki 1990, 1992; Hacker & Christie 1990) in combination with dynamic recrystallization (Dornbusch 1995).

Fig. 10. Local assignment of the textures. For the conventions of pole figure representation refer to Figure 5. The dip direction of the foliation is given in the clino-amphibole [100] pole figure, the dip direction of the lineation is indicated by an arrow.

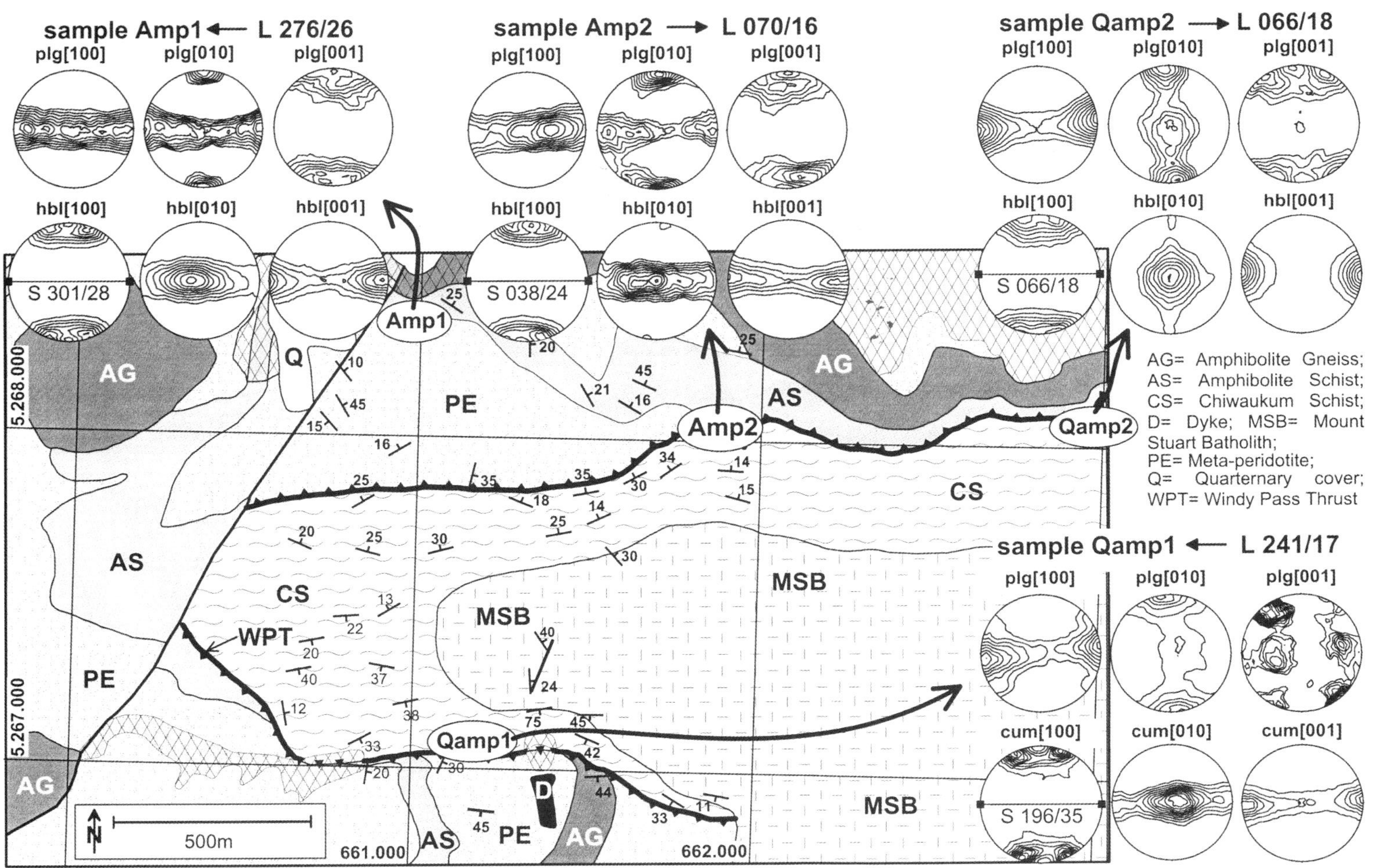
sample Amp1 ← L 276/26
plg[100]
plg[010]
plg[001]
hbl[100]
hbl[010]
hbl[001]
S 301/28
sample Amp2 → L 070/16
plg[100]
plg[010]
plg[001]
hbl[100]
hbl[010]
hbl[001]
S 038/24
sample Qamp2 → L 066/18
plg[100]
plg[010]
plg[001]
hbl[100]
hbl[010]
hbl[001]
S 066/18
AG= Amphibolite Gneiss;
AS= Amphibolite Schist;
CS= Chiwaukum Schist;
D= Dyke; MSB= Mount Stuart Batholith;
PE= Meta-peridotite;
Q= Quarternary cover;
WPT= Windy Pass Thrust
sample Qamp1 ← L 241/17
plg[100]
plg[010]
plg[001]
cum[100]
cum[010]
cum[001]
S 196/35
AG
AS
CS
MSB
PE
Q
D
WPT
Amp1
Amp2
Qamp1
Qamp2
5.268.000
5.267.000
661.000
662.000
500m
N

Table 2. *Hornblende and plagioclase peaks with potential significance for the QTA*

Mineral phase	h	k	l	d, Å	Relative intensity contribution, %	Bulk intensity, %
Hbl	0	1	1	4.9451	100.0	10.2
Hbl	0	4	0	4.5175	100.0	39.9
Hbl	0	2	1	4.4685	100.0	1.5
Plg	2	0	−1	4.0404	100.0	47.3
Plg	1	−1	1	3.8924	31.9	30.7
Hbl	−1	3	1	3.9058	47.0	
	0	3	1	3.9105	21.1	
Plg	1	1	1	3.7593	50.4	34.3
	1	−3	0	3.7447	49.6	
Plg	1	3	0	3.6433	65.4	13.1
	1	3	−1	3.6278	26.2	
Hbl	0	5	0	3.6140	8.4	
Plg	1	1	−2	3.4725	100.0	2.5
Hbl	1	3	1	3.3922	68.9	13.1
	0	4	1	3.3936	31.1	
Hbl	2	4	0	3.2909	100.0	4.3
Plg	0	0	2	3.1797	33.4	100.0
	2	0	−2	3.2024	28.8	
	0	4	0	3.2065	26.7	
	2	−2	0	3.2261	11.1	
Hbl	3	1	0	3.1533	57.6	29.3
	2	0	1	3.1232	5.3	
Plg	2	2	0	3.1395	37.1	
Plg	1	−3	1	3.0168	90.0	6.6
Hbl	−2	4	1	3.0315	10.0	
Plg	0	4	−1	2.9468	45.9	14.6
	0	2	−2	2.9311	29.2	
	2	2	−2	2.9150	24.9	
Hbl	1	5	1	2.7126	95.5	17.1
	−2	5	1	2.7079	7.5	
Hbl	0	6	1	2.5986	100.0	15.8
Hbl	−2	0	2	2.5654	85.7	17.6
	0	0	2	2.5705	14.3	
Hbl	2	6	1	2.1627	100.0	7.7

Relative intensity contributions of superposed reflections add to 100%, bulk intensities are normalized with respect to the strongest reflection (=100%) in the diffraction pattern.

(2) Crystal growth under differential stress (Schwerdtner 1964).
(3) Rigid body rotation (Ildefonse *et al.* 1990; Shelley 1994).

Gapais & Brun (1981) proposed a model for the development of texture types 2 and 3 (Fig. 8). They related the textures to the grain shape anisotropy of hornblende and the type of accommodated strain. Type 3 textures correspond to a prolate strain ellipsoid whereas type 2 textures refer to an oblate strain ellipsoid. Since our texture types and grain shape anisotropies can be correlated in the same manner, type 1 textures correspond to a plain strain ellipsoid whereas type 2 textures refer to an oblate strain ellipsoid. The conclusion seems to be justified that the clino-amphibole textures of type 1 and 2 represent different strain geometries ($\varepsilon_1 > \varepsilon_2 > \varepsilon_3$ and $\varepsilon_1 > \varepsilon_2 = \varepsilon_3$) rather than different deformation mechanisms.

Texture analyses of plagioclase described in the literature are difficult to compare with our study as data have been obtained by various methods and presented in different ways. In metamorphic rocks the lattice preferred orientation of plagioclase is usually not as strongly developed as that of other minerals (Siegesmund *et al.* 1989). Two plagioclase texture types are predominant. The first one is characterized by a single maximum of (010) poles perpendicular to the foliation, an *a*-axes great circle distribution and a small circle distribution of the *c*-axes around the foliation normal (Shelley 1977; Olsen & Kohlstedt 1985; Kruhl 1987; Ji & Mainprice 1988; Ague *et al.* 1990; Ullemeyer *et al.* 1994; Dornbusch 1995; Ullemeyer & Weber 1999). The *c*-axes small circle distributions show a submaximum parallel to the lineation. In metamorphic rocks the texture development is attributed to intracrystalline slip, predominantly by (010)[001], and additionally (010)[100], probably in combination with dynamic recrystallization (Ji & Mainprice 1988). An *a*-axes maximum perpendicular to the lineation results from slip on (010)[001], associated with shearing parallel to (010) and (001) (Olsen & Kohlstedt 1985; Kruhl 1987). A well-developed *a*-axes maximum parallel to the lineation was sometimes observed and attributed to mechanical rotation in unconsolidated tuffs in combination with mimetic crystallization (Shelley 1979). In gabbros even passive rotation might be a mechanism of texture formation, if an idiomorphic and anisotropic grainshape is assumed (Ague *et al.* 1990; Mainprice & Nicolas 1989).

The second type of plagioclase texture shows axial symmetric intensity distributions around an (001) pole maximum, which parallels the foliation pole (Kruhl 1987; Siegesmund *et al.* 1994; Ullemeyer & Weber 1999). Submaxima of the *a*-axes distribution may develop parallel to the lineation and submaxima of the *b*-axes distribution may develop perpendicular to the lineation (Siegesmund *et al.* 1994). From these textures, Kruhl (1987) inferred that (001)[100] and (001)[010] are the dominant slip systems at high metamorphic conditions. In combination with an *a*-axes maximum parallel to the lineation, Ullemeyer & Weber (1999) observed (010) and (001) pole maxima perpendicular to the foliation and submaxima of both poles normal to the lineation.

Our samples basically display the second type of texture with some differences in the intensity distributions. Assuming crystal plasticity as the dominant deformation mechanism and judging only from the pole figures, dominant (001)[100] slip and secondary (001)[010] or (010)[100] slip would be expected. However, in naturally deformed plagioclase the first two slip systems have not, so far, been clearly identified by TEM investigations (Siegesmund *et al.* 1994). Since idiomorphic grain shape is absent, a texture development of plagioclase by passive reorientation of grains can be excluded (Ague *et al.* 1990; Shelley 1979).

Although it is difficult to identify the active slip systems in plagioclase during texture development, the observed modifications of the fundamental texture type can be explained by different strain geometries in analogy to the clino-amphibole textures (Fig. 10). The well developed great circle distributions of the *a*- and *b*-axes parallel to the foliation (Figs 9c, d) can be related to an oblate strain ellipse, while plane strain is indicated by the unique maximum of the *a*-axes distribution parallel to the lineation and the great circle distribution of the *b*-axes around the lineation (Figs 9a, b).

The absence of preferred orientation of quartz in the quartz-bearing amphibolites remains enigmatic. Similar observations have been reported by Shelley (1989) and Wenk & Pannetier (1990). Shelley (1989) assumed obliteration of a weak pre-existing texture. Wenk & Pannetier (1990) found initial development and subsequent loss of quartz-preferred orientation during mylonitization of gabbro. This observation was interpreted such that grain size reduction during deformation hampered the crystal plastic deformation mechanisms in quartz. The same mechanism might also apply to the Chiwaukum Schists.

According to the homogeneous microstructure of most samples, even processes such as grain and phase boundary sliding may be relevant for the texture development. The contrasting deformation behaviour of the different minerals may also cause modifications of the ideal textures. However, the interaction of different mineral phases during deformation in polyphase rocks is still poorly understood. Therefore, any assumption considering these mechanisms would be hypothetical. Kruse & Stünitz (1999) describe a homogenous, fine-grained microstructure of amphibolites developed by granular flow. Since these microstructures are similar to our mylonitic amphibolites, crystal growth associated with granular flow could be a possible mechanism for the texture development of hornblende.

Geological significance of the results

The samples from the two tectonic units (Ingalls Complex in the hanging wall and Chiwaukum Schists in the footwall) display microfabrics, which are similar within the units but different for the two units (Fig. 10). Following the findings of Gapais & Brun (1981), a plain strain and oblate finite strain ellipsoid can be attributed to the Chiwaukum Schists and to the Ingalls Complex, respectively. The different types of straining may result from different strain paths in the individual tectonic units. In this case, it is remarkable that the youngest tectonic event, the Windy Pass thrusting, obviously did not change the microfabrics (samples Amp2, Qamp2) in the thrust zone itself. On the other hand, if we assume an overprinting of all four samples by the Windy Pass thrusting, the different texture types of the units are difficult to explain. This might be due to the different rheological behaviour of the two units. The textures of biotite schists from the Ingalls Complex, however, also indicate oblate strain (Gröger *et al.* 2000; Gröger 2001).

One of the most important results is the good agreement of the textures with the overall kinematics inferred from the structural analysis performed in the Windy pass area (Gröger 2001, Tischler 2001). Commonly, a NW–SE trending stretching lineation is described for the Cascades Crystalline Core as a major indicator for regional tectonic transport in a dextral strike-slip system (e.g. Brown & Talbot 1989). A competing model proposes compressional tectonics leading to NE or SW directed displacements (Paterson *et al.* 1994). This later model seems to fit best our textural and microstructural data as well as field observations, because structural geometries and the internal architecture of the Ingalls Complex suggest top-to-the NE/ENE nappe displacement. In addition, lattice preferred orientations of hornblende and plagioclase suggest similar kinematic directions trending NE/ENE. In contrast to the asymmetric large scale structures, the sense of shear is not recognizable from the texture due to the orthorhombic symmetry of the pole figures in respect to the foliation. Additional texture analyses are desirable in order to allow conclusions on the kinematics of the Cascades Crystalline Core.

Conclusions

This study demonstrates the feasibility of neutron diffraction for the texture analysis of polyphase rocks. Multiphase texture analyses are not a routine procedure (Wenk & Pannetier 1990) and samples require an individual strategy of data analysis. For amphibolites and even for quartz-bearing amphibolites, the high resolution of the neutron diffraction patterns gained at the SKAT texture diffractometer in Dubna provides pole figures that are suitable for 'Quantitative Texture Analyses' of hornblende and plagioclase. For the QTA of these low symmetric minerals, not only four or five pole figures but also additional experimental pole figures not used for the QTA had to be judged in order to obtain reliable results. The different mathematical methods of QTA proposed in recent years offer different approaches to determine the QTA. For testing the reliability of the texture results, however, our investigations show that it is not only sufficient to prove the compatibility of the experimental pole figures but also that a careful comparison of the experimental and recalculated pole figures is essential.

The microstructural and textural variations of the samples suggest a fabric development by intracrystalline slip and dynamic recrystallization of plagioclase and hornblende. For hornblende, the mechanisms crystal growth under differential stress and rigid body rotation may be relevant. Due to the limited number of laboratory experiments and numerical models of hornblende, plagioclase and polymineralic rocks in the literature, the mechanisms cannot be specified more precisely. However, the results allow us to draw the conclusion that the different texture types of hornblende and plagioclase are primarily controlled by different deformation modes and not by different deformation mechanisms. A major argument for this conclusion is the observation that in the individual samples plagioclase and hornblende show the development of similar textural features, i.e. a single grain or an axially symmetric intensity distribution. On the basis of the four samples, however, we cannot finally decide if the different fabrics, i.e. the suggested different strain paths, are the result of a different strain history or the consequence of different mineralogical compositions of the tectonic units. On the other hand, the study demonstrates that the methodological basis for continuing systematic investigations is now available. The results also show a clear variation of fabrics in the area studied, which is worthy of more systematic analysis. Moreover, microstructural and textural variations in amphibolites may not only be important for the kinematic interpretation but also for the determination of the anisotropic elastic properties of the earth's crust.

We gratefully acknowledge support by B. Miller and S. Paterson, who introduced us to the geology of the Cascades Crystalline Core of Washington. H. Lebit participated on their research project on the significance of mineral lineations in ductile deformed rocks (NSF: EAR-9614521). We are also grateful to J. Kruhl and W. Skrotzki for their constructive reviews. The manuscript benefited from discussions with C. Lüneburg and H.-R. Wenk. The neutron texture research of B. Leiss and K. Ullemeyer greatly benefits from the financial support of the German 'Bundesministerium für Bildung und Forschung' (Grants: 03-DUBGOE1, 03-DU0FRE). Technical and logistics support by the Frank Laboratory of Neutron Physics at Dubna/Russia is also acknowledged. H. Gröger thanks M. Tischler for common field work and fruitful discussions.

References

AGUE, D. M., WENK, H.-R. & WENK, E. 1990. Deformation and lattice orientations of plagioclase in gabbros from Central Australia. *Geophysical Monograph*, **56**, 173–186.

BIERMANN, C. & VAN ROERMUND, H. L. M. 1983. Defect structures in naturally deformed clinoamphiboles – a TEM study. *Tectonophysics*, **95**, 267–278.

BRAUN, G. 1994. Multi-channel X-ray measurements of textures with a standard texture goniometer. *In*: BUNGE, H.-J., SIEGESMUND, S., SKROTZKI, W. & WEBER, K. (eds) *Textures of Geological Materials*. DGM Press, 61–83.

BROWN, E. H. 1987. Structural geology and accretionary history of the Northwest Cascades system, Washington and British Columbia. *Geological Society of America Bulletin*, **99**, 201–214.

BROWN, E. H. & TALBOT, J. L. 1989. Orogen parallel extension in the North Cascades Crystalline Core, Washington. *Tectonics*, **8**(6), 1105–1114.

BUTLER, R. F., GEHREIS, G. E. & KODAMA, K. P. 2001. A moderate translation alternative to the Baja British Columbia hypothesis. *GSA Today*, **11**(6), 4–10.

CHATEIGNER, D., WENK, H. R. & PERNET, M. 1999. Orientation distributions of low symmetry polyphase materials using neutron diffraction data: application to a rock composed of quartz, biotite and felspar. *Textures and Microstructures*, **33**, 35–43.

COWAN, D. S., BRANDON, M. T. & GARVER, J. I. 1997. Geologic tests of hypotheses for large coastwise displacements – a critique illustrated by the Baja British Columbia Controversy. *American Journal of Science*, **297**, 117–173.

CUMBEST, R. J., DRURY, M. R., VAN ROERMUND, H. L. M. & SIMPSON, C. 1989. Dynamic recrystallization and chemical evolution of clinoamphibole from Senja, Norway. *Contributions to Mineralogy and Petrology*, **101**, 339–349.

DORNBUSCH, H.-J. 1995. Gefüge-, Mikrostruktur- und Texturuntersuchungen an Hochtemperatur-Scherzonen in granulitfaziellen Metabasiten der Ivrea-Zone. *Geotektonische Forschungen*, **83**, 1–94.

GAPAIS, D. & BRUN, J.-P. 1981. A comparison of mineral grain fabrics and finite strain in amphibolites from eastern Finland. *Canadian Journal of Earth Sciences*, **18**, 995–1003.

GRÖGER, H. R. 2001. *Teil I: Geologische Kartierung des Chiwaukum Schist Fensters an der 'Windy Pass' – Überschiebung im Cascades Crystalline Core, Washington, USA. Teil II: Gefüge- und Texturentwicklung amphibolitfazieller Gesteine an der 'Windy Pass' – Überschiebung in den nördlichen Cascade Mountains, WA, USA*. Diploma thesis, University of Göttingen, Germany.

GRÖGER, H. R., LEISS, B., ULLEMEYER, K. & LEBIT, H. 2000. Quantitative texture analyses of deformed amphibolites and biotite schists. *TSK 8, Conference Abstract, Terra Nostra 2000*, **5**, 22.

HACKER, B. R. & CHRISTIE, J. M. 1990. Brittle/ductile and plastic/cataclastic transitions in experimentally deformed and metamorphosed amphibolites. *Geophysical Monograph*, **56**, 127–147.

HELMING, K. & ESCHNER, T. 1990. A new approach to texture analysis of multiphase materials using a texture component model. *Crystalline Research and Technology*, **25**, 203–208.

ILDEFONSE, B., LARDEUX, J. M. & CARON, J. M. 1990. The behaviour of shape preferred orientation in metamorphic rocks: amphiboles und jadeites from the Monte Mucrone area, Sesia-Lanzo zone, Italian western Alps. *Journal of Structural Geology*, **12**, 1005–1012.

JI, S. & MAINPRICE, D. 1988. Natural deformation fabrics of plagioclase: implications for slip systems and seismic anisotropy. *Tectonophysics*, **147**, 145–163.

KALLEND, J. S. 1998. Determination of the orientation distribution from pole figure data. *In*: KOCKS, U. F., TOMÉ, C. N. & WENK, H.-R. (eds) *Texture and Anisotropy*. Cambridge University Press, Cambridge, 102–125.

KERN, H. & FAKIMI, M. 1975. Effect of fabric anisotropy on compressional wave propagation in various metamorphic rocks for the range of 20–750 °C at 2 kbars. *Tectonophysics*, **28**, 227–244.

KRUHL, J. H. 1987. Preferred lattice orientation of plagioclase from amphibolite and greenschist facies rocks near the insubric line (Western Alps). *Tectonophysics*, **135**, 233–242.

KRUHL, J. H. & HUNTEMANN, T. 1991. The structural state of the former lower continental crust in Calabria (S. Italy). *Geologische Rundschau*, **80/2**, 289–302.

KRUSE, R. & STÜNITZ, H. 1999. Deformation mechanisms and phase distribution in mafic high-temperature mylonites from the Jotun Nappe, southern Norway. *Tectonophysics*, **303**, 223–249.

LEBIT, H., LÜNEBURG, C. & CASEY, M. 1999. The geometry of folds and mineral lineations: Examples from the Cascade Crystalline Core (Washington, USA). EUG 10, *Journal of Conference Abstracts*, **4**, Nr. 1; Strasbourg, France.

LEISS, B., SIEGESMUND, S. & WEBER, K. 1999. Texture asymmetries as shear sense indicators in naturally deformed mono- and polyphase carbonates. *Textures and Microstructures*, **33**, 61–74.

LUZIN, V. V. & NIKOLAYEV, D. I. 1996. On the errors of the experimental pole figures. *Textures and Microstructures*, **25**, 121–128.

MAINPRICE, D. & NICOLAS, A. 1989. Development of shape and lattice preferred orientations: Application to the seismic anisotropy of the lower crust. *Journal of Structural Geology*, **11**, 391–398.

MATTHIES, S. & VINEL, G. W. 1982. An example demonstrating a new reproduction method of the ODF of texturized samples from pole figures. *Physica Status Solidi*, **112**, 115–120.

MILLER, R. B. 1980. *Structure, petrology, and emplacement of the ophiolitic Ingalls Complex, north-central Cascades, Washington*. PhD thesis, University of Washington.

MILLER, R. B. 1985. The ophiolitic Ingalls Complex, North-central Cascades, Washington. *Geological Society of America Bulletin*, **96**, 27–41.

MISCH, P. 1966. Tectonic evolution of the North Cascades of Washington State – A west Cordilleran case history. *In*: GUNNING, H. C. (ed.) *A symposium on the tectonic history and mineral deposits of the Western Cordillera in British Columbia and neighboring parts of the United States*. Canadian Institute of Mining and Metallurgy Special, **8**, 101–148.

MÜCKLICH, A. & KLIMANEK, P. 1994. Experimental errors in quantitative texture analysis from diffraction pole figures. *Materials Science Forum*, **157–162**, 275–286.

NIKOLAYEV, D. I. & ULLEMEYER, K. 1996. The effect of smoothing on ODF reproduction. *Textures and Microstructures*, **25**, 149–157.

OLSEN, T. S. & KOHLSTEDT, D. L. 1985. Natural deformation and recryztallization of some intermediate plagioclase feldspars. *Tectonophysics*, **111**, 107–131.

PATERSON, S. R., MILLER, R. B., ANDERSON, J. L., LUND, S., BENDIXEN, J., TAYLOR, N. & FINK, T. 1994. Emplacement and Evolution of the Mt. Stuart Batholith. *In*: SWANSON, D. A. & HAUGERUD, R. A. (eds) *Guides to Field Trips*. Geological Society of America Annual Meeting, Seattle, Washington.

PLUMMER, C. C. 1980. Dynamothermal contact metamorphism superposed on regional metamorphism in the pelitic rocks of the Chiwaukum Mountains area, Washington Cascades. *Geological Society of America Bulletin*, **91**, Part I: summary 386–388 and Part II, 1627–1668.

SCHWERDTNER, W. M. 1964. Preferred orientation of hornblende in a banded hornblende gneiss. *American Journal of Science*, **262**, 1212–1229.

SCHWERDTNER, W. M., SHEEHAM, P. M. & RUCKLIDGE, J. C. 1971. Variation in degree of hornblende grain alignment within two boudinage structures. *Canadian Journal of Earth Science*, **8**, 144–149.

SHELLEY, D. 1977. Plagioclase preferred orientation in Haast Schist, New Zealand. *Journal of Geology*, **85**, 635–644.

SHELLEY, D. 1979. Plagioclase preferred orientation, forshore groupe metasediments, Bluff, New Zealand. *Tectonophysics*, **58**, 279–290.

SHELLEY, D. 1989. Plagioclase and quartz preferred orientation in low-grade schist: the roles of primary growth and plastic deformation. *Journal of Structural Geology*, **11**, 1029–1037.

SHELLEY, D. 1994. Spider texture and amphibole preferred orientations. *Journal of Structural Geology*, **16**, 709–717.

SIEGESMUND, S., HELMING, K. & KRUSE, R. 1994. Complete texture analysis of a deformed amphibolite: comparision between neutron diffraction and U-stage data. *Journal of Structural Geology*, **16**, 131–142.

SIEGESMUND, S., TAKESHITA, T. & KERN, H. 1989. Anisotropy of Vp and Vs in an amphibolite of the deeper crust and its relationship to the mineralogical, microstructural and textural charakteristics of the rock. *Tectonophysics*, **157**, 25–38.

SKROTZKI, W. 1990. Microstructure in hornblende of a mylonitic amphibolite. *In*: KNIPE, R. J. & RUTTER, E. H. (eds) *Deformation mechanisms, rheology and tectonics*. Geological Society, London, Special Publications, **54**, 321–326.

SKROTZKI, W. 1992. Defect structures and deformation mechanisms in naturally deformed hornblende. *Physical Status Solidi*, **131**, 605–624.

TABOR, R. W., FRIZELL, V. A., BOOTH, D. D., WAITT, R. B., WHETTEN, J. T., ZARTMAN, R. E. 1993. *Geological map of the Skykomish River 30- by 60-minute quadrangle, Washington.* Miscellaneous Investigations Series, U. S. Geological Survey, Denver.

TISCHLER, M. 2001. *Kartierbericht zur Geologischen Karte der Windy Pass Region im Cascades Crystalline Core, Washington, USA*. Diploma thesis, University of Göttingen, Germany.

ULLEMEYER, K. & WEBER, K. 1999. Lattice preferred orientation as an indicator of a complex deformation history of rocks. *Textures and Microstructures*, **33**, 45–60.

ULLEMEYER, K., BRAUN, G., DAHMS, M., KRUHL, J. H., OLESEN, N. Ø. & SIEGESMUND, S. 2000. Texture analysis of muscovite-bearing quartzite: a comparison of some currently used techniques. *Journal of Structural Geology*, **22**, 1541–1557.

ULLEMEYER, K., HELMING, K. & SIEGESMUND, S. 1994. Quantitative texture analysis of plagioclase. *In*: BUNGE, H. J., SIEGESMUND, S., SKROTZKI, W. & WEBER, K. (eds) *Textures of geological materials.* DGM Press, Oberursel, 93-108.

ULLEMEYER, K., SPALTHOFF, P., HEINITZ, J., ISAKOV, N. N., NIKITIN, A. N. & WEBER, K. 1998. The SKAT texture diffractometer at the pulsed reactor IBR-2 at Dubna: experimental layout and first measurements. *Nuclear Instruments and Methods in Physics Research*, **A412**, 80–88.

WALITZI, E. M. & TROJER, F. J. 1965. Strukturuntersuchungen an einer Hornblende aus dem eklogitischen Gestein von Stramez, südliche Koralpe. *Tschermaks Mineralogische und Petrologische Mitteilungen*, **10**, 233–240.

WENK, E. 1936. Zur Genese der Bändergneise von Orno Huvud. *Bulletin of the Geological Insitution of the University of Uppsala*, **26**, 53–91.

WENK, H. R., BUNGE, H. J., JANSEN, E. & PANNETIER, J. 1986. Preferred orientation of plagioclase – neutron diffraction and U-stage data. *Tectonophysics*, **126**, 271–284.

WENK, H. R., CONT, L., XIE, L., LUTTEROTTI, L., RATSCHBACHER, L. & RICHARDSON, J. 2001. Rietveld texture analysis of Dabie Shan eclogite from TOF neutron diffraction spectra. *Journal of Applied Crystallography*, **34**, 442–453.

WENK, H. R., MATTHIES, S., DONOVAN, J. & CHATEIGNER, D. 1998. Beartex: a Windows-based program system for quantitative texture analysis. *Journal of Applied Crystallography*, **31**, 262–269.

WENK, H. R. & PANNETIER, J. 1990. Texture developed in deformed granodiorites from the Santa Rosa mylonite zone, southern California. *Journal of Structural Geology*, **12**, 177–184.

Quantitative texture analysis of glaucophanite deformed under eclogite facies conditions (Sesia-Lanzo Zone, Western Alps): comparison between X-ray and neutron diffraction analysis

M. ZUCALI[1], D. CHATEIGNER[2], M. DUGNANI[1], L. LUTTEROTTI[3] & B. OULADDIAF[4]

[1]*Dipartimento di Scienze della Terra, Università di Milano, Via Mangiagalli 34, I-20133 Milano, Italy (e-mail: Michele.Zucali@unimi.it)*

[2]*Laboratoire CRISMAT-ISMRA, bd. M. Juin, 14050 Caen, France*

[3]*Dipartimento di Ingegneria dei Materiali, Università degli Studi di Trento, Via Mesiano 77, I-38050 Trento, Italy*

[4]*Institut Laue-Langevin, Neutron for Science, Rue Jules Horowitz 6, BP 156, F-38042 Grenoble Cedex 9, France*

Abstract: X-ray and neutron diffraction techniques have been applied to quantitative texture analysis of a glaucophanite from the Sesia-Lanzo Zone (Western Italian Alps), naturally deformed under eclogite facies conditions. The comparison has been carried out in order to reveal the limits and problems of texture analysis related to strongly deformed polymineralic. Different methods of measuring and computing the orientation distribution function from diffraction data have been tested, in particular X-rays, direct peak integration, and neutron diffraction using Rietveld-texture analysis. Due to grain-size problems and heterogeneity of individual amphibole minerals, neutron radiation is shown to be the best probe for characterizing the whole rock: being more penetrative than conventional X-rays, a larger volume of the mineral aggregate is sampled, giving better statistics. However, results obtained by summing the corresponding individual spectra of at least three X-ray diffraction experiments on parallel slabs of the same specimen also give statistically valid, semi-quantitative results that reproduce the overall textures. The quantitative texture analysis shows the strong texture of the two generations of amphiboles (AmpI and AmpII), which are mainly characterized by $[001]^*$-directions at an angle of about 10° to the mineral lineation and by *(hk0)* planes describing girdles around the lineation. The texture is comparable to those described in the literature for amphibole deformed under different temperature and pressure conditions, and the pronounced asymmetry of the $[001]^*$ directions with respect to the mineral lineation is consistent with a non-coaxial component that occurs during the deformation.

In the recent past, quantitative texture analysis has been used in incredibly diverse applications in Earth Sciences (Wenk 1985; Bunge *et al.*1994; Kocks *et al.* 1998; Leiss *et al.* 2000). For instance in structural geology, tectonics, deformation processes and glaciology, texture analysis can be used to describe the anisotropy of fabrics (Baker *et al.* 1969; Baker & Wenk 1972; Gapais & Brun 1981; Bennet *et al.*1997; Leiss *et al.* 2000). In palaeontology, texture analysis brings new insights in to the phylogeny of molluscs and fossils (Chateigner *et al.* 2000). In geophysics, texture analysis aids interpretation of the anisotropic seismic wave propagation in the Earth's inner core (Wenk *et al.* 2000). Several techniques have been used to carry out texture analysis: the former optical U-stage has progressively been complimented by X-ray and neutron diffraction experiments, and more recently by the local electron back scattering diffraction approach (e.g., Prior *et al.*1999). In the case of X-ray and neutron diffraction, photographic films have rapidly been replaced by point detectors which have allowed the first real quantitative texture analyses by measuring diffraction pole figures. However, such detectors only measure one pole figure at a time and do not allow the separation of several textures in multiphase samples of the type commonly present in geological settings.

Bunge *et al.* (1982) proposed using neutrons to overcome the multiphase problem in texture analysis. Recently, Ricote & Chateigner (1999) used X-rays in which several pole figures were simultaneously acquired by use of a curved position sensitive detector. So far, however, quantitative texture studies of rocks composed of more than one phase of low crystal symmetry are rare

From: DE MEER, S., DRURY, M. R., DE BRESSER, J. H. P. & PENNOCK, G. M. (eds) 2002. *Deformation Mechanisms, Rheology and Tectonics: Current Status and Future Perspectives*. Geological Society, London, Special Publications, **200**, 239–253. 0305-8719/02/$15

(Siegesmund *et al.*1994; Wenk *et al.* 2000; Wenk *et al.* 2001).

Microstructural investigations, using optical microscopy, X-ray and neutron diffraction, scanning electron and transmission electron microscopy show that (100)[001] and (*hk0*)[001] are the most common slip systems in amphiboles (Table 1). Dynamic recrystallization (Cumbest *et al.* 1989), rigid body rotation (Ildefonse *et al.* 1990; Siegesmund *et al.* 1994) and cataclastic deformation (Nyman *et al.* 1992) may also occur. The most frequently recurrent amphibole textures consist of [001]-direction and (*hk*0) planes parallel to the stretching lineation (e.g., Schwerdtner 1964; Gapais & Brun 1981; Siegesmund *et al.* 1994). These studies also showed that the roles of intracrystalline plasticity, fracture, rigid body rotation and chemical forces

Table 1. *Deformation mechanisms in naturally and experimentally deformed amphiboles reported in the literature*

References	Deformation conditions	Observation techniques	*P–T* conditions	Mineral	Deformation mechanisms, slip systems and twin systems
Schwerdtner (1964); Schwerdtner *et al.* (1971)	NAT	OM		Hbl	stress-driven growth: [001] maxima perpendicular to {110} girdle
Riecker & Rooney (1969); Rooney *et al.* (1970)	EXP	OM-XR	400–600 °C; 5–15 kbar	Hbl	(100) (−101) twinning
Dollinger & Blacic (1975)	EXP/NAT	OM	AMP <800 °C >800 °C GS	Hbl Act	(100)[001] (100) variable directions (100)[001]
Kern & Fakhimi (1975)	EXP	XR	20–700 °C at 2kbar	Hbl	[110]//XZ planes of fabric and [001] deduced// lineation
Morrison 1976					
Gapais & Brun (1981)	NAT	XR	IP-LP/HT	Hbl	[001] // lineation
Biermann (1981)	NAT	—	—	—	(100) twinning
Biermann & Van Roermund, (1983)	NAT	OM TEM	450–600 °C 4–6 kbar	Clino-Amp	(hk0)[001]
Cumbest *et al.* (1989)	NAT	OM TEM	AMP	Hbl Clino-Amp	Dynamic re-crystallization (core-mantle microstructure)
Mainprice & Nicolas (1989)	NAT	OM-XR-TEM		hbl	[001]//lineation
Reynard *et al.* (1989)	NAT	TEM	ECL	Gln	(100)[001] {110}[001] (010)[100] {110}1/2⟨1−10⟩ (001)1/2⟨110⟩
Ildefonse *et al.* (1990)	NAT	OM	ECL	Gln	Rigid body rotation
Skrotzki (1990, 1992)	NAT	TEM	$T > 650$ °C	Hbl	(100)[001]
Krhul & Huntemann (1991)	NAT		GR	Hbl	[001]//lineation
Nyman *et al.* (1992)	NAT	OM SEM TEM	~500 °C 3–5 kbar	Hbl	Cataclastic deformation
Shelley, (1993, 1994; 1995)	NAT	OM	BS	Hbl	anisoptropic growth (or solution); [001]//lineation
Siegesmund *et al.* (1994)	NAT	OM ND	AMP	Hbl	Rigid body rotation or dynamic recrystallization
Berger & Stünitz (1996)	NAT	OM TEM	500–670 °C >4 kbar	Hbl	Rigid body rotation

NAT, naturally deformed; EXP, experimentally deformed; OM, optical microscopy; XR, X-ray; TEM, transmission electron microscopy; ND, neutron diffraction; AMP, amphibolite facies; GS, greenschist facies; BS, blueschist facies; ECL, eclogite facies; GR, granulite facies; Hbl, hornblende; Act, actinolite; Amp, amphibole; Gln, glaucophane.

during the deformation of amphiboles under different geological conditions, need more quantitative investigation.

Neutron diffraction is well known as the most reliable experimental technique for quantitative texture analysis (QTA) of naturally deformed rocks where statistics, in coarse-grained samples, is a big concern (Bouchez *et al.* 1979; Wenk *et al.* 1984; Kocks *et al.* 1998; Chateigner *et al.* 1999). X-ray diffraction suffers from poor statistics when the grain size is not very fine, as is the case in many geological situations. This work first investigates the possibility of using the X-ray diffraction method to achieve a better reliability at least for qualitative results. Secondly, we compare two different methodologies of analysis with respect to their applicability to more complex polymineralic rocks. Finally, amphibole textures obtained with X-ray and neutron diffraction techniques are compared with microstructural features and crystallographic orientations reported in literature.

Geological setting and sample description

The M26 sample is a glaucophanite, mainly constituted of winchitic amphiboles (≥97%) (Table 2). It crops out along the divide between Monte Mucrone and Mombarone, in the Eclogitic Micaschists Complex (EMC) of the Sesia-Lanzo Zone (Austroalpine domain – Western Alps, Italy) (Fig. 1). This zone consists of two main

Table 2. *Chemical analyses of M26 amphiboles*

	AmpI	AmpII	AmpIII
SiO_2	59.717	59.658	57.343
TiO_2	0.111	0.043	0.101
Al_2O_3	**12.112**	**11.946**	**2.009**
FeO	6.163	6.303	5.801
MnO	0.097	0.000	0.099
MgO	13.406	13.453	19.993
CaO	2.111	2.166	13.254
K_2O	0.064	0.000	0.075
Na_2O	**4.512**	**4.906**	**0.162**
Total	98.293	98.475	98.837
Si	7.945	8.035	7.875
Ti	0.008	0.008	0.008
Al	1.899	1.895	0.322
Fetot	0.690	0.707	0.669
Mn	0.008	0.000	0.008
Mg	**2.661**	**2.701**	**4.095**
Ca	**0.299**	**0.316**	**1.949**
K	0.008	0.000	0.017
Na	1.160	1.280	0.041

Stoichiometric ratios of elements based on 23 O. Fe^{tot} as Fe^{2+}.

tectonic units distinguished on the base of their lithological and metamorphic differences (e.g. Compagnoni 1977): the upper unit (II Zona Diorito-Kinzigitica) and the lower unit (Gneiss Minuti Complex and the EMC). The EMC shows a dominant Alpine imprint under eclogite facies conditions. The EMC consists of small lenses of biotite–garnet–Al silicates–metapelites (kinzigites), garnet–omphacite–glaucophane–metapelites, large omphacite–glaucophane–metaquartzdiorite bodies (metagranitoids and metaquartzdiorite of Monte Mars Complex, western part of the Monte Mucrone intrusion emplaced at 293 ± 1 Ma (Bussy *et al.* 1998)), lenses of metabasites (amphibole-bearing eclogites, eclogites and glaucophanites), pure and impure marbles, kyanite–chloritoid–garnet–quartzites, metre-size peridotitic lenses and andesitic dykes (Gosso 1977; Pognante *et al.* 1980; Passchier *et al.* 1981; Spalla *et al.* 1983; Williams & Compagnoni 1983; Lardeaux & Spalla 1991; Venturini *et al.* 1991). All lithologies, except Oligocene andesitic dykes (Dal Piaz *et al.* 1979; De Capitani *et al.* 1979; Beccaluva *et al.* 1983) show a penetrative structural and metamorphic Alpine re-equilibration under eclogite facies conditions. The age of the eclogitic metamorphism has been dated as Late Cretaceous–Early Paleocene (Inger *et al.* 1996; Duchene *et al.* 1997; Ruffet *et al.* 1997; Rubatto *et al.* 1999).

The structural framework of the EMC, along the Monte Mucrone–Mombarone divide, is the result of seven deformation phases (Pognante *et al.* 1980; Passchier *et al.* 1981; Williams & Compagnoni 1983). The most penetrative is D_2; it consists, within micaschists and gneisses, of a penetrative foliation (S2, 265°/40°) marked by eclogite facies minerals, as omphacite, blue-amphibole and phengite; in eclogites, amphibole-bearing eclogites and glaucophanites, D_2 mainly consists of an S_2 foliation, defined by a compositional layering associated with a mineral lineation (L_2, 250°/35°); the latter lineation (L_2) is defined by shape preferred orientation of omphacite and glaucophane. Subsequent deformation phases (D_3 and D_4) produced large-scale isoclinal and recumbent folds. Glaucophanites are metre-scale lenses or boudins within eclogitic micaschists and gneisses; they are characterized by centimetre to millimetre scale grain size variations. The macroscopic lineation (L_2) lies within the S_2 compositional layering and is marked by the shape preferred orientation of glaucophanes and by the preferred orientation of lenticular aggregates of glaucophanes (Fig. 2). No syn-D_2 fabric gradients occur at the mesoscopic scale within the glaucophanites and surrounding micaschists and gneisses. The

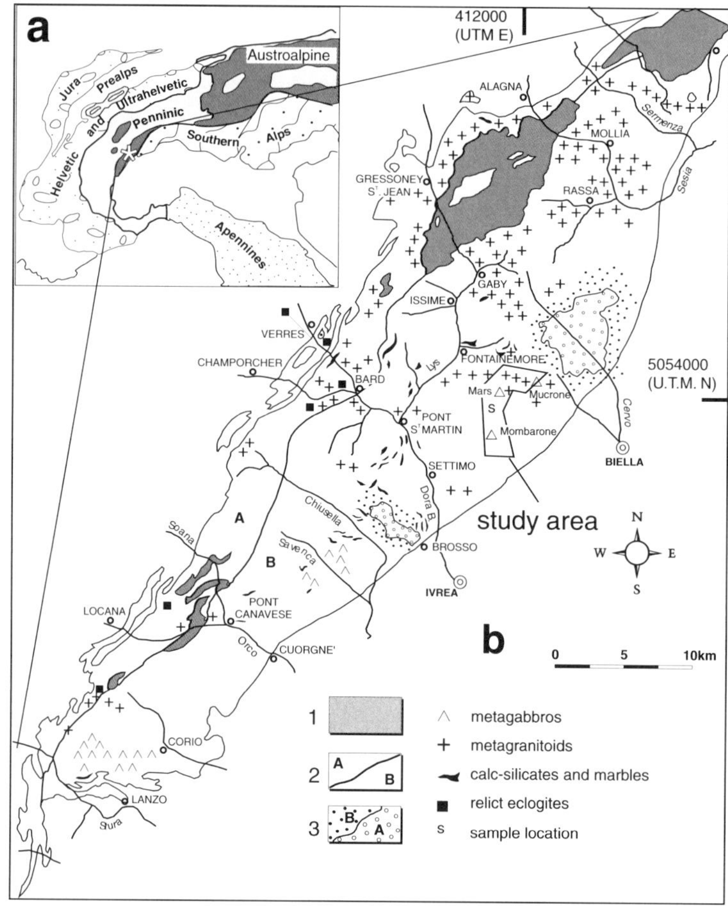

Fig. 1. (**a**) Tectonic outline of the Alpine Chain. A star identifies the Sesia-Lanzo Zone. (**b**) Simplified geological map of the Sesia-Lanzo Zone. 1, II Zona Diorito-Kinzigitica; 2A, Gneiss Minuti Complex (GMC); 2B, Eclogitic Micaschists Complex (EMC); 3A, non-metamorphic Tertiary intrusives; 3B, contact metamorphic aureoles.

glaucophanite has been sampled where minor overprinting by subsequent deformation and metamorphic transformations occurred (e.g. D_3 and D_4 folds).

Thin sections for optical microscopy were cut parallel to the mineral lineation (L_2) and perpendicular to the foliation S_2 (XZ plane in Fig. 2). Glaucophanite exhibits a compositional layering defined by alternating domains ($\leq$5 mm thick). Domains I have a lenticular shape; large AmpI occur as ellipsoidal grains (0.4–1 mm) within domains I, showing undulose extinction, deformation bands and in some places subgrains (Fig. 2). Domains I are discontinuous, and wrapped by domains II, parallel to S_2. AmpI grains do not have a shape preferred orientation with respect to S_2 and L_2. AmpI porphiroclasts, displaying undulose extinction and marginal subgrains, occur in the core of the domains I. The misorientation angle between adjacent subgrains is often higher than 5° (Fig. 2). Subgrains at the rims of the domains I display shape preferred orientation close to S_2 and L_2 directions. Domains II are defined by aggregates of AmpII. AmpII are mainly strain free with grain sizes <0.4 mm. Shape preferred orientation of AmpII is parallel to the mineral layering (S_2) (Fig. 2a) and the AmpII grains are similar in size to AmpI marginal subgrains. AmpI shows rutile, zircon, opaques and quartz inclusions. A third generation of amphibole (AmpIII) occurs locally at the grain boundaries between amphiboles (Fig. 2). Chlorite and white mica partially replace AmpI and AmpII or fill small fractures with quartz and carbonates. Garnet appears as small porphyroblasts ($\leq$1 mm) and forms less

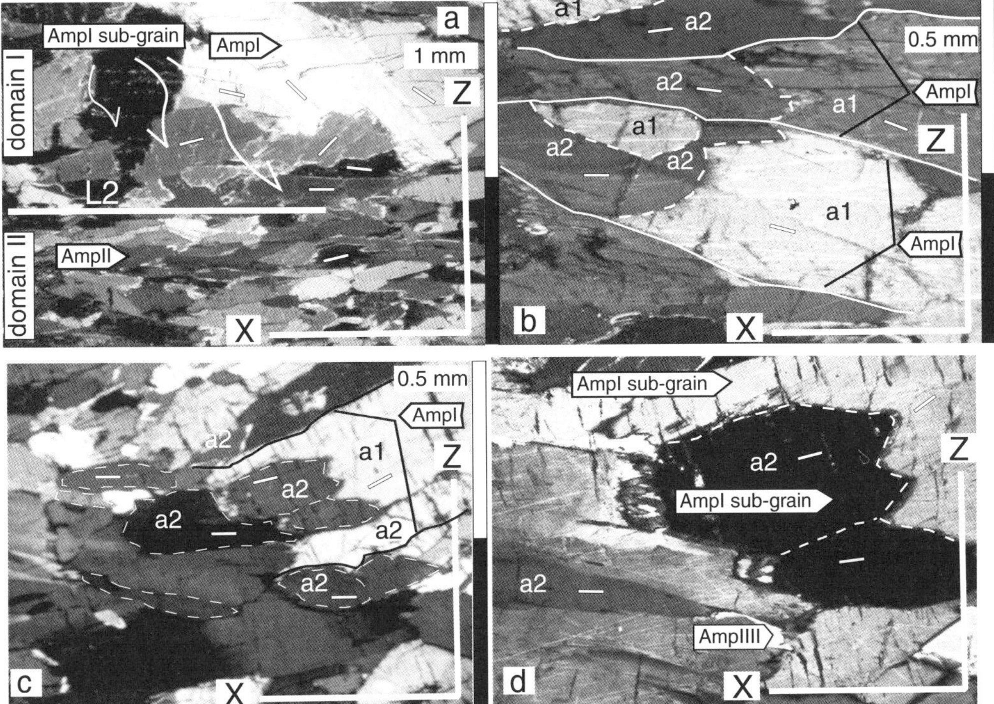

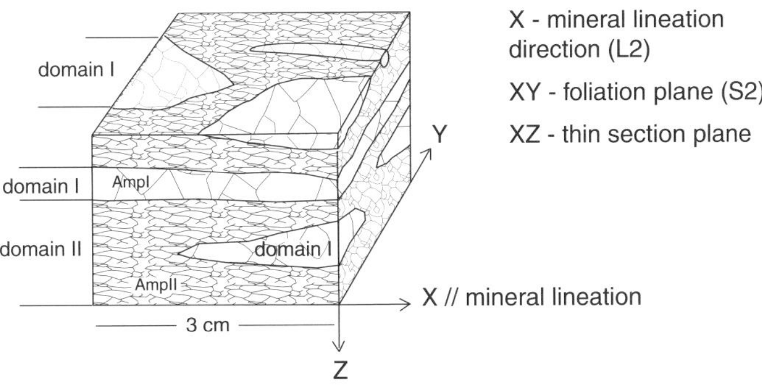

Fig. 2. (**a**) Alternating domains I and II, marking the mineralogical lineation (L_2). Domain I is defined by an elongated aggregate of AmpI with no shape preferred orientation; domain II is defined by shape preferred orientation of AmpII grains, elongated parallel to the mineral lineation (L_2). AmpI subgrains at the domain boundary show shape preferred orientation parallel to L_2. (**b**) Continuous white and black lines define the AmpI grain boundaries, while stippled white lines define AmpI subgrain boundaries; *a1* grains preserve shape and crystallographic orientation of AmpI; *a2* sub-grains show different degrees of misorientation with respect to AmpI crystal orientation. (**c**) and (**d**) AmpI, within domain I, with elongate strain free subgrains (*a2*); shape preferred orientations of subgrains mainly lie parallel to the X direction.

than 1% of the rock volume. Micro-fractures (2 mm thick), occurring at high angles (~90°) with respect to the layering, are mainly filled by fine aggregates of quartz, carbonates, diopside and white mica.

Mineral chemistry and quantitative diffraction analyses

Quantitative chemical analyses were performed on polished samples using an Applied Research

Laboratories electron microprobe fitted with six wavelength-dispersed spectrometers and a Tracor Northern Energy Dispersive Spectrometer 5600, using natural silicates as standards. An accelerating voltage of 15 kV, a sample current 20 nA and a beam current 300 nA were used. Matrix corrections were calculated using the ZAF procedure (Colby 1968). Table 2 shows representative mineral compositions of AmpI, AmpII and AmpIII. While AmpI and AmpII display the same chemical composition (winchite), AmpIII (actinolite) is characterized by a decrease in Al and Na content and an increase of Mg and Ca content.

A representative sample of M26 glaucophanite (1 cm^3) was powdered with an agate mortar and a powder diffraction pattern was collected using Cu Kα radiation, from 5° to 70°, in steps of 0.01° and 3 seconds per step. A quantitative phase analysis using the Rietveld method was performed using a richterite (Hawtorn *et al.* 1997) and a clinochlore (Smyth *et al.* 1981) as starting structural models. The atomic fractions were deduced from chemical analysis. The quantitative phase analysis confirmed that the total amount of chlorite and AmpIII is less than 5%. The other phases, observed using optical microscopy, are below the detection limits of the X-rays.

Texture measurements

Measuring single pole figures does not allow a quantitative texture analysis, even when measured completely; in order to compare between samples independently (e.g. porosity, stress/strain states, particle sizes and phase ratios) normalization of pole figures is needed (Bunge & Esling 1982). Normalization involves refining the orientation distribution function (ODF), which defines the texture strength and all components of the texture. The same single pole figure can be generated by many different ODFs, which means that interpretation of texture from a single pole figure is ambiguous.

X-ray diffraction

X-ray diffraction texture measurements were carried out using a Huber four-circle goniometer (closed eulerian cradle $+\theta$–2θ movements) mounted on an INEL X-ray generator, and Cu Kα wavelength radiation, monochromatized by an incident flat graphite monochromator. A curved position sensitive detector (PSD) with a 2θ resolution of 0.03° (INEL CPS-120) was used to acquire complete diffraction patterns at different positions of the sample in the 0–120° 2θ range. The flat samples were measured in reflection geometry. The incident angle on the sample ω, and the PSD position were chosen to maximize coverage in orientation space. The spectra were measured using a coverage from 0° to 355° in φ and from 0° to 70° in χ (Fig. 3) with incremental steps of 5° for both angles, which corresponds to $72 \times 15 = 1080$ measured 2θ patterns. Each pattern was acquired for 180 seconds.

We chose an X-ray beam size of 1×1 mm, collimated to provide sufficient resolution for peak separation and for an average grain size of 0.1–0.4 mm; larger AmpI grains (0.5–0.8 mm) correspond to less than 2% of the sample. The irradiated surface was increased by oscillations of the sample perpendicular to the lineation direction ($\pm$3 mm of amplitude) (Fig. 3). A further increase of the irradiated area could be provided by lowering ω, but such a procedure would give rise to more severe peak overlap due to a defocusing effect, particularly at high 2θ values, where much of the orientation information is located. Compared to a set-up with a point detector, this experimental design allows an optimized incident angle value of $\omega = 16°$. To further increase the grain statistics, three slabs of the same sample, cut perpendicular to the mineral layers and parallel to the mineral lineation, were measured. The respective raw intensities for each pole figure direction obtained from the three different experiments were summed before the ODF calculation.

Figure 3a shows the sum of 1080 diagrams of M26 used for the phase identification and peak indexing at the $\omega = 16°$ position, and Fig. 3b is an example of such a summed diagram for a lower incidence angle ($\omega = 5.3°$). Such diagrams are close to random powder patterns, but only approximately since the pole figures were incompletely measured. The summed diagram confirms that the AmpI and AmpII chemical compositions are similar. The increased overlap for the lower ω value can be clearly seen (Fig. 3b), leading to unreliable integration for 2θ values larger than 40°. For this reason only the experiments at $\omega = 16°$ have been used. Indexing of the diagrams used shows that at least twenty pole figures are theoretically usable for the ODF determination. Some peaks are however weak, closer to the background counts, and these would decrease the overall reliability of the ODF; therefore, only thirteen have been used in the analysis, removing peaks lower than 8% of the maximum.

To obtain these pole figures for texture analysis, data treatment was operated through direct

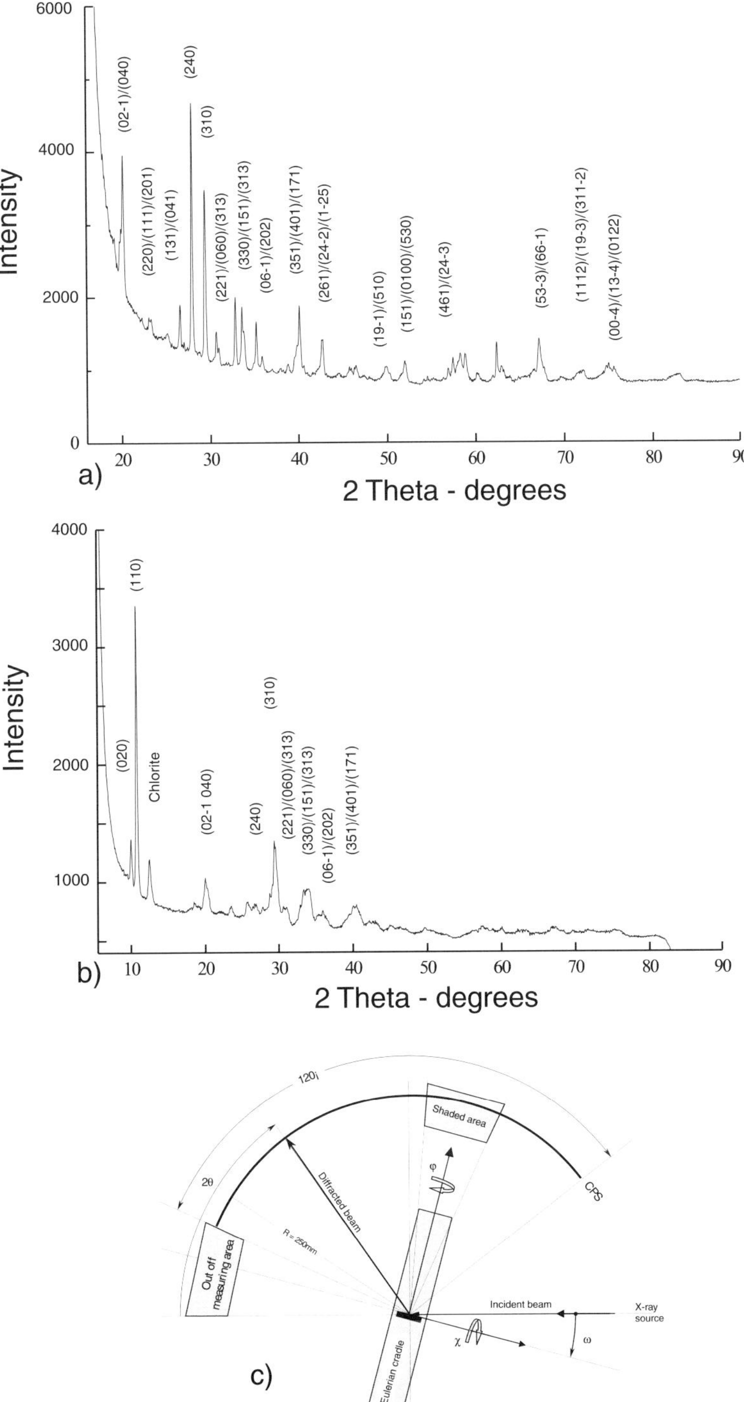

Fig. 3. Summed X-ray diffraction diagrams at 16° (**a**) and 5.3° (**b**) ω angles. (**c**) Instrument angles in reflection geometry.

Table 3. *Declared overlapping peaks for x-ray analysis, overlapping factors (%) and Chi ranges*

Miller indices			Declared overlaps (%)	min Chi	max Chi
1	1	1	30	0	70
2	2	0	20	0	70
2	0	1	50	0	70
1	3	1	30	0	70
0	4	1	70	0	70
2	2	1	20	0	70
0	6	0	1	0	70
3	1	5	79	0	70
1	5	1	40	0	70
3	3	0	40	0	70
3	1	3	20	0	70
0	6	−1	50	0	70
2	0	2	50	0	70
3	5	1	70	0	70
4	0	1	15	0	70
1	7	1	15	0	70
2	6	1	5	0	70
2	4	−2	70	0	70
1	−2	5	25	0	70
1	9	−1	50	5	70
5	1	0	50	5	70
1	5	1	15	5	70
0	10	0	5	5	70
5	3	0	80	5	70
4	6	1	50	10	70
2	4	−3	50	10	70
5	3	−3	30	15	70
6	6	−1	70	15	70
1	11	2	50	20	70
1	9	−3	30	20	70
3	11	−2	20	20	70
0	0	−4	10	20	70
1	3	−4	50	20	70
0	12	2	40	20	70

cyclic integration of the diffracted intensities and corrections for absorption using an INEL-LPEC software program (INEL 1986). The WIMV (Matthies & Vinel 1982) iterative method was subsequently used to compute the ODF from the selected pole figures. The quality of the result was assessed with reliability factors (RP0 and RP1, for global and above 1 m.r.d., or multiple of random distribution, values respectively). From the ODF, we can recalculate any pole figures, even those that are not available experimentally, and calculate parameters indicative of the texture strength: the texture index (F^2), and the texture entropy (S) (after Bunge 1982; Matthies 1991).

The different calculations have been carried out using the Berkeley Texture Package (BEARTEX, Wenk *et al.* 1998). Depending on the sample composition several diffraction peaks coming from different phases can be present at close 2θ positions. For X-ray data analysis where only the direct integration method was used to obtain pole figures, we only used single peaks, or peak overlaps coming from the same phase as a summed multi-pole figure. The overlaps can be treated by the WIMV method, assigning intensity contributions to each component of the multi-pole figure (weight in % in Table 3).

Neutron diffraction

Neutron diffraction experiments were carried out at the Institut Laue-Langevin (ILL, Grenoble, France) high flux reactor using the position-sensitive detector of the D1B beamline. The detector spans a 2θ range of 80° with a resolution of 0.2° and the wavelength used was 2.523 Å. The sample (1 cm^3) was mounted in a transmission Debye-Scherrer geometry and measured with the same scan grid as for X-ray experiments, but the pole figure coverage extended to 90° in χ, thanks to the low absorption of neutrons in this type of material. The ω, value was 10° and the counting time was 50 seconds per spectrum. A total of $72 \times 19 = 1368$ measured scans over the accessible 2θ range were made. Figure 4a shows the summed neutron diffraction pattern of the M26 sample for the 1368 scans. Since no defocusing effect occurs in transmission geometry and the full χ range is measured, such a diagram is closer to a powder diffraction pattern than an X-ray diffraction pattern, even when a blind area exists for pole figures at $\theta \neq \omega$. The neutron diffraction pattern shows an agreement between AmpI and AmpII chemical compositions. The smaller range of 2θ values, due to the larger wavelength used, gave fewer peaks. The MIMA function of Beartex, a subroutine based on the Minimal Pole Density Set (MPDS) criterion (Helming 1991), was used to compute the minimum requirement in experimental data to fulfil the condition that any ODF cell is determined by at least three points coming from the experimental pole figures. MIMA results show that the orientation space was completely covered by both the X-ray and neutron data sets, allowing recalculation of the quantitative textures using the WIMV algorithm.

Numerical integration is not suitable for reliable ODF determination because of the relatively low 2θ resolution, and either a Rietveld texture analysis or a peak fit separation (Howard & Preston 1989) should be chosen to resolve the overlaps. The latter is, however, less capable of separating exact and/or strong overlaps than

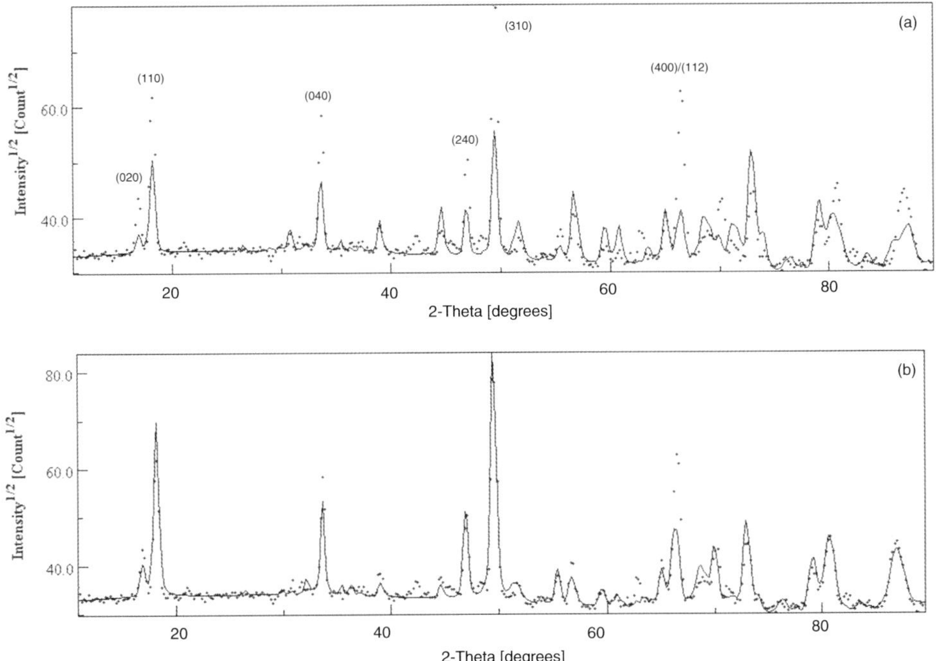

Fig. 4. (**a**) Neutron diffraction spectrum at $\varphi = 0°$ and $\chi = 60°$ after fitting by Maud, not considering texture. (**b**) Same spectrum as in (**a**) but considering texture. Crosses: measured data. Solid line: calculated pattern.

the Rietveld Texture analysis (Matthies *et al.* 1997). A Rietveld texture analysis (Lutterotti *et al.* 1997) was performed for all patterns with the software package MAUD (Lutterotti *et al.* 1999), considering amphibole as the only phase present. MAUD uses a Rietveld core routine to compute spectra and a so-called Le Bail algorithm (Matthies *et al.* 1997) to extract the differences between random and textured intensities for each computed peak. These spectra form the basis for computing the ODF using WIMV. Finally, the spectra were recomputed using the results from the ODF for the next Rietveld iteration step. Initially, lattice parameters (Table 4), a five-parameter background function, 2θ displacement of the peaks and the three-parameter Caglioti for the peak shape function, were simultaneously refined for five cycles assuming the sample was not textured. The discrepancy in fitting after these first steps suggests how strong the texture is (Fig. 4a). Then, the Le Bail-WIMV routines for the ODF computation were activated and texture was iteratively refined with crystal structure, yielding a better agreement between observed and calculated patterns (Fig. 4b). A first attempt to refine atomic positions as well as B_{iso} led to unusual values and these were replaced by values found in the literature for the next steps. Crystallographic results are compared with those of glaucophane (Comodi *et al.* 1991) and winchite (Ghose *et al.* 1986) in Table 4.

Table 4. *Lattice parameters for amphiboles*

Lattice parameters		Estimated standard deviation	Glaucophane (Comodi *et al.* 1991)	Winchite (Ghose *et al.* 1986)
a (Å)	9.5355	7.7517E-05	9.5310	9.7573
b (Å)	17.706	1.6669E-05	17.7590	17.9026
c (Å)	5.2823	7.1603E-05	5.3030	5.2886
beta (°)	103.78	8.1959E-05	103.59	103.81
Rw	0.1053	—		

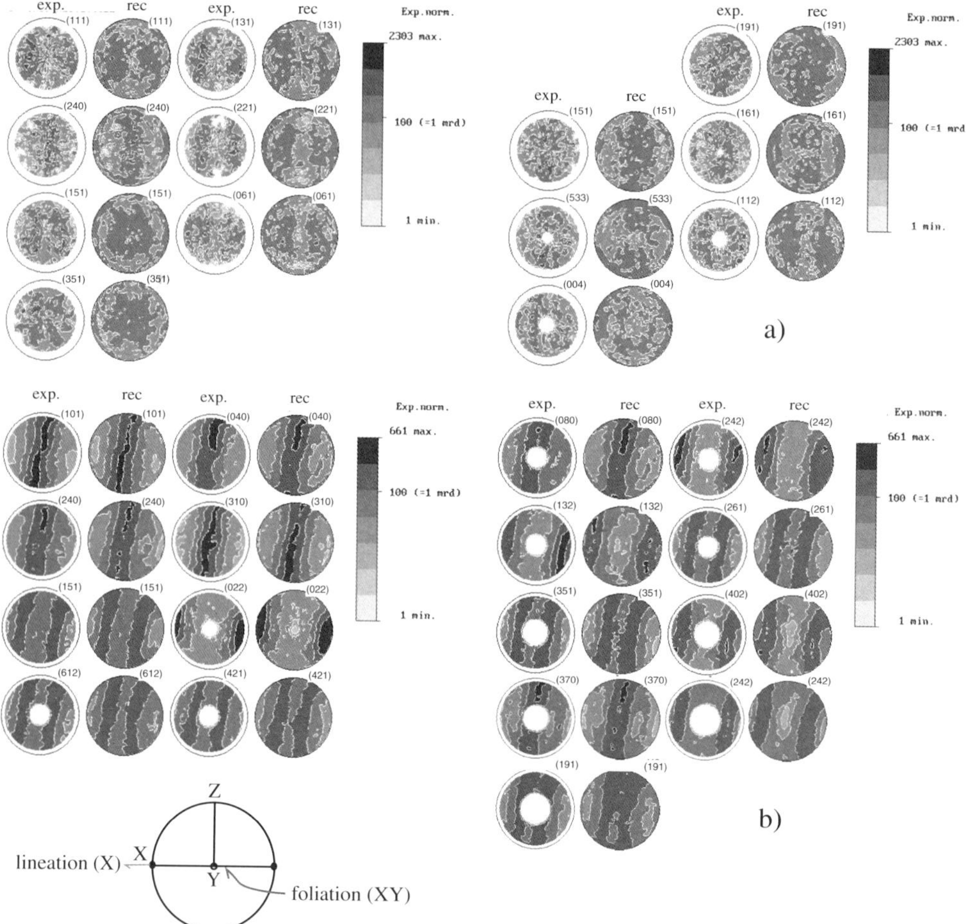

Fig. 5. Experimental normalized (exp.) and recalculated (rec.) X-ray (**a**) and neutron (**b**) pole figures for amphibole. Equal area projections. Logarithmic density scale. Intensity values are in m.r.d. ×100.

Results

Figure 5 shows the experimental normalized and recalculated pole figures from the X-ray and neutron analyses with respect to lineation direction and foliation plane. For X-rays, the high χ range was not completely covered and characteristic rings appear at unmeasured regions in recalculated pole figures (Fig. 5a). Such rings prove that, for the X-ray case, the MPDS is at the lower limit for the full ODF coverage.

X-ray pole figures (Fig. 5a) display symmetric and asymmetric girdles around mineral lineation for most pole figures (e.g. (111), (131), (240) and (221)). The quality assessment of the ODF is given by a comparison of the experimental and recalculated pole figures and by the RP0 and RP1 values in Table 5. We can clearly see in this case that the orientation density values are not as low as those revealed by neutrons. This is mainly due to the low grain statistics, which intrinsically force the WIMV algorithm to

Table 5. *Texture parameters after OD refinement from WIMV*

	X-ray pole figures	Neutron pole figures
Number of pole figures	13	17
OD minima (m.r.d.)	0.00	0.00
OD maxima (m.r.d.)	23.03	6.61
S	−2.91	−1.10
F^2 (m.r.d.^{2})	34.17	6.33
RP_0 averaged (%)	55.03	13.69
RP_1 averaged (%)	29.12	10.74

enhance the overall texture strength. However, these strong densities are obtained through an ODF refinement, the convergence of which ensures coherency between pole figures. A qualitative interpretation of the results is then made possible, with the grain-size problem affecting only the density values. Recalculated pole figures do not perfectly reproduce the experimental ones, as seen from the quite high RP factors. This is a consequence of the presence of some large grains. The strong amphibole texture is clearly shown by recalculating some special pole figures from the ODF focussing on principal crystal directions (Fig. 6a). The non-measurable [001]* directions are close to the lineation and [010]* and [100]* are mainly scattered within a plane perpendicular to the lineation direction. The [001]* directions make an angle lower than 10° with the lineation and display an angle of <15° with respect to XZ plane as shown by the asymmetry of maxima.

Similar features to those described above can also be observed in the neutron pole figures (Figs 5b and 6b). Here, experimental pole figures are smoother than those obtained with X-rays, and recalculated pole figures (Fig. 5b) closely reproduce the experimental ones, with corresponding lower RP values. Experimental and recalculated pole figures exhibit great and small circle distributions close or perpendicular to the lineation direction. As for the X-ray pole figures, a small angle between the lineation and [001]* directions is apparent, as shown by the low Miller indices pole figures (Fig. 6b). The [001]* directions are strongly oriented, with an angle slightly deviating from the lineation: approximately 10° and 15° with respect to the XZ plane. The [010]* and [100]* directions mainly describe maxima normal to the lineation direction.

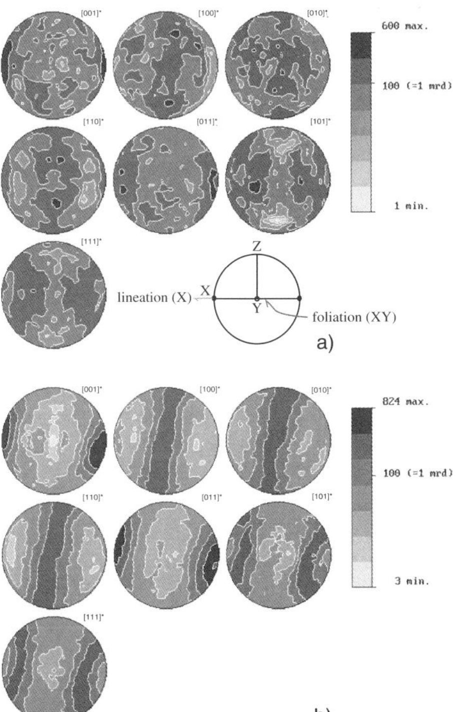

Fig. 6. X-ray (**a**) and neutron (**b**) pole figures recalculated with BEARTEX. Equal area projections. Logarithmic density scale. Intensity values are in m.r.d. ×100.

Discussion and conclusions

The M26 glaucophanite exhibits a strong texture, as shown by the pole density distribution of the amphibole and the texture indices reported in Table 5. The texture is comparable with those described in amphiboles deformed at different pressure and temperature conditions: the [001]* and [110]* directions mainly lie parallel and perpendicular to the lineation respectively (e.g., Schwerdtner 1964; Schwerdtner *et al.* 1971; Gapais & Brun 1981; Mainprice & Nicolas 1989; Kruhl & Huntemann 1991; Siegesmund *et al.* 1994), where the [110]* orientation is a function of the fabric components (Gapais & Brun 1981). In the M26 glaucophanites, [100]* and [010]* directions scatter within the YZ-plane of the strain ellipsoid, and [001]* directions lie at an angle with respect to the lineation and foliation. This can be interpreted as due to a dominant constrictional component of the finite strain (Gapais & Brun 1981). The asymmetry of the [001]* directions with respect to the fabric elements could be interpreted as having developed during a non-coaxial deformation (e.g., simple shear), as observed in other materials (Nicolas & Poirier 1976; Gapais & Cobbold 1987; Mainprice & Nicolas 1989; Law 1990; Wenk 1998). In the literature, the observed preferred orientations have been mainly interpreted as due to (100)[001] slip, rigid body rotation and oriented growth (see Table 1 for references). The crystallographic preferred orientations of the M26 sample suggest a strong influence of crystal shape, as shown by Ildefonse *et al.* (1990) for glaucophane within the Eclogitic Micaschists of the Sesia-Lanzo Zone and for hornblende in amphibolites where the weak matrix is constituted of plagioclase (Gapais & Brun 1981; Siegesmund *et al.* 1994). Our sample is matrix free, and domains II, mainly constituted by AmpII, wrapping around domains I. Furthermore, AmpI porphyroclasts show undulose to patchy extinction, deformation bands and subgrains (Fig. 2). Subgrains in the core of the domains I show neither undulose extinction nor shape preferred orientations. Subgrains at the rims of the domains I progressively show a tendency to be oriented parallel to the domains II. The grain size and preferred orientation of such subgrains is similar to those of the AmpII defining domains II and both are mainly strain free. These relationships account for a component of subgrain rotation during the development of S2 and for AmpII-rich domains II having mechanically behaved as a 'weak matrix' with respect to the domains I (Cumbest *et al.* 1989; Siegesmund *et al.* 1994).

The relationships between the meso–microstructures and crystallographic textures, described above, do not unequivocally discriminate the dominant deformation mechanism during the development of the glaucophanite preferred orientation. In general, the nature of the deformation mechanisms leading to amphibole textures still remains incompletely resolved (e.g. Stünitz 1989; Siegesmund *et al.* 1994). Aside from these limitations, the present work shows that similar crystallographic textures occur in amphiboles deformed under amphibolite and eclogite facies conditions, except for the pronounced asymmetries of $[001]^*$ directions with respect to the fabric elements. Further investigations on other lithologies of the Sesia-Lanzo Zone, such as eclogitic micaschists or eclogites, may provide more quantitative textural data of the deeply subducted slice of continental crust in the Sesia-Lanzo Zone. Such data can be used to study the mechanical behaviour of rock-forming minerals during the important natural process of subduction to mantle depths and their relationships with whole rock deformation mechanisms.

The comparison of the two techniques shows that X-ray data can produce semi-quantitative results, which reproduce the overall texture if a sufficient area of the sample, or different sections of the same sample, are scanned using similar conditions. Neutron results are statistically more reliable (Table 5). The reliability of the different computation techniques is confirmed by the textures obtained: both direct peak integration (X-ray) and Rietveld texture analysis (neutron) produced qualitatively similar results even when starting from raw intensities obtained through different acquisition techniques (Ullemeyer *et al.* 2000), and agree with the results reported in the literature. However, it can be seen that, besides the comparatively limited accessibility of neutron experiments to users, the much larger resolution of X-rays using a PSD gives a high potentiality in QTA of geological samples of half-millimetre sized grains. The relatively poor 2θ resolution of thermal neutrons makes the separation of peaks from several phases a hard task, even with a Rietveld-like technique. This can be demonstrated, as in this work, by the strong correlation existing between fitted parameters such as atomic positions or isotropic Debye-Waller factors and texture coefficients, causing the refinements to diverge when all the parameters are refined. No atomic structure variations between samples can be accessed under these conditions, and with the resolution used, this limits the neutron applicability. X-rays offer a better resolution, but their lower χ measurable range illustrates another limitation. The use of a Rietveld texture methodology overcomes this limitation by increasing the number of pole figures used in the QTA analysis, together with a better atomic structure definition than that which is accessible from the better resolution. This methodology requires a high number of diffraction diagrams and a diffraction apparatus equipped with a position sensitive detector or area detector that can realize such measurements in a reasonable experimental time. Indeed, if the grain statistic problem is alleviated by special measuring practices, such as larger oscillations than those used in the present work, X-ray diffraction using the direct integration method suggests that a complete quantitative description of textures and structures of strongly polyphasic rocks can be achieved by implementing the X-ray Rietveld texture analysis.

The authors greatly thank G. Gosso, M. I. Spalla and G. Artioli from Università di Milano. The manuscript was improved through the very constructive reviews of D. Gapais and B. Leiss. The work was funded by COF. MURST 40% 1999.

References

BAKER, D. W. & WENK, H. R. 1972. Preferred orientation in a low-symmetry quartz mylonite. *Journal of Geology*, **80**, 81–105.

BAKER, D. W., WENK, H. R. & CHRISTIE, J. M. 1969. X-ray analysis of preferred orientation in fine-grained quartz aggregates. *Journal of Geology*, **77**, 144–172.

BECCALUVA, L., BIGIOGGERO, B. *ET AL.* 1983. Post-collisional orogenic dyke magmatism in the Alps. *Memorie della Società Geoligica Italiana*, **26**, 341–359.

BENNET, K., WENK, H.-R., DURHAM, W. B., STERN, L. A. & KIRBY, S. H. 1997. Preferred crystallographic orientation in the ice I > II transformation and the flow of ice II. *Philosophical Magazine*, **A76**, 413–435.

BERGER, A. & STUNITZ, H. 1996. Deformation mechanism and reaction of hornblende: examples from the Bergell tonalite (Central Alps). *Tectonophysics*, **257**, 149–174.

BIERMANN, C. 1981. (100) deformation twins in naturally deformed amphiboles. *Nature*, **292**, 821–823.

BIERMANN, C. & VAN ROERMUND, H. L. M. 1983. Defect structures in naturally deformed amphiboles – a TEM study. *Tectonophysics*, **95**, 267–278.

BOUCHEZ, J. L., DERVIN, P., MARDON, J. P. & ENGLANDER, M. 1979. La diffraction neutronique appliquee a l'etude de l'orientation preferentielle de reseau dans les quartzites. *In*: NICOLAS, A., DAROT, M. & WILLAIME, C. (eds) *Mecanismes de Deformation des Mineraux et des Roches*. Bulletin de Mineralogie, **102**, 225–231.

BUNGE, H. J. 1982. *Texture analysis in Material Science – Mathematical methods*. Butterworths, London.

BUNGE, H. J. & ESLING, C. 1986. *Quantitative Texture Analysis*. DGM Verlag, Germany.

BUNGE, H. J., SIEGESMUND, S. & WEBER, K. 1994. *Textures of Geological Materials*. Oberusel. 399.

BUNGE, H. J., WENK, H. R. & PANNETIER, J. 1982. Neutron diffraction texture analysis using a 2q position sensitive detector. *Textures and Microstructures*, **5**, 153–170.

BUSSY, F., VENTURINI, G., HUNZIKER, J. & MARTINOTTI, G. 1998. U-Pb ages of magmatic rocks of the western Austroalpine Dent-Blanche-Sesia Unit. *Schweizerische Mineralogische und Petrographische Mitteilungen*, **78**, 163–168.

CHATEIGNER, D., HEDEGAARD, C. & WENK, H. R. 2000. Mollusc shell microstructures and crystallographic textures. *In*: LEISS, B., ULLEMEYER, K. & WEBER, K. (eds) *Textures and physical properties of rocks. Journal of Structural Geology. Special Issue*. **22**, 1723–1735.

CHATEIGNER, D., WENK, H. R. & PERNET, M. 1999. Orientation distributions of low symmetry polyphase materials using neutron diffraction data: application to a rock composed of quartz, biotite and feldspar. *Textures and Microstructures*, **33**, 35–43.

COLBY, J. W. 1968. Quantitative microprobe analysis of thin insulating films. *Advances in X-Ray Analysis*, **11**, 287–305.

COMODI, P., MELLINI, M., UNGARETTI, L. & ZANAZZI, P. F. 1991. Compressibility and high pressure structure refinement of tremolite, pargasite and glaucophane. *European Journal of Mineralogy*, **3**, 485–499.

COMPAGNONI, R. 1977. The Sesia-Lanzo zone: high-pressure low-temperature metamorphism in the Austroalpine continental margin. *Rendiconti della Società Italiana di Mineralogia e Petrologia*, **33**, 335–374.

CUMBEST, R. J., DRURY, M. R., VAN ROERMUND, H. L. M. & SIMPSON, C. 1989. Dynamic recrystallization and chemical evolution of clinoamphibole from Senja Norway. *Contributions to Mineralogy and Petrology*, **101**, 339–349.

DAL PIAZ, G. V., VENTURELLI, G. & SCOLARI, A. 1979. Calc-alkaline to ultrapotassic post-collisional volcanic activity in the internal northwestern Alps. *Memorie della Società Geologica Italiana, Padova*, **32**, 4–15.

DE CAPITANI, L., POTENZA FIORENTINI, M., MARCHI, A. & SELLA, M. 1979. Chemical and Tectonic contributions to the age and petrology of the Canavese and Sesia-Lanzo 'porphyrites'. *Atti Società Italiana di Scienze Naturali*, **120**, 151–179.

DOLLINGER, G. & BLACIC, J. D. 1975. Deformation mechanisms in experimentally and naturally deformed amphiboles. *Earth and Planetary Science Letters*, **26**, 409–416.

DUCHENE, S., BLICHERT, T. J., LUAIS, B., TELOUK, P., LARDEAUX, J. M. & ALBAREDE, F. 1997. The Lu-Hf dating of garnets and the ages of the Alpine high-pressure metamorphism. *Nature*, **387**, 586–589.

GAPAIS, D. & BRUN, J. P. 1981. A comparison of mineral grain fabrics and finite strain in amphibolites from eastern Finland. *Canadian Journal of Earth Sciences*, **18**, 995–1003.

GAPAIS, D. & COBBOLD, P. R. 1987. Slip system domains: 2, Kinematic aspects of fabric development in polycrystalline aggregates. *Tectonophysics*, **138**, 289–309.

GHOSE, S., KERSTEN, M., LANGER, K., ROSSI, G. & UNGARETTI, L. 1986. Crystal field spectra and Jahn Teller effect of Mn (super 34) in clinopyroxene and clinoamphiboles from India. *Physics and Chemistry of Minerals*, **13**, 291–305.

GOSSO, G. 1977. Metamorphic evolution and fold history in the eclogite micaschists of the upper Gressoney valley (Sesia-Lanzo zone, Western Alps). *Rendiconti della Società Italiana di Mineralogia e Petrologia*, **33**, 389–407.

HAWTORN, F. C., DELLA VENTURA, G., ROBERT, J.-L., WELCH, M. D., RAUDSEPP, M. & JENKINS, D. M. 1997. A Rietveld and infrared study of synthetic amphibole along the potassium-richterire tremolite join. *American Mineralogist*, **82**, 708–716.

HELMING, K. 1991. Minimal pole figures ranges for quantitative texture analysis. *Textures and Microstructures*, **19**, 45–54.

HOWARD, S. A. & PRESTON, K. D. 1989. Profile fitting of powder diffraction patterns. *Reviews in Mineralogy*, **20**, 217–275.

ILDEFONSE, B., LARDEAUX, J. M. & CARON, J. M. 1990. The behavior of shape preferred orientations in the metamorphic rocks: amphiboles and jadeites from the Monte Mucrone Area (Sesia-Lanzo Zone, Italian Western Alps). *Journal of Structural Geology*, **12**, 1005–1011.

INEL 1986. *Goman*. World Wide Web Address: http://www.inel.fr

INGER, S., RAMSBOTHAM, W., CLIFF, R. A. & REX, D. C. 1996. Metamorphic evolution of the Sesia–Lanzo Zone, Western Alps: time constraints from multi-system geochronology. *Contribution to Mineralogy and Petrology*, **126**, 152–168.

KERN, H. & FAKHIMI, M. 1975. Effect of fabric anisotropy on compressional wave propagation in various metamorphic rocks for the range 20–700°C at 2 kbars. *Tectonophysics*, **28**, 227–244.

KOCKS, F., TOME, C. & WENK, R. 1998. *Texture and Anisotropy*. Cambridge University Press, Cambridge.

KRUHL, J. H. & HUNTEMANN, T. 1991. The structural state of the former lower continental crust in Calabria. *Geologische Rundschau*, **80**, 289–302.

LARDEAUX, J. M. & SPALLA, M. I. 1991. From granulites to eclogites in the Sesia zone (Italian Western Alps): a record of the opening and closure of the Piedmont ocean. *Journal of Metamamorphic Geology*, **9**, 35–59.

LAW, R. D. 1990. Crustallographic fabrics: a selective review of their applications to research in structural geology. *In*: KNIPE, R. J. & RUTTER, E. H. (eds) *Deformation Mechanisms, Rheology and Tectonics*, Geological Society, London, Special Publications, **54**, 335–352.

LEISS, B., ULLEMEYER, K. & WEBER, K. 2000. *Textures and Physical Properties of Rocks*. Pergamon, Oxford, International.

LUTTEROTTI, L., MATTHIES, S. & WENK, H. R. 1999. MAUD (Material Analysis Using Diffraction): a user friendly Java program for Rietveld Texture Analysis and more. *In*: SZPUNAR, J. A. (ed.) *Proceedings of the Twelfth International Conference on Textures of Materials (ICOTOM-12)*, **2**, 1599.

LUTTEROTTI, L., MATTHIES, S., WENK, H. R., SCHULTZ, A. J. & RICHARDSON, J. J. W. 1997. Combined texture and structure analysis of deformed limestone from time-of-flight neutron diffraction spectra. *Journal of Applied Physics*, **81**, 594–600.

MAINPRICE, D. & NICOLAS, A. 1989. Development of shape and lattice preferred orientations: Application to the seismic anisotropy of the lower crust. *Journal of Structural Geology*, **11**, 391–398.

MATTHIES, S. 1991. On the principle of conditional ghost correction and its realization in existing correction concepts. *Textures and Microstructures*, **14–18**, 1–12.

MATTHIES, S. & VINEL, G. W. 1982. On the reproduction of the orientation distribution function of textured samples from reduced pole figures using the concept of ghost correction. *Physica Status Solidi (a)*, **112**, K111–114.

MATTHIES, S., LUTTEROTTI, L. & WENK, H. R. 1997. Advances in texture analysis from diffraction spectra. *Journal of Applied Crystallography*, **30**, 31–42.

MORRISON, S. D. J. 1976. Transmission electron microscopy of experimentally deformed hornblende. *American Mineralogist*, **61**, 272–280.

NICOLAS, A. & POIRIER, J. P. 1976. *Crystalline Plasticity and Solid State Flow in Metamorphic Rocks*. Wiley, London.

NYMAN, M. W., LAW, R. D. & SMELIK, E. A. 1992. Cataclastic deformation for the development of core mantle structures in amphibole. *Geology*, **20**, 455–458.

PASSCHIER, C. W., URAI, J. L., VAN LOON, J. & WILLIMAS, P. F. 1981. Structural geology of the Central Sesia-Lanzo Zone. *Geologie en Mijnbouw*, **60**, 497–507.

POGNANTE, U., COMPAGNONI, R. & GOSSO, G. 1980. Micro-mesostructural relationships in the continental eclogitic rocks of the Sesia-Lanzo zone: a record of a subduction cycle (Italian Western Alps). *Rendiconti della Società Italiana di Mineralogia e Petrologia*, **36**, 169–186.

PRIOR, D. J., BOYLE, A. P., *ET AL*. 1999. The application of electron backscatter diffraction and orientation contrast imaging in the SEM to textural problems in rocks. *American Mineralogist*, **84**, 1741–1759.

REYNARD, B., GILLET, P. & WILLAIME, C. 1989. Deformation mechanisms in naturally deformed glaucophanes; a TEM and HREM study. *European Journal of Mineralogy*, **1**, 611–624.

RICOTE, J. & CHATEIGNER, D. 1999. Quantitative texture analysis applied to the study of preferential orientations in ferroelectric thin films. *Boletín Sociedad Española de Ceramica y Vidrio*, **38**, 587–591.

RIECKER, R. E. & ROONEY, T. P. 1969. Water-induced weakening of hornblende and amphibolite. *Nature*, **224**, 1299.

ROONEY, T. P., RIECKER, R. E. & ROSS, M. 1970. Deformation twins in hornblende. *Science*, **169**, 173–175.

RUBATTO, D., GEBAUER, D. & COMPAGNONI, R. 1999. Dating of eclogite-facies zircons; the age of Alpine metamorphism in the Sesia-Lanzo Zone (Western Alps). *Earth and Planetary Science Letters*, **167**, 141–158.

RUFFET, G., GRUAU, G., BALLEVRE, M., FERAUD, G. & PHILIPPOT, P. 1997. Rb-Sr and (super 40) Ar-(super 39) Ar laser probe dating of high-pressure phengites from the Sesia Zone (Western Alps); underscoring of excess argon and new age constraints on the high-pressure metamorphism. *Chemical Geology*, **141**, 1–18.

SCHWERDTNER, W. M. 1964. Preferred orientation of hornblende in a banded hornblende gneiss. *American Journal of Science*, **262**, 1212–1229.

SCHWERDTNER, W. M., SHEEHAN, P. M. & RUCKLIDGE, J. C. 1971. Variation in degree of hornblende grain alignment within two boudinage structures. *Canadian Journal of Earth Sciences*, **8**, 144–149.

SHELLEY, D. 1993. *Igneous and Metamorphic Rocks under the Microscope*. Chapman and Hall, London.

SHELLEY, D. 1994. Spider texture and amphibole preferred orientation. *Journal of Structural Geology*, **16**, 709–717.

SHELLEY, D. 1995. Asymmetric shape preferred orientation as shear-sense indicators. *Journal of Structural Geology*, **17**, 509–518.

SIEGESMUND, S., HELMING, K. & KRUSE, R. 1994. Complete texture analysis of a deformed amphibolite: comparison between neutron, diffraction and U-stage data. *Journal of Structural Geology*, **16**, 131–142.

SKROTZKI, W. 1990. Microstructure in hornblende of a mylonite amphibolite. *In*: KNIPE, R. J. & RUTTER, E. H. (eds) *Deformation Mechanisms, Rheology and Tectonics*, Geological Society, London, Special Publications, **54**, 321–325.

SKROTZKI, W. 1992. Defect structures and deformation mechanisms in naturally deformed hornblende. Physica Status Solidi (a), 131, 605–624.

SMYTH, J. R., DYAR, M. D., MAY, H. M., BRICKER, O. P. & ACKER, J. G. 1981. Crystal structure refinement and Moessbauer spectroscopy of an ordered triclinic clinochlore. *Clays and Clay Min.*, **29**, 544–550.

SPALLA, M. I., DE MARIA, L., GOSSO, G., MILETTO, M. & POGNANTE, U. 1983. Deformazione e metamorfismo della Zona Sesia - Lanzo meridionale al contatto con la falda piemontese e con il massiccio di Lanzo, Alpi occidentali. *Memorie della Società Geologica Italiana, Padova*, **26**, 499–514.

STÜNITZ, H. 1989. *Partitioning of Metamorphism and Deformation in the Boundary Region of the 'Seconda Zona Diorito-Kinzigitica', Sesia Zone, Western Alps*. PhD Thesis, ETH, Zurich.

ULLEMEYER, K., BRAUN, G., DAHMS, M., KRUHL, J. H., OLESEN, N. O. & SIEGESMUND, S. 2000. Texture analysis of a muscovite-bearing quartzite; a comparison of some currently used techniques. *In*: LEISS, B., ULLEMEYER, K. & WEBER, K. (eds) *Textures and Physical Properties of Rocks. Journal of Structural Geology, Special Issue*, **22**, 1541–1577.

VENTURINI, G., MARTINOTTI, G. & HUNZIKER, J. C. 1991. The protoliths of the 'Eclogitic Micaschists in the lower Aosta Valley (Sesia-Lanzo zone, Western Alps). *Memorie della Società Geoligica Italiana*, **43**, 347–359.

WENK, H. R. 1985. *Preferred orientation in deformed metals and rocks; an introduction to modern texture analysis*. Academic Press, Orlando, FL, United States.

WENK, H. R. 1998. Plasticity modeling in minerals and rocks. *In*: KOCKS, U. F. TOMÉ, C. N. & WENK, H. R. (eds) *Texture and Anisotropy: Preferred Orientations in Polycrystals and their Effect on Material Properties*, 676.

WENK, H. R., CONT, L., LUTTEROTTI, L., RATSCHBACHER, L. & RICHARDSON, J. 2001. Rietveld texture analysis of Dabie Shan eclogite from TOF neutron diffraction spectra. *Journal of Applied Crystallography*, **34**, 442–453.

WENK, H. R., KERN, H., SCHAFER, W. & WILL, G. 1984. Comparison of x-ray and neutron diffraction in textures analysis of carbonate rocks. *Journal of Structural Geology*, **6**, 687–692.

WENK, H. R., MATTHIES, S., DONOVAN, J. & CHATEIGNER, D. 1998. Beartex: a Windows-based program system for quantitative texture analysis. *Journal of Applied Crystallography*, **31**, 262–269.

WENK, H. R., MATTHIES, S., HEMLEY, R. J., MAO, H. K. & SHU, J. 2000. The plastic deformation of iron at pressures of the Earth's inner core. *Nature*, **45**, 1044–1047.

WILLIAMS, P. F. & COMPAGNONI, R. 1983. Deformation and metamorphism in the Bard area of the Sesia-Lanzo zone, Western Alps, during subductionand uplift. *Journal of Metamamorphic Geology*, **1**, 117–140.

Stress and deformation in subduction zones: insight from the record of exhumed metamorphic rocks

BERNHARD STÖCKHERT

Institut für Geologie, Mineralogie und Geophysik, Sonderforschungsbereich 526, Ruhr-Universität Bochum, D-44780 Bochum, Germany (e-mail: bernhard.stoeckhert@ruhr-uni-bochum.de)

Abstract: High pressure (HP) and ultrahigh (UHP) metamorphic rocks are exhumed from subduction zones at high rates on the order of plate velocity (cm/year). Their structural and microstructural record provides insight into conditions and physical state along the plate interface in subduction zones to depths of >100 km. Amazingly, many identified (U)HP metamorphic rocks appear not to be significantly deformed at (U)HP conditions, despite their history within a high strain rate mega-shearzone. Other (U)HP metamorphic rocks seem to be deformed exclusively by dissolution–precipitation creep. Indications of deformation by dislocation creep are lacking, apart from omphacite in some eclogites. Available flow laws for dislocation creep (extrapolated to low natural strain rates, which is equivalent to no deformation on the time scales of subduction and exhumation, i.e., 1 to 10 Ma) pose an upper bound to the magnitude of stress as a function of temperature along the trajectory followed by the rock. Although the record of exhumed (U)HP metamorphic rocks may only be representative of specific types or evolutionary stages of subduction zones, for such cases it implies: (1) strongly localized deformation; (2) predominance of dissolution–precipitation creep and fluid-assisted granular flow in the shear zones, suggesting Newtonian behaviour; (3) low magnitude of differential stress; which (4) is on the order of the stress drop inferred for earthquakes; and (5) negligible shear heating. These findings are easily reconciled with exhumation by forced flow in a low viscosity subduction channel prior to collision, implying effective decoupling between the plates.

An unexpected wealth of high-pressure (HP) and previously unknown ultrahigh-pressure (UHP) metamorphic rocks has been identified within the last two decades (Harley & Carswell 1995; Schreyer 1995; Coleman & Wang 1995; Ernst & Liou 1999; Liou 1999; Ernst 1999). The characteristic low temperature/pressure ratios require that these rocks have been exhumed from subduction zones. Thus their record provides valuable insight into the physical state at depth and the trajectories followed by these rocks, which can be integrated into the results of geophysical field studies, laboratory experiments, and numerical simulations (e.g. Peacock 1996; Hacker & Peacock 1995; Ernst & Peacock 1996; Hacker 1996) to address fundamental questions concerning subduction zones. These questions comprise the degree of localization of deformation in interplate shear zones, the magnitude of shear stress, the nature and position of subduction zone seismicity, the interaction between deformation and phase transformations, the role of devolatilization, the pore fluid pressure and transport properties and their variation in space and time, and finally the typical particle trajectories followed by rocks during progressive subduction and exhumation. The latter can only be recorded to a limited extent by the present structure of mountain belts, due to the general lack of complete information. The purpose of the present paper is to highlight the significance of the record of (U)HP metamorphic rocks, beyond that of occurrence of uncommon mineral assemblages or derived very high pressures, and to emphasize the opportunities presented by the available information. As most of the peculiar features have not received appropriate attention by researchers so far, it is hoped that this paper will stimulate more investigations focussed on the objective assessment of the (micro)structural record of exhumed rocks, and its interpretation in terms of subduction zone rheology and the general geodynamics of convergent plate boundaries.

The record of metamorphic rocks exhumed from subduction zones

Pressure–temperature paths

Pressure–temperature (P–T) paths are available for a great number of HP- and UHP-metamorphic terranes. Most reveal significant cooling during decompression (e.g. Harley & Carswell 1995; Carswell & Zhang 1999; Ernst

From: DE MEER, S., DRURY, M. R., DE BRESSER, J. H. P. & PENNOCK, G. M. (eds) 2002. *Deformation Mechanisms, Rheology and Tectonics: Current Status and Future Perspectives*. Geological Society, London, Special Publications, **200**, 255–274. 0305-8719/02/$15 © The Geological Society of London.

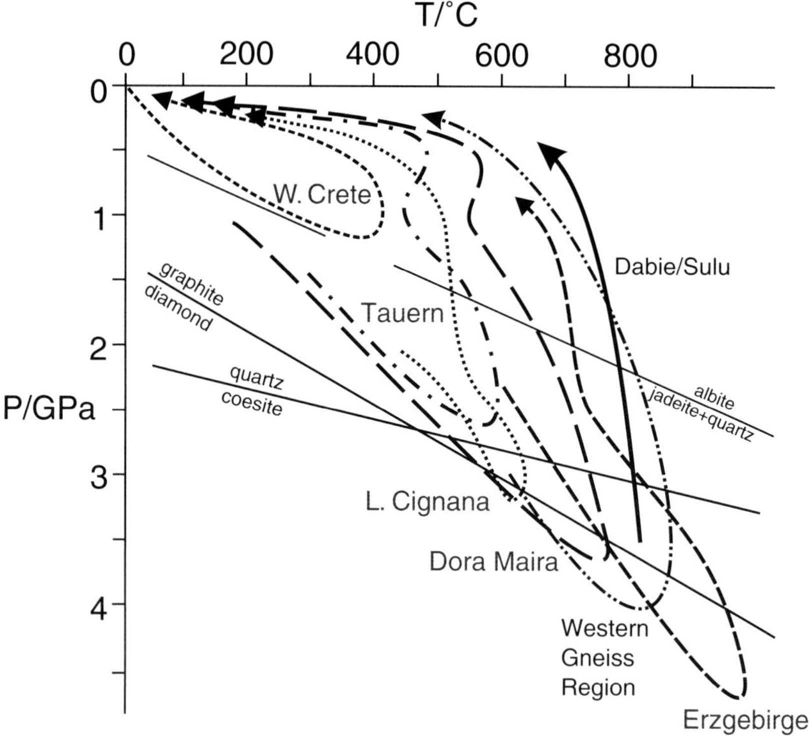

Fig. 1. Selected *P–T* paths of HP and UHP metamorphic rocks: Erzgebirge (Massonne 1999), Dora Maira (Chopin & Schertl 1999), Lago Cignana (Reinecke 1998), Tauern (Stöckhert *et al.* 1997), Western Gneiss Region, Norway (Terry *et al.* 2000*a*, *b*), Dabie/Sulu, China (Wallis *et al.* 1999), Crete (Küster & Stöckhert 1997). Note that the shape of *P–T* paths is always subject to large uncertainties and that a (U)HP metamorphic terrain cannot be characterized by a single loop. Nevertheless, the inferred general form is similar for most

1999), although a number of exceptions occur with a significant portion showing near-isothermal decompression (e.g. Carswell *et al.* 2000) or a more complex shape with significant changes in slope (Reinecke 1998). Some examples are displayed in Fig. 1. In general, most *P–T* paths are presented without consideration of error bars, which can be very large for thermobarometry. Apart from the uncertainties in the position of phase equilibria and the formulation of the usually applied geothermobarometers, it is very difficult to assess local equilibrium within a phase assemblage in a natural rock, and in fact equilibrium may not be attained at all in many (U)HP metamorphic mineral assemblages. Furthermore, a subduction-related high pressure metamorphic complex is neither characterized by a single type of *P–T* loop, nor is a single *P–T* trajectory sufficient to derive or to test tectonic models (Gerya *et al.* 2001). Notwithstanding these restrictions, the available information appears to be sufficiently robust and the points made in this contribution are not seriously affected by the uncertainties.

Time constraints

The results of isotopic dating with a variety of methods indicate that exhumation of continental UHP metamorphic rocks of the Dora Maira Massif (Gebauer *et al.* 1997; Rubatto & Hermann 2001), and UHP metamorphic oceanic slices in the Piemonte Zone (Amato *et al.* 1999), both tectonic units of the western Alps, took place at rates on the order of plate velocity. Very high rates of burial and heating are also indicated by preservation of steep compositional gradients in zoned minerals and around inclusions (Perchuk *et al.* 1999; Perchuk & Philippot 1997, 2000). As the geochronological data can only provide average rates, and the time brackets may cover late stages of slower cooling and decompression after reaching normal crustal depth, the actual exhumation rates may be even higher than the minimum rates suggested by these studies. For older UHP metamorphic terranes, such as the Triassic Dabie Sulu belt in Central China (e.g. Hacker *et al.* 1997, 2000), the Paleozoic Kokchetav massif in Kazachstan (e.g. Kaneko *et al.* 2000; Maruyama & Parkinson 2000), the Caledonian Western Gneiss Region in Norway (e.g. Terry *et al.* 2000*b*), and the Variscan Erzgebirge in Germany (e.g. Massonne 1999), larger age uncertainties preclude precise estimates on the exhumation rates, although the available data appear to reconcile with exhumation to normal crustal depths within less than 10 Ma.

The record of deformation at (U)HP conditions

Apart from some structural and microstructural observations reported from the Dora Maira Massif (e.g. Wheeler 1991; Henry *et al.* 1993; Michard *et al.* 1993, 1997), the UHP-metamorphic belt in China (e.g. Wallis *et al.* 1997, 1999; Hacker *et al.* 2000; Faure *et al.* 1999), and the Kokchetav massif in Kazachstan (e.g. Yamamoto *et al.* 2000), the deformation-related fabrics of UHP-metamorphic rocks have received much less attention when compared to the numerous studies on phase relations. In particular, a quantitative assessment of the deformation mechanisms and the inferred boundary conditions of deformation at great depth, and of the changing state of stress along the particle trajectory (e.g. van der Klauw *et al.* 1997), is not available for UHP metamorphic units. For less deeply buried HP-metamorphic rocks there are numerous (micro)structural studies, but sufficiently detailed documentation and objective interpretation is similarly scarce. In part, this may be a consequence of the rather inconspicuous (micro)structural record, and in particular of the apparent absence of high-strain shear zones commonly observed in other crustal units. In fact, the record of (U)HP metamorphic rocks appears to be rather systematic and intriguing for the given tectonic setting. Most (U)HP-metamorphic rocks, where they are not affected by the common late-stage overprint at normal crustal depth under greenschist to amphibolite facies metamorphic conditions, reveal a common picture: many are not notably deformed at HP or UHP-conditions at all, or the recorded deformation took place by poorly specified deformation mechanisms other than disclocation creep.

For instance, for the coesite-bearing pyrope quartzites (Chopin *et al.* 1991; Schertl *et al.* 1991; Compagnoni *et al.* 1995) of the Dora Maira Massif, western Alps, Michard *et al.* (1993, 1995) describe a preferred orientation of minerals overgrown by the large pyrope poikiloblasts. Also, these authors present evidence of strain shadows to the large pyrope crystals, implying a weak deformation at UHP metamorphic conditions. Apart from the presumably inherited compositional layering and related preferred orientation of minerals grown during UHP metamophism, there is no further record of pervasive deformation in the pyrope quartzites.

A particulary impressive rock body in the Dora Maira Massif is the UHP metamorphic Brossasco granite (Biino & Compagnoni 1992; Bruno *et al.* 2001). This km-sized granite body, intersected by late greenschist-facies shear zones, has remained undistorted during burial to >100 km depth and subsequent exhumation, with the original magmatic fabric perfectly preserved (Fig. 2). In view of the ease with which granite bodies elsewhere acquire a gneissic fabric, the undistorted UHP metamorphic Brossasco granite is taken to represent a stress gauge that indicates that the maximum differential stress as a function of temperature has remained too low to drive any significant deformation along its trajectory in the subduction zone (Stöckhert & Renner 1998; Renner *et al.* 2001). Similar undeformed UHP metamorphic granites (Hirajima *et al.* 1993; Wallis *et al.* 1999) and gabbros (Zhang & Liou 1997) have been reported from the Dabie-Sulu UHP belt in China. In general, UHP metamorphic rocks appear either weakly deformed at depth, with a foliation and mineral preferred orientation pattern inherited from an earlier stage and overgrown by the UHP mineral assemblage, or essentially undeformed, as in rocks with preserved primary magmatic fabrics.

In contrast to the quartz- or coesite-rich rocks described above, eclogites derived from precursors of basaltic composition frequently reveal a marked foliation and a pronounced shape (SPO) and crystallographic (CPO) preferred orientation of omphacite, even in rocks that underwent HP metamorphism at rather low temperatures (e.g. Buatier *et al.* 1991; Philippot & van Roermund 1992; Godard & van Roermund 1995; Piepenbreier & Stöckhert 2001). The CPO is attributed to deformation by dislocation creep and is consistent with the established glide systems in clinopyroxene (e.g. Mauler *et al.* 2000). However, at least in some cases, the combined SPO and CPO of omphacite in eclogites could also be the result of oriented nucleation and anisotropic growth. Also, many eclogites lack an obvious SPO or CPO of omphacite and

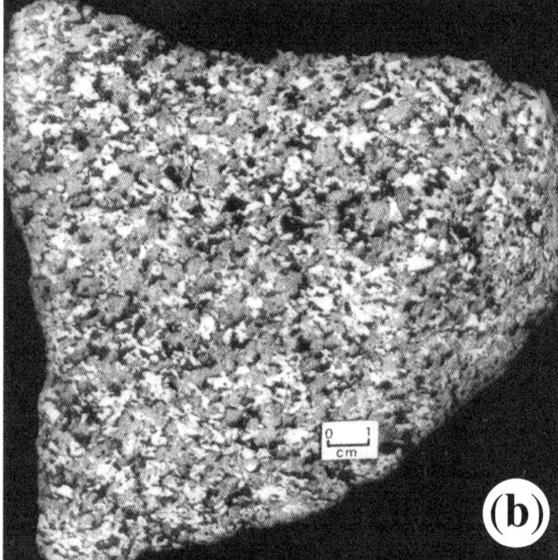

Fig. 2. Structure and microfabric of the UHP metamorphic Brossasco granite from the Dora Maira massif, western Alps. Undistorted magmatic fabric (**a**, **b**) and microstructure (**c**) revealing fine-grained quartz aggregates with foam structure. This quartz has grown at the expense of coesite, that in turn had formed from magmatic quartz during UHP metamorphism. Biotite rimmed by coronitic garnet, plagioclase decomposed to fine-grained aggregates of breakdown phases.

even primary magmatic structures (e.g. undistorted pillows, Bearth 1959) are preserved in places. Finally, it should be stressed that the occasional record of deformation of omphacite by dislocation creep should not be mistaken as an indication that this deformation regime is relevant for the subduction shear zone in general.

Microstructures and deformation mechanisms

Evidence for deformation by dissolution precipitation creep is mainly restricted to low to intermediate grade HP metamorphic rocks. An unequivocal microstructural record for this deformation mechanism has not been described from higher grade (U)HP metamorphic rocks. On the other hand, evidence for significant deformation by dislocation creep is lacking. In particular microstructures of major constituents indicating dynamic recrystallization or recovery and a pronounced CPO (apart from the forementioned omphacite fabrics in some eclogites) are rarely observed. Three representative examples (dealt with elsewhere in more detail) are given below.

(1) Delicate fossils in carbonate rocks of Crete, Greece, buried to 35 km depth, were not distorted when overgrown by coarse-grained metamorphic aragonite (Fig. 3a). Also, they remained undistorted during folding and boudinage at depth, as syn-HP-deformation was essentially by dissolution–precipitation creep not affecting the interior of large grains (Stöckhert *et al.* 1999). The grain-size sensitivity of dissolution–precipitation creep is reflected by boudinage of the coarse-grained layers within homogeneously deformed fine-grained aragonite marbles, revealing a microstructure characteristic for dissolution–precipitation creep (Fig. 3b) and lacking a CPO (Wachmann & Stöckhert, unpublished data). In siliciclastic rocks of the same series, progressive deformation at HP metamorphic conditions ($c.\ T = 400 \pm 50\,^{\circ}C$, $P = 1 \pm 0.3$ GPa) was essentially by dissolution–precipitation creep with enhanced dissolution at mica–quartz interfaces in poly-phase phyllites (Fig. 3c), and stress concentration in pure (nearly single phase) quartzites and quartz veins deforming by dislocation creep (Schwarz & Stöckhert 1996; Stöckhert *et al.* 1999).

(2) Garnet poikiloblasts (Fig. 4a) in metasediments from the Tauern Window, eastern Alps, crystallized during HP metamorphism at the maximum depth of burial of *c.* 70 km ($T \approx 600\,^{\circ}C$ at $P \approx 2.5$ GPa) have overgrown

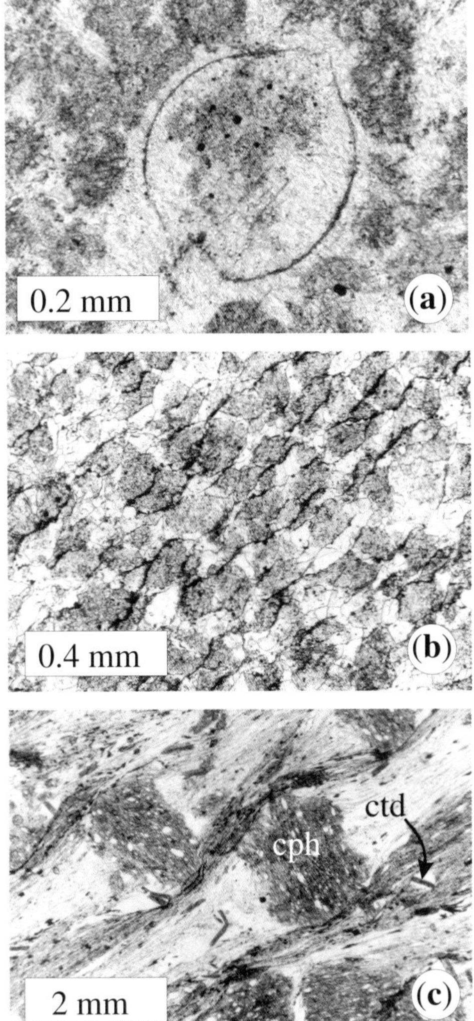

Fig. 3. Microstructures of low-temperature HP metamorphic rocks from Crete (see Stöckhert *et al.* 1999 for details). (**a**) Coarse-grained aragonite marble boudin with undistorted fossil shell overgrown by metamorphic aragonite, precluding pervasive deformation of this carbonate rock during burial to >25 km depth (crossed polarizors). (**b**) Fine-grained aragonite marble homogeneously deformed by dissolution–precipitation creep in the stability field of aragonite, with unsoluble opaque particles enriched at interfaces normal to shortening direction and clear overgrowth at interfaces normal to the stretching direction (crossed polarizors). (**c**) Phyllite deformed by dissolution–precipitation creep. Mg-carpholite (cph) poikiloblasts have overgrown a foliation defined by mica and truncated clastic quartz grains, with strain shadows and strain caps at carpholite and chloritoid (ctd) revealing ongoing deformation by dissolution–precipitation creep. There is no evidence for crystal plastic deformation of quartz.

quartz aggregates with a typical foam microstructure (Fig. 4c; Stöckhert *et al.* 1997) and random crystallographic orientation (Fig. 4b; Eismann & Stöckhert, unpublished data). The lack of a quartz CPO as preserved inside the garnet in the well-foliated schist indicates that dislocation creep was not activated during progressive burial to 70 km depth. Also, the foam structure controlled by interfacial free energy shows that stress was too low to drive deformation by dislocation creep at the stage when the rock became detached from the downgoing plate and started to return towards the surface.

(3) The classical pyrope quartzites of the Dora Maira Massif, western Alps reveal a coarse-grained (0.3–0.5 mm) foam structure of quartz (Fig. 4d). The quartz formed at the expense of coesite in an early stage of exhumation (Chopin *et al.* 1991) does not reveal a CPO (Lämmerhirt & Stöckhert, unpublished data). The preservation of a conspicuous palisade-type SPO pattern of elongated quartz grains (e.g. Michard *et al.* 1995) related to the coesite–quartz phase transformation, the undisturbed interfacial free energy controlled foam structure, and the lack of CPO preclude any post-UHP deformation by dislocation creep during exhumation.

For these examples, the microstructural record implies that the levels of differential stress (as a function of temperature) required to drive deformation of calcite/aragonite or quartz by dislocation creep were generally not reached at HP-conditions, and during most of the burial and early exhumation history. Likewise, the undistorted UHP metamorphic granites indicate that the same holds true for the stage of UHP metamorphism in the stability field of coesite.

Experimental constraints on flow strength

The ultimate flow strength of rocks as a function of temperature and strain rate is limited by dislocation creep (e.g. Ranalli 1995; Evans and Kohlstedt 1995). Thus, for rocks not deformed at (U)HP conditions at all, or exclusively deformed by mechanisms like dissolution–precipitation creep without activation of crystal plastic processes, an upper bound to differential stress at (U)HP conditions can be obtained by extrapolation of flow laws for dislocation creep to a strain rate of $10^{-15}\ s^{-1}$, which is equivalent to no notable deformation on the time scales of subduction and exhumation (1 to 10 Ma). As a first approximation, the ultimate strength of many polyphase crustal rocks buried in subduction zones and undergoing (U)HP metamorphism is controlled by the predominating

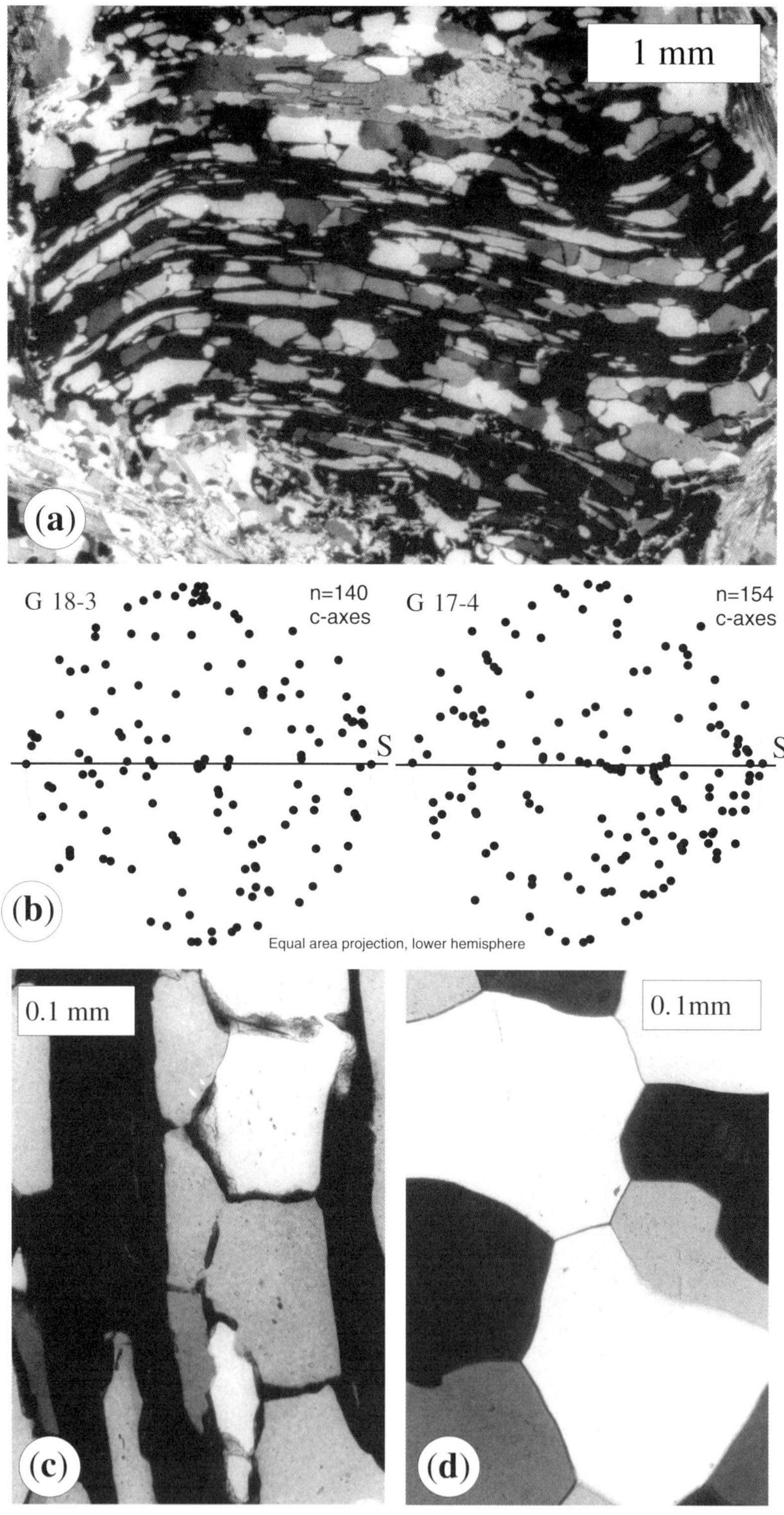
1 mm
(a)
G 18-3
n=140
c-axes
G 17-4
n=154
c-axes
S
S
(b)
Equal area projection, lower hemisphere
0.1 mm
0.1mm
(c)
(d)

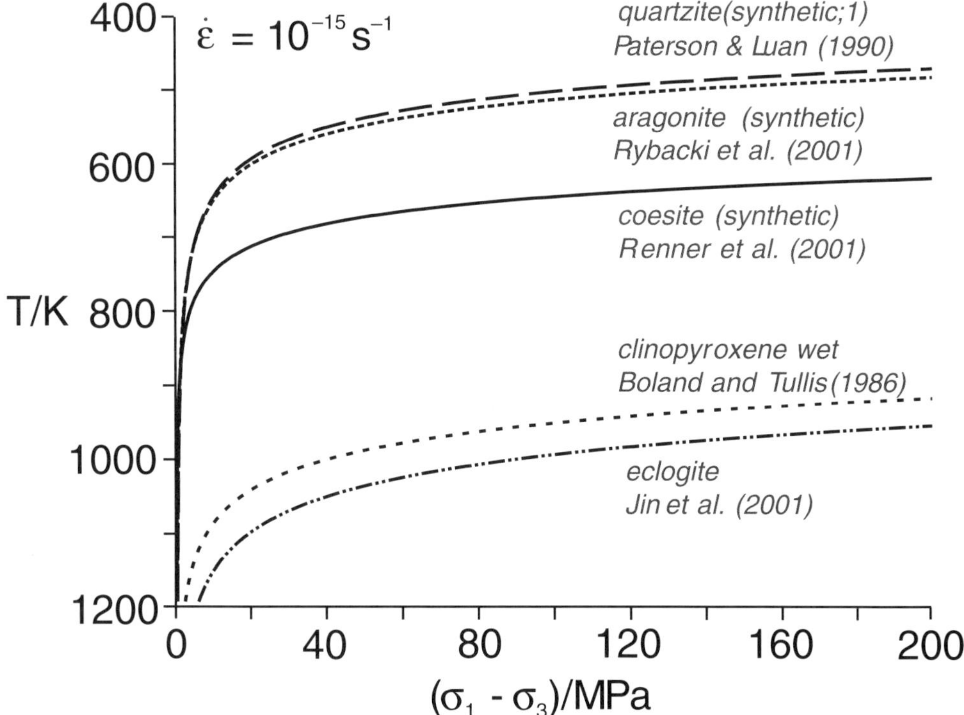

Fig. 5. Strength as a function of temperature for rock-forming minerals relevant at HP and UHP metamorphic conditions, based on experimental flow laws for dislocation creep (power law, $d\varepsilon/dt = A\exp(-Q/RT)\sigma^n$, with $d\varepsilon/dt$ denoting the strain rate in s^{-1}, A is the pre-exponential factor given as $MPa^{-n}\,s^{-1}$, Q is the activation energy in kJ/mole, n is the stress exponent, T is the absolute temperature in K, and R is the gas constant). The parameters for quartz are from Paterson & Luan (1990; $Q = 135$ kJ/mol, $n = 3.1$, $\log A = -7.19$ [$MPa^{-n}n^{-1}$]), for aragonite from Rybacki *et al.* (in prep.; $Q = 246$ kJ/mol, $n = 5.25$, $\log A = 0.52$ [$MPa^{-n}n^{-1}$]), for coesite from Renner *et al.* (2001; $Q = 257$ kJ/mol, $n = 3.1$, $\log A = -3.16$ [$MPa^{-n}n^{-1}$]), for clinopyroxene from Boland & Tullis (1986; $Q = 490$ kJ/mol, $n = 3.3$, $\log A = 5.17$ [$MPa^{-n}n^{-1}$]), and for eclogite from Jin *et al.* (2001; $Q = 480$ kJ/mol, $n = 3.43$, $\log A = 3.3$ [$MPa^{-n}n^{-1}$]. For discussion of the experimental techniques, data reduction, and uncertainties the reader is referred to the original papers. The strain rate of $10^{-15}\,s^{-1}$ is equivalent to no significant deformation on the time scales inherent to burial and exhumation of (U)HP metamorphic crustal slices, i.e. 1 to 10 Ma.

constituent, i.e. omphacite for basaltic composition, quartz or coesite for granitic composition, and aragonite for carbonate rocks.

An experimental flow law for dislocation creep of quartz has been provided by, among others, Paterson & Luan (1990), for coesite by Renner *et al.* (2001), for aragonite by Rybacki *et al.* (in prep.), and for omphacite with additional garnet, i.e. eclogite, by Jin *et al.* (2001). The graphical representation of these flow laws extrapolated to a geological strain rate of $10^{-15}\,s^{-1}$ is depicted in Fig. 5. Unfortunately, the crucial test of the reliability of these flow laws and their extrapolation (i.e. the comparison with natural

Fig. 4. Interfacial free energy controlled quartz microstructure in HP metamorphic rocks. (**a**) Poikiloblastic garnet in micaschist from the Eclogite Zone of the Tauern Window, eastern Alps (see Stöckhert *et al.* 1997 for details). (**b**) Two examples for the absence of a CPO (*c*-axes) of quartz overgrown by poikiloblastic garnet at HP metamorphism, as shown in (**a**) and (**c**), indicating that dislocation creep of quartz was not activated in the foliated schist during burial to about 70 km (Eismann & Stöckhert, unpublished data). (**c**) Closer view showing interfacial free energy controlled quartz microstructure (foam structure) overgrown and preserved by garnet shown in a) during HP metamorphism at *c.* 600 °C and 2.5 GPa. (**d**) Foam structure of quartz that has formed at the expense of coesite in UHP metamorphic pyrope quartzite (Chopin *et al.* 1991, Schertl *et al.* 1991) from the Dora Maira massif, western Alps. There is no indication of crystal plastic deformation of quartz during exhumation.

microstructures in rocks with well-constrained metamorphic conditions, e.g. Stöckhert *et al.* 1999; Hirth *et al.* 2001) is impossible in the case of coesite and aragonite for two reasons. First, these phases are readily transformed to quartz and calcite, respectively, upon exhumation (Mosenfelder & Bohlen 1997; Liu & Yund 1993) and syn-(U)HP microstructures are erased. Second, based on observations on natural (U)HP metamorphic rocks, it remains doubtful whether stress levels sufficient for deformation by dislocation creep are ever typically reached at natural (U)HP conditions, and consequently it is doubtful whether the microstructures required for comparison are even formed in nature. In contrast to coesite and aragonite, omphacite microstructures have been reported from some eclogites, reflecting deformation by dislocation creep, even at temperatures below 500 °C (Piepenbreier & Stöckhert 2001). Deformation by dislocation creep at such low temperatures is in conflict with the extrapolation of experimental data for diopside (e.g. Boland & Tullis 1986), that invariably predict an unrealistically high flow strength for $T < 700\,^{\circ}\mathrm{C}$ at low geological strain rates (Fig. 5). Even the eclogite flow law proposed by Jin *et al.* (2001) does not allow deformation at reasonable stress levels at temperatures below 700 °C (Fig. 5). Based on the natural record it is therefore proposed that sodic pyroxenes may have a significantly lower flow strength compared to diopside. This prediction is supported by new experimental results on synthetic jadeite aggregates (Orzol *et al.* 2001), and awaits further verification and extension to the omphacite solid solution series. As pointed out by Jin *et al.* (2001) and Piepenbreier & Stöckhert (2001), the flow strength of eclogite is of particular interest, as metamorphosed basaltic oceanic crust is expected to form a continuous layer between both lithospheric plates in subduction zones, with the ultimate strength of eclogite thus posing an upper bound to interplate shear stress.

Discussion

Localization of deformation

Rocks that remained weakly deformed or undeformed at (U)HP-conditions indicate a high degree of localization of deformation within the subduction zone. Assuming a simple situation without return flow, a cumulative thickness of interplate shear zone(s) of 2 km, and a convergence rate of only 1 cm/year, strain rates would be on the order of $10^{-13}\,s^{-1}$. Higher convergence rates and smaller cumulative thickness of the shear zones, as suggested by the huge volumes of undeformed or weakly deformed (U)HP metamorphic rocks, require correspondingly higher strain rates, maybe even of the order of $10^{-11}\,s^{-1}$, which are well above the rates usually considered as relevant for long-term tectonic processes. Also, more complicated kinematic patterns, including return flow in a subduction channel as a feasible mechanism for rapid exhumation of (U)HP metamorphic rocks (see below), imply higher strain rates.

In any case, the postulated shear zones are obviously difficult to identify in the field. Indeed, the present author is not aware of any descriptions and detailed microstructural analyses of high-strain shear zones unequivocally developed at UHP conditions in the literature. There may be several reasons for this shortcoming. First, it is possible that the shear zones developed during progressive burial at HP and UHP metamorphic conditions are obliterated by progressive metamorphism. Second, they may be continuously active during burial and exhumation, or repeatedly reactivated, with localization possibly controlled by the bulk composition of a weak rock type. In this case the microstructures developed at higher metamorphic grade would become replaced by those developed later at lower grade during exhumation, with the high grade fabrics being erased. Third, it appears possible that the (micro)structural record of high strain deformation at (U)HP conditions is unspecific, due to specific deformation mechanisms predominating at HP and in particular at UHP metamorphic conditions. In view of the apparent lack of strongly deformed rocks in crustal slices rapidly exhumed from a deep-reaching mega-shearzone, this third possibility is particularly attractive and deserves careful evaluation.

Upper bound on flow stress

Undeformed rocks and/or the unequivocal microstructural record of rocks deformed exclusively by dissolution–precipitation creep prove that stresses (compared to temperatures) were too low to drive dislocation creep along the entire trajectory. As such, these rocks serve as a kind of 'flight recorder', a stress gauge that would have recorded episodes of sufficient stress by corresponding permanent strain and the respective microfabrics indicative of deformation by dislocation creep. Taking a strain rate on the order of $10^{-15}\,s^{-1}$ as equivalent to no notable deformation on the time scales of subduction and exhumation (1 to 10 Ma), an upper bound to stress as a function of temperature is imposed by extrapolation (subject to the uncertainties outlined above) of the flow laws for dislocation creep of quartz/coesite or aragonite to this strain rate (Fig. 5).

As a first approximation, the kinematic framework of a subduction zone suggests a layered structure (e.g. Ji & Zhao 1994; Kirby *et al.* 1996). Also, the isotherms are closely spaced and subparallel to the subduction shear zone (e.g. Peacock 1996). Evidently, interplate deformation should then be localized in the weakest layer of this sandwich structure, the strength of which in turn limits the average shear stress and the degree of interplate coupling. If this is true, the preservation and exhumation of undeformed rocks, combined with the bounds imposed by extrapolation of the experimental flow laws, i.e. the 'flight recorder concept' outlined above, indicates that the interplate shear zones must be very weak, with a very low effective viscosity of the respective materials. The same holds true for rocks deformed exclusively by mechanisms other than dislocation creep. Evidently, this conclusion is rather robust, as the eventually identified microstructural record of deformation by dislocation creep may simply reflect a temporary local stress concentration, related to spatial heterogeneity of the crust (e.g. Schwarz & Stöckhert 1996), or to instantaneous stress redistribution during seismic events (e.g. Taylor *et al.* 1996). In contrast, the failure of the exhumed stress gauge to indicate any deformation by dislocation creep along its entire particle path is clearly more significant, as it precludes attainment of the required stress level (as a function of temperature) throughout the entire burial and exhumation history.

For continental material, with a significant volumetric proportion of quartz respectively coesite, the flow laws depicted in Fig. 5 pose an upper bound to differential stresses of <10 MPa at the typical temperatures of HP-metamorphism (using the quartz flow law of Paterson & Luan, 1990), and barely a few MPa at the typical temperatures of UHP-metamorphism, using the flow law for coesite (Renner *et al.* 2001). Likewise, the flow law for aragonite (Rybacki *et al.*, in prep.) indicates that in the case of the aragonite marbles exposed on Crete, which remained undeformed or exclusively deformed by dissolution–precipitation creep, differential stress never exceeded a few MPa at the peak temperatures of $400 \pm 50\,^{\circ}C$.

Clearly the extrapolation of experimental flow laws of rock-forming minerals to typical geological strain rates is subject to considerable uncertainty (e.g. Paterson 1987). As the quoted uncertainties in the experimental data depend on the mode of correction and data reduction, which in turn depend on the understanding of the processes in the experimental sample assembly, the reader is referred to the details given in the original papers. In the case of quartz, comparison of the predictions of experimental flow laws with both the microstructural and thermometric record of exhumed rocks on one hand (Stöckhert *et al.* 1999; Hirth *et al.* 2001), and with the depth distribution of intracontinental earthquakes (Chen & Molnar 1983; Meissner & Strehlau 1982) on the other hand, supports the validity of the extrapolation. Furthermore, the flow law for coesite displayed in Fig. 5 represents an upper bound to strength (Renner *et al.* 2001). The upper bounds on the magnitude of stress in subduction zones proposed in this paper are based on these flow laws for quartz and coesite, respectively. Notwithstanding the remaining uncertainty, it should be born in mind that unequivocal microstructural evidence of deformation in the dislocation creep regime is common in metamorphic rocks from a broad variety of tectonic settings. The typical absence of such microstructures in HP and UHP metamorphic rocks indicates comparatively low stress levels at the given temperatures – independent of the problems inherent in the extrapolation of experimental flow laws.

Comparison with information on stress in present day subduction zones

The magnitude of shear stress in subduction zones has been inferred from seismological information (e.g. Choy & Boatwright 1995), from heat flow data (e.g. von Herzen *et al.* 2001; Hyndman & Wang 1993), and theoretical models (e.g. Molnar & England 1990; Wang & He 1999). The typical values proposed by these studies span two orders of magnitude, ranging from close to zero (e.g. Wang & He 1999; Wang 2000) to several tens of MPa (e.g. von Herzen *et al.* 2001), or even 100 MPa (e.g. Molnar & England 1990). Clearly, the stress field can be highly heterogeneous on various length scales, and thus a 'typical value' for a subduction zone may be a spurious number. However, the fact that rocks have been buried and exhumed without being subject to sufficient stress to drive deformation of their major constituents by dislocation creep at any time is consistent with those results indicating a very low shear stress, at least at elevated temperatures beyond the seismogenic zone (Tichelaar & Ruff 1993; Ruff & Tichelaar 1996).

A seismological quantity derived from radiation spectra of earthquakes (e.g. Scholz 1990; Stacey 1992) is the stress drop. Although the physical significance of the calculated values is not entirely clear and the quantitative result is model-dependent (Beresnev 2001), stress drop is generally accepted to range between about 1–10 MPa, which is the same order as the upper bound to differential stress in subduction zones posed by the microstructural record of exhumed (U)HP metamorphic rocks combined with experimental flow laws. In this case, with information so far restricted to the depth range $<c.$ 150 km, the stress drop for subduction zone earthquakes should be close to total and the stored energy should be almost quantitatively radiated as seismic waves, with a very high efficiency (e.g. Stacey 1992). Negligible remaining stress would be consistent with lubrication by a fluid and effective decoupling along the fault plane due to zero effective pressure.

Suspected deformation mechanisms

At present, predictions of the potential deformation mechanisms in the low viscosity shear zones are speculative. At any rate, volatiles are carried down in subduction zones in huge amounts and liberated as a fluid phase during dehydration reactions. High pressures strongly enhance the solubility of rock-forming minerals in aqueous solutions (e.g. Manning 1994; Shen & Keppler 1997) resulting in an extraordinarily high concentration of solutes at conditions of HP and UHP metamorphism (e.g. Scambelluri *et al.* 1998; Scambelluri & Philippot 2001). In view of the continuous fluid supply, likely deformation mechanisms (apart from the possibility of brittle failure at very low effective pressure $\sigma_{ii}/3 - P_{fluid}$) could be dissolution–precipitation creep and grain boundary sliding with grain shape accomodation by dissolution and precipitation in the presence of an interconnected aqueous fluid or a hydrous melt (Fig. 6). Constitutive equations to predict the behaviour of materials undergoing deformation by fluid or melt assisted granular flow, with a melt fraction below the critical value (e.g. Rosenberg 2001) for flow to be concentrated in the melt itself, have been proposed by Paterson (1995, 2001). In this model, strain rate is a function of effective grain or aggregate size d (decreasing proportional to $1/d^2$) and linearly dependent on stress (Newtonian behaviour), controlled by either viscosity of the flowing melt, rate of diffusion within the melt, or rate of reaction at the solid–melt interfaces (Paterson 2001). Experimental studies on partially molten mantle rocks in the diffusion creep regime have shown that the strain rate is increased by a factor of 25 in the presence of 7% of melt (Hirth & Kohlstedt 1995).

Microstructures indicative of such a deformation mechanism operative at high temperatures in the presence of a supercritical fluid at (U)HP metamorphic conditions have not been identified in natural rocks so far, and, unfortunately, the respective microstructural record cannot be expected to be unequivocal. Any synkinematic fabrics are likely to be obliterated by grain growth and mineral reactions when deformation is accompanied or followed by annealing.

For the pressure and temperature realm of UHP metamorphism, the recent experimental identification of supercritical fluids (Shen & Keppler 1997; Bureau & Keppler 1999) in aqueous silicate systems has opened new perspectives. At these conditions, there is no coexistence between a low density/low viscosity fluid and a high density/high viscosity silicate melt. Supercritical fluids with intermediate properties are formed. Inclusions of these fluids, now crystallized to yield a complex mineral assemblage including diamond, have been identified in garnet from UHP metamorphic gneisses from the Erzgebirge, Germany (Stöckhert *et al.* 2001). Experimental investigations on the distribution of such fluids at typical UHP metamorphic conditions ($T = 700\,^{\circ}C$ at $P = 3.5\,GPa$, melt fraction about 10%) have shown (Mönicke *et al.* 2001) that the fluid-filled interstices are characterized by a high ratio γ_{ss}/γ_{sl} (with γ_{ss} denoting the interfacial free energy between crystals, and γ_{sl} that between the fluid and a crystal) and by predominantly rational (low-Miller index) interfaces, resulting in a highly irregular and complex shape of the

Fig. 6. Example from Franciscan HP–LT metamorphic siliciclastic series (Pacheco Pass), visualizing the contrast in effective viscosity between single phase material (folded quartz vein) undergoing deformation by dislocation creep and polyphase material (phyllite) undergoing deformation by dissolution–precipitation creep. (**a**) and (**b**) are optical micrographs, (**c**) and (**d**) are SEM images of the phyllite and the quartz vein, respectively. Note that, as a first approximation, the grain size is identical in both materials. The fold shape suggests a viscosity contrast of two to three orders of magnitude (e.g. Ramsay & Huber 1987).

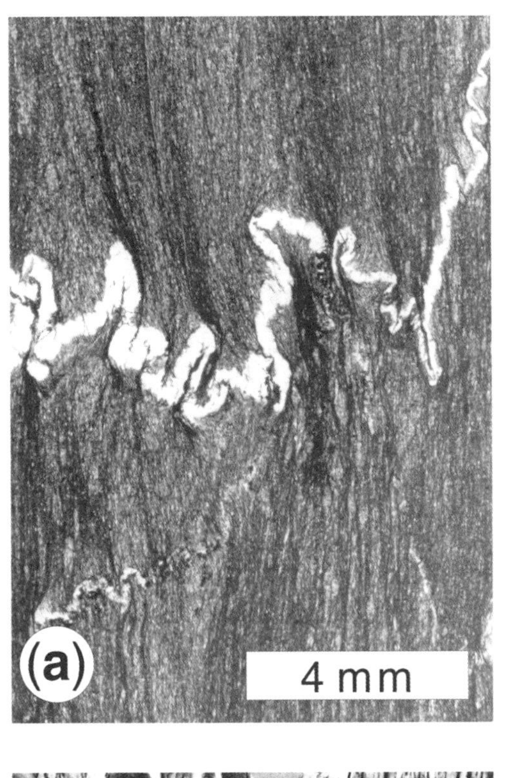
(a)
4 mm

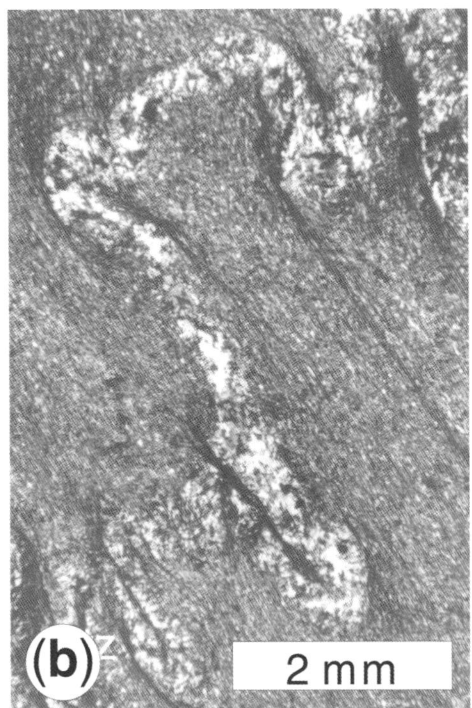
(b)
2 mm

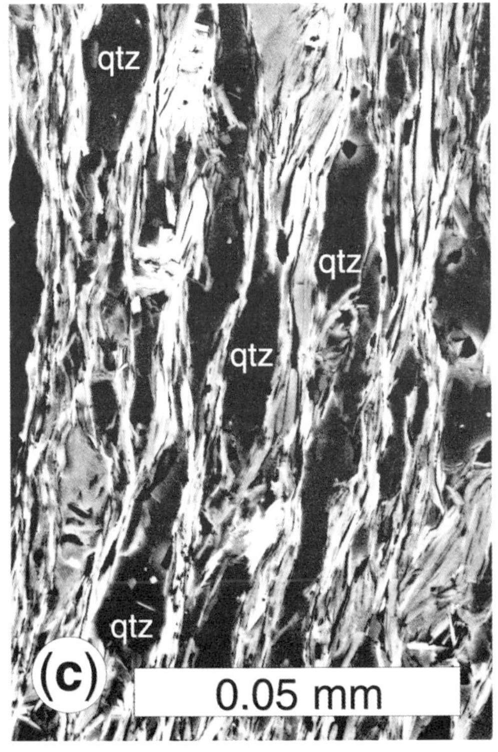
qtz
qtz
qtz
qtz
(c)
0.05 mm

(d)
0.05 mm

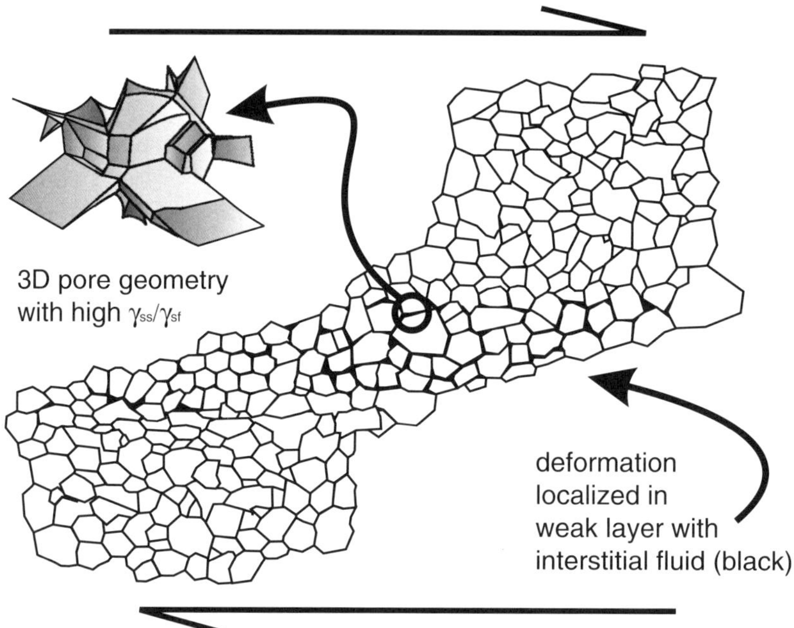

Fig. 7. Sketch visualizing the proposed mechanism to account for localized deformation and low effective viscosity at UHP metamorphic conditions: fluid assisted granular flow, controlled by the quantity and distribution of (supercritical) fluid in different rock types. The characteristic shape of the fluid-filled interstitials is based on experimental studies at UHP metamorphic conditions (Mönicke *et al.* 2001). Stress becomes concentrated at the edges of the wedge-shaped offshoots.

fluid-filled interstices with a low volume/interface area ratio (Laporte & Provost 2000). Owing to their shape, the interstices are interconnected in three dimensions, but the solid portions of the rock are polycrystalline and polyphase aggregates, as considered in the model proposed by Paterson (2001). The further exploration of the composition, reaction kinetics, density, viscosity and pore geometry of these supercritical fluids for a variety of host rocks at (U)HP conditions will be an important task for the near future, as the rheology of deeply subducted crust and in particular the localization of deformation into narrow shear zones possibly related to specific rock compositions (Fig. 7) may crucially depend on the properties and distribution of the interstitial fluids.

For both of the interrelated deformation mechanisms inferred above, i.e. dissolution–precipitation creep and fluid assisted granular flow, grain-size-sensitive behaviour and a linear (Newtonian) dependence of strain rate on stress are predicted (e.g. Rutter 1983). Complications arise because grain-size sensitivity is obviously not sufficient to describe the influence of microstructure on dissolution–precipitation creep. Instead, the portion and nature of the interfaces between unlike minerals controls viscosity (Schwarz & Stöckhert 1996; Stöckhert *et al.* 1999), as also indicated by experimental studies (e.g. Hickman & Evans 1995; Bos *et al.* 1999). This is illustrated here using an example of a very low grade HP-metamorphic phyllite from the Pacheco Pass area in the Franciscan Mélange (e.g. Ernst 1993), where an early formed quartz vein became folded during progressive deformation of the phyllitic matrix (Fig. 6). The microstructure of the monophase quartz vein indicates deformation by dislocation creep, with a recrystallized grain size of *c.* 0.005–0.01 mm. In contrast, the phyllitic matrix indicates deformation exclusively by dissolution–precipitation creep, with clastic quartz grains truncated at their interphase boundary with mica flakes, but not deformed by crystal plastic processes and not recrystallized. The diameter of the flattened clastic quartz grains in the phyllite compares to the recrystallized grain size in the quartz vein. The shape of the folded vein suggests a high contrast in effective viscosity, which may be two to three orders of magnitude (e.g. Ramsay & Huber 1987). This example illustrates that a polyphase rock capable of undergoing deformation by dissolution–precipitation creep has a much lower strength compared to a single-phase quartz vein, notwithstanding the similar grain size. Constitutive equations relevant to natural rock deformation and crustal properties must take this effect into account (e.g. Wheeler 1992), with a purely grain-size sensitive term being insufficient. Furthermore, significant stress concentration can occur in specific layers, the microstructural record of which will therefore not reflect the typical average stress level.

At present, the effective viscosities of polyphase rocks undergoing deformation by dissolution–precipitation creep and fluid-assisted granular flow are poorly constrained by experimental results. From the viscosity contrast with material that was deformed by dislocation creep and for which experimental flow laws are available, effective viscosities on the order of 10^{19} Pa s or below have been suggested for low grade HP-metamorphic phyllites on Crete deformed by dissolution–precipitation creep (Stöckhert *et al.* 1999). This order compares well to the viscosities proposed by Shreve and Cloos (1986) and Cloos and Shreve (1988*a*, *b*) for sedimentary material in subduction channels.

For a given grain size *d*, increasing temperature will decrease the effective viscosity. However, grain size in metamorphic rocks generally increases with increasing temperature, and the increase in viscosity of a material undergoing deformation by dissolution–precipitation creep or fluid-assisted granular flow is proportional to d^2 or d^3. As such, the effect of increasing temperature on viscosity may be counterbalanced to some extent by increasing grain size, and viscosity changes with depth may be moderate.

Implications for the kinematic pattern and tectonic models

In combination with the *P–T* paths, the high exhumation rates pose upper bounds on the size of the exhumed bodies and thus stimulate questions concerning the kinematics and driving forces for exhumation, and in particular the inherent rheological aspects. There are a number of concurrent exhumation models (see Platt 1993 for a review). The shape of the *P–T* paths in combination with the constraints on the relevant time-scale discussed above, i.e. exhumation rates corresponding to plate velocity, and are best reconciled with the corner flow model (e.g. Cloos & Shreve 1988*a*, *b*). This concept is also consistent with the limited size of many (U)HP metamorphic rock bodies and their intimate association with metamorphic rocks revealing markedly different *P–T* conditions or *P–T* paths (e.g. Chopin *et al.* 1991; Henry *et al.* 1993; Eide 1995; Dong *et al.* 1998; Cong *et al.* 1999; Chopin & Schertl 1999; Terry *et al.* 2000*a*). One possible exception is the very extensive UHP metamorphic area in the Dabie/Sulu region in China which may constitute a coherent continental slice (Hacker *et al.* 2000; Ye *et al.* 2000*a*, *b*), contrasting with the small slices identified elsewhere. Chemenda *et al.* (1995) have proposed a model for the exhumation of coherent continental crust driven by buoyancy. However, as the ultimate strength of continental crust at UHP metamorphic conditions should be controlled by coesite, Stöckhert and Renner (1998) and Renner *et al.* (2001) have questioned this concept, because at typical UHP metamorphic conditions the crust should be too weak to withstand the buoyancy forces.

Exhumation of small and internally undeformed (U)HP metamorphic slices by forced flow in a subduction channel narrowing with depth requires a low viscosity matrix, which has not so far been unequivocally identified. Possible reasons for this have been considered above in the discussion of the nature of the shear zones, which in their entity can be considered as equivalent to what is referred to as matrix here.

It is possible that – at least to some extent – the low viscosity matrix is represented by serpentinite derived from the progressively hydrating hanging wall of the subduction channel (Peacock 1993; Peacock & Hyndman 1999; Gerya *et al.* 2001). Serpentinite and peridotite are intimately associated with most occurrences of smaller (U)HP metamorphic slices in mélange type associations (e.g. Little *et al.* 1993; Blake *et al.* 1995; Parkinson 1996; Guillot *et al.* 2000) and serpentine mud diapirism rising from up to 25 km depth with incorporated blueschist facies blocks has been observed in the active Mariana forearc (Fryer *et al.* 1999).

Three scenarios for subduction zones with a low viscosity subduction channel are displayed in Fig. 8. Figure 8a illustrates the simple situation with continuous subduction and no return flow, which is probably irrelevant when considering the record of rapidly exhumed HP and UHP metamorphic rocks. In this case, exhumation of HP and UHP rocks could only be achieved after collision. Figure 8b introduces return flow in a narrowing subduction channel, with exhumed material forming an orogenic wedge, following the principle proposed by Cloos (1982) for the Franciscan complex in California. Figure 8c shows a scenario derived from the model shown in Fig. 8b, but introduces an important modification. Here, forced flow in the low viscosity

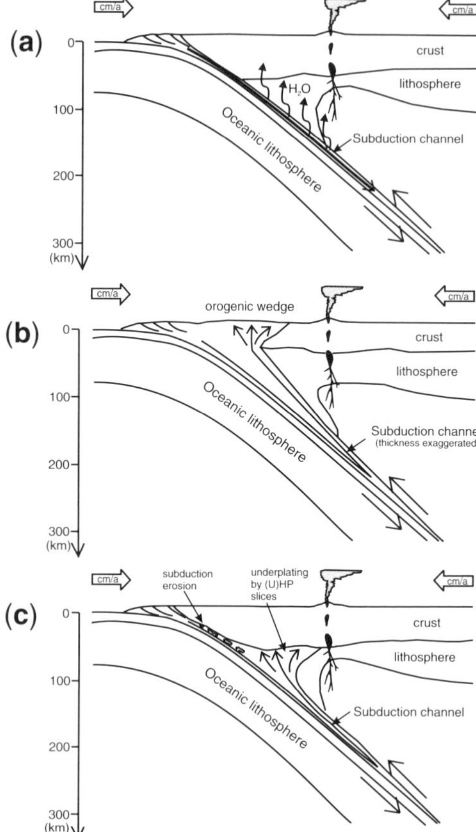

Fig. 8. Three scenarios for subduction zones, based on the subduction channel concept (not to scale). (**a**) Subduction flux greatly exceeds return flux; no exhumation of (U)HP metamorphic rocks driven by forced flow. (**b**) Subduction channel with exhumation by forced flow into an accretionary prism. (c) Subduction channel with forced flow and exhumation to a deep crustal level, from which the rocks may be further exhumed after continental collision at a later stage, and thus after a prolonged history with P, T depending on their successive position in the underplated volume. In contrast to model (**b**), model (**c**) allows subduction erosion to take place concomitantly with exhumation. Also, it allows long-term storage of former (U)HP metamorphic material at a deep crustal level, with the record of (U)HP metamorphism being selectively erased, depending on rock type and presumably availability of fluids.

subduction channel (controlled by rheology and thus by realization of the boundary conditions required to activate the specific deformation mechanisms envisaged above) is restricted to the *P T* realm of HP and UHP metamorphism. Material delivered by return flow is stored at the base of the crust of the upper plate, eventually losing its (U)HP memory with time, but with a fair chance of becoming exhumed after collision. This scenario reconciles with the evidence that UHP metamorphism seems to happen prior to collision (e.g. Eide & Liou 2000), and not as a consequence of collision. Note that the scenario in Fig. 8c takes into account the possibility of subduction erosion (von Huene & Scholl 1991), with material removed from the front of the upper plate carried down into the subduction channel. This implies that UHP metamorphic continental crust is not necessarily part of the downgoing plate, but can be derived from an active continental margin during continuous subduction of oceanic lithosphere. This scenario appears to be consistent with petrological, geochronological and (micro)structural information, and may also reconcile with the depth distribution of subduction thrust earthquakes (e.g. Ruff & Tichelaar 1996; Peacock & Hyndman 1999).

Are these conditions valid for subduction zones in general?

All conclusions drawn here are subject to the general validity of the record of the exhumed (U)HP-metamorphic rocks. In fact, it may be strongly biased, since exhumation may be highly selective and restricted to specific types of subduction zones or episodes of reorganization. It is possible that exhumation during plate convergence is facilitated at retreating margins undergoing subduction roll back (e.g. Royden 1993). Notably, exhumed high-pressure metamorphic rocks are merely absent along the active continental margin of the Andes. There, corresponding to the scenario illustrated in Fig. 8c, HP and UHP metamorphic rocks may be stored at deeper crustal levels of the upper plate (and eventually losing their memory) while subduction still continues. These rocks may reach the surface only after collision and underplating by continental crust, as in most part of the Alps.

Notwithstanding the uncertainties about their general validity, the features discussed in this paper indicate that, at least under certain conditions, extensive volumes of rock can be buried at convergent plate boundaries to depths well exceeding 100 km, and then return to normal crustal levels within a time span of 1 to 10 Ma, without undergoing any significant deformation. This means that they were not subject, at any time during their burial and exhumation history, to the level of differential stress (as a function of temperature) required for deformation by dislocation creep, or in cases even for any deformation by whatsoever mechanism.

Summary and conclusions

Amazingly, we can learn a lot about stress and deformation in subduction zones from undeformed rocks, or from rocks that underwent deformation by dissolution–precipitation creep or fluid-assisted granular flow, a mechanism for which appropriate constitutive equations and experimental calibrations allowing application to natural conditions are not yet available. Following the principle of a 'flight recorder', exhumed rocks are considered as a stress gauge that would provide an appropriate signal, if subject to a sufficient stress (as a function of temperature, and for appropriate time span) to cause a permanent deformation. Although deformation by dislocation creep seems to be only exceptional in subduction zones, the experimental flow laws for this deformation regime can be applied to provide an upper bound to stress.

If the microstructural record of exhumed (U)HP metamorphic rocks is of general significance for subduction zones, and if we accept the validity of experimental flow laws and extrapolation of laboratory data as feasible, a tentative summary and account of implications of these findings looks as follows.

(1) The entire absence of deformation observed in some (U)HP metamorphic rocks, and the absence of deformation during exhumation prior to reaching a normal crustal depth in many, is much more significant than any record of deformation identified so far.

(2) Deformation must be highly localized, although the microstructural record of the shear zones appears to be incomplete, either because it is not specific and hence not recognized, or obliterated by later annealing and progressive reactions.

(3) A well-recorded deformation mechanism in lower grade HP metamorphic rocks is dissolution–precipitation creep. At greater depth, supercritical fluids at partially or completely wetted grain and solid phase boundaries are suspected to facilitate fluid assisted granular flow and thus to control crustal strength.

(4) Both deformation mechanisms imply Newtonian behaviour, with the effect of increasing temperature on viscosity counterbalanced to some extent by increasing grain size.

(5) As a large-scale model, corner flow in a narrowing subduction channel appears feasible for these low viscosities, allowing exhumation of internally undeformed small slices with plate velocity and cooling during decompression.

(6) Interplate shear stress is expected to be very low in subduction zones (a few MPa or less) and thus of the same order as the stress drop inferred for seismic events.

(7) The previous point implies that the stress drop for earthquakes may be total rather than partial, and that the efficiency of seismic radiation is high.

(8) Finally, any significant contribution of shear heating to the thermal budget of subduction zones is unlikely.

In the present context there appear to be two research goals still outstanding. The first is to develop a better understanding of dissolution–precipitation creep and fluid-assisted granular flow in complex polyphase materials. This includes derivation of constitutive equations and experimental calibration, which are needed as input for more realistic numerical simulations, and a more complete and objective assessment of deformation at depth in (U)HP terranes. In the case of evidence for deformation by crystal plastic processes, the possibilities of local stress concentration related to spatial heterogeneity, and of short-term deformation related to rapid synseismic loading should be carefully explored. The second goal is to improve the integration of geophysical field studies with laboratory experiments, numerical simulations and, last but not least, the record of natural rocks, as here (in contrast to Hutton's famous statement) the past may be the key to the present.

This paper summarizes some basic concepts and ideas developed over the last decade, with continuous benefit from discussions with many colleagues, in particular J. Renner, E. Rybacki, R. Wirth, G. Dresen, W. Schreyer, H.-J. Massonne, H.-P. Harjes, to name just a few. The organizers of the DRT 2001 meeting at Nordwijkerhout are thanked for the invitation to present these results and thoughts as a keynote. S. Thomson is thanked for correcting the English. The reviews by Harry Green and by an anonymous referee, who urged me to add an additional note of caution concerning the extrapolation of experimental flow laws, are gratefully acknowledged. Financial support has been provided by the Deutsche Forschungsgemeinschaft within the scope of the Research Group 'High Pressure Metamorphism in Nature and Experiment' and within the Collaborative Research Center SFB 526 'Rheology of the Earth – from the Upper Crust into the Subduction Zone'.

References

AMATO, J. M., JOHNSON C. M., BAUMGARTNER, L. P. & BEARD, B. L. 1999. Rapid exhumation of the Zermatt-Saas ophiolite deduced from high-precision Sm-Nd and Rb-Sr geochronology. *Earth and Planetary Science Letters*, **171**, 425–438.

BEARTH, P. 1959. Über Ekogite, Glaukophanschiefer und metamorphe Pillowlaven. *Schweizer Mineralogisch-Petrologische Mitteilungen*, **39**, 267–286.

BERESNEV, I. A. 2001. What we can and cannot learn about earthquake sources from the spectra of seismic waves. *Bulletin of the Seismological Society of America*, **91**, 397–400.

BIINO, G. & COMPAGNONI, R. 1992. Very-high pressure metamorphism of the Brossasco coronite metagranite, southern Dora-Maira massif, western Alps. *Schweizerische Mineralogische und Petrographische Mitteilungen*, **72**, 347–363.

BLAKE JR., M. C., MOORE, D. E. & JAYKO A. S. 1995. The role of serpentinite melanges in the unroofing of UHPM rocks: an example from the Western Alps of Italy. *In*: COLEMAN, R. G. & WANG, X. (eds) *Ultrahigh Pressure Metamorphism*. Cambridge University Press, Cambridge, 182–205.

BOLAND, J. N. & TULLIS, T. E. 1986. Deformation behavior of wet and dry clinopyroxenite in the brittle to ductile transition region. *Geophysical Monograph*, **36**, 35–49.

BOS, B., PEACH, C. J. & SPIERS, C. J. 1999. Frictional-viscous flow of simulated fault gouge caused by the combined effects of phyllosilicates and pressure solution. *Tectonophysics*, **327**, 173–194.

BRUNO, M., COMPAGNONI, R. & RUBBO, M. 2001. The ultra-high pressure coronitic and pseudomorphous reactions in a metagranodiorite from the Brossasco-Isasca Unit, Dora-Maira Massif, western Italian Alps: a petrographic study and equilibrium thermodynamic modelling. *Journal of Metamorphic Geology*, **19**, 33–43.

BUATIER, M., VAN ROERMUND, H., DRURY, M. R. & LARDEAUX, J. M. 1991. Deformation and recrystallization mechanisms in naturally deformed omphacites from the Sesia-Lanzo zone; geophysical consequences. *Tectonophysics*, **195**, 11–27.

BUREAU, H. & KEPPLER, H. 1999. Complete miscibility between silicate melts and hydrous fluids in the upper mantle: experimental evidence and geochemical implications. *Earth and Planetary Science Letters*, **165**, 187–196.

CARSWELL, D. A. & ZHANG, R. Y. 1999. Petrographic characteristics and metamorphic evolution of ultrahigh-pressure eclogites in plate-collision belts. *International Geology Review*, **41**, 781–798.

CARSWELL, D. A., WILSON, R. N. & ZHAI, M. 2000. Metamorphic evolution, mineral chemistry and thermobarometry of schists and orthogneisses hosting ultra-high pressure eclogites in the Dabieshan of Central China. *Lithos*, **52**, 121–155.

CHEMENDA, A. I., MATTAUER, M., MALAVIEILLE, J. & BOKUN, A. N. 1995. A mechanism for syn-collisional rock exhumation and associated normal faulting: results from physical modelling. *Earth Planetary Science Letters*, **132**, 225–232.

CHEN, W. P. & MOLNAR, P. 1983. Focal depths of intracontinental and intraplate earthquakes and their implications for the thermal and mechanical properties of the lithosphere. *Journal of Geophysical Research*, **88**, 4183–4214.

CHOPIN, C. & SCHERTL, H. P. 1999. The UHP Unit in the Dora-Maira-Massif, Western Alps. *International Geology Review*, **41**, 765–780.

CHOPIN, C., HENRY, C. & MICHARD, A. 1991. Geology and petrology of the coesite-bearing terrain, Dora Maira massif, western Alps. *European Journal of Mineralogy*, **3**, 263–291.

CHOY, G. L. & BOATWRIGHT, J. L. 1995. Global patterns of radiated seismic energy and apparent stress. *Journal of Geophysical Research*, **100**, 18205–18228.

CLOOS, M. 1982. Flow melanges: numerical modeling and geologic constraints on their origin in the Franciscan subduction complex, California. *Geological Society of America Bulletin*, **93**, 330–345.

CLOOS, M. & SHREVE, R. L. 1988*a*. Subduction-channel model of prism accretion, melange formation, sediment subduction, and subduction erosion at convergence plate margins: 1. background and description. *Pure and Apllied Geophysics*, **128**, 455–500.

CLOOS, M. & SHREVE, R. L. 1988*b*. Subduction-channel model of prism accretion, melange formation, sediment subduction, subduction erosion at convergence plate margins: 2. implications and discussion. *Pure and Applied Geophysics*, **128**, 501–545.

COLEMAN, R. G. & WANG, X. 1995. Overview of the geology and tectonics of UHPM. *In*: COLEMAN, R. G. & WANG, X. (eds) *Ultrahigh Pressure Metamorphism*. Cambridge University Press, Cambridge, 1–32.

COMPAGNONI, R., HIRAJIMA, T. & CHOPIN, C. 1995. Ultra-high-pressure metamorphic rocks in the Western Alps. *In*: COLEMAN, R. G. & WANG, X. (eds) *Ultrahigh Pressure Metamorphism*. Cambridge University Press, Cambridge, 206–243.

CONG, B., WANG, Q. & ZHAI, M. 1999. New data regarding hotly debated topics concerning UHP metamorphism of the Dabie-Sulu belt, East-Central China. *International Geology Review*, **41**, 827–835.

DONG, S., CHEN, J. . & HUANG, D. 1998. Differential exhumation of tectonic units and ultahigh-pressure metamorphic rocks in the Dabie Mountains, China. *The Island Arc*, **7**, 174–183.

EIDE, E. A. 1995. A model for the tectonic history of HP and UHPM regions in east central China. *In*: COLEMAN, R. G. & WANG, X. (eds) *Ultrahigh Pressure Metamorphism*. Cambridge University Press, Cambridge, 391–426.

EIDE, E. A. & LIOU, J. G. 2000. High-pressure blueschists and eclogites in Hongan: a framework for addressing the evolution of high- and ultra-high-pressure rocks in central China. *Lithos*, **52**, 1–22.

ERNST, W. G. 1993. Metamorphism of Franciscan tectonostratigraphic assemblage, Pacheco Pass area, east-central Diablo Range, California. Coast Ranges. *Geological Society of America Bulletin*, **105**, 618–636.

ERNST, W. G. 1999. Metamorphism, partial preservation, and exhumation of ultrahigh-pressure belts. *The Island Arc*, **8**, 125–153.

ERNST, W. G. & LIOU, J. G. 1999. Overview of UHP metamorphism and tectonics in well-studied collisional orogens. *International Geology Review*, **41**, 477–493.

ERNST, W. G. & PEACOCK, S. M. 1996. A thermotectonic model for preservation of ultrahigh-pressure phases in metamorphosed continetal crust. *In*: BEBOUT, G. E., SCHOLL, D. W., KIRBY, S. H. & PLATT, J. P. (eds) *Subduction Top to Bottom*. American Geophysical Union, Washington, 171–178.

EVANS, B. & KOHLSTEDT, D. L. 1995. Rheology of rocks. *In*: AHRENS, T. J. (ed.) *Rock Physics & Phase Relations – A Handbook of Physical Constants*. American Geophysical Union, Washington, 148–165.

FAURE, M., LIN, W., SHU, L., SUN, Y. & SCHÄRER, U. 1999. Tectonics of the Dabieshan (eastern China) and possible exhumation mechanism of ultra high-pressure rocks. *Terra Nova*, **11**, 251–258.

FRYER, P., WHEAT, C. G. & MOTTL, M. J. 1999. Mariana blueschist mud volcanism: Implications for conditions within the subduction zone. *Geology*, **27**, 103–106.

GEBAUER, D., SCHERTL, H.-P., BRIX, M. & SCHEYER, W. 1997. 35 Ma old ultrahigh-pressure metamorphism and evidence for very rapid exhumation in the Dora Maira Massif, Western Alps. *Lithos*, **41**, 5–24.

GERYA, T. V., STÖCKHERT, B., MARESCH, W. V. & PERCHUK, A. L. 2001. Exhumation dynamics of high-pressure rocks in a hydrated mantle wedge: constraints from 2D numerical experiments. *Beihefte zum European Journal of Mineralogy*, **13**, 64.

GODARD, G. & VAN ROERMUND, H. 1995. Deformation-induced clinopyroxene fabeics from eclogites. *Journal of Structural Geology*, **17**, 1425–1443.

GUILLOT, S., HATTORI, K. H. & DE SIGOYER, J. 2000. Mantle wedge serpentinization and exhumation of eclogites: insights from eastern Ladakh, northwest Himalaya. *Geology*, **28**, 199–201.

HACKER, B. 1996. Eclogite formation and the rheology, buoyancy, seismicity and H_2O content of oceanic crust. *In*: BEBOUT, G. E., SCHOLL, D. W., KIRBY, S. H. & PLATT, J. P. (eds) *Subduction Top to Bottom*. American Geophysical Union, Washington, 337–346.

HACKER, B. R. & PEACOCK S. M. 1995. Creation, preservation, and exhumation of UHPM. *In*: COLEMAN, R. G. & WANG, X. (eds) *Ultrahigh Pressure Metamorphism*. Cambridge University Press, Cambridge, 159–181.

HACKER, B. R., RATSCHBACHER, L., WEBB, L. & SHUWEN, D. 1997. What brought them up? Exhumation of the Dabie Shan ultrahigh-pressure rocks. *Geology*, **23**, 743–746.

HACKER, B. R., RATSCHBACHER, L. *ET AL*. 2000. Exhumaton of ultrahigh-pressure continental crust in east central China: Late Triassic–Early Jurassic tectonic unroofing. *Journal of Geophysical Research*, **105**, 13339–13364.

HARLEY, S. L. & CARSWELL, D. A. 1995. Ultradeep crustal metamorphism: a prospective view. *Journal of Geophysical Research*, **100**, 8367–8380.

HENRY, C., MICHARD, A. & CHOPIN, C. 1993. Geometry and structural evolution of ultra-high-pressure and high-pressure rocks from the Dora Maira massif, western Alps, Italy. *Journal of Structural Geology*, **15**, 965–981.

HICKMAN, S. & EVANS, B. 1995. Kinetics of pressure solution at halite–silica interfaces and intergranular clay films. *Journal of Geophysical Research*, **100**, 13113–13132.

HIRAJIMA, T., WALLIS, S. R., ZHAI, M. & YE, K. 1993. Eclogitized metagranitoid from the Su-Lu ultra-high pressure province, eastern China. *Proceedings of the Japan Academy*, **69**, Series B, 249–254.

HIRTH, G. & KOHLSTEDT, D. L. 1995. Experimental constraints on the dynamics of the partially molten upper mantle: deformation in the diffusion creep regime. *Journal of Geophysical Research*, **100**, 1981–2001.

HIRTH, G., TEYSSIER, C. & DUNLAP, W. J. 2001. An evaluation of quartzite flow laws based on comparisons between experimentally and naturally deformed rocks. *International Journal of Earth Sciences*, **90**, 77–87.

HYNDMAN, R. D. & WANG, K. 1993. Thermal constraints on the zone of major thrust earthquake failure: the Cascadia subduction zone. *Journal of Geophysical Research*, **98**, 2039–2060.

JI, S. & ZHAO, P. 1994. Layered rheological structure of subducting oceanic lithosphere. *Earth and Planetary Science Letters*, **124**, 75–94.

JIN, Z. M., ZHANG, J., GREEN, H. W. & JIN, S. 2001. Eclogite rheology: implications for subducted lithosphere. *Geology*, **29**, 667–670.

KANEKO, Y., MARUYAMA, S., *ET AL*. 2000. Geology of the Kikchetav UHP-HP metamorphic belt, Northern Kazakhstan. *The Island Arc*, **9**, 264–283.

KIRBY, S., ENGDAHL, E. R. & DENLINGER, R. 1996. Intermediate-depth intraslab earthquakes and arc volcanism as physical expressions of crustal and uppermost mantle metamorphism in subducted slabs. *In*: BEBOUT, G. E., SCHOLL, D. W., KIRBY, S. H. & PLATT, J. P. (eds) *Subduction Top to Bottom*. American Geophysical Union, Washington, 195–214.

KÜSTER, M. & STÖCKHERT, B. 1997. Density changes of fluid inclusions in high-pressure low-temperature metamorphic rocks from Crete: A thermobarometric approach based on the creep strength of the host minerals. *Lithos*, **41**, 151–167.

LAPORTE, D. & PROVOST, A. 2000. Equilibrium geometry of a fluid phase in a polycrystalline aggregate with anisotropic surface energies: dry grain boundaries. *Journal of Geophysical Research*, **105**, 25937–25953.

LIOU, J. G. 1999. Petrotectonic summary of less intensively studied UHP regions. *International Geology Review*, **41**, 571–586.

LITTLE, T. A., HOLCOMBE, R. J. & SLIWA, R. 1993. Structural evidence für extensional exhumation of Blueschist-bearing serpentinite matrix melange, New England Orogen, Southeast Queensland, Australia. *Tectonics*, **12**, 536–549.

LIU, M. & YUND, R. A. 1993. Transformation kinetics of polycrystalline aragonite to calcite – new experimental data, modeling, and implications. *Contributions to Mineralogy and Petrology*, **114**, 465–478.

MANNING, C. E. 1994. The solubility of quartz in H_2O in the lower crust and upper mantle. *Geochimica et Cosmochimica Acta*, **58**, 4831–4839.

MARUYAMA, S. & PARKINSON, C. D. 2000. Overview of the geology, petrology and tectonic framework of the high-pressure–ultrahigh-pressure metamorphic belt of the Kokchetav Massif, Kazakhstan. *The Island Arc*, **9**, 439–455.

MASSONNE, H.-J. 1999. A new occurrence of microdiamonds in quartzofeldspathic rocks of the Saxonian Erzgebirge, Germany, and their metamorphic evolution. *Proceedings of the VIIth International Kimberlite Conference: Capetown*, **2**, 533–539.

MAULER, A., BYSTRICKY, M., KUNZE, K. & MACKWELL, S. 2000. Microstructures and lattice preferred orientations in experimentally deformed clinopyroxene aggregates. *Journal of Structural Geology*, **22**, 1633–1648.

MEISSNER, R. & STREHLAU, J. 1982. Limits of stresses in continental crusts and their relation to the depth-frequency distribution of shallow earthquakes. *Tectonics*, **1**, 73–89.

MICHARD, A., HENRY, C. & CHOPIN, C. 1995. Structures in UHPM rocks: a case study from the Alps. *In*: COLEMAN, R. G. & WANG, X. (eds) *Ultrahigh Pressure Metamorphism*. Cambridge University Press, Cambridge, 132–158.

MICHARD, D., CHOPIN, C. & HENRY, C. 1993. Compression versus extension in the exhumation of the Dora-Maira coesite-bearing unit, Western Alps, Italy. *Tectonophysics*, **221**, 173–193.

MOLNAR, P. & ENGLAND, P. 1990. Temperatures, heat flux, and frictional stress near major thrust faults. *Journal of Geophysical Research*, **95**, 4833–4856.

MÖNICKE, A., BURCHARD, M., DUYSTER, J., MARESCH, W. V., RÖLLER, K. & STÖCKHERT, B. 2001. Experimental studies on fluid distribution and properties at ultra-high pressure metamorphic conditions: Implications for the flow strength of deeply subducted crust. *Beihefte zum European Journal of Mineralogy*, **13**, 126.

MOSENFELDER, J. L. & BOHLEN, S. R. 1997. Kinetics of the coesite to quartz transformation. *Earth and Planetary Science Letters*, **153**, 133–147.

ORZOL, J., STÖCKHERT, B. & RUMMEL, F. 2001. Experimental deformation of synthetic polycrystalline jadeite aggregates. *EOS*, **82**, F1145.

PARKINSON, C. D. 1996. The origin and significance of metamorphosed tectonic blocks in mélanges. *Terra Nova*, **8**, 312–323.

PATERSON, M. S. 1995. A theory for granular flow accomodated by material transfer via an intergranular fluid. *Tectonophysics*, **245**, 135–151.

PATERSON, M. S. 1987. Problems in the extrapolation of laboratory rheological data. *Tectonophysics*, **133**, 33–43.

PATERSON, M. S. 2001. A granular flow theory for the defomation of partially molten rock. *Tectonophysics*, **335**, 51–61.

PATERSON, M. S. & LUAN 1990. Quartzite rheology under geological conditions. *In*: KNIPE, R. J. & RUTTER, E. H. (eds) *Deformation Mechanisms, Rheology and Tectonics*. Geological Society, London, Special Publications, **54**, 299–307.

PEACOCK, S. M. 1993. Large-scale hydration of the lithosphere above subducting slabs. *Chemical Geology*, **108**, 49–59.

PEACOCK, S. M. 1996. Thermal and petrologic structure of subduction zones (overview). *In*: BEBOUT, G. E., SCHOLL, D. W., KIRBY, S. H. & PLATT, J. P. (eds) *Subduction Top to Bottom*. American Geophysical Union, Washington, 119–134.

PEACOCK, S. M. & HYNDMAN, R. D. 1999. Hydrous minerals in the mantle wedge and the maximum depth of subduction thrust earthquakes. *Geophysical Research Letters*, **26**, 2517–2520.

PERCHUK, A. & PHILIPPOT, P. 1997. Rapid cooling and exhumation of eclogitic rocks from the Great Caucasus, Russia. *Journal of Metamorphic Geology*, **15**, 299–310.

PERCHUK, A. L. & PHILIPPOT, P. 2000. Geospeedometry and time scales of high-pressure metamorphism. *International Geology Review*, **42**, 207–223.

PERCHUK, A., PHILIPPOT, P., ERDMER, P. & FIALIN, M. 1999. Rates of thermal equilibration at the onset of subduction deduced from diffusion modeling of eclogitic garnets, Yukon-Tanana terrane, Canada. *Geology*, **27**, 531–534.

PHILIPPOT, P. & VAN ROERMUND, H. L. M. 1992. Deformation processes in eclogitic rocks: evidence for the rheological delamination of the oceanic crust in deeper levels of subduction zones. *Journal of Structural Geology*, **14**, 1059–1077.

PIEPENBREIER, D. & STÖCKHERT, B. 2001. Plastic flow of omphacite in eclogites at temperatures below 500 °C – implications for interplate coupling in subduction zones. *International Journal of Earth Sciences*, **90**, 197–210.

PLATT, J. P. 1993. Exhumation of high-pressure rocks: a review of concepts and processes. *Terra Nova*, **5**, 119–133.

RAMSAY, J. G. & HUBER, M. I. 1987. *The Techniques of Modern Structural Geology Volume 2: Folds and Fractures*. Academic Press, London.

RANALLI, G. 1995. *Rheology of the Earth*. Chapman & Hall, London.

REINECKE, T. 1998. Prograde high- to ultrahigh-pressure metamorphism and exhumation of oceanic sediments at Lago di Cignana, Zermatt-Saas Zone, Western Alps. *Lithos*, **42**, 147–189.

RENNER, J., STÖCKHERT, B., ZERBIAN, A., RÖLLER, K. & RUMMEL, F. 2001. An experimental study into the rheology of synthetic polycrystalline coesite aggregates. *Journal of Geophysical Research*, **106**, 19411–19429.

ROSENBERG, C. L. 2001. Deformation of partially molten granite: a review and comparison of experimental and natural case studies. *International Journal of Earth Sciences*, **90**, 60–76.

ROYDEN L. H. 1993. The tectonic expression slab pull at continental convergent boundaries. *Tectonics*, **12**, 303–325.

RUBATTO, D. & HERMANN, J. 2001. Exhumation as fast as subduction. *Geology*, **29**, 3–6.

RUFF, L. J. & TICHELAAR, B. W. 1996. What controls the seismogenic plate interface in subduction zones? *In*: BEBOUT, G. E., SCHOLL, D. W., KIRBY, S. H. & PLATT, J. P. (eds) *Subduction Top to Bottom*. American Geophysical Union, Washington, 105–112.

RUTTER, E. H. 1983. Pressure solution in nature, theory and experiment. *Journal of the Geological Society London*, **140**, 725–740.

SCAMBELLURI, M. & PHILIPPOT, P. 2001. Deep fluids in subduction zones. *Lithos*, **55**, 213–227.

SCAMBELLURI, M., PENNACCHIONI, G., & PHILIPPOT, P. 1998. Salt-rich aqueous fluids formed during eclogitization of metabasites in the Alpine continental crust (Austroalpine Mt. Emilius unit, Italian western Alps). *Lithos*, **43**, 151–167.

SCHERTL, H.-P., SCHREYER, W. & CHOPIN, C. 1991. The pyrope-coesite rocks and their country rocks at Parigi, Dora Maira massif, western Alps: detailed petrography, mineral chemistry, and P-T path. *Contributions to Mineralogy and Petrology*, **108**, 1–21.

SCHOLZ, C. H. 1990. *The Mechanics of Earthquakes and Faulting*. Cambridge University Press, Cambridge.

SCHREYER, W. 1995. Ultradeep metamorphic rocks: the retrospective viewpoint. *Journal of Geophysical Research*, **100**, 8353–8366.

SCHWARZ, S. & STÖCKHERT, B. 1996. Pressure solution in siliciclastic HP-LT metamorphic rocks – constraints on the state of stress in deep levels of accretionary complexes. *Tectonophysics*, **255**, 203–209.

SHEN, A. H. & KEPPLER, H. 1997. Direct observation of complete miscibility in the albite-H_2O system. *Nature*, **385**, 710–712.

SHREVE, R. L. & CLOOS, M. 1986. Dynamics of sediment subduction, melange formation, and prism accretion. *Journal of Geophysical Research*, **91**, 10229–10245.

STACEY, F. D. 1992. *Physics of the Earth* (third edition). Brookfield Press, Kenmore.

STÖCKHERT, B. & RENNER, J. 1998. Rheology of crustal rocks at ultrahigh pressure – when continents collide. *In*: HACKER, B. R. & LIOU, J. G. (eds) *Geodynamics and Geochemistry of Ultrahigh-Pressure Rocks*. Kluwer, Amsterdam, 57–95.

STÖCKHERT, B., BRIX, M. R., KLEINSCHRODT, R., HURFORD, A. J. & WIRTH, R. 1999. Thermochronometry and microstructures of quartz – a comparison with experimental flow laws and predictions on the temperature of the brittle–plastic transition. *Journal of Structural Geology*, **21**, 351–369.

STÖCKHERT, B., DUYSTER, J., TREPMANN, C. & MASSONNE, H.-J. 2001. Microdiamond daughter crystals precipitated from supercritical COH + silicate fluids included in garnet, Erzgebirge, Germany. *Geology*, **29**, 391–394.

STÖCKHERT, B., MASSONNE, H. & NOWLAN, E. U. 1997. Low differential stress during high-pressure metamorphism: the microstructural record of a metapelite from the Eclogite Zone, Tauern Window, Eastern Alps. *Lithos*, **41**, 103–118.

STÖCKHERT, B., WACHMANN, M., KÜSTER, M. & BIMMERMANN, S. 1999. Low effective viscosity during high pressure metamorphism due to dissolution precipitation creep: the record of HP-LT metamorphic carbonates and siliciclastic rocks from Crete. *Tectonophysics*, **303**, 299–319.

TAYLOR, M. A. J., ZHENG, G., RICE, J. R., STUART, W. D. & DMOWSKA, R. 1996. Cyclic stressing and seismicity at strongly coupled subduction zone. *Journal of Geophysical Research*, **101**, 8363–8381.

TERRY, M. P., ROBINSON, P., HAMILTON, M. A. & JERCINOVIC, M. J. 2000*b*. Monazite geochronology of UHP and HP metamorphism, deformation, and exhumation, Nordoyane, Western Gneiss Region, Norway. *American Mineralogist*, **85**, 1651–1664.

TERRY, M. P., ROBINSON, P. & KROGH RAVNA, E. J. 2000*a*. Kyanite eclogite thermobarometry and evidence for thrusting of UHP over HP metamorphic rocks, Nordoyane, Western Gneiss Region, Norway. *American Mineralogist*, **85**, 1637–1650.

TICHELAAR, B. W. & RUFF, L. J. 1993. Depth of seismic coupling along subduction zone. *Journal of Geophysical Research*, **98**, 2017–2037.

VAN DER KLAUW, S. N, REINECKE, T. & STÖCKHERT, B. 1997. Exhumation of ultrahigh-pressure metamorphic oceanic crust from Lago di Cignana, Piemontese zone, western Alps: the structural record in metabasites. *Lithos*, **41**, 79–102.

VON HERZEN, R., RUPPEL, C., MOLNAR, P., NETTLES, M., NAGIHARA, S. & EKSTRÖM, G. 2001. A constraint on the shear stress at the Pacific-Australian plate boundary from heat flow and seismicity at the Kermadec forearc. *Journal of Geophysical Research*, **106**, 6817–6833.

VON HUENE, R. & SCHOLL, D. W. 1991. Observations at convergent margins concerning sediment sub duction, subduction erosion, and the growth of continental crust. *Reviews of Geophysics*, **29**, 279–316.

WALLIS, S., ENAMI, M. & BANNO, S. 1999. The Sulu UHP Terrane: A Review of the Petrology and Structural Geology. *International Geology Review*, **41**, 906–920.

WALLIS, S. R., ISHIWATARI, A., *ET AL.* 1997. Occurrence and field relationships of ultrahigh-pressure metagranitoid and coesite eclogite in Su-Lu terrane, eastern China. *Journal of the Geological Society*, **154**, 45–54.

WANG, K. 2000. Stress-strain ‘paradox’, plate coupling, and forearc seismicity at the Cascadia and Nankai subduction zones. *Tectonophysics*, **319**, 321–338.

WANG, K. L. & HE, J. 1999. Mechanics of low-stress forearcs: Nankai and Cascadia. *Journal of Geophysical Research*, **104**, 15191–15205.

WHEELER, J. 1991. Structural evolution of a subducted continental sliver: the northern Dora Maira massif, Italian Alps. *Journal of the Geological Society*, **148**, 1101–1113.

WHEELER, J. 1992. Importance of pressure solution and coble creep in the deformation of polymineralic rock. *Journal of Geophysical Research*, **97**, 4579–4586.

YAMAMOTO, H., ISHIKAWA, M., ANMA, R. & KANKEO, Y. 2000. Kinematic analysis of ultrahigh-pressure–high-pressure metamorphic rocks in the Chaglinka-Kulet area of the Kokchetav Massif, Kazakhstan. *The Island Arc*, **9**, 304–316.

YE, K., CONG, B. & YE, D. 2000*a*. The possible subduction of continental material to depths greater than 200 km. *Nature*, **407**, 734–736.

YE, K., YAO, Y., KATYAMA, I., CONG, B., WANG, Q. & MARUYAMA, S. 2000*b*. Large areal extent of ultrahigh-pressure metamorphism in the Sulu ultrahigh-pressure terrane of East China: new implications from coesite and omphacite inclusions in zircon of granitic gneiss. *Lithos*, **52**, 157–164.

ZHANG, R. Y. & LIOU, J. G. 1997. Partial transformation of gabbro to coesite-bearing eclogite from Yangkou, the Sulu terrane, eastern China. *Journal of Metamorphic Geology*, **15**, 183–202.

Non-linear feedback loops in the rheology of cooling-crystallizing felsic magma and heating-melting felsic rock

JEAN-PIERRE BURG[1] & JEAN-LOUIS VIGNERESSE[2]

[1] *Geologisches Institut, University of Zurich and ETH-Zentrum, Sonneggstrasse 5, CH-8006 Zürich, Switzerland (e-mail: jean-pierre.burg@ethz.ch)*
[2] *CREGU, UMR CNRS 7566 G2R, BP 23, F-54501 Vandoeuvre Cedex, France (e-mail: jean-louis.vigneresse@g2r.uhp-nancy.fr)*

Abstract: At least six major parameters control the rheology of partially molten systems: melt content, rate of melt production, reaction to strain of the solid component, reaction to strain of the molten component, temperature and chemical composition of the source rocks. We examine their interactions to understand the rheology of partly molten rocks and partly crystallized magmas. In particular, this paper focuses on the rheology in the transitional domains between two pairs of thresholds that bracket a transitional regime between solid state and fluid behaviour during melting and crystallization, respectively. We review related information and point out non-linear effects that develop during heating of melting rocks and cooling of crystallizing magmas. Owing to the non-linear interactions, positive or negative feedback loops accelerate or damp the system. Melt in migmatite experiences shear-softening which, along with strain partitioning, facilitates melt segregation. Conversely, the increasing number of rigid crystals during cooling increases the suspension viscosity (shear hardening), which soon inhibits magma movement. These effects reinforce the asymmetry between solid-to-melt and melt-to-solid transitions. They severely contradict the concept of one rheological critical melt percentage valid for both melting and crystallization transitions. Fabric acquisition competes with nucleation and crystal growth, thus leading to hysteresis of the stress–strain rate curves. Implications for field observations include horizontal magma segregation, magma extraction and successive magma intrusions.

Introduction

Complex interactions between solid and molten phases along with applied stress and resulting strain in partially molten rocks pose challenging problems for students of rheological processes. For example, compaction models of magma extraction (McKenzie 1984) do not explain discontinuous and irregular melt segregation observed in natural shear zones, and more sophisticated mathematical formulations fail to explain how melt is extracted in a volume large enough to feed a granitic pluton (Brown *et al.* 1995; Petford *et al.* 1993). Experimental studies of partial melting in rocks have chiefly aimed at estimating melt-producing reactions for a given magma type (e.g. Clemens & Vielzeuf 1987; Patiño-Douce & Beard 1995) but the small sample sizes are not conducive to assessing mechanical processes during melting. In contrast, experimental deformation of melting rocks constrains mechanisms that accommodate the bulk sample deformation (Rushmer 1995; Rutter & Neuman 1995; Kohlstedt *et al.* 2000). From all issued studies, the mechanisms and the amount of melt segregation appear to be extremely sensitive to boundary conditions.

Magma segregation in the continental crust has been approached through the study of migmatites. These metamorphic rocks have been partly molten and the crystallized melt forms leucosomes associated with residual components, the melanosome (Mehnert 1968; Ashworth 1985; Brown 1994). Microstructures are clues to deformation mechanisms active during crystallization (Vernon 2000; Rosenberg 2001), but often disappear with further melting. Another approach to understanding the behaviour of partially molten rocks has involved fabric studies of plutonic rocks (Bouchez 1997). However, plutonic fabrics mostly record the waning strain history that obliterated earlier strain patterns. Therefore, they do not give clues to the physical state of the magma during segregation and subsequent ascent from the source.

Partially molten crustal rocks (migmatites and crystallizing felsic magmas, here simply referred as PMR) are two-phase systems relevant to investigations of complex behaviours. One phase is a mobile viscous fluid (migmatitic or residual melt) and the second is the stiff visco-plastic or rigid material (restitic matter or crystals). Taken as solid–liquid systems, attempts to get rheological information from experimental and theoretical studies of concentrated suspensions are frustrating for their application to melting and crystallization (reviews in:

From: DE MEER, S., DRURY, M. R., DE BRESSER, J. H. P. & PENNOCK, G. M. (eds) 2002. *Deformation Mechanisms, Rheology and Tectonics: Current Status and Future Perspectives*. Geological Society, London, Special Publications, **200**, 275–292. 0305-8719/02/$15 © The Geological Society of London.

Adler *et al.* 1990; Liu & Masliyah 1996). In effect, these studies consider successive static situations, each valid for a given particle concentration. However, the continuously changing fraction of the fluid phase with respect to the solid matrix during melting or crystallization of rocks alters the transport property of the whole system. Connection of melt pockets during progressive partial melting builds up channels through which magma can flow (Laporte & Watson 1995). Magma crystallization, on the other hand, builds a crystalline framework through which applied stresses can be transmitted. Therefore, PMRs are transitional stages that cannot be directly compared to suspensions, mushes or slurries in which the solid phase content has no time variation (Fowler 1987).

Besides temperature and chemical composition of the source, four major physical effects interact and cause non-linear behaviour of PMR. These effects concern: (1) material and stress transfer; (2) strain partitioning between the solid and fluid phases; (3) melting and crystallization rates; and (4) non-linear interactions between each system component, each also presenting non-linear behaviours. The non-linear interactions explain the inadequacy of classical models for rocks undergoing state transitions (Barboza & Bergantz 1998; Renner *et al.* 2000).

This paper first reviews key findings pertinent to the rheology of PMR (see also Miller *et al.* 1988; Vigneresse *et al.* 1991; Nicolas 1992; Renner *et al.* 2000; Rosenberg 2001). Aiming to investigate the implications of non-linear effects on the bulk behaviour of PMR, we examine the non-linear responses of each component of the melt–solid system to stress. We explore these responses under increasing and decreasing temperatures, thus simulating evolutions of melting rocks and crystallizing magmas, respectively. During melting, positive feedback loops internally amplify and enhance horizontal melt segregation. Conversely, melt segregates in crystallizing magmas under action of negative (counteractive) loops. The discontinuous and irregular motion of melt is controlled by its rheological contrast with the solid phase and affects the subsequent amount, periodicity and composition of extracted melt. We were unable to derive a unique equation or 'model' that would fit all melt-involving states, but we suggest several ways of tracking the behaviour of PMR. The present paper focusses on felsic rocks because the authors know many examples of migmatites and plutons. The general concepts should also apply to other rock types, with the restriction that thresholds and viscosity have different values, and stress effects have different amplitudes.

Matter and stress transfer thresholds

Structural studies of PMR often refer to Arzi's (1978) rheological critical melt percentage (RCMP). The viscosity of a dilute suspension of rigid particles in a viscous fluid increases with the number of particles up to a threshold (the RCMP) beyond which aggregates of solid particles lock the system. This concept has been applied to melting rocks by inferring process symmetry (Van der Molen & Paterson 1979). However, a single and symmetric RCMP for melting and crystallization does not match geological observation (Vigneresse *et al.* 1996). We suggest using different thresholds for migmatite and crystallizing magma.

Liquid versus rigid percolation thresholds

Melting and crystallization are transitional states whose rheology can be approached with the percolation theory (Stauffer 1985), which addresses the transport properties of a given network. During partial melting, melt connection along grain boundaries corresponds to a bond-percolation. The bond is the linking film or tube permitting melt movement from pocket to pocket. Conversely, grain attachment during crystallization is a site-percolation, a term that refers to the creation of a continuum of touching particles, each one being located at a site. In bond and site-percolations, the volume of material ensuring connection differs. The bond-percolation threshold is significantly smaller than the site-percolation threshold because it is easier to build links along continuous grain boundaries than attach particles of finite size (Stauffer 1985). For melting rocks and crystallizing magmas, these situations have been termed the liquid percolation threshold (LPT) and rigid percolation threshold (RPT), respectively (Vigneresse *et al.* 1996; Fig. 1). In felsic magmas, measured and calculated values of 8 volume% of melt for the LPT and 55 volume% of solid for the RPT vary by a few % according to the actual mineralogy (Vigneresse *et al.* 1996). Both thresholds should show up in the rheological parameters since the shear modulus becomes null when stresses are transmitted through the melt.

Melt escape and particle locking thresholds

The cohesive strength of migmatites disappears at about 20–25 volume% melt, after connection of melt pockets. The melt escape threshold

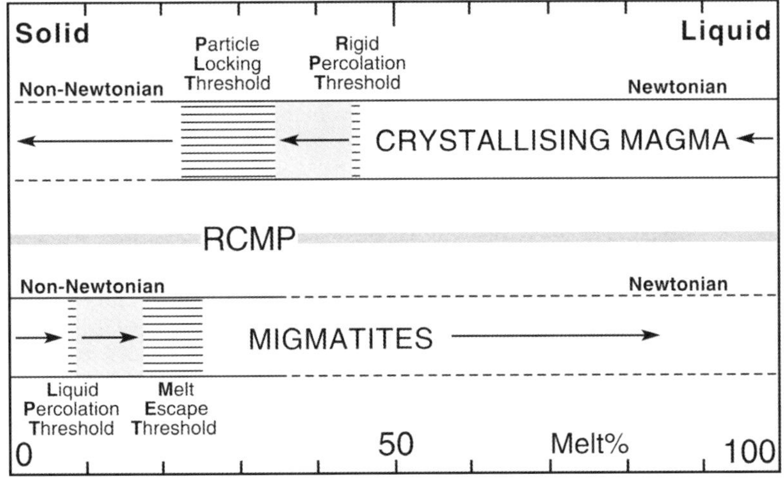

Fig. 1. Thresholds (hatched) and behaviour of PMR during the transition from solid to liquid (bottom) and from liquid to solid (top). The range of RCMP (Arzi 1978) is given for comparison. We are particularly concerned with the shaded intervals between thresholds.

(MET) marks the loss of cohesion of the source rock during melting (Fig. 1). In crystallizing magmas, close packing of phenocrysts locks the solid framework at much higher crystal proportions (70–75 volume%) than the onset of interactions between solid particles, which results in a loose packing (Vigneresse *et al.* 1996). The particle locking threshold (PLT) corresponds to total locking. The exact threshold values depend on external factors such as composition and applied stress.

Non-linear melting and crystallization rates

Melting in source rocks

Melt content and production are non-linear in space and time because melting rocks include several minerals that react differently to temperature variations (Fig. 2).

Incipient melting is governed by the free energy of crystals and thus varies from grain to grain (Laporte & Watson 1995). The lower free energy a mineral pair presents, the easier it is for melting to occur, such as at triple (in 2D) or quadruple (in 3D) junctions. Minerals and fluids may be randomly spread in rocks, but a given assemblage of minerals with similar melting capacities will be irregularly distributed. For instance, a rock with 30% quartz in randomly dispersed small grains does not have the same melt production, at a given temperature, as a rock of same composition with quartz concentrated in a single domain. Accordingly, melting is heterogeneously distributed.

Experimental melting has shown that production of more melt than the MET only occurs when water-saturated pelitic, biotite-muscovite- or biotite-bearing rocks leave the stability field of biotite, above 820 °C (Vielzeuf & Holloway 1988; Patiño-Douce & Johnston 1991; Gardien

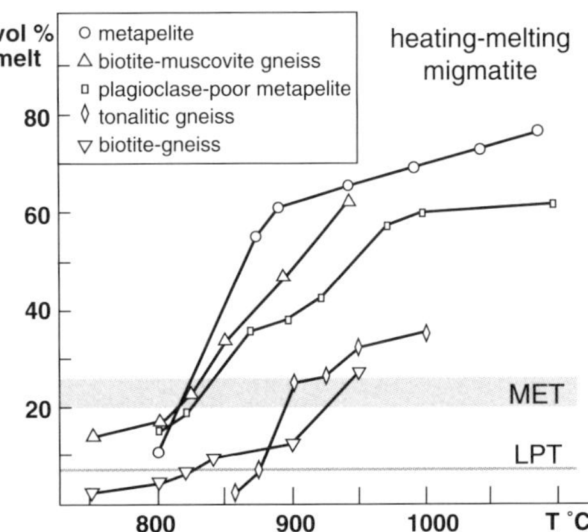

Fig. 2. Melt fraction as a function of temperature for experimental melting of crustal derived sources. Metapelite from Vielzeuf & Holloway (1988), biotite–muscovite gneiss and biotite–gneiss from Gardien *et al.* (1995), plagioclase-poor metapelite from Patiño-Douce & Johnston (1991) and tonalite from Rutter & Wyllie (1988). LPT and MET thresholds as in Fig. 1. Steps in the non-linear melting curves indicate a new mineral participating in melting.

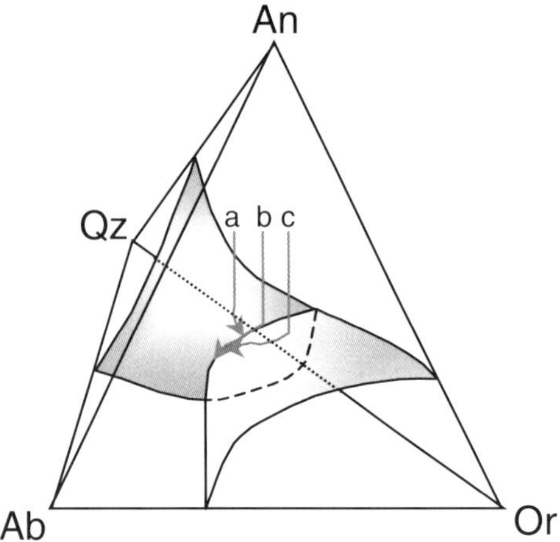

Fig. 3. Anorthite–albite–orthoclase–quartz (An–Ab–Or–Qz) system with its cotectic surfaces (shaded). Three sources with close compositions (**a**, **b** and **c**) follow different cotectic slopes. As a result, their final compositions (arrows) are different.

et al. 1995; Patiño-Douce & Beard 1995). For a given bulk composition, the increase in melt content is not rate constant with temperature but varies as different mineral-out reactions are overstepped (e.g. Vielzeuf & Holloway 1988). Around 800 °C, heating of about 75 °C suffices to metasediments to go from the LPT to the MET, yielding about 25% melt. Conversely, above 850 °C, further heating by about 200 °C is required to increase melt by 25%. These variations are reflected in melt production curves, which are concave downward for metasediments and concave upward for metaigneous rocks (Fig. 2).

Rocks exhibit, on a small (centimetre) scale, compositional variations to which their bulk response to temperature, including melting, is very sensitive (Fig. 3). Thermodynamic data constrain the melting temperature of any given mineral, but eutectic behaviour buffers melting as soon as this mineral is mixed with other phases (Johannes & Holtz 1996). In the system quartz–orthoclase–albite–anorthite–H_2O, the cotectic surfaces of each eutectic sub-system plunge toward a minimum (Johannes & Holtz 1996). The starting composition on the cotectic surfaces and the orientation of the slope on these surfaces control the chemical evolution of the melt (Fig. 3). Consequently, the amount of melt changes considerably for small variations in temperature and chemical composition of the source rock. Introduction of a fluid further lowers the melting temperature of minerals (Mehnert 1968; Johannes & Holtz 1996). Therefore, water generated by the dehydration of hydrous minerals is immediately consumed for melting. Since fluids in rocks are heterogeneously distributed, melt production in migmatites is non-linear in space and time, even if a linear temperature distribution is assumed.

Crystallization of magmas

The bulk crystallization rate in magma chambers varies spatially because it depends on heat exchange between magma and host rocks. In particular, the crystallization front of a pluton advances irregularly towards its core from tortuous margins where a crystal suspension rapidly becomes a mush and ultimately forms a rigid envelope. This envelope insulates the residual magma (Marsh 1996) resulting in an unequally distributed growth rate.

Besides, magma crystallization is essentially non-linear with temperature. Changes in the slope of melt fraction curves (Fig. 4) mark the appearance or disappearance of mineral phases.

Strain partitioning

The contrasting rheology between coexisting phases (new crystals versus residual melt or restitic crystals versus new melt) in PMR induces strain partitioning at the grain scale (solid grain/adjacent viscous medium) and crustal scale (non-molten country rocks/partially molten body). Partitioning is fundamental because it increases the vorticity, hence the non-coaxial component of deformation within the weak melt (Vigneresse *et al.* 1996; Vigneresse & Tikoff 1999). This non-coaxial component addresses two important rheological issues: viscosity and shear softening or hardening (i.e. shear thinning or shear thickening, as they are called in the literature on physics of materials).

Strain partitioning during partial melting

The melt film between mineral grains triggers strain partitioning as soon as stress is applied because the melt viscosity is several orders of magnitude lower than that of the grains. Indeed, a weak phase within a deforming matrix is a major cause of strain localization (Burg & Wilson 1987; Olgaard 1990; Bazant 1988; Herwegh *et al.* 1997). Strain rate in the melt is thus faster than in the solid matrix, hence allowing melt redistribution (Dell'Angelo

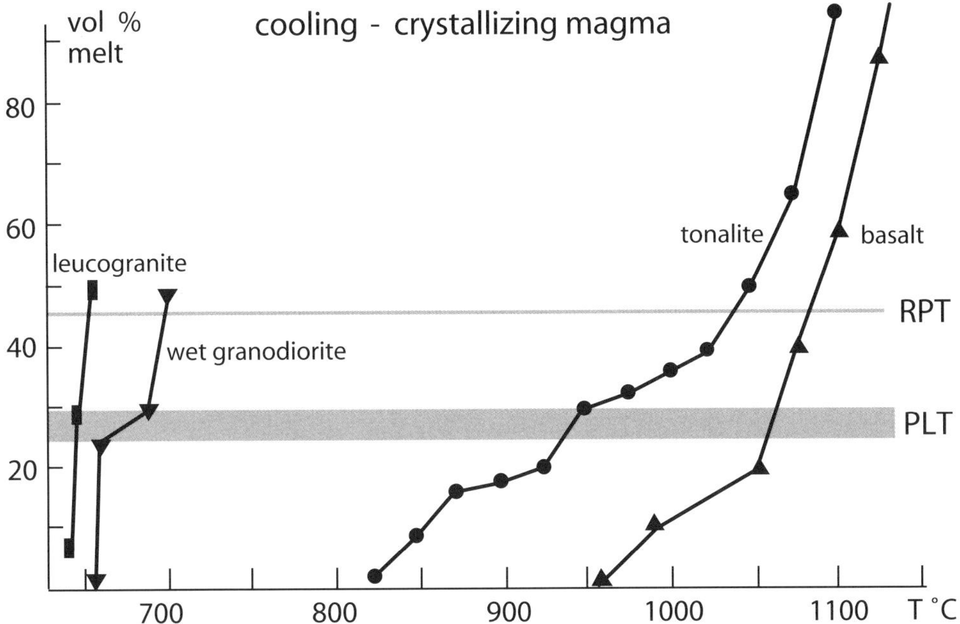

Fig. 4. Melt fraction with respect to temperature for experimental crystallization of different magma compositions. Basalt from Wright & Okamura (1977), leucogranite and wet granodiorite computed from Bouchez *et al.* (1992). RPT and PLT as in Fig. 2. Breaks in the non-linear crystallization curves indicate that a new mineral phase enters or leaves the system.

& Tullis 1988; Bagdassarov *et al.* 1996). Observations on migmatites suggest that melts, represented by leucosomes, tend to be channelled into shear zones (Brown *et al.* 1999). Occurrence of shear zones contributes to bulk strain softening of the rock mass; channelling in these deformation sites involves changes in grain size and/or water content that modify the strain-stress curve. An important consequence is the reduction of the bulk viscosity during deformation to accommodate fast shear strain localized into the melt. A lower viscosity takes up more strain into the melt whose displacement accelerates, which in turn induces further non-coaxial deformation. Shear softening is important at strain rates faster than $10^{-4.5}\,s^{-1}$ for magma compositions that range from nephelinite to rhyolite (Webb & Dingwell 1990). As a corollary, syn-deformation melting has fundamental rheological consequences since it leads to faster extraction of less viscous melt.

Strain partitioning during magma crystallization

Strain partitioning mainly concerns the strong crystals during magma crystallization. Crystal-poor magmas are viscous media in which shear-hardening results from crystals interacting while they rotate (Jeffery 1922; Barnes 1989; Jezek *et al.* 1994; Arbaret *et al.* 1997; Hoffman 1998). At higher crystal contents, crystals form clusters that behave as a single particle with a new aspect ratio, and thus a different rotation rate (Tikoff & Teyssier 1994; Joseph *et al.* 1994). Further crystallization results in a non-cohesive skeleton of solid particles whose rigidity is sufficient to sustain external constrictive but not extensive stress (Guyon *et al.* 1990). In addition, the skeleton cannot transmit shear stress because particles are not tied to each other. While crystals aggregate, dilatancy becomes necessary in crystal-rich magmas because crystals must disconnect to override and pass each other before rearranging in denser packing (Barnes 1989; Bashir & Goddard 1991). The entire system becomes constrained once all crystals touch each other, which marks closed packing. At higher crystal concentration, increased contacts over-constrain the system and the ability of crystal to move is reduced by several orders of magnitude (Moukarzel 1999). In summary, particle rotation, strain partitioning and dilatancy impart to non-cohesive aggregates some strength but a permanently evolving rheology.

Rheology of PMR

Rheology is expressed either in a stress/strain (σ/ε) curve whose instantaneous slope is the shear modulus, or in a stress/strain rate $\sigma/\dot{\varepsilon}$ curve whose instantaneous slope is viscosity:

$$\eta = \sigma/\dot{\varepsilon} \qquad (1)$$

(e.g. Ranalli 1995). In both cases, the instantaneous slope refers to either the linear slope if material is Newtonian or to the local property if the ($\sigma/\dot{\varepsilon}$) curve is non-linear. We first consider solid matrix and melt separately, and then try to understand how they can influence bulk rheology when they coexist in a PMR.

Rheology of solid state rocks in sub-melting conditions

Partial melting of crustal rocks takes place at 700–1000 °C and 500–900 MPa (Mehnert 1968; Ashworth 1985). The solid-state rheology of rock-forming minerals under these pressure-temperature conditions is typically written:

$$\dot{\varepsilon} = A\sigma^{n}\exp(-Q/RT) \qquad (2)$$

where A is a grain-size sensitive constant, n the power-law coefficient, Q the activation energy, R the perfect gas constant and T the absolute temperature (Kirby & Kronenberg 1987; Carter & Tsenn 1987; Wilks & Carter 1990).

We selected experimental calibrations on granite, amphibolite and felsic granulite because they probably represent the bulk upper, intermediate and lower crusts, respectively (Table 1). To infer values of the equivalent viscosity, we first computed the differential stress (σ) with respect to temperature from 700 to 1000 °C and strain rate from 10^{-16} to $10^{-6}\,s^{-1}$, which are parameters relevant to natural melting conditions (Fig. 5). In this window, stress at a constant strain rate decreases with temperature for all rock types. It is nearly identical for amphibolite and granulite, and overlaps the high strain rate

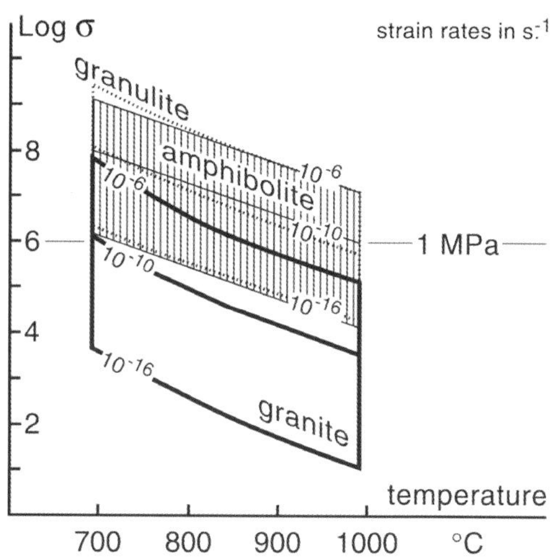

Fig. 5. Differential stress–temperature log diagram computed from the power law rheology for granite, amphibolite and granulite (source values in Table 1). For each rock, differential stress is computed for fast (10^{-6}), moderate (10^{-10}) and slow ($10^{-16}\,s^{-1}$) strain rates. Temperatures between 700 and 1000 °C span the range of melting temperatures of crustal rocks.

stress in granite. We will use the amphibolite parameters to model the behaviour of mid-crustal source rocks.

Melt rheology

The computed viscosity of molten silicate (Bottinga & Weill 1972; McBirney & Murase 1984; Baker 1998; Petford & Clemens 1999) matches experimental measurements (Ryan & Blevins 1987; Ryerson *et al.* 1988; Spera *et al.* 1988), which themselves fit models for granitic melts (Goto *et al.* 1997). Viscosity values of felsic melt, supposed to be Newtonian, range from 10^6 to 10^8 Pa s. We considered an Arrhenian variation with temperature following the law:

$$\eta = A_n\exp(-E_0/RT) \qquad (3)$$

in which A_n is a constant and E_0 the activation energy of the system (Spera *et al.* 1988). Heating from 800 to 900 °C corresponds to one order of magnitude reduction of the viscosity. This is negligible compared to the range of viscosity variation following changes of stress intensity (Fig. 5). Therefore a Newtonian approximation is valid enough between 700 and 1000 °C and at strain rates $<10^{-4.5}\,s^{-1}$. We adopted a stress-independent viscosity ranging from 10^6 to

Table 1. *Parameters used in model of crustal source material for migmatites*

Layer	log(A) $MPa^{-n}\,s^{-1}$	Qc $kJ\,mol^{-1}$	n
Granodiorite	1.5	212	2.4
Amphibolite	−9.11	244	3.7
Felsic granulite	−4.82	243	3.1

(From Kirby & Kronenberg 1987; Wilks & Carter 1990).

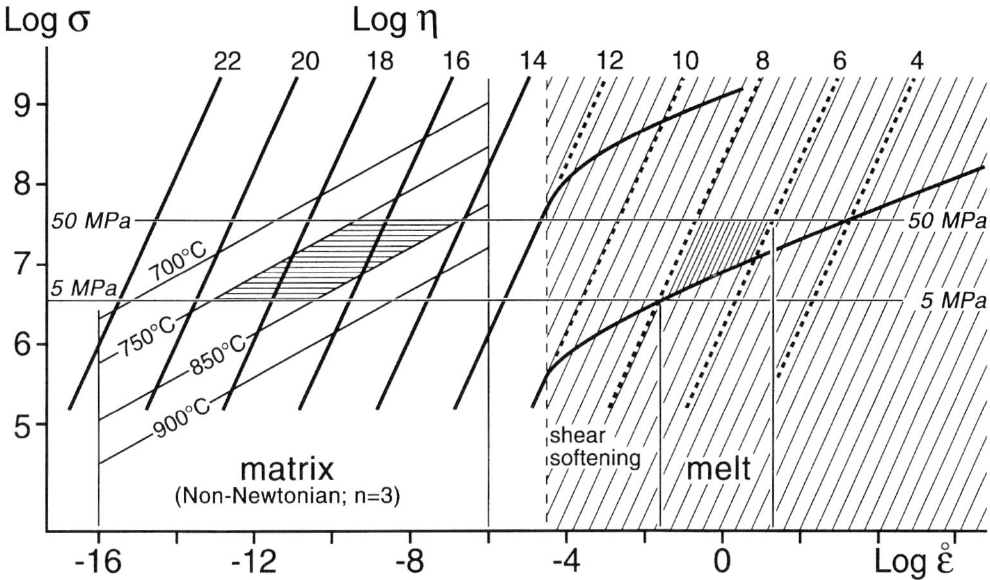

Fig. 6. Stress–strain rate diagram for PMR. Stress values were extracted from Fig. 5 for the 700–900 °C temperature range. Viscosity is computed in a Newtonian mode as the ratio of stress to strain rate. Restricting temperatures to 750–850 °C and differential stress to 5–50 MPa yields the viscosity of the matrix (left hatched box, 10^{14} to 10^{20} Pa s). Melt viscosity (right shaded box) is between 10^{6} to 10^{8} Pa s, corresponding to strain rate faster than 10^{-1} s^{-1} at 5–50 MPa stresses. At strain rate faster than $10^{-4.5}$ s^{-1} (dashed vertical line and hatched domain) shear softening results in a break in the viscosity curves (Webb & Dingwell 1990). The contrasted viscosity between matrix and melt, and their non-linear difference, rules out mere averaging of matrix and melt viscosities.

10^{8} Pa s. In a linear relationship between strain rate and a stress of about 10^{7} Pa (10 MPa), these viscosities correspond to strain rates of 10^{-1} and 10^{1} s^{-1}, respectively.

However, shear softening reduces viscosity by 2 orders of magnitude at strain rates higher than $10^{-4.5}$ s^{-1} (Webb & Dingwell 1990). This effect is responsible for the break in the slope of viscosity isovalues (Fig. 6) and departure from the Newtonian behaviour (Dingwell *et al.* 1996).

Rheology of migmatites during melting

To express the temperature/strain rate relationship in terms of viscosity, we extract from Fig. 5 the stress values obtained between 700 and 900 °C, each value corresponding to one strain rate through viscosity (Eq. 1). We obtain a parallelogram that contains every possible stress versus strain rate point of migmatites in the considered temperature range (Fig. 6). The parallelogram covers the first-order rheology of crustal rocks undergoing melting because it integrates both the rheology of crustal materials (Eq. 2) and the rheology of silicate melts (Webb & Dingwell 1990). The lines of equivalent Newtonian viscosity are deduced from the stress and strain rate coordinates. Those passing through the parallelogram field range from 10^{22} to 10^{13} Pa s.

In a second step, we restrict temperatures to between 750 and 850 °C, the incipient melting temperature of a granitic protolith and the dehydration melting temperature of biotite, respectively (Johannes & Holtz 1996). We also constrain the differential stress between 5 and 50 MPa, considering that higher figures would produce fracturing in the matrix and lower ones would produce extremely slow strain rates ($<10^{-14}$ s^{-1}). The corresponding lines bound a small parallelogram that spans strain rates from 10^{-14} to 10^{-7} s^{-1} (Fig. 6). The equivalent viscosity of melt-containing rocks lies now between 10^{20} and 10^{14} Pa s. By contrast, the viscosity of the small amount of melt remaining in the matrix, below the LPT, is 10^{6} to 10^{8} Pa s (Petford & Clemens 1999).

It is now possible to discuss the consequences on the rheology of progressively melting migmatites. As long as the rock contains little melt, crystal grains of the matrix remain attached to each other. Stress applied to one grain is transmitted to its neighbours, up to some correlation length that marks the range of crystal interaction. In the melt, interaction and chemical exchange are possible only by diffusion over a shorter correlation length than stress in the matrix. Thus, the bulk rheology of low-melt migmatites is nearly that of the matrix. Migmatite bodies showing structural orientations similar to those unmolten country rocks support this proposition (Weber *et al.* 1985; Barbey *et al.* 1990; Sawyer 1991). However, this bulk behaviour does not explain local and independent melt mobility indicated by leucosomes cutting regional structures (e.g. Hopgood 1999), which we will comment on later.

Viscosity and rheology of crystallizing magma

The growing number of crystals and dilatancy increase the viscosity until particle aggregation inhibits movement in the crystallizing magma. Previous studies investigating how these and other processes interact and affect viscosity refer to the Einstein (1906) and Roscoe (1952) expression:

$$\eta = \eta_0(1 - \Phi/\Phi_{max})^{-n}. \quad (4)$$

This power law equation implies that the viscosity (η) of a suspension depends on the volume fraction (Φ) compared to the maximum possible concentration (Φ_{max}) of solid particles in a fluid of viscosity (η_0). Since (η) increases with particle concentration, and hence with cooling, magma in course of crystallization should show shear hardening. In fact, the complex evolution of viscosity must be considered under various situations. (1) At low shear rate, the exponent n is around 2.5, the system behaves as a dispersed suspension and shear-hardening occurs for a given effective volume fraction while particles start clustering (Joseph *et al.* 1994). (2) At faster strain rates, a more complex exponent ($n\Phi_{max}$), equal to about 1.8, replaces n (Krieger & Dougherty 1959). The suspension of particle clusters must have a lower viscosity than at low strain rate and undergoes rheofluidization (the viscosity increases less than it should under a constant strain rate, Quemada 1998). (3) When a significant number of crystals have formed, packing fixes viscosity at a plateau value and the magma behaves as a granular medium with an intrinsic strength.

We first consider the latter case. For a dry or wet granular medium the static friction angle (ϕ) relates tangential (τ) to normal (σ) stress:

$$\tau = \sigma \tan \phi \quad (5)$$

where $\tan \phi$ is a coefficient that depends on friction at grain contacts (Scholz 1990; Marone 1998). It commonly ranges between 0.85 and 0.65, resulting in a friction angle of about 30° (Chester 1995; Blanpied *et al.* 1995). In dynamic conditions, the friction coefficient varies with the displacement velocity, the slip history and the sample porosity (Marone 1998). In addition, dilatancy implies that a new angle φ is defined, which sums the particle/particle friction angle and the parameters that a particle must overcome to slide freely over adjacent particles before close packing.

If magma becomes too sluggish, two mechanisms compete.

(1) Particle aggregation lowers the melt/crystal surface energy, which is the difference between twice the solid/liquid surface energy and the solid/solid surface energy (Laporte & Watson 1995). The aggregation kinetic is limited by the collision frequency between the particles, which directly depends on both the particle concentration (Φ) and the shear rate.

(2) Particle growth tends to minimize the melt/crystal surface energy by narrowing the size distribution of aggregates and smoothing surfaces (Voorhees 1985). It is diffusion limited with a characteristic length of about the individual particle equivalent radius. The characteristic diffusion time is about 1000 s for about 1 mm big crystals and chemical diffusivity of about $10^{-9}\ cm^2\ s^{-1}$ around 900 °C (Johannes & Holtz 1996). Consequently, textural coarsening is likely to take place in magma that cannot flow over long distances owing to lack of space or fast cooling.

Experiments with up to 30–40% crystals (Lejeune & Richet 1995) are consistent with Equation 4. They imitate the large-scale behaviour of a crystallizing magma whose viscosity increases with decreasing temperature. The viscosity of a leucogranitic magma close to its solidus increases by about one order of magnitude for <50 °C cooling. However, this bulk response does not apply to small-scale behaviour and does not take into account dilatancy (Onoda & Liniger 1990).

Strict application of Equation 4 states that viscosity increases with the solid concentration up to the maximum packing (Φ_{max}) (Adler *et al.* 1990; Liu & Masliyah 1996). Close to Φ_{max}, particles are in contact. Clusters then deform only above some critical stress level, but are destroyed if the strain rate is fast and the associated stronger flow rearranges particles in closer packing.

Developing the power function of the Einstein-Roscoe equation in series indicates that crystallizing magma has a yield stress and should be considered under a constant strain rate description:

$$\eta = \eta_0(1 - \Phi/\Phi_{max})^{-n} \approx \eta_0(1 + n\Phi/\Phi_{max}). \quad (6)$$

Since high strain rate leads to a strong readjustment of the packing (Onoda & Liniger 1990; Rogers *et al.* 1994), we assume that Φ_{max} depends on some strain rate $\dot{\varepsilon}_0$, which writes:

$$\Phi_{max} = \Phi_0\dot{\varepsilon}_0 \quad (7)$$

leading to

$$\eta = \eta_0 + \eta_0\Phi/\Phi_0\dot{\varepsilon}_0. \quad (8)$$

Multiplying both sides by strain rate, and putting all second right hand members in a constant $\alpha(\Phi_0, \dot{\varepsilon}_0)$, yields the equivalent equation:

$$\sigma = \sigma_f + \eta_0\Phi\alpha(\Phi_0, \dot{\varepsilon}_0)_{\dot{\varepsilon}}. \quad (9)$$

This equation describes a Bingham flow with a yield stress (σ_f) (Adler *et al.* 1990).

Yield stress is important for low applied stress and hence low strain rate. The few experiments on basaltic or dacitic lavas measured yield strength values of 17–78 Pa (Gauthier 1973; Pinkerton & Norton 1995; Cashman *et al.* 1999). The yield strength of granitic magma is not known, but evidence of Bingham behaviour is recognised from phenocryst fabrics varying from the border to the centre of small veins (Shaw 1965; Komar 1976). Strain is irregularly distributed in space and time. It depends on the transmission of stress from the ambient stress field, and on friction between particles. The yield strength decreases with crystallization, thus explaining crystal accumulation at rims of granitic intrusions (Pons *et al.* 1995; Brun *et al.* 1990).

Behaviour loops

Combined strain partitioning and non-linear rheology produce feedback loops that bolster the non-linearity of PMR. Positive feedback loops amplify processes and are effective during partial melting. Conversely, negative loops over-dampen the rheology of crystallizing magma.

Positive feedback loops in melting

Interactions between melt-production rate, strain partitioning and shear softening act as positive feedback loops enhancing melt segregation (Fig. 7). Heating produces more melt and lessens the melt viscosity; a lowered viscosity partitions more strain into the melt, which increases strain rate; shear softening diminishes

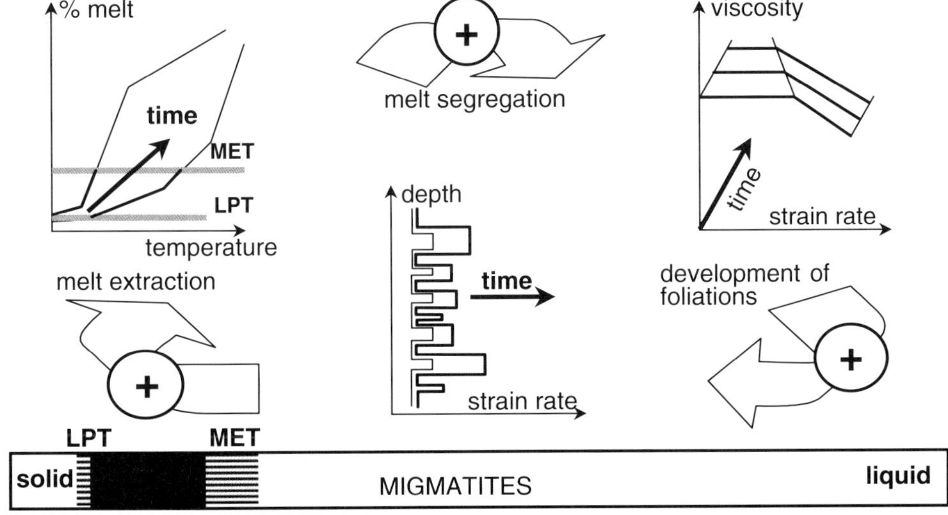

Fig. 7. Interacting non-linear processes in migmatite during partial melting. (Top left) non-linear melting produces more melt with temperature and time during melting. (Top right) viscosity decreases with strain rate and shear-softening takes place in the melt. (Middle) distribution of molten material with depth is highly anisotropic. All processes combine in positive feedback loops (circled + between two arrows) that accelerate melt segregation. The curves are for a granitic magma composition, as shown in Fig. 2.

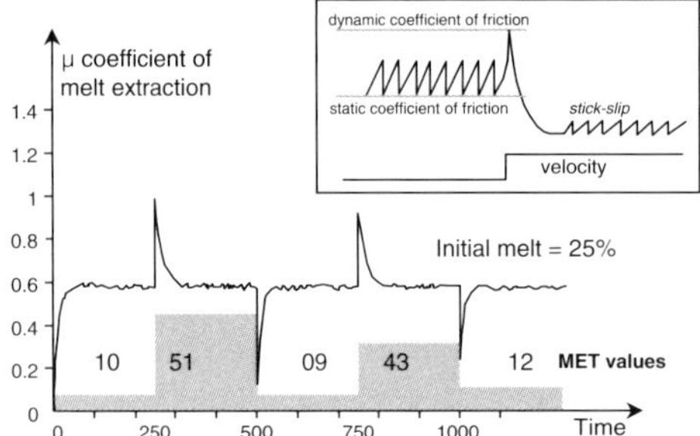

Fig. 8. Erratic, stick-slip-like evolution of a numerical model of melt extraction starting with 25% melt content (Vigneresse & Burg 2000). (μ) is the ratio of extracted melt percentage versus the initial melt percentage available in the matrix. Various theoretical MET values (in grey) induce transient variations of μ followed by a return to a smooth extraction rate.

further the already low melt viscosity. This process combination favours melt extraction.

The bulk viscosity of matrix–melt systems remains controlled by the relaxation of the matrix when melt percolates through it. The bulk viscosity should decrease with increasing amount of melt, hence with temperature, because the melt viscosity is lower than that of the matrix. Once there is enough melt to break matrix cohesion, the bulk viscosity becomes nearly the melt viscosity.

To examine the influence of deformation on the feedback loops, we modelled the quantity of melt extracted under coaxial and non-coaxial deformation (Fig. 8). In the absence of tectonic deformation, heat controls melt production and the only applied force is gravity. The system is static and magma extraction stops rapidly after extraction due to coaxial compaction has removed the overabundant magma. Magma is easily extracted when shear deformation is applied, although the extracted volume is erratic in time (Fig. 8), an extraction behaviour that has been observed in analogue models (Barraud *et al.* 2001).

Negative feedback loops in magma crystallization

In cooling magmas, interactions between shear hardening and strain partitioning generate negative feedback loops that ultimately choke the system (Fig. 9). With cooling, the growing proportion of crystals adds stiffness to the magma while melt viscosity increases. A robust estimate, using Equation 3, indicates that viscosity of a leucogranitic magma increases by two orders of magnitude (only one with Arrhenian viscosity) when the crystal content increases from 50% to 66%. This roughly corresponds to only 20 °C

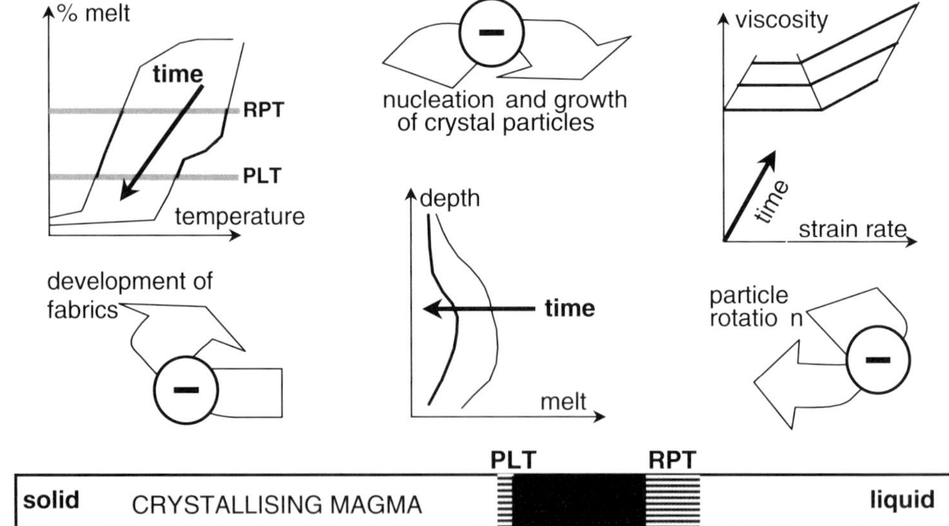

Fig. 9. Interacting non-linear processes during magma crystallization. (Top left) non-linear crystallization results in less residual melt fraction with time. (Top right) viscosity increases because of shear-hardening due to particle concentration. (Middle) melt concentrates towards the core of plutons because of border effects. All processes combine in negative feedback loops that choke and ultimately stop melt circulation. The curves are for a granitic magma composition, as shown in Fig. 4.

cooling (Fig. 4). The combination of cooling and increasing number of crystals raises viscosity by 3 orders of magnitude. Under constant applied stress, the strain rate is correspondingly reduced by 3 orders of magnitude.

This feedback loop counteracts strain partitioning (Fig. 9) and thus reduces magma mobility, slowing down the magma flow. With the onset of crystallization, magma becomes less able to internally reorganize and will cool down with little further movement. Correspondingly, magnetic susceptibility in felsic magmas reflects the last strain increment (Bouchez 1997). Although transposition to strain is not direct, anisotropy values, commonly lower than 3%, do not reflect high strain rates. Rather, they indicate that magma is transported *en masse* with little internal strain after it has upwelled to upper crustal levels.

Discussion

Rheological differences between melting and crystallizing PMR

Formulation of bulk properties of composite systems commonly uses arithmetic or geometric average values of each material property. For instance a system composed of two materials with viscosity η_1 and η_2 has an arithmetic mean viscosity η_a:

$$\eta_a = \Phi\eta_1 + (1 - \Phi)\eta_2 \tag{10}$$

and a geometric mean viscosity η_g,

$$1/\eta_g = \Phi/\eta_1 + (1 - \Phi)/\eta_2 \tag{11}$$

where Φ is the percentage of the phase with subscript 1. These equations are the Voigt and Reuss approximations, respectively (Guéguen & Palciauskas 1990). They correspond to a system under constant strain rate (Eq. 10) or under constant stress (Eq. 11). Focusing on the rheological differences within PMR, we can approach state equations for melting and crystallization transitions with some intrinsic assumptions.

For melting, we start from the matrix rheology and the melt viscosity calculated above. Implicitly, we assume that the two-phase material is treated at constant stress and thus strain partitioning exists between melt and matrix. For crystallization, we take viscosity from Equation 4, thus formulating the system under constant strain rate.

We use tonalite and pelite as extreme rock types for which melt content in function of temperature is known (Vielzeuf & Holloway 1988; Rutter & Wyllie 1988). Computing an average viscosity for both rock types, we obtain different curves for melting and crystallization transitions. These curves do not show the superposition anticipated by the RCMP concept (Fig. 10).

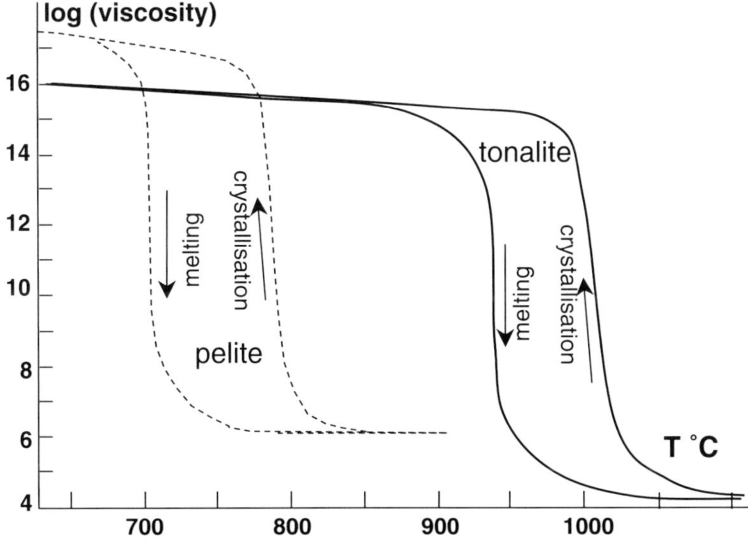

Fig. 10. Viscosity curves computed for a pelite (dashed) and a tonalite (solid line). Melt content computed from Figs 2 and 4. Rapid variations of the viscosity reflect the threshold values that depend on the melt fraction. The rheologically significant thresholds and associated drastic changes in viscosity occur at different temperatures for melting and crystallization. The shift between the melting and crystallization curves is similar to hysteresis in memory dependent materials.

They trace a cycle suggesting hysteresis, and thus a different mode of energy loss during melting and crystallization transitions (Brokate & Sprekels 1996). The large viscosity change, around 700 °C for pelite and at 950 °C for tonalite, reflects the liquid percolation threshold during melting. The viscosity change around 775 °C (pelite) and 1025 °C (tonalite) during crystallization reflects the rigid percolation threshold.

Continuous versus discontinuous motion of one phase in PMR

We have mentioned secant leucosomes indicating expulsion of melt from migmatites. Similarly, magmatic fabrics provide evidence for bulk laminar flow at slow strain rates (Bouchez 1997; Paterson *et al.* 1998). However, metre-scale disturbances of the larger scale fabrics and heterogeneous enclave patterns (Tobisch *et al.* 1997) show that the bulk flow is locally perturbated. Depending on the scale, the description of PMR thus relates either to a liquid, or to a viscous solid, or both in any proportion. Besides scale lengths, the time span for melting or crystallization to take place is very different.

Melt segregation through strain partitioning in migmatites drives the system from a stage of non-connected and heterogeneously distributed melt below the MET to a stage of connected melt films focussing strain. The evolution occurs at slow melting rate and slow strain rate because the system is governed by the matrix rheology. The discontinuous and repetitive melt motion obtained using Lagrangian modelling is similar to stick-slip motion (Fig. 8). It represents an alternating switch from matrix-dominated to melt-dominated rheology. In crystallizing magma, melt escape obeys the discontinuous motion of the solid crystalline framework. Regions of smaller friction coefficient concentrate deformation into a smaller volume, inducing a feedback loop that localizes strain (Barnes 1989; Marone 1998) and local extensional zones act as sinks for the residual melt.

Geological relevance

Magma extraction

We argue that magma extraction results from both melting (temperature effect) and rheology (tectonic stress and intrinsic property of the melt).

In crustal rocks, melt is essentially generated from melting of muscovite, quartz and plagioclase (Thompson 1982). The resulting leucosomes lack ferromagnesian minerals (e.g. amphibole, biotite). Therefore, melting quickly followed by melt ascent should not provide magma of granitic composition. En-route assimilation of ferromagnesian minerals from intruded country rocks is unrealistic in terms of volume. Consequently, the migmatite leucosome–melanosome system must undergo further melting that involves ferromagnesian minerals to produce a granitic melt (Sawyer 1998). For example, breakdown of biotite liberates water and increases the melt volume, which allows the MET, above which melt may escape (Petford *et al.* 2000) to be overcome.

Competition between horizontal and vertical segregation has an influence on melt productivity. According to Le Chatelier's principle, fast melt production slows down the melting rate. Conversely, fast melt extraction increases melt productivity. However, if the melt is extracted upwards, it rapidly becomes disconnected from its source, which becomes depleted in material adequate for melting and cannot produce more melt. If melt segregates horizontally and remains close to its source, a low melting rate may produce a large amount of melt with a homogeneous composition. We conclude with Sawyer (1994) that migmatites crystallized with a high melt fraction denote low ambient deformation, unable to quickly extract significant melt volumes. On the contrary, source regions with melt productivity high enough to produce magma and feed granitic plutons should have been submitted to intense deformation. The remaining source region would comprise restitic rocks, devoid of melt (i.e. granulites).

Equilibrium versus disequilibrium compositions

Strain partitioning induced by horizontal tectonic stresses enhances the horizontal separation between melt and residual phases which is further reinforced by the melt flow channelled between unmolten layers (Vigneresse & Burg 2000). Horizontal segregation has two major consequences. First, melt may chemically re-equilibrate because it remains for some time at constant pressure and temperature conditions while concentrating at some level. Second, upward flow of felsic magma cannot be reduced to a pervasive percolating fluid through a more or less porous crust because melt is collected before extraction in dykes (Petford *et al.* 2000).

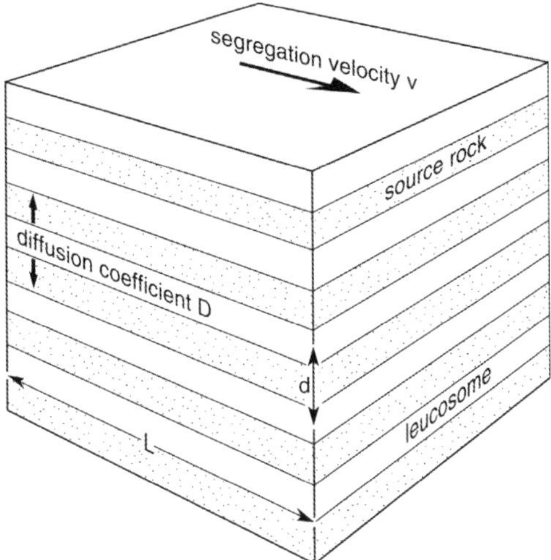

Fig. 11. Schematic model of PMR of length L in which melt circulates with a velocity v along small conduits separated by a distance d. To be in chemical equilibrium with its matrix, diffusion Ds of an element s of the melt must be faster than the time for melt to segregate.

Melt production and segregation influences the equilibrium textures. For instance, if both melting and melt extraction rates are fast, melt can hardly re-equilibrate with its matrix (Sawyer 1994). A robust estimate of the influence of melt circulation in a migmatite can be obtained using the Péclet number, which defines the ratio of diffusion versus advection effects (Crank 1975). For this purpose, we consider chemical diffusion of elements that are present within the leucosome.

A migmatitic section of length L is idealized by a series of leucosome veins separated by a distance d and in which melt segregates with a velocity v (Fig. 11). The time (t_{adv}) for melt to segregate is given by:

$$t_{adv} = L/v. \quad (12)$$

By comparison, a rough estimate of the time (t_{diff}) for an element (s) to diffuse is given by:

$$t_{diff} = d^2/D_s \quad (13)$$

where D_s is the diffusion coefficient of the considered element.

The Péclet number (Pe) is the inverted ratio between the two time quantities:

$$Pe = vd^2/D_sL. \quad (14)$$

Disequilibrium occurs when advection (mass transport) is faster than diffusion, or if Pe is larger than 10 (Ottino 1989). In migmatites, we adopt a length L of about 1 m and vein spacing about 0.1 m, which results in a $d^2/L \approx 0.01$. In granitic magmas, tracer diffusivities range from 10^{-16} to $10^{-7}\,cm^2\,s^{-1}$ (Jambon 1982). Slow diffusion of Si and Al controls diffusion of non-alkaline elements that have chemical diffusivity values of about $10^{-9}\,cm^2\,s^{-1}$ around 900 °C (Carmichael *et al.* 1974). Excepting Li and Na, all other cations have diffusion coefficients smaller than $10^{-7}\,cm^2\,s^{-1}$ at about 900 °C (Johannes & Holtz 1996). Whereas diffusion time can easily be computed, advection time is not directly measurable and must be estimated. Calculated viscosities yield strain rates faster than $10^{-4}\,s^{-1}$ in melt whilst strain rates range from 10^{-14} to $10^{-8}\,s^{-1}$ in the solid matrix (Fig. 6). If one assumes that strain rates for material to move by about 1 m (the section length L) convert to velocity, Péclet numbers of *c.* 10^7 for melt and below 10^3 for the matrix are obtained. Those Péclet numbers indicate that advection largely dominates over diffusion. Therefore, melt segregation by melt movement within the matrix leads to melt disequilibrium. It can be argued that segregated melt remaining in the matrix can re-equilibrate if the time interval between deformation pulses is long (about 1 Ma) as under a bulk strain rate of $10^{-10}\,s^{-1}$. This time interval is very long compared to that between migmatite melting and upwelling (about 1 Ma in total, Vanderhaeghe & Teyssier 1997). Accordingly, we expect that most migmatite melts will show disequilibrium composition (see also Sawyer 1991).

Fabrics in imbricated magma intrusions

Interactions between PMR include imbricated magma intrusions from the injection of melt into another melt to the intrusion of a melt into a solid rock (Bédard 1993; Fernandez & Gasquet 1994; Hallot *et al.* 1996; Fernandez *et al.* 1997). Clearly, different structures should document different crystallization states between the RPT and the PLT of the older, intruded magma.

When two melt batches interact, mixing leads to chemical homogenization. New magma dismembers the fabric of the intruded, loosely connected framework of particles. Similar new-fabric patterns in both magmas integrate concordant contacts (Fig. 12). The two magmas may partly exchange mobile elements and a profile across a facies (paragenesis) change displays a smooth variation of those elements. A more crystallized, older, magma batch begins to sustain stress and resist to new intrusion, maintaining its bulk fabric. However, fabric variations close to contacts reflect mechanical interaction

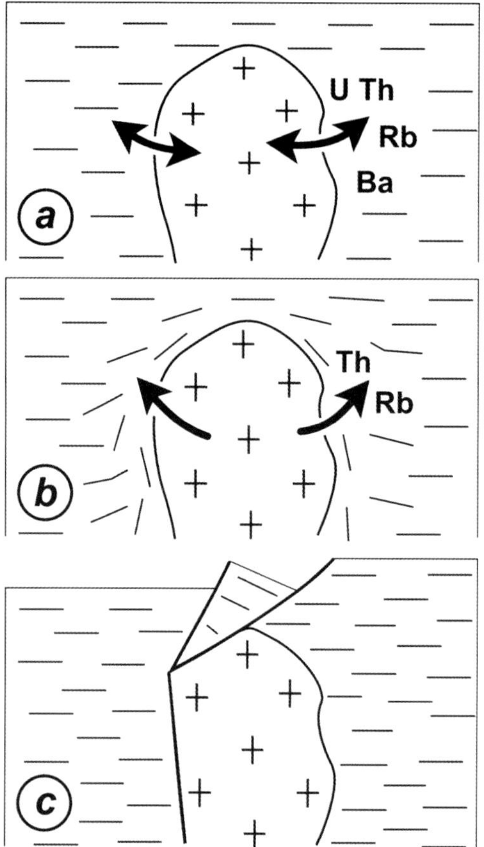

Fig. 12. New intrusion in an already settled magma. (**a**) The new intrusion takes place when the former magma is not yet totally crystallized. Chemical rehomogenisation of the residual liquid occurs. (**b**) The new magma intrudes a yet partly crystallized magma. Flattening occurs at the contacts, segregating incompatible elements from the residual liquid as reported in the granitic massif of Cabeza de Araya (Vigneresse & Bouchez 1997). (**c**) In the case of a fast rate intrusion, brittle fracturing may take place within the not yet crystallized magma. No chemical exchange occurs.

(Vigneresse & Bouchez 1997). Exchange of incompatible elements is limited, but compaction of the earlier magma expels residual fluids so that near-contact enrichment in Th, Rb, Ba may be expected (Fig. 12). Sharply discordant fabrics indicate that new magma breaks and disrupts a former, consolidated magmatic body. No large-scale chemical exchange should then take place.

Shear zones in a partly crystallized magma compare to proto-faults that disrupt the fabric of a non-consolodated framework of crystals (Brun *et al.* 1990; Pons *et al.* 1995; Smith 1996; John & Stunitz 1997). They indicate that fracturing is the response of PMR to a new, high strain rate intrusion (Dingwell 1997).

Conclusion

We have examined the relationships that occur in migmatites between melting, melt segregation and deformation in one hand, and in felsic magmas between crystallization, deformation and crystal interactions on the other hand.

Solid-to-suspension transitions during melting and crystallization are not symmetrical. Two thresholds that depend on the respective percentage of melt and solid phase bound three rheological stages. For melting, a transitional rheology takes place between 8 and 20–25% melt. During crystallization, the transitional rheology takes place between 50 to 25% melt. In the transitional domain, the rheology of the two-phase material presents bulk shear softening (reduction of viscosity with shear rate) during melting and shear hardening (increase of viscosity with shear rate) during crystallization.

Similarities of deformation during meting and crystallization include strain partitioning and non-linearity in time (mostly due to temperature variation) and space (due to variability of chemical composition, including fluids). Non-linearity is responsible for feedback loops. They are positive during melting and negative during crystallization. In both cases, the loops modify melt extraction. As a consequence, strain-partitioning-driven, horizontal melt segregation due to non-coaxial deformation overcomes buoyancy and matrix-compaction-driven melt segregation. Conversely, during crystallization, melt segregates only into local dilatant sinks and the negative feedback loops damp large-scale melt segregation.

Stick-slip-like melt motion in PMR results from the coexistence of melt and solid phase components with contrasted rheologies. The motion is discontinuous in space and time, leading to numerous leucosome generations in migmatites. In a crystallizing magma, the negative feedback loops counteract this effect and restrict the appearance of proto-faults to the early-crystallized rims of plutons. The cycle of melting/crystallization shows hysteresis in transport properties, amongst which viscosity is a specific example.

This paper results from a long maturation starting from a presentation at the 1991 GAC/MAC Meeting in Toronto with P. Barbey (U. Nancy) and M. Cuney (Cregu, Nancy). Later on, B. Tikoff (U. Madison), while staying in France, introduced one author (J.-L. Vigneresse) to strain partitioning. A workshop at

ETH Zurich later revealed the importance of non-linear rheology and feedback loops. Numerous fruitful discussions came out from poster or informal presentations at meetings and conferences, attendance at which was funded by Cregu, Nancy. We acknowledge E. Sawyer and L. Kriegsman, and later T. Blenkinsop and Y. Park for their constructive reviews and encouragement. Finally, P. Lagrange (Cregu) provided his skill for most figures and slides.

References

ADLER, P., NADIM, A. & BRENNER, H. 1990. Rheological models of suspensions. *Advances in Chemical Engineering*, **15**, 1–67.

ARBARET, L., DIOT, H., BOUCHEZ, J. L., LESPINASSE, P. & SAINT BLANQUAT, M. DE 1997. Analogue 3D simple shear experiments of magmatic biotite subfabrics. *In*: BOUCHEZ, J.-L., HUTTON, D. H. W. & STEPHENS, W. E. (eds) *Granite: from segregation of melt to emplacement fabrics*. Kluwer, Dordrecht, 129–143.

ARZI, A. A. 1978. Critical phenomena in the rheology of partially melted rocks. *Tectonophysics*, **44**, 173–184.

ASHWORTH, J. R. 1985. *Migmatites*. Blackie & Sons, Glasgow.

BAGDASSAROV, N. S., DORFMAN, A. M. & DINGWELL, D. B. 1996. Modelling of melt segregation processes by high-temperature centrifuging of partially molten granites – I. Melt extraction by compaction and deformation. *Geophysical Journal International*, **127**, 616–626.

BAKER, D. R. 1998. Granitic melt viscosity and dike formation. *Journal of Structural Geology*, **20**, 1395–1404.

BARBEY, P., MACAUDIÈRE, J. & NZENTI, J. P. 1990. High-pressure dehydration melting of metapelites. Evidence from the migmatites of Yaoundé (Cameroon). *Journal of Petrology*, **31**, 401–427.

BARBOZA, S. A. & BERGANTZ, G. W. 1998. Rheological transitions and the progress of melting of crustal rocks. *Earth and Planetary Science Letters*, **158**, 19–29.

BARNES, H. A. 1989. Shear thickening ('Dilatancy') in suspensions of nonaggregating solid particles dispersed in Newtonian liquids. *Journal of Rheology*, **33**, 329–366.

BARRAUD, J., GARDIEN, V., ALLEMAND, P. & GRANDJEAN, P. 2001. Analog modelling of melt segregation and migration during deformation. *Physics and Chemistry of the Earth*, **A26**, 317–323.

BASHIR, Y. M. & GODDARD, J. D. 1991. A novel simulation method for the quasi-static mechanics of granular assemblages. *Journal of Rheology*, **35**, 849–885.

BAZANT, Z. P. 1988. Softening instability: Part I – Localization into a planar band. *Journal of Applied Mechanics*, **55**, 517–522.

BÉDARD, L. P. 1993. Injections multiples de magmas dans un conduit nourricier: implications sur le remplissage des plutons et l'extraction des magmas. *Canadian Journal of Earth Science*, **30**, 124–131.

BLANPIED, M. L., LOCKNER, D. A. & BYERLEE, J. D. 1995. Frictional slip of granite at hydrothermal conditions. *Journal of Geophysical Research*, **B100**, 13045–13064.

BOTTINGA, Y. & WEILL, D. F. 1972. The viscosity of magmatic silicate liquids: a model for calculation. *American Journal of Science*, **272**, 438–475.

BOUCHEZ, J. L. 1997. Granite is never isotropic: an introduction to AMS studies of granites. *In*: BOUCHEZ, J. L., HUTTON, D. H. W. & STEPHENS, W. E. (eds) *Granite: from segregation of melt to emplacement fabrics*. Kluwer, Dordrecht, 95–112.

BOUCHEZ, J. L., DELAS, C., GLEIZES, G., NÉDÉLEC, A. & CUNEY, M. 1992. Submagmatic microfractures in granites. *Geology*, **20**, 35–38.

BROKATE, M. & SPREKELS, J. 1996. *Hysteresis and phase transitions*. Applied Mathematical Sciences, **121**, Springer Verlag, Heidelberg.

BROWN, M. 1994. The generation, segregation, ascent and emplacement of granite magma: the migmatite-to-crustally-derived granite connection in thickened orogens. *Earth Science Reviews*, **36**, 83–130.

BROWN, M., AVERKIN, Y. A., MCLELLAN, E. L. & SAWYER, E. 1995. Melt segregation in migmatites. *Journal of Geophysical Research*, **B100**, 15655–15679.

BROWN, M. A., BROWN, M., CARLSON, W. D. & DENISON, C. 1999. Topology of syntectonic melt-flow networks in the deep crust: Inferences from three-dimensional images of leucosome geometry in migmatites. *American Mineralogist*, **84**, 1793–1818.

BRUN, J. P., GAPAIS, D., COGNÉ, J. P., LEDRU, P. & VIGNERESSE, J. L. 1990. The Flamanville granite (NW France): an unequivocal example of an expanding pluton. *Geological Journal*, **25**, 271–286.

BURG, J. P. & WILSON, C. J. L. 1987. Deformation of two-phase systems with contrasting rheology. *Tectonophysics*, **135**, 199–205.

CARMICHAEL, I. S. E., TURNER, F. J. & VERHOOGEN, J. 1974. *Igneous Petrology*. McGraw-Hill, New York.

CARTER, N. L. & TSENN, M. T. 1987. Flow properties of continental lithosphere. *Tectonophysics*, **136**, 27–63.

CASHMAN, K. V., THORNBER, C. & KAUAHIKKAUA, J. P. 1999. Cooling and crystallization of lava in open channels, and the transition of Pahoehoe lava to Aa. *Bulletin Volcanologique*, **61**, 306–323.

CHESTER, F. M. 1995. A rheologic model for wet crust applied to strike-slip faults. *Journal of Geophysical Research*, **B100**, 13033–13044.

CLEMENS, J. D. & VIELZEUF, D. 1987. Constraints on melting and magma production in the crust. *Earth and Planetary Science Letters*, **86**, 287–306.

CRANK, J. 1975. *The mathematics of diffusion*. Oxford University Press, Oxford.

DELL'ANGELO, L. N. & TULLIS, J. 1988. Experimental deformation of partially melted granitic aggregates. *Journal of Metamorphic Geology*, **6**, 495–615.

DINGWELL, D. B. 1997. The brittle-ductile transition in high level granitic magmas: material constraints. *Journal of Petrology*, **38**, 1635–1644.

DINGWELL, D. B., HESS, K. U. & KNOCHE, R. 1996. Granite and granitic pegmatite melts: volumes and viscosities. *Transactions of the Royal Society of Edinburgh, Earth Sciences*, **87**, 65–72.

EINSTEIN, A. 1906. Eine neue Bestimmung der Molekuldimensionnen. *Annales de Physique*, **19**, 289–306.

FERNANDEZ, A. & GASQUET, D. 1994. Relative rheological evolution of chemically contrasted coeval magmas: example of the Tichka plutonic complex (Morocco). *Contributions to Mineralogy and Petrology*, **116**, 316–326.

FERNANDEZ, C., CASTRO, A., DE LA ROSA, J. D. & MORENTO-VENTAS, I. 1997. Rheological aspects of magma transport inferred from rock structures. *In*: BOUCHEZ, J. L., HUTTON, D. H. W. & STEPHENS, W. E. (eds) *Granite: from segregation of melt to emplacement fabrics*. Kluwer, Dordrecht, 75–91.

FOWLER, A. C. 1987. Theories of mushy zones: applications to alloy solidification, magma transport, frost heave and igneous intrusions. *In*: LOPER, D. E. (ed) *Structure and dynamics of partially solidified systems*, NATO ASI Series E125. Martinus Nijhoff, Dordrecht, 159–199.

GARDIEN, V., THOMPSON, A. B., GRUJIC, D. & ULMER, P. 1995. Experimental melting of biotite + plagioclase + quartz ± muscovite assemblages and implications for crustal melting. *Journal of Geophysical Research*, **B100**, 15581–15591.

GAUTHIER, F. 1973. Field and laboratory studies of the rheology of Mount Etna lavas. *Philosophical Transactions of the Royal Society of London*, **274**, 83–98.

GOTO, A., OSHIMA, H. & NISHINA, Y. 1997. Empirical method of calculating the viscosity of peraluminous silicate melts at high temperature. *Journal of Volcanology and Geothermal Research*, **76**, 319–327.

GUÉGUEN, Y. & PALCIAUSKAS, V. 1992. *Introduction à la physique des roches*. Hermann, Paris.

GUYON, E., ROUX, S., HANSEN, A., BIDEAU, D., TROADEC, J. P. & CRAPO, H. 1990. Non-local and non-linear problems in the mechanics of disordered systems: application to granular media and rigidity problems. *Reports on Progress in Physics*, **53**, 373–419.

HALLOT, E., DAVY, P., BRÉMOND D'ARS, J. DE, AUVRAY, B., MARTIN, H. & VAN DAMME, H. 1996. Non-Newtonian effects during injection in partially crystallised magmas. *Journal of Volcanology and Geothermal Research*, **71**, 31–44.

HERWEGH, M., HANDY, M. R. & HEILBRONNER, R. 1997. Temperature- and strain-dependent microfabric evolution in monomineralic mylonite: evidence from in situ deformation of norcamphor. *Tectonophysic*, **280**, 83–106.

HOFFMAN, R. L. 1998. Explanation for the cause of shear thickening in concentrated colloidal suspsensions. *Journal of Rheology*, **42**, 111–123.

HOPGOOD, A. M. 1999. *Determination of structural successions in migmatites and gneisses*. Kluwer Academic Publishers, Dordrecht.

JAMBON, A. 1982. Tracer diffusion in granitic melts: experimental results for Na, K, Rb, Cs, Ca, Sr, Ba, Ce, Eu to 1300°C and model of calculation. *Journal of Geophysical Research*, **B87**, 797–810.

JEFFERY, J. B. 1922. The motion of ellipsoidal particles immersed in viscous fluid. *Proceedings of the Royal Society of London*, **A102**, 161–179.

JEZEK, J., MELKA, R., SCHULMANN, K. & VENERA, Z. 1994. The behaviour of rigid triaxial ellipsoidal particles in viscous flows – modeling of fabric evolution in a multiparticle system. *Tectonophysics*, **229**, 165–180.

JOHANNES, W. & HOLTZ, F. 1996. *Petrogenesis and experimental petrology of granitic rocks*. Springer Verlag, Heidelberg.

JOHN, B. E. & STUNITZ, H. 1997. Magmatic fracturing and small-scale melt segregation during pluton emplacement: evidence from the Adamello massif (Italy). *In*: BOUCHEZ, J. L., HUTTON, D. H. W. & STEPHENS, W. E. (eds) *Granite: from segregation of melt to emplacement fabrics*. Kluwer, Dordrecht.

JOSEPH, D. D., LIU, Y. J., POLETTO, M. & FENG, J. 1994. Aggregation and dispersion of spheres falling in viscoelastic liquids. *Journal of Non-Newtonian Fluid Mechanics*, **54**, 45–86.

KIRBY, S. H. & KRONENBERG, A. K. 1987. Rheology of the lithosphere: selected topics. *Reviews of Geophysics*, **25**, 1219–1244.

KOHLSTEDT, D. L., BAI, Q., WANG, Z. C. & MAI, S. 2000. Rheology of partially molten rocks. *In*: BAGDASSAROV, N., LAPORTE, D. & THOMPSON, A. B. (eds) *Physics and chemistry of partially molten rocks*. Kluwer, Dordrecht.

KOMAR, P. D. 1976. Phenocryst interactions and the velocity profile of magma flowing through dikes and sills. *Geological Society of America Bulletin*, **87**, 1336–1342.

KRIEGER, I. M. & DOUGHERTY, T. J. 1959. A mechanism for non Newtonian flow in suspensions of rigid spheres. *Transactions of the Society of Rheology*, **3**, 137–152.

LAPORTE, D. & WATSON, E. B. 1995. Experimental and theoretical constraints on melt distribution in crustal sources: the effect of crystalline anisotropy on melt interconnectivity. *Chemical Geology*, **124**, 161–184.

LEJEUNE, A. M. & RICHET, P. 1995. Rheology of crystal-bearing silicate melts: an experimental study of high viscosities. *Journal of Geophysical Research*, **B100**, 4215–4229.

LIU, S. & MASLIYAH, J. H. 1996. Rheology of suspensions. *In*: SCHRAMM, L. L. (ed) *Suspensions: fundamentals and applications in the petroleum industry*. American Chemical Society, Washington, Advances in Chemistry Series **251**, 107–176.

MARONE, C. 1998. Laboratory-derived friction laws and their application to seismic faulting. *Annual Reviews in Earth and Planetary Sciences*, **26**, 643–696.

MARSH, B. D. 1996. Solidification fronts and magmatic evolution. *Mineralogical Magazine*, **60**, 5–40.

McBirney, A. R. & Murase, T. 1984. Rheological properties of magmas. *Annual Reviews of Earth and Planetary Science*, **12**, 337–357.

McKenzie, D. 1984. The generation and compaction of partially molten rocks. *Journal of Petrology*, **25**, 713–765.

Mehnert, K. R. 1968. *Migmatites and the origin of granitic rocks*. Elsevier, Amsterdam.

Miller, C. F., Watson, E. B. & Harrison, T. M. 1988. Perspectives on the source, segregation and transport of granitoid magmas. *Transactions of the Royal Society of Edinburgh, Earth Sciences*, **79**, 135–156.

Moukarzel, C. F. 1999. Granular matter instability: a structural rigidity point of view. *In*: Thorpe, M. F. & Duxbury, P. M. (eds) *Rigidity Theory and Applications*. Plenum, New York, 125–142.

Nicolas, A. 1992. Kinematics in magmatic rocks with special reference to gabbros. *Journal of Petrology*, **33**, 891–915.

Onoda, G. Y. & Liniger, E. G. 1990. Random loose packing of uniform spheres and the dilatancy onset. *Physical Review Letters*, **64**, 2727–2730.

Olgaard, D. L. 1990. The role of second phase in localizing deformation. *In*: Knipe, R. J. & Rutter, E. H. (eds) *Deformation Mechanisms, Rheology and Tectonics*. Geological Society, London, Special Publications, **54**, 175–181.

Ottino, J. M. 1989. *The kinematics of mixing: stretching, chaos and transport*. Cambridge Univesity Press, Cambridge.

Paterson, S. R., Fowler, T. K., Schmidt, K. L., Yoshinobu, A. S., Yuan, E. S. & Miller, R. B. 1998. Interpreting magmatic fabric patterns in plutons. *Lithos*, **44**, 53–82.

Patiño-Douce, A. E. & Beard, J. S. 1995. Dehydration melting of biotite gneiss and quartz amphibolite from 3 to 15 kbar. *Journal of Petrology*, **96**, 707–738.

Patiño-Douce, A. E. & Johnston, A. D. 1991. Phase equilibria and melt productivity in the pelitic system: implications for the origin of peraluminous granitoids and aluminous granites. *Contributions to Mineralogy and Petrology*, **107**, 202–218.

Petford, N. & Clemens, J. D. 1999. Granitic melt viscosity and silicic magma dynamics in contrasting tectonic settings. *Journal of the Geological Society of London*, **156**, 1057–1060.

Petford, N., Cruden, A. R., McCaffrey, K. J. W. & Vigneresse, J. L. 2000. Granite magma formation, transport and emplacement in the Earth's crust. *Nature*, **408**, 669–673.

Petford, N., Kerr, R. C. & Lister, J. R. 1993. Dike transport of granitoid magmas. *Geology*, **21**, 845–848.

Pinkerton, H. & Norton, G. 1995. Rheological properties of basaltic lavas at subsolidus temperatures: laboratory and field measurements on lavas from Mount Etna. *Journal of Volcanology and Geothermal Research*, **68**, 307–323.

Pons, J., Barbey, P., Dupuis, D. & Léger, J. M. 1995. Mechanism of pluton emplacement and structural evolution of a 2.1 Ga juvenile continental crust: the Birimian of southwestern Niger. *Precambrian Research*, **70**, 281–301.

Quemada, D. 1998. Rheological modelling of complex fluids: II – Shear thickening behavior due to shear induced flocculation. *European Physical Journal of Applied Physics*, **2**, 175–181.

Ranalli, G. 1995. *Rheology of the Earth*. Chapman & Hall, London.

Renner, J., Evans, B. & Hirth, G. 2000. On the rheologically critical melt fraction. *Earth and Planetary Science Letters*, **181**, 585–594.

Rogers, C. D. F., Dijkstra, T. A. & Smalley, T. A. 1994. Particle packing from an Earthscience viewpoint. *Earth-Science Reviews*, **36**, 59–82.

Roscoe, R. 1952. The viscosity of suspensions of rigid spheres. *British Journal of Applied Physics*, **3**, 267–269.

Rosenberg, C. L. 2001. Deformation of partially-molten granite: a review and comparison of experimental and natural case studies. *International Journal of Earth Sciences*, **90**, 60–76.

Rushmer, T. 1995. An experimental deformation study of partially molten amphibolite. Application to low fraction melt segregation. *Journal of Geophysical Research*, **B100**, 15681–15695.

Rutter, E. H. & Neumann, D. H. K. 1995. Experimental deformation of partially molten Westerly granite under fluid-absent conditions with implications for the extraction of granitic magma. *Journal of Geophysical Research*, **B100**, 15697–15715.

Rutter, J. M. & Wyllie, P. J. 1988. Melting of vapour absent tonalite at 10 kb to simulate dehydrating melting in the deep crust. *Nature*, **331**, 159–160.

Ryan, M. P. & Blevins, J. Y. K. 1987. *The viscosity of synthetic and natural melts and glasses at high temperatures and 1 bar (10^5 Pascals) pressure and at higher temperature*. U.S. Geological Survey Bulletin, **1764**.

Ryerson, F. J., Weed, H. C. & Piwinskii, A. J. 1988. Rheology of subliquidus magmas. 1. Picritic compositions. *Journal of Geophysical Research*, **B93**, 3421–3436.

Sawyer, E. W. 1991. Disequilibrium melting and the rate of melt residuum separation of mafic rocks from the Grenville Front, Québec. *Journal of Petrology*, **32**, 701–738.

Sawyer, E. W. 1994. Melt segregation in the continental crust. *Geology*, **22**, 1019–1022.

Sawyer, E. W. 1998. Formation and evolution of granite magmas during crustal reworking: the significance of diatexites. *Journal of Petrology*, **39**, 1147–1167.

Scholz, C. S. 1990. *The mechanics of earthquakes and faulting*. Cambridge University Press, New York.

Shaw, H. R. 1965. Comments on the viscosity, crystal settling and convection in granitic magmas. *American Journal of Science*, **263**, 120–152.

Smith, J. V. 1996. Ductile-brittle transition structures in the basal shear zone of a rhyolite lava flow, eastern Australia. *Journal of Volcanology and Geothermal Research*, **72**, 217–223.

SPERA, F. J., BORGIA, A., STRIMPLE, J. & FEIGENSON, M. 1988. Rheology of melts and magmatic suspensions. 1. Design and calibration of concentric cylinder viscometer with application to rhyolitic magma. *Journal of Geophysical Research*, **B93**, 10273–10294.

STAUFFER, D. 1985. *Introduction to percolation theory*. Taylor & Francis, London.

THOMPSON, A. B. 1982. Dehydration melting of pelitic rocks and the generation of H_2O-undersaturated granitic liquids. *American Journal of Science*, **282**, 1567–1595.

TIKOFF, B. & TEYSSIER, C. 1994. Strain and fabric analyses based on porphyroclast interaction. *Journal of Structural Geology*, **16**, 477–491.

TOBISCH, O. T., MCNULTY, B. A. & VERNON, R. H. 1997. Microgranitoid enclave swarms in granitic plutons, Central Sierra Nevada, California. *Lithos*, **40**, 321–339.

VANDERHAEGHE, O. & TEYSSIER, C. 1997. Formation of the Shuswap metamorphic core complex during late-orogenic collapse of the Canadian Cordillera: role of ductile thinning and partial melting of the mid- to lower crust. *Geodinamica Acta*, **10**, 41–58.

VAN DER MOLEN, I. & PATERSON, M. S. 1979. Experimental deformation of partially-melted granite. *Contributions to Mineralogy and Petrology*, **70**, 299–318.

VERNON, R. H. 2000. Review of microstructural evidence of magmatic and solid-state flow. *Electronic Geoscience*, **5**.

VIELZEUF, D. & HOLLOWAY, J. R. 1988. Experimental determination of the fluid-absent melting reactions in the pelitic system. Consequences for crustal differentiation. *Contributions to Mineralogy and Petrology*, **98**, 257–276.

VIGNERESSE, J. L. & BOUCHEZ, J. L. 1997. Successive granitic magma batches during pluton emplacement: the case of Cabeza de Araya (Spain). *Journal of Petrology*, **38**, 1767–1776.

VIGNERESSE, J. L. & BURG, J. P. 2000. Continuous versus discontinuous melt segregation in migmatites : insights from a cellular automaton model. *Terra Nova*, **12**, 188–192.

VIGNERESSE, J. L. & TIKOFF, B. 1999. Strain partitioning during partial melting and crystallising felsic magmas. *Tectonophysics*, **312**, 117–132.

VIGNERESSE, J. L., BARBEY, P. & CUNEY, M. 1996. Rheological transitions during partial melting and crystallisation with application to felsic magma segregation and transfer. *Journal of Petrology*, **37**, 1579–1600.

VIGNERESSE, J. L., CUNEY, M. & BARBEY, P. 1991. Deformation assisted crustal melt segregation and transfer. *Geological Association of Canada/Mineralogical Association of Canada Abstract*, **16**, A128.

VOORHEES, P. W. 1985. The theory of Ostwald ripening. *Journal of Statistical Physics*, **38**, 231–252.

WEBB, S. L. & DINGWELL, D. B. 1990. Non-Newtonian rheology of igneous melts at high stresses and strain rates: experimental results for rhyolite, andesite, basalt and nephelinite. *Journal of Geophysical Research*, **B95**, 15695–15701.

WEBER, C., BARBEY, P., CUNEY, M. & MARTIN, H. 1985. Trace element behaviour during migmatization. Evidence for a complex melt-residuum-fluid interaction in the St. Malo migmatite dome (France). *Contributions to Mineralogy and Petrology*, **90**, 52–62.

WILKS, K. R. & CARTER, N. L. 1990. Rheology of some continental lower crustal rocks. *Tectonophysics*, **182**, 57–77.

WRIGHT T. L. & OKAMURA, R. T. 1977. *Cooling and crystallisation of tholeiitic basalt, 1965 Makahopuhi Lava Lake, Hawaii*. US Geological Survey Professional Paper, **1004**.

Do calcite rocks obey the power-law creep equation?

JÖRG RENNER[1,2] & BRIAN EVANS[1]

[1] *Department of Earth, Atmospheric and Planetary Sciences, Massachusetts Institute of Technology, Cambridge, MA 02139, USA (e-mail: brievans@mit.edu)*

[2] *Present address: Institute for Geology, Mineralogy, and Geophysics, Ruhr-University, D-44780 Bochum, Germany*

Abstract: The power-law creep equation, $\dot{\varepsilon} \propto \sigma^n \exp(-Q/RT)$, is commonly used to relate strain rate, $\dot{\varepsilon}$, stress, σ, and temperature, T, for thermally activated dislocation creep in rocks. When triaxial deformation experiments on marble and limestone samples are performed at temperatures of 400–1050 °C, to strains <0.2, and with strain rates between 10^{-3} and $10^{-7}\,s^{-1}$, the variations in strength among different rocks at nominally identical conditions are much larger than the experimental uncertainty. During dislocation creep, the strengths of various limestones and marbles decrease with increasing grain size, similar to the Hall-Petch effect in metals. The stress sensitivity of strain rate, $n' = \partial \ln \dot{\varepsilon}/\partial \ln \sigma$, and the temperature sensitivity of strain rate, $Q' = -R\partial \ln \dot{\varepsilon}/\partial(1/T)$, differ greatly for the various calcite aggregates. There is a systematic dependence of n' and Q' on stress, grain size, and perhaps, temperature, and there is no interval in stress where n' is constant. Thus, the steady-state power-law equation is an inadequate description of dislocation creep in calcite rocks. To improve the constitutive law, it may be necessary to include at least one additional state variable that scales with grain size.

Under elevated pressure, over a broad range of temperatures, calcite deforms by mechanisms involving crystalline defects, including: dislocation glide, cross-slip, and climb; mechanical twinning; self-diffusion; and grain boundary sliding (Heard & Raleigh 1972; Rutter 1974; Schmid 1976). Strain is often accommodated by a combination of these mechanisms, but strength is usually thought to be dominated by one or two steps that are rate limiting (Frost & Ashby 1982). One important element in understanding tectonic processes is the specification of a constitutive relation that links the strength of a rock to a few thermodynamic variables, material constants, and state or structure parameters (Carter & Tsenn 1987; Paterson 1987). If both the thermodynamic parameters and state parameters remain constant, then the constitutive relation is said to be steady state; but if some important element of the state or structure varies during deformation, then the mechanical properties will also vary (Argon 1975). In that case, rules for the evolution of the state and structure become important (Stouffer & Dame 1996). Both steady-state (see reviews by Kohlstedt *et al.* 1995; de Bresser *et al.* 2002) and state-variable laws (Covey-Crump 1994, 1998) have been proposed for calcite rocks deformed at various conditions of temperature and confining pressure.

Background

Steady-state constitutive laws

Steady-state descriptions of the mechanical creep behaviour of calcite are often divided heuristically into three major regimes (Rutter 1974; Schmid *et al.* 1977, 1980; Walker *et al.* 1990). At low temperatures and high stresses, the strength is only modestly dependent on strain rate, and the deformation mechanism is often called low-temperature plasticity, or power-law breakdown. At higher temperatures, creep involves dislocation climb and cross-slip (de Bresser & Spiers 1990, 1993). At the highest temperatures, self-diffusion and/or boundary sliding dominate (Schmid *et al.* 1977; Walker *et al.* 1990).

In the intermediate regime, the most commonly used flow law is the power-law creep equation:

$$\dot{\varepsilon} = \dot{\varepsilon}_{pl,0}\sigma^n \exp\left(-\frac{Q_{pl}}{RT}\right) \tag{1}$$

where: $\dot{\varepsilon}$ is the strain rate; $\dot{\varepsilon}_{pl,0}$ is a pre-exponential factor perhaps containing a dependence on chemical fugacity and temperature; σ is the differential stress; n is an index of the stress sensitivity of strain rate, often called the stress exponent; Q_{pl} is the activation enthalpy for creep; R is the gas constant; and T is the absolute temperature

From: DE MEER, S., DRURY, M. R., DE BRESSER, J. H. P. & PENNOCK, G. M. (eds) 2002. *Deformation Mechanisms, Rheology and Tectonics: Current Status and Future Perspectives*. Geological Society, London, Special Publications, **200**, 293–307. 0305-8719/02/$15 © The Geological Society of London.

(Kohlstedt *et al.* 1995). If this equation is valid, then there will be regions in temperature and stress-space where n and Q_{pl} are relatively constant. In this view, strain rate should be independent of grain size. Alternative descriptions also exist. For example, de Bresser & Spiers (1990) have suggested that deformation in this regime might be limited by cross-slip.

For deformation by diffusion creep and grain boundary sliding (Schmid *et al.* 1977, 1980; Walker *et al.* 1990), the constitutive law is commonly given as:

$$\dot{\varepsilon} = \dot{\varepsilon}_{diff,0} \frac{\sigma}{d^m} \exp\left(-\frac{Q_{diff}}{RT}\right) \quad (2)$$

where $\dot{\varepsilon}_{diff,0}$ may depend on the chemical fugacity of major or impurity components, d is the grain size, m is an index of the sensitivity of strain rate to grain size and is 2 or 3 depending on the dominant diffusion path, and Q_{diff} is the activation energy for lattice or grain boundary diffusion.

State variable approach

At low homologous temperature (i.e. $T/T_m < 0.4$, where T_m is the melting temperature), strength often changes during deformation, and a history-dependent constitutive relation is needed. One way to insert history dependence is to include one or more variables that describe some elements of the state or structure that are important for mechanical behaviour and change during deformation (Stouffer & Dame 1996). Such variables might be identified explicitly, e.g. dislocation geometries, subgrain dimensions, precipitate orientation, crystallographic orientations, or other features of the structure (Stone 1991). Alternatively, they might be defined implicitly from macroscopic behaviour associated with the development of the microstructure (Hart 1970). At temperatures less than 600 °C, Carrara marble continues to work harden up to strains of 0.2. Using Hart's (1970, 1976) technique to define an implicit state variable, Covey-Crump (1998) found that the state variable increases with strain, but that the rate of increase becomes smaller at higher strains and at higher temperatures. Above 600 °C, the mechanical state may achieve a steady state at strains of 0.15 or higher, but the technique does not have high enough resolution to prove that steady state exists. At higher temperatures (730–930 °C), recent experiments, also on Carrara marble, using the torsion technique (Pieri *et al.* 2001) suggest that the evolution of strength may continue up to shear strains as large as 2.5.

Mechanical testing of calcite rocks

High quality mechanical data exist for a variety of calcite rocks including Yule marble, Solnhofen limestone, Carrara marble, and synthetic marble, as does detailed information regarding dislocation processes in single crystals (Turner *et al.* 1954; Wang *et al.* 1996; de Bresser & Spiers 1997). Some success has been achieved in relating single crystal deformation behaviour to the generation of preferred orientation in polycrystals (Takeshita *et al.* 1987; Tome *et al.* 1991; Lebensohn *et al.* 1998). Consequently, it is very instructive to compare the mechanical data for various limestones and marbles. Significantly, when the power-law creep description is used, the parameters determined for each rock differ greatly (Evans & Kohlstedt 1995). Most tests have been duplicated by several laboratories and, apparently, differences do not owe to experimental uncertainties.

The uncertainty in the rheology is of great concern when extrapolating flow laws to geological conditions, reconstructing palaeo-deformation conditions from microstructure imprints, and predicting the strength contrast in polymineralic aggregates. Extrapolated strain rates vary by more than 3 orders of magnitude for given stress and temperature (Evans & Kohlstedt 1995; Brodie & Rutter 2000; de Bresser *et al.* 2002). Thus, there is actually considerable contention regarding which constitutive relation to use for calcite rocks. Should the relation be steady state or should it contain a state variable? Or, is a combination of the two required? If steady state, which formulation should be used, and under what conditions is each relation valid? If state-variable formulations are used, then can the state variable be defined explicitly?

In this paper, we examine the applicability of the commonly used power-law creep equation and conclude that this paradigm does not adequately represent the current data for strains less than 0.20, even at relatively high temperatures (700–1000 °C). During dislocation creep at different fixed stresses, the stress sensitivity of strain rate varies, as does the temperature sensitivity of strain rate. In the intermediate stress/strain-rate regime of dislocation creep, the strength of calcite rocks depends inversely on grain size, although the data are not sufficient to distinguish the exact functional dependence. None of these variations are predicted by the power-law creep equation. An improved constitutive law must include at least one additional parameter that scales with grain size.

Data and results

Variability in the strength of different calcite rocks is apparent even when they are tested under common conditions of strain rate and temperature (Fig. 1). Even within a single block of material, the strength may vary by about 20%. In addition to the explicit state variables mentioned above, variations in strength can arise from differences in grain size, accessory minerals, porosity, or chemical impurities. The data plotted here come from low-strain tests ($\varepsilon < 0.2$) on Carrara marble, Solnhofen limestone, Yule marble, and synthetic limestones made from hot-isostatically pressed calcite (see Table 1 for sources). We have not considered the effect of mechanical twinning on strength, because that process requires stresses above 50–100 MPa in the coarse-grained marbles and more than 200–300 MPa in the finer-grained natural limestone and synthetic marbles (Rowe & Rutter 1990). Moreover, twins are seldom seen in the quenched samples.

Stress sensitivity

To evaluate the power law, we re-examined literature data for both single crystals and polycrystalline aggregates and approximated

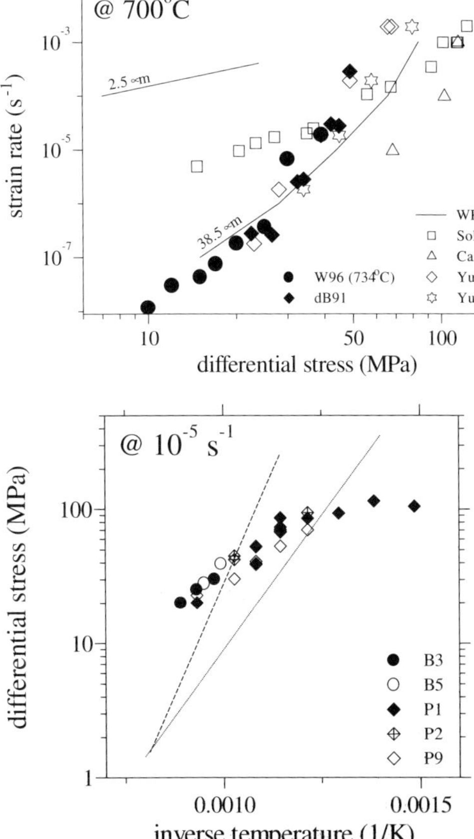

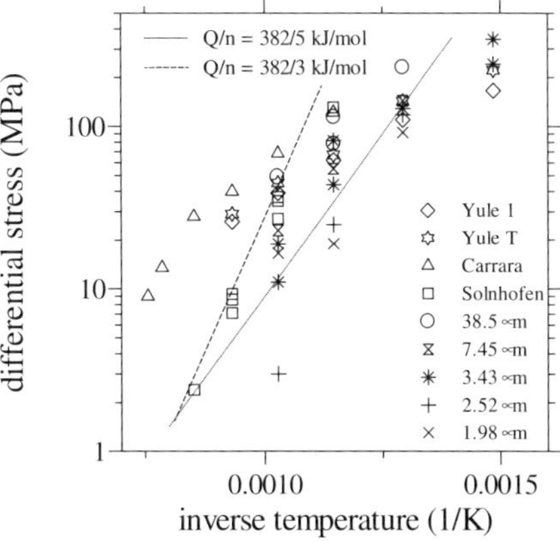

Fig. 1. Results of deformation experiments on single-crystal calcite, natural and synthetic marbles, and natural limestones. (**a**) Strain rate versus differential stress at a temperature of 700 °C. (**b**) and (**c**) Stress versus temperature for polycrystalline and single crystal samples at a strain rate of $\sim 10^{-5}$ s^{-1}. For data labelling and grain sizes, see Table 1. The lines in (**a**) represent results of relaxation experiments. The creep tests by Wang *et al.* (1996), W96, represent axial strains of less than 1%. Constant strain-rate tests were performed to maximum axial strains between about 5 and 25% and are represented by the stress at 10%. Most often, differential stress remained constant over the range of strain explored. The strength of polycrystals varies considerably at a given temperature (**b**). In contrast, the single crystal data show only moderate variation (**c**) despite a range of Mn impurity content covering about 2 orders in magnitude up to 600 ppm. Thus, grain boundaries seem to play a more important role than impurities. The lines in (**b**) and (**c**) show activation terms calculated according to $\sigma \propto \exp(Q/nRT)$ (see the power law, Eq. 1) using activation energies determined for Ca self-diffusion in calcite (Farver & Yund 1996; Fisler & Cygan 1999) and stress exponents as expected for regular power-law creep, i.e. 3.5 to 5. In addition to the obvious discrepancy to these anticipated power-law activation parameters, neither the bulk of observations nor an individual data set seem to be in accord with a constant activation parameter Q/n.

Table 1. *Sources of data on deformation of calcite single crystals, limestones, and marbles*

Material	Reference	Legend in Figures	Technique*	$d^{\dagger}$ (mm)
Single crystal, P1	de Bresser (1991); de Bresser & Spiers (1990)	dB91	Constant strain rate	7 ± 3
Single crystal, P2	de Bresser (1991); de Bresser & Spiers (1990)	dB91	Constant strain rate	7 ± 3
Single crystal, P9	de Bresser (1991); de Bresser & Spiers (1990)	dB91	Constant strain rate	7 ± 3
Single crystal, B3	Wang *et al.* (1996), their 4a	W96	Creep	4 ± 2
Single crystal, B5	Wang *et al.* (1996), their 4b	W96	Creep	4 ± 2
Yule marble, 1	Heard (1963); Heard & Raleigh (1972)	Y1	Constant strain rate	0.35 ± 0.15
Yule marble, T	Heard (1963); Heard & Raleigh (1972)	YT	Constant strain rate	0.35 ± 0.15
Carrara marble	Schmid *et al.* (1980)	CM	Constant strain rate, relaxation	0.15 ± 0.05
Solnhofen limestone	Schmid (1976); Schmid *et al.* (1977)	SL	Constant strain rate, relaxation	0.005 ± 0.002
Synthetic	Bruhn *et al.* (1999)	BOD99	Constant strain rate	0.008–0.012
Synthetic	Walker *et al.* (1990)	WRB90	Constant strain rate, relaxation	0.002–0.040

*Heard (1963) and Heard & Raleigh (1972) performed extension tests with the maximum compressive stress parallel (1-samples) and perpendicular (T-samples) to the natural shape preferred orientation. All other tests were performed in compression. For the limestone and the marbles each data point entering the flow law determination corresponds to an individual sample while Wang *et al.* (1996) determined all data on a single sample. de Bresser (1991) combined stepping tests and testing of several samples from the same parent crystal.

†Sample size for single crystals and grain size for polycrystalline aggregates.

the stress sensitivity of strain rate ($n' \equiv \partial \ln \dot{\varepsilon} / \partial \ln \sigma|_T$) as the quotient of differences:

$$n' \simeq \left. \frac{\ln(\dot{\varepsilon}_1/\dot{\varepsilon}_2)}{\ln(\sigma_1/\sigma_2)} \right|_T . \quad (3)$$

Stress and strain rate data come from constant strain rate and creep tests (Fig. 1; Table 1). In the range of conditions considered here, tests most often reached apparent steady state. The stress at 10% strain was consistently chosen in case of constant strain-rate tests. Rather than performing linear regressions at constant temperature, we calculated n' from pairs of experiments at a particular temperature (labelled 1 and 2 in Eq. 3), that were adjacent in terms of strain rate. Thus, n' could be analysed as a function of stress (Fig. 2). The stress sensitivity is plotted versus the arithmetic mean of the respective stress interval. The width of the stress intervals is about ±20% of the arithmetic mean.

If the power law (Eq. 1) applies, stress sensitivity of strain rate should be constant, i.e. $n' = n$. However, for each of the natural and synthetic calcite materials n' increases monotonically with increasing stress (Fig. 2a). Data from relaxation tests on synthetic marbles, Solnhofen limestone (Walker *et al.* 1990), and Carrara marble (Schmid *et al.* 1980) also indicate that the stress sensitivity changes continuously with stress. Exponents from single-crystal data are upper bounds to those observed for polycrystalline aggregates at a given stress, whereas exponents for Solnhofen limestone (grain size about 5–10 μm) are lower bounds (Fig. 2a). Representing the various calcite materials by their respective crystal or grain size (Table 1) reveals a systematic increase of the stress sensitivity with increasing grain size (Fig. 2b). Temperature has a subordinate effect, but for a given suite of experiments there is a tendency for n' to increase with increasing temperature (Fig. 2b).

Temperature sensitivity

The data compilation in Figures 1b and 1c demonstrates that simple activation terms ($\sigma \propto \exp(Q/nRT)$, as suggested by the power law, Eq. 1) cannot account for the temperature dependence of the strength of the various calcite rocks. Similar to the procedure for the stress sensitivity, we therefore estimated the temperature sensitivity of strain rate ($Q' \equiv -R \partial \ln \dot{\varepsilon} / \partial (1/T)|_\sigma$), as:

$$Q' \simeq R \frac{\ln(\dot{\varepsilon}_1/\dot{\varepsilon}_2)}{(1/T_2 - 1/T_1)|_\sigma} \quad (4)$$

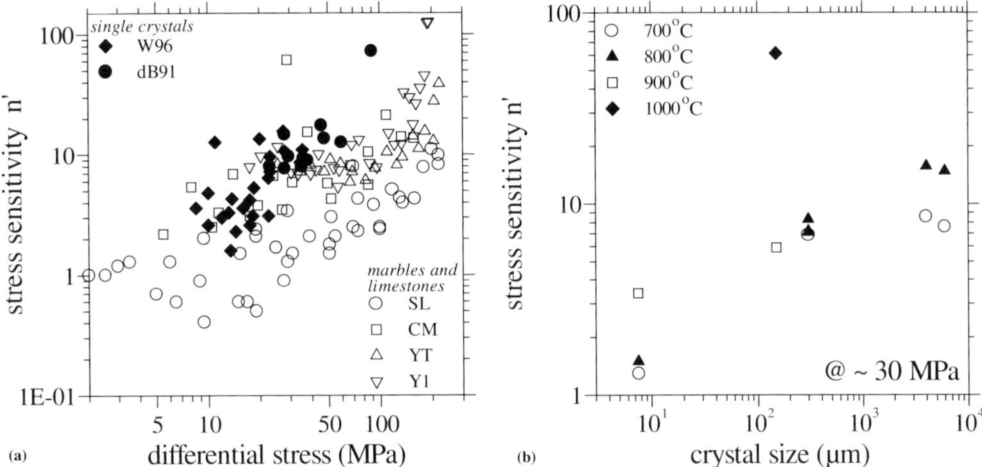

Fig. 2. Apparent stress sensitivity for calcite aggregates and single crystals. For data labelling and grain sizes see Table 1. (**a**) Stress sensitivity of strain rate, n', versus differential stress for single crystal calcite and polycrystalline limestones and marbles. Each value is determined from two tests at the same temperature and separated by a small stress interval. The stress sensitivity is plotted versus the arithmetic mean of the respective stress interval. The width of the stress intervals is about ±20% of the arithmetic mean. (**b**) Stress sensitivity versus crystal/grain size at a differential stress of about 30 MPa.

for different rocks, by evaluating pairs of tests at nearly identical stress, but at two different temperatures (labelled 1 and 2). We identify the results by the average of the two temperatures T_1 and T_2 in the legend of Fig. 3. The temperature sensitivities formally correspond to apparent activation energies and vary between about 100 and almost 1000 kJ/mol depending on the rock that is tested. Values for the fine-grained synthetic rocks tested at low temperatures ($d \sim 10\,\mu$m, Bruhn *et al.* 1999) are smaller than those for natural marbles and limestones. There is a tendency for Q' to increase with stress for a given rock. Both the increase and the absolute values tend to be larger for rocks with larger grain sizes and/or tests at higher temperatures.

When necessary, we linearly interpolated strain rate/strength pairs to obtain strain rates at a particular strength. Because the sensitivities are local gradients obtained from a limited set of observations, they are subject to uncertainty. However, it is quite unlikely that the systematic variations could simply be artefacts of the procedure. In particular, the inferred temperature dependence of the stress sensitivity and the stress dependence of the temperature sensitivity are in accord with a Maxwell-type relation between second derivatives of strain rate:

$$\partial n'/\partial T = (T^2\sigma/R)\partial Q'/\partial\sigma.$$

Discussion

Activation energy

The micromechanical models underlying the power law equation often assume that deformation rate is limited by self-diffusion of a constituent phase. In metals, activation energies measured in creep at higher temperatures frequently agree well with activation energies for self-diffusion (Poirier 1978). Diffusion kinetics in minerals are much more complicated and can be affected by extrinsic factors, including substitutional impurities and fluid-phase fugacity, as well as intrinsic parameters, such as total pressure and temperature (see Table 2). At ambient pressure, either Ca or C is the slowest diffusing constituent in calcite (Fisler & Cygan 1999). Under these conditions the diffusion of Ca seems to be dominated by extrinsic vacancies (Kronenberg *et al.* 1984), whereas carbon may diffuse as carbonate ions (Mirwald 1979; Fisler & Cygan 1999; Labotka *et al.* 2000).

Despite the lack of certainty about the rate-controlling species for self-diffusion it is interesting to compare activation energies for self-diffusion and creep. Flow-law parameters for calcite aggregates from various studies in the literature are collected in de Bresser *et al.* (2002). The apparent activation energies for different rocks under similar conditions vary considerably. A particular problem in the

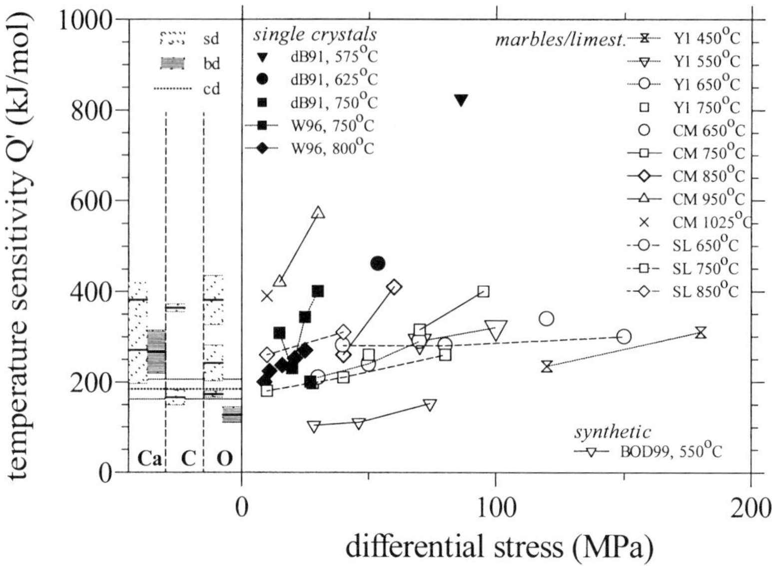

Fig. 3. Temperature sensitivity of strain rate, Q', versus differential stress. Temperatures given in the legends represent the mean of the temperature intervals of Q' determination. On the left side of the plot, we present the activation energies determined from self diffusion (sd) and grain boundary diffusion (gb) of the constituents of calcite (Ca, C, O) as listed in Table 2. Also given is the estimate for core diffusion (cd) of an unspecified vacancy (Liu & Evans 1997). Activation energies from deformation and diffusion experiments cover roughly the same range of values. The activation energies from deformation experiments seem to increase with increasing stress, increasing temperature, and increasing grain size.

comparison of diffusion and creep experiments arises because of the variation in chemical conditions among the experiments. The diffusion experiments that were performed dry under a CO_2 atmosphere close to ambient pressure are directly applicable to the unconfined deformation of single crystals. However, the polycrystalline calcite rocks were tested at elevated confining pressure with mostly unspecified/uncontrolled conditions for water or CO_2 pressures (but see Rutter 1974). The diffusivities of both C and O in calcite in a pure CO_2 atmosphere decrease significantly when pressure is increased from ambient to 100 MPa. Also, the activation energies decrease by a factor of 2 for the same pressure change (Labotka *et al.* 2000). The presence of water enhances O diffusion but leaves the activation energy unaltered.

In experiments designed to measure the kinetics of dislocation recovery, Liu & Evans (1997) determined an activation energy of approximately 185 kJ mol^{-1} for core diffusion of unspecified vacancies that would be pertinent if that process controlled deformation rate. At low stresses diffusion creep or grain-boundary sliding might accommodate the deformation. Then, instead of transport through the grain interiors, transport on grain boundaries might be rate controlling (for example, for Coble-creep, $m = 3$ in Eq. 2). The grain boundary diffusion of oxygen is orders of magnitude faster than that of calcium (Farver & Yund 1996, 1998). The latter is apparently insensitive to pressure. The kinetics of carbon transport on the boundary have not been explored so far. The agreement between activation energies for deformation of Solnhofen limestone at a confining pressure of 300 MPa (Schmid *et al.* 1977, 1980; 214 kJ/mol) and for deformation of synthetic marbles deformed at ambient pressure (Freund, pers. comm., 2001; 299 ± 59 kJ/mol), and activation energies determined for Ca boundary diffusion (Table 2) is fair. The activation energy determined for grain-size-sensitive creep of synthetic marbles deformed at 200 MPa (Walker *et al.* 1990; 190 kJ/mol) is smaller, but is probably still within the combined experimental uncertainties.

Changes in the diffusion kinetics of the rate-controlling step for high-temperature creep are likely to contribute to the variability of the temperature sensitivity of strain rate of deformation presented in Fig. 3, particularly at the transition from deformation controlled by dislocation activity to that controlled by grain-boundary processes.

Table 2. *Activation energies for diffusion of calcite constituents*

Element	Activation energy, self Diffusion (kJ/mol)	Conditions & Reference	Migrating species	Activation energy, boundary diffusion (kJ/mol)	Conditions & Reference
Ca	271 ± 47	Dry, 1 atm CO_2 (Fisler & Cygan 1999)	Extrinsic Ca vacancies	267 ± 47	Dry (Farver & Yund 1996)
	382 ± 37	Dry, 1 atm CO_2 (Farver & Yund 1996)			
C	166 ± 16	Dry, 100 MPa CO_2 (Labotka *et al.* 2000)	Carbonate ions CO_3^{2-} (Fisler & Cygan 1999; Mirwald 1979)		
	364 ± 8	Wet (Kronenberg *et al.* 1984); Dry (Anderson 1969)			
O	242 ± 39	Dry, 100 MPa CO_2 (Labotka *et al.* 2000)	Intrinsic vacancies	127 ± 17	Wet (Farver & Yund 1998)
	c. 381	Dry 1 atm (Anderson 1969)			
	173 ± 6	Hydrothermal 10–350 MPa (Farver 1994)	H_2O molecule		

However, even when restricting the analysis to conditions that indicate dislocation creep, i.e. stress exponents in excess of 3, we find that the temperature sensitivity of strain rate increases rather systematically with stress, temperature, and grain size. Thus, the determined temperature sensitivities are unlikely to correspond to activation energies of a diffusion process, as suggested if dislocation climb were rate controlling.

Grain-size dependence of dislocation creep

In Fig. 4 we present a synopsis of the strength and strain rate data for various calcite aggregates tested at 700 °C. Notice that strength decreases with decreasing grain size at high temperatures and low stresses, which is characteristic of diffusion creep, whereas strength increases with decreasing grain size at lower temperatures and higher stresses. In the high temperature/low stress regime, stress exponents below 2 are observed, which is also consistent with a significant contribution of diffusion/grain boundary sliding creep. Wang (1994) and de Bresser *et al.* (2001) discuss a combination of grain-size-sensitive diffusion/grain-boundary sliding and grain-size-insensitive power-law dislocation creep. The stress and temperature sensitivities of such macroscopic composite laws do vary but only between the parameters corresponding to the two mechanisms considered. If grain growth occurs, the determination of constitutive parameters may be biased depending upon the specific experimental procedures. For example, temperature sensitivity (apparent activation energy) will be underestimated if determined from tests on samples with initial grain size systematically increasing with increasing deformation temperature. However, such composite laws cannot explain the inversion in grain-size dependence or the continuous increase in stress exponent with increasing stress (Fig. 2a), especially when those exponents do not lie within the range of 'normal' power law exponents, i.e. 3 to 5.

The switch in grain-size dependence can be seen in deformation tests on a limestone (Schmid & Paterson 1977) that is composed of very fine-grained ooids (1–2 μm) in a sparitic matrix (*c.* 200 μm). Strain analysis (Schmid & Paterson 1977) indicates that the ooids are strong relative to the matrix at low temperatures, but that strain concentrates in the ooids at higher

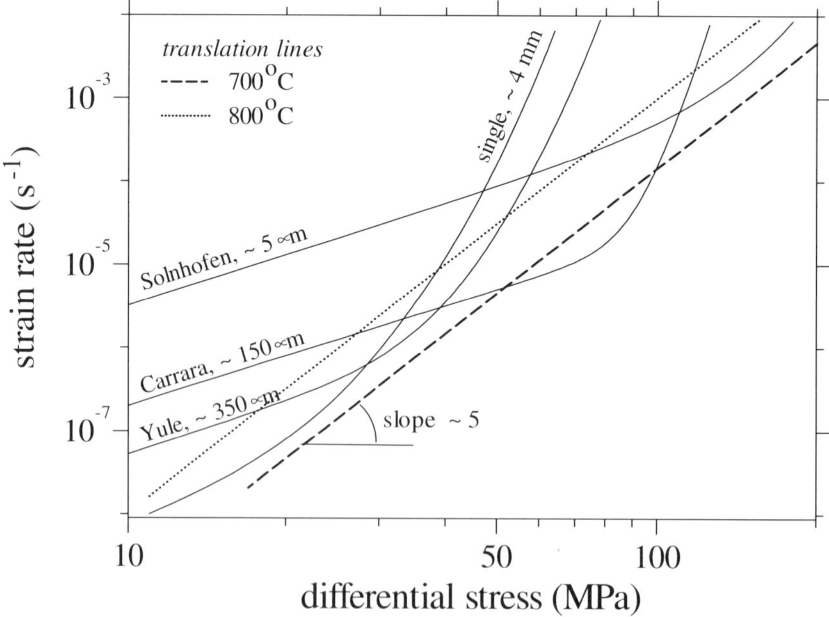

Fig. 4. Trends of strain rate versus stress for single and polycrystalline calcite rocks simplified and extrapolated based on Figure 1a. The effect of grain size changes from a diffusion creep relation at low stresses to a Hall-Petch relation at higher stresses. The individual curves may result from the translation of a master curve along the dashed line (Stone 1991). The translation line is related to the kinetics of the deformation process on the lattice scale. The state variable either is directly related to grain size or scales with it.

temperatures. Similarly, the oolitic limestone is weaker than Solnhofen limestone at low temperatures and high stresses, but stronger at higher temperatures and low stresses. The reverse is true for its strength in comparison to that of Carrara marble.

At lower temperatures, the yield stress of metals often exhibits an inverse dependence on the square root of grain size (Hall 1951). This dependence, called the Hall-Petch relation, could arise from interactions between dislocations piled-up at grain boundaries and is usually thought to occur when recovery processes are not rapid enough to relieve internal stresses. For marble tested near room temperature, the stress at which inelastic strain occurs also decreases with increasing $\sqrt{d}$ (Brace 1961; Olsson 1974; Wong *et al.* 1996). Because the 'yield strength' under those conditions is pressure dependent, and because volumetric-strain measurements at room temperature indicate dilation, it is probable that deformation in this regime involves the cataclastic propagation of initial flaws that scale with the grain size (Fredrich *et al.* 1990).

Both mechanisms for Hall-Petch type relations, the first involving dislocation pile-ups, and the second involving crack extension, would be expected to operate at fairly low temperatures. Thus, it is surprising to encounter grain-size dependence of strength in marble deformed at temperatures and pressures near to the transition from dislocation creep to diffusion flow. However, such grain-boundary strengthening of steady-state creep deformation has been envisaged if, for example, dislocation sources in the grain boundaries exhibit a critical stress for activation independent of grain size (Hirth 1972). It seems that, similar to the requirements for metallic alloys (Lasalmonie & Strudel 1986), describing the constitutive behaviour for calcite under these conditions also needs to include an explicit state variable that scales with grain size.

Alternative steady-state flow laws for dislocation creep

Below, we consider three classes of constitutive equations for dislocation-controlled creep that provide descriptions alternative to the power law (Eq. 1), including jogged-screw dislocation models, cross-slip models, and dislocation-glide models. Despite their differing micromechanical motivations, they are formally similar; each involves an exponential stress relation that predicts increasing stress sensitivity of strain rate with increasing stress (see Fig. 2 and Table 3). However, none of them includes a grain-size dependence nor do they include an internal state variable. The screw dislocation model may be written (Jonas 1969):

$$\dot{\varepsilon}_{ex} = A_{ex} \exp\left(\frac{\sigma}{B_{ex}}\right) \exp\left(-\frac{Q}{RT}\right)$$

$$\text{with } B_{ex} = \text{const or} B_{ex} = \frac{RT}{V} \tag{5}$$

where V denotes the activation volume, and A_{ex} is a pre-exponential factor that may be weakly temperature dependent. Descriptions for creep controlled by cross-slip typically have coupled temperature and stress dependencies (e.g. de Bresser & Spiers 1993):

$$\dot{\varepsilon}_{cs-c} = A_{cs-c}\sigma^{p} \exp\left\{-\frac{Q_{cs}}{RT}[\ln(\sigma_0/\mu_0) + \ln \mu]\right\}$$

$$\text{with } p = 2 + \frac{Q_{cs}}{RT} \tag{6}$$

and

$$\dot{\varepsilon}_{cs-d} = A_{cs-d}\sigma^2 \exp\left\{-\frac{Q_{cs}}{RT}(1 - b_{cs-d}\sigma)\right\} \tag{7}$$

where Equation (6) describes control by constriction before a cross slip event and Equation (7) describes control by dissociation on the new glide plane, respectively. The parameters, μ and μ_0, denote the shear modulus at temperature T and at absolute zero, respectively; σ_0 is the flow stress at absolute zero; A_{cs-c}, and A_{cs-d} are pre-exponential factors for constriction control and dissociation control respectively. Q is an activation energy related to some diffusion process; Q_{cs} is a thermal activation parameter that also depends on the elastic properties of the material, the Burgers vector and the stacking fault energy of the dislocations. Note that the exponential law (Eq. 5) and the laws for cross-slip controlled by dissociation (Eqs 6, 7) are indistinguishable in the high-stress limit. The constriction case predicts that the stress sensitivity $n' = p$ decreases with increasing temperature (Table 3) in contradiction to observations (Fig. 2b).

A third alternative is to assume creep to be controlled by dislocation glide against a lattice resistance (Frost & Ashby 1982). The central parameter of such a glide model is the stress necessary to initiate glide without thermal activation, called the Peierls stress, σ_P:

$$\dot{\varepsilon}_{cs-P} = A_{cs-P}\sigma^2 \exp\left(\frac{\sigma}{\sigma_P}\right) \exp\left(-\frac{Q_P}{RT}\right) \tag{8}$$

Table 3. *Analytical stress sensitivities n' and temperature sensitivities Q'**

Flow law	Eq.	$n' \equiv \partial \ln \dot{\varepsilon} / \partial \ln \sigma \vert_{T,d}$	$\sigma_a \uparrow$	$T \uparrow$	$d \uparrow$	$Q' = -R \partial \ln \dot{\varepsilon} / \partial (1/T) \vert_{\sigma,d}$	$\sigma_a \uparrow$	$T \uparrow$	$d \uparrow$
power law	(1)	$n \frac{\sigma_a - q\sigma_i}{\sigma}$	$\downarrow^{\dagger}$	$\downarrow^{\ddagger}$	$\downarrow$	$Q_{pl}\left(1 + n \frac{\sigma_i Q_k}{\sigma Q_{pl}}\right)$	$\downarrow$	$\downarrow$	$\downarrow$
exponential law	(5)								
$B_{ex} = const$		$\frac{\sigma_a - q\sigma_i}{B_{ex}}$	$\uparrow$	$\uparrow^{\ddagger}$	$\uparrow$	$Q_{ex}\left(1 + \frac{\sigma_i Q_k}{B_{ex} Q_{ex}}\right)$	$\uparrow$	$\downarrow$	$\downarrow$
$B_{ex} = RT/V$		$\frac{V(\sigma_a - q\sigma_i)}{RT}$	$\uparrow$	$\downarrow\uparrow$	$\uparrow$	$Q_{ex}\left(1 + \frac{\sigma_i V Q_k}{RT Q_{ex}}\right) - V\sigma$	$\downarrow\uparrow$	$\downarrow$	$\downarrow$
cross-slip									
constriction	(6)	$\left(2 + \frac{Q_{cs}}{RT}\right) \frac{\sigma_a - q\sigma_i}{\sigma}$	$\downarrow^{\dagger}$	$\downarrow$	$\downarrow$	$Q_{cs}\left[\ln \frac{\sigma_0}{\mu_0} - \ln \frac{\sigma}{\mu} + \frac{Q_k}{Q_{cs}}\left(2 + \frac{Q_{cs}}{RT}\right) \frac{\sigma_i}{\sigma}\right]$	$\downarrow$	$\downarrow$	$\downarrow$
dissociation	(7)	$2 \frac{\sigma_a - q\sigma_i}{\sigma} + \frac{Q_{cs} b_{cs-d}}{RT} (\sigma_a - q\sigma_i)$	$\downarrow\uparrow$	$\downarrow\uparrow$	$\downarrow\uparrow$	$Q_{cs}\left[2 \frac{Q_k}{Q_{cs}} \frac{\sigma_i}{\sigma} + b_{cs-d}\left(\sigma_i \frac{Q_k}{RT} - \sigma\right) + 1\right]$	$\downarrow\uparrow$	$\downarrow$	$\downarrow\uparrow$
Peierls law	(8)	$\left(2 + \frac{\sigma}{\sigma_P}\right) \frac{\sigma_a - q\sigma_i}{\sigma}$	$\uparrow$	$\uparrow$	$\uparrow$	$Q_P - Q_k\left(\frac{\sigma_i}{\sigma} + \frac{\sigma_i}{\sigma_P}\right)$	$\downarrow\uparrow$	$\downarrow\uparrow$	$\downarrow\uparrow$
$\sigma_P = (\Sigma_{P,0} + K_d d^{-m}) \cdot (T_m - T)$		$2 + \frac{\sigma}{\sigma_P}$	$\uparrow$	$\uparrow$	$\uparrow$	$Q_P + \frac{\sigma}{\sigma_P} \frac{RT^2}{T_m - T}$	$\uparrow$	$\uparrow$	$\uparrow$
		experimental observations (Fig. 2)	$\uparrow$	$(\uparrow)$	$\uparrow$	experimental observations (Fig. 3)	$\uparrow$	$\uparrow$	$\uparrow$

*In the steady state relations we replaced stress by $\sigma = \sigma_a - \sigma_i = \sigma_a - \sigma_a^q k_0 \exp(Q_k/RT) d^{-m}$. For the Peierls law we also investigated a grain size and temperature dependence of the Peierls term. Arrows indicate the response of each parameter to an increase in applied stress, temperature, and grain size.

†For $q \neq 1$, otherwise independent of applied stress.

‡For $Q_k \neq 0$, otherwise independent of temperature.

Equation (8) is formally equivalent to the cross-slip equation for the dissociation case with

$$b_{cs-d} = \frac{RT}{Q_{cs}\sigma_P}$$

whereas in the high-stress limit $B_{ex} = \sigma_P$.

Internal stresses and grain size

The motion of mobile dislocations in polycrystalline aggregates can be opposed by an internal stress field resulting from the presence of other dislocations, subgrain boundaries, and dislocation tangles, or by interactions of the mobile dislocations with grain boundaries, precipitates, and other obstacles (see for example Stouffer & Dame 1996). For metals, this internal stress or backstress, σ_i, is often large enough to reduce the local force acting on a particular dislocation to values substantially below that owing to the applied stress, σ_a, if it were acting alone. Variations in the internal stresses can cause deviations from the direct proportionality between the applied stress and the square of the dislocation density (Mecking & Kocks 1981) and can alter the force necessary for the dislocations to overcome discrete obstacles or lattice frictional forces (Davies *et al.* 1973; Kocks *et al.* 1975; Gibeling & Nix 1980; Li & Langdon 1999). Failure to consider the backstress may yield values of n' that are unexpectedly high or that change with stress. Similarly, apparent activation energies may appear to change with temperature (Jonas 1969; Solomon & Nix 1970; Dennison *et al.* 1978). In the following, we explicitly denote the applied stress as σ_a. Then, the average local stress is $\sigma_a - \sigma_i$. Assuming that the average local stress is the important parameter causing dislocation motion, we can replace σ by $\sigma_a - \sigma_i$ in the power law (Eq. 1). This substitution leads to stress exponents that decrease with increasing stress as long as the internal stress is positive and constant or positively correlated with the applied stress (Table 3).

For calcite, de Bresser (1996) posited that dislocation densities are not determined by stress alone, but that they vary with grain size at a given stress. He suggested that total dislocation density was given by $\sigma = a\rho_t^{0.5} - bd^{-0.5}$, a relation that seems opposite to the Hall-Petch effect. However, de Bresser (1991) argued that the total dislocation density is comprised of geometrically necessary and statistically stored dislocations, such that $\rho_t = \rho_g + \rho_s$. After a few percent of straining, the geometrically necessary dislocations are assumed to have a constant density, independent of stress, but inversely proportional to grain size, i.e. $\rho_g \propto d^{-1}$. If the total dislocation density is a function of stress then the statistically stored dislocations that accommodate deformation, depend on stress and grain size in a complex manner (de Bresser 1991):

$$\rho_s = c_1\sigma^2 + c_2\sigma d^{-0.5} - c_3 d^{-1}. \quad (9)$$

The cross-slip equations (Eqs 6 and 7) can be modified by incorporating this relation:

$$\dot{\varepsilon}_{cs-cg} = (A_{cs-cg}\sigma^2 + B_{cs-cg}\sigma d^{-0.5} - C_{cs-cg}d^{-1}) \times \sigma^{p-2}\exp\left\{-\frac{Q_{cs}[\ln(\sigma_0/\mu_0) + \ln\mu]}{RT}\right\} \quad (10)$$

$$\dot{\varepsilon}_{cs-dg} = (A_{cs-dg}\sigma^2 + B_{ds-cg}\sigma d^{-0.5} - C_{ds-cg}d^{-1}) \times \exp\left\{-\frac{Q_{cs}(1 - b_{cs-dg}\sigma)}{RT}\right\} \quad (11)$$

where A_{cs}, B_{cs}, and C_{cs} are constants. The stress dependence is more complicated than the simple Hall-Petch relation, but if $C_{cs-cg} > B_{cs-cg}\sigma d^{0.5}/2$, strain rates increase with increasing grain size as required. Note that there will be restrictions on the constants, so that negative strain rates do not result from decreasing grain size.

More generally, specifying the internal stress as $\sigma_i = \sigma_a^q k d^{-m}$ with $k = k_0\exp(Q_k/RT)$ and exponents $0 \le q \le 1$ and $0.5 \le m \le 1$ we can account for the notion that internal stresses depend on applied stress (Argon & Takeuchi 1981; Argon 1996), temperature (Dennison *et al.* 1978; Dunand & Jansen 1997; Li & Langdon 1999), and grain size (Li & Chou 1970; Lasalmonie & Strudel 1986). When $\sigma = \sigma_a - \sigma_i$ is used in either the power law (Eq. 1), the exponential law (Eq. 5), or the constriction controlled cross-slip (Eq. 6), and if the internal stress is grain-size dependent as specified above, then Hall-Petch type relations between applied stress (strength) and grain size will result ($\sigma_a = S_0 + k_{HP}d^{-m}$). The cross-slip equation for the dissociation case (Eq. 7) and the Peierls law (Eq. 8) cannot be rearranged simply, but $\partial\ln\dot{\varepsilon}_{cs-d}/\partial d > 0$ as required for decreasing strength with increasing grain size. Analytical expressions for stress sensitivity, $n' \equiv \partial\ln\dot{\varepsilon}/\partial\ln\sigma$, and temperature sensitivity of strain rate, $Q' \equiv -R\partial\ln\dot{\varepsilon}/\partial(1/T)$, become complex when the flow laws are modified by an internal stress (see Table 3). Upon modification by an internal stress term, Equations 5 and 8 potentially satisfy most of the experimental observations, but other models with similar mathematical form might also do as well. Most of the equations fail to predict an increase in activation energy with increasing temperature. Also, the inferred dependencies of n' and Q' on grain size do not seem to be modelled correctly.

As another variation of an effect of internal stresses, we consider resistance against glide to be composed of the intrinsic Peierls stress and a backstress, for example generated by dislocation pileups in front of grain boundaries, such that the Peierls term in Equation (8) obeys $\sigma_P = \sigma_{P,0} + k_d d^{-m}$. Accounting for the expected proportionality between both the intrinsic Peierls stress and the magnitude of the pile-up contribution, and the shear modulus (Hirth & Lothe 1982; Li & Chou 1970), we introduce a temperature dependence such that $\sigma_P = (\Sigma_{P,0} + K_d d^{-m})(T_m - T)$. This temperature relation accounts for the physical limits of the shear modulus, which has a finite value at $T = 0$ and vanishes at the melting temperature T_m. Then Equation 8 predicts variations in the stress sensitivity of strain rate and activation energy that agree with the experimental data (Table 3, Figs 2 and 3).

Internal state or structure

The results of a micromechanical approach by Stone (1991) show intriguing resemblance to the observations summarized in Fig. 4. Stone's composite flow law is developed from constitutive equations for two dominant mechanisms on the lattice scale and incorporates statistical variations in subgrain size, D. In this model, athermal yielding follows a simplified Hall-Petch relation between local stress and subgrain size ($\sigma_{l,at} = A_{at}D^{-r}$). Thermally activated migration of subgrain walls is supposed to operate simultaneously obeying a diffusion flow law

$$\dot{\varepsilon}_{l,t} = A_t \exp\left(-\frac{Q_t}{RT}\right)\sigma_{l,t}^{\zeta} D^{-\psi}.$$

Local stresses, σ_l, peak at the critical subgrain size D_0, such that subgrains smaller than D_0 deform by the thermal mechanism, while larger subgrains deform by the athermal mechanism. Uniform strain rate among subgrains is assumed when summing the local stresses to gain the strength of the aggregate. The isostructural strain-rate sensitivity of strength, $\nu = 1/n'|_D$, for a constant average subgrain size (Stone 1991, his equation 11) increases from 0 to $1/\zeta$ with decreasing strain rate, a prediction that mimics the observed increase of the stress sensitivity with increasing stress. At given stress, the stress sensitivity ($1/\nu$) decreases with decreasing average subgrain size. Consequently, the temperature sensitivity of strain rate at constant structure, $Q' \propto \nu$, increases with stress and grain size.

The relations among apparent flow-law parameters and experimental conditions predicted by Stone (1991), are similar to those found for calcite aggregates, if subgrain size and grain size are simply correlated and if the deformation experiments qualify as 'isostructural'. The importance of the effect of structure and its evolution on stress sensitivity has been emphasized and investigated analytically by Jonas (1969). Dislocations interact with the elements of the initial structure, subgrain and grain boundaries. Eventually, structural steady state may be reached where the distributions of dislocation, subgrain boundaries, and grain boundaries attain a dynamic equilibrium, leading to piezometric relations. As discussed above, for calcite rocks relaxation, creep, and constant strain rate tests all yield the same dependence of stress sensitivity on stress, temperature, and grain size (Fig. 2), despite the fact that the strains characteristic of these methods differ substantially. This strain independence may indicate that, even though the dislocation density may adjust to the imposed stress state almost instantaneously (de Bresser 1996), the pertinent microstructure, e.g. subgrain and/or grain-size distribution does not significantly change over the range of investigated strains (cf. Stone 1991).

Conclusions

The calcite materials previously tested in laboratory experiments differ in structure and apparently maintain these differences through small-strain (<20%) deformation experiments. Representing the structural state of the various calcite materials by their crystal size successfully rationalizes their differences in strength. In detail, the stress and temperature sensitivities of strain rate during dislocation creep of calcite rocks and single crystals are systematically related to stress, temperature, and grain size. Unexpectedly, grain size is an important parameter, even when deformation is accommodated by dislocation processes. A conventional power-law equation cannot account for these observations. By including an internal stress or a complex lattice resistance for dislocation glide and an exponential stress dependence of strain rate, a constitutive law motivated by a single dislocation mechanism may be constructed that agrees with observations over a wide range in experimental conditions. There is also qualitative agreement between the experimental observations and another micromechanical approach that combines subgrain-size dependent hardening and diffusion creep processes accounting for statistical variations in the structure (Stone 1991).

On the basis of the previously published data from various calcite materials, we cannot favour either of these approaches to a constitutive equation. Singling out a particular dislocation mechanism has the merit of constraining an important endmember of the deformation processes operating in calcite materials. The state-variable approach (Stone 1991) is attractive due to its potential to cover the whole spectrum of high-temperature plasticity including diffusion creep by a single constitutive relation. Grain size may be only a poor substitute for the true state variable, in particular, if their correlation evolves during deformation. Besides subgrain size or dislocation density, impurity and second phase content are candidates. Initial grain size could be a result of variations in the latter parameters, which are known to affect grain growth kinetics. The structure or state variable does not seem to evolve to a steady-state value during the standard small-strain experiments, but it appears to change very slowly with strain. Future work has to consider the obvious changes in grain size associated with dynamic recrystallization during extensive straining. Many detailed mechanical tests and microstructure observations are necessary before the appropriate internal variable can be explicitly identified.

Funding was provided by NSF grants EAR9627863 and 9903307. J. Renner gratefully acknowledges support from the Alexander von Humboldt foundation. P. Chopra, S. Covey-Crump, M. Drury, and H. de Bresser provided insightful and constructive reviews that helped us to improve the manuscript.

References

ANDERSON, T. F. 1969. Self-diffusion of carbon and oxygen in calcite by isotope exchange with carbon dioxide. *Journal of Geophysical Research*, **74**, 3918–3932.

ARGON, A. S. 1975. Physical basis of constitutive equations for inelastic deformation. *In*: ARGON, A. S. (ed) *Constitutive Equations in Plasticity*. M.I.T. Press, Cambridge, MA., 1–22.

ARGON, A. S. 1996. Mechanical properties of single phase crystalline media: Deformation in the presence of diffusion. *In*: CAHN, R. W. & HAASEN, P. (eds) *Physical Metallurgy*. Elsevier, Amsterdam.

ARGON, A. S. & TAKEUCHI, S. 1981. Internal stresses in power-law creep. *Acta Metallurgica*, **29**, 1877–1884.

BRACE, W. F. 1961. Dependence of fracture strength of rocks on grain size. *Bulletin of the Pennsylvania State University Mineral Industries Experiment Station, (Proceedings of 4th US Symposium on Rock Mechanics)*, **76**, 99–103.

BRODIE, K. H. & RUTTER, E. H. 2000. Deformation mechanisms and rheology: why marble is weaker than quartzite. *Journal of the Geological Society, London*, **157**, 1093–1096.

BRUHN, D. F., OLGAARD, D. L. & DELL'ANGELO, L. N. 1999. Evidence for enhanced deformation in two-phase rocks: experiments on the rheology of calcite–anhydrite aggregates. *Journal of Geophysical Research*, **104**, 707–724.

CARTER, N. L. & TSENN, M. C. 1987 Flow properties of the continental lithosphere. *Tectonophysics*, **136**, 27–63.

COVEY-CRUMP, S. J. 1994. The application of Hart's state variable description of inelastic deformation to Carrara marble at $T < 450\,^{\circ}C$. *Journal of Geophysical Research*, **99**, 19793–19808.

COVEY-CRUMP, S. J. 1998. Evolution of mechanical state in Carrara Marble during deformation at 400 °C to 700 °C. *Journal of Geophysical Research*, **103**, 29781–29794.

DAVIES, P. W., NELMES, G., WILLIAMS, K. R. & WILSHIRE, B. 1973. Stress-change experiments during high-temperature creep of copper, iron, and zinc. *Metals Science Journal*, **7**, 87–92.

DE BRESSER, J. H. P. 1991. *Intracrystalline deformation of calcite*. PhD thesis. Rijksuniversiteit, Utrecht.

DE BRESSER, J. H. P. 1996. Steady state dislocation densities in experimentally deformed calcite materials; single crystals versus polycrystals. *Journal of Geophysical Research*, **101**, 22189–22201.

DE BRESSER, J. H. P. & SPIERS, C. J. 1990. High-temperature deformation of calcite single crystals by r^+ and f^+ slip. *In*: KNIPE, R. J. & RUTTER, E. H. (eds) *Deformation Mechanisms, Rheology and Tectonics*. Geological Society, London, Special Publications, **54**, 285–298.

DE BRESSER, J. H. P. & SPIERS C. J. 1993. Slip systems in calcite single crystals deformed at 300–800 °C. *Journal of Geophysical Research*, **98**, 6397–6409.

DE BRESSER, J. H. P. & SPIERS, C. J. 1997. Strength characteristics of the r, f, and c slip systems in calcite. *Tectonophysics*, **272**, 1–23.

DE BRESSER, J. H. P., EVANS, B. & RENNER, J. 2002. On estimating the strength of calcite under natural conditions. *In*: DE MEER, S., DRURY, M. R., DE BRESSER, J. H. P. & PENNECK, G. M. (eds) *Deformation Mechanisms, Rheology and Tectonics: Current Status and Future Perspectives*. Geological Society, London, Special Publications, **200**, 309–329.

DE BRESSER, J. H. P., TER HEEGE, J. H. & SPIERS, C. J. 2001. Grain size reduction by dynamic recrystallization: Can it result in major rheological weakening? *International Journal of Earth Sciences, (Geologisches Rundschau)*, **90**, 28–45.

DENNISON, J. P., HOLMES, P. D. & WILSHIRE, B. 1978. The creep and fracture behavior of the cast, nickel-based superalloy, IN100. *Materials Science and Engineering*, **33**, 35–47.

DUNAND, D. C. & JANSEN, A. M. 1997. Creep of metals containing high volume fractions of unshearable dispersoids – Part I. Modeling the effect of dislocation pile-ups upon the detachment threshold stress. *Acta Materialia*, **45**, 4569–4581.

EVANS, B. & KOHLSTEDT, D. L. 1995. Rheology of rocks. *In*: AHRENS, T. J. (ed) *Rock Physics & Phase Relations, Handbook of Physical Properties of Rocks*. American Geophysical Union, Washington, D. C., pp 148–165.

FARVER, J. R. 1994. Oxygen self-diffusion in calcite: Dependence on temperature and water fugacity. *Earth and Planetary Science Letters*, **121**, 575–587.

FARVER, J. R. & YUND, R. A. 1996. Volume and grain boundary diffusion of calcium in natural and hot-pressed calcite aggregates. *Contributions to Mineralogy and Petrology*, **123**, 77–91.

FARVER, J. R. & YUND R. A. 1998. Oxygen grain boundary diffusion in natural and hot-pressed calcite aggregates. *Earth and Planetary Science Letters*, **161**, 189–200.

FISLER, D. K. & CYGAN R. T. 1999. Diffusion of Ca and Mg in calcite. *American Mineralologist*, **84**, 1392–1399.

FREDRICH, J. T., EVANS, B. & WONG, T. F. 1990. Effect of grain size on brittle and semibrittle strength: Implications for micromechanical modelling of failure in compression. *Journal of Geophysical Research*, **95**, 10907–10920.

FROST, H. J. & ASHBY, M. F. 1982. *Deformation-Mechanism Maps: The Plasticity and Creep of Metals and Ceramics*. Pergamon, Tarrytown, N.Y.

GIBELING, J. C. & NIX, W. D. 1980. A numerical study of long range internal stresses associated with subgrain boundaries. *Acta Metallurgica*, **28**, 1743–1752.

HALL, E. O. 1951. *The Deformation and Aging of Mild Steel, III, Discussion of Results*. Proceeding of the Physical Society, London, **64B**, 747–753.

HART, E. W. 1970. A phenomenological theory for plastic deformation of polycrystalline materials. *Acta Metallurgica*, **18**, 599–610.

HART, E. W. 1976. Constitutive equations for the non-elastic deformation of metals. *Journal of Engineering Materials Technology*, **98**, 193–202.

HEARD, H. C. 1963. Effect of large changes in strain rate in the experimental deformation of Yule Marble. *Journal of Geology*, **71**, 162–195.

HEARD, H. C. & RALEIGH, C. B. 1972. Steady-state Flow in Marble at 500° to 800 °C. *Geological Society of America Bulletin*, **83**, 935–956.

HIRTH, J. P. 1972. The influence of grain boundaries on mechanical properties. *Metallurgical Transactions*, **3**, 3047–3067.

HIRTH, J. P. & LOTHE, J. 1982. *Theory of Dislocations*. Krieger, Malabar, Fl.

JONAS, J. J. 1969. The back stress in high temperature deformation. *Acta Metallurgica*, **17**, 397–405.

KOCKS, U. F., ARGON, A. S. & ASHBY, M. F. 1975. Thermodynamics and kinetics of slip. *In*: CHALMERS, B., CHRISTIAN, J. W. & MASSALSKI, T. B. (eds) *Progress in Materials Sciences*, Vol 19. Pergamon Press, Oxford.

KOHLSTEDT, D. L., EVANS, B. & MACKWELL, S. M. 1995. Strength of the lithosphere: Constraints imposed by laboratory experiments. *Journal of Geophysical Research*, **100**, 17587–17603.

KRONENBERG, A. K., YUND, R. A. & GILETTI, B. J. 1984. Carbon and oxygen diffusion in calcite: effects of Mn content and P_{H_2O}. *Physics and Chemistry of Minerals*, **11**, 101–112.

LABOTKA, T. C., COLE, D. R. & RICIPUTI, L. R. 2000. Diffusion of C and O in calcite at 100 MPa. *American Mineralogist*, **85**, 488–494.

LASALMONIE, A. & STRUDEL, J. L. 1986. Influence of grain size on the mechanical behaviour of some high strength materials. *Journal of Material Science*, **21**, 1837–1852.

LEBENSOHN, R. A., WENK. H.-R. & TOME, C. N. 1998. Modelling deformation and recrystallization textures in calcite. *Acta Materialia*, **46**, 2683–2693.

LI, J. C. M. & CHOU, Y. T. 1970. The role of dislocations in the flow stress grain size relationships. *Metallurgical Transactions*, **1**, 1145–1159.

LI, Y. & LANGDON, T. G. 1999. A unified interpretation of threshold stresses in the creep and high strain rate superplasticity of metal matrix composites. *Acta Materialia*, **47**, 3395–3403.

LIU, M. & EVANS, B. 1997. Dislocation recovery kinetics in single-crystal calcite. *Journal of Geophysical Research*, **102**, 24801–24809.

MECKING, H. & KOCKS, U. F. 1981. Kinetics of flow and strain-hardening. *Acta Metallurgica*, **29**, 1865–1875.

MIRWALD, P. W. 1979. Electrical conductivity of calcite between 300 and 1200 °C at a CO_2 pressure of 40 bars. *Physics and Chemistry of Minerals*, **4**, 291–297.

OLSSON, W. A. 1974. Grain size dependence of yield stress in marble. *Journal Geophysical Research*, **79**, 4859–4862.

PATERSON, M. S. 1987. Problems in the extrapolation of laboratory rheological data. *Tectonophysics*, **133**, 33–43.

PIERI, M., BURLINI, L., KUNZE, K., STRETTON, I. & OLGAARD, D. L. 2001. Rheological and microstructural evolution of Carrara Marble with high shear strain: results from high temperature torsion experiments. *Journal of Structural Geology* **23**, 1393–1413.

POIRIER, J. P. 1978. Is power-law creep diffusion controlled? *Acta Metallurgica*, **27**, 401–403.

ROWE, K. J. & RUTTER, E. H. 1990. Palaeostress estimation using calcite twinning: experimental calibration and application to nature. *Journal of Structural Geology*, **12**, 1–17.

RUTTER, E. H. 1974. The influence of temperature, strain rate and interstitial water in the experimental deformation of calcite rocks. *Tectonophysics*, **22**, 311–334.

SCHMID, S. M. 1976. Rheological evidence for changes in the deformation mechanism of Solnhofen limestone towards low stress. *Tectonophysics*, **31**, T21–T28.

SCHMID, S. M. & PATERSON, M. S. 1977. Strain analysis in an experimentally deformed oolitic limestone. *In*: SAXENA, S. K. & BHATTACHARAJI, S. (eds) *Energetics of Geological Processes*. Springer, NY, 67–93.

SCHMID, S. M., BOLAND, J. N. & PATERSON, M. S. 1977. Superplastic flow in finegrained limestone. *Tectonophysics*, **43**, 257–291.

SCHMID, S. M., PATERSON M. S. & BOLAND, J. N. 1980. High temperature flow and dynamic recrystallization in Carrara marble. *Tectonophysics*, **65**, 245–280.

SOLOMON, A. A. & NIX, W. D. 1970. Interpretation of high temperature plastic deformation in terms of measured effective stresses. *Acta Metallurgica*, **18**, 863–876.

STONE, D. S. 1991. Scaling laws in dislocation creep. *Acta Metallurgica et Materialia*, **39**, 599–608.

STOUFFER, D. D. & DAME, L. T. 1996. *Inelastic Deformation of Metals: Models, Mechanical Properties and Metallurgy*. John Wiley & Sons, New York.

TAKESHITA, T., TOME, C., WENK, H. R. & KOCKS, U. F. 1987. Single-crystal yield surface for trigonal lattices; application to texture transitions in calcite polycrystals. *Journal of Geophysical Research*, **92**, 12917–12930.

TOME, C. N., WENK, H. R., CANOVA, G. R. & KOCKS, U. F. 1991. Simulations of texture development in calcite; comparison of polycrystal plasticity theories. *Journal of Geophysical Research*, **96**, 11865–11875.

TURNER, F. J., GRIGGS, D. T. & HEARD, H. C. 1954. Experimental deformation of calcite crystals. *Geological Society of America Bulletin*, **65**, 883–933.

WALKER, A. N., RUTTER, E. H. & BRODIE, K. H. 1990. Experimental study of grain-size sensitive flow of synthetic, hot-pressed calcite rocks. *In*: KNIPE, R. J. & RUTTER, E. H. (eds) *Deformation Mechanisms, Rheology and Tectonics*, Geological Society, London, Special Publications, **54**, 259–284.

WANG, J. N. 1994. The effect of grain size distribution on the rheological behavior of polycrystalline materials. *Journal of Structural Geology*, **16**, 961–970.

WANG, Z.-C., BAI, Q., DRESEN, G., WIRTH, R. & EVANS, B. 1996. High temperature deformation of calcite single crystals. *Journal of Geophysical Research*, **101**, 377–20, 390.

WONG, R. H. C., CHAU, K. T. & WANG, P. 1996. Microcracking and grain size effect in Yuen Long Marbles. *International Journal of Rock Mechanics and Mining Sciences & Geomechanics Abstracts, 1996*, **33**, 479–485.

On estimating the strength of calcite rocks under natural conditions

J. H. P. DE BRESSER[1], B. EVANS[2] & J. RENNER[2,3]

[1] *HPT-laboratory, Faculty of Earth Sciences, Utrecht University, P.O. Box 80.021, 3508 TA Utrecht, the Netherlands (e-mail: J.H.P.deBresser@geo.uu.nl)*
[2] *Department of Earth, Atmospheric and Planetary Sciences, Massachusetts Institute of Technology, Cambridge, MA 02139, USA*
[3] *Present address: Institute for Geology, Mineralogy, and Geophysics, Ruhr-Universität, D-44780 Bochum, Germany*

Abstract: Field studies of calcite mylonites often document microstructures produced by dislocation creep. In contrast, flow laws derived from experiments predict that calcite rocks should deform mostly by diffusion creep during tectonic processes. To investigate this apparent discrepancy, we compare stresses estimated by microstructural piezometers to those obtained by extrapolation of experimentally derived flow laws. Considering shear zones from different geological settings, a clear trend is observed of increasing recrystallized grain size with increasing temperature. However, there is a large spread in grain size and associated stress. Because separate flow laws have been defined for various different marbles and limestones, the strengths predicted for a given set of conditions differ significantly. The stress estimates based on the piezometers and strength extrapolated from the various experimentally derived dislocation creep flow laws agree qualitatively, but no single flow law predicts all the palaeostress estimates. Even if experimental data are disregarded, the field observations are not consistent with a hypothetical law for Coble creep; they are consistent with a power law for dislocation creep, but only if the material constants are different from those currently determined in laboratory experiments.

Calcite is one of the best-studied geological materials. Owing to their importance in the tectonic evolution of mountain ranges, the microstructure of many calcite rocks have been investigated, including those deformed at low metamorphic grades (Burkhard 1990; Kennedy & Logan 1997; Pfiffner 1982) and at intermediate to high-grade conditions (Behrmann 1983; Busch & van der Pluijm 1995; Heitzmann 1987; Bestmann *et al.* 2000; Molli *et al.* 2000; Ulrich *et al.* 2002). The stress under which these rocks were deformed can be estimated by a piezometric relation between applied stress and, for example, the size of the dynamically recrystallized grains produced (Twiss 1977; Drury & Urai 1990).

The strength of rocks can also be estimated by identifying the mechanism of deformation, then placing constraints on strain rate, temperature and pressure, and finally, applying a constitutive law for the appropriate mechanism. Strain is often easily measured in field studies and can be combined with absolute or relative age dating to estimate average strain rates. For crustal deformation zones, rates on the order of 10^{-12}–10^{-15} s^{-1} have been suggested (e.g. Pfiffner & Ramsay 1982). If geothermometry and barometry are used to constrain the thermodynamic conditions, then the strength of naturally deformed rocks can be estimated by extrapolating a particular flow law.

How do the estimates of palaeostress and strength compare? Are they consistent? To shed light on these questions, we compare estimates from the microstructures in several shear zones with strength predictions from most of the current constitutive relations. Although there is qualitative agreement between the two methods, there are also some puzzling inconsistencies.

Shear zones and dynamic recrystallization

Rocks from shear zones appear to be a natural choice for examination because the large strains involved may lead to diagnostic 'steady state' microstructures. Besides the assumption (1) that strain increases significantly towards the centre of the shear zone, it is also usually implicitly or explicitly assumed (2) that the rock directly outside the shear zone, i.e. the protolith, represents the shear zone material in its undeformed state and (3) that progressive localization is caused by strain weakening of the material in the shear zone. For this discussion the assumption that the grain size developed within the shear zone records the stress conditions at some point in time during rock deformation is also important.

Each of the above assumptions needs to be examined in each shear zone. For example, a

From: DE MEER, S., DRURY, M. R., DE BRESSER, J. H. P. & PENNOCK, G. M. (eds) 2002. *Deformation Mechanisms, Rheology and Tectonics: Current Status and Future Perspectives*. Geological Society, London, Special Publications, **200**, 309–329. 0305-8719/02/$15 © The Geological Society of London.

Table 1. *Results of field studies on naturally deformed calcite rocks as presented in the literature*[*]

Location	Reference	Rock classification[†]	Type[‡]	Deformation temp. [°C]	Grain size [μm]	Recrystallization mechanism§	Deformation mechanism‖	Stress¶ [MPa]
Bancroft shear zone, Canada	Busch & Van der Pluijm (1995); Van der Pluijm (1991)	Protomylonite	A	475 (±50)	47 (±16)[I,a]	Rot	Disloc	27 (±8)
		Protomylonite	A	475 (±50)	67 (±29)[I,a]	Rot	Disloc + GSS	20 (±6)
		Coarse mylonite	B	475 (±50)	51 (±18)[I,a]	Rot	Disloc	26 (±7)
		SC-mylonite	B	475 (±50)	38 (±13)[I,a]	Rot (+Migr?)	Disloc	33 (±9)
		Ultramylonite	C	475 (±50)	30 (±9)[I,a]	Rot	Disloc	41 (±12)
		Fine ultramylonite	C	475 (±50)	21 (±5)[I,a]	Rot	Disloc	56 (±12)
Alpi Apuane, Carrara, Italy	Molli *et al.* (2000)	High strain, type B1	B	370 (±25)	150–200[II]	Migr.	Disloc	24–31
		High strain, type B2	B	340 (±25)	40–50[II]	Rot (+Migr?)	Disloc	26–32
Glarus overthrust, Switzerland	Pfiffner (1982)	Footwall	A	350 (±50)	7.5 (±2.5)[III,b]	Rot + Migr	Disloc	138–448
							$\rho = 5.7 \times 10^{13}\ m^{-2}$	208 (±49)
		Mylonite	B	350 (±50)	8.0 (±2.5)[III,b]	Rot + Migr	Disloc + GSS	130–423
	Briegel & Goetze (1978)						$\rho = 4.4 \times 10^{13}\ m^{-2}$	143 (±39)
Infrahelvetic, Switzerland	Pfiffner (1982)	Slight/moderat. deformed	A	350 (±50)	3.7 (±0.4)[III,b]	Rot + Migr	Disloc	257–840
							$\rho = 5.7 \times 10^{13}\ m^{-2}$	208 (±49)
		Highly deformed limestone	B	350 (±50)	9.8 (±2.0)[III,b]	Rot + Migr	Disloc + GSS	109–353
							$\rho = 5.4 \times 10^{13}\ m^{-2}$	201 (±47)
Thassos, Greece	Bestmann *et al.* (2000)	Protomylonite	A	600 (±20)	1000–3000[II]	Migr	Disloc	2–6
		Protomylonite	A	325 (±25)	10–50[II]	Rot	Disloc	32–58
		Ultramylonite	C	325 (±50)	14[II]	Rot	Disloc	80 (±22)
Sierra Alhamilla, Spain	Behrmann (1983)	Fine grained marble	A	300	52[V]	Rot + Migr	Disloc	22–71
		Fine grained marble	A	300	45[V]	Rot + Migr	Disloc + GSS	29–91
		Calcite-mylonite	B	300	26[V]	Rot + Migr	Disloc	46–148
		Calcite-mylonite	B	300	19[V]	Rot + Migr	Disloc	61–196
		Calcite-mylonite	B	300	15[V]	Rot + Migr	Disloc	75–242
		Calcite-dolomite mylonite	B	300	10[V]	Rot + Migr	Disloc	107–347
Alpine 'root' zone, Switzerland	Heitzmann (1987)	High grade marble	A	675 (±25)	500–1000[IV]	Migr	Disloc	6–11
		Protomylonite	A	300	20–50[IV]	Rot + Migr	Disloc	36–114
		Mylonite	B	300	14–28[IV]	Rot + Migr	Disloc	54–172
		Ultramylonite	C	300	5–10[IV]	Rot + Migr	GSS	138—448[**]
Doldenhorn nappe, Switzerland	Burkhard (1990)	Mylonite	B	330	7.9 (±4.2)[V,a]	Rot + Migr	Disloc + GSS?	132–428
		Mylonite	B	340	7.6 (±4.1)[V,a]	Rot + Migr	Disloc + GSS?	136–443
Heidnischbiel		Mylonite	B	350	69 (±48)[V,a]	Rot + Migr	Disloc + GSS?	20–62
Diablerets nappe		Thrust mylonite	B	300	4.3 (±2.0)[V,a]	Rot + Migr	Disloc? + GSS	225–735
Gellihorn nappe		Hinge of large fold	A	180	3.0 (±1.5)[V,a]	Rot + Migr	Disloc? + GSS	309–1012
Retzligl		Fault mylonite	B	160	1[V]	Rot	Disloc	813 (±227)
Tutt shear zone, Wales, UK	De Bresser (1991)	Mylonite	B	200 (±25)	1.3 (±0.7)[V]	Rot + Migr	Disloc	645–2131
							$\rho = 2.5 \times 10^{13}\ m^{-2}$	125 (±25)

McConnell thrust, Canada	Kennedy & Logan (1997)	Mylonite	B	140 (±25)	1 [IV]	Rot + Migr	Disloc + brittle	813–2692
Bohemian Massif, CZ Parautochthon	Ulrich *et al.* (2002)	Type 1	A	275 (±25)	160 (±55) [VI,c]	Migr	Disloc	29 (±10)
Parautochthon		Type 3	C	275 (±25)	36 (±25) [VI,c]	Rot	GSS	36 (±10)**
Lower Nappe		Type 2	B	440 (±10)	365 (±170) [VI,c]	Rot + Migr	Disloc + GSS	5–14
Upper Nappe		Type 1	A	530 (±30)	685 (±380) [VI,c]	Migr	Disloc	8 (±3)
Upper Nappe		Type 2	B	530 (±30)	360 (±220) [VI,c]	Rot + Migr	Disloc + GSS	5–14
Upper Nappe		Type 3	C	530 (±30)	170 (±80) [VI,c]	Rot	GSS	9 (±2)**
Internal Zone		Type 1	A	625 (±15)	1085 (±620) [VI,c]	Migr	Disloc	5 (±2)
Internal Zone		Type 3	C	625 (±15)	100 (±45) [VI,c]	Rot	GSS	14 (±4)**
Naxos, Greece	Schenk (2000, MSc report)	Marble unit + pegmatite	A	700 (±25)	700—1000 [IV]	Rot	Disloc	2–3
Naxos, Greece††	Covey-Crump & Rutter (1989); Urai *et al.* (1990)	High grade marble	D	290–720	200—1800 [III]	Migr	Disloc	3–20

* The data and conclusions presented in the various studies were taken at face value; only if not presented by the authors themselves, we interpreted the mechanisms involved on the basis of the described microstructures.

† Nomenclature as used by the various authors.

‡ Classification this study.

§ Rot: rotation recrystallization; Migr: migration recrystallization.

|| Disloc: dislocation creep, with dislocation density ρ if measured; GSS: grain-size-sensitive (diffusion, grain boundary sliding) creep.

Grain measurement technique:

I The long dimension of each grain (caliper dimension).

II Diameter of equivalent circle (from grain area measurement using quantitative image analysis).

III Line intercept method.

IV Not explicitly stated.

V Arithmetic mean of long and short dimensions.

VI The Feret dimension (caliper dimension) corrected by $4/\pi$.

a Mean value and standard deviation of individual samples, b Average of multiple samples and range, c Average of multiple samples and average standard deviation.

¶ Palaeostress estimated using piezometric relations, errors calculated using standard errors of equations (see Appendix). Lower and upper bounds correspond to maximum and minimum grain size or rotation and migration recrystallization, respectively.

** Stress during recrystallization event, not necessarily corresponding to stress during GSS creep.

†† The increase in grain size with temperature observed in these rocks was originally attributed to normal grain growth (i.e. static annealing; Covey-Crump & Rutter 1989), but was reinterpreted as dynamic migration recrystallization by Urai *et al.* (1990), see also Rutter (1995).

number of natural shear zones are probably related to inherited layer-parallel heterogeneities, e.g. the Glarus overthrust, Switzerland (Pfiffner 1982), or the Hunter Valley thrust, USA (Kennedy & Logan 1998). At least for these shear zones, the assumption that the protolith represents the shear zone material in its undeformed state may not be valid. In these cases, flow laws derived from experiments on samples resembling the protolith, rather than the material composing the fault, are probably of limited applicability.

Much work has centered on relating strain localization in shear zones to some process for material weakening (for example, see discussions by Poirier 1980; White 1979*a*; White *et al.* 1985; Evans & Wong 1985; Hobbs *et al.* 1990; Rutter 1995, 1999). Candidate processes include: weakening resulting from a switch in deformation mechanism from dislocation creep to grain-size-sensitive diffusion creep; softening caused by the production of new grains free of internal strain via cyclic dynamic recrystallization (Tullis & Yund 1985; Urai *et al.* 1986; Rutter 1995; Peach *et al.* 2001); weakening resulting from development of a crystallographic lattice preferred orientation (e.g. Pieri *et al.* 2001); and thermal softening owing to shear heating. Before continuing, it is important to recall that material weakening is not actually a necessary prerequisite for the initiation of localization, especially if the deformation mechanisms involved do not conserve volume. Such mechanisms include those that are dilatant (e.g., brittle cracking), but also include solution transfer processes and diffusion mechanisms (Hobbs *et al.* 1990).

Naturally deformed calcite rocks almost universally show evidence of dislocation creep, using microstructural criteria such as shape preferred orientation (SPO), crystallographic or lattice preferred orientation (LPO), and undulatory extinction in cross-polarized light. Further, dynamic recrystallization, as suggested by sutured grain boundaries and/or progressive subgrain rotation, is common at all temperatures $T > 150\,^{\circ}\mathrm{C}$. In fact, many deformed calcite rocks show evidence for both rotation and migration recrystallization in the same sample.

We collected grain-size measurements given in the literature for calcite rocks in several shear

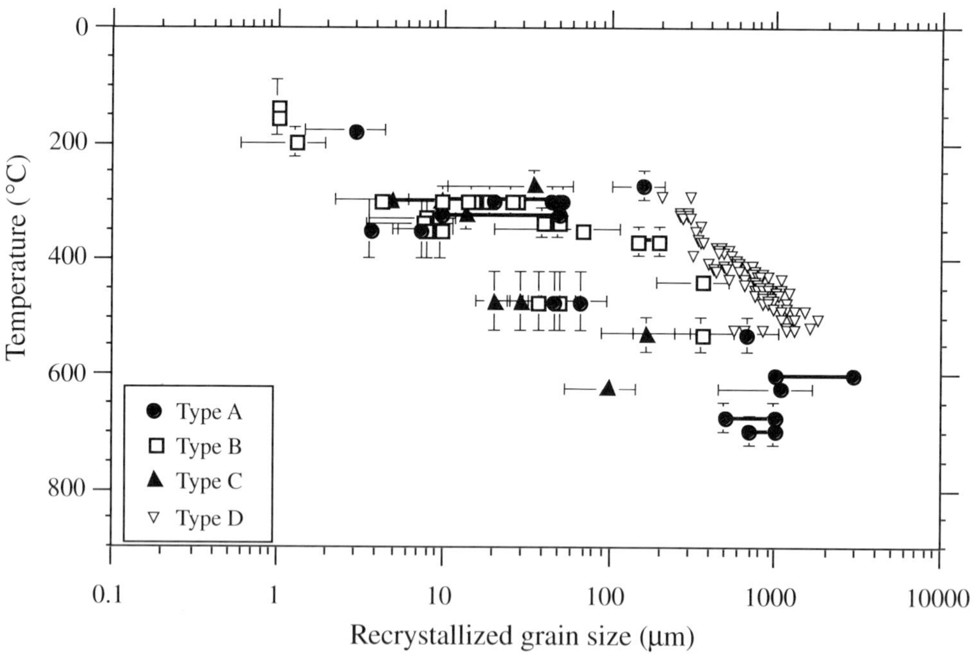

Fig. 1. Recrystallized grain size of some naturally deformed calcite rocks as a function of temperature (see Table 1). Type A, protomylonites (low/moderate strain); Type B, mylonites (high strain); Type C, ultramylonites; Type D, Naxos marbles. The grouping of data according to type A to D suggests a trend of decreasing recrystallized grain size from low-strain protomylonites towards the ultra-mylonites often found in the centre of a shear zone. Some controversy extends to the Naxos rocks; while Covey-Crump & Rutter (1989) asserted that the grain structure of those rocks has been established by static grain growth, Urai *et al.* (1990) suggested that the microstructure resulted from a deformation event.

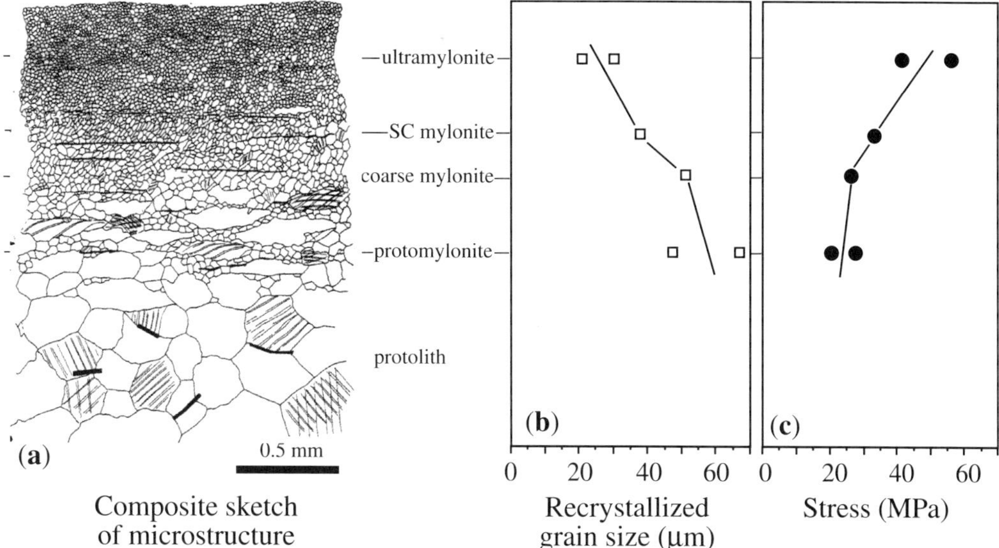

Fig. 2. Microstructure of the Bancroft shear zone, Ontario, Canada (see Table 1 and Van der Pluijm 1991). (**a**) Composite sketch of the microstructure (reprinted from *Journal of Structural Geology*, **13**, 1125–1135, with the kind permission of Elsevier Science). (**b**) Grain size as a function of position in the shear zone. (**c**) Estimated palaeostresses assuming recrystallization by rotation mechanism (Table 1), resulting in an apparent palaeostress recorded in the protomylonite (20 MPa) that is smaller than that recorded by the ultramylonite in the centre of the shear zone (55 MPa).

zones (Table 1). Data from a particular study were included if the study provided an estimate for the recrystallized grain size, constrained the deformation temperature, and presented evidence allowing an interpretation of the deformation and recrystallization mechanism (Table 1; Fig. 1). In most studies, a distinction was made between protomylonites from the boundary of the deformation zone and mylonites from the centre. A few studies used a systematic classification scheme based on the volume percentage of recrystallized (new) grains as compared to porphyroclasts (old grains) (see for example Passchier & Trouw, 1996; protomylonite, mylonite s.s., ultramylonite). To avoid conflicting terminology, we refer to all low/moderate strain shear zone rocks and protomylonites as 'type A', all mylonites and 'high-strain rocks' as 'type B', and the few ultramylonites as 'type C'. We also included data from a set of marbles from Naxos. The origin of the grain structure of the Naxos rocks is contested. Covey-Crump & Rutter (1989) posited that coarsening of the grain structure occurred by normal grain growth, hence these data may not be dynamically recrystallized at all, but may simply reflect the grain size achieved during a prolonged heat treatment. However, Urai *et al.* (1990) suggested that the microstructure of at least some of the Naxos rocks resulted from dynamic recrystallization during an orogenic event. We will evaluate this difference in interpretation, and we have plotted the data as 'type D' to clearly distinguish them from the shear zone data.

Within the set of field studies included in our analysis, the recrystallized grain size generally increases with the inferred temperature of deformation (Fig. 1). At a given temperature, the recrystallized grain size appears smaller if the rock is strained more (type B versus type A) and/or if more of the rock volume is recrystallized (type C versus type A). An example of this trend, from an individual shear zone, is given in Fig. 2. In this shear zone, the mean recrystallized grain size progressively decreases going from the protomylonite to the ultramylonitic part of the sample (after Van der Pluijm 1991; Busch & Van der Pluijm 1995).

Estimates of strength and palaeostresses

Palaeostresses

We applied two conventional methods of microstructural palaeo-piezometry, namely the recrystallized grain-size method and the dislocation density method (see Appendix). Calcite twinning piezometry was not used (Rowe & Rutter 1990; Laurent *et al.* 2000), since this method works well only for weakly deformed cover rocks, and is believed to substantially overestimate stresses in the highly strained samples from shear zones (Ferrill 1998). Whenever the recrystallization mechanism was identified, we used the appropriate piezometer, i.e. for migration or rotation recrystallization (see Appendix for details). When evidence was reported for both mechanisms, we calculated two stress values, and present a stress-connecting bar rather than a single datum (Fig. 3).

The use of a recrystallized grain-size piezometer is associated with a number of uncertainties that are both theoretical and practical (White 1979*b*; De Bresser *et al.* 2001; see also Appendix). In addition to the challenge of identifying the appropriate mechanism, there are questions about the influence of temperature, solid-solution impurities and second phases on the processes of dynamic recrystallization. One practical problem arises in the specification of grain size, especially when comparing one study with another. Several size parameters are commonly used, including mean intercept length, diameter of a circle of equivalent area, and Feret dimension, as well as others. In some cases the exact procedure used to obtain an average recrystallized grain size was not clear. Conversion from one scale factor to another is not completely straightforward, particularly if stereological corrections have been applied. Thus, in Table 1, we have tried to identify the general method used, but have not attempted to convert from one size measure to another. On this account, variability in average grain size of factors up to 3 might accrue.

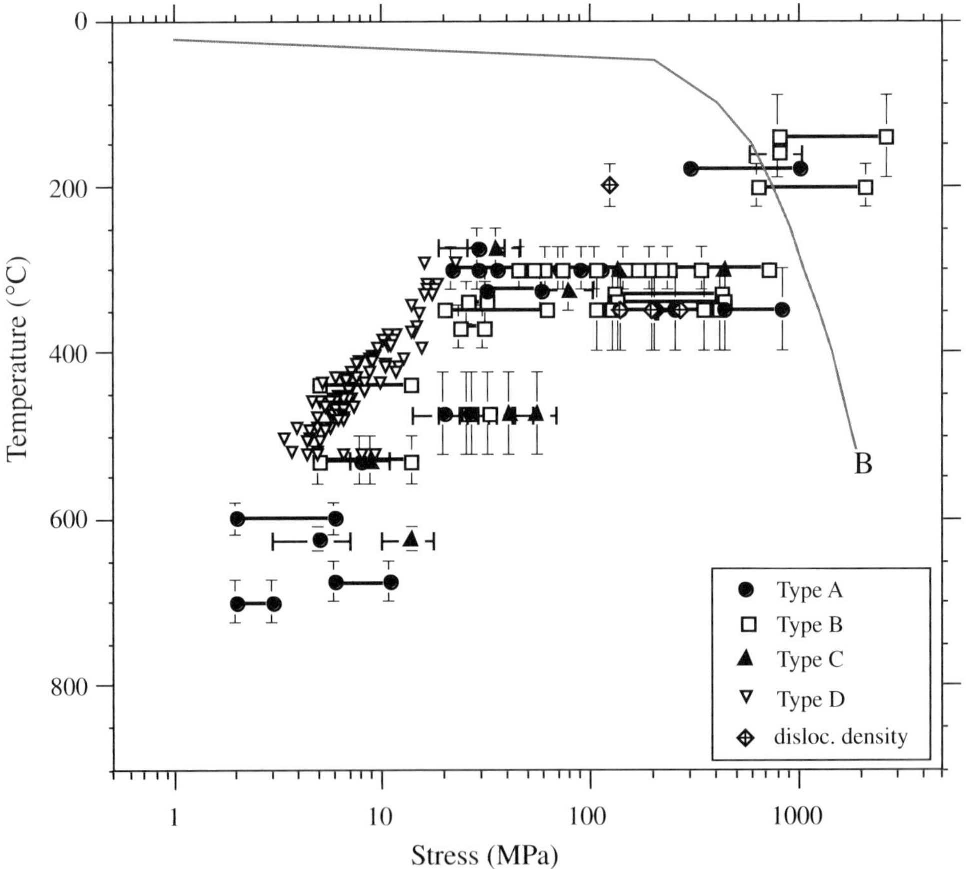

Fig. 3. Palaeostresses calculated from the recrystallized grain size data of Fig. 1 using the piezometers given in the Appendix. In cases where the authors specified a particular recrystallization mechanism we applied the appropriate piezometer. When rotation as well as migration recrystallization were observed we calculated both palaeostress values and represent them by means of a connecting bar. B refers to Byerlee's rule (1978) for frictional sliding (cf. Kohlstedt *et al.* 1995). Type A–D as in Fig. 1.

Stresses calculated from the recrystallized grain size increase with decreasing temperature, from about 2 MPa at $T = 700\,^{\circ}\mathrm{C}$, up to 500 MPa at $T = 300\,^{\circ}\mathrm{C}$, and above 1000 MPa at $T = 150\,^{\circ}\mathrm{C}$ (Fig. 3). The latter exceed strengths calculated from Byerlee's rule for frictional sliding. At $T = 300$–$350\,^{\circ}\mathrm{C}$, calculated palaeostress varies by more than one order of magnitude, even if one excepts the Naxos rocks (Type D). Low palaeostresses generally correlate with weakly deformed rocks (Type A, e.g. protomylonites) while higher stresses are found for rocks that presumably are highly strained (Type C, i.e. ultramylonites). On occasion the recrystallized grain size decreases towards the centre of a shear zone (see Table 1, Fig. 2 and discussion below). The data on dislocation densities are too sparse to derive a trend, but estimated values for stress roughly agree with values from the grain size piezometer.

Strengths extrapolated from experimental flow laws

Deformation tests have been carried out on a variety of calcite rocks, under a wide range of conditions (for a recent compilation see Renner & Evans 2002). At temperatures above about 500 °C, intracrystalline plastic mechanisms play an important role at laboratory strain rates. Work on single crystals has identified the relevant slip systems (Griggs 1938; Turner *et al.* 1954; Paterson & Turner 1970; Braillon & Serughetti 1976; De Bresser & Spiers 1997) and quantified their kinetics (De Bresser & Spiers 1990, 1993; Wang *et al.* 1996). Experiments on natural and synthetic polycrystalline aggregates with different grain sizes show similarities to the single crystal behavior at moderate temperatures and high stresses. However, grain boundary sliding or diffusive transport on grain boundaries dominates the deformation at high temperatures and/or low stresses (Schmid *et al.* 1977; Walker *et al.* 1990; see also Fig. 3 of Renner & Evans 2002). Flow laws derived from experiments on different calcite rocks show substantial variation (Table 2). The data seem robust and the variations are not experimental artifacts, but the exact causes of the variations are not known. Thus, there is reason to question the generality of the experimentally derived constitutive equations, but in the following, we will use these flow laws without reservations.

The experimentally derived dislocation creep laws (Table 2) were used to calculate strain rate contours at $10^{-12}\,\mathrm{s}^{-1}$ (Fig. 4). Stresses predicted by the various laws diverge at almost all temperatures and are as high as 1000 MPa at temperatures between 100 and 150 °C, at the chosen strain rate. At these low temperatures, only the flow law for Carrara marble from De Bresser (Table 2) predicts stresses below 200 MPa. Grain-size-sensitive (GSS) creep laws were determined for Solnhofen limestone (Schmid *et al.* 1977) and synthetic marble (Walker *et al.* 1990). The two materials are similar in grain size and porosity, but they are probably different in solid solution impurities, accessory mineral content, and perhaps other microstructural parameters as well. The two GSS laws (Table 2) agree reasonably well for aggregates with 1–10 μm grain size but deviate from each other at larger grain sizes. The discrepancy at large grain sizes owes to the difference in the grain size exponent, which, for Solnhofen limestone, is not well constrained.

Even assuming the form of the flow laws applies to natural deformation conditions, the flow law constants are, of course, experimentally derived and are to some degree uncertain. The variance in prediction increases as extrapolations in temperature and strain rate increase. For example, using the quoted uncertainties in the flow law parameters, the stress predicted by the flow law for Carrara marble (see Table 2) is uncertain by at factor of 2–4 at 300–400 °C and natural strain rates (see also Fig. 6).

Despite the discrepancies between the predictions of the flow laws in Table 2, one systematic result emerges. The flow laws predict natural deformation of calcite rocks to be dominated by GSS mechanisms for a large range in temperature (Fig. 4). Using the creep law by Walker *et al.* (1990), fine-grained calcite aggregates (~1 μm) should deform by GSS mechanisms at temperatures above 200 °C, and coarse-grained calcite aggregates (~1000 μm) at temperatures above 350–400 °C. In practice, these results imply that GSS mechanisms will dominate in the majority of natural shear zones, especially where grain size is small.

Comparison of inferred palaeostresses and extrapolated flow laws

A comparison of palaeostresses inferred from the record of natural rocks and strength derived from extrapolation of flow laws requires specification of at least two constitutive variables, namely strain rate and temperature. In the following, we first discuss the microstructure of rocks from a particular field example, the Sesia zone (Western Alps), to analyse stress–strain rate relations for natural deformation at a

Table 2. *Flow law parameters for calcite single crystals, limestones, and marbles*

Material; grain size*	Reference	ln(A) ($MPa^{-n}\ \mu m^{m}\ s^{-1}$)	n	Q (kJ/mol)	m	B (MPa)	Process†	Stress range (MPa)	Label in Figures
Single crystal P2	De Bresser & Spiers (1990),	−8.8	11.5	362	–	–	disl.	20–115	dBS90
7 ± 3 mm	De Bresser (1991)								
Single crystals	Wang *et al.* (1996)	–	3–4	200–400	–	–	disl.	<30	
4 ± 2 mm									
Yule marble, 1;	Heard (1963),	18.0	–	260	–	9.1	disl.	>110	HR72
350 ± 150 mm	Heard & Raleigh (1972)	−9.0	8.3	260	–	–		<110	
Yule marble, T;	Heard (1963),	16.1	–	239	–	13.8	disl.	>140	
350 ± 150 mm	Heard & Raleigh (1972)	−8.3	7.7	256	–	–		<140	
Carrara marble;	Schmid *et al.* (1980)	7.1	7.6	418	–	–	disl.	20–100	SPB80-1
150 ± 50 mm		18.4	4.2	427	–	–		<20	SPB80-2
Carrara marble;	Rutter (1974),	13.4	–	260	–	11.4	disl.	>100	R74
150 ± 50 mm	Covey-Crump (1998)								
Carrara marble;	De Bresser (in press)	19.5	2	423	–	‡	disl.	7–55	dB
150 ± 50 mm									
Carrara marble	Renner *et al.* (in prep.)	16.1	2	373	–	§	disl.	‖	RES
150 ± 50 mm									
Solnhofen	Schmid (1976),	−0.3	–	197	–	16.0	disl.	>190	
limestone;	Schmid *et al.* (1977)	7.8	4.7	298	–	–	disl.	<190	
5–10 µm		10.0 §	1.7	214	–	–	GSS	<100	SBP77
		(13.5, 15.3)			(2, 3)¶				
synthetic;	Walker *et al.* (1990)	4.6	3.3	190	1.3	–	trans.	<250	
2–40 µm		11.3	1.7	190	1.9	–	GSS	<25	WRB90

For a general flow law $\dot{\varepsilon} = A\sigma^n \exp(\sigma/B)\exp(-Q/RT)d^{-m}$: boils down to a conventional power law if B is not included.
* Sample size for single crystals.
† disl.: dislocation creep; GSS: grain-size-sensitive creep; trans.: transitional.
‡ $B = \dfrac{RT}{cQ}$ with $c = 3.8 \times 10^{-3}$.
§ $B = B_d(T_m - T)$ with $B_d = 14.0\,\mathrm{MPa\,K^{-1}}$ and $T_m = 1600\,\mathrm{K}$.
‖ All data by Schmid *et al.* (1980).
¶ Pre-exponential factor is valid for a grain size of 5.9 µm; the factors in parenthesis are calculated assuming $m = 2$ or 3 (see p. 43 and Fig. 3 of Schmid *et al.* (1977)).

Heard (1963) and Heard & Raleigh (1972) performed extension tests with the maximum compressive stress parallel (1-samples) and perpendicular (T-samples) to the natural shape preferred orientation of Yule marble. All other tests were performed in compression. For the limestone and the marbles each data point entering the flow law parameter calculation corresponds to an individual sample while Wang *et al.* (1996) and De Bresser (1991) combined stepping tests and testing of several samples from the same parent crystal.

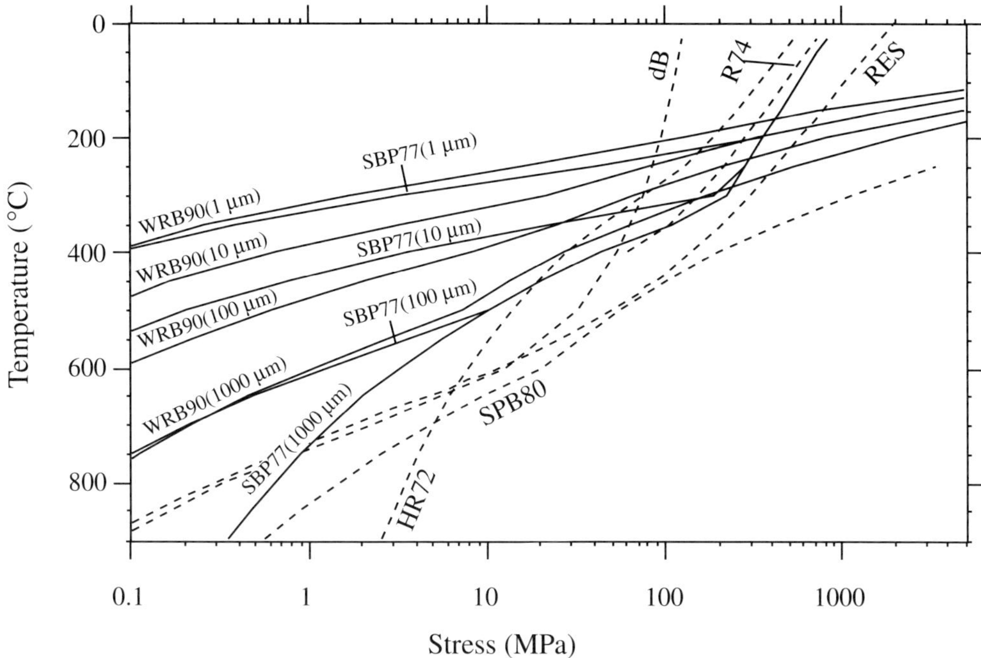

Fig. 4. Extrapolation of experimentally derived flow laws for various calcite aggregates (Table 2) to a strain rate of $10^{-12}\,s^{-1}$. Solid lines represent grain-size-sensitive (diffusion) creep laws, while dashed lines refer to dislocation creep laws. Grain-size-sensitive creep strengths are given for average grain sizes of 1, 10, 100 and 1000 μm.

given temperature. Then, we compare the entire set of palaeostresses derived from the field studies to flow laws extrapolated to a restricted range of strain rate. It is emphasized that recrystallized grain size and dislocation density piezometers are intrinsically related to dislocation creep. Nevertheless, it is possible that the deformation of a studied rock material has been accommodated by a combination of dislocation creep and a second mechanism that did not result in an obvious imprint on the microstructure. If this is the case, a comparison of palaeostresses determined applying a piezometer with iso-strain rate contours predicted by dislocation creep laws remains valid (within about one order of magnitude in strain rate) provided that both mechanisms contribute more or less equally to the bulk deformation. One candidate for the second mechanism is GSS (diffusion) creep. Thus, the palaeostresses are also compared to iso-strain rate contours predicted by flow laws for GSS creep.

The Sesia zone

The microstructure of a shear zone from the Sesia zone, Western Alps (Fig. 5) is used to compare the stresses estimated from the piezometers with the strength predicted by the various dislocation creep flow laws and the two GSS laws (Table 2), at a fixed temperature (Fig. 6). According to the GSS flow law for synthetic calcite rocks (Walker *et al.* 1990), the protolith with a grain size in the order of 1000 μm (Fig. 5a) can accommodate strain rates up to almost $10^{-13}\,s^{-1}$ by GSS flow before reaching the palaeostresses of roughly 100 MPa predicted by the piezometers to hold for the ultramylonite with a recrystallized grain size of $d \sim 20$ μm (Figs 5a and 6a). In contrast, combined with the absence of deformation features in the protolith, the GSS flow law for Solnhofen limestone (Schmid *et al.* 1977) postulates that the protolith was essentially static (strain rate as low as $10^{-17}\,s^{-1}$: see Fig. 6a). Grain size reduction by dynamic recrystallization owing to a switch to dislocation creep (caused by an increase in tectonic stress and strain rate) seems in good accord with the extrapolations of dislocation creep flow laws for Yule and Carrara marble (HR72 and dB in Fig. 6, respectively: see Table 2). Within the range of considered palaeostresses, the dislocation creep flow laws for Carrara marble by Schmid *et al.* (1980) predict strain rates that appear too slow to achieve the large strain associated with the shear zone.

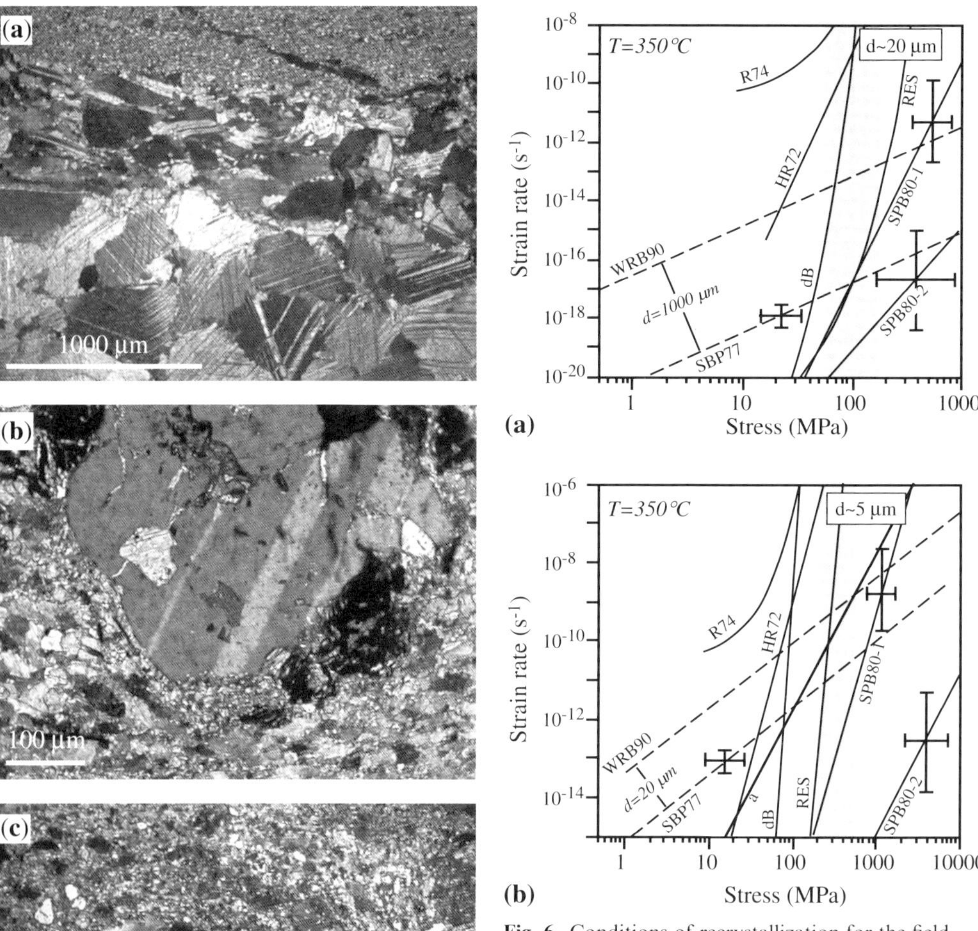

Fig. 5. Optical photomicrographs of the microstructure from the Sesia zone (Western Alps). The shear zone was likely to have been active at a depth of about 30 km (*c.* 1 GPa) and a temperature of 300–400 °C. The protolithic marble was heated to a maximum temperature of about 700–750 °C at *c.* 0.7 GPa. (**a**) Transition from the protolith (lower part of the microstructure, $d \approx 1000\,\mu m$) to mylonite (top part, $d \approx 20\,\mu m$). This transition occurs over a few centimetres or even less. (**b**) Plagioclase clasts embedded in the mylonite. Note the rim of very fine ($d \approx 5\,\mu m$) recrystallized grains around the clast. (**c**) Quartz clasts embedded in the mylonite.

Fig. 6. Conditions of recrystallization for the field example from the Sesia zone. Strain rate–stress conditions were calculated for a temperature of 350 °C using the flow laws given in Table 2. Error bars for SBP77 and SPB80-1/2 are representative for the complete isotherm, and give an indication of the uncertainties in stress and strain rate if laboratory flow laws are extrapolated to natural conditions. The errors were calculated making use of the standard errors of the fitting parameters of the respective flow laws. The ranges of palaeostresses (shaded areas) are based on the piezometric relations presented in the Appendix. (**a**) Recrystallization of the protolith ($d \approx 1000\,\mu m$) to an ultramylonite with $d \approx 20\,\mu m$. Isotherms for grain-size-sensitive creep (dashed lines) calculated for $d \approx 1000\,\mu m$. (**b**) Recrystallization of the ultramylonite to grains of $d \approx 5\,\mu m$ (Fig. 5b). Isotherms for grain size sensitive creep (dashed lines) calculated for $d \approx 20\,\mu m$. The line labeled 'qtz' represents the extrapolation of the quartzite flow law proposed by Hirth *et al.* (2001).

Palaeostress could be underestimated if the ultramylonite grains were enlarged by static growth during uplift. Strain rates would also be underestimated if the deformation temperature were underestimated.

An alternative approach to *in situ* strength determination is to use a scale of relative competence (Ramsay 1983). Brodie & Rutter (2000) recently concluded that field evidence generally indicates that calcite rocks appear weaker than quartz rocks when deformed together. By comparing strengths extrapolated from flow laws for calcite and quartzite, those authors suggested that, in nature, calcite rocks should deform dominantly by GSS mechanisms (cf. Fig. 4). The rocks from the Sesia zone provide an opportunity to examine the competence contrast between the mylonitic calcite matrix and several types of embedded clasts. The recrystallized rims around undeformed feldspar clasts indicate load transfer into this much stronger phase (Fig. 5b). The small grain size (<5 μm) observed in the recrystallized rims indicates stresses of 200–640 MPa, if Rutter's (1995) piezometers are used (see Appendix). In comparison to the GSS flow law as given by Walker *et al.* (1990), the mylonite had to flow around the strong obstacle at almost laboratory strain rates to achieve a switch to dislocation creep (Fig. 6b). The two flow laws that agreed with the recrystallization conditions for the mylonite with $d \sim 20\,\mu m$ (dB and HR72: Fig. 6a) hardly allow for sufficient strength to reach the estimated palaeostresses (Fig. 6b). While the quartz clasts are probably stronger than the calcite matrix, the strength contrast between the mylonite and the quartz clasts does not appear to be as pronounced as that around the feldspar clasts (Figs 5b and 5c). In fact, in this case, a recrystallization rim is absent. The quartz clast appears to have been elongated, perhaps by plastic flow, since the clast shows patchy undulatory extinction in polarized light. Recently, a set of dislocation creep parameters have been suggested that are in accord with experimental and natural observations on quartzites (Stöckhert *et al.* 1999; Hirth *et al.* 2001). The applicability of this quartzite flow law to an isolated clast is certainly questionable, yet we note that it actually predicts flow stresses that are close to the palaeostresses required to recrystallize the ultramylonite to a grain size of 5 μm (Fig. 6b).

The above examples from the Sesia zone demonstrate that palaeo-piezometry and extrapolation of laboratory derived flow laws can put constraints on the conditions during natural deformation, but that uncertainties arise related to the question of which particular flow law to use in the analysis.

Palaeostress estimates and extrapolated strengths

In Figs 7a–f, palaeostress estimates from piezometers applied to calcite rocks from a wide range of natural shear zones (Table 1) are compared with strength contours extrapolated from the experimentally determined flow laws (Table 2). Curves are given for strain rates ranging from 10^{-11} to $10^{-15}\,s^{-1}$. Some strength contours lie close to part of the piezometric data, but none of the extrapolated flow laws fits all the natural data. Notice that the spread in palaeostresses at given T is generally larger than can be generated by variations of strain rate within the range given. Rates slower than $10^{-15}\,s^{-1}$ would not generate a substantial amount of strain within the time available. If strain rates are allowed to vary by 5 orders of magnitude, the stress exponent of a standard power law, $n = \partial \ln \dot{\varepsilon} / \partial \ln \sigma$, has to be about 3.5 to explain the variation in stresses of about one and a half order of magnitudes found for a temperature of about 300 °C ($n = \ln(\dot{\varepsilon}_{max}/\dot{\varepsilon}_{min}) / \ln(\sigma_{max}/\sigma_{min}) = 5/1.5$). As before, errors in the determination of temperature could also be responsible for some of the spread in palaeostresses.

Discussion

Variations in recrystallized grain size in individual shear zones

The microstructure in the Bancroft zone, Canada, shows an interesting transition (Fig. 2 – Van der Pluijm 1991; Busch & Van der Pluijm 1995). The relative proportion of recrystallized grains increases towards the centre of the shear zone, an observation taken to indicate progressive strain localization. In addition, the size of the recrystallized grains decreases towards the centre (Fig. 2b). Several scenarios might allow different grain sizes to be established at different locations in the shear zone. The basic assumption of many of these scenarios is that stress changes occur during the progressive straining of the rocks, and that the different grain sizes record different stress values. Alternatively, abnormally small grain sizes may be maintained via inhibition of grain growth via Zener pinning of grain boundaries owing to second phase particles (Evans *et al.* 2001). In that case, variations in grain size cannot simply be related

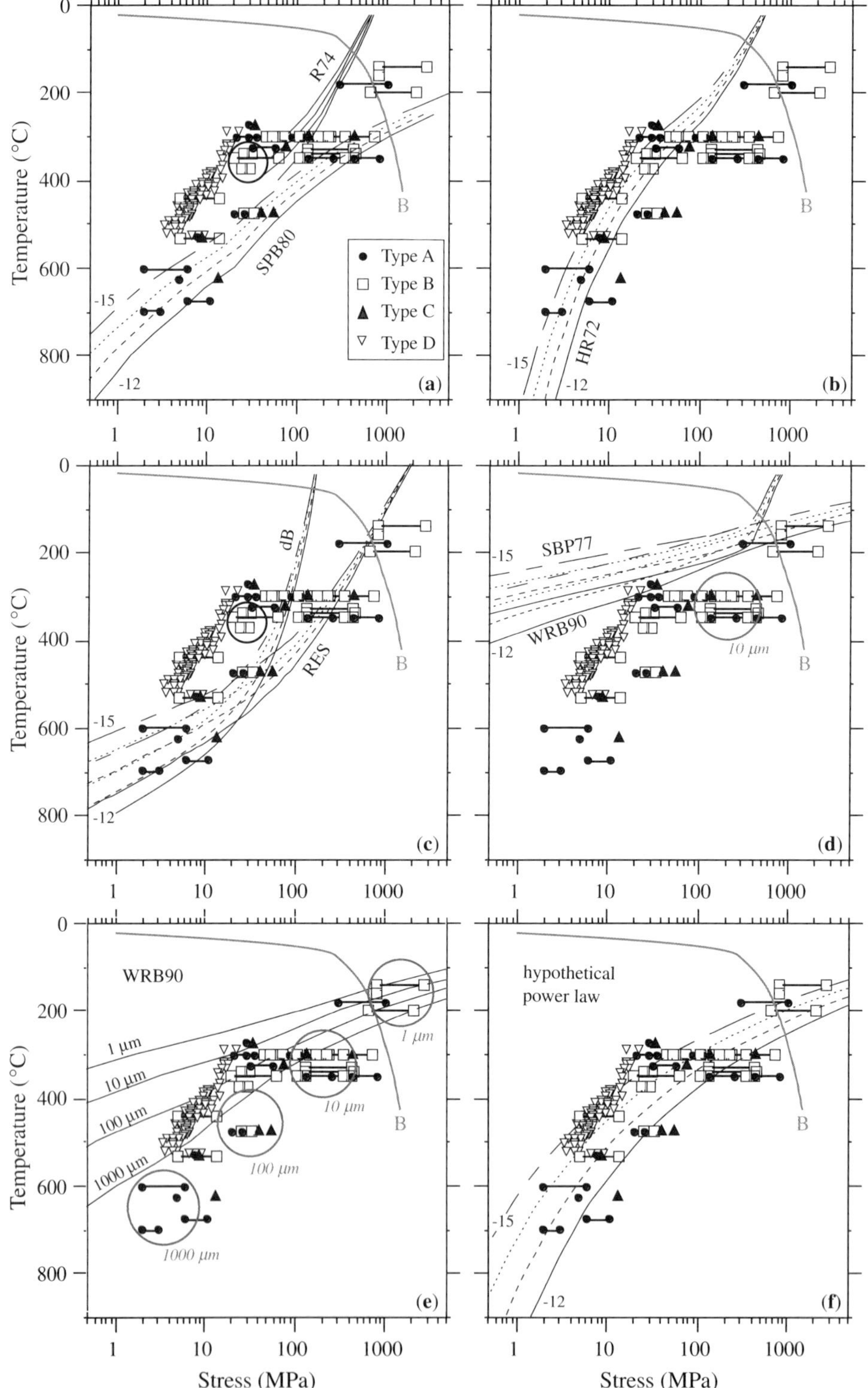
Temperature (°C)
Stress (MPa)
R74
SPB80
HR72
dB
RES
SBP77
WRB90
hypothetical power law
B
-15
-12
1 μm
10 μm
100 μm
1000 μm
Type A
Type B
Type C
Type D
(a)
(b)
(c)
(d)
(e)
(f)

to changes in stress during the evolution of the shear zone.

Assuming that the recrystallized grain sizes do record the stress, then the inferred palaeostresses increase from shear zone boundary to centre (see Fig. 2c). It may be that the individual grain sizes are created during different stages of the formation of the shear zone (White 1979*b*). A similar grain structure was developed in a shear zone formed during compression testing of magnesium (White *et al.* 1985). As the shear zone developed in that material, the sample showed overall weakening, accompanied by recrystallization in two stages. The first stage occurred during the stress peak as fine-grained material was formed within the shear zone. The second occurred as the shear zone broadened outwards, and the overall stress fell, when a second generation of larger grains formed. Apparently, grain growth in the initial finer grains was inhibited so that those grains did not grow to equilibrate with the lower stress. It also seems conceivable that the reverse scenario is possible, namely that outward broadening of a hardened core could occur. Alternatively, the recrystallization mechanism might change towards the centre or the parameters in the piezometric relation may evolve with strain.

Variations in recrystallized grain size within individual shear zones and their host rocks do not always occur. In a well-known example, the Glarus overthrust (Switzerland), the protolith (footwall) rocks and the calc-mylonites do not differ in either dislocation densities or in the recrystallized grain sizes (Briegel & Goetze 1978; Pfiffner 1982 – see also Table 1). Does this microstructure suggest that palaeostress was more or less constant through the complex? Recently, Österling (pers. comm.) performed a thorough analysis of 3D grain size distributions across natural shear zones in marbles near Carrara, Italy. He also found that recrystallized grain size remained constant across the zone, thus corroborating the suggestions made for the Glarus example.

The grain structure around clasts in the Sesia zone provide an example of yet another cause of variability in recrystallized grain size. Using palaeo-piezometers and finite-element calculations, Kenkmann & Dresen (1998) analysed the stress gradients around porphyroclasts in a retrograde amphibolite facies shear zone in the Ivrea zone. Those authors concluded that stress concentrations in the order of 1.7–2.0 existed in the vicinity of the clasts. The magnitude of the stress concentrations probably depended on the nature of the clast–matrix coupling and the effective viscosity contrast of the two minerals.

Complications in the comparison between extrapolated strength and estimated stress

The strongest trend in the microstructural data is a general increase in recrystallized grain size with temperature (Fig. 1). Such a trend agrees with the notion of thermally activated plastic flow, since the increase in size corresponds to a decrease in palaeostress (Fig. 3). The apparent effect of strain on recrystallized grain size could result from variations in strain rate, but could also be related to poor understanding of the evolution of microstructure in shear zones or to the kinetics of recrystallization and grain growth. Importantly, the strength extrapolations from the flow laws do not match the palaeostresses estimated from the natural microstructures. A particular example of this mismatch is formed by the data on Carrara marble (see Figs 7a and 7c). The recrystallized grain size piezometer for calcite rocks (Appendix) is calibrated on the basis of experiments on this specific material. Yet,

Fig. 7. Comparison between palaeostresses of naturally deformed calcite rocks as a function of temperature (Fig. 3, Table 1) and strain rate contours predicted by the extrapolation of experimentally derived flow laws (Table 2). Stresses constrained by Byerlee's rule (1978) for frictional sliding are represented by a solid line labeled B. Strain rates 10^{-12}, 10^{-13}, 10^{-14}, 10^{-15} s^{-1}, except for (**e**) where all lines represent 10^{-12} s^{-1}. Type A–D as in Fig. 1. (**a**) Exponential and power laws for Carrara marble (Rutter 1974; Schmid *et al.* 1980). The palaeostress estimates for the natural deformation experienced by Carrara marble are highlighted (circle). At low temperature, the power laws yield flow stresses up to an order of magnitude higher than estimated on the basis of palaeo-piezometry. (**b**) Exponential and power law for Yule marble (Heard 1963; Heard & Raleigh 1972). (**c**) Exponential laws with stress dependent activation energies derived from new experiments on Carrara marble (dB and RES – see Table 2). Naturally deformed Carrara marble is highlighted. The RES flow law fits the low temperature data better. (**d**) Grain-size-sensitive creep laws for synthetic marble with a grain size $d = 10\,\mu m$ (Walker *et al.* 1990) and Solnhofen limestone with $d = 5$–$10\,\mu m$ (Schmid *et al.* 1977). Natural deformed calcite rocks with $d \approx 10\,\mu m$ are highlighted. (**e**) Grain-size-sensitive creep of synthetic marble (Walker *et al.* 1990) for $d = 1$–$1000\,\mu m$. To facilitate the comparison, palaeostress estimates for natural rocks are grouped according to grain size. Within the range of strain rates chosen, the grain-size-sensitive laws invariably predict stresses that are too low compared to the palaeostresses. (**f**) Hypothetical power law with $n \approx 3.5$, $Q \approx 180$ kJ/mol and $A_{pl} \approx 10^{-5\pm2}$ $MPa^{-3.5}$ s^{-1}.

palaeo-stress estimates for natural Carrara marble (deformed at $T \sim 350\,^{\circ}C$ – Figs 7a and 7c) are lower than values calculated using flow laws for the same Carrara marble as used to determine the piezometric relation. It is important, thus, to remember that there are potential complications in both estimated stress and extrapolated strength. Specifically, there are uncertainties in: (1) the interpretation of microstructures of exhumed rocks; (2) the formulation of piezometers; and (3) the applicability of flow laws. These uncertainties are discussed below in some detail.

Several factors might cause the exhumed microstructure to provide misleading information about the conditions during the main episode of straining. For example, the interpretation of natural microstructures could be complicated by: (a) a more striking appearance of dislocation creep features compared to other mechanisms which were active but did not leave much microstructural imprint; (b) the preservation of the imprint of a dislocation creep period that either survived the accommodation of bulk deformation by another mechanism or reflects a late stage overprint; (c) complex histories of stress, temperature, and metamorphic reactions; and (d) overprinting of the recrystallized grain size by normal grain growth.

Some static growth during uplift may occur, particularly in (fine-grained) ultramylonites. Returning to the Naxos data, it is interesting to notice that stresses inferred from the piezometer are systematically low compared with the remainder of the data. Hence, even if the microstructure may indeed be interpreted as the result of a deformation event (Urai *et al.* 1990), the strain rate was very low (e.g. Fig. 7b). However, increases in grain size owing to grain growth (cf. Covey-Crump & Rutter 1989) could easily lead to an underestimation of palaeostresses. If this is the case, the true palaeostresses would be closer to realistic strain rates resulting from the extrapolations of the flow laws. Nevertheless, even if we ignore the data from Naxos, inferred palaeostresses vary by 1–2 orders of magnitude at 300–400 °C.

The piezometric relations themselves have considerable uncertainty. The relations are usually purely empirical relationships with no direct connection to the mechanical behaviour (i.e. the flow law). Some microphysical modeling, however, suggests that direct relations between piezometers and flow laws do exist (e.g., Edward *et al.* 1982; Derby & Ashby 1987; De Bresser *et al.* 1998, 2001). It is possible that the grain size versus stress relation contains a temperature dependence and/or that the grain size exponent of the piezometric relation and the power law exponent of the flow law are directly related. Consistency between stress estimates from piezometers and flow laws is then warranted, and mismatches, such as exemplified by the data on Carrara marble, can be avoided.

The flow laws that describe the laboratory behaviour of a particular calcite material might also be inappropriate for extrapolation. The flow laws are currently formulated on the basis of low-strain experiments (but see Pieri *et al.* 2001) and neglect other important variables such as impurity content or secondary solid phases dispersed through the matrix, both of which are ubiquitous in nature (e.g., Herwegh & Jenni 2001). In the case of calcite, Mg is one solute that might be particularly important. Considering the significant strength contrast between calcite and dolomite in laboratory experiments, it might be expected that solid solution of magnesium would increase the strength of calcite. However, the extrapolated stresses already tend to overestimate the palaeostresses, and thus the discrepancy would be enlarged. Finally, little effort has been invested to discover a flow law that unifies the contrasting laboratory observations on different calcite rocks. Potential candidates for such a law are discussed by Renner & Evans (2002). Here, we follow an alternative approach of postulating flow laws on theoretical grounds and testing them against the data derived from the field studies.

Inversion using simple theoretical flow laws

Experiments contain practical limitations to the range of temperature and strain rates that can be investigated. Within these restricted ranges, rocks may deform by a complicated mix of mechanisms, and it is possible that the end-member mechanisms cannot be investigated under practical conditions. For example, the experiments of Walker *et al.* (1990) appear to show deformation by mixed grain-size-sensitive flow and dislocation creep (Table 2, transitional law).

As an alternative to the laboratory flow laws, one might therefore consider a constitutive law for diffusion creep on a purely theoretical basis, i.e. Newtonian Coble creep, controlled by grain boundary diffusion:

$$\dot{\varepsilon}_{cc} = A_{cc}\sigma d^{-m} \exp(-Q_{bd}/RT) \qquad (1)$$

where $\dot{\varepsilon}_{cc}$ is the strain rate for Coble creep ($m = 3$), σ is the differential stress, d is the grain size, Q_{bd} is the activation energy for grain boundary diffusion, R is the gas constant, and T is the absolute temperature. Further, assume that (a) such a flow law applies to fine-grained natural rocks as are found in shear zones, and (b) the pre-exponential factor A_{cc} is rather insensitive to variations in impurity content and chemical fugacities. Then, the grain size of the rocks that actively contributed to the accommodation of tectonic deformation at the inferred temperature would follow:

$$d = (A_{cc}\sigma/\dot{\varepsilon}_{cc})^{1/m} \exp(-Q_{bd}/mRT). \quad (2)$$

On the basis of calcium grain boundary diffusion experiments, Farver & Yund (1996) determined $Q_{bd} = 267 \pm 47$ kJ/mol. The factor A_{cc} can in principle be calculated using theoretical constraints (Coble 1963), but uncertainties are large. The factor can also be estimated from recent deformation experiments on synthetic fine-grained marbles (Freund, pers. comm. 2001; Herwegh, pers. comm. 2001). Despite large differences in the chemical composition of the synthetic marbles used in these two studies, there is little variation in the pre-exponential factor ($A_{cc} = 4.4 \times 10^8$ μm^3 s/MPa if $Q_{bd} = 267$ kJ/mol). Taking reasonable values for stress and strain rate conditions of $\sigma = 1$–100 MPa at $\dot{\varepsilon}_{cc} = 10^{-15}$–$10^{-12}$ s^{-1} yields values for $\sigma/\dot{\varepsilon}_{cc}$ (i.e. viscosities) of 10^{15}–10^{14} MPa s. Employing these constraints on Q_{bd}, A_{cc} and $\sigma/\dot{\varepsilon}_{cc}$, Equation 2 was used to calculate hypothetical grain-size–temperature contours. These are shown in Fig. 8 together with the recrystallized grain size and deformation temperature data from the selected field studies (Table 1). Figure 8 shows that, in general, the observed grain sizes are too large for a given temperature to be in accord with the calculations using a hypothetical Coble creep flow equation. To come to a better agreement, substantially larger stresses and/or smaller strain rates are needed, probably resulting in unrealistic values. Note, however, that the above approach becomes invalid in case

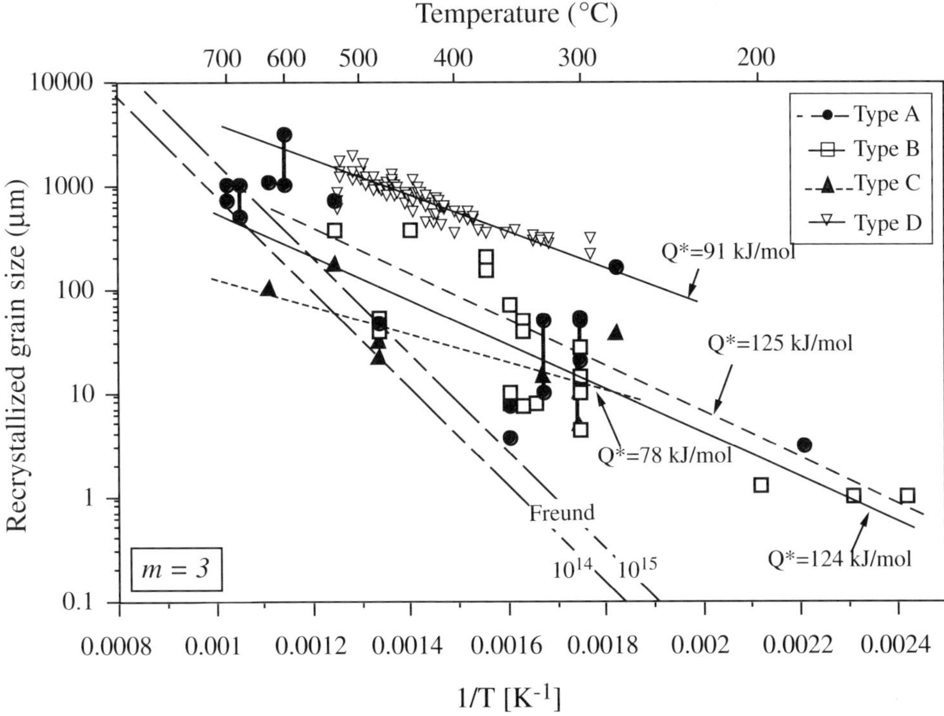

Fig. 8. The recrystallized grain size data (Table 1, Fig. 1) versus inverse deformation temperature. The lines labeled 'Freund' represent the predictions based on a hypothetical Newtonian creep law for grain boundary diffusion (Coble) creep (Eqs 1 and 2), using $Q_{bd} = 267$ kJ/mol (Farver & Yund 1886) and a pre-exponential factor of 4.4×10^8 μm^3 s/MPa derived on the basis of deformation experiments on fine-grained marbles (Freund, pers. comm. 2001). Numbers refer to ratios $\sigma/\dot{\varepsilon}_{cc}$ (in MPa s). Best fit lines for the individual data were determined using Eqs 2 and 4, with $Q^* = Q_{bd}$ if Eq. 2 is applied assuming a constant ratio $\sigma/\dot{\varepsilon}_{cc}$, and $Q^* = (Q_{sd}/n) - Q_{bd}$ if Eq. 4 is used.

substantial grain growth has occurred after deformation of the natural rocks.

As an alternative to the hypothetical Coble creep law (Eq. 1), one may also attempt to constrain a hypothetical (grain size insensitive) power-law:

$$\dot{\varepsilon}_{pl} = A_{pl}\sigma^n \exp(-Q_{sd}/RT) \qquad (3)$$

where Q_{sd} is the activation energy for volume diffusion. According to theoretical micromechanical models the stress exponent should range between 3 and 5 when dislocation climb is rate limiting. The activation energy Q_{sd} should coincide with the activation energy for lattice (self) diffusion of the slowest species in calcite. The spread of palaeostresses at about 300 °C requires a stress exponent n of about 3.5 (or lower) to be explained by differences in strain rate (Fig. 7). Ca is probably the slowest species for lattice diffusion in calcite and has an activation of 382 kJ/mol (Farver & Yund 1996). If the natural observations are assumed to be given by a power law, we can use the inferred palaeo-conditions to constrain the pre-exponential factor. An estimate for a representative flow stress of 100 MPa at a strain rate of $10^{-14\pm1}$ s^{-1} and a temperature of 300 °C (Fig. 7) yields $A_{pl} \approx 10^{14\pm3}$ $MPa^{-3.5}$ s^{-1}. For this approach to be valid the experimentally observed strength of calcite aggregates has to be less than that predicted by the hypothetical power law, i.e. at experimental conditions a more efficient deformation mechanism dominates. For the parameters given above, this is not the case (Fig. 9). However, Fisler & Cygan (1999) recently found a lower activation energy for the self-diffusion of Ca in calcite (almost identical to the boundary diffusion energy of Farver & Yund (1996) (see also Table 2 in Renner & Evans 2002). Also, the activation energy for dislocation recovery kinetics determined by Liu & Evans (1997) is lower than the value for Q_{sd} of Farver & Yund (1996). Applying the Liu & Evans value (180 kJ/mol) as a lower bound, a pre-exponential factor of $A_{pl} \approx 10^{-5\pm2}$ $MPa^{-3.5}$ s^{-1} results. Then, the predicted power law strengths are indeed larger than the experimental observations (Fig. 9) and also fit the field data well (Fig. 7f).

If it exists, such a hypothetical power law would probably not be accessible in the laboratory. Thus, it is extremely important to identify the deformation mechanism active at laboratory conditions and to describe it with a unified constitutive equation. The predictons of this equation for natural conditions can then be compared to the hypothetical power law to infer the rate-controlling mechanism at geological conditions.

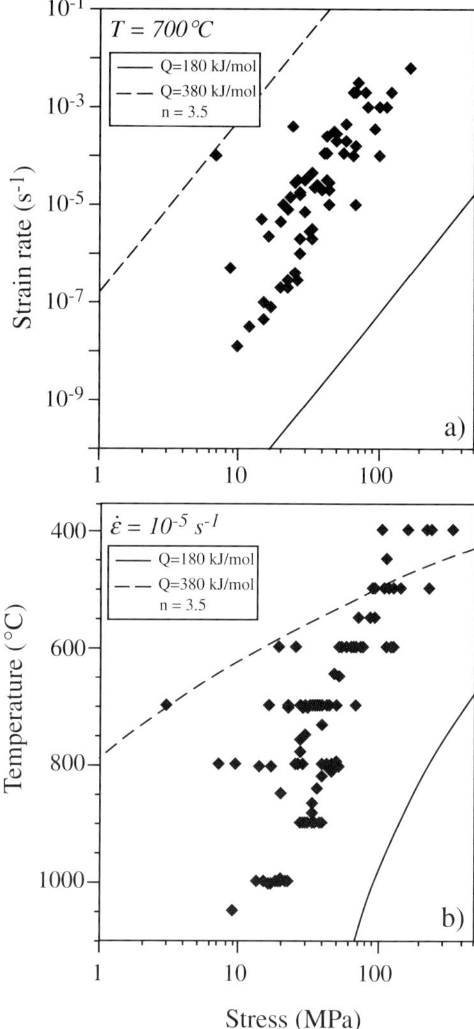

Fig. 9. Predictions of a hypothetical power law with a pre-exponential factor calculated to agree to the field observations (see text and Fig. 7e) in comparison to experimental measurements. These measurements comprise results from constant strain rate tests, creep tests and relaxation tests, and form the input data for the flow laws given in Table 2. (**a**) Strain rate versus stress at T = 700 °C. (**b**) Temperature versus stress at laboratory strain rate 10^{-5} s^{-1}.

Recrystallization weakening and the rheological boundary hypothesis

One common hypothesis for weakening in shear zones is that dynamic recrystallization results in a switch in mechanisms, e.g. from dislocation creep to diffusion creep. In calcite rocks, neither large-strain simple-shear experiments (Schmid *et al.* 1987), nor extension experiments (Rutter 1995) or torsion experiments (Pieri *et al.* 2001) have resulted in a switch from dislocation creep to grain-size sensitive flow that was accompanied by large strength drops. If grain boundary migration is not inhibited by pores or secondary phases, grain growth may counterbalance progressive grain refinement by dynamic recrystallization. In the balanced state, stresses might tend to follow the rheological boundary between the dislocation creep field and the diffusion creep field (De Bresser *et al.* 1998, 2001). Considering the strong evidence for dynamic recrystallization associated with dislocation creep, we could assume that the mylonites from shear zones actually reached the steady state proposed by this rheological boundary hypothesis. Then, the dynamic, steady-state grain size would obey the relation $d = C\sigma^{-(n-1)/m}\exp[(Q_{sd} - Q_d)/mRT]$ where n, m and Q_{sd} have the same meaning as in Equations 1–3, and Q_d represents the activation energy for diffusion creep (i.e. $Q_d = Q_{bd}$ for Coble creep). In fact, this boundary equation is an example of a piezometric relation derived under consideration of the flow behavior, thus ensuring consistency between palaeostress estimates from piezometric relations and strength estimates from extrapolated flow laws. In the case of Nabarro-Herring diffusion creep contributing to the overall creep rate (i.e. $m = 2$ and $Q_d = Q_{sd}$), iso-grain-size lines are also iso-stress lines in Fig. 8 (no temperature dependence).

Using the rate equations, i.e. Equations 1 and 3, the boundary equation can be rewritten for deformation at constant strain rate but varying temperature, resulting in:

$$d = E\exp[((Q_{sd}/n) - Q_{bd})/mRT] \quad (4)$$

where E is a constant. Assuming that the GSS component of the deformation is governed by Coble creep ($m = 3$), and taking $Q_{sd} = 382 \pm 37\,\mathrm{kJ/mol}$ and $Q_{bd} = 267 \pm 47\,\mathrm{kJ/mol}$ (Farver & Yund 1986), the factor $((Q_{sd}/n) - Q_{bd})$ takes values of $140 \pm 60\,\mathrm{kJ/mol}$ for $n = 3$ and $190 \pm 55\,\mathrm{kJ/mol}$ for $n = 5$. The value obtained for $n = 3$ is in reasonably good agreement with the values representing the slopes of the best fit lines for the individual data sets in Fig. 8, lending some support to the rheological boundary hypothesis. However, the correspondence becomes less good at higher n-value. It must be concluded that, at present, the set of field observations is not sufficiently well constrained to allow any definite conclusions regarding the validity of the boundary equations for calcite under natural conditions.

Conclusions and suggestions for future work

Current flow laws for creep of calcite agree qualitatively with estimates of the strength obtained from microstructure under some natural conditions of temperature and strain rate, but no single flow law is consistent with all the estimates from all the shear zones. The discrepancy may arise, in part, from the complex mechanical history recorded by the microstructure, as has been discussed by previous authors. For example, various elements of the microstructure may have recorded deformation conditions that changed with time or position. Further, the grain size may not have achieved a steady-state value, owing either to pinning by second phase particles or because of continuing evolution with strain. Some of the variability in stress estimates from one study to another may also owe to different methods of measurement of average grain-size, or to uncertainty in the identification of the recrystallization mechanism.

When extrapolated to natural strain rates, the experimentally derived flow laws suggest that natural calcite rocks should deform mostly by grain-size-sensitive (GSS) diffusion creep (Brodie & Rutter 2000). In contrast, field studies almost always show evidence of dislocation creep. In principle, the boundary hypothesis (De Bresser *et al.* 2001) provides an explanation for the predominance of dislocation creep recorded by the microstructure. However, this approach does not explain the initiation and progression of localization. More has to be learned about the mechanical boundary conditions of shear zones before the recrystallization weakening hypothesis can be fully evaluated. In this regard, progress in the comparison of experimental and field work may come from determining grain-size distributions rather than average values (Ter Heege 2002; Ter Heege *et al.* 2002). The understanding of piezometers, their relation to flow laws, and their dependence on parameters such as strain and temperature is an additional area for further work. Relative competence contrasts appear to provide a viable alternative where piezometers cannot be applied, but more observations with better specified boundary conditions are needed. In the long run, it would be desirable to establish an internally consistent set of flow laws for key minerals.

The authors thank E. Rutter and M. Drury for helpful and insightful comments, and an anonymous reviewer for his/her input. B. Stöckhert generously allowed us access to his incredible collection of thin sections from the Sesia zone. Financial support came from the US National Science Foundation, EAR-9903307 and EAR-0087533 (BE) and from the Alexander von

Humboldt Foundation (JR). At the time of the DRT meeting in the Netherlands, J. Renner enjoyed being a visiting scientist at the GeoForschungsZentrum Potsdam.

Appendix

We follow the conventional subdivision of the recrystallization mechanisms into two end-members, rotation and migration recrystallization (e.g., Drury & Urai 1990). The first involves the formation of new grains by progressive misorientation of subgrains. In the second, a reworked microstructure develops as a result of grain boundary migration owing to gradients in deformation energy that cause grain dissection, grain coalescence and development of new grains from grain boundary bulges (e.g., Means 1989).

The recrystallized grain size (d, in μm) – stress (σ, in MPa) piezometers were adopted from experimental calibrations on Carrara marble (Rutter 1995):

$$\text{Rotation recrystallization: } \sigma = 10^{2.91 \pm 0.12} d^{-0.88} \quad (5)$$

$$\text{Migration recrystallization: } \sigma = 10^{3.43 \pm 0.15} d^{-0.89} \quad (6)$$

These equations are modifications of earlier relations for rotation and migration recrystallization (Schmid *et al.* 1980; Friedman & Higgs 1981). Note that the distinction between rotation and migration mechanisms was made largely on the basis of grain size (Rutter 1995). All equiaxed grains of the same size as subgrains were assumed to have developed by rotation recrystallization, whereas large grains were assumed to have developed by grain boundary migration. This approach was justified on the basis of key-microstructures that indicated that subgrains can be seen to progressively generate new grains by rotation, and subsequently act as nuclei for migration. Thus, the application of the piezometers (Eqs 5 and 6) to naturally deformed rocks requires the recrystallization mechanism to be specified. In practice, many rocks show evidence for both mechanisms (see Table 1), but quoted recrystallized grain sizes invariably have not been split according to the two basic mechanisms. For these rocks, Eqs 5 and 6 can only provide upper and lower bounds for the palaeostress.

The dependence of dislocation density (ρ, in m^{-2}) on stress (σ, in MPa) has been calibrated on the basis of calcite single crystal experiments (De Bresser 1996):

$$\sigma = 10^{-6.12 \pm 0.86} \rho^{0.62 \pm 0.07} \quad (7)$$

Dislocation densities measured in experimentally deformed polycrystals match the single crystal data at high stress ($\sigma > 40$ MPa), but deviate from these at lower stress. This deviation appears to be more pronounced if the polycrystal has a smaller grain size. We did not include this effect here because of the uncertainties associated with it.

References

BEHRMANN, J. H. 1983. Microstructure and fabric transitions in calcite tectonites from the Sierra Alhamilla (Spain). *Geologische Rundschau*, **72**, 605–618.

BESTMANN, M., KUNZE, K. & MATTHEWS, A. 2000. Evolution of a calcite marble shear zone complex on Thassos Island, Greece: microstructural and textural fabrics and their kinematic significance. *Journal of Structural Geology*, **22**, 1789–1807.

BRAILLON, P. & SERUGHETTI, J. 1976. Deformation plastique de monocristaux de calcite en compression suivant ⟨001⟩. *Physica Status Solidi (a)*, **36**, 637–646.

BRIEGEL, U. & GOETZE, C. 1978. Estimates of differential stress recorded in the dislocation structure of Lochseiten Limestone (Switzerland). *Tectonophysics*, **48**, 61–76.

BRODIE, K. H. & RUTTER, E. H. 2000. Deformation mechanisms and rheology; why marble is weaker than quartzite. *Journal of the Geological Society, London*, **157**, 1093–1096.

BURKHARD, M. 1990. Ductile deformation mechanisms in micritic limestones naturally deformed at low temperatures (150–350 °C). *In*: KNIPE, R. J. & RUTTER, E. H. (eds) *Deformation mechanisms, rheology and tectonics*. Geological Society, London, Special Publications, **54**, 241–257.

BUSCH, J. P. & VAN DER PLUIJM, B. A. 1995. Calcite textures, microstructures and rheological properties of marble mylonites in the Bancroft shear zone, Ontario, Canada. *Journal of Structural Geology*, **17**, 677–688.

BYERLEE, J. D. 1978. Friction of rocks. *Pure and Applied Geophysics*, **116**, 615–626.

COBLE, R. L. 1963. A model for boundary diffusion controlled creep in polycrystalline materials. *Journal of Applied Physics*, **34**, 1679–1682.

COVEY-CRUMP, S. J. 1998. Evolution of mechanical state in Carrara Marble during deformation at 400 °C to 700 °C. *Journal of Geophysical Research*, **103**, 29781–29794.

COVEY-CRUMP, S. J. & RUTTER, E. H. 1989. Thermally-induced grain growth of calcite marbles on Naxos Islands, Greece. *Contributions to Mineralogy and Petrology*, **101**, 69–86.

DE BRESSER, J. H. P. 1991. Intracrystalline deformation of calcite. *Geologica Ultraiectina*, **79** (PhD thesis, Utrecht University, the Netherlands).

De Bresser, J. H. P. 1996. Steady state dislocation densities in experimentally deformed calcite materials; single crystals versus polycrystals. *Journal of Geophysical Research*, **101**(10), 22189–22201.

De Bresser, J. H. P. On the mechanism of dislocation creep of calcite at high temperature: inferences from experimentally measured pressure sensitivity and strain rate sensitivity of flow stress. *Journal of Geophysical Research*, in press.

De Bresser, J. H. P. & Spiers, C. J. 1990. High-temperature deformation of calcite single crystals by r^+ and f^+ slip). *In*: Knipe, R. J. & Rutter, E. H. (eds) *Deformation mechanisms, rheology and tectonics*. Geological Society, London, Special Publications, **54**, 285–298.

De Bresser, J. H. P. & Spiers, C. J. 1993. Slip systems in calcite single crystals deformed at 300–800 °C. *Journal of Geophysical Research*, **98**, 6397–6409.

De Bresser, J. H. P. & Spiers, C. J. 1997. Strength characteristics of the r, f, and c slip systems in calcite. *Tectonophysics*, **272**, 1–23.

De Bresser, J. H. P., Peach, C. J., Reijs, J. P. J. & Spiers, C. J. 1998. On dynamic recrystallization during solid state flow; effects of stress and temperature. *Geophysical Research Letters*, **25**, 3457–3460.

De Bresser, J. H. P., Ter Heege, J. H. & Spiers, C. J. 2001. Grain size reduction by dynamic recrystallization: can it result in major rheological weakening? *International Journal of Earth Sciences (Geologische Rundschau)*, **90**, 28–45.

Derby, B. & Ashby, M. F. 1987. On dynamic recrystallization. *Scripta Metallurgica*, **21**, 879–884.

Drury, M. R. & Urai, J. L. 1990. Deformation-related recrystallization processes. *Tectonophysics*, **172**, 235–253.

Edward, G. H., Etheridge, M. A. & Hobbs, B. E. 1982. On the stress dependence of subgrain size. *Textures and Microstructures*, **5**, 127–152.

Evans, B. & Wong, T.-F. 1985. Shear localization in rocks induced by tectonic deformation. *In*: Bazant, Z. P. & Evanston, I. L. (eds) *Mechanics of geomaterials; rocks, concretes, soils*, 189–210.

Evans, B., Renner, J. & Hirth, G. 2001. A few remarks on the kinetics of static grain growth in rocks. *International Journal of Earth Sciences (Geologische Rundschau)*, **90**, 88–103.

Farver, J. R. & Yund, R. A. 1996. Volume and grain boundary diffusion of calcium in natural and hot-pressed calcite aggregates. *Contributions to Mineralogy and Petrology*, **123**, 77–91.

Ferrill, D. A. 1998. Critical re-evaluation of differential stress estimates from calcite twins in coarse-grained limestone. *Tectonophysics*, **285**, 77–86.

Fisler, D. K. & Cygan, R. T. 1999. Diffusion of Ca and Mg in calcite. *American Mineralogist*, **84**, 1392–1399.

Friedman, M. & Higgs, N. G. 1981. Calcite fabrics in experimental shear zones. *In*: Carter, N. L., Friedman, M., Logan, J. M. & Stearns, D. W. (eds) *Mechanical behavior of crustal rocks*. The Handin volume. Geophysical Monograph, AGU, **24**, 11–27.

Griggs, D. T. 1938. Deformation of single calcite crystals under high confining pressures. *American Mineralogist*, **23**, 28–33.

Heard, H. C. 1963. Effect of large changes in strain rate in the experimental deformation of Yule Marble. *Journal of Geology*, **71**, 162–195.

Heard, H. C. & Raleigh, C. B. 1972. Steady-state flow in marble at 500 to 800 °C. Geological Society of America, Bulletin, **83**, 935–956.

Heitzmann, P. 1987. Calcite mylonites in the central Alpine 'root zone'. *Tectonophysics*, **135**, 207–215.

Herwegh, M. & Jenni, A. 2001. Granular flow in polymineralic rocks bearing sheet silicates: new evidence from natural examples. *Tectonophysics*, **332**, 309–320.

Hirth, G., Teyssier, C. & Dunlap, W. J. 2001. An evaluation of quartzite flow laws based on comparisons between experimentally and naturally deformed rocks. *International Journal of Earth Sciences (Geologische Rundschau)*, **90**, 77–87.

Hobbs, B. E., Mühlhaus, H.-B. & Ord, A. 1990. Instability, softening and localization of deformation. *In*: Knipe, R. J. & Rutter, E. H. (eds) *Deformation mechanisms, rheology and tectonics*. Geological Society, London, Special Publications, **54**, 143–165.

Kenkmann, T. & Dresen, G. 1998. Stress gradients around porphyroclasts: palaeopiezometric estimates and numerical modeling. *Journal of Structural Geology*, **20**, 163–173.

Kennedy, L. A. & Logan, J. M. 1997. The role of veining and dissolution in the evolution of fine-grained mylonites; the McConnell Thrust, Alberta. *Journal of Structural Geology*, **19**, 785–797.

Kennedy, L. A. & Logan, J. M. 1998. Microstructures of cataclasites in a limestone-on-shale thrust fault; implications for low-temperature recrystallization of calcite. *Tectonophysics*, **295**, 167–186.

Kohlstedt, D. L., Evans, B. & Mackwell, S. J. 1995. Strength of the lithosphere: constraints imposed by laboratory experiments. *Journal of Geophysical Research*, **100**, 17587–17602.

Laurent, P., Kern, H. & Lacombe, O. 2000. Determination of deviatoric stress tensors based on inversion of calcite twin data from experimentally deformed monophase samples; Part II, Axial and triaxial stress experiments. *Tectonophysics*, **327**, 131–148.

Liu, M. & Evans, B. 1997. Dislocation recovery kinetics in single-crystal calcite. *Journal of Geophysical Research*, **102**, 24801–24809.

Means, W. D. 1989. Synkinematic microscopy of transparent polycrystals. *Journal of Structural Geology*, **11**, 163–174.

Molli, G., Conti, P., Giorgettic, G., Meccheri, M. & Österling, N. 2000. Microfabric study on the deformational and thermal history of the Alp Apuane marbles (Carrara marbles), Italy. *Journal of Structural Geology*, **22**, 1809–1825.

Passchier, C. W. & Trouw, R. A. J. 1996. *Microtectonics*. Springer, Berlin.

PATERSON, M. S. & TURNER, F. J. 1970. Experimental deformation of constrained crystals of calcite in extension. *In*: PAULITSCH, P. (ed.) *Experimental and natural rock deformation*. Darmstadt, Germany.

PEACH, C. J., SPIERS, C. J. & TRIMBY, P. W. 2001. Effect of confining pressure on dilatation, recrystallization, and flow of rock salt at 150°C. *Journal of Geophysical Research*, **106**, 13315–13328.

PFIFFNER, O. A. 1982. Deformation mechanisms and flow regimes in limestone from the Helvetic zone of the Swiss Alps. *Journal of Structural Geology*, **4**, 429–444.

PFIFFNER, O. A. & RAMSAY, J. G. 1982. Constraints on geological strain rates; arguments from finite strain states of naturally deformed rocks. *Journal of Geophysical Research*, **87**, 311–321.

PIERI, M., BURLINI, L., KUNZE, K., STRETTON, I. & OLGAARD, D. L. 2001. Rheological and micro structural evolution of Carrara Marble with high shear strain; results from high temperature torsion experiments. *Journal of Structural Geology*, **23**, 1393–1413.

POIRIER, J. P. 1980. Shear localization and shear instability in materials in the ductile field. *Journal of Structural Geology*, **2**, 135–142.

RAMSAY, J. G. 1983. Rock ductility and its influence on the development of tectonic structures in mountain belts. *In*: HSUE, K. J. (ed) *Mountain building processes*. Academic Press, NY.

RENNER, J. & EVANS, B. 2002. Do calcite rocks obey the power-law creep equation? *In*: DE MEER, S., DRURY, M. R., DE BESSER, J. H. P. & PENNOCK, G. M. (eds) *Deformation Mechanisms, Rheology and Tectonics: Current Status and Future Perspectives*, Geological Society, London, Special Publications, **200**, 293–307.

ROWE, K. J. & RUTTER, E. H. 1990. Palaeostress estimation using calcite twinning: experimental calibration and application to nature. *Journal of Structural Geology*, **12**, 1–17.

RUTTER, E. H. 1974. The influence of temperature, strain rate and interstitial water in the experimental deformation of calcite rocks. *Tectonophysics*, **22**, 311–334.

RUTTER, E. H. 1995. Experimental study of the influence of stress, temperature, and strain on the dynamic recrystallization of Carrara Marble. *Journal of Geophysical Research*, **100**, 24651–24663.

RUTTER, E. H. 1999. On the relationship between the formation of shear zones and the form of the flow law for rocks undergoing dynamic recrystallization. *Tectonophysics*, **303**, 147–158.

SCHMID, S. M. 1976. Rheological evidence for changes in the deformation mechanism of Solenhofen Limestone towards low stresses. *Tectonophysics*, **31**, T21–T28.

SCHMID, S. M., BOLAND, J. N. & PATERSON, M. S. 1977. Superplastic flow in finegrained limestone. *Tectonophysics*, **43**, 257–291.

SCHMID, S. M., PANOZZO, R. & BAUER, S. 1987. Simple shear experiments on calcite rocks; rheology and microfabric. *Journal of Structural Geology*, **9**, 747–778.

SCHMID, S. M., PATERSON, M. S. & BOLAND, J. N. 1980. High temperature flow and dynamic recrystallization in Carrara Marble. *Tectonophysics*, **65**, 245–280.

STÖCKHERT, B., BRIX, M. R., KLEINSCHRODT, R., HURFORD, A. J. & WIRTH, R. 1999. Thermochronometry and microstructures of quartz – a comparison with experimental flow laws and predictions on the temperature of the brittle–plastic transition. *Journal of Structural Geology*, **21**, 351–369.

TER HEEGE, J. H. 2002. Relationship between dynamic recrystallization, grain size distribution and rheology. *Geologica Ultraiectina*, **218** (PhD thesis, Utrecht University, the Netherlands).

TER HEEGE, J. H., DE BRESSER, J. H. P. & SPIERS, C. J. 2002. The influence of dynamic recrystallization on the grain size distribution and rheological behaviour of Carrara marble deformed in axial compression. *In*: DE MEER, S., DRURY, M. R., DE BESSER, J. H. P. & PENNOCK, G. M. (eds) *Deformation Mechanisms, Rheology and Tectonics: Current Status and Future Perspectives*, Geological Society, London, Special Publications, **200**, 331–353.

TULLIS, J. & YUND, R. A. 1985. Dynamic recrystallization of feldspar: a mechanism for ductile shear zone formation. *Geology*, **13**, 238–241.

TURNER, F. J., GRIGGS, D. T. & HEARD, H. C. 1954. Experimental deformation of calcite crystals. *Geological Society of America, Bulletin*, **65**, 883–933.

TWISS, R. J. 1977. Theory and applicability of a recrystallised grain size paleopiezometer. *Pure and Applied Geophysics*, **115**, 227–244.

ULRICH, S., SCHULMANN, K. & CASEY, M. 2002. Microstructural evolution and rheological behaviour of marbles deformed at different crustal levels. *Journal of Structural Geology*, **24**, 979–995.

URAI, J. L., MEANS, W. D. & LISTER, G. S. 1986. Dynamic recrystallization of minerals. *In*: HOBBS, B. E. & HEARD, H. C. (eds) *Mineral and rock deformation: laboratory studies*. Geophysical Monograph, AGU, **36**, 161–199.

URAI, J. L., SCHUILING, R. D. & JANSEN, J. B. H. 1990. Alpine deformation on Naxos (Greece). *In*: KNIPE, R. J. & RUTTER, E. H. (eds) *Deformation mechanisms, rheology and tectonics*. Geological Society, London, Special Publications, **54**, 509–522.

VAN DER PLUIJM, B. A. 1991. Marble mylonites in the Bancroft shear zone, Ontario, Canada; microstructures and deformation mechanisms. *Journal of Structural Geology*, **13**, 1125–1135.

WALKER, A. N., RUTTER, E. H. & BRODIE, K. H. 1990. Experimental study of grain-size sensitive flow of synthetic, hot-pressed calcite rocks. *In*: KNIPE, R. J. & RUTTER, E. H. (eds) *Deformation mechanisms, rheology and tectonics*. Geological Society, London, Special Publications, **54**, 259–284.

WANG, Z. C., BAI, Q., DRESEN, G., WIRTH, R. & EVANS, B. 1996. High-temperature deformation of calcite single crystals. *Journal of Geophysical Research*, **101**, 20377–20390.

WHITE, S. H. 1979*a*. Grain and sub-grain variations across a mylonite zone. *Contributions to mineralogy and petrology*, **70**, 193–202.

WHITE, S. H. 1979*b*. Difficulties associated with paleostress estimates. *Bulletin de Mineralogie*, **102**, 210–215.

WHITE, S. H., DRURY, M. R., ION, S. E. & HUMPHREYS, F. J. 1985. Large strain deformation studies using polycrystalline magnesium as a rock analogue. Part 1: Grain size paleopiezometry in mylonite zones. *Physics of the Earth and Planetary Interiors*, **40**, 201–207.

The influence of dynamic recrystallization on the grain size distribution and rheological behaviour of Carrara marble deformed in axial compression

J. H. TER HEEGE, J. H. P. DE BRESSER AND C. J. SPIERS

High Pressure and Temperature Laboratory, Faculty of Earth Sciences, Utrecht University, PO Box 80021, 3508 TA Utrecht, The Netherlands (email: jtheege@geo.uu.nl)

Abstract: Strain localization and associated rheological weakening are often attributed to grain size reduction resulting from dynamic recrystallization. Most studies investigating rheological changes due to dynamic recrystallization regard recrystallized grain size as a single value that is uniquely related to stress during steady-state deformation. However, rock materials invariably exhibit a grain size distribution with distribution parameters that may be altered by dynamic recrystallization during transient deformation, affecting the rheological behaviour. This study aims to investigate the effect of deformation conditions on the evolution of grain size distribution and rheological behaviour during dynamic recrystallization in the approach to steady state. To study this, we have deformed Carrara marble to natural strains of 0.15–0.90 in axial compression at temperatures of 700–990 °C, stresses of 15–65 MPa, strain rates of 3.0×10^{-6}–4.9×10^{-4} s^{-1} and a confining pressure of 150 or 300 MPa, and analysed the grain size distribution of each sample. The results show that during dynamic recrystallization, grain size distributions evolve by a competition between grain growth and grain size reduction. The relative importance of grain-size-sensitive creep increases as the average grain size is reduced with strain. Minor weakening is observed, which is probably insufficient to cause strain localization in nature.

It is commonly observed that deformation in natural rocks tends to localize in relatively narrow zones. Localized deformation is often found in calcite rocks that are believed to act as weak rocks accommodating large-scale displacements along major thrusts in mountain belts, such as the Glarus thrust in the Alps (Schmid 1975). One of the key issues in rock deformation studies is the cause of the rheological weakening necessary for strain localization. It has been frequently proposed that dynamic recrystallization could produce the rheological weakening required for strain localization by inducing sufficient grain size reduction to cause a switch in deformation mechanism from dislocation to diffusion creep (e.g. White *et al.* 1980; Rutter & Brodie 1988; Karato & Wu 1993). However, this requires that the reduced grain size remains constant during diffusion creep, which will only occur if grain growth is prevented, for example if inhibited by second phases (e.g. Olgaard 1990). For single phase materials, grain growth during diffusion creep may drive deformation back to the boundary between the diffusion and dislocation creep fields, where dynamic recrystallization and grain size reduction associated with dislocation creep become important again. This notion has led to the hypothesis that for single phase materials, the recrystallized grain size tends to organize itself in the boundary between the dislocation and diffusion creep field during steady-state deformation (De Bresser *et al.* 1998, 2001). This boundary hypothesis model, like most models describing dynamic recrystallization, treats recrystallized grain size as a single value and implicitly assumes steady-state deformation. However, most rock materials exhibit some grain size distribution with distribution parameters, such as median and standard deviation, which may be altered during dynamic recrystallization. These alterations can be expected to cause changes in rheological behaviour (Freeman & Ferguson 1986; Ter Heege *et al.* 2000). In addition, recent torsion experiments on rock materials such as anhydrite (Stretton & Olgaard 1997), olivine (Bystricky *et al.* 2000) and calcite (Paterson & Olgaard 2000; Pieri *et al.* 2001*a*, *b*) have revealed that high strains are required to reach microstructural and mechanical steady state. Therefore, models describing the rheology of rock materials under natural conditions, which may have been deformed to a range of strains, must account for the evolution of grain size distribution and rheological behaviour during transient deformation in the approach to steady state. This requires a detailed understanding of the influence of dynamic recrystallization on microstructural

From: DE MEER, S., DRURY, M. R., DE BRESSER, J. H. P. & PENNOCK, G. M. (eds) 2002. *Deformation Mechanisms, Rheology and Tectonics: Current Status and Future Perspectives*. Geological Society, London, Special Publications, **200**, 331–353. 0305-8719/02/$15

evolution and mechanical behaviour. So far, only very few experimental deformation studies have been carried out that systematically contribute to this understanding (Covey-Crump 1998; Bystricky *et al.* 2000; Pieri *et al.* 2001*a*). The broad aim of this study is to provide a systematic investigation that helps to fill this gap, concentrating on the important rock-forming mineral, calcite.

Calcite is one of the best-studied materials in rock deformation and experiments on polycrystalline calcite aggregates, such as marble, date back to the early days of experimental rock-deformation studies (Griggs 1936; Griggs & Miller 1951). Since these early days, the mechanical behaviour of calcite rocks has been explored by numerous studies, which have demonstrated that different calcite rocks at a wide range of conditions dominantly deform by intracrystalline plastic flow accompanied by dynamic recrystallization, in particular at temperatures above ~600 °C (e.g. Yule marble: Heard & Raleigh 1972, Solnhofen limestone: Rutter 1974; Schmid *et al.* 1977; and Carrara marble: Rutter 1974, Schmid *et al.* 1980, Covey-Crump 1998). Only at small grain size and low stresses has grain-size-sensitive flow been observed in calcite rocks, presumably controlled by grain boundary sliding accommodated by dislocation or diffusional processes (Solnhofen limestone: Schmid *et al.* 1977; hot pressed synthetic calcite aggregates: Walker *et al.* 1990). Except for some extensional tests (Rutter 1995, 1998), in these studies samples have been deformed in axial compression, which limited the amount of strain that could be attained. Higher strains have been reached by means of simple shear experiments on a diagonal saw cut assembly containing calcite rock slices (Kern & Wenk 1983; Schmid *et al.* 1987) and more recently by high-strain torsion experiments (Casey *et al.* 1998; Pieri *et al.* 2001*a*, *b*). Axial compression tests are often used to make inferences on steady-state recrystallized grain size and rheological behaviour of calcite rocks, despite the limited strain reached. In view of the large strains required to reach true steady state (shear strain well over 10: Pieri *et al.* 2001*a*) compared to the strains attainable in axial compression (equivalent to shear strains of ~0.7), it is questionable if this approach is justified. Within the broad aim defined above, the specific objectives of this study are systematically to investigate: (1) the evolution of the grain size distribution and rheological behaviour of Carrara marble with increasing strain as a function of temperature during transient deformation; and (2) the influence of stress and temperature on the grain size distribution that has developed at selected strains.

To investigate the above, cylindrical samples of Carrara marble were deformed to natural strains of 0.15–0.90 in axial compression at temperatures in the range of 700–990 °C, stresses of 15–65 MPa, strain rates of 3.0×10^{-6}–$4.9 \times 10^{-4}\ s^{-1}$ and a confining pressure of 150 or 300 MPa. Microstructural analysis of undeformed and deformed samples was conducted, applying quantitative image analysis techniques. Compared with previous studies on calcite deformed in axial compression, this research has the following new elements.

(1) Instead of determining a single value representing the average grain size, complete grain size distributions have been analysed. This approach provides important additional information on the evolution of differently sized grains during recrystallization.

(2) The evolution of grain size distribution and rheological behaviour in the approach to steady state gives insight into the processes altering the grain size distribution, both during transient and steady state deformation.

(3) The influence of deformation conditions on the relative importance of processes altering the microstructure and their effect on the characteristics of the grain size distribution is investigated by analysis of samples deformed to selected strains at a range of stresses, temperatures and strain rates.

Experimental procedure

Material and sample preparation

The marble investigated in this study originates from a block of Carrara marble ('Lorano Bianco' type) that has been selected as a laboratory standard for deformation experiments on marble. The structural elements, chemical composition and material properties of the marble are well characterized and the structure, stratigraphy and tectonic setting of the source of the block in the area near Carrara in the Apuane Alps in Italy have been well documented (Molli *et al.* 2000; Pieri *et al.* 2001*a*). Microscopic analysis in combination with chemical analysis of minor and trace elements using X-ray fluorescence (XRF) and inductively coupled plasma (ICP) spectroscopy (Table 1) revealed that the marble consists of ~99% pure calcite with few grains of muscovite, quartz, dolomite and graphite.

Samples were cored in an arbitrary but constant direction from the block of Carrara marble. Sample ends were machined and

Table 1. *Chemical analysis of Carrara marble ('Lorano Bianco' type)*

	XRF* (wt.%)	ICP† (wt.%)		XRF‡ (ppm)	ICP§ (ppm)		XRF‡ (ppm)	ICP§ (ppm)
CaO	56.86	55.11	Mg	—	2250	Zr	<6	27
MgO	0.85	0.82	Sr	122	97	P	—	6
SiO_2	—	0.07	Mn	—	27	Cu	5	<1
Al_2O_3	—	0.05	Ti	—	24	K	—	4
P_2O_5	0.07	trace	Nd	<14	—	Pb	<4	—
Fe_2O_3	0.07	0.01	V	<12	<0.8	Nb	3	—
Na_2O	0.04	trace	Fe	—	12	Ni	3	—
K_2O	0.02	trace	Ba	11	—	Y	3	1.3
TiO_2	0.008	trace	La	10	—	As	<3	—
MnO	0.002	trace	Cr	<10	27	Ga	<3	—
LOI‖	44.00	42.83	Sn	<9	1.5	U	<3	—
			Th	<9	—	Rb	<2	—
Total	101.92	98.88	Sc	8	—	Zn	1	—
			Ce	—	7	Be	—	<0.2

*XRF on fused glass bead – this study. †ICP (average) – Pieri *et al.* (2001*a*). ‡XRF on pressed powder pellets – this study. §ICP-Rutter (1995). ‖LOI – Lost On Ignition.

polished in such a way that they were parallel (to within 0.02 mm) to each other and perpendicular ($\leq 1°$) to the sample axis. Measurements of grain aspect ratios (see Results below) revealed that the starting material has a weak shape preferred orientation. Samples were found to have been cored in a direction of ~20° to the longest axis of grains in the starting material. Samples with a length of 25, 16 and 11 mm and a diameter of 9.9, 8.0 and 8.0 mm were used for the experiments to ~0.15, ~0.30, ~0.40 natural strain respectively. The samples, together with Al_2O_3 ceramic spacers for the samples shorter than 25 mm, were jacketed in a 0.2 mm thick copper tube with a copper cup at each end, which were welded together creating a gas-tight copper capsule.

Experiments, data acquisition and processing

The samples were deformed in axial compression by constant displacement rate tests in a constant volume, internally heated, argon gas medium apparatus, consisting of a water cooled 1 GPa pressure vessel with a three-zone Kanthal-AF wire furnace. The vessel was mounted in a horizontally placed Instron 1362 servo-controlled testing machine. Piston displacement was controlled by an Instron actuator and measured externally with a linear variable differential transformer (LVDT, 2 µm resolution). Sample shortening was calculated by correcting piston displacement for apparatus distortion. Axial force was applied to the sample by displacement of a dynamically sealed deformation piston against a stationary reaction piston. Axial force was measured both internally, using a Heard-type internal force gauge (100 kN full scale, ~20 N resolution), and externally, using an Instron 100 kN load cell (accurate to 0.1% of the full scale output or 0.5% of the indicated load, whichever is greater). Confining pressure was measured using a strain-gauge-type pressure transducer (10^5 Pa resolution). In general, pressure remained constant within 2 MPa during the experiments, but in some overnight experiments small variations in pressure (<5 MPa) occurred. Stress on the sample was calculated from the internal axial force, corrected for pressure variations and copper jacket strength, assuming constant volume and homogeneous deformation. Temperature was measured using three Pt/10%Rh (type S) thermocouples located next to the capsule wall. The temperature gradient along the sample was determined to be in the order of 1 or 2 °C (see McDonnell 1997; McDonnell *et al.* 1999 for a detailed description of the apparatus). At the end of each experiment the furnace was immediately shut down, resulting in rapid cooling of the sample (1000 °C to 100 °C in ~13 min). All measurements are estimated to be accurate to within 5%. However, taking into account inaccuracies caused by slight heterogeneous straining of the samples due to piston friction, the error in stress, strain and strain rate is estimated to be <10%.

Samples were deformed to natural strains of 0.15–0.40 at strain rates of 3.0×10^{-6}–$4.9 \times 10^{-4}\ s^{-1}$, temperatures of 700–990 °C and a confining pressure of 150 or 300 MPa, resulting in flow stresses in the range 14.6–65.0 MPa (Table 2). One sample was deformed to a natural strain of ~0.90 in three subsequent tests on the same sample (36LM900/0.90). This required removing the sample twice from the gas

Table 2. *Deformation and microstructural characteristics*

Test[*]	Stress at peak [MPa]	Strain at peak	Strain rate at peak [s^{-1}]	Flow stress[†] [MPa]	Strain rate[†] [s^{-1}]	$T^{\dagger}$ [°C]	$P^{\dagger}$ [MPa]	Nat. strain	Exp. dur. [hr]	Weak. (peak-end)[‡] [%]	Median grain size [μm]	Av. grain size [μm]	Av. grain size log-space [μm]	Stand. dev. (log d)	Min.-max. grain size [μm]	Skewness (log d)	Kurtosis (log d)	Av. grain aspect ratio[§]	No. of grains	Min. RX area [%]
Starting material	—	—	—	—	—	—	—	—	—	—	35	68	41 (1.62)	0.43	11-405	0.39	−1.02	1.16	1363	—
Start 1 hr at 900 °C	—	—	—	—	—	—	—	—	—	—	47	78	52 (1.72)	0.40	11-385	0.11	−1.15	1.10	1155	—
15LM950	24.9	0.03	2.7×10^{-6}	14.6	4.2×10^{-6}	962	150	0.446	39.07	41	42	54	42 (1.62)	0.31	9-265	−0.06	−0.58	1.33	1797	38
23LM900	25.4	0.19	2.9×10^{-6}	21.8	3.9×10^{-6}	906	303	0.424	38.12	14	29	36	29 (1.46)	0.28	9-300	0.24	−0.21	1.31	3962	50
23LM990	28.1	0.13	3.5×10^{-5}	25.0	4.6×10^{-5}	992	299	0.386	2.89	11	32	48	35 (1.54)	0.33	8-450	0.55	0.04	1.34	1669	20
36LM830/0.15	38.5	0.15	3.0×10^{-6}	36.7	3.0×10^{-6}	829	301	0.170	17.41	5	58	85	60 (1.77)	0.38	10-502	0.02	−0.72	1.37	969	—
36LM830/0.30	43.9	0.19	3.2×10^{-6}	39.6	3.4×10^{-6}	830	299	0.291	27.95	10	52	80	52 (1.72)	0.41	10-614	0.06	−0.85	1.29	1114	9
36LM830	43.4	0.13	2.8×10^{-6}	38.7	4.0×10^{-6}	835	299	0.446	40.09	11	45	62	48 (1.68)	0.31	10-467	0.18	−0.48	1.46	2202	22
36LM900/0.15	37.4	0.01	3.1×10^{-5}	34.1	3.7×10^{-5}	903	305	0.158	1.70	9	80	100	69 (1.84)	0.41	10-479	−0.35	−0.86	1.32	805	—
36LM900/0.15r[‖]	37.8	0.11	3.5×10^{-5}	37.4	3.5×10^{-5}	899	301	0.141	1.42	1	—	—	—	—	—	—	—	—	—	—
36LM900/0.30	35.6	0.02	3.2×10^{-5}	32.9	4.0×10^{-5}	901	303	0.269	2.14	8	71	97	67 (1.82)	0.40	10-481	−0.29	−0.78	1.29	802	—
36LM900/0.30	heated for *c.* 1 hr at 900 °C										35	66	43 (1.63)	0.40	11-435	0.37	−0.92	1.38	1462	—
36LM900	39.9	0.16	3.7×10^{-5}	36.3	4.5×10^{-5}	898	299	0.399	2.98	9	68	85	63 (1.80)	0.35	11-481	−0.20	−0.57	1.59	1090	15
36LM900/0.90a[¶]	33.9	0.16	3.5×10^{-5}	33.5	3.5×10^{-5}	900	299	0.161	1.37	—	—	—	—	—	—	—	—	—	—	—
36LM900/0.90b[¶]	—	—	—	31.9	4.6×10^{-5}	900	302	0.299	2.07	—	—	—	—	—	—	—	—	—	—	—
36LM900/0.90c[¶]	—	—	—	26.2	7.1×10^{-5}	902	306	0.437	2.20	23	22	29	24 (1.37)	0.26	6-288	0.69	0.79	1.40	2737	36
36LM950/0.15	36.7	0.06	3.3×10^{-4}	35.2	3.6×10^{-4}	960	302	0.157	0.08	4	103	118	90 (1.95)	0.35	9-372	−0.65	−0.12	1.30	565	—
36LM950/0.30	38.5	0.07	3.3×10^{-4}	32.5	4.1×10^{-4}	957	300	0.298	0.23	16	72	90	69 (1.84)	0.33	10-434	−0.22	−0.54	1.30	1127	8
36LM950	42.6	0.15	3.6×10^{-4}	31.6	4.9×10^{-4}	949	301	0.464	0.33	26	42	58	42 (1.62)	0.35	10-371	0.06	−0.58	1.51	2189	19
50LM730	58.5	0.18	3.0×10^{-6}	52.2	3.9×10^{-6}	726	298	0.436	39.56	11	27	39	30 (1.47)	0.30	6-243	0.65	0.04	1.36	1416	31
50LM730r[‖]	59.4	0.15	2.9×10^{-6}	52.1	3.9×10^{-6}	731	298	0.435	39.40	12	—	—	—	—	—	—	—	—	—	—
50LM780	55.6	0.13	3.5×10^{-5}	47.3	4.7×10^{-5}	776	298	0.429	3.21	15	32	42	33 (1.52)	0.30	8-306	0.14	−0.53	1.42	3094	52
50LM830	62.6	0.13	3.5×10^{-4}	53.3	4.7×10^{-4}	835	297	0.438	0.33	15	41	50	40 (1.61)	0.29	10-326	−0.01	−0.52	1.45	3588	46
65LM700	73.5[**]	0.23[**]	3.1×10^{-6}	65.0	3.8×10^{-6}	698	299	0.431	39.44	12[**]	20	31	23 (1.36)	0.29	5-294	1.08	1.26	1.40	1871	30

[*] For convenience in comparing tests, stress (first number) and temperature (second number) are quoted in the test numbers. For samples deformed to natural strains less or more than ~0.4, the last number in the name gives the approximate natural strain.
[†] Flow stress, temperature (T), strain rate and pressure (P) averaged over a strain of 0.01 at the end of the test are quoted.
[‡] Weakening is calculated by $((\sigma_{peak} - \sigma_{end})/\sigma_{peak}) \times 100\%$ with σ_{peak} and σ_{end} the flow stress at the peak or at the end of the experiment.
[§] Average measured grain aspect ratio perpendicular to the compression direction is quoted.
[‖] Samples broke during removal from gas apparatus or jacket, no grain size distributions determined. Tests are quoted to give an indication of the reproducibility of the mechanical data.
[¶] Three-step experiment, sample removed from apparatus and re-polished and re-jacketed after each step (twice in total). In total a natural strain of 0.897 was reached. See text for details.
[**] Value is influenced by dip in stress-strain curve due to day-night fluctuation of the pressure vessel cooling-water temperature affecting the internal load cell signal. Peak stress is likely to be slightly higher (~74 MPa) and strain at peak lower (~0.16) if temperature fluctuations of cooling-water are corrected for.

apparatus and jacket and, after re-polishing and re-jacketing, inserting the sample into the apparatus again and heating to 900 °C before restoring the load.

Microstructural analysis

Microstructural analysis was carried out on ultrathin (~5–10 µm) optical sections using a Leica DMRX light microscope, equipped with a Sony DXC-950P colour video camera. For grain size analysis, a mosaic covering an area of 3–12 mm^2 in the middle of the sample was made of colour photographs taken in transmitted light at 5× magnification under crossed-polarized light. Grains in the photographs were traced manually on transparencies. Quantitative analysis of the area (A_{grain}) and perimeter (P_{grain}) of individual grains, the total grain boundary area (A_{gb}) (i.e. the total area made up by the traced boundary in 2D section) and 22.5° ferets (i.e. the orthogonal distance between a pair of parallel tangents to the grain at angles of 22.5° intervals) was conducted on the scanned transparencies using the image analysing program Leica QWin Pro Version 2.2 (1997). The thickness of the manually traced grain boundary in the 2D section was added to the area of individual grains using

$$A_{cor} = A_{grain} + (P_{grain}/P_{tot}) \times A_{gb} \quad (1)$$

where P_{tot} is the total measured grain perimeter and A_{cor} is the corrected area of individual grains in 2D section. Grain size is presented as equivalent circular diameters (ECD), calculated using the corrected grain area A_{cor} without any (stereological) correction for sectioning effects. With this method, the number of grains in the lowest class (detection limit: 5–9 µm, equivalent to 1 pixel size) tends to be overestimated, because all grains that are visible optically but have a size below the detection limit of the image analysis program are incorporated in the lowest class. For all samples, the logarithmic ECD frequency distribution (referred to as grain size distribution) and the distribution of area fraction occupied by a given ECD class (referred to as area distribution) is given in a single histogram.

Delta-histograms have been constructed by subtracting two histograms with the same subdivision of class intervals. Delta-histograms show differences in ECD frequency or area fraction of grains between samples and can be used to investigate evolution of the grain size distribution with strain or during heating. When logarithmic grain size histograms of two samples deformed to different strains are used to construct a delta-histogram, grain areas A_{cor} have been corrected to account for out-of-plane strain, which causes a reduction of grain area in 2D section unrelated to grain size reduction due to the recrystallization process. In axial symmetric straining, the three principal stretches s_1, s_2 and s_3 (with s_3 parallel to the compression direction) are related by:

$$s_1 = s_2 = \sqrt{1/s_3} \quad (2)$$

if the volume of the deformed sample is constant. This means that a spherical grain with diameter d with maximum area in thin section of $\frac{1}{4}\pi d^2$ flattens to an ellipsoid with diameter $d_1 = d_2 > d_3$ and a maximum area in thin section of:

$$\tfrac{1}{4}\pi d_1 d_3 = \tfrac{1}{4}\pi(ds_1)(ds_3) = \tfrac{1}{4}\pi d^2\sqrt{s_3}. \quad (3)$$

Hence, A_{cor} needs to be corrected by a factor $\sqrt{s_3}$ to estimate the true influence of dynamic recrystallization on 2D microstructural modification. An estimate of the minimum amount of recrystallized area as seen in 2D section (A_{Rxmin}) has been obtained by summing the negative area fraction differences (= area reduction) of the grain size classes, i.e. the area made up by the grains in the distribution that are reduced in size. A_{Rxmin} is a minimum estimate since growth of (recrystallized) grains results in an increase of the area fraction difference in the grain size classes of the delta-histograms. This means area reduction due to dynamic recrystallization in these classes is underestimated. Grain aspect ratios were calculated by dividing the feret measured perpendicular to the compression direction by the feret measured parallel to the compression direction.

Results

Mechanical data

Table 2 lists test conditions and characteristics of all performed experiments. Note that test numbers are used with approximate stress, temperature and final natural strain (if not ~0.4) as the first, second and third numbers, respectively. Selected, representative stress (σ) versus strain (ε) curves are depicted in Fig. 1. The flow stresses increase with increasing strain rate and decreasing temperature. Most curves show a peak stress or plateau in flow stress at $\varepsilon = 0.1$–0.2 followed by weakening at higher strains. Experiments 15LM950, 36LM900/0.15 and 36LM900/0.30 weaken after a peak at a lower value of strain. However, the duplication experiment 36LM900/0.15r, which shows a peak stress at

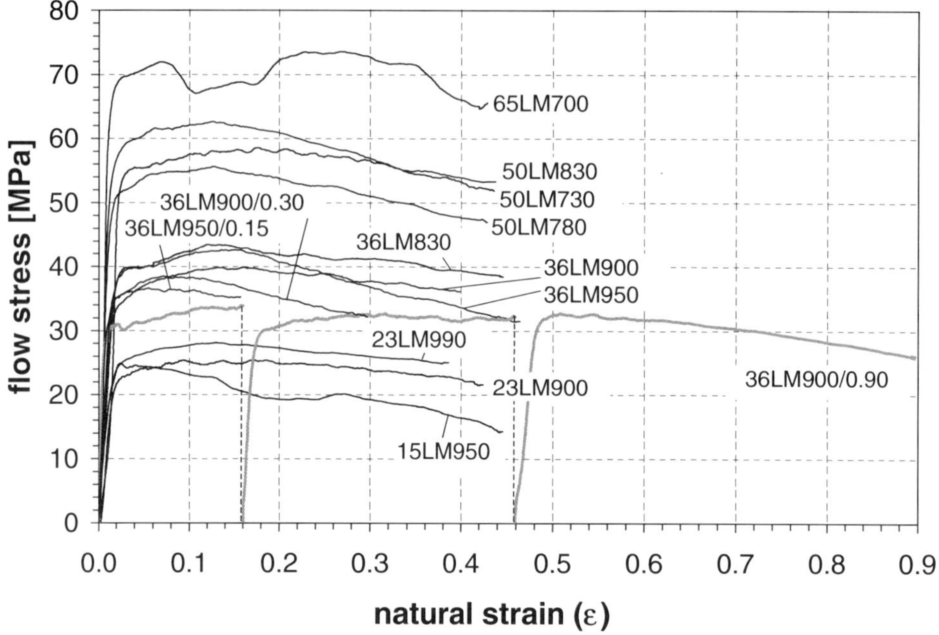

Fig. 1. Typical stress–strain curves for Carrara Marble deformed in axial compression showing the influence of stress and temperature on the flow stress. Sample 36LM900/0.90 was deformed to a strain of 0.9 in three subsequent steps (thick lines). The sudden dip in flow stress in experiment 65LM700 is due to day–night fluctuation of the pressure vessel cooling-water temperature, affecting the internal load cell signal. This effect was corrected in all other overnight experiments. The experimental conditions and results of all tests are depicted in Table 2.

$\varepsilon = 0.11$, indicates that peaks at lower strains are an exception rather than a general feature. The amount of weakening tends to increase with increasing temperature. Experiment 36LM900/0.90 (23% total weakening) shows that weakening continues at least up to $\varepsilon = 0.90$. As a consequence, the observed weakening in the other samples, deformed to lower strains, should be regarded as a minimum value. True steady state was not achieved in any of the experiments. Duplication of selected experiments (i.e. 36LM900/0.15r and 50LM730r) indicates that flow stress at a given ε can be reproduced within ~10% (Table 2). Flow stresses, which are the result of test conditions chosen on the basis of a previous study on the same material (De Bresser 2002), are grouped into five sets of broadly constant stress, namely 15, 23.4 ± 1.6, 35.6 ± 4.0, 50.3 ± 3.0 and 65 MPa (Table 2).

Qualitative microstructural observations

Microscopic observations of the Carrara Marble starting material (Fig. 2a) indicated that grain

Fig. 2. Overview (left) and details (right) of typical microstructures of undeformed (**a**) and deformed (**b**–**h**) Carrara marble. (**a**) Straight twins, straight or gently curved grain boundaries and triple point junctions intersecting at 120° in the starting material. (**b**) Progressive rotation of subgrains leading to the nucleation of new grains in sample 23LM990. (**c**) Grain flattening and irregular grain boundaries and shapes in sample 36LM950. (**d**) Irregular grain boundaries due to (sub)grain boundary migration in grains nucleated by subgrain rotation in sample 23LM900. Subvertical stripes are related to sectioning or polishing. (**e**) Large flattened grains with small, recrystallized grains at the grain boundaries forming a core-mantle structure in sample 65LM700. (**f**) Bulge migration recrystallization with nucleation of new grains at grain boundary bulges caused by grain boundary migration in sample 50LM780. (**g**) Extensive grain boundary migration leading to recovery in grains and regular grains with gently curved grain boundaries in sample 15LM950. (**h**) Grain size reduction by the formation of new grain boundaries dissecting large parts of a relict grain and by grain boundary migration of adjacent grains consuming a relict grain in sample 36LM950. Compression direction vertical in (**c**), (**e**) and (**g**).

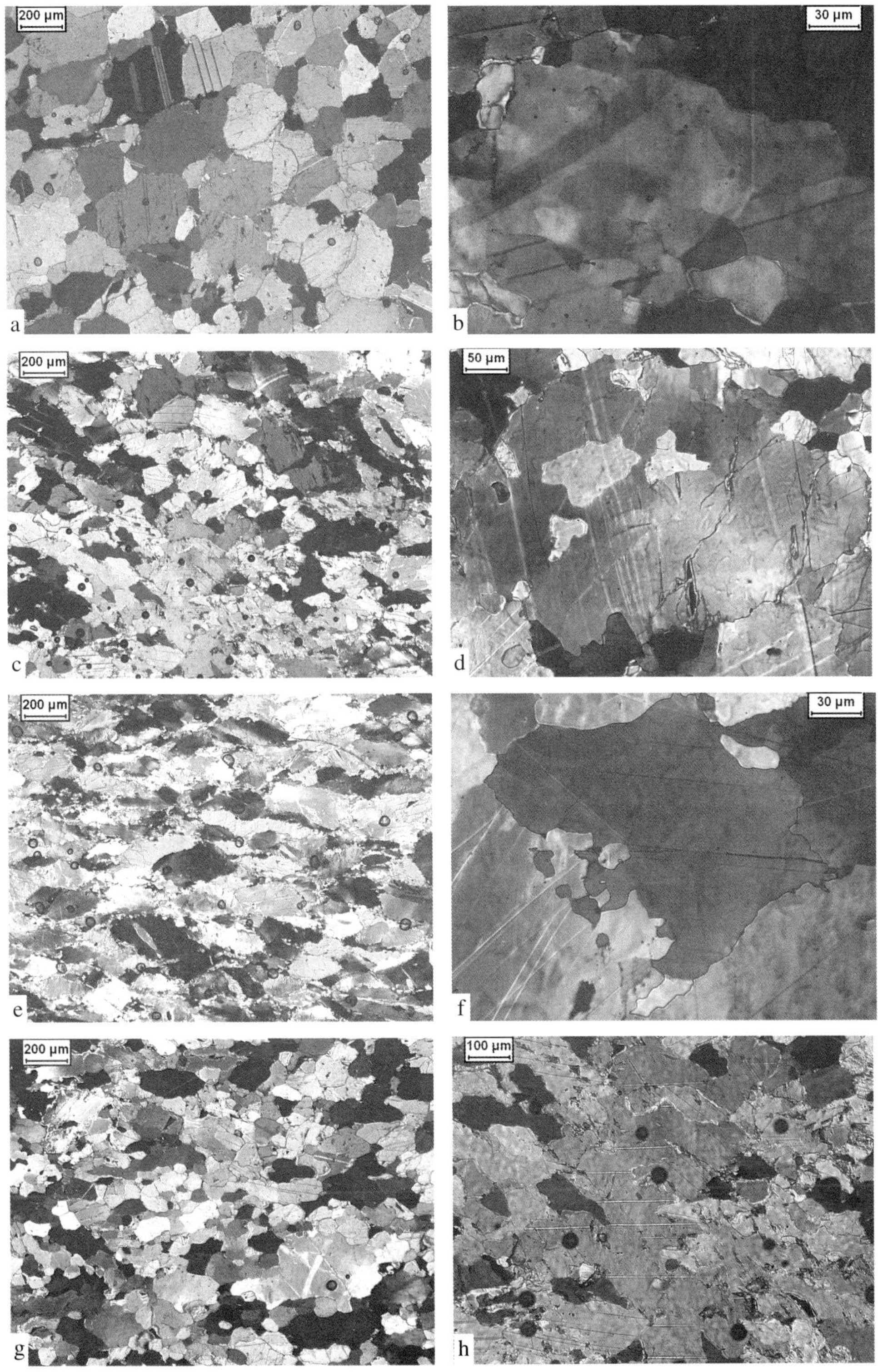
200 µm
a
30 µm
b
200 µm
c
50 µm
d
200 µm
e
30 µm
f
200 µm
g
100 µm
h

size was fairly homogeneous in the sample. The grains had a weak shape preferred orientation, but lacked indications of intracrystalline deformation, such as undulose extinction, deformation lamellae or lattice preferred orientation. Within individual grains, thin twins with straight boundaries were recognized. Grain boundaries were generally straight or only gently curved, and preferentially intersected at triple point junctions with angles of $\sim 120^{\circ}$, although 90° junctions were also present. Qualitatively, the microstructure of the starting material did not change significantly after heating for approximately 1 hr at 900 °C (equivalent to the time needed to reach a stable temperature in the experiments).

The influence of strain on the microstructure is illustrated by samples 36LM830/0.15 to 36LM950 (Table 2), deformed to strains of 0.15–0.90 in the temperature range 830–950 °C at a near-constant stress of 35.6 ± 4.0 MPa. The samples deformed to a strain of *c.* 0.15 show very minor grain flattening and an increase in grain size. At strains above *c.* 0.15, grain shapes become more irregular and grains become more flattened with increasing strain, with the long axis perpendicular to the compression direction (Fig. 2c). In addition, grain boundaries become more irregular and curved, and intersection of grain boundaries in triple junctions at an angle of *c.* 120° is less common. Following the nomenclature described in Passchier & Trouw (1996), the grain aggregates change from inequigranular-polygonal in the starting material to seriate-interlobate after deformation to a strain of *c.* 0.4. With increasing strain, deformation bands, undulose extinction and deformation lamellae become more apparent. Thin twins with straight boundaries remain present as internal grain features, but thick twins with serrated twin boundaries tend to occur more frequently with increasing strain. All the above features are also apparent in sample 36LM900/0.90 deformed in three subsequent steps to a total strain of *c.* 0.9, but grains are more flattened. In addition, grain size is smaller and clusters of very small equiaxed grains are present. Qualitatively, the microstructure of 36LM900/0.30 did not change significantly after heating for about 1 hr at 900 °C.

In all samples deformed to a strain of *c.* 0.4 (Figs 2b–h), grain size is reduced with respect to the starting material. Grain size reduction is most apparent in the samples deformed at high stress and low temperature (e.g. 50LM730 and 65LM700), where a core-mantle structure has developed consisting of large flattened grains with smaller grains at the grain boundaries (Fig. 2e). Three distinctive types of microstructural features associated with grain size reduction could be identified:

(1) Subgrains in individual grains or occasionally near twins. In addition, grains similar in size to the subgrains are present within larger grains or at grain boundaries (Fig. 2b). These grains locally form clusters of equiaxed grains with gently curved grain boundaries. Some of the grains inside larger grains have irregular shapes and grain boundaries (Fig. 2d).
(2) Irregular grain boundaries forming bulges. Near such boundaries, grains that are similar in size and shape as the bulges are present (Fig. 2f).
(3) Large relict grains that can be recognized by pervasive calcite twins running through several adjacent grains (Fig. 2h). The relict grains are dissected by grain boundaries across larger parts of the grains or by large grains or arrays of small grains cross-cutting the relict grains.

Microstructures of type (1) and (2) have been found in all samples and tend to occur more frequently with increasing stress and decreasing temperature. Both types are most apparent in 65LM700 (Fig. 2e), but are less developed in 15LM950 where grain boundaries are mostly gently curved without bulges, and internal grain structures do not show many subgrains (Fig. 2g). Other features accompanying type (1) and (2) microstructures, such as deformation bands, undulose extinction and deformation lamellae, occur more frequently with increasing stress and decreasing temperature and are most apparent in sample 65LM700 (Fig. 2e). Microstructure type (3) is most apparent in samples deformed at relatively high temperature and low stress where relict grains can be identified, such as 23LM990 and 36LM950 (Fig. 2h). Thin twins with straight boundaries as well as thick twins with irregular boundaries are present in all samples. With increasing stress, the larger grains become more flattened with the long axis perpendicular to the compression direction and grain boundaries tend to be more irregular and aligned. At a relatively low stress (*c.* 23 and *c.* 36 MPa) and high temperature ($T = 830$–990 °C), the microstructure does not seem to vary significantly with increasing temperature. At relatively high stress (*c.* 50 MPa) and low temperature ($T =$ 700–830 °C), the microstructure changes from a core-mantle type structure to a microstructure with intensely flattened interlobate to amoeboid grains.

Quantitative microstructural results

All histograms with ECD frequency and area fraction distributions per individual grain size (ECD) class are depicted in Figs 3 and 4. In addition, Fig. 5 shows delta-histograms illustrating the difference in ECD frequency and area fraction of grains for deformed samples, for the starting material heated for about 1 hr at 900 °C or the sample re-heated for about 1 hr at 900 °C after deformation. Statistical descriptors characterizing the grain size distributions are given in Table 2. Standard deviation, skewness and kurtosis (of log d) are plotted in Fig. 6. Figure 7 shows plots of the median and average grain size versus strain or temperature.

Starting material
The grain size distribution of the starting material is positively skewed, platykurtic

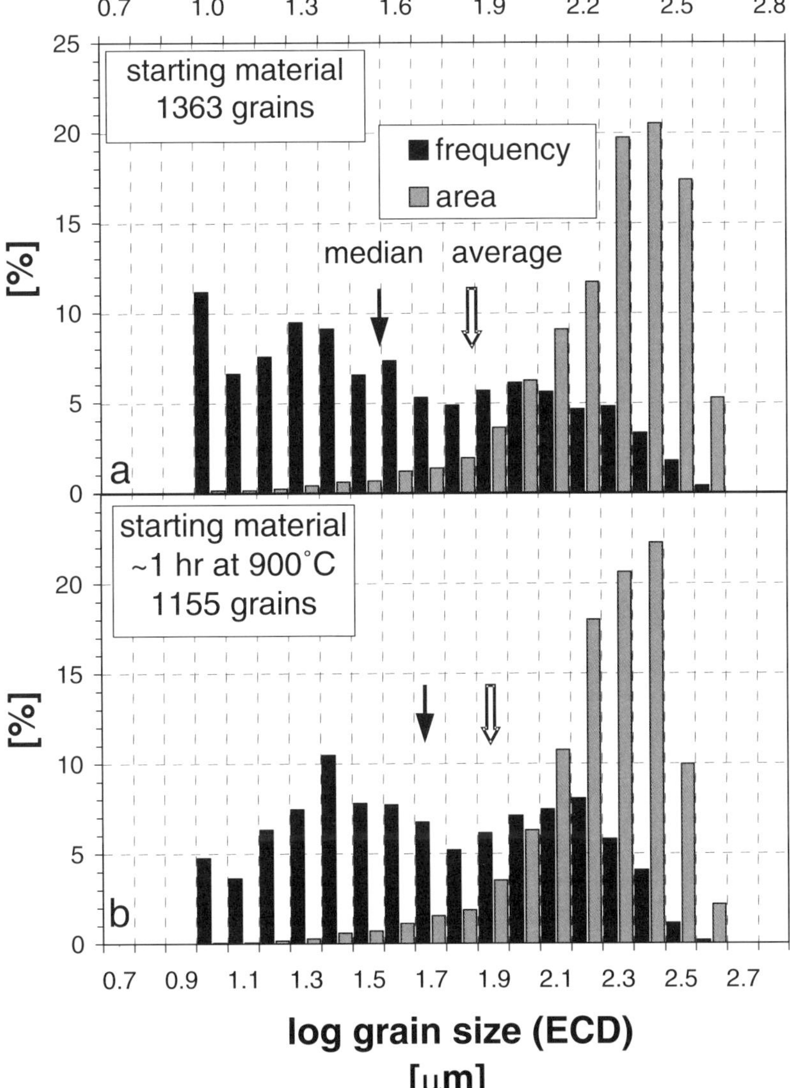

Fig. 3. Logarithmic grain size (equivalent circular diameter-ECD) distributions showing ECD frequency (black bars) and area fraction (grey bars) of the undeformed samples. Sample number, median (closed arrow) and average (open arrow) grain size are indicated. (**a**) Carrara marble starting material. (**b**) Starting material heated for ~1 hr at 900 °C. See Table 2 and Fig. 6 for statistical descriptors and other details on the experiments.

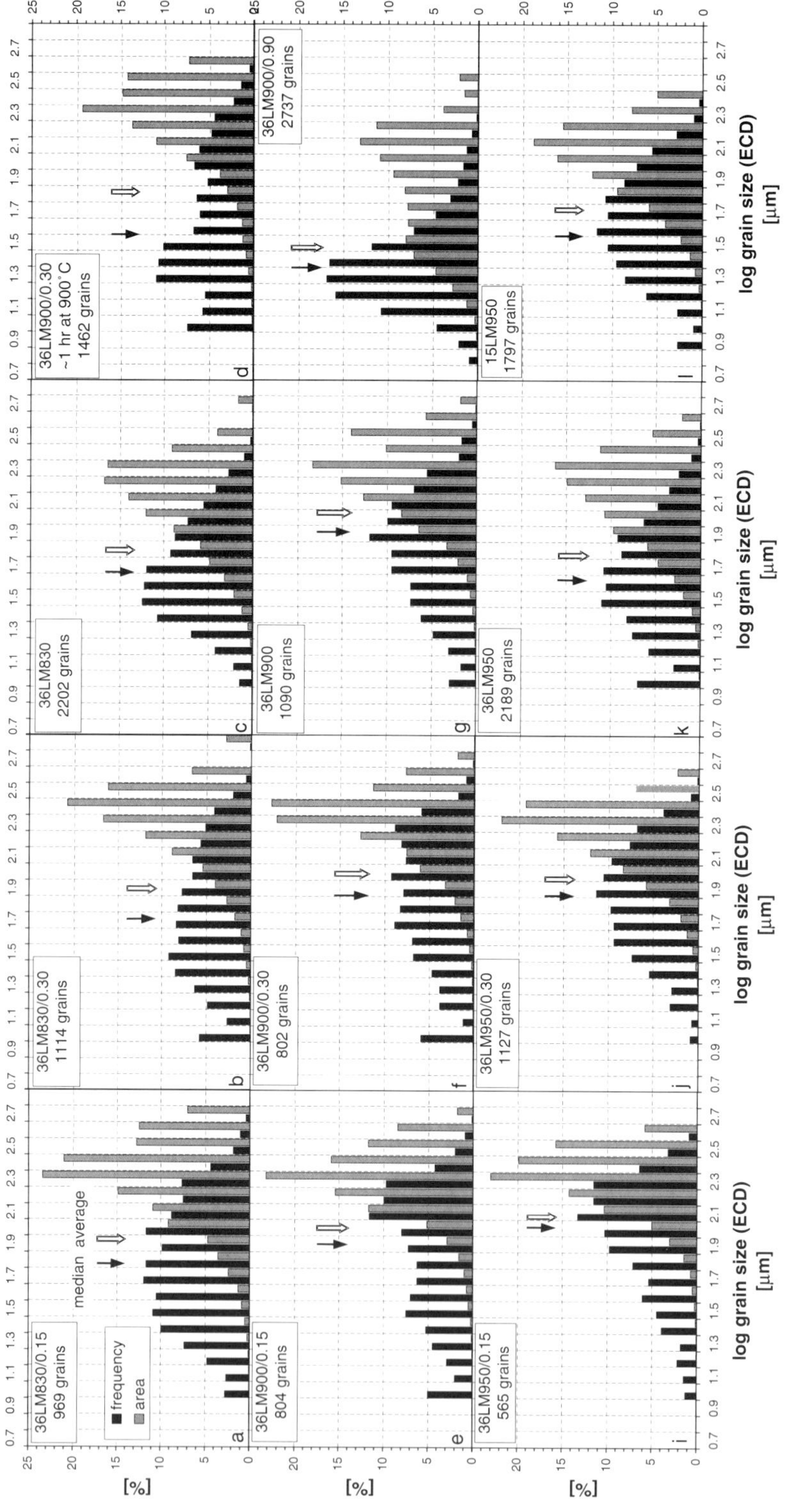
36LM830/0.15
969 grains
frequency
area
median
average
36LM830/0.30
1114 grains
36LM830
2202 grains
36LM900/0.30
~1 hr at 900°C
1462 grains
36LM900/0.15
804 grains
36LM900/0.30
802 grains
36LM900
1090 grains
36LM900/0.90
2737 grains
36LM950/0.15
565 grains
36LM950/0.30
1127 grains
36LM950
2189 grains
15LM950
1797 grains
a
b
c
d
e
f
g
h
i
j
k
l
[%]
log grain size (ECD)
[µm]

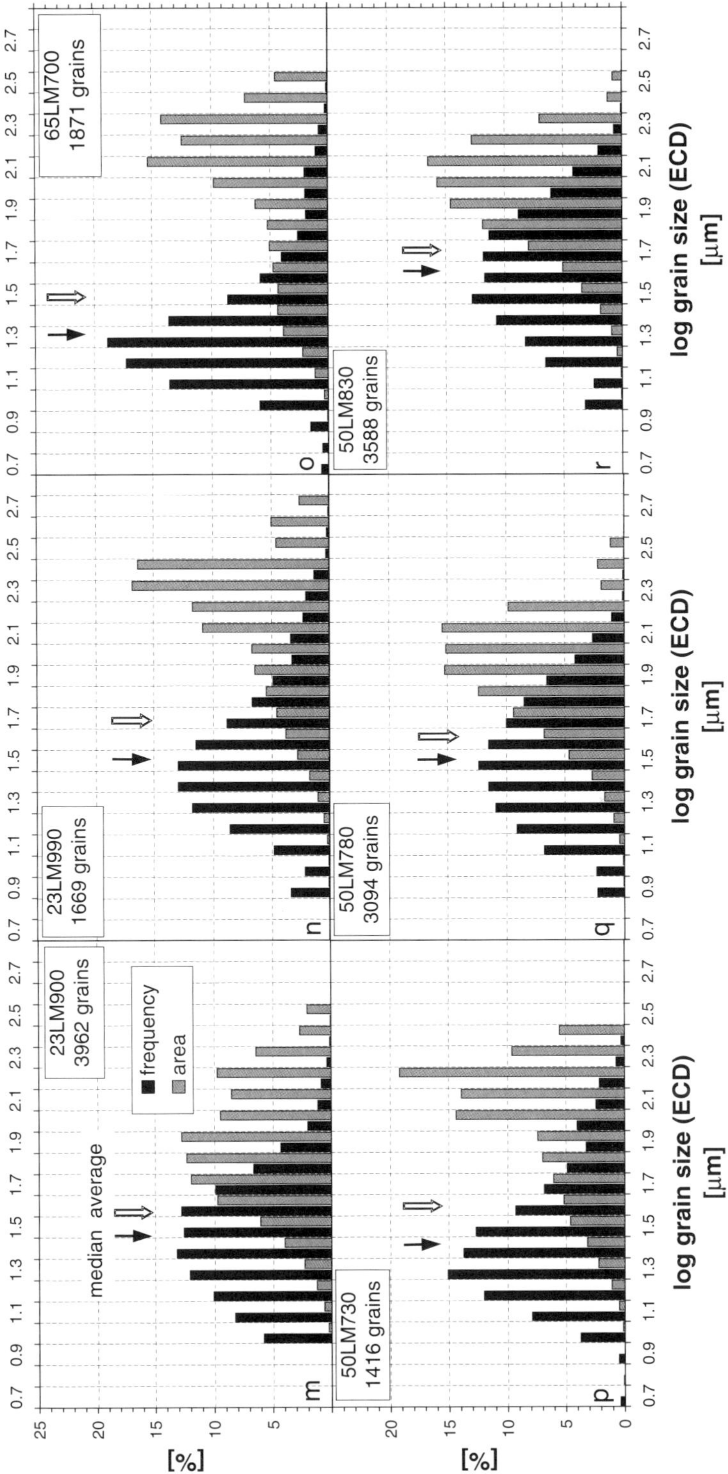

Fig. 4. Logarithmic grain size (equivalent circular diameter-ECD) distributions showing ECD frequency (black bars) and area fraction (grey bars) of the deformed samples. Sample number, median (closed arrow) and average (open arrow) grain size are indicated. Where possible, histograms are ordered so that strain is increasing from left to right at constant stress and temperature (a–c, e–h, i–k) or temperature is increasing from left to right at constant stress and strain (m–n, p–r). See Table 2 and Fig. 6 for statistical descriptors and other details on the experiments.

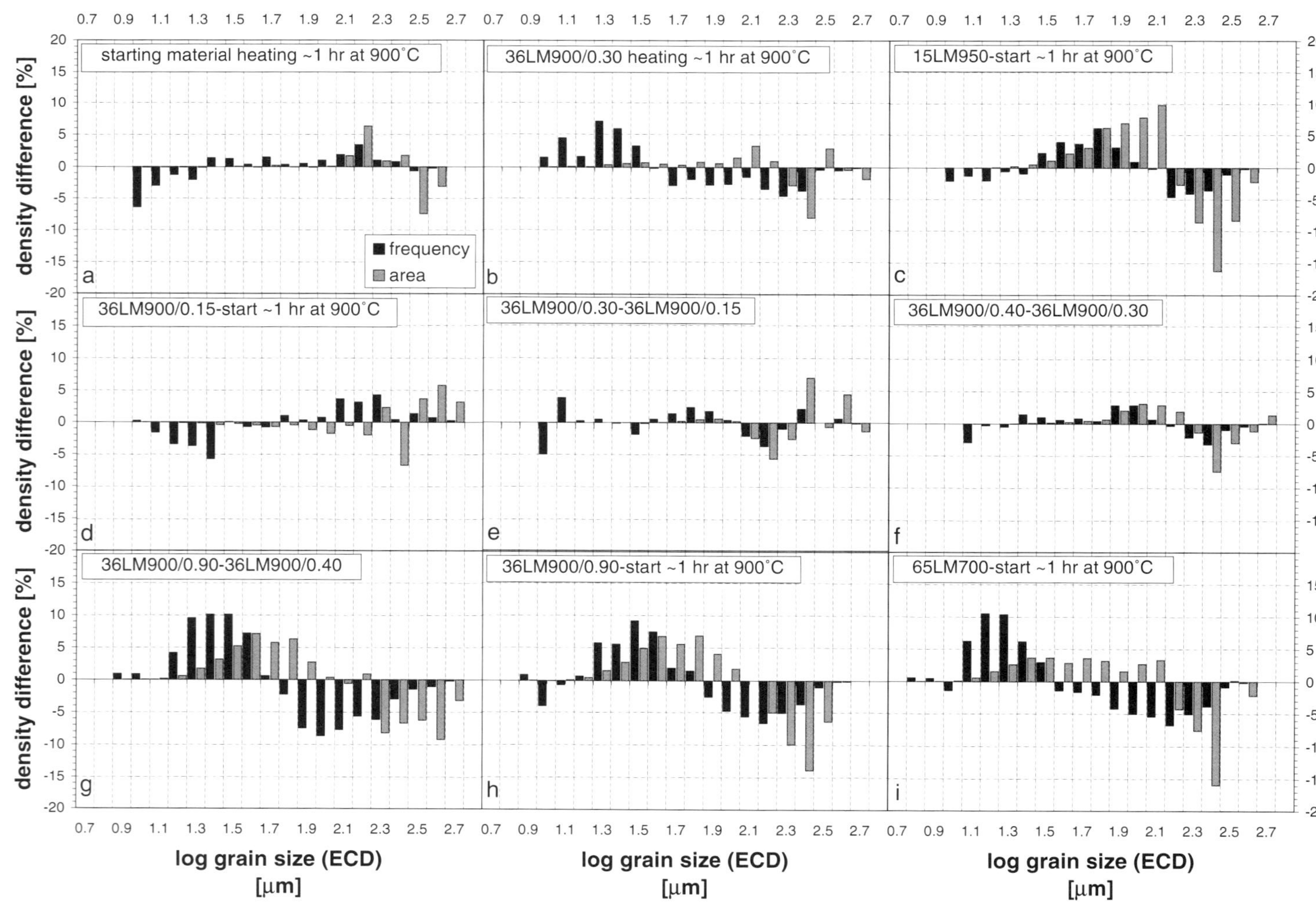

starting material heating ~1 hr at 900°C
36LM900/0.30 heating ~1 hr at 900°C
15LM950-start ~1 hr at 900°C
frequency
area
a
b
c
36LM900/0.15-start ~1 hr at 900°C
36LM900/0.30-36LM900/0.15
36LM900/0.40-36LM900/0.30
d
e
f
36LM900/0.90-36LM900/0.40
36LM900/0.90-start ~1 hr at 900°C
65LM700-start ~1 hr at 900°C
g
h
i
density difference [%]
log grain size (ECD) [μm]

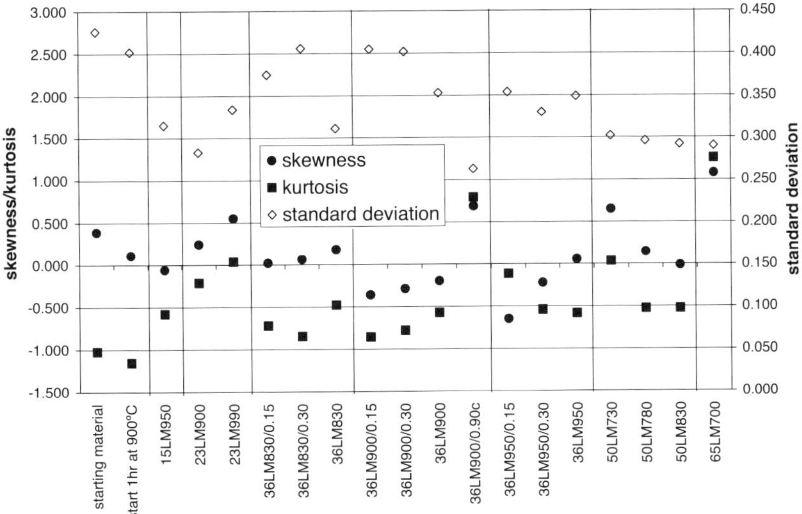

Fig. 6. Plot of the measured statistical descriptors, i.e. standard deviation, skewness and kurtosis (of log d) for undeformed and deformed samples of Carrara marble. For an ideal lognormal distribution, skewness and kurtosis are zero.

(negative kurtosis) to bimodal with modes at 20 and 100 µm (Fig. 3a, Table 2). Although the starting material originates from the same block of Carrara Marble as used by Pieri *et al.* (2001*a*, *b*), the grain size distribution is different from that reported by those authors, who found a unimodal grain size distribution with a mode at 178 µm. This may be due to natural variations in the grain size distribution of the Carrara marble (Molli *et al.* 2000), or an artefact of the different approaches to grain size analysis used. Note that grain sizes smaller than 30 µm have not been analysed by Pieri *et al.* (2001*a*, *b*). A bimodal distribution thus might have been overlooked, since the bimodal character of the starting material used in this study becomes apparent only at grain sizes smaller than 45 µm. The area distribution is unimodal with 90% of the area in section taken up by grains larger than 100 µm and a peak at 251 µm. The grains have an average aspect ratio of 1.30 with the long axis of the grains at an angle of *c.* 20° with respect to the sample axis. If measured perpendicular to the sample axis, this shape preferred orientation results in an average aspect ratio of 1.16 (Table 2). The starting material remained bimodal during heating for ~1 hr at 900 °C (equivalent to the time needed to reach a stable temperature in the experiments), although the two modes shifted to slightly higher values of 25 and 158 µm (Fig. 3b). The delta-histogram (Fig. 5a) shows that classes with grains of 22–282 µm increase in frequency and area fraction, while frequency or area fraction decrease in the other classes. This implies that the modification of microstructure during heating was caused by growth of grains smaller than 22 µm, reducing the grain size of grains larger than 282 µm (possibly due to a higher surface energy of the small grains). The overall result is an increase in the average and median grain size (Table 2). The average grain aspect ratio slightly decreased during heating (Table 2).

Deformed material

Table 2 shows that the median grain size and the average of log d are generally close for the deformed samples. In addition, the skewness and kurtosis of the logarithmic grain size distributions of all deformed samples show values close to zero. These statistical descriptors indicate that the grain size distributions of the deformed samples are approximately lognormal,

Fig. 5. Delta-histograms showing differences in ECD frequency (black bars) and area fraction (grey bars) calculated from the logarithmic grain size distributions of Figs 3 and 4 (corrected for out-of-plain strain where necessary). The delta-histograms show changes in ECD frequency and area fraction due: (**a**) to heating of the starting material; (**b**) re-heating of 36LM900/0.30; (**d**–**h**) progressive deformation; and (**c**–**i**) deformation at different conditions.

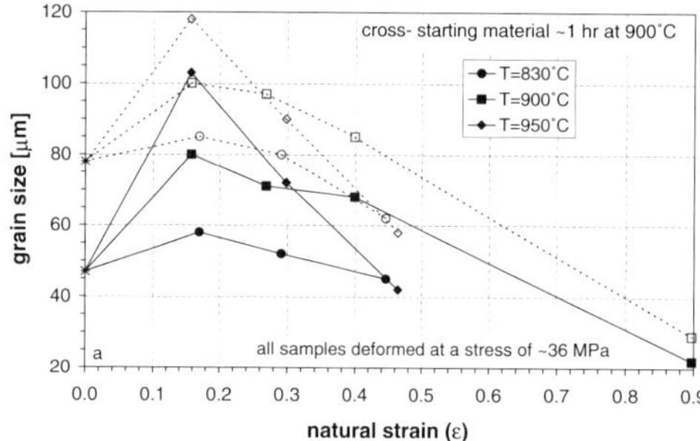

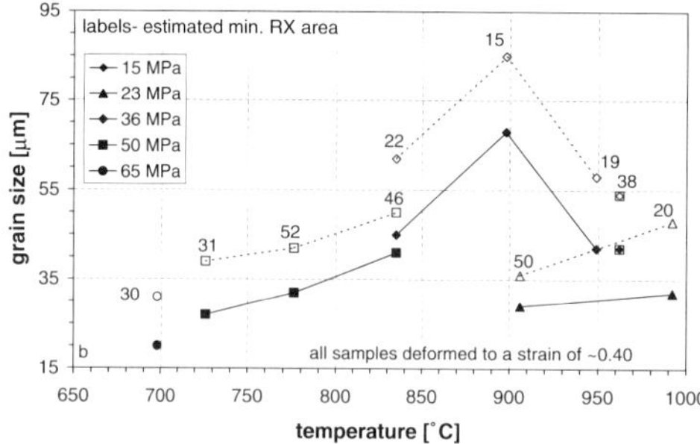

Fig. 7. Changes of median (open symbols) and average (closed symbols) grain size (**a**) with strain for $T = 830$—$950\,^{\circ}$C and (**b**) with temperature at a stress of 15–65 MPa.

since for an ideal lognormal distribution the median and average of log d coincide, and skewness and kurtosis are zero. However, none of the grain size distributions are exactly lognormal and most distributions are slightly platykurtic and positively skewed at the higher strains. Some trends of mode, median, average, standard deviation and skewness with strain, stress and temperature can be identified. These will be discussed below.

The area distributions of all deformed samples are negatively skewed with a mode at 71–282 μm, always higher than the mode of the grain size distribution.

Influence of strain

The evolution of the grain size distributions with strain has been investigated for strains of 0.15–0.90 at a stress of 35.6 ± 4.0 MPa and temperatures of 830–950 °C (Figs 4a–c, e–h, i–k and Table 2). The grain size distributions have changed significantly with respect to the starting material heated for about 1 hr at 900 °C (Fig. 3b), even at strains as small as 0.15. At $T = 830$ and 950 °C, the distributions at 0.15 strain are roughly unimodal (Fig. 4a, i), while the distribution of the sample deformed at $T = 900\,^{\circ}$C is bimodal (Fig. 4e). Skewness is close to zero or negative and kurtosis is negative. With increasing strain, all grain size distributions become unimodal. The skewness of the distributions increases with strain, changing from negative to positive for $T > 830\,^{\circ}$C (Fig. 6). The standard deviation of the grain size distribution decreases with increasing strain (Table 2, Fig. 6).

Sample 36LM900/0.90 was deformed to a strain of *c*. 0.90 in three successive steps (Table 2). The possibility that the heating period in between loading steps modified the microstructure has been assessed by heating sample 36LM900/0.30 to 900 °C for ~1 hr, which is equivalent to the time required to reach a stable temperature. The grain size distribution (Fig. 4d) and delta-histogram of the re-heated sample compared to 36LM900/0.30 (Fig. 5b) show that heating resulted in a grain size distribution that is bimodal, with more grains in the classes below 35 μm than in the distribution before heating. However, the flow stress after reloading agrees well with the flow stress at the end of the previous step (Fig. 1). Hence, heating between subsequent steps resulted in some microstructural modification, but did not significantly affect the mechanical behaviour. The growth of very small grains at best resulted in a slight over-estimation of the number of grains in the classes below 35 μm in the distribution of 36LM900/0.90 (Fig. 4h).

The evolution of the grain size distribution with strain at a stress of 35.6 ± 4.0 MPa and temperatures 830–950 °C is illustrated by the delta-histograms shown in Figs 5d–h for samples deformed at 900 °C. The delta-histogram in Fig. 5d shows that for sample 36LM900/0.15 the frequency of grains smaller than 28 μm decreases substantially, while the frequency of grains with sizes above 56 μm increases compared to the starting material heated for about 1 hr at 900 °C. This indicates that during deformation to a strain of *c*. 0.15, the grain size distribution predominantly changes by the growth of small grains, which causes removal of grains from the smaller grain size classes and an increase of grains in the larger classes. This results in an increase in median and average grain size at a strain of *c*. 0.15 with respect to the heated starting material for all investigated temperatures (Fig. 7a). The delta-histograms (Figs 5e–g) show that as strain increases from *c*. 0.15 to 0.90 at 900 °C, the frequency of grains in the classes with the smallest grain sizes (below 14 μm) remains approximately constant, while the frequency of grains in the classes with the largest grains decreases and the frequency of grains in between the smallest and largest size increases. The smallest grain size class, in which the frequency of grains decreases, becomes smaller (from 112 μm to 56 μm) with increasing strain. This indicates that with ongoing deformation, the grain size distribution predominantly changes by the progressive reduction of the larger grains in the distribution, which causes removal of grains from the larger grain size classes and an increase of grains in smaller classes. However, the frequency of the classes with the smallest grains of the distributions remains approximately constant with strain, indicating that grains are removed from these classes by grain growth during deformation, compensating the grains added to these classes by nucleation of new grains (cf. Fig. 2f). Most of the change in frequency that has occurred after a strain of *c.* 0.90 (Fig. 5h) takes place in the strain interval 0.40–0.90 (Fig. 5g). Microstructural alteration during deformation results in an overall decrease of the median and average grain size with strain for all investigated temperatures.

The minimum recrystallized area increases with increasing strain for $T = 830–950\,^{\circ}\mathrm{C}$ (Table 2). Average grain aspect ratios increase with strain (Table 2), which is in agreement with the qualitative observation that grains progressively flatten, with their long axis at an angle of more than 45° to the compression direction.

Influence of temperature and stress

The influence of temperature and stress on the grain size distribution has been investigated for samples deformed to a strain of *c.* 0.4 (at this strain, samples clearly exhibit recrystallization-related microstructures), selecting samples deformed at approximately the same stress and different temperatures or approximately the same temperature and different stresses (Figs 4c, g, k, l, m–r, Table 2). There are no trends of standard deviation, skewness or kurtosis with changing temperature or stress that are consistent for all investigated conditions (fig. 6).

Median and average grain size increase with increasing temperature at all stresses, except at a stress of ~36 MPa (Fig. 7b). At this stress, the median and average grain size increase with increasing temperature for $T = 830–900\,^{\circ}\mathrm{C}$ and then decrease for $T = 900–950\,^{\circ}\mathrm{C}$. It is recalled that the median and average grain size are determined from grain size distributions of partially (i.e. incompletely) recrystallized samples (minimum estimated recrystallized area in the samples shown by the labels in Fig. 7b). The data thus do not allow conclusions with respect to a correlation between median or average recrystallized grain size and stress.

Delta-histograms (Fig. 5c, i) show that the ECD frequency and area fraction of the classes with the largest grains decrease, while the classes with the smallest grains remain approximately constant and the classes in-between increase in frequency and area fraction. As mentioned earlier, this indicates that the distributions have changed with respect to the heated starting material by grain size reduction of the larger grains and grain growth of the smallest grains in the distribution.

Discussion

Deformation mechanisms

Microstructural observations show widespread evidence for crystal plastic deformation in all deformed samples, such as deformation bands, grain flattening, undulose extinction, subgrains and deformation lamellae in the larger grains (Fig. 2). In addition, equiaxed grains as small as *c.* 10 μm are present, which does not indicate intracrystalline plasticity. According to Walker *et al.* (1990), grains of ~10 μm size can be expected to deform by grain-size-sensitive mechanisms under the conditions investigated in this study. With increasing strain, a larger contribution of grain-size-sensitive flow to the overall creep rate may be expected, as a greater number of small grains are progressively formed. The microstructures of samples deformed to a low strain of *c.* 0.15 do not show widespread evidence for the nucleation of small grains, and grain growth rather than reduction dominates up to this strain. All stress peaks occur at a strain below *c.* 0.20, mostly around a strain of *c.* 0.15. It is therefore inferred that deformation at the stress peaks is strongly dominated by dislocation creep mechanisms, and that grain-size-sensitive mechanisms only become significant at the higher strains.

In Fig. 8, peak stresses are plotted against strain rate for different temperatures. The data show a high strain rate sensitivity of the stress, which is in general agreement with previous studies on the mechanical behaviour of marble at high temperature (Rutter 1974; Schmid *et al.* 1980; Covey-Crump 1998; Pieri *et al.* 2001*a*). Our peak stress data were obtained for a limited range of conditions, which hampers the drawing of definite conclusions regarding the rate-controlling mechanism during dislocation creep on the basis of the mechanical data. No microstructural evidence that allows direct identification of the rate controlling mechanism is available. However, the high strain rate sensitivity of stress observed in this study and the additional observation in previous studies that the strain

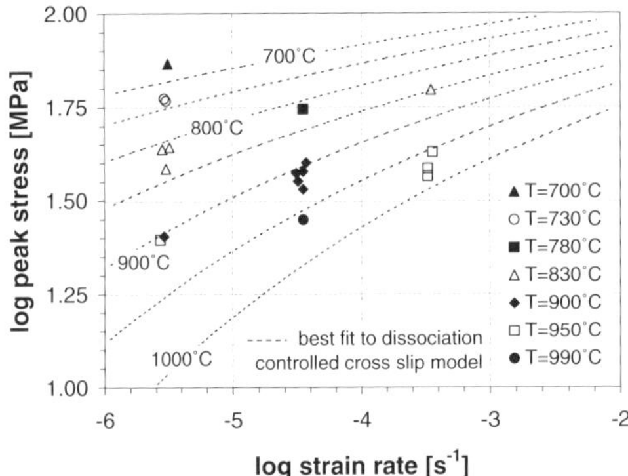

Fig. 8. Plot of log peak stress versus log strain rate for all experiments. Isotherms for a best fit of the data to a theoretical model for flow by dissociation-controlled cross-slip of dislocations are indicated.

rate sensitivity changes with temperature and/or stress (Rutter 1974; Schmid *et al.* 1980; Pieri *et al.* 2001*a*) indicate that the mechanical behaviour of marble at high temperature cannot be accurately described by a single power law creep equation of the type derived for conventional climb-controlled dislocation creep (with a constant stress exponent *n*, see Renner & Evans 2002). Rather, it has been suggested that deformation of marble under conditions similar to those investigated in the present study is controlled by cross slip of dislocations (De Bresser & Spiers 1990). A detailed evaluation of mechanical data for calcite against dislocation cross-slip models is beyond the scope of this paper, but is the subject of a separate study that includes new deformation experiments on Carrara marble (see De Bresser 2002). In this separate study, cross slip is presented as a mechanism capable of dislocation recovery (as is dislocation climb), that thus prevents continuous work hardening (Poirier 1976; De Bresser 1991). Models for dislocation cross slip all start from dissociated dislocations, forming stacking faults, which need to be constricted to allow the cross-slip step. This is followed by (re-)dissociation of the dislocation and continued glide in the cross-slip plane. The dissociation step is thought to control the rate of dislocation cross slip of coarse grained marble at intermediate to high temperature, and hence the creep rate. For present purposes, we rely on this conclusion, and we infer that the peak stress data represent nominally pure dislocation creep and can be best described by a rate equation for dissociation-controlled cross-slip creep.

We have fitted our mechanical data to the dissociation-controlled cross-slip model of Escaig (1968), in which creep rate ($\dot{\varepsilon}_{cs}$) is related to flow stress (σ) by:

$$\dot{\varepsilon}_{cs} = K\left(\frac{\sigma}{\mu}\right)^2 \exp\left[-\frac{\Delta U_{cs}\left(1-\frac{\alpha b \sigma}{\gamma}\right)}{kT}\right] \quad (4)$$

where γ is the stacking fault energy (in J/m^2), α is a geometrical constant, b is the Burgers vector (taken as $b = 6.37 \times 10^{-10}$ m), k is the gas constant ($k = 1.381 \times 10^{-23}$ J/K) and K is a dimensionless rate constant. ΔU_{cs} represents the activation energy at zero applied stress (in J) described by:

$$\Delta U_{cs} = \left(\frac{\mu b^3}{1859(\gamma/\mu b)}\right)\left(\ln\left[\frac{2\sqrt{3}}{16\pi(\gamma/\mu b)}\right]\right)^{0.5} \quad (5)$$

where μ is the shear modulus. Best fit of (Equation 4) to the peak stress data, applying a non-linear least squares regression method, yields LOG(K) $= 8.98 \pm 1.19$, $\gamma = 0.17 \pm 0.02$ J/m^2 and $\alpha = 1.41 \pm 0.15$ (we used $\delta\mu/\delta T = -9.7$ and $\mu = 34180$ Pa at 0 K and 300 MPa). Isotherms for cross-slip-controlled dislocation creep have been included in Fig. 8 using Equations 4 and 5 with best fit values for K, γ and α.

Evolution of flow stress with strain

The microstructure of Carrara marble evolves with strain and this evolution appears to be associated with a decrease in flow stress. Under the investigated conditions, weakening following the peak in flow stress may be explained by:

(1) Geometric weakening due the development of a lattice-preferred orientation (LPO), in particular by progressive alignment of easy slip directions.
(2) Weakening due to the recovery of dislocations by grain boundary migration consuming grains with high dislocation densities, which leads to a reduction in dislocation density faster than recovery by dislocation climb or cross slip.
(3) Weakening due to an increase of the importance of grain-size-sensitive flow mechanisms as a result of the nucleation of small grains by dynamic recrystallization.

Texture weakening by alignment of easy slip systems parallel to the shear direction has been proposed as an explanation for the limited rheological weakening (~10%) observed in high-strain torsion tests (Pieri *et al.* 2001*a*). However, this cannot explain the rheological weakening observed in experiments deformed in axial compression, because the easy slip systems tend to rotate to an orientation perpendicular to the compression direction, effectively making it progressively harder to continue glide of dislocations along the slip planes.

Weakening due to recovery of dislocations by grain boundary migration is put forward by Rutter (1998) to explain the weakening observed in axial compression or extension tests on Carrara marble. Although grain boundary migration is widespread in the samples deformed at high temperature, grain boundary migration is less active at temperatures below *c.* 730 °C and sample scale microstructural alteration by the consumption of grains with high dislocation densities has not been observed at these temperatures. However, strain weakening at $T = 700$–730 °C (experiments 50LM730, 65LM700) is of a similar magnitude to that observed at higher temperature, which indicates that the weakening cannot be controlled by grain boundary migration.

We will now evaluate the possibility that the observed strain weakening after a strain of *c.* 0.15 is the result of an increase in the contribution of grain size sensitive deformation mechanisms to the overall creep rate, due to the nucleation of small grains by dynamic recrystallization. Comparison of the stress–strain curves with microstructural observations reveals that there is a direct correlation between the onset of strain weakening and nucleation of small grains by dynamic recrystallization after a strain of *c.* 0.15. Up to the peak in stress, grains predominantly grow and flatten without significant nucleation of new grains. In order to quantify the contribution of grain-size-sensitive flow of grains in the distribution to the overall rheological behaviour of Carrara marble, a combined grain-size-sensitive grain-size-insensitive flow law is required. In addition, this flow law needs to be able to account for variations in the grain size distribution (cf. Ter Heege *et al.* 2000). For grain size insensitive flow in Carrara marble, the flow law for dissociation controlled cross slip, given in Equations 4 and 5, is used. Studies of grain-size-sensitive deformation in calcite aggregates (Schmid *et al.* 1977; Walker *et al.* 1990) indicate that the creep rate for grain-size-sensitive flow ($\dot{\varepsilon}_{gss}$) can be related to flow stress (σ) by:

$$\dot{\varepsilon}_{gss} = A\sigma^n d^m \exp\left[-\frac{H}{RT}\right] \quad (6)$$

where H is the apparent activation enthalpy for grain-size-sensitive flow (in kJ/mol), R is the gas constant (in kJ/mol), d is grain size (in μm), A is a rate constant (in $\mathrm{Mpa}^{-n}\,\mu\mathrm{m}^{-m}\,\mathrm{s}^{-1}$, for σ in MPa) and m, n are the empirically-derived grain size and stress exponent, respectively. Both studies suggest that grain-size-sensitive flow of calcite rocks occurs by grain boundary sliding accommodated by a combination of dislocation and diffusional processes. The most extensive study was performed by Walker *et al.* (1990) on synthetic hot-pressed calcite aggregates. They found that $A = 10^{4.93}\,\mathrm{Mpa}^{-n}\,\mu\mathrm{m}^{-m}\,\mathrm{s}^{-1}$, $H = 190$ kJ/mol, $m = -1.87$ and $n = 1.67$.

In our analysis, we adopt the following assumptions.

(1) In an individual grain size class, grain-size-insensitive creep is dominated by dissociation-controlled cross slip with a creep rate ($\dot{\varepsilon}_{cs}$), described by Equations 4 and 5, and grain-size-sensitive creep is dominated by grain boundary sliding with a creep rate ($\dot{\varepsilon}_{gss}$) described by (Equation 6).

(2) Both mechanisms contribute independently to the total creep rate ($\dot{\varepsilon}_i$) of grains in an individual grain size class (i), given by $\dot{\varepsilon}_i = \dot{\varepsilon}_{cs} + \dot{\varepsilon}_{gss}$.

(3) The grain size distribution at the peak stress (at $\varepsilon = 0.1$–0.2) is similar to the distribution analysed at a strain of ~0.15.

(4) Either stress is uniform in the sample with the bulk flow stress (σ) equal to local stresses of grains (σ_i) in individual grain size classes ($\sigma = \sigma_1 = \sigma_2 = \cdots = \sigma_i$), or strain rate is uniform with the bulk strain rate ($\dot{\varepsilon}$) equal to local strain rates of grains in individual grain size classes ($\dot{\varepsilon} = \dot{\varepsilon}_1 = \dot{\varepsilon}_2 = \cdots = \dot{\varepsilon}_i$). The uniform stress and uniform strain rate assumptions give upper and lower bounds for the rate of deformation in polycrystalline aggregates (Freeman and Ferguson 1986; Wang 1994; Ter Heege *et al.* 2000). If stress is uniform, the bulk strain rate for Carrara marble with a grain size distribution consisting of j classes can be written as the volume average of the strain rate of grains in individual classes with a volume fraction v_i (Raj & Ghosh 1981; Freeman and Ferguson 1986; Ter Heege *et al.* 2000):

$$\dot{\varepsilon} = \dot{\varepsilon}_1 v_1 + \dot{\varepsilon}_2 v_2 + \cdots + \dot{\varepsilon}_j v_j = \sum_{i=1}^{i=j} \dot{\varepsilon}_i v_i. \quad (7)$$

Or, if strain rate is uniform, the bulk stress can be written as:

$$\sigma = \sigma_1 v_1 + \sigma_2 v_2 + \cdots + \sigma_j v_j = \sum_{i=1}^{i=j} \sigma_i v_i \quad (8)$$

with $v_1 + v_2 + \cdots + v_j = \sum_{i=1}^{j} v_i = 1$.

Volume fractions of individual grain size classes were calculated using the StripStar computer program (Heilbronner & Bruhn 1998).

The program uses the so-called Schwartz–Saltykov method to calculate volume fractions of individual classes of a 3D grain size distribution from frequencies of grain size distributions based on 2D grain sections, assuming grains are spherical (Underwood 1968). Bulk flow stresses can be determined directly from Equation 8 using Equations 4 to 6 and the imposed strain rate in the experiments or flow stress can be chosen in Equations 4 to 6 so that the bulk strain rate from Equation 7 equals the imposed strain rate in the experiments. In this way, changes in bulk flow stress can be calculated from changes in grain size distribution for both the uniform stress and the uniform strain rate assumption. Reliable calculations of strain weakening due to changes in grain size distribution can be made at 830 °C, using samples 36LM830/0.15 and 36LM830, and at 900 °C, using samples 36LM900/0.15 and 36LM900/0.90. The results are presented in Fig. 9.

Taking into account all uncertainties, notably regarding the applicability of the grain size sensitive flow law to Carrara marble, as well as experimental reproducibility and sample variability, the calculated strain weakening is close to the observed weakening for both uniform stress and strain rate. At this stage, this should not be regarded as undisputable evidence that the strain weakening observed in our experiments is due to an increased contribution of grain-size-sensitive flow as a result of changes in grain size distribution. It is merely meant to illustrate that changes in grain size distribution can explain the observed weakening and should be considered as a serious alternative for other weakening mechanisms like the alignment of easy slip systems (Pieri *et al.* 2001*a*) or recovery by grain boundary migration (Rutter 1998).

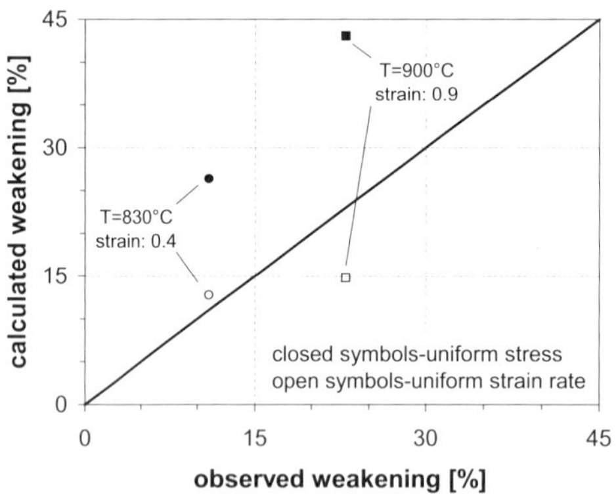

Fig. 9. Strain weakening of the flow stress relative to the peak stress, calculated using changes in grain size distribution and associated increase of the importance of grain size sensitive flow with increasing strain, plotted against the strain weakening observed in experiments at 830 and 900 °C. Calculated strain weakening, assuming uniform stress or strain rate (giving upper and lower bounds), and the observed weakening are in reasonably good agreement for these experiments as the points fall near the line where calculated and observed weakening are similar.

Evolution of dynamically recrystallizing microstructure

The systematic investigation of the effect of deformation of Carrara marble on the microstructure at different strains, temperatures and stresses carried out in this study, provides a picture of the microstructural evolution during transient deformation in the approach to steady state. We make three important inferences based on the microstructural observations, regarding: (1) the effect of heating of the starting material prior to deformation; (2) processes altering the microstructure during deformation; and (3) the evolution of (recrystallized) grain size as a function of strain.

First, heating of the starting material for about 1 hr at 900 °C resulted in an increase of the average and median grain size due to minor growth of small grains at the expense of larger grains in the grain size distribution. Because the grain size distributions of the deformed samples are compared to the heated starting material rather than to that of the original block, observed changes in distributions are truly syn-deformational.

Second, grain size distributions continuously change during deformation by a combination of growth of the smallest grains and refinement of the larger grains in the distribution. Grain growth predominantly occurred by migration of boundaries of small relatively strain-free grains into larger grains. Based on the three distinctive types of microstructural features identified in the results, we infer that grain size reduction in marble occurred by: (1) the formation of new grains by progressive subgrain rotation (Fig. 2b); (2) nucleation of grains behind grain boundary bulges (Fig. 2f); and (3) dissection of grains by formation of new grain boundaries across larger parts of old grains or by cross-cutting of (migrated) new grains (Fig. 2h). Grain growth and grain size reduction are competing processes during dynamic recrystallization. The relative importance of both processes is dependent on the deformation conditions, as evidenced by comparing the samples deformed to strains of ~0.40. At relatively low temperatures (below ~730 °C, in samples 50LM730 and 65LM700), core–mantle microstructures (Fig. 2e) develop and grain size distributions are positively skewed and leptokurtic (Figs. 4o–p). At relatively high temperatures microstructures with interlocking (e.g. sample 36LM950, Fig. 2c) or even polygonal shaped grains (e.g. sample 15LM950, Fig. 2g) develop and grain size distributions are closer to log-normal (Fig. 4c, g, k, m-n, q–r, 6). This indicates that at high temperatures relatively rapid grain boundary migration prevents the development of core–mantle structures by rapid growth of nucleated grains (mantle), consuming larger older grains with higher dislocation density (core). From this, it is concluded that the relative kinetics of grain growth by grain boundary migration and grain refinement during dynamic recrystallization determines the characteristics of the grain size distribution.

Third, the delta-histograms show that with ongoing deformation, the classes with positive frequencies progressively belong to smaller grain sizes (compare Figs 5e–g). Thus not only did the complete (old + new grains) grain size distribution continuously evolve to high strain, but also the recrystallized grain size distribution did not reach steady state. It follows that recrystallized grain size–stress (piezometric) relations based on low strain (*c.* 0.15) microstructures need to be regarded as characteristic of transient deformation, and hence are not generally applicable. At best the recrystallized grain size obtained from the delta-histograms gives an upper limit of the recrystallized grain size at steady state. In addition, analysis of nucleated small (sub-grain size) grains will give a lower limit, as recrystallized grains grow by grain boundary migration after nucleation. These grains will be the smallest in the grain size distribution that develops during steady state as a result of a competition between grain growth and grain size reduction. It is emphasized that individual grains can undergo several cycles of nucleation and growth before a steady state microstructure has developed. It also means that it will be problematic to calibrate separate piezometric relations for rotation and migration recrystallization (cf. Rutter 1995) as grains nucleated from subgrains will continuously grow by grain boundary migration and grains are continuously refined by a combination of subgrain rotation and nucleation behind bulges resulting from grain boundary migration.

Comparison with high-strain torsion experiments

Recently, the same type of Carrara marble ('Lorano Bianco' type) has been deformed by Pieri *et al.* (2001*a*, *b*) to high shear strains in torsion, at conditions that fall in the range of conditions investigated in this study. These authors claim that at a shear strain of 1 (equivalent to a natural strain of *c.* 0.5), recrystallized grains make up only *c.* 5% of the area (in 2D section) of the sample. The delta-histograms presented in this study show that already at a strain of *c.* 0.4, a higher percentage of the area (at least 15–52%) of the sample has recrystallized at all conditions (Table 2, Fig. 5c, i). Pieri *et al.* (2001*a*) determined the recrystallized area fraction on the basis of the area fraction made up by new nucleated grains with a size approximately similar to the final recrystallized grain size at high strains (shear strain of *c.* 11). Their estimate should be regarded as a minimum because grains are not always reduced to this size, and grains may grow after nucleation during ongoing dynamic recrystallization. In this study, we determined the recrystallized area fraction from the delta-histograms, indicating the alteration of the area distribution with respect to the starting material. The delta-histograms (Fig. 5d–h) of this study and the grain size distributions analysed by Pieri *et al.* (2001*a*) show that the grain size distributions continue to change up to very high strains (unattainable in axial compression) and that grain size is progressively reduced. As mentioned before, this means that the microstructural observations in the present study are indicative of transient deformation. However, the results can be used to investigate the microphysical processes altering the microstructure during dynamic recrystallization and their effect on grain size distribution. In addition, inferences on the development of a steady-state microstructure can be made as it is likely that that the processes active during transient deformation operate during steady-state deformation as well.

The amount of weakening observed in the experiments conducted in this study (9–41%) is comparable or higher than the weakening observed by Pieri *et al.* (2001*a*) in their torsion tests, and of similar magnitude as observed by Rutter (1995, 1998). It should be noted that part of the observed weakening in the samples deformed to strains larger than *c.* 0.3 may be an artefact of the data processing as deformation in the samples becomes increasingly heterogeneous with increasing strain. Pieri *et al.* (2001*a*) claimed that mechanical steady state was only reached at high strain ($\gamma > 5$, equivalent to a natural strain of *c.* 1.6), which is clearly not reached in this study. The weakening observed in the axial compression tests in this study cannot be explained by the progressive alignment of easy slip directions, as argued above. This means that either the weakening in the torsion tests of Pieri *et al.* was not caused by the progressive alignment of easy slip directions, in contrast to their suggestion, or that the weakening in the two studies occurred by different mechanisms due to differences in the deformation geometry. However, a mechanism different from alignment of easy slip directions is more likely, as it is unclear to us how deformation geometry can affect the weakening mechanism.

Geodynamical implications

The recent high-strain torsion tests on anhydrite (Stretton & Olgaard 1997), calcite (Casey *et al.* 1998; Pieri *et al.* 2001*a*) and olivine rocks (Bystricky *et al.* 2000) indicated that steady-state deformation may be reached only at high strains (Paterson & Olgaard 2000), e.g. at shear strains above 10 for microstructural steady state in Carrara marble (Pieri *et al.* 2001*a*, *b*). This implies that transient rheological behaviour associated with evolving microstructure is important in nature, where rocks deformed to a wide range of strains occur. In practice, it will be difficult to recognize steady state from microstructures of natural rocks, as the results obtained in this study indicate that grain size continuously decreases with ongoing deformation. However, the present study shows that it is possible to calculate the evolution of flow stress during transient deformation from the evolution of grain size distribution, provided that the individual grain-size-sensitive and grain-size-insensitive rate equations are well constrained for the material under consideration. If further research supports these findings, the method will provide better constraints for transient deformation of natural rocks evolving towards steady state, because changes in grain size distribution due to dynamic recrystallization are accounted for. A microphysical model that describes the evolution of the grain size distribution during dynamic recrystallization is required.

In the approach to steady state, grain size distributions of experimentally deformed Carrara marble evolve by a combination of grain size reduction and grain growth acting as competing processes. Both processes need to balance in order to develop a steady-state microstructure. This implies that the grain size distribution at steady state depends on the relative kinetics of grain growth and grain size reduction, determined by the deformation conditions. Hence, the steady-state grain size distribution can be expected to be a characteristic of true steady state with distribution parameters that are uniquely related to stress and temperature. Our data suggest that the grain size distributions are close to lognormal, although some deviations from an ideal lognormal distribution are present. Considering that the deviations bear no systematic relation with stress or temperature, we infer that in general the distributions are best described by a lognormal grain size distribution. This corroborates models that predict the development of a steady-state grain size distribution during dynamic recrystallization based on a competition between nucleation and growth (e.g. Shimizu 1999). If the distributions at steady state are approximately lognormal, they can be uniquely described by the median and standard deviation (Aitchison & Brown 1957), which can then be regarded as microstructural state variables. The development of a unique steady-state grain size distribution during dynamic recrystallization by a competition between grain size reduction of large grains deforming by grain-size-insensitive mechanisms, and growth of small grains deforming by grain-size-sensitive mechanisms, is consistent with the boundary hypothesis model, proposed by De Bresser *et al.* (1998, 2001).

On the basis of the results presented in this paper, some inferences can be made on the causes for rheological weakening and strain localization in Carrara marble. The results indicate that extensive rheological weakening by a complete switch in deformation mechanism due to grain size reduction associated with dynamic recrystallization is not possible, because at steady state deformation will occur by a combination of grain-size-sensitive and grain-size-insensitive creep mechanisms. Instead, dynamic recrystallization alters the grain size distribution, causing limited weakening due to an increase in the relative contribution of grain-size-sensitive creep mechanisms. It is therefore concluded that in single phase materials, such as Carrara marble, the degree of weakening caused by dynamical recrystallization is probably insufficient to produce significant strain localization (Braun *et al.* 1999; Rutter 1999). This means that in order to explain strain localization, often observed in natural marble rocks, other weakening mechanisms should be considered (see White *et al.* 1980), such as brittle processes (Bos & Spiers 2001), or grain growth should be inhibited, for example by the presence of second phases (Olgaard 1990).

Summary and conclusions

Cylindrical samples of Carrara marble were deformed to natural strains of 0.15–0.90 in axial compression at temperatures in the range of 700–990 °C, stresses of 15–65 MPa, strain rates of 3.0×10^{-6}–$4.9 \times 10^{-4}\ s^{-1}$ and a confining pressure of 150 or 300 MPa. The aim of the experiments was to investigate the evolution of grain size distribution and rheological behaviour as a result of dynamic recrystallization during transient deformation in the approach to steady state. We conclude the following.

(1) Under the deformation conditions investigated, Carrara marble showed ductile flow behaviour with high strain rate sensitivity of flow stress, if compared at fixed strains. The observed mechanical behaviour is in agreement with the results of previous axial compression experiments on Carrara marble for which cross-slip-controlled dislocation creep was inferred to be the dominant deformation mechanism. Microstructural evidence was found for dislocation creep and dynamic recrystallization by grain boundary migration and progressive subgrain rotation.

(2) In general, individual stress–strain curves show a peak stress followed by continuous weakening to natural strains of at least 0.4 at all investigated conditions and up to ~0.9 at 900 °C, without reaching true steady state.

(3) Measurements of grain sizes on traced micrographs show an evolution of the grain size distribution from bimodal at the start to a positively skewed or roughly lognormal distribution with increasing strain. At strains above *c.* 0.15, the median and average grain size decreases with increasing strain. The recrystallized area, as seen in 2D sections, ranged from a minimum of 8% to 52%, depending on deformation conditions. The grain size distributions evolved during dynamic recrystallization by a competition of grain growth due to grain boundary migration and grain size reduction due to progressive subgrain rotation, dissection of grains by new grains or grain boundaries or nucleation of new grains behind grain boundary bulges caused by grain boundary migration.

(4) Geometric ('texture') weakening resulting from alignment of easy slip systems, which has been proposed to occur in high-temperature torsion tests on Carrara marble, cannot explain the weakening as observed in the current experiments because of the different (axial symmetric) deformation geometry. Further, since rates of grain boundary migration were slow in some of the experiments showing strain weakening, it is unlikely that the observed weakening is due to selected removal of dislocations by grain boundary migration.

(5) We infer that peak stress and subsequent weakening behaviour observed in torsion and compression experiments was probably due to a change in relative importance of grain-size-sensitive (diffusion) creep with respect to grain-size-insensitive (dislocation) creep as the number of small grains in the distribution increased with strain. This is quantitatively supported by calculations of weakening resulting from changes in grain size distributions based on a composite flow law, combining grain boundary sliding accommodated by both diffusional and dislocation processes and cross-slip-controlled dislocation creep. This type of behaviour is consistent with the boundary hypothesis model proposed by De Bresser *et al.* (1998, 2001) and implies that both steady state and transient grain size distribution parameters will be uniquely related to stress and temperature. Also consistent with the framework of the boundary hypothesis is the fact that the degree of weakening, observed both in our experiments and in torsion tests, is minor and probably insufficient to produce significant strain localization.

(6) If the present conclusions on Carrara marble apply to other materials, a model describing the evolution of the grain size distribution during dynamic recrystallization would offer a way of tracking rheological behaviour in nature during deformation from transient to steady state. The data presented in this paper may form a starting point for formulating such a model.

P. van Krieken and G. Kastelein are thanked for technical support. We gratefully acknowledge constructive comments from G. Lloyd, J. Newman and G. Pennock.

References

Aitchison, J. & Brown, J. A. C. 1957. *The lognormal distribution, with special reference to its use in economics*. Cambridge University Press.

Bos, B. & Spiers, C. J. 2001. Experimental investigation into the microstructural and mechanical evolution of phyllosilicate-bearing fault rock under conditions favouring pressure solution. *Journal of Structural Geology*, **23**, 1187-1202.

Braun, J., Chéry, J., Poliakov, A., Mainprice, D., Vauchez, A., Tomassi, A. & Daignières, M. 1999. A simple parameterization of strain localization in the ductile regime due to grain size reduction: a case study for olivine. *Journal of Geophysical Research*, **104**, 25167-25181.

Bystricky, M., Kunze, K., Burlini, L. & Burg, J.-P. 2000. High shear strain of olivine aggregates: rheological and seismic consequences. *Science*, **290**, 1564–1567.

Casey, M., Kunze, K. & Olgaard, D. L. 1998. Texture of Solnhofen limestone deformed to high strains in torsion. *Journal of Structural Geology*, **20**, 255–267.

COVEY-CRUMP, S. J. 1998. Evolution of mechanical state in Carrara marble during deformation at 400° to 700 °C. *Journal of Geophysical Research*, **103**, 29781–29794.

DE BRESSER, J. H. P. 1991. Intracrystalline deformation of calcite. PhD thesis, Utrecht University. *Geologica Ultraiectina*, **79**.

DE BRESSER, J. H. P. 2002. On the mechanism of dislocation creep of calcite at high temperature: Inferences from experimentally measured pressure sensitivity and strain rate sensitivity of flow stress. *Journal of Geophysical Research*, in press.

DE BRESSER, J. H. P. & SPIERS, C. J. 1990. High-temperature deformation of calcite single crystals by r^+ and f^+ slip. *In:* Knipe, R. J. & Rutter, E. H. (eds) *Deformation Mechanisms, Rheology and Tectonics*. Geological Society, London, Special Publications, **54**, 285–298.

DE BRESSER, J. H. P., PEACH, C. J., REIJS, J. P. J. & SPIERS, C. J. 1998. On dynamic recrystallization during solid state flow: effects of stress and temperature. *Geophysical Research Letters*, **25**, 3457–3460.

DE BRESSER, J. H. P., TER HEEGE, J. H. & SPIERS, C. J. 2001. Grain size reduction by dynamic recrystallization: can it result in major rheological weakening? *International Journal of Earth Sciences*, **90**, 28–45.

ESCAIG, B. 1968. Sur le glissement dévié des dislocations dans la structure cubique à faces centrées. *Journal de Physique*, **29**, 225–239.

FREEMAN, B. & FERGUSON, C. C. 1986. Deformation mechanism maps and micromechanics of rocks with distributed grain sizes. *Journal of Geophysical Research*, **91**, 3849–3860.

GRIGGS, D. T. 1936. Deformation of rocks under high confining pressures–I. Experiments at room temperature. *Journal of Geology*, **44**, 541–577.

GRIGGS, D. T. & MILLER, W. B. 1951. Deformation of Yule marble: Part I. Compression and extension experiments on dry Yule marble at 10000 atmospheres confining pressure, room temperature. *Geological Society of America Bulletin*, **62**, 853–862.

HEARD, H. C. & RALEIGH, C. B. 1972. Steady state flow in marble at 500° to 800 °C. *Geological Society of America Bulletin*, **83**, 935–956.

HEILBRONNER, R. & BRUHN, D. 1998. The influence of three-dimensional grain size on the rheology of polyphase rocks. *Journal of Structural Geology*, **20**, 695–705.

KARATO, S. & WU, P. 1993. Rheology of the upper mantle. *Science*, **260**, 771–778.

KERN, H. & WENK, H.-R. 1983. Calcite texture development in experimentally induced ductile shear zones. *Contributions to Mineralogy and Petrology*, **83**, 231–236.

MCDONNELL, R. D. 1997. *Deformation of fine-grained synthetic peridotite under wet conditions*. PhD thesis, Utrecht University. *Geologica Ultraiectina*, **152**.

MCDONNELL, R. D., PEACH, C. J. & SPIERS, C. J. 1999. Flow behavior of fine-grained synthetic dunite in the presence of 0.5 wt% H_2O. *Journal of Geophysical Research*, **104**, 17823–17845.

MOLLI, G., CONTI, P., GIORGETTI, G., MECCHERI, M. & OESTERLING, N. 2000. Microfabric study on the deformational and thermal history of the Alpi Apuane marbles (Carrara marbles), Italy. *Journal of Structural Geology*, **22**, 1809–1825.

OLGAARD, D. L. 1990. The role of second phase in localizing deformation. *In:* KNIPE, R. J. & RUTTER, E. H. (eds) *Deformation Mechanisms, Rheology and Tectonics*. Geological Society, London, Special Publications, **54**, 175–181.

PASSCHIER, C. W. & TROUW, R. A. J. 1996. *Microtectonics*. Springer-Verlag, Heidelberg.

PATERSON, M. S. & OLGAARD, D. L. 2000. Rock deformation tests to large shear strains in torsion. *Journal of Structural Geology*, **22**, 1341–1358.

PIERI, M., BURLINI, L., KUNZE, K., STRETTON, I. & OLGAARD, D. L. 2001*a*. Rheological and microstructural evolution of Carrara marble with high shear strain: results from high temperature torsion experiments. *Journal of Structural Geology*, **23**, 1393–1413.

PIERI, M., KUNZE, K., BURLINI, L., STRETTON, I. C., OLGAARD, D. L., BURG, J.-P. & WENK, H.-R. 2001*b*. Texture development of calcite by deformation and dynamic recrystallization at 1000 K during torsion experiments of marble to large strains. *Tectonophysics*, **330**, 119–140.

POIRIER, J. P. 1976. On the symmetrical role of cross-slip of screw dislocations and climb of edge dislocations as recovery processes controlling high-temperature creep. *Revue de Physique Appliquee*, **11**, 731–738.

RAJ, R. & GHOSH, A. K. 1981. Micromechanical modelling of creep using distributed parameters. *Acta Metallurgica*, **29**, 283–292.

RENNER, J. & EVANS, B. 2002. Do calcite rocks obey the power-law creep equation? *In*: DE MEER, S., DRURY, M. R., DE BRESSER, J. H. P. & PENNOCK, G. M. (eds) *Deformation Mechanisms, Rheology and Tectonics: Current Status and Future Perspectives*. Geological Society, London, Special Publications, **200**, 293–307.

RUTTER, E. H. 1974. The influence of temperature, strain-rate and interstitial water in the experimental deformation of calcite rocks. *Tectonophysics*, **22**, 311–334.

RUTTER, E. H. 1995. Experimental study of the influence of stress, temperature, and strain on the dynamic recrystallization of Carrarra marble. *Journal of Geophysical Research*, **100**, 24651–24663.

RUTTER, E. H. 1998. Use of extension testing to investigate the influence of finite strain on the rheological behaviour of marble. *Journal of Structural Geology*, **20**, 243–254.

RUTTER, E. H. 1999. On the relationship between the formation of shear zones and the form of the flow law for rocks undergoing dynamic recrystallization. *Tectonophysics*, **303**, 147–158.

RUTTER, E. H. & BRODIE, K. H. 1988. The role of tectonic grain size reduction in the rheological stratification of the lithosphere. *Geologische Rundschau*, **77**, 295–308.

SCHMID, S. M. 1975. The Glarus Overthrust: field evidence and mechanical model. *Eclogae Geologicae Helvetiae*, **68**, 247–280.

SCHMID, S. M., BOLAND, J. N. & PATERSON, M. S. 1977. Superplastic flow in fine grained limestone. *Tectonophysics*, **43**, 257–291.

SCHMID, S. M., PANOZZO, R. & BAUER, S. 1987. Simple shear experiments on calcite rocks: Rheology and microfabric. *Journal of Structural Geology*, **9**, 747–778.

SCHMID, S. M., PATERSON, M. S. & BOLAND, J. N. 1980. High temperature flow and dynamic recrystallization in Carrara marble. *Tectonophysics*, **65**, 245–280.

SHIMIZU, I. 1999. A stochastic model of grain size distribution during dynamic recrystallization. *Philosophical Magazine A*, **79**, 1217–1231.

STRETTON, I. C. & OLGAARD, D. L. 1997. A transition in deformation mechanism through recrystallization: evidence from high strain, high temperature torsion experiments. *EOS Transactions of the American Geophysical Union*, **78**, F723.

TER HEEGE, J. H., DE BRESSER, J. H. P. & SPIERS, C. J. 2000. The effect of grain size distribution on rheology. *EOS Transactions of the American Geophysical Union*, **81**, F1208-F1209.

UNDERWOOD, E. E. 1968. Particle size distribution. *In:* DEHOFF, R. T. & RHINES, F. N. (eds) *Quantitative Microscopy*. McGraw-Hill Series in Materials Science and Engineering, 149–200.

WALKER, A. N., RUTTER, E. H. & BRODIE, K. H. 1990. Experimental study of grain-size sensitive flow of synthetic, hot-pressed calcite rocks. *In:* Knipe, R. J. & Rutter, E. H. (eds) *Deformation Mechanisms, Rheology and Tectonics*. Geological Society, London, Special Publications, **54**, 259–284.

WANG, J. N. 1994. The effect of grain size distribution on the rheological behavior of polycrystalline materials. *Journal of Structural Geology*, **16**, 961–970.

WHITE, S. H., BURROWS, S. E., CARRERAS, J., SHAW, N. D. & HUMPHREYS, F. J. 1980. On mylonites in ductile shear zones. *Journal of Structural Geology*, **2**, 175–187.

Deformation of the continental lithosphere: Insights from brittle–ductile models

JEAN-PIERRE BRUN

Géosciences Rennes UMR 6118 CNRS/Rennes 1 University, 35042 Rennes Cedex, France (e-mail: Jean-Pierre.Brun@univ-rennes1.fr)

Abstract: 2D deformation experiments on multilayer models of a brittle–ductile lithosphere are reviewed. The experimental method consists of simulating simplified strength profiles which incorporate brittle (frictional) and ductile (viscous) rheologies with gravity forces. A selection of models built with sand and silicone putties to represent brittle and ductile lithosphere layers, respectively, is used to illustrate the effects of variations in strength profiles on deformation patterns. Models of extension first consider lithosphere necking and the development of narrow rifts, with application to continental rifts and passive margins, and, second, lithosphere spreading with application to the development of wide rifts and core complexes. Models of compression compare sandbox-type and brittle–ductile multilayer-type experiments. Results are applied to mountain belt formation and, in particular, to the Pyrenees and the western Alps. Both extensional and compressional experiments demonstrate that the presence/absence of a sub-Moho brittle mantle and the coupling/decoupling between brittle and ductile layers play a dominant role on localized versus distributed deformation, at lithosphere scale.

Since the early work of Goetze & Evans (1979), strength profiles have become a basic tool for the study of lithosphere mechanics. Even if a large number of uncertainties remains, strength profiles substantiate the so-called 'brittle–ductile' layering of the lithosphere (see review by Ranalli 1997). During the last fifteen years, laboratory experiments on sand–silicone multilayer models have been used to test the mechanical behaviour of a large spectrum of simplified strength profiles, with application to various types of lithosphere-scale deformation. A selection of 2D experiments, in extension and compression, is used here to discuss the effects of mantle rheological layering and of coupling between brittle and ductile layers on lithosphere deformation.

Modelling principles and techniques

The experimental method and techniques developed at Geosciences Rennes to study the deformation of layered brittle–ductile systems operating at crustal or lithosphere scales consist of simulating simplified strength profiles which incorporate brittle (frictional) and ductile (viscous) rheologies with gravity forces (Figs 1a and b). Scaling relationships between the prototype and the model are obtained by keeping the average strength of the ductile layers correctly scaled with respect to the strength of the brittle layers and the gravity forces.

Scaling

For a small-scale model to be representative of a natural example (a prototype), a dynamic similarity in terms of distribution of stresses, rheologies and densities between model and prototype is required (Hubbert 1937; Ramberg 1981).

In the equation of dynamics:

$$(\partial\sigma_{ij}/\partial X_{ij}) + \rho(g - (\partial^2\varepsilon_{ij}/\partial t^2)) = 0, \qquad (i, j = 1, 2, 3) \quad (1)$$

where σ_{ij} are the components of stress, ε_{ij} the components of deformation, X_{ij} space coordinates, ρ density, g acceleration of gravity and t time. Any modification of the length scale multiplies the first term by $\sigma^*(L^*)^{-1}$, the second by ρ^*g^*, and the third by $\rho^*\varepsilon^*(t^*)^{-2}$, where the exponent (*) refers to model/prototype ratios (e.g. $\sigma^* = \sigma_m/\sigma_p$).

Because Equation 1 must be respected, we obtain the two following conditions:

$$\sigma^* = \rho^* g^* L^*, \quad (2)$$

$$\varepsilon^* = g^*(t^*)^2. \quad (3)$$

Inertial forces being negligible in geological processes (Hubbert 1937; Ramberg 1981), only Equation 2 must be verified.

Brittle–ductile models

In our experiments, carried out under normal gravity, the gravity ratio is $g^* = 1$. The densities of model materials range between 1,100 and 1,400 kg/m^3 and those of rocks between 2,300 and 3,000 kg/m^3. Because model and prototype densities are in the same order of magnitude,

From: DE MEER, S., DRURY, M. R., DE BRESSER, J. H. P. & PENNOCK, G. M. (eds) 2002. *Deformation Mechanisms, Rheology and Tectonics: Current Status and Future Perspectives.* Geological Society, London, Special Publications, **200**, 355–370. 0305-8719/02/$15

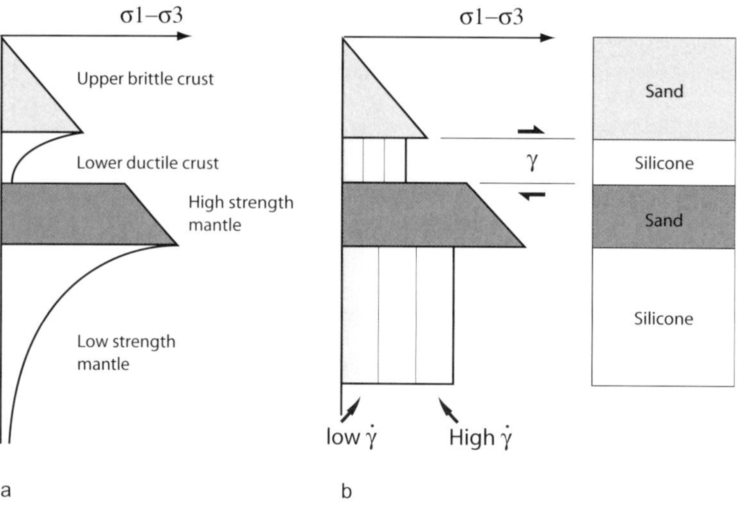

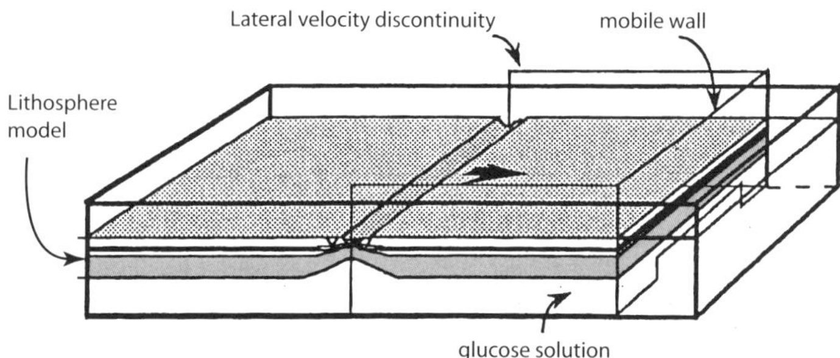

Fig. 1. Principles of modelling of brittle–ductile systems. (**a**) Strength profile of natural system, i.e. prototype. (**b**) Multilayer model made of sand and silicone putties and corresponding strength profiles for varying shear strain rates $\dot{\gamma}$. (**c**) Apparatus used to deform a sand–silicone multilayer floating above a glucose solution. See text for further explanation.

the density ratio is $\rho^* \approx 1$. Therefore Equation 2 simplifies to:

$$\sigma^* \approx L^*. \quad (4)$$

In other words, the ratios of stresses and lengths must be nearly equal.

To model brittle layers with a Mohr–Coulomb type behaviour:

$$\tau = C + (\tan\Phi)\sigma \quad (5)$$

where τ is the shear stress, C the cohesion in MPa, Φ the angle of friction, and σ the normal stress, we use Fontainebleau sand with density $\rho \approx 1{,}400\,\mathrm{kg/m^3}$ and whose angle of friction Φ is in the range 30–33°, without significant cohesion C. According to Byerlee (1978), most crustal rocks are characterized by $C = 50\,\mathrm{MPa}$ and $\Phi = 31°$. The length–scale ratio being $L^* \approx 10^{-6}$, the cohesion of the sand should be around 5.10^{-5} Pa. This value is negligible compared with the maximum differential stresses in models which, for $\Phi = 30°$, are given in extension by:

$$\sigma_1 - \sigma_3 = 2/3\rho g z \quad (6)$$

and in compression by:

$$\sigma_1 - \sigma_3 = 2\rho g z \quad (7)$$

where σ_1 and σ_3 are the maximum and minimum principal stresses and z is the depth.

To represent ductile layers, we use silicone putties with variable Newtonian viscosities (from 10^3 to 10^4 Pa s) and variable densities (from 1,100 to 1,600 kg/m^3). According to the mean strain rate applied to the models, the layers of silicone putty can have extremely variable strengths given by:

$$\sigma_1 - \sigma_3 = \eta\dot{\varepsilon} \text{ or } \eta\dot{\gamma} \quad (8)$$

where η is the viscosity and $\dot{\varepsilon}$ the strain rate. The shear strain rate $\dot{\gamma}$ is used as an approximation for experiments in which deformation in ductile layers is close to simple shear. Therefore, for given relative thicknesses of sand and silicone putty, a wide range of strength profiles can be simulated for different applied strain rates (Fig. 1b).

Experimental procedure

In lithosphere experiments, models made of sand and silicone layers float above a low-viscosity syrup representing the asthenosphere. The basal boundary is nearly slip-free allowing models to respond isostatically to gravity forces induced by variations in thickness. Displacements are applied at a constant rate along the lateral borders of models (Fig. 1c) or models are allowed to spread under their own weight, giving a displacement rate which continuously decreases with time. In models designed to represent crustal or upper-crustal sub-systems, displacements are applied on a rigid basal plate at constant rate, using moving plastic sheets. In spreading-type experiments, nearly basal free-slip is obtained with a liquid soap film coating the basal boundary.

Uses and limitations of brittle–ductile models

As indicated above, multilayers of sand and silicone putties can be deformed at various strain rates to test the deformation response of a wide range of strength profiles, in extension or compression. The technique has been applied to a large number of tectonic situations, at various scales. The present paper reviews a selection of 2D experiments pertinent for the understanding of the lithosphere rheological layering and the tectonic consequences. However, a number of 3D applications have also been made (e.g. continental indentation: Davy & Cobbold 1988; lateral extrusion in collision: Ratschbacher *et al.* 1991; subduction: Faccenna *et al.* 1996; oblique subduction: Pinet & Cobbold 1992; Pubellier & Cobbold 1996; oblique rifting: Tron & Brun 1991; block rotation in rifting: Souriot & Brun 1992). Sand–silicone multilayers have also been used as laboratory models to study the mechanics of faulting and to test new theoretical aspects (e.g. fractal nature: Davy *et al.* 1990*b*; localization and fault growth: Davy *et al.* 1995).

The technique has, however, some inherent limitations. To keep the models tractable in the laboratory, their size is necessarily limited. This imposes a level of resolution according to the size of the geological problem concerned. Models are unable to incorporate rheological changes due to temperature variations during deformation. The measurement of dynamic parameters at any point in models (e.g. stresses, displacements, or strain rates) is not possible. Numerical models which do not suffer from these limitations are, on the other hand, less adapted to simulation of faulting and to 3D modelling. Consequently, a combination of analogue and numerical modelling is desirable for many problems.

Lithosphere extension

Continental lithosphere extension occurs in various tectonic environments. From a mechanical point of view, a major difference exists between localized rifting within a normal thickness crust (30–40 km) and widespread extension of thickened crust (Brun & Choukroune 1983), called narrow and wide rifting respectively (Buck 1991). Narrow rifting starts with continental rifts (e.g. Rhine, Baikal, or Ethiopian rifts) and ends with development of passive margins and with continental break up (Fig. 2). Wide rifting occurs during (e.g. Tibet or Aegean) or after the cessation of convergence (e.g. Basin and Range of the Western United States). Analogue modelling of lithosphere extension, presented in the two next sections, substantiates that the difference between narrow rift and wide rift results from two distinct modes of mechanical instability: necking versus spreading (Brun 1999).

Necking and narrow rifts

In necking experiments, constant rate displacements are applied to the lithosphere model at lateral velocity discontinuities to stimulate a localized necking (Fig. 1c). The behaviour of the lithosphere with various rheological layers can therefore be tested. Experiments which have been carried out (Allemand 1990; Allemand

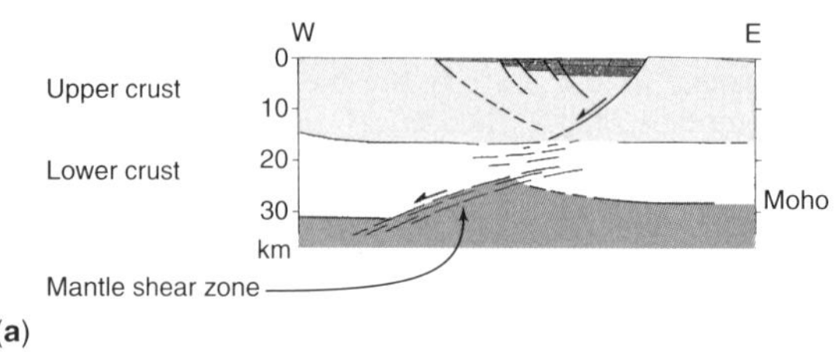

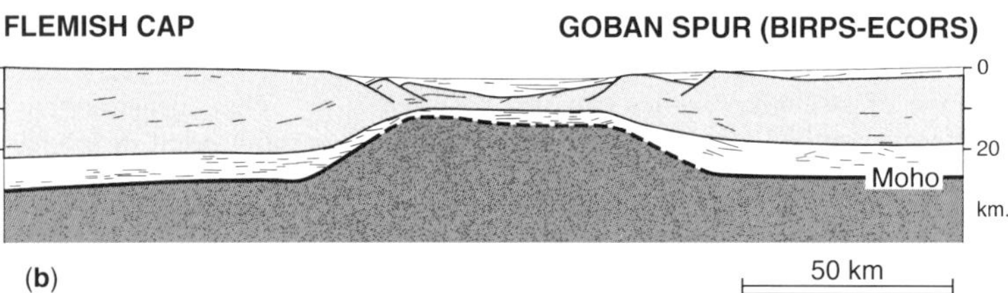

Fig. 2. Natural examples of narrow rifts. (**a**) Section of the Rhine graben from the DEKORP–ECORS deep seismic line (modified after Brun *et al.* 1992). The existence of a mantle shear zone suggests a high strength and localizing sub-Moho mantle. Faults mark the upper brittle crust and thinning of the lower crust suggests a ductile deformation. (**b**) Flemish Cap–Goban Spur section from deep seismic lines showing the geometry of conjugate passive margins in North Atlantic at the time of continental break up (modified after Keen *et al.* 1989).

et al. 1989; Allemand & Brun 1991; Beslier 1991; Brun & Beslier 1996) have led to two important outcomes.

First, the presence/absence of a high-strength brittle layer in the upper part of the lithosphere mantle plays a major role. In models where the mantle is entirely ductile, the strength peak is located at the base of the upper crust (e.g. three-layer-type model B/D/D; Fig. 3a). In such models, a single-rift structure develops, directly connected to the lateral velocity discontinuities. In four-layer models with a brittle sub-Moho mantle, the strength peak is located in the lithospheric mantle (e.g. four-layer-type model B/D/B/D; Fig. 3b). A localized zone of necking connected to the lateral velocity discontinuities develops first in the high-strength mantle layer. Extension is transmitted to the upper crust through ductile flow of the lower crust, and generally several rifts are produced.

Second, when a high-strength sub-Moho mantle is present, the tectonic response of the model is also strongly controlled by the coupling between brittle and ductile layers (Fig. 4). For given viscosities of the ductile layers, the mechanical behaviour of the multilayer model depends on the strain rate. At low strain rate (low B/D coupling; Fig. 4a), deformation remains localized in brittle layers, resulting in a single rift in the upper crust and a single zone of necking in the sub-Moho mantle. At higher strain rates (medium and strong B/D coupling; Figs 4b and c), deformation becomes distributed and the width of the deformed zone increases. The brittle sub-Moho mantle is affected by several zones of necking, of which the number increases with increasing strain rate and which define a boudinage-type structure at lithosphere

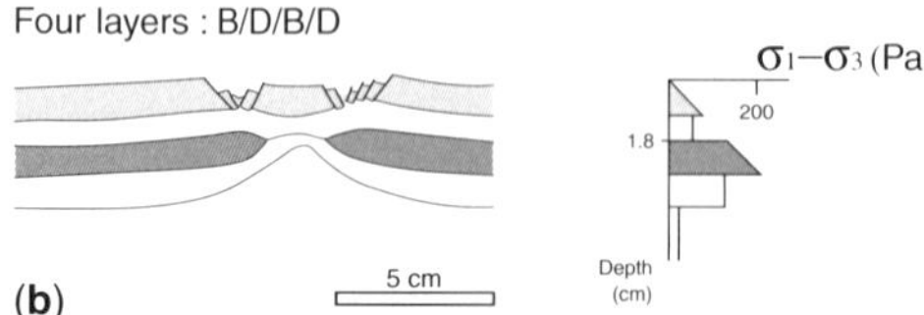

Fig. 3. Narrow rift-type extension of sand–silicone multilayers without (**a**) and with (**b**) a brittle layer representing a high-strength sub-Moho mantle (modified after Allemand *et al.* 1989). See text for further explanation.

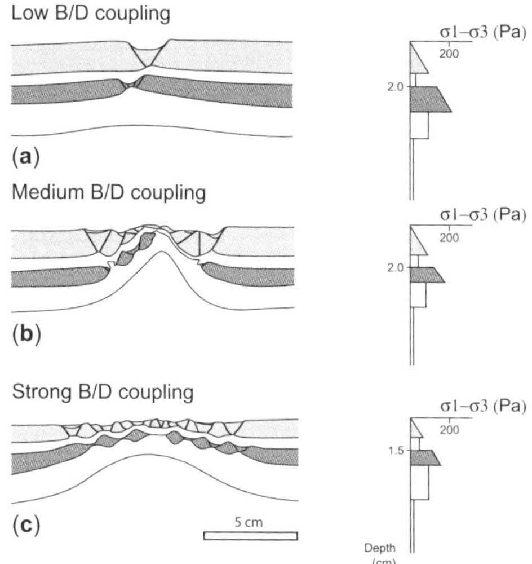

Fig. 4. Effects of brittle–ductile coupling on the necking of four-layer-type models (modified after Beslier 1991).

scale. With increasing deformation, in medium B/D coupling type models (Fig. 4b), stretching concentrates between some boudins, allowing the lower ductile lithospheric mantle to rise up and eventually reach the surface. Note that this mantle exhumation process does not require any type of detachment fault cross-cutting the whole lithosphere (for details see Brun & Beslier 1996). Moreover, experiments run for various types of brittle–ductile coupling sustain that the whole lithosphere stretches according to a nearly pure shear fashion, despite the internally heterogeneous deformation involving local shear zones.

These models also illustrate the particular role of the lower ductile crust in lithosphere stretching. At low strain rates for a given viscosity, the lower crust acts as a decoupling level (i.e. a décollement) between the upper crust and the lithospheric mantle. Conversely, at high strain rates the ductile strength of the lower crust increases and the upper crust becomes more strongly coupled with the lithospheric mantle. Comparable effects can result from variations of viscosity at constant strain rate. As viscosity is strongly temperature dependent, the strength of ductile layers decreases with heating and increases with cooling.

Consequently, if a temperature change occurs during rifting the viscosity of the lower crust also changes, leading to a variation in ductile strength and in brittle–ductile coupling (see discussion on Viking graben extension in Brun & Tron 1993).

Spreading and wide rifts/core complexes

In spreading experiments, a two-layer-type model is left flowing under its own weight above a rigid horizontal plate (Fig. 5) (Faugère & Brun 1984; Brun *et al.* 1985) or above a low viscosity fluid (Gautier *et al.* 2000; Martinod *et al.* 2000). Deformation is driven by gravitational forces. The ductile layer flows horizontally, allowing the brittle layer to extend. A major difference in comparison with necking experiments is that, from the early stages of deformation onwards, faulting affects most of the model length. In necking experiments faulting affects only a part of the model, the width of which at initiation depends on brittle–ductile coupling.

In models with homogeneous viscosity of the ductile layer, the envelope of the brittle–ductile interface remains nearly flat during deformation (Fig. 5). Normal faulting accommodating extension in the brittle layer defines tilted blocks or horsts and grabens. At high strain rate, the tilted block mode is favoured (Brun 1999) and the dominant fault vergence and sense of block tilting are controlled by the sense of shear at the brittle-ductile interface (Brun *et al.* 1985). The experiment shown in Fig. 5 shows blocks

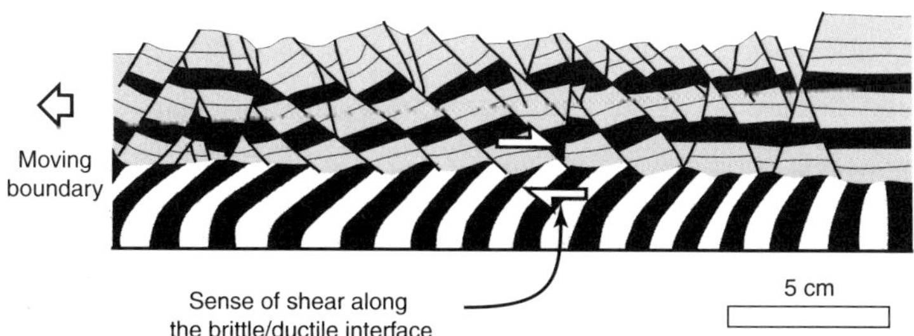

Fig. 5. Wide-rift-type extension in a two-layer brittle–ductile model (modified after Faugère & Brun 1984).

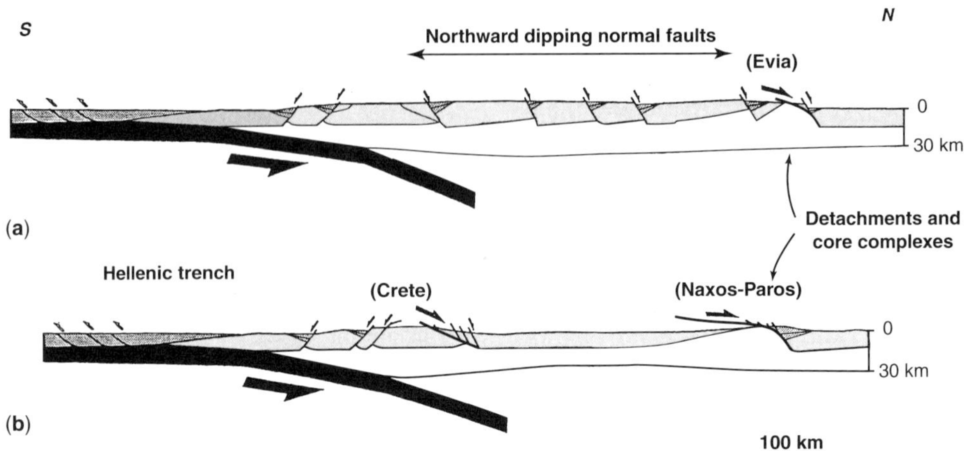

Fig. 6. Regional-scale cross sections showing the pattern of wide rifting in the south Aegean (modified after Jolivet *et al.* 1994). (**a**) Section across the Peloponeses showing N-dipping normal faults and crustal-scale tilted blocks. (**b**) Section in the central part of the south Aegean Sea passing through the Naxos-Paros core complexes.

tilted toward the moving boundary. The relative sense of displacement on normal faults between blocks is identical to the sense of shear at the brittle–ductile interface.

The kinematic pattern of block tilting observed in experiments (Fig. 5) can be compared with the regional-scale pattern of tilted blocks observed in the south Aegean. A N–S cross-section across the Peloponneses (Fig. 6a) displays, along more than 200 km, a series of master normal faults all dipping northward and delimiting crustal-scale blocks tilted toward the south. Aegean extension is due to the southward migration of the Hellenic trench (i.e. subduction rollback: Le Pichon & Angelier 1981), which allows gravity spreading of a previously thickened lithosphere (Gautier *et al.* 2000). Therefore the model and the Aegean geometry display similar kinematic relationships between faulting and block tilting in the extended domain and displacement of the domain boundary.

In models where a low viscosity anomaly is placed in the ductile layer immediately below the brittle layer, the envelope of the brittle–ductile interface does not remain flat during deformation (Fig. 7a) and leads to the formation of core complexes (Brun *et al.* 1994). Even for low viscosity contrast (i.e. one order of magnitude) between the local anomaly and the surrounding ductile layer, stretching of the brittle layer localizes above the low viscosity anomaly (see Fig. 7a). Subsequent localized thinning of the brittle layer is compensated by an uprise of ductile material. Above and around the domal uprise of ductile layers, deformation of the brittle layer is governed by a main detachment fault and a listric accommodation fault (Fig. 7b). The detachment fault and its upward convex shape result from the progressive coalescence and rotation to horizontal of initially steeply dipping normal faults, as proposed by Buck (1988) and Wernicke & Axen (1988). Strong block rotation in the detachment footwall is accommodated by a major listric fault. This process, demonstrated in experiments, is at variance with early models of core complexes which postulated or argued an initial flat-lying attitude of detachment faults (Wernicke 1985).

The theoretical analysis of Buck (1991) identified three modes of lithosphere extension: narrow rifts, wide rifts, and core complexes. Analogue modelling suggests that core complexes, instead of being a separate mode of extension, can be considered as local perturbations within the wide rift mode. An illustration of this is given by the wide-rift-type Aegean extension with coexisting tilted blocks and core complexes (Fig. 6), which developed during the same deformation event (Gautier *et al.* 2000).

Lithosphere thrusting

Lithosphere-scale thrusting is the primary mechanism of mountain belt building. The structure of mountain belts is mostly known through surface geology and deep reflection seismics. However, the role of the mantle part of the lithosphere is still poorly understood. Deciphering the role of lithospheric mantle in the processes of

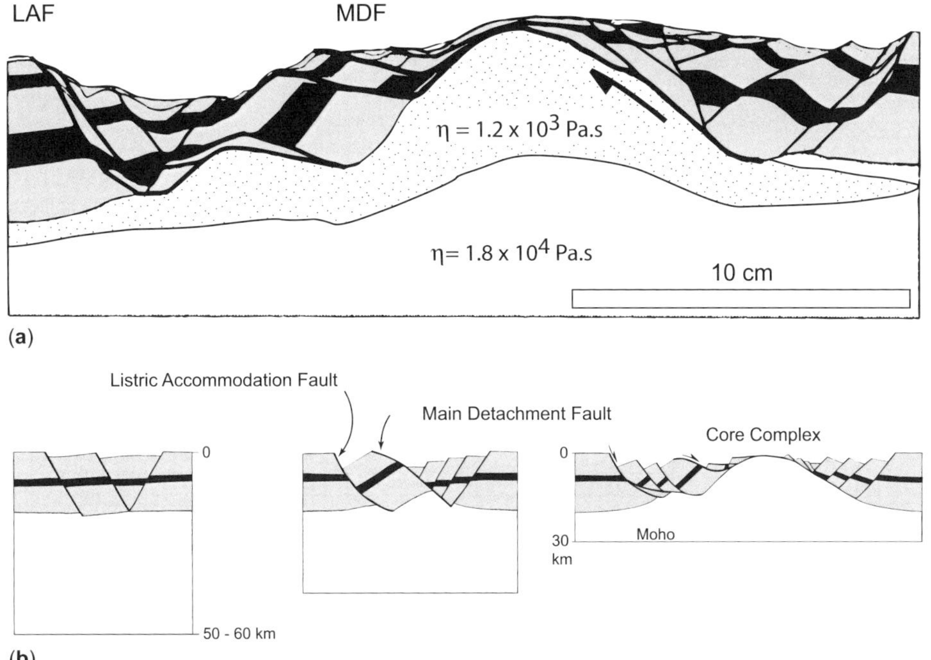

Fig. 7. Core complex-type deformation in a two-layer model (modified after Brun *et al.* 1994). (**a**) Model cross section. Viscosity (η) of the stippled part of the ductile layer is an order of magnitude lower than the rest of the ductile layer. LAF: Listric accommodation fault. MDF: Main detachment fault. Note the strong tilting of brittle blocks below the MDF. (**b**) Sketch of model evolution adapted to continental crust scale.

lithosphere shortening is therefore an important target for which analogue and numerical modelling techniques will be useful tools to test mechanical hypotheses.

Sandbox-type models

Since Hubbert (1951), sandbox experiments have been widely used to model geological faulting. In those of Malavieille (1984), applied to mountain belt formation, the whole continental crust was represented by a single sand layer lying partly above a plastic sheet which was able to slide through a slit in an underlying rigid substratum representing the lithospheric mantle (Fig. 8). When pulling on the plastic sheet, the sand layer was shortened at this basal velocity discontinuity. This gave birth to conjugate thrust faults defining a nearly symmetrical pop-up structure which became asymmetrical with increasing displacement. Shortening was accommodated on one side of the pop-up by a permanent fault zone which remained active during the whole experiment. A sequence of faults with short lifetimes (here called transient faults) developed on the opposite side. Each of these faults accommodated a small amount of shortening until it lost its connection with the basal velocity discontinuity. It then became inactive, and was incorporated in the pop-up and passively transported on top of the permanent fault zone. A new transient fault initiated and the process reiterated, leading to an asymmetric growth of the pop-up. The 'pop-up stage' ended and gave place to a 'thrust wedge stage' when the transient faults started developing at increasing distances from the basal velocity discontinuity, using the basal surface (i.e. the Moho) as a décollement.

Numerical models by Willet *et al.* (1993), Beaumont *et al.* (1996) or Ellis & Beaumont (1999) which also considered the entire continental crust as made of frictional materials, provided results rather similar to those of sandbox experiments (Fig. 9). The whole crust was involved in an asymmetric pop-up structure whose constitutive faults were connected to a velocity discontinuity (S on Fig. 9) located along the Moho at the junction between overriding and subducted plates.

Such models with purely frictional properties successfully explain a number of characteristics

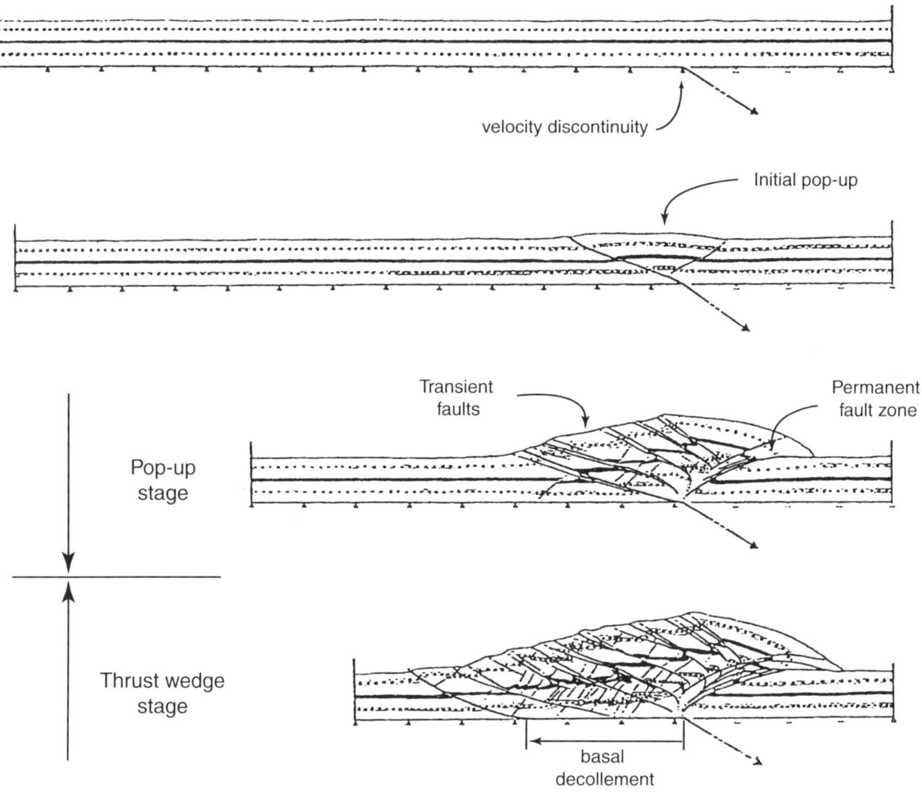

Fig. 8. Sandbox-type experiments simulating the development of a mountain belt at crustal-scale (modified after Malavieille 1984). The transition between an early pop-up stage and a late thrust wedge stage is introduced for the purpose of the present review. See text for further explanation.

of shallow thrust wedge systems (Davis *et al.* 1983). However, their application to mountain belt formation (e.g. Malavieille 1984; Willet *et al.* (1993); Beaumont *et al.* 1996) faces some difficulties. The use of a local velocity discontinuity at the base of the deforming layer (i.e. the plastic sheet sliding through a slit in the rigid substratum in sandbox experiments or the point 'S' in numerical models) strongly predetermines the deformation pattern. The crust whose behaviour is entirely frictional must decouple along the Moho and, consequently, thickening never involves the lithospheric mantle. From a geological point of view, this is not necessarily realistic. Modifications were later introduced, in analogue models (e.g. Gutscher *et al.* 1998) and in numerical models (e.g. Pfiffner *et al.* 2000), to allow a lower part of the crust to be subducted, and to take into account a power-law type rheology of the lower crust (Beaumont & Quinlan 1994; Willet 1999; Pfiffner *et al.* 2000). However, despite these modifications, in these models the lithospheric mantle remains undeformable, preventing the possibility of crust–mantle imbrication.

Brittle–ductile models

In experiments made to study thrust systems within a brittle–ductile lithosphere, a constant rate of shortening is applied at lateral velocity discontinuities to sand–silicone multilayers floating on top of a low viscosity fluid representing the asthenosphere. The procedure is the same as for lithosphere extension (Fig. 1c), except that boundary displacements here result in shortening. The behaviour of various types of rheological layering of the lithosphere have been tested in 2D by Davy *et al.* (1990*a*); Davy & Cobbold (1991); Martinod (1991); Burg *et al.* (1994); Martinod & Davy (1994) and Davy *et al.* (1995).

In models that do not involve a sand layer representing a high-strength sub-Moho mantle (two layers B/D and three layers B/D/D;

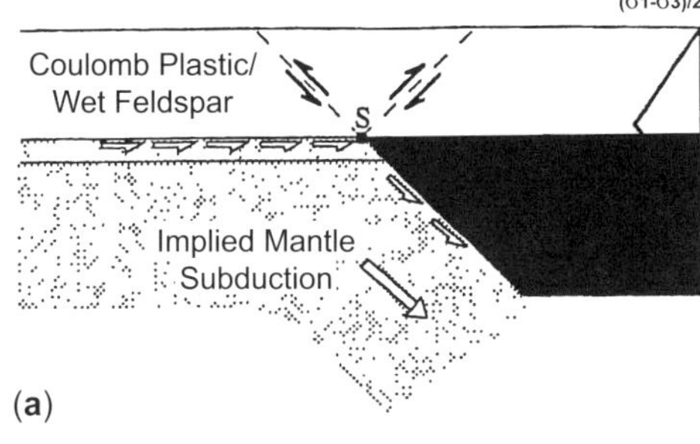

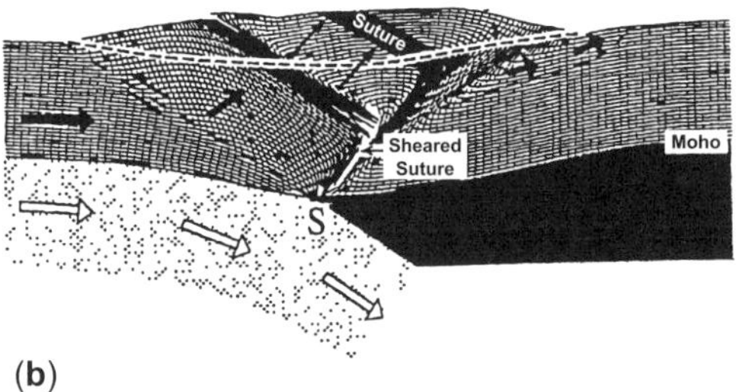

Fig. 9. Numerical model of crustal-scale deformation in Alpine-type orogens (modified after Beaumont *et al.* 1996).

Figs.10a and b), the whole lithosphere tends to thicken rather homogeneously. The upper brittle crust is affected by thrusts with variable vergence in a rather large domain, the width of which increases with increasing shortening. Even when the ductile mantle is stronger than the ductile crust (three layers B/D/D; Fig. 10b), no strain localization develops at lithosphere scale. Such types of models, where highest strengths are located in the upper crust, produce wide domains of distributed deformation, which do not really compare with mountain belts of Phanerozoic age. They could, however, have important implications in domains of high geothermal gradients (e.g. if mantle delamination occurs) and, for similar reasons, in Archean or lower Proterozoic orogenies.

Models where the sub-Moho mantle is made of sand (four layers B/D/B/D; Fig. 10c) show a different type of deformation. A localized zone of thrusting cross-cutting the whole lithosphere develops at early stages. The underthrusted unit is subducted into the asthenosphere. At the end of the experiment presented here, a wedge of crustal material has been carried down to more than three times the initial crustal thickness. In nature, this would mean more than 100 km. In terms of dynamic evolution, two stages appear in the development of thrusting in the upper crust. After a first stage of deformation, localized at the mergence of the lithosphere-scale thrust zone, thrusting migrates outward, but the underlying lithospheric mantle remains undeformed. This two-stage evolution, first localized and then propagating laterally in the upper crust, may be compared with the pop-up/thrust wedge evolution observed in sandbox models. It is the lower ductile crust which acts here as a décollement between the deformed upper crust and the undeformed mantle.

Buckling experiments of four-layer-type models (B/D/B/D) (Martinod 1991), have been carried out with moderate amounts of shortening (Fig. 11a and b). Despite the application of an uniaxial displacement, the experiments display 3D deformation effects within the lithosphere mantle such as lateral duplications (Fig. 11a) or vergence variability (Fig. 11b) of mantle thrusts. In these experiments, the upper crust was strongly decoupled from the lithospheric mantle. The upper crust deformation resulted in pop-up and pop-down structures which show no simple relationships with the sub-Moho

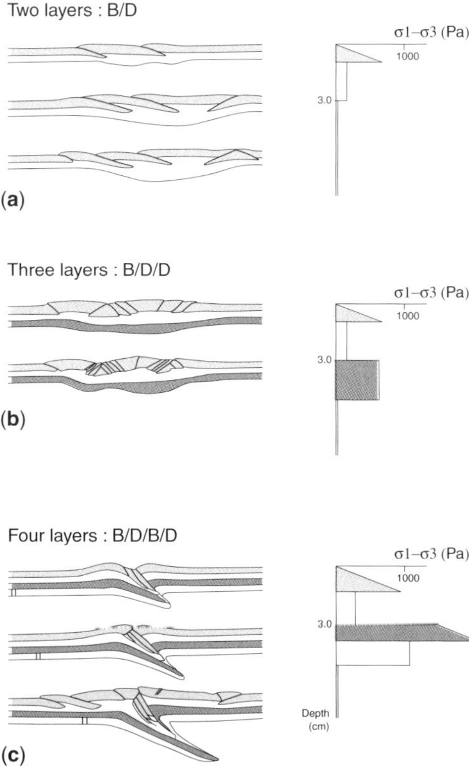

Fig. 10. Brittle–ductile models of lithosphere scale thrusting (modified after Davy & Cobbold 1991). See text for further explanation.

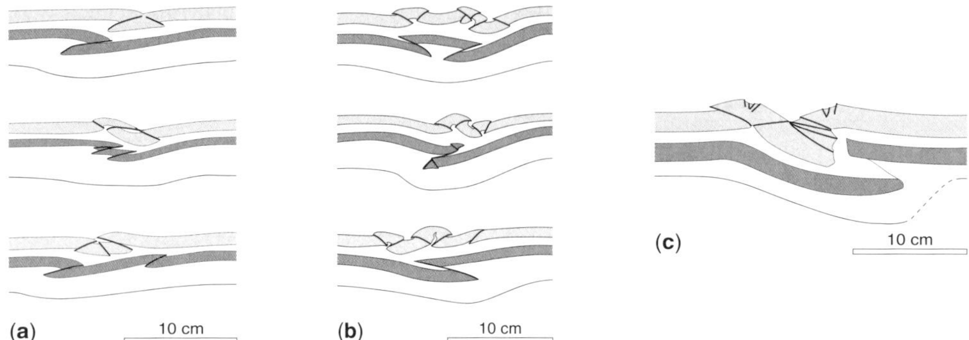

Fig. 11. Patterns of upper crust and mantle thrusting in four-layer-type models. (**a**) and (**b**). Serial sections showing geometrical variations of thrusts along strike in two different models (modified after Martinod 1991). (**c**) Model showing a strong decoupling along the lower ductile crust (modified after Davy & Cobbold 1991).

mantle thrust zones. The lower crust acted here as an efficient décollement zone. This contrasts with more strongly coupled models in which the asymmetry of upper crust deformation reflects the asymmetry of mantle thrusting (Figs 10c and 11c).

The Pyrenean mountain belt as imaged by deep reflection seismics (Fig. 12) consists of a central domain of brittle upper crust uplifted between thrust faults that dip northward in the south and southward in the north. The amount of horizontal shortening and the erosion are moderate enough to allow a restoration of the initial crustal-scale geometry (Roure *et al.* 1989; Choukroune *et al.* 1990). The Pyrenees therefore offer an example of a collisional belt with moderate deformation on the large scale. Even if the initial geometry is obviously complicated by a previous extensional event, the thrust system is comparable to the four-layer model B/D/B/D of Fig. 10c. As in the model, all thrusts are connected to the mantle thrust zone and the axial zone defines an asymmetric pop-up at the crustal scale. These observations support the notion of a high strength and localizing sub-Moho mantle (Davy *et al.* 1990*a*). The restoration (upper case in Fig. 12) suggests an amount of shortening in the order of 100 km and, to the south, major thrust faults cut the whole upper crust and flatten into the lower crust. As observed in the four-layer model of Fig. 10c, the evolution of the Pyrenean thrust system seems to have reached the stage at which deformation starts to migrate outward using the lower crust as a décollement.

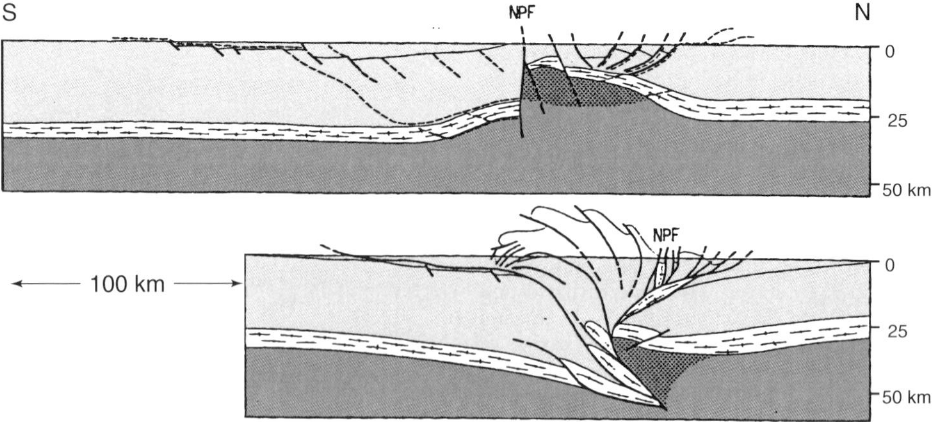

Fig. 12. Structure of the Eastern Pyrenees from the ECORS seismic line and crustal-scale balanced cross section (modified after Choukroune *et al.* 1990). NPF: North Pyrenean Fault.

The Western Alps, as imaged by deep reflection seismics (Fig. 13a), allow two different types of interpretation, at lithosphere scale (Figs 13b and c). If deep reflections labelled M1, M2, M3 are all interpreted as Moho reflections, two mantle thrust zones must be considered (Nicolas *et al.* 1990). Such an interpretation, already put forward by Ménard & Thouvenot (1987), is supported by gravity modelling (Bayer *et al.* 1989) and numerical modelling (Burov *et al.* 1999). If reflection M2 does not correspond to a Moho reflection, the whole Alpine deformation pattern, from the Jura mountains to the Po plain, is related to a single mantle thrust (Roure *et al.* 1996; Stampfli & Marchant 1997; Schmid & Kissling 2000). In this second type of interpretation, compatible with gravity modelling by Rey *et al.* (1990), M2 is interpreted as the interface between lower and upper European crust (Schmid & Kissling 2000). Further to the east, the profile NRP 20 East (Schmid *et al.* 1996) displays a less debatable configuration with a single mantle thrust zone, inherited from the previous subduction. Such a lateral variation in the deformation pattern is more probably related to the arcuate shape of the western Alps (see discussion by Schmid & Kissling 2000).

In the interpretation considering two mantle thrusts (Fig. 13b), a first stage of shortening (Fig. 14), accommodated by displacement along the mantle thrust 1 (subduction zone), lead to the imbrication of the Pennine nappes and their incorporation into a crustal-scale structure of pop-up type. A second stage of shortening started with the initiation of the mantle thrust 2 and the thrusting of the external crystalline massives. This interpretation has two important implications. First, the ductile lower crust must have acted as a decoupling layer (i.e. a décollement) between the upper brittle crust and the high-strength sub-Moho mantle. Second, the initiation of mantle thrust 2 should have resulted from an increase in shear strength within mantle thrust zone 1. This could be due to an increasing contact between the high-strength layers of the overriding and subducting mantle. In the second interpretation (Fig. 13c), the mantle thrust zone corresponds to the previous subduction. The lower crust of the subducting plate, the top of which is the reflector M2, was duplicated along a flat-lying shear zone, of more than 50 km. Such a lower crust duplication has two important implications. First, the lower crust is not a weak layer acting as a ductile décollement, but must have a rheological behaviour allowing shear strain localization. Second, decoupling between the lower crust and the mantle occurs directly along the Moho. For the above reasons, the 'two mantle thrust' interpretation is more compatible with four-layer sand–silicone models (Figs. 10 and 11). Conversely, the 'single mantle thrust' interpretation better fits with sandbox type models (Figs. 8 and 9). Choosing between the two types of models obviously needs further mechanical consideration, in particular, concerning the rheology of the lower crust and lithospheric mantle.

Discussion/conclusion

The selection of experiments reviewed here shows that two parameters exert a strong influence on lithosphere-scale deformation patterns, namely mantle rheology and brittle–ductile coupling. The rheological layering of the continental lithosphere is, first of all, dependent on the composition of crust and mantle, including fluids, and thermal state. In addition, deformation parameters, in particular strain rates, are extremely important. The variations of each of these basic parameters, their wide range of possible combinations, and the number of unknowns and inevitable approximations make the problem rather intractable. One can, however, make first-order approximations and separate 'hot lithospheres' with highest strength located in the brittle upper crust from 'cold lithospheres' with highest strength in the sub-Moho lithospheric mantle. This is basically the approximation made in analogue experiments, in which various types of simplified strength profiles are used for materials deformed in compression or extension at different strain rates.

Role of mantle rheology

A high-strength sub-Moho mantle tends to localize deformation at the lithosphere scale. In extension, the high-strength layer develops a necking-type instability which, depending on the applied strain rate, produces one local necking or a boudinage pattern. This localized zone of stretching in the mantle further controls the necking at the lithospheric scale. Crustal stretching can reach high values in the centre of the necking zone, allowing the mantle to exhume. In compression, the high-strength mantle layer is offset at an early stage by a thrust which further cuts the whole lithosphere allowing the subduction of the underthrusted unit.

Conversely, a low-strength ductile mantle tends to distribute the deformation at the lithospheric scale. In extension no zone of necking

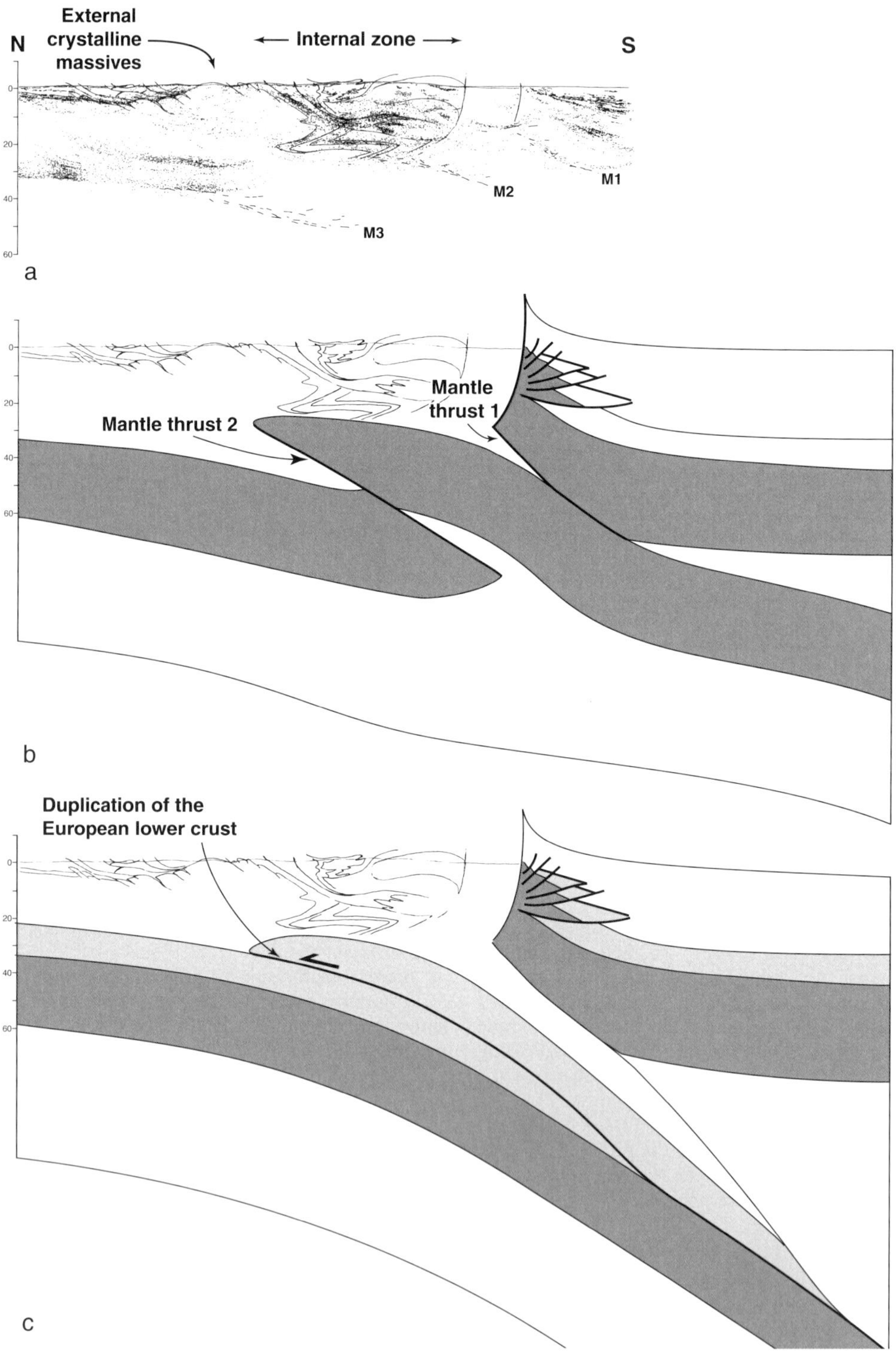
WESTERN ALPS (ECORS / CROP)
N
External crystalline massives
Internal zone
S
M1
M2
M3
a
Mantle thrust 2
Mantle thrust 1
b
Duplication of the European lower crust
c
0
20
40
60

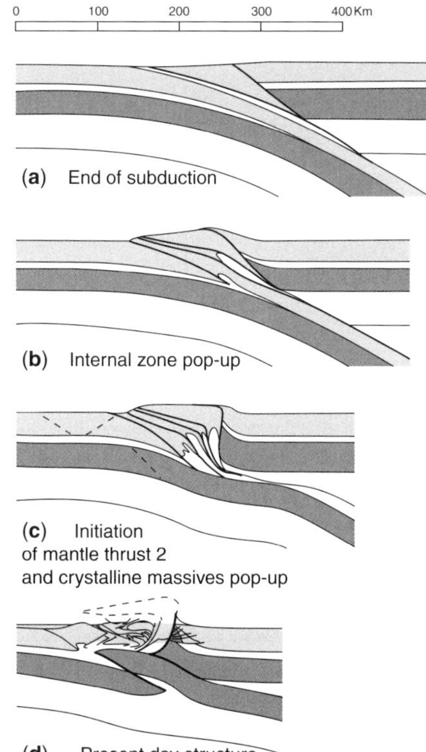

Fig. 14. A possible evolution of the Western Alps, involving two mantle thrusts (see Fig. 13b). (**a**) Geometry of the system at the end of subduction. (**b**) and (**c**) Evolution of the internal zone crustal-scale pop up during displacement along mantle thrust 1. This stage ends up with initiation of a new thrust in the sub-Moho mantle and of an upper crust pop up giving the external crystalline massives. (**d**) Present-day structure. Displacement along mantle thrust 2 is responsible for uplifting of internal zone crust and mantle.

develops. In compression, the lithosphere thickens rather homogeneously and no thrust transects the lithosphere.

From a geological point of view, four-layer models of the lithosphere with a high-strength sub-Moho mantle, i.e. cold lithosphere, better apply to tectonic environments of Phanerozoïc age: continental rifts (Allemand *et al.* 1989; Allemand & Brun 1991); volcanic and non-volcanic passive margins (Brun & Beslier 1996; Callot *et al.* 2000); and continental collision (Davy & Cobbold 1988; Ratschbacher *et al.* 1991). Two- or three-layer models with no high-strength brittle layer in the mantle, i.e. hot lithosphere, apply to more specific tectonic environments of phanerozoic age such as wide rifts of Basin and Range or Aegean type (Gautier *et al.* 2000; Martinod *et al.* 2000), including core complexes (Brun *et al.* 1994). In this case, an originally cold lithosphere becomes a hot lithosphere after thickening and thermal relaxation. This evolution is accompanied by a strong decrease of the bulk lithospheric strength which allows extension by gravity collapse (Brun 1999). More generally, hot lithosphere models should apply to the Archean and early Proterozoic, where no blueschist and only rare eclogites occur, suggesting the absence of lithosphere-scale thrusting. This is in agreement with a drastic change of P–T conditions at 700–800 Ma, after which abnormally high P–T metamorphism began at subduction zones (Maruyama *et al.* 1996).

Strain localization and brittle-ductile coupling

Using two-layer brittle–ductile models with boundary conditions allowing the development of strike-slip faults, Davy *et al.* (1995) showed that a transition between localized and distributed deformation occurs when the strength of the brittle layers is 5–10 times larger than the strength of the ductile layers.

The experiments presented here show that coupling between brittle and ductile layers is especially important in four-layer-type lithospheres with a high strength and strongly localizing sub-Moho mantle. Viscosities being constant, in a given set of experiments, the strength of ductile layers and the coupling between brittle and ductile layers vary as a function of strain rate. Low strain rates decrease the lower crust strength which then can act as a décollement between the upper crust and the mantle. Conversely, high strain rates increase the lower crust strength which then couples the upper crust and the mantle. Necking experiments show that low brittle–ductile coupling favours strain localization at the lithospheric scale and that with increasing coupling deformation becomes distributed.

Fig. 13. Structure of the Western Alps after the ECORS-CROP seismic line. (**a**) Seismic reflectors and contours of geological units (modified after Stampfli & Marchant 1997). Note the three deep reflections M1, M2, M3. Interpretations at lithosphere scale involving two (**b**) and one (**c**) mantle trusts. The dark grey layer represents the high strength sub-Moho mantle. The pale grey layer represents the European lower crust according to the interpretation of Schmid & Kissling (2000). See text for explanation.

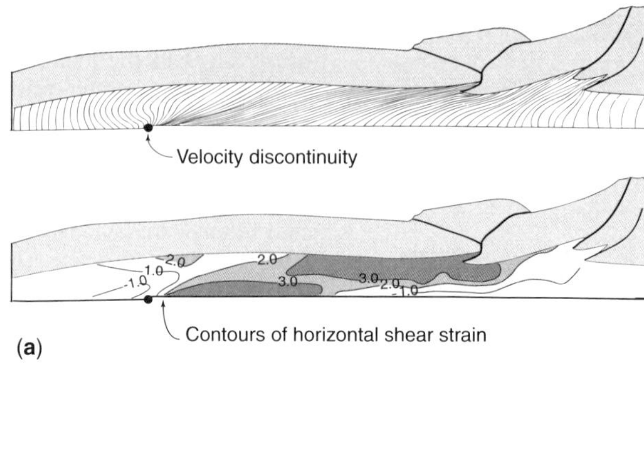

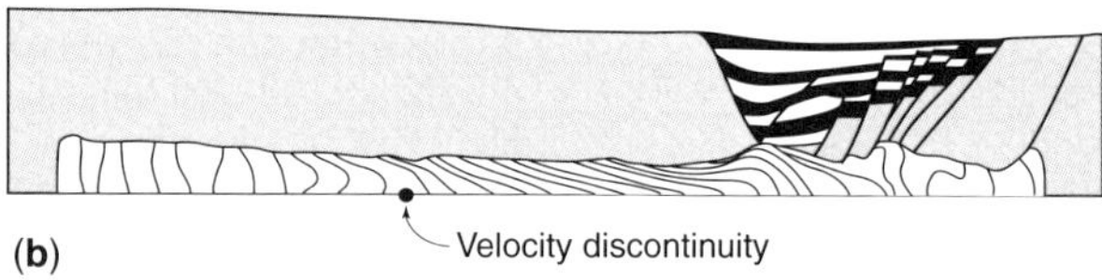

Fig. 15. Two-layer models deformed in compression (**a**) and extension (**b**) along a basal velocity discontinuity. Distortion of passive vertical markers in the ductile layer outlines a high strain zone joining the basal velocity discontinuity to the fault zone in the brittle layer.

Kinematic localization in ductile layers

The conclusions above about localized versus distributed deformation, at the scale of brittle–ductile systems, emphasize that a particular mode of localization commonly develops in brittle–ductile experiments, but inside ductile layers. The two experiments shown in Fig. 15 are two-layer brittle–ductile models deformed in an apparatus which applies a compression (Fig. 15a) or an extension (Fig. 15b) along a velocity discontinuity at the base of the ductile layer. In both experiments, the distortion of passive vertical markers in the ductile layer outlines the existence of a zone of high shear strain oblique to the layer boundaries. The zone joins the basal velocity discontinuity to the base of thrusts (Fig. 15a) or normal faults (Fig. 15b) developing in the brittle layer. This type of strain localization is not dependant on the intrinsic rheological properties as the ductile layer is made of a Newtonian silicone putty. It is a pure kinematic effect in which the deformation zone acts as a shortcut between localized displacements applied at upper and lower layer boundaries.

In four-layer lithosphere models, similar kinematic localization occurs in the lower crust layer, connecting zones of faulting in the upper crust and sub-Moho mantle (e.g. Figs 4 and 5 in Brun & Beslier 1994). This process is likely to occur in the continental lithosphere. This could possibly explain some large-scale low-angle shear zones in the ductile lower crust more effectively than other processes.

I would like to thank the organizers of the DRT 2001 meeting for their invitation to present a review of analogue modelling of brittle–ductile systems, applied to lithosphere deformation. Since the first trials with E. Faugère, twenty years ago, the method and the techniques have been considerably improved, in particular due to the contribution of P. Cobbold and P. Davy and their students, and the permanent technical assistance of J.-J. Kermarrec. Special thanks are due to D. Gapais for hours of discussion when preparing the present paper and his careful check of the manuscript, as well as to X. Fort for the design of the figures. I would also like to thank the reviewers M. Handy and D. Niewland and the editor H. de Bresser for extremely useful remarks and suggestions for improvement of the text.

References

Allemand, P. 1990. Approche expérimentale de la mécanique du rifting continental. *Mémoires de Géosciences Rennes*, **38**.

Allemand, P. & Brun, J.-P. 1991. Relation between continental rift width and rheological layering of lithosphere: a working hypothesis. *Tectonophysics*, **188**, 63–70.

ALLEMAND, P., BRUN, J.-P., DAVY, P. & VAN DEN DRIESSCHE, J. 1989. Symétrie et asymétrie des rifts et mécanismes d'amincissement de la lithosphère. *Bulletin de la Société Géologique France*, **8**, 445–451.

BAYER, R., CAROZZO, M. T., LANZA, R., MILETTO, M. & REY, D. 1989. Gravity modelling along the ECORS–CROP vertical seismic reflection profile through the western Alps. *Tectonophysics*, **162**, 203–218.

BEAUMONT, C. & QUINLAN, G. 1994. A geodynamic framework for interpreting crustal-scale seismic-reflectivity patterns in compressional orogens. *Geophysical Journal International*, **116**, 754–783.

BEAUMONT, C., ELLIS, S., HAMILTON, J. & FULLSACK, P. 1996. Mechanical model for subduction-collision tectonics of Alpine-type compressional orogens. *Geology*, **24**, 675–678.

BESLIER, M.-O. 1991. Formation des marges passives et remontée du manteau: modelisation experimentale et exemple de la marge de la Galice. *Mémoires de Géosciences Rennes*, **45**.

BRUN, J.-P. 1999. Narrow rifts versus wide rifts: inferences for the mechanics of rifting from laboratory experiments. *Philosophical Transactions of the Royal Society of London*, **A357**, 695–710.

BRUN, J.-P. & BESLIER, M.-O. 1996. Mantle exhumation at passive margins. *Earth and Planetary Sciences Letters*, **142**,161–173.

BRUN, J.-P. & CHOUKROUNE, P. 1983. Normal faulting, block tilting and decollement in a stretched crust. *Tectonics*, **2**, 345–356.

BRUN, J.-P. AND TRON V. 1993. Development of the North Viking Graben: inferences from laboratory modelling. *Sedimentary Geology*, **86**, 31–51.

BRUN, J.-P., CHOUKROUNE, P. & FAUGÈRE, E. 1985. Les discontinuités significatives de l'amincissement crustal: application aux marges passives. *Bulletin de la Société Géologique de France*, **8**, 139–144.

BRUN, J.-P., GUTSCHER, M. A. & DEKORP–ECORS TEAMS 1992. Deep crustal structure of the Rhine Graben from DEKORP–ECORS seismic reflection data: a summary. *Tectonophysics*, **208**, 139–147.

BRUN, J.-P., SOKOUTIS, D. & VAN DEN DRIESSCHE, J. 1994. Analogue modelling of detachment fault systems. *Geology*, **22**, 319–322.

BUCK, W. R. 1988. Flexural rotation of normal faults. *Tectonics*, **7**, 959–973.

BUCK, W. R. 1991. Modes of continental lithospheric extension. *Journal of Geophysical Research*, **96**, 20161–20178.

BURG, J.-P., DAVY, P. & MARTINOD, J. 1994. Shortening of analogue models of the continental lithosphere: New hypotheses for the formation of the Tibetan Plateau. *Tectonics*, **13**, 475–483.

BUROV, E., PODLADCHKOV, Y., GRANDJEAN, G. & BURG, J.-P. 1999. Thermo-mechanical approach to validation of deep crustal and lithsopheric structures inferred from multidisciplinary data: application to the Western and Northern Alps. *Terra Nova*, **11**, 124–131.

BYERLEE, J. 1978. Friction of rocks. *Pure and Applied Geophysics*, **116**, 615–626.

CALLOT, J.-P., GRIGNÉ, C., GEOFFROY, L. & BRUN, J.-P. 2000. Development of volcanic passive margins: two-dimensional laboratory models. *Tectonics*, **20**, 148–159.

CHOUKROUNE, P., ROURE, F., PINET, B. & ECORS PYRENEES TEAM 1990. Main results of the ECORS Pyrenees profile. *Tectonophysics*, **173**, 411–423.

DAVIS, D., SUPPE, J. & DAHLEN, A. 1983. Mechanics of fold-and-thrust belts and accretionary wedges. *Journal of Geophysical Research*, **88**, 1153–1172.

DAVY, P. & COBBOLD, P. R. 1988. Indentation tectonics in nature and experiment. 1. Experiments scaled for gravity. *Bulletin of the Geological Institutions of Uppsala*, **14**, 129–141.

DAVY, P. & COBBOLD, P. R. 1991. Experiments on shortening of a 4-layer continental lithosphere, *Tectonophysics*, **188**, 1–25.

DAVY, P., CHOUKROUNE, P. & SUZANNE, P. 1990*a*. Hypothèses de déformation de la lithosphère appliquées à la formation des Pyrénées. *Bulletin de la Société Géologique de France*, **8**, 219–228.

DAVY, P., HANSEN, A., BONNET, E. & ZHANG, S. Z. 1995. Localisation and fault growth in layered brittle–ductile systems for deformations of the continental lithosphere. *Journal of Geophysical Research*, **100**, 6281–6294.

DAVY, P., SORNETTE, A. & SORNETTE, D. 1990*b*. Some consequences of a proposed fractal nature of continental faulting, *Nature*, **348**, 56–58.

ELLIS, S. & BEAUMONT, C. 1999. Models of convergent boundary tectonics: implications for the interpretation of Lithoprobe data. *Canadian Journal of Earth Sciences*, **36**, 1711–1741.

FACCENNA, C., DAVY, P., BRUN, J.-P., FUNICIELLO, R., GIARDINI, D., MATTEI, M. & NALPAS, T. 1996. The dynamics of back-arc extension: an experimental approach to the opening of the Tyrrhenian Sea. *Geophysical Journal International*, **126**, 781–795.

FAUGÈRE, E. & BRUN, J.-P. 1984. Modélisation expérimentale de la distension continentale. *Comptes Rendus de l'Académie des Sciences de Paris*, **299**, 365–370.

GAUTIER, P., BRUN, J.-P., MORICEAU, R., SOKOUTIS, D., MARTINOD, J. & JOLIVET, L. 2000. Aegean extension: timing, kinematics and causes. *Tectonophysics*, **315**, 31–72.

GOETZE, C. & EVANS, B. 1979. Stress and temperature in the bending lithosphere as constrained by experimental rock mechanics. *Journal of the Royal Astronomical Society*, **59**, 463–478.

GUTSCHER, M. A., KUKOWSKY, N. & MALAVIEILLE, J. 1998. Material transfer in accretionary wedges from analysis of a systematic series of analog experiments. *Journal of Structural Geology*, **20**, 406–416.

HUBBERT, K. 1937. Theory of scale models as applied to the study of geological structures. *Geological Society of America Bulletin*, **48**, 1459–1520.

HUBBERT, K. 1951. Mechanical basis for certain familiar geologic structures. *Geological Society of America Bulletin*, **62**, 355–372.

KEEN, C., PEDDY, C., DE VOOGD B. & MATTHEWS, D. 1989. Conjugate margins of Canada and Europe: results from deep reflection profiling. *Geology*, **17**, 173–176.

JOLIVET, L., BRUN, J.-P., GAUTIER, P., LALLEMAND, S. & PATRIAT, M. 1994. 3D kinematics in the Aegean from the Early Miocene to Present, insights from the ductile crust. *Bulletin de la Société Géologique de France*, **165**, 195–209.

LE PICHON, X. & ANGELIER, J. 1981. The Aegean Sea. *Philosophical Transactions of the Royal Society of London*, **A300**, 357–372.

MALAVIEILLE, J. 1984. Modélisation expérimentale des chevauchements imbriqués: Application aux chaines de montagnes. *Bulletin de la Société Géologique de France*, **7**, 129–138.

MARTINOD, J. 1991. Instabilités périodiques de la lithosphere (Flambage, boudinage) en compression et en extension. *Mémoire et Documents du CAESS*, **44**.

MARTINOD, J. & DAVY, P. 1994. Periodic instabilities during compression of the lithosphere. 2. Analogue experiments. *Journal of Geophysical Research*, **99**, 12057–12069.

MARTINOD, J., HATZFELD, D., BRUN, J.-P., DAVY, P. & GAUTIER, P. 2000. Continental collision, gravity spreading, and kinematics of Aegea and Anatolia. *Tectonics*, **10**, 290–299.

MARUYAMA, S., LIOU, J. G. & TERABAYASHI, M. 1996. Blueschists and eclogites of the world and their exhumation. *International Geology Review*, **38**, 485–594.

MÉNARD, G. & THOUVENOT, F. 1987. Coupes équilibrées crustales: méthodologie et applications aux Alpes occidentales. *Geodynamica Acta*, **1**, 35–45.

NICOLAS, A., HIRN, A., NICHOLICH, R. & POLINO, R. 1990. Lithosphere wedging in the Western Alps inferred from the ECORS–CROP traverse. *Geology*, **18**, 587–590.

PFIFFNER, A., ELLIS, S. & BEAUMONT, C. 2000. Collision tectonics in the Swiss Alps from geodynamic modelling. *Tectonics*, **19**, 1065–1094.

PINET, N. & COBBOLD, P. R. 1992. Experimental insights into the partitioning of motion within zones of oblique subduction. *Tectonophysics*, **206**, 371–388.

PUBELLIER, M. & COBBOLD, P. R. 1996. Analogue models for the transpressional docking of volcanic arcs in the Western Pacific. *Tectonophysics*, **253**, 33–52.

RAMBERG, H. 1981. *Gravity Deformation and the Earth Crust*. Academic Press, London and New York.

RANALLI, G. 1997. *Rheology of the lithosphere in space and time*. Geological Society, London, Special Publications, **121**, 19–37.

RATSCHBACHER, L., MERLE, O., DAVY, P. & COBBOLD, P. R. 1991. Lateral extrusion in the Eastern Alps, Part 1: Boundary conditions and experimental experiments scaled for gravity. *Tectonics*, **10**, 245–256.

REY, D., QUARTA. T., MOUGE, P., MILETTO, M., LANZA, R., GALDEANO, A., CARROZZO, M. T., BAYER, R. & ARMANDO, E. 1990. Gravity and aeromagnetic maps of the Western Alps: contribution to the knowledge of the deep structures along the ECORS–CROP seismic profile. *In*: ROURE, F., HEITZMAN, P. & POLINO, R. (eds) *Deep Structure of the Alps*. Mémoire de la Société Géologique de France, **156**, 107–121.

ROURE, F., CHOUKROUNE, P. & POLINO, R. 1996. *Deep seismic reflection data and new insights on the bulk geometry of mountain ranges*. Compte Rendus de l'Académie des Sciences, Paris, Série IIa, **322**, 345–359.

ROURE, F., CHOUKROUNE, P. & ECORS PYRENEAN TEAM 1989. ECORS deep seismic data and balanced cross sections: geologic constraints on the evolution of the Pyrenees. *Tectonics*, **8**, 41–50.

SCHMID, S. M. & KISSLING, E. 2000. The arc of the western Alps in the light of geophysical data on deep crustal structure. *Tectonics*, **19**, 62–85.

SCHMID, S. M., PFIFFNER, O. A, FROITZHELM, G., SCHÖNBORN, G. & KISSLING, E. 1996. Geophysical-geological transect and tectonic evolution of the Swiss-Italian Alps. *Tectonics*, **15**, 1036–1064.

SOURIOT, T. & BRUN, J.-P. 1992. Faulting and block rotation in the Afar triangle, East Africa: the Danakyl 'crank-arm' model. *Geology*, **20**, 911–914.

STAMPFLI, G. M. & MARCHANT, R. H. 1997. Geodynamic evolution of the Tethyan margins of the Western Alps. *In*: PFIFFNER, O. A., LEHNER, P., HEITZMANN, P., MUELLER, S. & STECK, A. (eds) *Deep Structure of the Swiss Alps: Results of NRP 20*. Birkhäuser, Boston, 223–240.

TRON, V. & BRUN, J.-P. 1991. Experiments on oblique rifting in brittle-ductile systems. *Tectonophysics*, **188**, 71–84.

WERNICKE, B. 1985. Uniform-sense normal simple shear of the continental lithosphere. *Canadian Journal of Earth Sciences*, **22**, 108–125.

WERNICKE, B. & AXEN, G. J. 1988 On the role of isostasy in the evolution of normal fault systems. *Geology*, **16**, 848–851.

WILLET, S. D. 1999. Rheological dependence of extension in wedge models of convergent orogens. *Tectonophysics*, **305**, 419–435.

WILLET, S. D., BEAUMONT, C. & FULLSACK, P. 1993. Mechanical model for the tectonics of doubly vergent compressional orogens. *Geology*, **21**, 371–374.

Evidence for steady fault-accommodated strain in the High Himalaya: progressive fault rotation of the southern Tibet detachment system in NW Bhutan

G. WIESMAYR[1], M. A. EDWARDS[1,2], M. MEYER[1], W. S. F. KIDD[3], D. LEBER[1], H. HÄUSLER[1] & D. WANGDA[4]

[1] *Institut für Geologie, Universität Wien, Austria (e-mail: medwards@gmx.net)*
[2] *Previous address: Asian Tectonics Research Unit, Institut für Geologie, TU-Bergakademie Freiberg, Germany*
[3] *Department of Geological Sciences, State University of New York at Albany, Albany NY 12222, USA*
[4] *Geological Survey of Bhutan, Ministry of Trade and Industry, Thimphu, Bhutan*

Abstract: We present fault analyses from the exhumed middle crustal slab of the High Himalaya in eastern Lunana in NW Bhutan. Fault planes from within two-mica, tourmaline-bearing leucogranites, leucogranitic rocks and migmatites indicate a complex brittle fault pattern with two distinct fault groups. A first group of faults (D_1) characterized by chlorite, quartz and tourmaline slickenfibres is mainly defined by steeply SSE-dipping oblique-slip normal faults, and by shallowly NNW-dipping normal faults. A second, younger group of faults (D_2) characterized by cataclasis products comprises strike-slip faults displaying conjugate patterns and E- and W-dipping conjugate normal faults, all which indicate E–W extension. Cross-cutting relationships amongst the D_1 fault group demonstrate that progressively steeper members of the fault group become younger within the NNW-dipping faults and become older within the SSE-dipping faults. These are all post-dated by the D_2 fault group. The D_1 fault group indicates that the slab experienced ongoing NNW–SSE extension (i.e. flow) via brittle fault accommodation, contemporaneous with fault rotation. This may reflect rotation of the entire upper orogen due to movement over deeply located major ramp structures formed by out-of-sequence thrusting (Kakhtang Thrust) within the High Himalayan Slab of the Bhutan Himalaya.

The evolution of deformation–accommodation processes in exhuming middle crust within mountain belts spans a history from the onset of orogenesis, through deep burial and exhumation, to final exposure and quiescence. Processes range from true high temperature intracrystalline plasticity through the poorly-understood realm or 'boundary' of frictional–viscous creep, to wholesale brittle and neotectonic processes. In this short contribution, we examine an area of relatively rapid orogenic convergence and exhumation; the eastern Himalaya. We examine the role of brittle processes in the upper part of the exhuming middle crust as part of the orogenic wedge. Here we find evidence for later-stage exhumation in the form of continued deformation accommodated by dense brittle fault networks.

Background

The Himalayan mountain belt is the Earth's largest and arguably most spectacular example of continent–continent collision. The arc is over 2000 km in length and represents the accommodation of convergence since India first collided with Asia at *c.* 65 to 55 Ma (e.g. Patriat & Achache 1984; Rowley 1998). Rates of shortening and exhumation of the Himalayan orogen are relatively rapid; present day convergence is *c.* 1–2 cm a^{-1} (Bilham *et al.* 1997) and probably has been in this range at least since the late Pliocene (Powers *et al.* 1998). Younger denudation of the Himalaya is on the order of 1 km Ma^{-1} (Einsele *et al.* 1996), and exhumation rates over the last 10 Ma are as high as 5 mm a^{-1} (Schneider *et al.* 1999). Taken together, these data imply that orogenic strain rates in the Himalaya are relatively rapid. This has relevance for our study of brittle-mechanism continued deformation of the exhuming middle crust in the High Himalaya.

Since collision began, several hundred kilometres of India have been subducted or underthrust beneath Tibet to depth of 10s to possibly greater than 100 km (e.g. Hauck *et al.* 1998; Owens & Zandt 1997; Guillot *et al.* 1997; O'Brien *et al.* 2001). This is, for the most part, expressed within rocks of amphibolite to lower granulite

From: DE MEER, S., DRURY, M. R., DE BRESSER, J. H. P. & PENNOCK, G. M. (eds) 2002. *Deformation Mechanisms, Rheology and Tectonics: Current Status and Future Perspectives*. Geological Society, London, Special Publications, **200**, 371–386. 0305-8719/02/$15

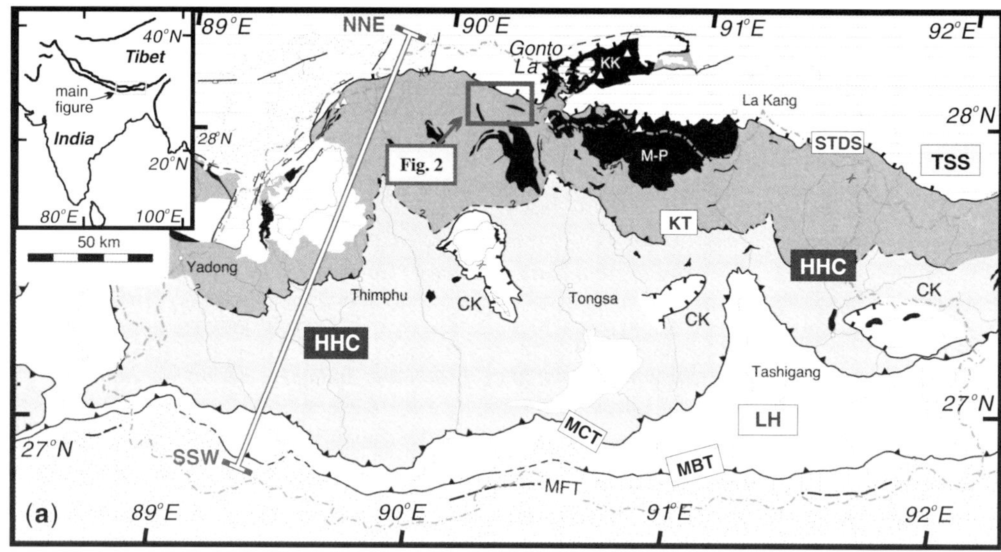

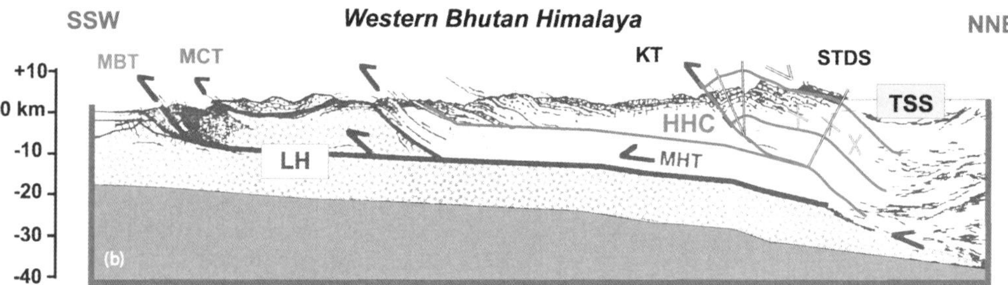

Fig. 1. (**a**) Overview geological map of Bhutan (after Gansser 1983; Edwards *et al.* 1999; Grujic *et al.* 2002. NNE–SSW is line of section shown in (**b**). Major structures: MFT, Main Frontal Thrust; MBT, Main Boundary Thrust; MCT, Main Central Thrust; CK, Klippe of Chekka Series; KT, Kakthang Thrust; STDS, Southern Tibet Detachment System. Main litho-units: LH, Lesser Himalaya; HHC, Higher Himalayan Crystalline; TSS, Tethyan Sedimentary Sequence; black shading, big Himalayan leucogranites: KK, Khula Kangri Pluton, MP, Monlakarchung-Pasalum Pluton. Other names are towns/villages. Inset shows location of main figure within Himalayan orogen. (**b**) Crustal-scale cross-section through western Bhutan Himalaya incorporating geological data from Gansser (1964) fit to interpreted reflection seismic data of Hauck *et al.* (1998). Additional interpretive crustal-scale frontal ramp and fault bend folding related to out-of-sequence thrusting on Kakhtang Thrust is proposed (see discussion).

facies that form a 5–40 km thick crystalline slab or wedge, known as the High Himalayan Crystalline (HHC: Fig. 1a and b). This wedge is bound below and above by two major tectonic discontinuities: the Main Central Thrust (MCT) and the Southern Tibet Detachment System (STDS), respectively. The MCT zone represents a several kilometer thick zone of high-strain rocks associated with an inverted metamorphic field gradient (see Grasemann & Vannay 1999 for a comprehensive review). The STDS is a system of gently N-dipping normal fault zones which typically juxtapose greenschist grade rocks over upper amphibolite to granulite facies rocks (e.g. Burg *et al.* 1983; Edwards *et al.* 1996).

The HHC in the Bhutan High Himalaya (Fig. 1) has a notably greater structural thickness than to the west in Sikkim and Nepal (Gansser 1983). This is associated with a major out-of-sequence thrust (the Kakhtang Thrust: Grujic *et al.* 1996), and is probably related to the multi-stage history of the STDS in this area (Edwards *et al.* 1996) and to the curiously young (Late Miocene) plutonism (Edwards & Harrison 1997). The major fabric of the HHC in the Bhutan High Himalaya is defined by foliation in metapelites, migmatites and high strain zones, and is for the most part gently N-dipping. Locally, however, the HHC is domed, and gently antiformally folded to S-dipping.

Eastern Lunana

The data we report here are from the eastern Lunana area near the Tibetan frontier (Fig. 2). The area affords striking topography and full outcrop when not obscured by glaciers and Quaternary deposits. Our measurements are from the HHC of the STDS zone immediately below the Tethyan sequences – in the footwall of the (southward projection of the) Gonto La detachment. The Gonto La detachment is the name given to the main low-angle normal fault of the STDS from the nearby Khula Kangri area. Southward projection of the most southerly exposure ('detachment' in Fig. 2a) shows that it is well above the present erosion surface (Fig. 2b; cf. Fig. 4 of Edwards *et al.* 1999). Overwhelmingly, the rocks present in the eastern Lunana area are two-mica, tourmaline-bearing leucogranites intruding leucogranitic and migmatitic host rocks; metapelites are conspicuous by their absence. The effects of plutonism and associated high-temperature deformation at very low relative viscosities have led to an abundance of chaotically-textured rocks made up of the interaction of intrusive bodies and migmatitic hosts. This has, almost everywhere, obscured any pre-existing foliation or planar-type regular fabric in the migmatites and orthogneisses.

Post-dating this plutonism and low viscosity deformation is a dense network of faults, the object of our study (Fig. 3). Exposed portions of the majority of fault surfaces are a few to

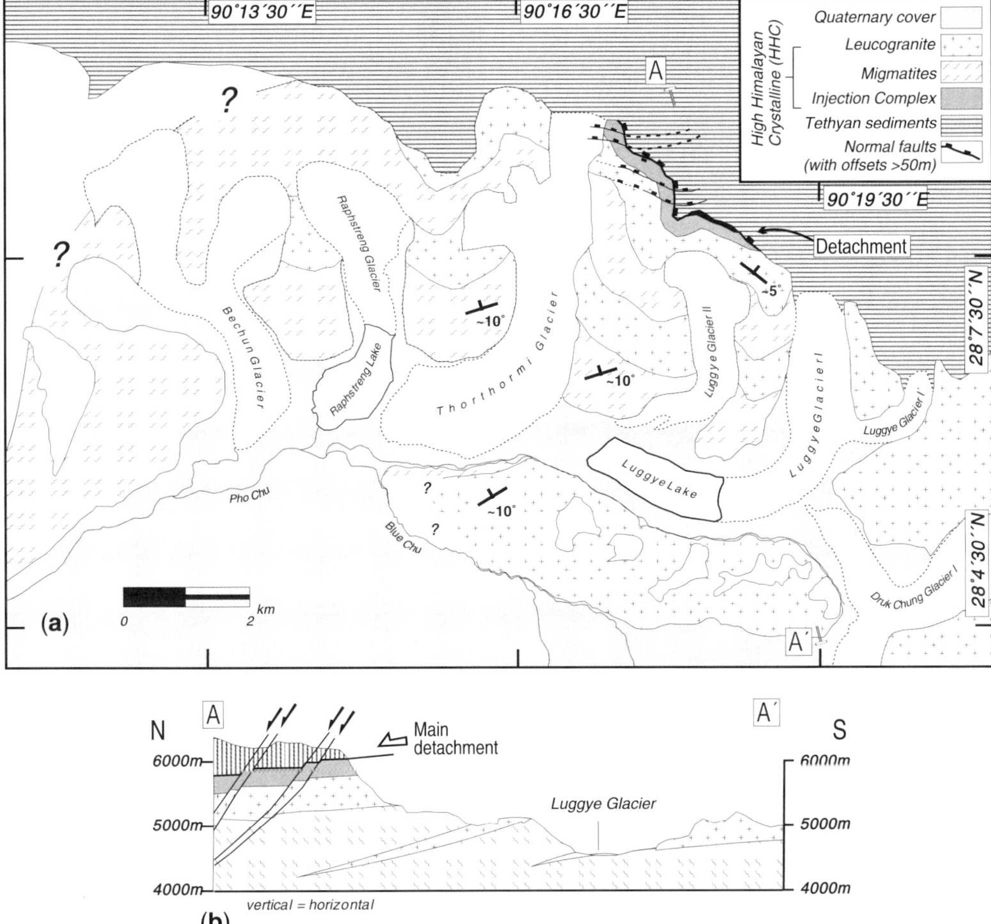

Fig. 2. Geological map (**a**) and cross-section (**b**) for eastern Lunana (after Gansser 1983; Wiesmayr, unpublished data). Legend in (**b**) as for (**a**). Map interpolated from processed remote satellite (IRS-1D) data.

Fig. 3. Field photographs of representative fault surfaces. (**a**) Tabular pegmatite body (arrowed) post-dating main leucogranite plutonism and migmatisation; note orientation is parallel to nearby D_1a faults with chlorite-slickenfibre-coated surface. Hammer for scale. (**b**) Cross-cutting relationship within D_1a fault sub-group; fault and slickenfibre orientations are shown in lower hemisphere equal area projection. (**c**) Gently (*c.* 20°) NW-dipping normal faults. (**d**) Close-up of outcrop lower fault surface shown in (**c**). Quartz slickenfibres are NW-dipping, offsteps and pluck marks indicate normal sense of movement; missing block has been displaced NNW (looking SSE). (**e**) NE–SW trending set of vertical and NW-dipping fault surfaces in ridge south of Pho Chhu Valley (*c.* 200 m high, viewer is facing SW).

tens of metres in length, overall well-developed, and essentially planar. Faults that are present within the south wall of the Table Mountain chain in Lunana are several hundreds of metres in exposure. Surfaces are typically either sites for mineralization (tourmaline, chlorite and quartz expressed as slickenfibres) or cataclasis (various types of gouge, crush and damage zones). Some very young faults are hosts for soft sediment dyke infill. It is noteworthy that the faults do not splay, sole or diminish into, nor are they strained along with, and nor do they provide pathways for, the magmatic material that is their crystalline host rock. Locally, pegmatite bodies (all of which post-date the main leucogranite plutonism and migmatization) are found in orientations parallel to faults (Fig. 3a). The various 'low temperature' mineral deposition upon fault surfaces, and the presence of pegmatites are evidence for young fluid flow. The widespread occurrence of thermally elevated waters (e.g. 'hot springs') in the Bhutanese Highlands is further evidence for young fluid flow.

Fault analysis

Throughout the eastern Lunana area, we measured: fault size; fault orientation; sense of displacement(s); and 'overprinting' relationships if more than one phase of displacement was found preserved, or if one fault was observed cutting another (Fig. 3b). Faults include fractures of all sizes where displacement was recognisable. Fault size was qualitatively assessed from lateral extent of the displacement surface. Sense of displacement criteria include: growth direction

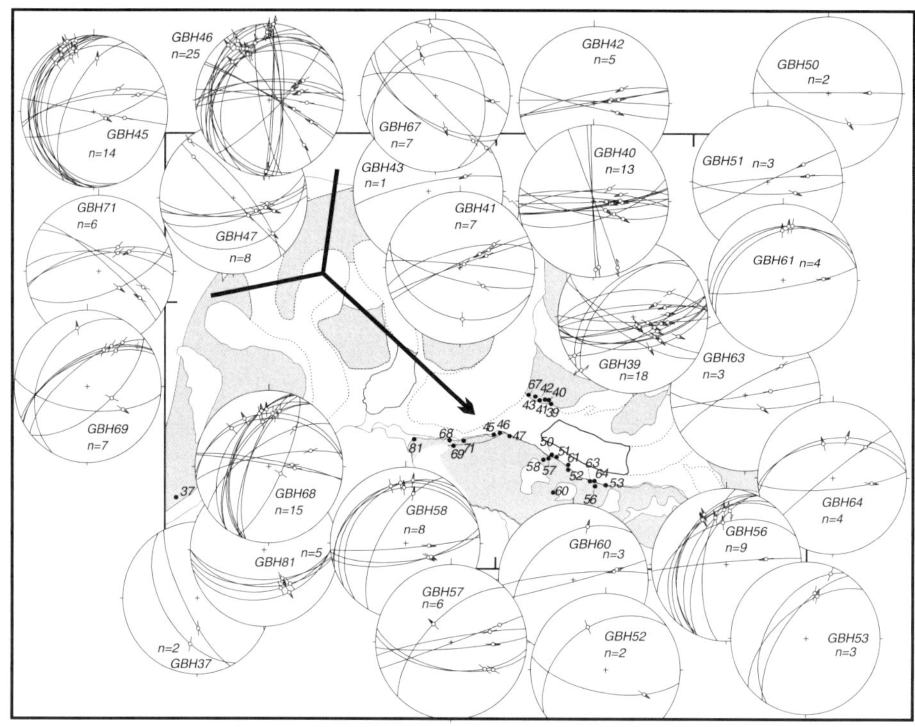
GBH46
n=25
GBH45
n=14
GBH67
n=7
GBH43
n=1
GBH42
n=5
GBH40
n=13
GBH50
n=2
GBH51 n=3
GBH71
n=6
GBH47
n=8
GBH41
n=7
GBH61 n=4
GBH39
n=18
GBH63
n=3
GBH69
n=7
GBH68
n=15
GBH58
n=8
GBH64
n=4
GBH56
n=9
GBH60
n=3
GBH81 n=5
GBH57
n=6
GBH52
n=2
GBH53
n=3
n=2
GBH37

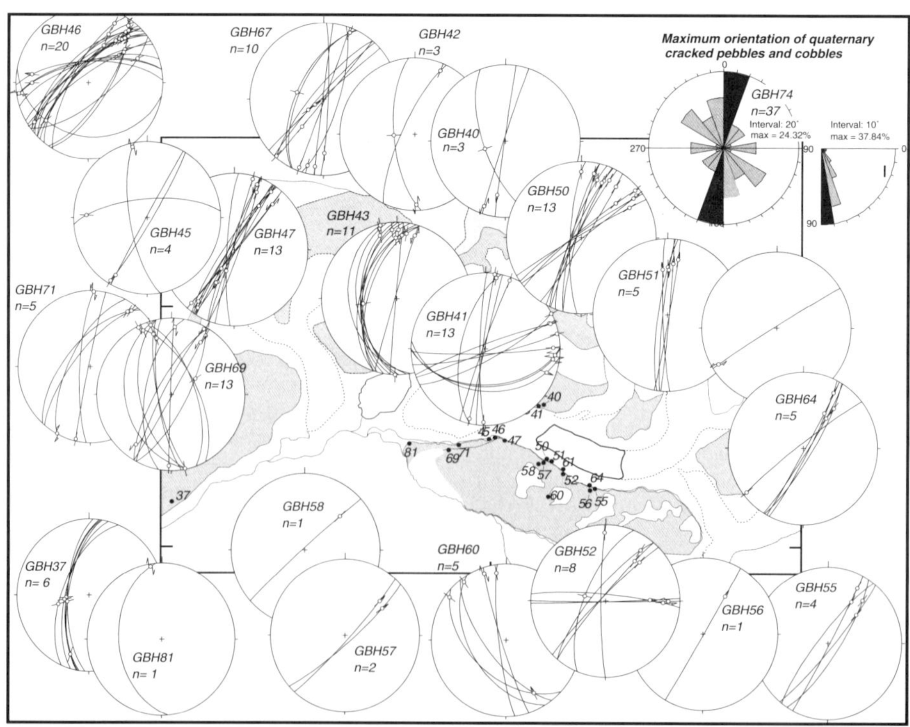
GBH46
n=20
GBH67
n=10
GBH42
n=3
Maximum orientation of quaternary cracked pebbles and cobbles
GBH74
n=37
Interval: 20°
max = 24.32%
Interval: 10°
max = 37.84%
GBH40
n=3
GBH50
n=13
GBH45
n=4
GBH47
n=13
GBH43
n=11
GBH71
n=5
GBH51
n=5
GBH41
n=13
GBH69
n=13
GBH64
n=5
GBH58
n=1
GBH37
n= 6
GBH60
n=5
GBH52
n=8
GBH55
n=4
GBH56
n=1
GBH81
n= 1
GBH57
n=2

of mineral fibers (typically slickenfibres composed of tourmaline, chlorite or quartz: Figs 3c and d); polarity of stylolites; steps or pluck marks on the fault surface; Riedel shears; and offset or aligned grains in gouge (Hancock 1985; Petit 1987).

In the field, it became apparent that faults incorporated either mineralization or cataclasis products (including soft sedimentary dyking) but nowhere were both found together. Moreover, in about fifteen instances mineralization faults were cross-cut by cataclasis product faults, but nowhere was the opposite relationship found. The relative chronology provided by this cross-cutting relationship indicates the two fault groups were generated by different events separated in time. Additionally, it is likely that the mineralization occurred at deeper crustal levels than the cataclasis, and this is consistent with the generation of the cataclasis product faults being later than the mineralization faults in the history of the steadily exhuming upper portion of the HHC. Based upon this interpretation of two separate events, we term the mineralization and cataclasis faults D_1 and D_2 respectively.

For each field locality, the fault data that we have discriminated into our D_1 and D_2 groups are displayed on Figs 4a and b, respectively. In the following section we attempt to test the robustness of this fault group discrimination.

Palaeostress and palaeostrain calculation attempts

Overall deformation has been assumed to be coaxial and offset along the faults is relatively small with respect to the spacing of faults. In this case, strain axes should coincide with stress axes. Therefore, we used kinematic as well as dynamic calculations of the stress tensor to crash test our fault group discrimination (Fig. 5). We used the numerical-dynamic analysis (NDA) Method (Spang 1972; Sperner 1996), the P–T-axes Method (Turner 1953), the Direct Inversion Method (Angelier & Goguel 1979; Sperner *et al.* 1993) and the Dihedra Method (Angelier & Mechler 1977).

There are significant limitations in using palaeostress techniques in rocks such as granites, migmatites and gneisses where there is no originally horizontal datum such as bedding. In such lithologies there is no operation to separate faults which pre- and post-date folding/tilting or rotation, in contrast to layered sedimentary rocks (e.g. Sperner *et al.* 1993). Despite these caveats, applying a palaeostress calculation can provide an identification of fault groups that were all formed under similar regimes.

We initially used non-discriminated fault data (i.e. D_1 plus D_2) from six field localities (GBH 40, 41, 46, 47, 49 & 71) whose data densities are large enough to be statistically valid for the calculation concerned (Fig. 5). In all cases, there is a striking inconsistency amongst the six localities. Moreover, the dramatic differences between the results of the four methods for each given locality provides further indication of the incompatibility of the D_1 group with the D_2 group. Note that for the P/T-axes Method and the Dihedra Method, the results are shown numerically, but not graphically. Applying each of the methods to D_1 alone and D_2 alone was more successful: D_2 fault data gave consistent results amongst the outcrops and amongst the four methods for the total summed D_2 data (D_2_all). We feel confident that D_2 is a real discrete event and that the grouping of the cataclasis product faults into a single 'event' is justified. D_1 however did not provide consistent results, neither summed nor as separate field localities, despite the visually elegant distribution of the total summed D_1 data (compare goodness of visual grouping of both D_1 _all and D_2_all with the entire D_1 plus D_2 dataset in the lower part of Fig. 5). Despite this, there are a number of key relationships within the D_1 fault group that can explain the apparent complexity.

D_1 faults

Within the investigated area, the faults that belong to D_1 form a dense network of fractures on the outcrop scale, typically displaying conspicuous black chlorite-slickenfibres and

Fig. 4. Brittle fault data measurements for each field locality. Circles are lower-hemisphere equal area projections with measured fault plane data (see text). Great circles are fault planes with accompanying lineations shown as poles and slip directions as lines. Slip direction indicated by arrow head (high confidence) or no arrow head (low confidence). GBH*x* corresponds to numbered black dot on map (field locality). *n* is number of data. Outlines of glaciers and grey shades are to allow easier cross-referencing with Fig. 2, and do not denote specific lithologies (also see Fig. 2 for lats/longs etc). (**a**) D_1 fault group (NNW–SSE extension). (**b**) D_2 fault group (conjugate strike-slip sets and E–W dipping conjugate normal faults indicating younger E–W extension), including rose diagram illustrating orientation trends of cracks in Quaternary morainal cobbles and pebbles.

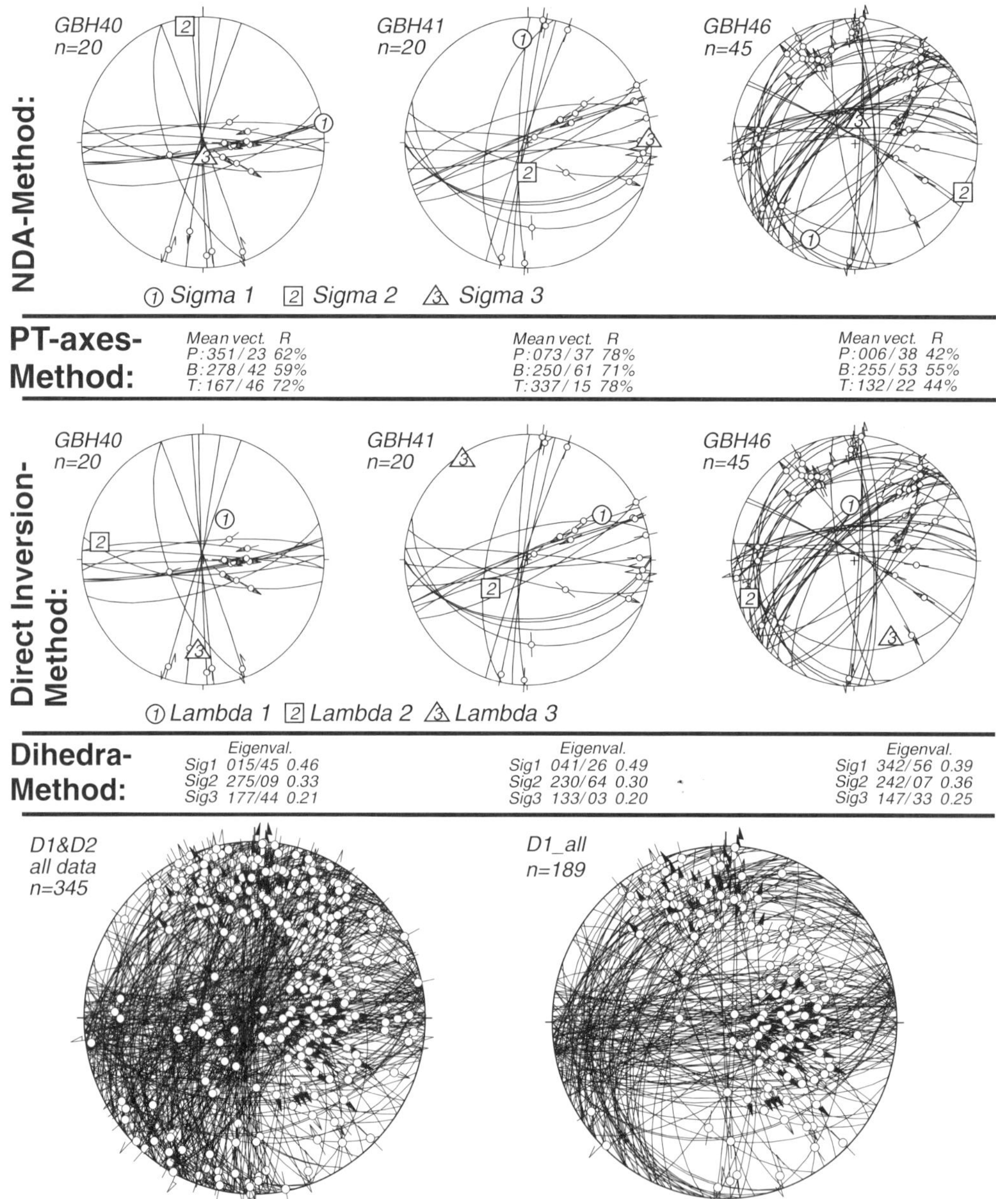

Fig. 5. Attempted palaeostress calculations using four different methods on D_1 and D_2 data sets for the six localities with densest data (upper portion). Lower portion shows total summed data for all faults (D_1 and D_2) separately. Only D_2_all gave consistent results and only these are therefore shown for all palaeostress calculation methods.

subordinate quartz-slickenfibres, both forming pronounced slickenlines. Figure 4a displays the D_1 group fault data for each field locality. Within the D_1 fault group, two sub-groups are visually apparent. Figure 6a (upper) shows a plot of all the D_1 data and a summary of the main fault orientations and lineations schematically plotted for clarity. The trends in fault orientation form two main sub-groups that we term D_1a and D_1b.

The D_1a sub-group comprises ENE–WSW trending fault planes dipping between 60–80° to the SSE, to vertical, and then to steeply NNW-dipping. Slickenside lineations vary

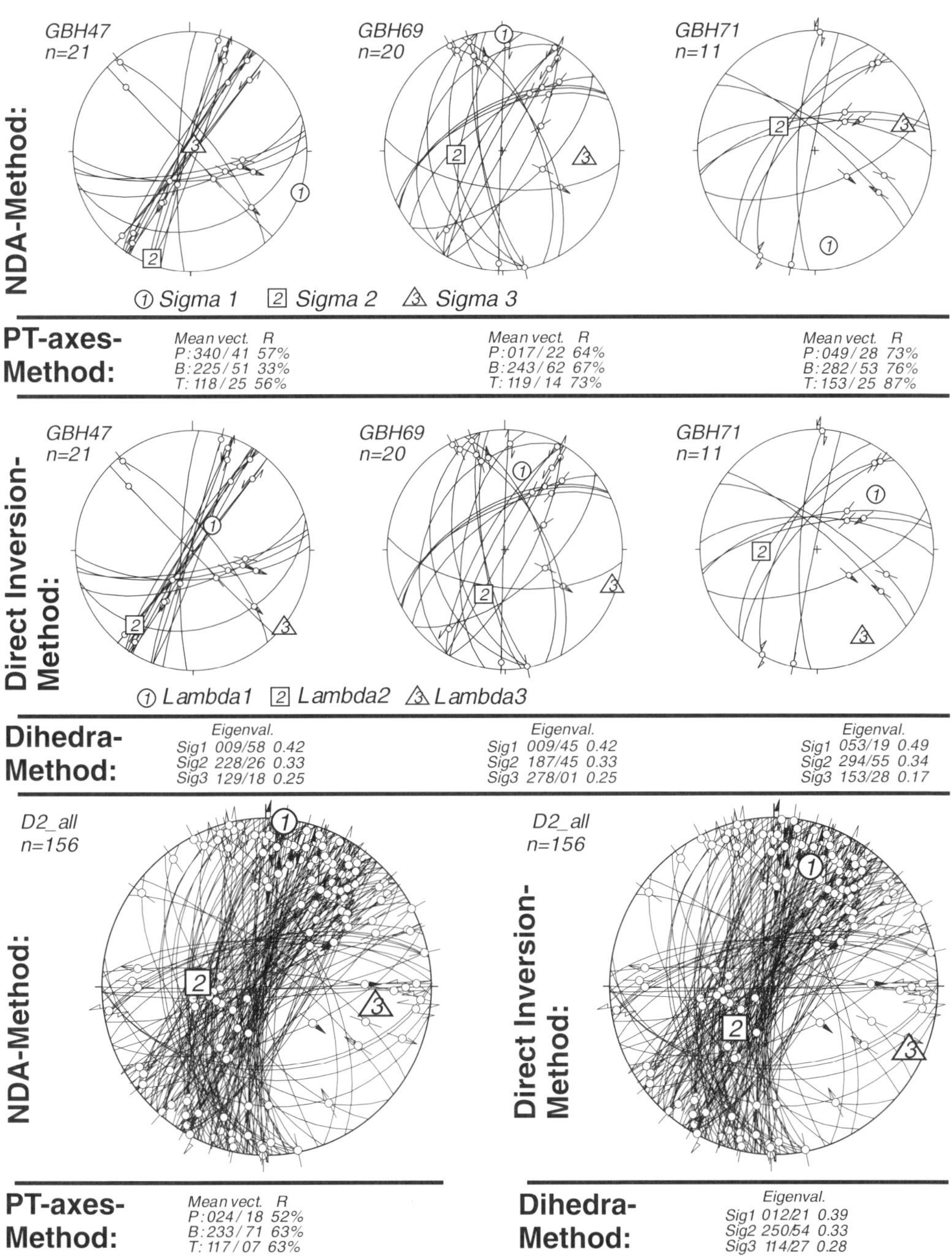

gradually with fault dip, ranging from SSE-plunging to progressively more easterly plunging with increasing fault dip. The SSE-dipping fault planes show normal and oblique slip, while for the NNW-dipping fault planes the shear sense changes to reverse displacement, thereby forming oblique thrusts (in the present orientation). Key cross-cutting fault relationships and overprinting lineations within the D_1a sub-group (Fig. 3b and Table 1) reveal that more steeply dipping faults are progressively older. The gradual change in fault orientations and displacement together with the relative chronology provided by this cross-cutting relationship is suggestive of a

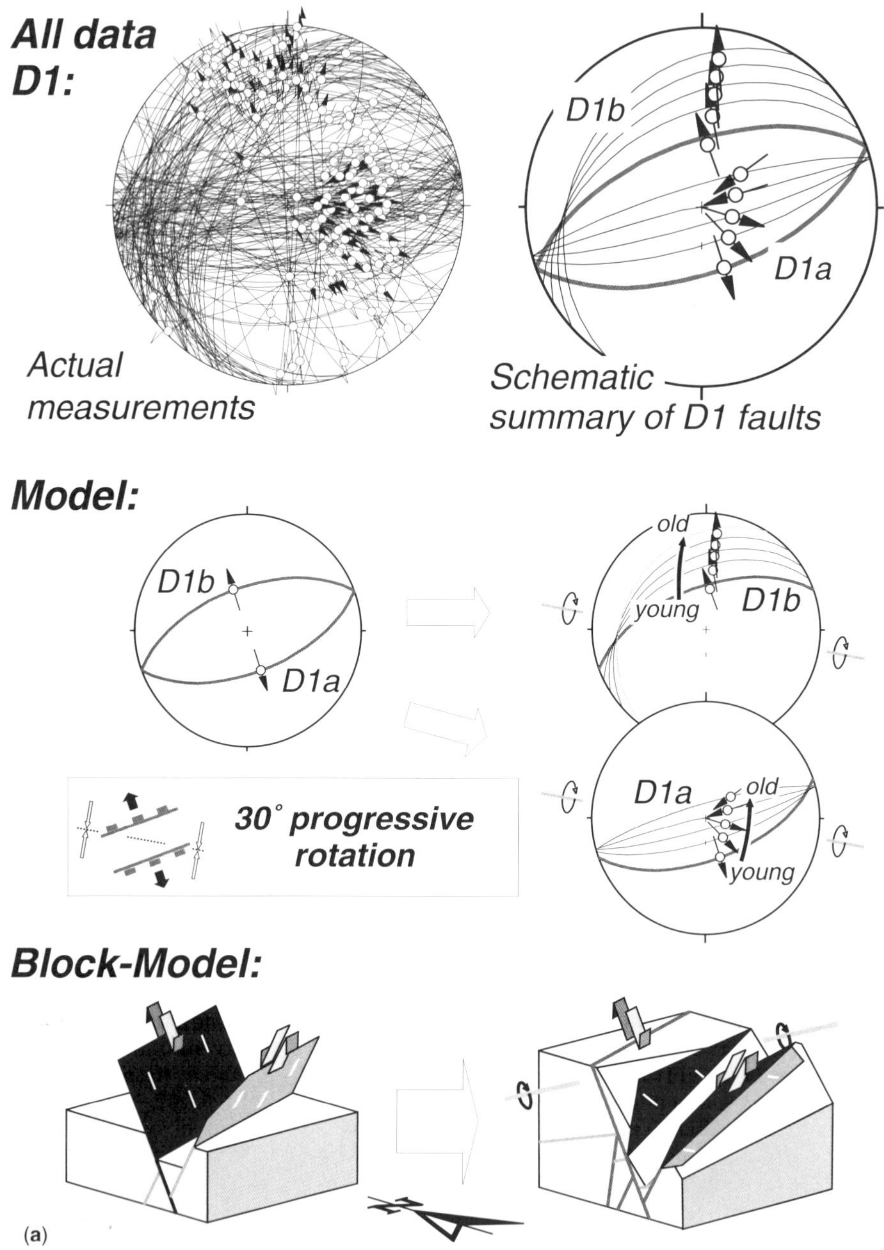

Fig. 6. (**a**) Interpretation of D_1 data and schematic model of progressive fault development. Upper portion: all D_1 data are plotted together and shown alongside a schematically displayed summary of the data illustrating how the D_1a and D_1b sub-groups are distributed. Middle portion: interpretative model of how original upright conjugate fault pairs and lineations rotate and new faults progressively form during a 30° rotation about an axis of 100/00 to give observed present-day distribution. Lower portion: 3D block model diagram of conjugate pair rotation. (**b**) General diagram of how the D_2 fault group was generated by simultaneous strike-slip and normal faulting caused by E–W extension and N–S shortening (see Fig. 5 for palaeostress results).

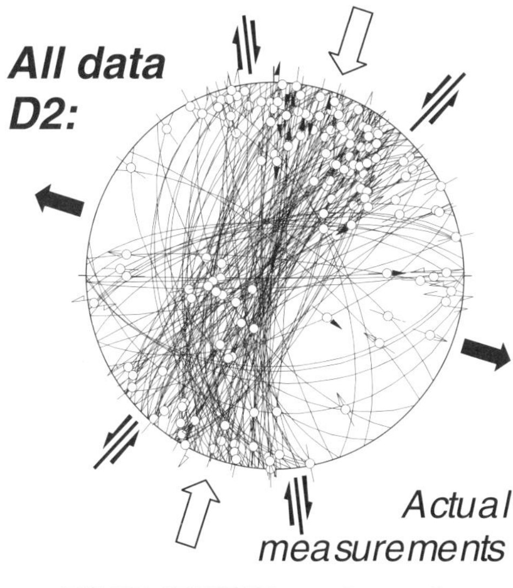

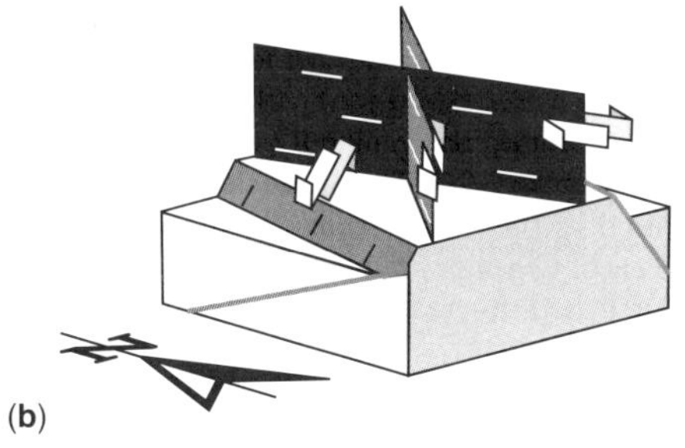

(b)

Table 1. *Relative age constraints for $D_1 a$ faults*

Outcrop	Fault	Lineation	Relative age
Examples of cross-cutting within $D_1 a$-faults			
GBH 39	355/78	070/64	old
	164/81	157/80	young
GBH 60	355/83	080/44	old
	165/88	078/40	young
Examples of overprinting lineations on $D_1 a$-faults			
GBH 39	181/84	112/73	young
		083/52	old
	164/81	157/80	young
		085/64	old
	165/78	150/75	young
		080/57	old
GBH 47	162/67	138/66	young
		108/52	old

rotation. In this scenario, the present steeply NNW-dipping oblique thrusts formed as SSE-dipping normal faults followed by *c.* 25–30° of rotation. In places, late stage, tourmaline-bearing pegmatitic veins and tabular bodies cross-cut the leucogranitic host gneisses. They are conspicuously parallel to the D_1a fault set (Fig. 3a). In the vicinity of the pegmatites, the faults are associated with stretched tourmaline needles that are aligned parallel to chlorite and quartz-slickenside lineations.

The D_1b sub-group consists of 60–10° NNW-dipping normal faults. Slickenside lineations change progressively from NNW- to N-plunging with decreasing fault dip. The second sub-group contains the same types of slickenfibres mineralization as the first sub-group, with chlorite and quartz linear fibres demonstrating normal sense of movement. Fault displacement is constrained by slickenfibres to be a few centimetres, reaching several tens of metres at a maximum, as shown by offset markers (cf. Fig. 2b).

D_2 faults

The D_2 group post-dates the D_1 group (see above). The faults are characterized by cataclasites formed within narrow brittle shear zones that offset D_1 faults. Figure 4b displays, for each field locality, the fault data that we have assigned to our D_2 suite. The faults mainly comprise: (1) NE–SW- to N–S-trending sinistral strike-slip faults; (2) NNW–SSE-trending dextral strike slip-faults; and (3) W-dipping normal faults (Figs 4b and 6b).

The NE–SW sinistral strike-slip faults comprise weakly to very well-developed subhorizontal lineations formed by scratching and gouging of the fault surface. The sinistral sense of displacement is also constrained by offsets of older D_1 faults. Subordinate mineralization of quartz as well as calcite occurs in fault parallel veins. The NNW–SSE dextral strike-slip faults display weak subhorizontal striations, and senses of movement are again additionally constrained by offsets of older D_1 faults. The W-, and E-dipping faults are those most commonly associated with cataclasites and are less abundant than the sinistral and dextral strike-slip faults. The overall orientation and homogeneous distribution of the faults is obvious in the remotely observed and interpolated fault traces illustrated in Fig. 7.

In addition to the cataclasis product faults of D_2 a few Quaternary or Recent features are present. Ten or so examples of steeply W-dipping, *c.* N–S trending soft-sediment fractures

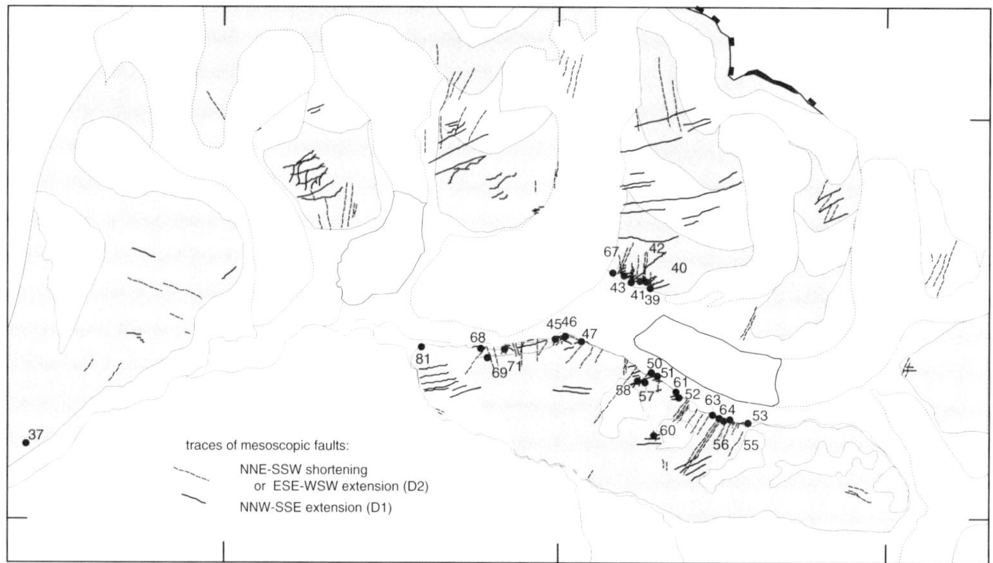

Fig. 7. Remotely sensed extrapolation of mapped and remotely observed faults throughout eastern Lunana area assisted by processed remote satellite (IRS-1D) data. Outline map from Fig. 2 employed as in Fig. 4.

infilled with sand dyke structures are present in Lunana, as well as fracture sets of partly-consolidated morainal cobbles and pebbles. Figure 4b includes frequency diagram plots of the dips and trends of the cobbles and pebbles. Based upon the orientation trends of both these features, we include them as part of the D_2 event and note that this suggests a continuation of D_2 up to the present.

Interpretation

The D_2 event appears to be straightforward. As indicated by the palaeostress calculation results, the D_2 faults are representative of a WNW–ESE extension. Figure 6b shows the basic orientation of faults forming under such a regime.

Many of the D_1 fault data show a clear progressive relationship between age and fault dip; older faults are successively more steeply dipping within the SSE-dipping D_1a fault population. Although we did not directly observe cross-cutting relationships in the D_1b sub-group, the range of fault dips is most readily interpreted as a corresponding progression of successively forming and rotating faults. This points to a straightforward scenario whereby D_1a and D_1b faults form as conjugate sets of normal faults with an acute angle of *c.* 60–40°, a result of NNW–SSE-directed extension as faults form. We infer that the faults have dynamically evolved and undergone a rotation of *c.* 25–30° about a slightly oblique, horizontal rotation axis (100/00). Simultaneously, new sets of conjugate faults have developed in the original NNW–SSE direction, leading to the scattering orientation of faults and slickenside lineations after 25–30° of finite rotation. This is schematically illustrated in the middle part of Figure 6a. This model nicely explains why steeper SSE-dipping or overturned normal faults are progressively older.

To explain the progressive rotation that we recognize in the D_1 group, we propose that the entire section of rocks examined in the field area rotated while faults were formed. Figure 8 shows a model with three time slices ($t = 3$, 2, and 1) to illustrate the approximate sequence of events that have operated in our hypothesis. Note that the three parts of the figure do not depict discrete events but snapshots of a continuous process. At the starting point ($t = 3$) the entire upper portion of the HHC slab, together with the (southern extension of the) Gonto La detachment and the overlying Tethyan rocks were dipping *c.* north at a steeper angle than is observed today. The fundamental assumption of this model is that the maximum principal stress axis ($\sigma1$) at the Earth's surface is subvertical and $\sigma2$ and $\sigma3$ (the intermediate and minimum principal stress axes) are subhorizontal in the extensional regime (allowing an insignificant topographic gradient, Burchfiel & Royden 1985). Consequently, conjugate normal faults form with a *c.* 60–70° dip. At

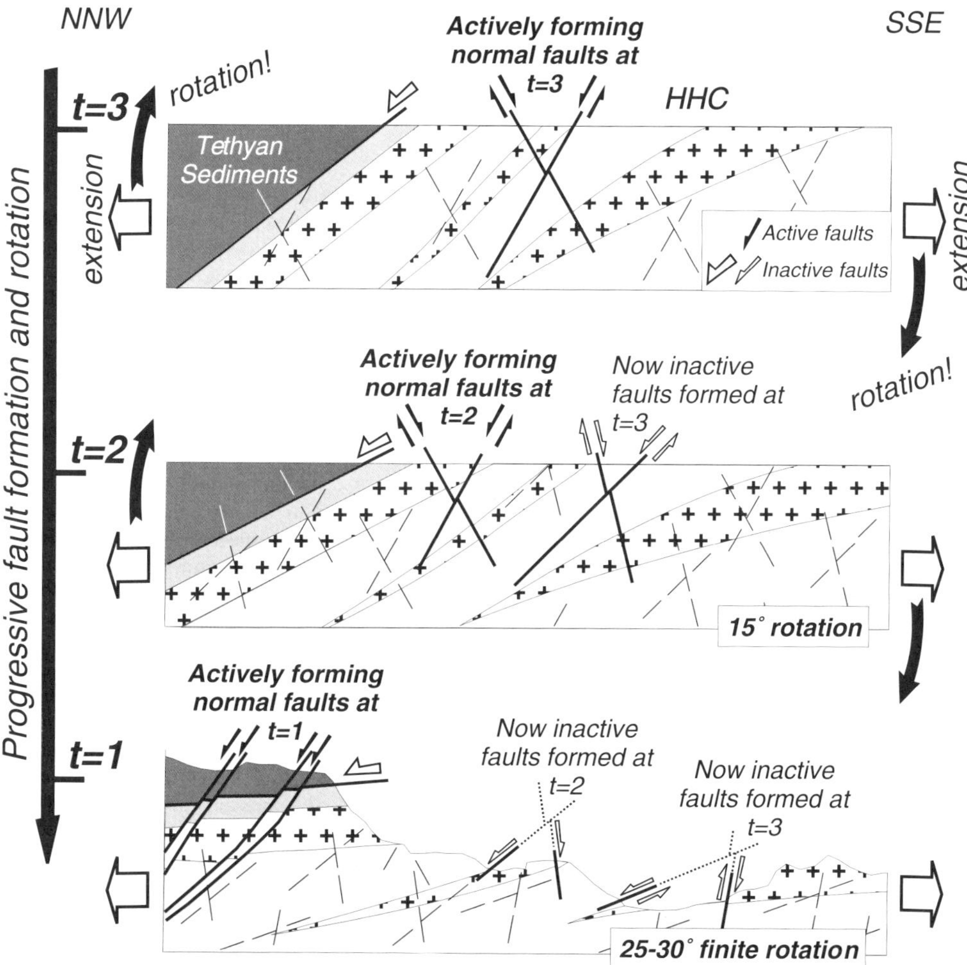

Fig. 8. Model to explain observed succession of rotated faults associated with D_1 group. Entire upper slab rotates to a total of 30°. Three stages $t = 3$, 2, and 1 are time slices of a continuous process and are not discrete events. $t = 3$: within the High Himalayan Crystalline (HHC) and the Tethyan Sediments conjugate normal faults form. $t = 2$: after 15° rotation faults formed at $t = 3$ have rotated and are now typically inactive, while new faults form. $t = 1$: after further rotation with new faults forming in the optimum orientation. Main detachment attains current *c.* dip 10° towards NNW. For clarity, only three orientations of conjugate fault pairs are selected.

$t = 3$, the upper portion of the slab is in a general state of extension (broad expansive flow). Movement along the Gonto La detachment is minimal, however, and there is no further accumulation of intra-grain strain in the granite mylonite. Only local extensional shear bands and internal minor normal faulting in the detachment hanging wall occur (see descriptions in Edwards *et al.* 1996). This point in time also marks the onset of the deformation realm where our mapped normal faults become the principal strain accommodation mechanism. After some time, $t = 2$, the slab is shown after approximately half the entire rotation (*c.* 15°). Faults formed at $t = 3$ have now rotated and are mostly inactive, and are becoming cut by newly forming faults in the optimum orientation. The final time slice shown, $t = 1$, represents the end of the rotation and present-day general geometry of the High Himalayan slab in eastern Lunana after a total of 25–30° foreland-directed finite rotation. Note that this is not t = 0 (the present day); the D_2 event post-dates $t = 1$. The STDS, now gently, *c.* 10° N-dipping, has repeatedly been cut by the later set of NNW-dipping D_1 normal faults.

We do not think that there has been significant slip on the faults after formation. We do not observe large fault offsets, and we speculate that cyclical fluid flux governs failure (hence faulting) to a large extent where periodic increases in hydrostatic pressure and thereby a sudden lowering of the effective stress occurs, leading to fresh faulting in favourable Mohr-Coulomb orientations being preferred rather than renewed slip on less favorably oriented pre-existing faults. Repeated cycling of fluid playing an instrumental role in rock failure is well documented from the Himalaya (Craw *et al.* 1997) and elsewhere (Fletcher 1998; Streit & Cox 2000). Note that we do not observe older faults reactivated in the opposite sense, but we do sometimes observe a normal-sense reactivation on rotated normal faults. This is consistent with the total indicated rotation for the slab being *c.* 25–30° – significantly less than the amount required to re-orient a fault into the region of opposite-sense reactivation.

Discussion

D_1 *deformation*

In terms of a mechanism to allow rotation of the entire thickness of the HHC in this area, we speculate that there is a crustal-scale ramp at several kilometres depth in this part of the Himalaya (Fig. 1b, originally from Gansser 1964 and Hauck *et al.* 1998). In this figure we have included the Kakhtang Thrust, a crustal structure that incorporates a frontal ramp in the main Himalayan thrust (Edwards *et al.* 1996; Grujic *et al.* 1996). This provides a mechanism for rotation of the HHC in this area. We allow fault bend folding of the entire upper slab as out-of-sequence displacement along the Kakhtang Thrust is accommodated. The progressive transport of the overthrusting slab along the Kakhtang structure results in a clear gradual rotation. It is important to note that only a small portion of the HHC is considered here: i.e. the area above the Kakhtang Thrust that is restricted to the western Bhutan High Himalaya. Moreover, we stress that this is a general suggestion to allow the upper HHC to rotate.

The three light grey crosses in Fig. 1b correspond to the fate of one conjugate fault pair initially formed on the lower ramp at three separate stages ($t = 3$, 2, and 1) during the advance of the slab as it is transported up and over the ramp of the Kakhtang Thrust. Subsequent newly forming faults are not shown. Note that the region of extension is not large. It can either be the general realm above a neutral surface of the slab travelling through the fault bend fold hinge, or a general region of extension in the upper portion of the slab synchronous with lower level contraction, as shown by wedge extrusion models for the HHC (e.g. Vannay & Grasemann 2001 and references therein).

Our suggested rotation of the entire HHC is not, we feel, unreasonable or without precedent. Edwards *et al.* (1996) discussed the likelihood of a 20–30° rotation of the Gonto La detachment based upon the presence of a cross-cutting steep normal fault (the Dzong Chu Fault) all along the northern portions of the Khula Kangri area. Other supporting evidence in the form of palaeomagnetic data has been reported from both the Spiti (NW India) and from Manaslu (Nepal), for example, where Schill *et al.* (2001, 2002) obtained data for a corresponding rotation of the upper HHC in this area. Deep seismic reflection data from Hauck *et al.* (1998) lend further support; these authors have shown evidence for large crustal scale features at depth below southern Tibet. Earthquake studies have additionally shown the occurrence of seismic slip deeper than 40 km in this area (Ekstrom 1987).

Our model is inconsistent with a channel–flow model to explain the thermotectonic evolution of the HHC (Grujic *et al.* 2002). Such a model requires that the present-day dips, and thus angles between, the MHT and the STDS have remained unchanged since times of high-temperature ductile flow of the HHC slab.

D_2 *deformation*

The E–W extension of D_2 is consistent with widespread observations throughout southern Tibet and the Himalaya of young E–W expansion (e.g., the graben systems of Armijo *et al.* 1986). It has been debated, however, whether E–W extension is contemporaneous with, or post-dates, the N–S extension associated with the STDS. Our data support the latter view or, at a minimum, require that the history of E–W extension in southern Tibet is protracted.

Conclusion

Fault surface slip data from the High Himalaya in eastern Lunana indicate two deformation products. A NNW-directed extension of D_1 is indicated by steeply SSE-dipping oblique-slip, and shallowly NNW-dipping, normal faults, while a (younger) E–W extension of D_2 is indicated by E-, and W-dipping conjugate pattern strike-slip faults. Progressive cross-cutting relationships amongst the D_1 fault group indicate a rotation of the local upper Himalaya. We speculate that this is due to a crustal-scale ramp and extension in the upper part of the HHC related to late reverse faulting along the Kakhtang Thrust. The D_2 E–W extension is considered to be part of the extension ubiquitous throughout southern Tibet.

The work of the Austrian party was funded within the Austrian–Bhutanese Glacial Lake Outburst Flood (GLOF) Mitigation Project, funded by the Austrian Federal Ministry of Foreign Affairs. We thank all the members of the Bhutan expedition of the University of Vienna, the University of Agricultural Sciences, Vienna and the Geological Survey of Bhutan for assistance and open discussion of ideas and data. Part of the writing was completed while G. Wiesmayr was visiting Freiberg as a DAAD scholar. We are indebted to B. Grasemann for speedy assistance on several occasions and to R. Lisle and an anonymous reviewer for very constructive reviews.

References

ANGELIER, J. & GOGUEL, J. 1979. Sur une méthode simple de détermination des axes pricipaux des constraintes pour une population de failles. *Comptes Rendus de l'Academie des Sciences, Paris*, **288**, 307–310.

ANGELIER, J. & MECHLER, P. 1977. Sur une méthode graphique de recherche des constraintes principales également utilisable en tectonique et en séismologie: la methode des dièdres droits. *Bulletin de la Société Géologique de France*, **7**, 1309–1318.

ARMIJO, R., TAPPONNIER, P., MERCIER, J. L. & TONGLIN, H. 1986. Quaternary extension in southern Tibet: Field observations and tectonic implications. *Journal of Geophysical Research*, **91**, 13803–13872.

BILHAM, R., LARSON, K., FREYMÜLLER, J. & MEMBERS, P. I. 1997. GPS measurements of present-day convergence across the Nepal Himalaya. *Nature*, **386**, 61–64.

BURCHFIEL, B. C. & ROYDEN L. H. 1985. North–south extension within the convergent Himalayan region. *Geology*, **13**, 679–682.

BURG, J. P., PROUS, F., TAPPONNIER, P. & MING, C. G. 1983. Deformation phase and tectonic evolution of the Lhasa block. *Eclogae Geologicae Helveticae*, **76**, 643–665.

CRAW, D., CHAMBERLAIN, C. P., ZEITLER, P. K. & KOONS, P. O. 1997. Geochemistry of a dry steam geothermal zone formed during rapid uplift of Nanga Parbat, northern Pakistan. *Chemical Geology*, **142**, 11–22.

ECKSTROM, G. A. 1987. *A Broad Band Method of Earthquake Analysis*. PhD thesis, Harvard University.

EDWARDS, M. A. & HARRISON T. M. 1997. When did the roof collapse? Late Miocene N–S extension in the High Himalaya revealed by Th-Pb monazite dating of the Khula Kangri granite. *Geology*, **25**, 543–546.

EDWARDS, M. A., KIDD, W. S. F., LI, J., YUE, Y. & CLARK, M. 1996. Multi-stage development of the southern Tibet detachment system near Khula Kangri. New data from Gonto La. *Tectonophysics*, **260**, 1–19.

EDWARDS, M. A., PÊCHER, A., KIDD, W. S. F., BURCHFIEL, B. C. & ROYDEN, L. H. 1999. The Southern Tibet Detachment System (STDS) at Khula Kangri, Eastern Himalaya: a large area, shallow detachment stretching into Bhutan? *Journal of Geology*, **107**, 623–631.

EINSELE, G., RATSCHBACHER, L. & WETZEL, A. 1996. The Himalaya-Bengal fan denudation-accumulation system during the past 20 Ma. *Journal of Geology*, **104**, 163–184.

FLETCHER, C. 1998. Effects of pressure solution and fluid migration on initiation of shear zones and faults. *Tectonophysics*, **295**, 139–165.

GANSSER, A. 1964. *The Geology of the Himalayas*. John Wiley, London.

GANSSER, A. 1983. *Geology of the Bhutan Himalayas*. Birkhauser Verlag, Basel.

GRASEMANN, B. & VANNAY, J. C. 1999. Flow controlled inverted metamorphism in shear zones. *Journal of Structural Geology*, **21**, 743–750.

GRUJIC, D., CASEY, M., DAVIDSON, C., HOLLISTER, L. S., KÜNDIG, R., PAVLIS, T. & SCHMID, S. 1996. Ductile extrusion of the Higher Himalayan Crystalline in Bhutan: evidence from quartz microfabrics. *Tectonophysics*, **260**, 21–43.

GRUJIC, D., HOLLISTER, L. S. & PARISH, R. R. 2002. Himalayan metamorphic sequence as an orogenic channel: insight from the kingdom of Bhutan. *Earth and Planetary Science Letters*, **198**, 177–191.

GUILLOT, S., DE SIGOYER, J., LARDEAUX, J. M. & MASCLE, G. 1997. Eclogitic metasediments from the Tso Morari area (Ladakh, Himalaya): evidence for continental subduction during India-Asia convergence. *Contributions to Mineralogy and Petrology*, **128**, 197–212.

HANCOCK, P. L. 1985. Brittle microtectonics: principles and practice. *Journal of Structural Geology*, **7**, 437–457.

HAUCK, M. L., NELSON, K. D., BROWN, L. D., ZHAO, W. & ROSS, A. R. 1998. Crustal structure of the Himalayan orogen at *c.* 90° east longitude from Project INDEPTH deep reflection profiles. *Tectonics*, **17**, 481–500.

O'BRIEN, P. J., ZOTOV, N., LAW, R., KHAN, M. A. & JAN, M. Q. 2001. Coesite in Himalayan eclogite and implications for models of India-Asia collision. *Geology*, **29**, 435–438.

OWENS, T. J. & ZANDT, G. 1997. Implications of crustal property variations for models of Tibetan plateau evolution. *Nature*, **387**, 37–42.

PATRIAT, P. & ACHACHE, J. 1984. India-Eurasia collision chronology has implications for crustal shortening and driving mechanism of plates. *Nature*, **311**, 615–621.

PETIT, J. P. 1987. Criteria for the sense of movement on fault surfaces in brittle rocks. *Journal of Structural Geology*, **9**, 597–608.

POWERS, P. M., LILLIE, R. J. & YEATS, R. S. 1998. Structure and shortening of the Kangra and Dehra Dun re-entrants, Sub- Himalaya, India. *GSA Bulletin*, **110**, 1010–1027.

ROWLEY, D. B. 1998. Minimum age of initiation of collision between India and Asia north of Everest based upon the subsidence history of the Zhepure Mountain section. *Journal of Geology*, **106**, 229–235.

SCHILL, E., APPEL, E., GAUTAM, P. & DIETRICH, P. 2002. Thermo-tectonic history of the Tethyan Himalayas deduced from palaeomagnetic record of metacarbonates from central Nepal (Shiar Khola). *Journal of Asian Earth Science*, **20**, 203–210.

SCHILL, E., APPEL, E., ZEH, O., SINGH, V. K. & GAUTAM, P. 2001. Coupling of late-orogenic tectonics and secondary pyrrhotite remanences: towards a separation of different rotation processes and quantification of rotational underthrusting in the western Himalayas (N-India). *Tectonophysics*, **337**, 1–21.

SCHNEIDER, D. A., EDWARDS, M. A., KIDD, W. S. F., KHAN, A. M., SEEBER, L. & ZEITLER, P. K. 1999. Tectonics of Nanga Parbat, western Himalaya: synkinematic plutonism within the doubly vergent shear zones of a crustal-scale pop-up structure. *Geology*, **27**, 999–1002.

SPANG, J. H. 1972. Numerical method for dynamic analysis of calcite twin lamellae. *Geological Society of America Bulletin*, **83**, 467–472.

SPERNER, B. 1996. *Computer programs for the finematic analysis of brittle deformation structures and the ertiary tectonic evolution of the western Carpathians (Slovakia)*. Tübinger Geowiss, Arbeiten, A27.

SPERNER, B., RATSCHBACHER, L. & OTT, R. 1993. A Turbo Pascal program package for graphical presentation and reduced stress tensor calculation. *Computers and Geoscience*, **19**, 1361–1388.

STREIT, J. E. & COX, F. E. 2000. Asperity interactions during creep of simulated faults at hydrothermal conditions. *Geology*, **28**, 231–234.

TURNER, F. J. 1953. Nature and dynamic interpretation of deformation lamellae in calcite of three marbles. *American Journal of Science*, **251**, 276–298.

VANNAY, J. C. & GRASEMANN, B. 2001. Himalayan inverted metamorphism and syn-convergence extension as a consequence of general shear extrusion. *Geological Magazine*, **138**, 253–276.

Strain localization by fracturing and reaction weakening – a mechanism for initiating exhumation of subcontinental mantle beneath rifted margins

M. R. HANDY[1] & H. STÜNITZ[2]

[1] *Geowissenschaften, Freie Universität Berlin, D-12249 Berlin, Germany*
[2] *Geologisch-Paläontologisches Institut, Universität Basel, CH-4056 Basel, Switzerland*

Abstract: Rift-related strain localization in spinel lherzolite from an exhumed passive continental margin in the Southern Alps involved two stages. (1) Critical fracturing coincided with heterogeneous nucleation of plagioclase, olivine, and hornblende aggregates to form discrete, ultrafine-grained (0.5–0.6 μm) shear zones oriented at high angles to the pre-existing foliation in the host rock. The syntectonic replacement of spinel lherzolite by lower pressure, plagioclase–hornblende lherzolite documents extensional exhumation under high temperature (700–900 °C) conditions accompanied by limited fluid infiltration. Deformation involved a combination of dislocation creep (ol) and diffusion-accommodated viscous granular flow (plag, ol, hbl aggregates). (2) Hydrous deformation at lower temperatures (200–400 °C) involved the formation of serpentine–chlorite mylonite and cataclasite along discrete, anastomozing shear zones oriented at low angles to the pre-existing foliation. Both stages involved drastic weakening, particularly once the shear zones coalesced subparallel to the extensional shearing plane. The top of the lithospheric mantle was initially strong, but is inferred to have become weaker than both the underlying mantle and the overlying mafic lower crust. The interconnection of such strong-then-weak delamination zones to form trans-lithospheric extensional shear zones accelerated rifting and led to the exhumation of subcontinental mantle during the late stages of continental breakup.

Mantle rocks are exhumed beneath non-coaxial extensional shear zones in rifted continental margins (Lemoine *et al.* 1987; Lister *et al.* 1991) and slow-spreading ocean ridges adjacent to ocean transform faults (Karsons 1991; Tuscholke & Lin 1994). In both settings, deformation is localized at the lithospheric scale, such that extensional strain is transferred from breakaway normal faults bounding asymmetrical rift basins in the upper crust down to mylonitic shear zones in the lower crust and mantle (Wernicke 1985; Lemoine *et al.* 1987). The geometry of these faults within the crust is partly constrained from structural studies of exposed crustal sections (Brodie & Rutter 1987*a*; Handy 1987) and reflection seismological experiments (McGeary & Warner 1985; Reston 1987; Keen *et al.* 1991), but the mechanisms triggering their nucleation and growth at depth, particularly within the mantle, remain enigmatic. The inaccessibility of active mantle shear zones coupled with their poor resolution in reflection seismological profiles obviously limit our ability to use large-scale geometry at depth as a reliable guide to mantle rheology.

Most models of strain localization harbour the assumption that strain localizes either within pre-existing weak lithologies of a compositionally and rheologically stratified lithosphere (e.g. Ranalli & Murphy 1987) or where a weak mechanical phase nucleates within a stronger lithology (e.g. Kirby 1985). In this paper, we term these two types of weakness, respectively, 'inherited weakness' and 'induced weakness'. Although there is consensus that localization results in bulk weakening once weak zones coalesce, some experimental and theoretical studies show that the onset of localization can also involve hardening if deformation is dilatant or involves a dilational component (Hobbs *et al.* 1990). This raises the possibility that large-scale shear zones initiate not in weak layers of the lithosphere, but in strong or preferentially stressed layers where dilational processes like cataclasis and metamorphic phase transformation are favoured.

To date, evidence for this hypothesis is equivocal. Microstructural studies of upper mantle rocks exposed at the surface reveal that most strain in the upper mantle is accommodated by viscous creep mechanisms (Drury *et al.* 1991; Newman *et al.* 1999; Furusho & Kanagawa 1999). Although fracturing has been identified as a potential localization mechanism in upper mantle rocks (Handy 1989; Vissers *et al.* 1997) and brittle precursors of mylonitic shear zones are ubiquitous in crustal rocks (Mitra 1978; Dixon & Williams 1983; FitzGerald & Stünitz 1993; Evans 1991; Wibberley 1999; Wintsch *et al.* 1995), fractures formed at high temperatures just

From: DE MEER, S., DRURY, M. R., DE BRESSER, J. H. P. & PENNOCK, G. M. (eds) 2002. *Deformation Mechanisms, Rheology and Tectonics: Current Status and Future Perspectives*. Geological Society, London, Special Publications, **200**, 387–407. 0305-8719/02/$15

prior to and/or during mylonitization are rarely preserved. Most fractures are associated with late mylonitic or post-mylonitic deformation rather than with the onset of mylonitization. Strain localization is therefore usually attributed to progressive dynamic recrystallization, in some cases enhanced by a transition to grain-size-sensitive creep mechanisms like diffusion-accommodated viscous granular flow (Vissers *et al.* 1995; de Bresser *et al.* 2001). The interconnection of such shear zones is believed to weaken the lithosphere during rifting (Handy 1994; Vissers *et al.* 1995), although the extent of this weakening is debatable.

In this paper, we present evidence that extensional shear zones in ultramafic rocks of the former rifted Apulian continental margin (southern Alps) nucleated in the upper mantle as dilatant shear fractures under fluid-deficient conditions. These fractures are shown to have been nuclei for discrete shear zones containing ultrafine-grained aggregates whose syntectonic mineral assemblage documents decompression during extensional exhumation. Based on an analysis of grain size, distribution and shape characteristics, we discuss the probable deformation mechanisms in these aggregates and the implications thereof for strain-dependent changes in the rheology of the lithospheric mantle in non-volcanic, rifted continental margins. In the final section, we integrate these findings with previous work to propose a qualitative mechanical model for distal parts of the rifted Apulian continental margin.

Geological setting

Structures and metamorphism related to Mesozoic, Tethyan rifting are preserved in several circum-Mediterranean mountain belts, where lower continental crustal and upper mantle rocks were exhumed in the footwall of large extensional faults prior to their incorporation within these Tertiary orogens (Drury *et al.* 1991; Vissers *et al.* 1995 and references therein). These pre-orogenic structures are also well preserved and accessible in the Ivrea–Verbano Zone, located in the westernmost part of the Southern Alps (Fig. 1a). The shear zones investigated in this paper transect the Balmuccia ultramafic body, one of several ultramafic bodies within the northwestern margin of the Ivrea–Verbano Zone.

The Ivrea–Verbano Zone is what was originally the deepest part of a fragmented piece of Paleozoic continental crust, the Ivrea crustal section (Fig. 1b). This crust was attenuated first during Early Permian transtensional tectonics, but especially during Early Mesozoic rifting (Handy & Zingg 1991). Shallower to intermediate levels of this rifted continental crust are exposed in adjacent units to the SE (Strona–Ceneri Zone) and NW (Sesia Zone, Fig. 1). Tertiary tectonics, primarily related to transpression along the Insubric Line (Fig. 1), verticalized the Ivrea–Verbano Zone, essentially exposing a cross section of the intermediate to lower continental crust at the surface (Zingg *et al.* 1990; Handy *et al.* 1999 and references therein). In the Ivrea crustal section, this Tertiary deformation was brittle and did not destroy the penetrative pre-Tertiary (Early Mesozoic and earlier) fabrics and mineral assemblages (Schmid *et al.* 1989).

The Balmuccia ultramafic body (Fig. 2) comprises mostly peridotite (spinel lherzolite) with subordinate pyroxenite bands (e.g. Rivalenti *et al.* 1981). It forms the intrusive base of a layered mafic complex (the Mafic Complex, Fig. 1) that itself intruded pre-Variscan mafic rocks and Carboniferous metasediments some 300–320 Ma (age criteria in Handy *et al.* 1999; Vavra *et al.* 1999). The ultramafic rocks therefore intruded at, or just above, the late Variscan (pre-Permian) crust-mantle boundary (Boudier *et al.* 1984), which is constrained by geobarometric studies to have occupied a depth of about 34–44 km at the time of intrusion (Shervais 1979; Rivalenti *et al.* 1981, 1984; Sinigoi *et al.* 1994). Magmatic way-up criteria indicate that the presently vertical magmatic banding in the Balmuccia ultramafic body (Fig. 2) was subhorizontal at the time of intrusion (Rivalenti *et al.* 1975).

The shear zones described below truncate, and therefore clearly post-date the magmatic to high-grade subsolidus structures related to Variscan orogenesis (Figs 3, 4). These latter structures include a subvertical compositional banding defined by cumulate pyroxenite layers (Fig. 4a) and a penetrative schistosity (oriented about 275°/85°, Fig. 2) locally containing a spinel mineral lineation plunging 50°–60° N. This schistosity is defined by pyroxene and spinel

Fig. 1. Geology of the Ivrea–Verbano Zone. (**a**) Map showing ultramafic bodies named after nearby towns (Balmuccia, Baldissero, Finero, Premosello) within the Mafic Complex. Inset map shows location within Alpine chain, western Europe. (**b**) Cross section of the western Central Alps and Southern Alps showing location of the Ivrea crustal section (modified from Schmid & Kissling 2000). Trace of cross section in (**a**) corresponds to section in (**b**) marked by arrow labelled 'Ivrea crustal cross section'.

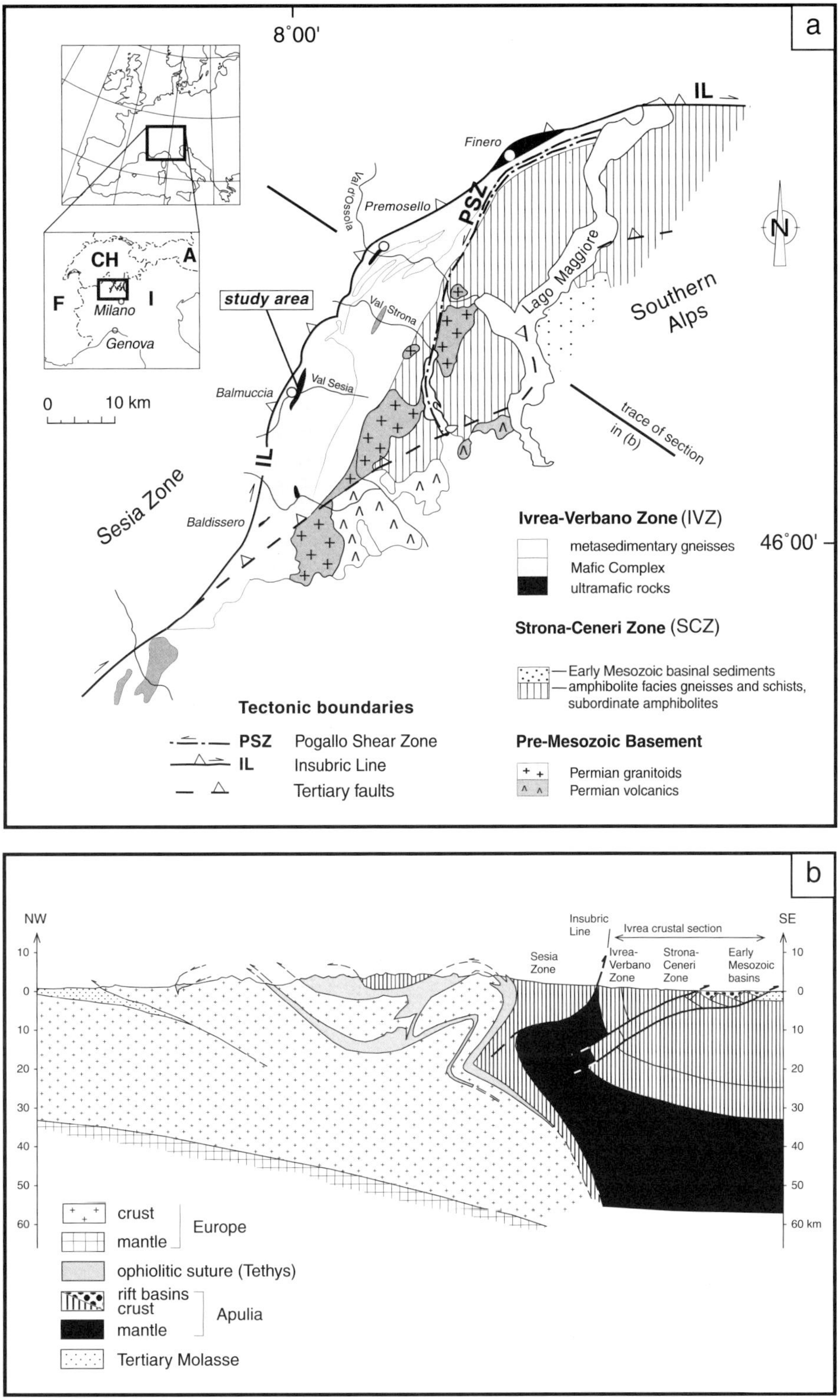

a
8°00'
46°00'
IL
Finero
Premosello
Val d'Ossola
PSZ
Lago Maggiore
Southern Alps
N
CH
A
F
I
Milano
Genova
study area
Val Strona
Val Sesia
Balmuccia
0 10 km
trace of section in (b)
Sesia Zone
Baldissero
Ivrea-Verbano Zone (IVZ)
metasedimentary gneisses
Mafic Complex
ultramafic rocks
Strona-Ceneri Zone (SCZ)
Early Mesozoic basinal sediments
amphibolite facies gneisses and schists, subordinate amphibolites
Tectonic boundaries
PSZ Pogallo Shear Zone
IL Insubric Line
Tertiary faults
Pre-Mesozoic Basement
Permian granitoids
Permian volcanics

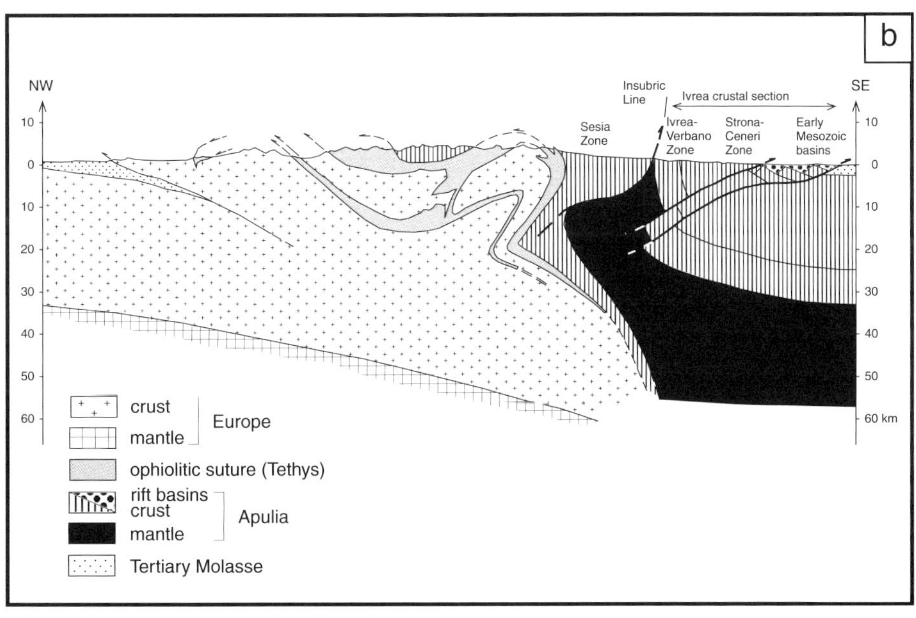
b
NW
SE
Insubric Line
Ivrea crustal section
Sesia Zone
Ivrea-Verbano Zone
Strona-Ceneri Zone
Early Mesozoic basins
10
0
10
20
30
40
50
60
60 km
crust
mantle
Europe
ophiolitic suture (Tethys)
rift basins
crust
mantle
Apulia
Tertiary Molasse

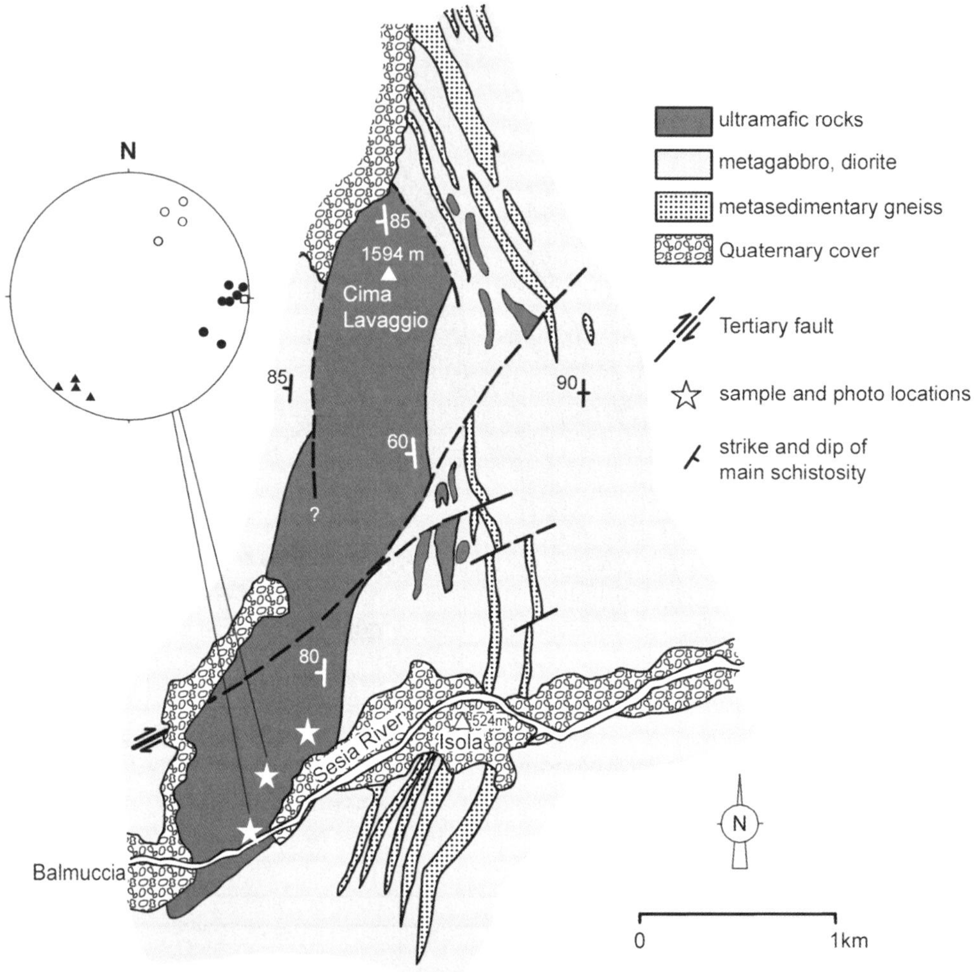

Fig. 2. Geological map of the Balmuccia ultramafic body and surrounding rocks within the Ivrea–Verbano Zone (simplified from Rivalenti *et al.* 1984; Quick *et al.* 1994). Equal area projection (lower hemisphere) with poles to foliation of type 1 shear zones (closed triangles) and type 2 shear zones (closed circles) with respect to the average orientation of the pre-existing compositional banding and granulite facies schistosity (open square) in a quarry of the Balmuccia ultramafic complex. Open circles are poles to cracks at the ends of type 2 shear zones.

grains aligned subparallel, or locally at low angles, to the compositional banding (Fig. 4). The schistosity is axial planar to tight to isoclinal folds that, in other parts of the Ivrea–Verbano Zone (Steck & Tièche 1976; Kruhl & Voll 1978/79), have been attributed to heterogeneous non-coaxial shearing under amphibolite to granulite facies conditions in Early Permian time (Handy & Zingg 1991). All of these early structures are associated with coarse-grained (1–2 mm) annealed microfabrics (Fig. 5a, Garuti & Friolo 1978/79) that characterize the protoliths adjacent to the mylonitic shear zones investigated here.

Rift-related structures in the Ivrea–Verbano Zone include the shear zones described below and have been mapped over an area of several hundred square kilometers (Brodie & Rutter 1987*a*; Handy 1987; Handy & Zingg 1991). Anastomozing, amphibolite to granulite facies, mylonitic shear zones accommodated extension parallel to the main foliation which follows the arc of the Ivrea–Verbano Zone. These shear zones are more common in the northeastern part of the Ivrea–Verbano Zone, where thermo-barometric data indicate the greatest amount of crustal thinning (Handy *et al.* 1999). The most

Fig. 3. Shear zones in lherzolite of the Balmuccia ultramafic body. (**a**) Type 2 shear zones appear as elongate cracks (arrows) and have conjugate orientations. The dark surface to the right is the serpentine-coated surface of such a shear zone. (**b**) Termination of a type 2 shear zone marked by discordant splays and late pseudotachylite (match tip). Type 1 shear zones (arrows) are truncated by the type 2 shear zone. Match for scale is 5 cm long.

prominent Early Mesozoic extensional structure is the Pogallo Shear Zone (PSZ in Fig. 1a), a 1–3 km wide mylonitic belt which accommodated sinistral, noncoaxial shear under retrograde amphibolite to greenschist facies conditions within quartz-rich metasediments. We shall discuss the significance of these structures in the final section, but first analyse their temporal equivalents in the ultramafic bodies.

Structural geology of the shear zones

Shear zones in the Balmuccia peridotite are strikingly narrow (Fig. 3). They comprise intricate strands of planar and arcuate, crack-like shear zones ranging in width from less than a mm to a few cm at most. Most of our observations were made in a quarry just above the road between the towns of Balmuccia and Bottorno, flanked by the northern bank of the Sesia river (locations indicated by stars in Fig. 2). It is unfortunate that the quarry has been active intermittently over the years, so that some of the outcrops reported in this paper no longer exist.

Two types of shear zone can be discerned (Figs. 3a, b). (1) Black shear zones form individually or as conjugate pairs which are truncated by later shear zones (Fig. 3b) and oriented at 45° to 60° to the pre-existing compositional banding (Fig. 4a). These shear zones contain microstructural evidence for viscous creep of olivine and pyroxenes, as described below. (2) Dark green to white, serpentine- and chlorite-bearing shear zones form long (one to several meters, Fig. 3a), generally planar strands that strike approximately N–S and dip subvertically (Fig. 2). These type 2 shear zones contain a very fine grained mylonitic foliation that is defined by aligned serpentine and chlorite grains. At their ends, these shear zones arc at high angles to the pre-existing foliation (Figs 2, 3b), often terminating as horsetail splays several cm to dm from the host strand.

Type 2 shear zones truncate and overprint type 1 (Fig. 3b), and are therefore clearly younger than the latter. Indeed, the high grade assemblages in the type 1 shear zones are rarely well

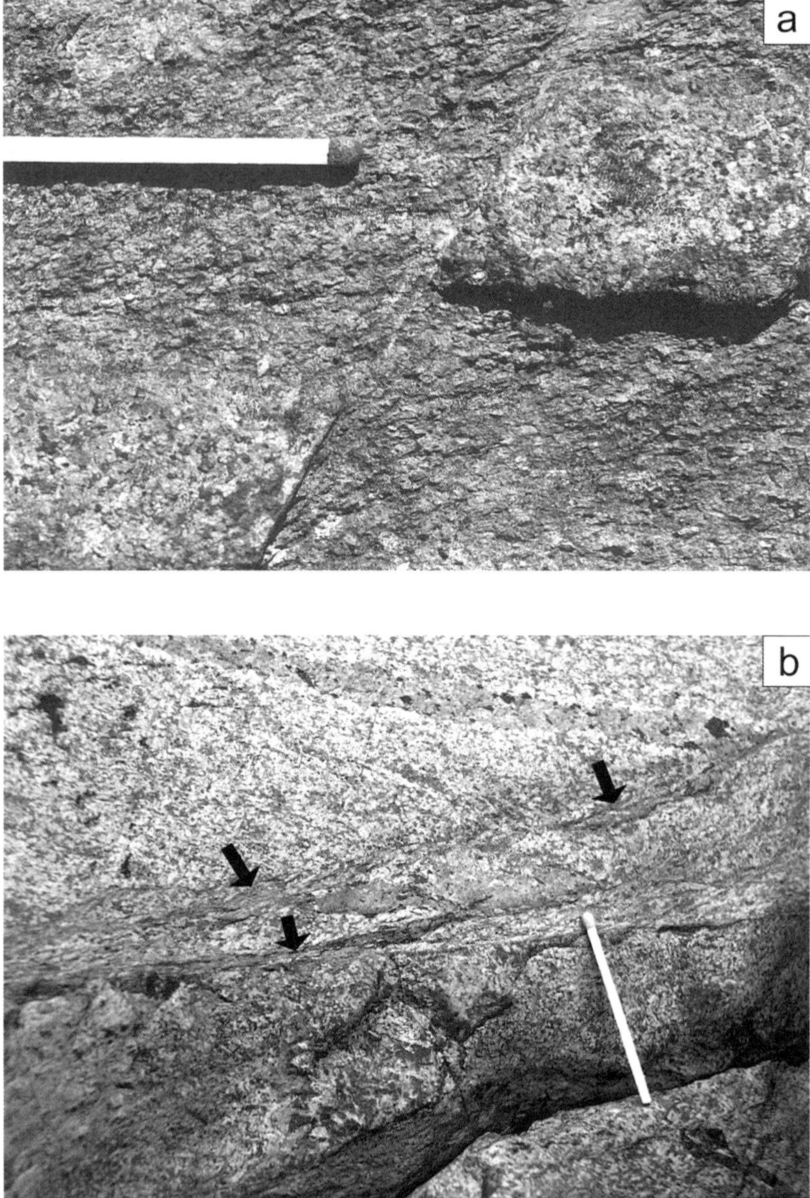

Fig. 4. Type 1 shear zones. (**a**) Shear zone offsets cumulate pyroxenite layer. Match for scale is 5 cm long. (**b**) Shear zones defined by closely spaced mylonitic foliation (arrows) truncates granulite facies schistosity and is related to vein-type pseudotachylite (above match tip).

preserved, as they usually show various degrees of brittle overprinting. This overprint takes the form of tectonic breccia or locally even pseudotachylite and ultracataclasite.

The displacement along both types of shear zone rarely exceeds a few centimeters (Fig. 4a), although the cumulative displacement along several strands making up type 2 shear zones is often a meter or more. The shear zones are spatially associated with vein-type pseudotachylites (Fig. 4b), some of which contain mineralogical evidence for formation under granulite facies conditions (Obata & Karato 1995). Most pseudotachylites, however, cut both types of shear zones and are clearly related to late, discordant cracks.

The type 1 shear zones and associated pseudotachylites have received considerable attention over the past twenty-five years (Garuti & Friolo 1978/79; Skrotzki *et al.* 1990; Obata & Karato 1995; Jin *et al.* 1998), and similar occurrences have been described from other ultramafic bodies further to the northeast in the Ivrea–Verbano Zone, especially in the Finero body (Kruhl & Voll 1978/79; Brodie 1980; Handy 1989) and the Premosello body (Rutter & Brodie 1988). Debate has centered on both the age of the shear zones (Variscan, Early Permian or Jurassic?) and on the mechanisms of strain localization leading to their growth: reaction-enhanced weakening (Brodie 1980); geometric softening of olivine (Obata & Karato 1995); dynamic recrystallization coupled with a transition to diffusion-accommodated grain-boundary sliding in olivine (Rutter & Brodie 1988; Jin *et al.* 1998); or a combination of these mechanisms (Handy 1989) have been proposed as possibilities. We first address the issue of localization before turning to the age of mylonitic shearing and the implications thereof for lithospheric extensional faults.

Microstructures and mineral assemblages

High temperature

To help resolve the debate on the mechanisms of strain localization in the upper mantle, we examined a remarkably well-preserved example of a type 1 shear zone in a spinel lherzolite (Fig. 4a). This shear zone is only 0.5 mm wide and offsets a magmatic pyroxenite band by 3.8 cm. As this magmatic band is oriented at 60° to the shear zone, the finite strain determined from the width and displacement is $\gamma = (380\,\text{mm}/0.5\,\text{mm})\cot 60^\circ = 75$. Despite the minor displacement, this shear strain is almost an order of magnitude greater than the shear strains determined across larger natural shear zones (e.g. Grocott & Watterson 1980) or experimental shear zones in high pressure experiments (e.g. Schmid *et al.* 1987; Paterson & Olgaard 2000). Our sample is therefore a good natural laboratory for studying high-strain microstructures in upper mantle rock.

Regarded in thin section, the type 1 shear zone cuts both the foliation and individual grains in the spinel lherzolite protolith along sharp, uneven boundaries (Fig. 5b). The shear zone contains a very fine grained matrix that envelops rounded clasts of olivine (ol), clinopyroxene (cpx), orthopyroxene (opx) and spinel (sp) (Fig. 5b). The clasts are obviously derived from the lherzolitic and pyroxenitic protoliths.

The matrix has a weak foliation defined by microscopic laminae that are oriented parallel to subparallel to the shear zone boundaries. Locally, these laminae are buckled and form intrafoliational folds. The laminae have contrasting grain sizes and, as observed in backscattered electron (BSE) images (Fig. 6), also have different compositions. Fine-grained (10–30 μm) laminae of pure olivine (ol) and of pure clinopyroxene (cpx) alternate with even finer grained (0.5–6 μm) layers of a mixture of plagioclase (plag), hornblende (hbl), olivine (ol), clinopyroxene (cpx), and orthopyroxene (opx; Fig. 6a–d). The mixed layers extend from, or envelop, spinel and cpx porphyroclasts (Fig. 6b, c). In the transmission electron microscope (TEM), the phases in the mixed layers have a grain size as small as 0.2 μm. The small grains are usually dislocation free (Fig. 7a). The shapes of the individual phases do not vary systematically from grain to grain. This invariance of grain shape, the lack of dislocations, the very small grain size, and the dispersed distribution of these phases within the aggregate are all attributes which suggest that deformation involved viscous grain boundary sliding, probably accommodated by diffusional mass transfer along grain boundaries.

The laminae of pure ol and, more rarely, of pure cpx grains in the mylonitic matrix of type 1 shear zones occur in the vicinity of ol and cpx porphyroclasts. The matrix grains range in diameter from 10 to 30 μm and have formed by progressive dynamic recrystallization of these porphyroclasts. Olivine grains in these matrix laminae have moderate dislocation densities ($\sim 10^{14}\,\text{m}^{-2}$) and show well-organized subgrain boundaries (Fig. 7b). This TEM observation, together with core mantle structures observed in the optical microscope, are diagnostic of subgrain rotation recrystallization (Poirier & Nicolas 1975). The matrix grains with high angle boundaries are generally subequant but locally have lobate grain boundaries, indicating that some grain boundary migration recrystallization also occurred.

The ultrafine-grained plag–hbl–cpx–opx–ol layers display a weak shape preferred orientation subparallel to the shear zone boundary defined by the margins of the large cpx and spinel grains (Fig. 6c). In most bands, however, the grains are equant and have rather straight boundaries. Some hbl/cpx phase boundaries are coherent, stepped boundaries along the *b*- and *a*-axes of both minerals (Fig. 7c). The heterogeneous distribution of grains and presence of

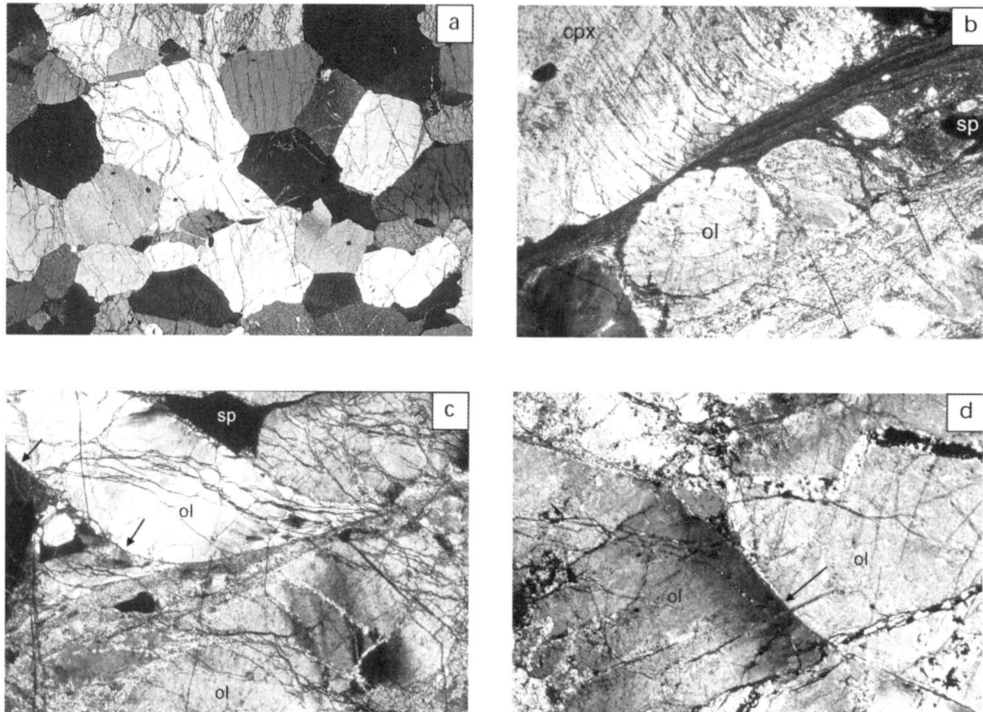

Fig. 5. Microstructures of a type 1 shear zone and its host rock, as observed in a polarizing light microscope with crossed nichols in the XZ fabric plane. (**a**) Protogranular microstructure of the host spinel lherzolite (frame length = 11.5 mm). (**b**) Thin mylonitic band at edge of shear zone in Fig. 4a. Note truncated twins in clinopyroxene (cpx), rounded clasts of olivine (ol), and spinel (sp) in ultrafine grained matrix of olivine, plagioclase, and hornblende (frame length = 1.35 mm). (**c**) Mylonitic microstructure in shear zone depicted in Fig. 4a. Arrows point to one of several arcuate cracks emanating from the side of the shear zone (frame length = 3.4 mm). (**d**) Close up of tip of crack in olivine grain indicated by arrows in (**c**), (frame length = 1.35 mm). Note undulose extinction adjacent to the crack.

coherent boundaries indicate that clinopyroxene and pargasitic hornblende coexisted stably in these layers.

Thus, the ol and cpx within the microscopic laminae of the mylonitic matrix are interpreted to have formed by different mechanisms. Pure ol and cpx layers (10–30 μm) derived from the dynamic recrystallization of ol and cpx porphyroclasts, whereas very small ol and cpx grains (0.5–6 μm) in plag–hbl–cpx–opx–ol layers were engendered by synkinematic mineral reactions.

The size and shape of the clasts within the matrix vary with their mineralogy. Spinel porphyroclasts are rounded and/or elongate and boudinaged, and are replaced by phase mixtures at their boundaries (Figs. 6b, c). Cpx and opx clasts are generally angular at the edges of the shear zone, rounded within the shear zone and have tails of dynamically recrystallized grains extending from their ends (Fig. 5b). Their internal microstructures are therefore interpreted to reflect the initial stages of deformation and localization, as discussed below.

Arcuate cracks emanate from the margins of the mylonitic matrix and terminate as inter- and transgranular fractures in the host rock (Figs. 5c, d). These cracks are unrelated to the late cracks mentioned above, as they are truncated by the latter and contain olivine and brown, pargasitic hornblende (Figs 6, 7). They therefore formed under the same high-grade conditions as the mylonitic matrix. The terminations of the cracks are oriented synthetically with respect to the bulk sinistral displacement along the shear zone boundaries. Displacement parallel to the crack boundaries increases from the crack tips (Fig. 5d) to their confluence with the mylonitic shear zone (Fig. 5c). Taken together, these observations indicate that the fractures opened as extensional shear fractures prior to and/or during viscous creep in the mylonitic matrix.

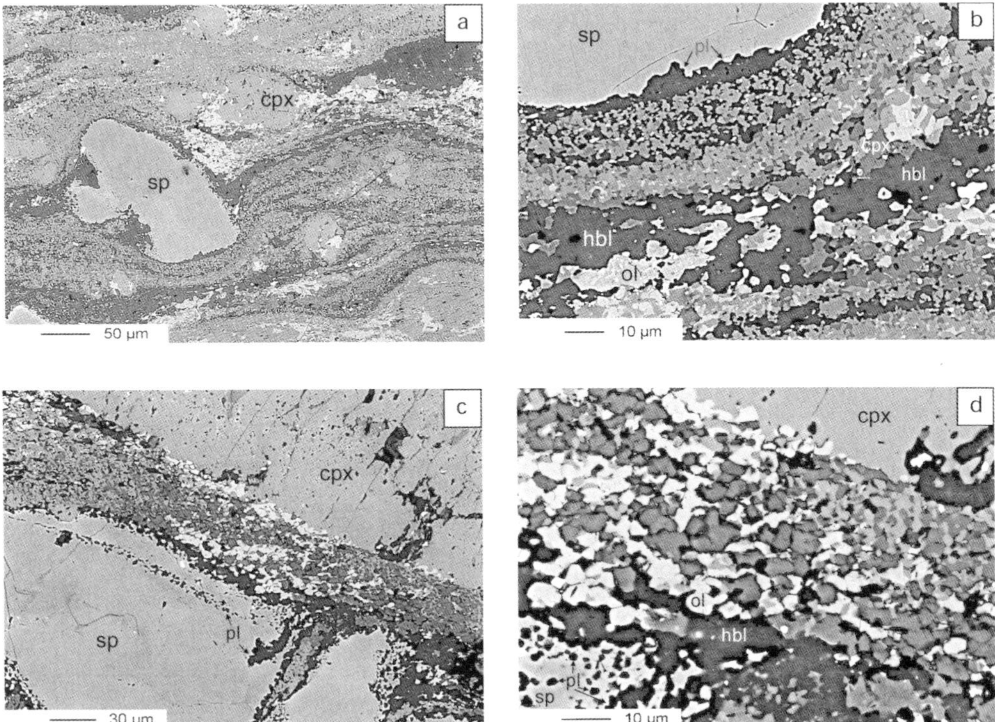

Fig. 6. Backscatter electron (BSE) images of mylonite in a type 1 shear zone. (**a**) Porphyroclasts of spinel (sp) and clinopyroxene (cpx) embedded in thin laminae of mixed phase layers consisting of olivine, plagioclase, orthopyroxene, clinopyroxene, and hornblende. (**b**) Detail of (**a**): boundary of a spinel grain (sp), where plagioclase (dark, pl), clinopyroxene (medium grey, cpx), and hornblende (slightly darker grey, hbl) form. The three new phases are well mixed. (**c**) A spinel grain (lower left, sp) and a clinopyroxene grain (upper right, cpx). Between these two grains is a thin mixed phase layer of plagioclase (pl), hornblende (hbl), olivine (ol), and clinopyroxene (cpx). (**d**) Detail of (**c**), showing the distribution of phases. plagioclase (dark, pl), olivine (lightest grey, ol), hornblende (darker grey, hbl), and clinopyroxene (lighter grey, cpx).

They appear to have propagated critically as inferred from their inter- and transgranular geometry, which is diagnostic of brittle intergranular and transgranular creep fracture (Gandhi & Ashby 1979).

Some fractures in the type 1 shear zones are kinematically related to small pull-apart structures that contain a glassy, former melt phase (Fig. 4b, see also Obata & Karato 1995). However, we do not think that deformation in the fine-grained mylonitic matrix described above involved melting because the thin layers of very fine-grained reaction products (plag, ol, hbl, cpx) alternate with monomineralic layers of dynamically recrystallized olivine. Dynamic recrystallization of olivine is indicative of lower strain rates than the seismic rates required to form pseudotachylite (e.g. Obata & Karato 1995), especially at the high temperatures of the deformation. The mylonitization in the matrix may have occurred during post-seismic creep, as described below.

Low temperature

The microstructure of type 2 shear zones is typically cataclastic, with fragments of altered, high-grade minerals floating in a ultrafine-grained, locally foliated matrix of serpentine and chlorite (Fig. 8). Evidence of this hydrous, greenschist facies overprint is also evident in type 1 shear zones, with transgranular serpentine-filled cracks cutting across the high grade matrix and porphyroclasts (Fig. 5c). Even in the mylonitic matrix of the type 1 shear zones, a fine-grained mixture of chlorite + magnetite ± serpentine forms pseudomorphs of former small spinel grains (Fig. 7d). These pseudomorphs indicate that the retrograde phase

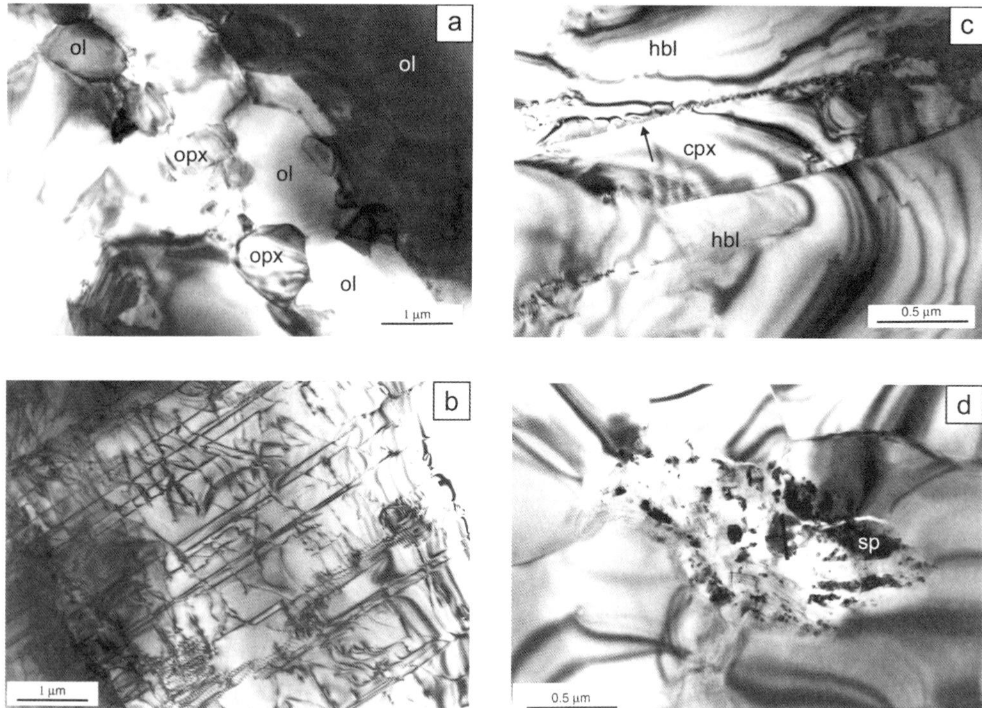

Fig. 7. Transmission electron microscope (TEM) images of mylonite in type 1 shear zone. (**a**) Olivine and opx grains in mixed phase laminae derived from spinel and cpx porphyroclasts. The very fine grains are dislocation-free and have equant shapes. (**b**) Dislocations within an olivine grain from a dynamically recrystallized monophase olivine layer (average diameter 20–50 μm) in the tail of an olivine porphyroclast. Note the well-organized subgrain walls. (**c**) View parallel to the *c*-axis of a clinopyroxene grain (oriented east-west in center) between two pargasitic hornblende grains. The upper boundary of the cpx grain (arrow) is coherent along the shared *a* and *b* crystallographic directions of cpx and hornblende. Mottled contrast along this boundary reflects steps in the coherent boundary. (**d**) Pseudomorphous replacement of spinel (sp) by an aggregate of serpentine, chlorite and magnetite. Small spinel aggregates are found in some of the pseudomorphous aggregates.

mixture grew after the high temperature deformation in type 1 shear zones.

Pressure–temperature conditions of type 1 shear zones

The BSE images of the very fine plag–hbl–cpx–ol layers in Figure 6 indicate that these phases all formed from the breakdown of spinel according to the discontinuous reaction:

$$sp + cpx + opx + H_2O = an + hbl + ol.$$

This syntectonic reaction and phase assemblage provides good constraints on the P–T conditions of deformation (Fig. 9). The spinel-lherzolite to plagioclase-lherzolite transition in the CFMASH (CaO–FeO–MgO–Al_2O_3–SiO_2–H_2O) system occurs at pressures between 400–750 MPa at temperatures of 700–1000 °C (Gasparik 1987; Bucher & Frey 1994; Furusho & Kanagawa 1999). This transition shifts to higher pressures at lower anorthite contents of the plagioclase, but there is almost no shift with pressure for anorthite contents between An_{80} to An_{100} (Gasparik 1987). EDS spectra of plagioclase grains in our samples taken with the SEM (the grains are too small to be analysed with the electron microprobe) indicate almost no Na and Si/Al ratios of approximately one, so that the anorthite content of the plagioclase is probably $\geq An_{90}$. Thus, pressures for the type 1 shear zones can be approximated with the CFMASH system, and the maximum pressure of the plagioclase lherzolite mylonite is about 600 MPa at 1000 °C (Fig. 9).

Chlorite was not stable during deformation, and therefore temperatures must have exceeded about 750 °C for the given low pressure range (chl-out curve, Fig. 9). The presence of pargasitic

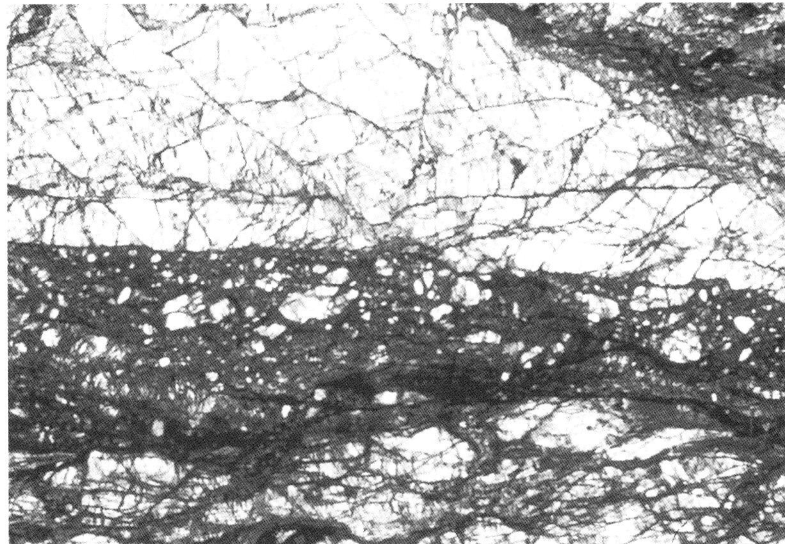

Fig. 8. Cataclastic microstructure of a type 2 shear zone. Note the angular, fractured clasts of partly serpentinized olivine (large clast at top, small clasts in matrix) and a nearly opaque matrix comprising mainly ultrafine grained serpentine and chlorite. View in the XZ fabric plane. Length of micrograph is 3.4 mm.

hornblende constrains the upper temperature limit of the stability field to an extent dependent on the bulk composition of the rock. The two experimental amphibole dehydration curves correspond to MORB pyrolite compositions (Niida & Green 1999, amph-out (2) curve in Fig. 9) and to depleted mantle compositions (Wallace & Green 1991, amph-out (1) curve in Fig. 9). The Balmuccia composition lies between these two compositions (Sinigoi *et al.* 1994), so that the upper temperature limit of deformation was somewhere between 900 and 1000 °C. Deformation in type 1 shear zones therefore occurred within a maximum range of 750–1000 °C at pressures of 400–600 MPa (shaded area in Fig. 9). Similar pressures were inferred for shear zones in the Balmuccia peridotite by Obata (1976) and Walter & Presnall (1994).

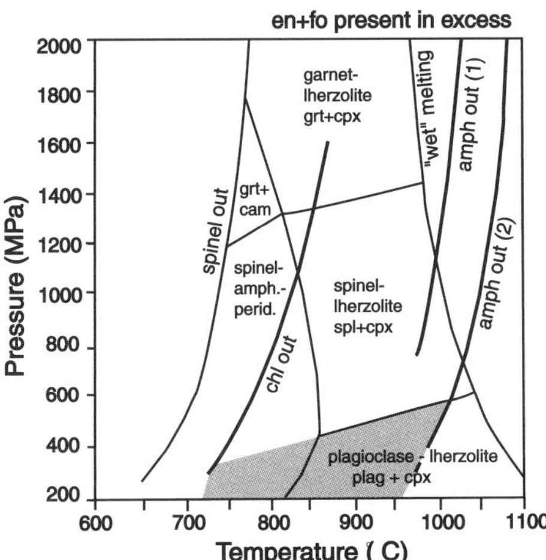

Fig. 9. P–T diagram for ultramafic rocks with the inferred pressure and temperature fields for the high-temperature shear deformation during Early Mesozoic rifting of the upper mantle. Stability fields and curves for stability of spinel and chlorite after Bucher & Frey (1994), experimentally determined hornblende stability curves from Wallace & Green (1991) (amph-out, 1) and Niida & Green (1999), (amph-out, 2). The probable maximum pressure and temperature range for the high temperature deformation of type 1 shear zones is indicated by grey shading.

The P–T conditions of strain localization within type 1 shear zones are consistent with conditions inferred for syntectonic microstructures in the pure olivine and clinopyroxene layers described above. Dynamic recrystallization of olivine in these layers constrains the temperature of deformation to have been at least 600–700 °C based on the extrapolation of laboratory flow laws for grain-size-insensitive power law creep of hydrous olivine to geological strain rates of 10^{-11}–$10^{-14}\,s^{-1}$ (Handy & Zingg 1991, references therein). The 10–30 μm diameter of dynamically recrystallized olivine grains from the coarser grained laminae of the mylonitic matrix is consistent with differential stresses ranging from 100–200 MPa according to the laboratory grain size piezometer of Van der Wal *et al.* (1993): $D = 0.015\Delta\sigma^{-1.33}$, where D and $\Delta\sigma$ are, respectively, the average grain diameter (in m) and differential stress (in MPa). These high differential stresses are only rough estimates but are consistent with the microstructural evidence that deformation in the sample occurred at the brittle-to-viscous transition.

The growth of pargasitic hornblende at the expense of clinopyroxene and in the presence of anhydrous reactants (sp, ol, cpx, opx) indicates that the reaction consumed water. The amount of water cannot have been great, however, given the small size and minor volume proportion of the hornblende grains.

The geodynamic significance of this syntectonic reaction is that strain localization in the type 1 shear zones was associated with a significant pressure drop. The spinel lherzolite that had equilibrated at a depth of about 34–44 km prior to localized shearing (Shervais 1979; Rivalenti *et al.* 1981, 1984; Sinigoi *et al.* 1994) was partly replaced by a plagioclase–hornblende lherzolite at shallower depths. Subsequently, the type 2 shear zones were active under hydrous greenschist facies conditions, probably at temperatures below 400 °C, in view of the stable mineral assemblage described in the previous section (Trommsdorff & Evans 1974; Trommsdorff 1983; Bucher & Frey 1994). Based on this information, a range of possible depths during deformation can be estimated from isotopic mineral cooling ages in the Ivrea–Verbano Zone and from a knowledge of geotherms in attenuating lithosphere. Accordingly, temperatures at the base of the attenuating Ivrea–Verbano Zone decreased from 800 °C to below 300 °C by Early to Middle Jurassic time (Handy & Zingg 1991 and references therein). At an assumed transient geothermal gradient of 30 °C/km during extensional shearing (e.g. Chapman 1986), these temperatures correspond to a depth range of 10–25 km for localized shearing in the Balmuccia peridotite.

Age of the shear zones

Neither the shear zones nor the pseudotachylites in the Balmuccia peridotite have been isotopically dated, partly because dating such small volumes of heterogeneous, fine grained aggregates with current techniques poses a formidable analytical challenge and is likely to generate ambiguous numbers. For now at least, the age of the type 1 and 2 shear zones can be constrained by considering the temperature estimates above in the context of published isotopic mineral cooling ages in the Ivrea–Verbano Zone.

The high temperature, type 1 shear zones are overprinted by, and therefore certainly older than the type 2 shear zones, but younger than the penetrative foliation and annealed microstructure in the spinel lherzolite host rocks. The main foliation and annealing in the host rocks are dated at 300–320 Ma by analogy with similar sub-solidus, high-grade structures in mafic rocks and metasediments in the Ivrea–Verbano Zone (Handy *et al.* 1999). Type 1 shear zones associated with high-grade pseudotachylites therefore post-date Variscan orogenesis. An Early Mesozoic (Late Triassic to Early Jurassic?) age for the type 1 shear zones and related pseudotachylites is likely in light of similar high-grade metamorphic conditions for mylonitic shear zones in the northeastern part of the Ivrea–Verbano Zone. There, the shear zones attenuate, and therefore post-date, isobars associated with 270–290 Ma magmatism and metamorphism (Handy *et al.* 1999). Furthermore, the petrological evidence in the type 1 shear zones for decompression and exhumation accords well with structural and petrological evidence in metabasic and pelitic rocks of the Ivrea–Verbano Zone for rapid cooling and E–W directed, extensional mylonitic shearing some 180–230 Ma (Handy & Zingg 1991). This extensional shearing is clearly related to the formation of Latest Triassic to Early to Middle Jurassic rift basins in the upper crust of the Southern Alps (Handy 1987).

The ≤400 °C temperatures for the type 2 shear zones correspond roughly to closing temperatures for isotopic systems in biotite (K-Ar, Rb-Sr, *c.* 300 °C) and muscovite (K-Ar, *c.* 350 °C), all of which yield 160–220 Ma cooling ages in the Ivrea–Verbano Zone (Zingg *et al.* 1990 and references therein). Only near the greenschist facies mylonites of the Insubric Line do biotite ages locally fall below 160 Ma, presumably due to incipient chloritization associated with Tertiary deformation (Handy & Zingg 1991). Thus, the published mica ages in the Ivrea–Verbano Zone indicate that since 160 Ma, temperatures in the lower crust and upper mantle of the Southern Alps never exceeded 300–350 °C. Most deformation in the type 2 shear zones is therefore inferred to have occurred in temporal continuity with type 1 shear zones during Early Mesozoic rifting. However, given the proximity of the Balmuccia ultramafic body to the Tertiary Insubric Line (Fig. 1a) and the brittle overprint of many type 1 and 2 shear zones, we cannot rule out that some type 2 shear zones were reactivated during Tertiary time. Indeed, with the exception of pseudotachylites related to type 1 shear zones, most pseudotachylites in the Ivrea–Verbano Zone are interpreted to have formed in response to

Oligo-Miocene, oblique backthrusting along the Insubric Line (Zingg *et al.* 1990; Techmer *et al.* 1993).

To summarize, both type 1 and type 2 shear zones are believed to have formed during a single deformational event related to rifting in Early Mesozoic time. However, Tertiary (Alpine) reactivation of some type 2 shear zones cannot be ruled out.

Discussion

Micromechanisms of strain localization

We interpret the type 1 and 2 shear zones in the Balmuccia ultramafic body to represent two successive stages of strain localization during rifting of the continental lithosphere of the Southern Alps in Early Mesozoic time. During stage 1, extensional shear fracturing under granulite facies conditions triggered the nucleation of ultrafine-grained mylonitic shear zones. This strain-dependent brittle-to-viscous transition is inferred to have occurred rapidly and in the presence of only minor amounts of fluid. The total amount of fluid introduced to the shear zones and contained in pargasite is estimated to be of the order of 0.1 wt% H_2O. The transition involved the spontaneous nucleation of lower pressure minerals (plag, hbl) along cracks and dilatant grain boundaries of the higher pressure phases (sp, cpx). Critical crack growth was therefore intimately related to the nucleation of minerals with larger molar volumes than the higher pressure phases which they replaced.

During stage 2, strain concentrated within very long, narrow shear zones at low angles to the pre-existing foliation and subparallel to the extensional shearing plane. As discussed in the next section, these presently subvertical shear zones were originally subhorizontal to moderately dipping during Early Mesozoic extension. Localization involved the breakdown of olivine, pyroxene and hornblende to form the hydrous phases serpentine and chlorite in the fine-grained mylonitic matrix as well as in parts of the undeformed rock adjacent to the mylonites. This testifies to a significant influx of fluids during the latter stages of extensional deformation. Like the type 1 shear zones before, strain localization in the type 2 shear zones was closely related to dilatancy and syntectonic phase transformations in the presence of a hydrous fluid phase.

Figure 10 shows a possible strength evolution for the two stages of strain localization outlined above. This evolution is speculative, based as it is on a correlation of microstructures in our samples with experimental microstructures and rheologies obtained at high temperatures and laboratory strain rates (Paterson 1987). Nevertheless, it serves as a qualitative guide to the strain-dependent behaviour of the upper mantle during rifting.

The stress–strain curve in this figure pertains to the entire volume of rock affected by deformation (i.e. the host lherzolite plus fractures and

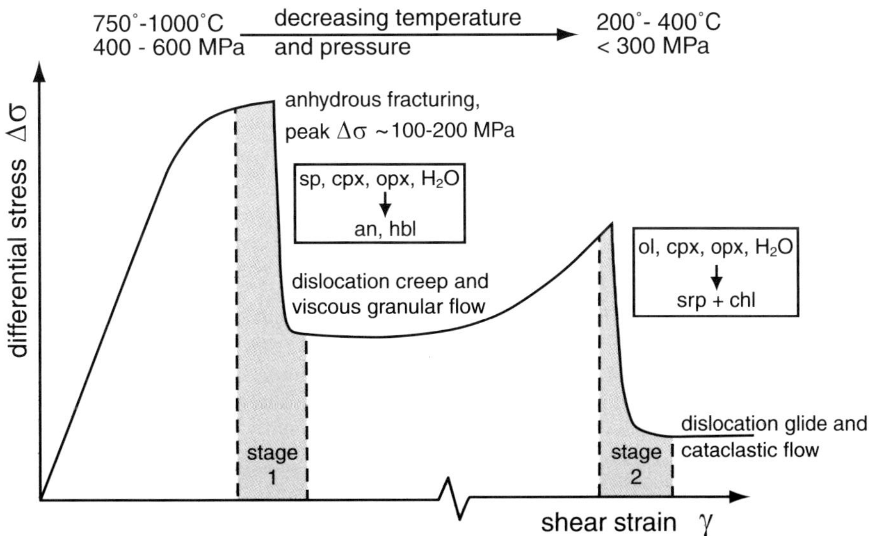

Fig. 10. Schematic strength versus strain evolution of lherzolite in the Balmuccia ultramafic body during Early Mesozoic rifting inferred from microstructures (see text).

shear zones). An initial increase in stress just prior to fracturing is inferred from traces of low strain intracrystalline plasticity adjacent to fractures in the olivine grains of the wall rock (Fig. 5d). Peak strength then coincided approximately with fracturing during the formation of the type 1 shear zones (Fig. 10). Fracturing marked the onset of the spinel- to plagioclase-lherzolite transition, which is limited to the immediate vicinity of the fractures. The fracturing increased permeability which in turn enhanced fluid infiltration. The interconnection of these fractures lined with very small strain-free grains drastically weakened the deforming rock, indicated in Figure 10 by a marked strength drop. This strength drop is inferred from numerous experimental and theoretical studies showing that significant weakening is induced by dynamic recrystallization (Zeuch 1982, 1983; Tullis & Yund 1985; see de Bresser *et al.* 2001 for a recent discussion) and, more importantly, by a transition to diffusion or reaction-accommodated viscous granular flow (e.g. White & Knipe 1978; Stünitz & Tullis 2001). These mechanisms were detected in the mylonitic laminae of the type 1 shear zones, and have been recognized in gabbroic mylonites (Brodie & Rutter 1985; Stünitz 1993; Kruse & Stünitz 1999) as well as in peridotitic mylonites (Boullier & Guegen 1975; Newman *et al.* 1999; Furusho & Kanagawa 1999). The high differential stress range (100–200 MPa) estimated from the small sizes of dynamically recrystallized olivine grains was probably transient, and marked a short period of rapid slip parallel to the shear zone boundaries during and/or immediately after macroscopic failure at peak strength. The growth of the mylonitic matrix may have stabilized the deformation at a stress significantly lower than the peak strength recorded by the observed grain size of dynamically recrystallized olivine (Fig. 10).

It is interesting to note that strain partitioning within very fine-grained mylonitic microstructures like those observed in our samples has been produced in the laboratory at similarly high temperatures (900 °C) and high strain rates ($\gamma = 5 \times 10^{-5}\ s^{-1}$: Stünitz *et al.* 1999) in similar lithologies. This lends credence to the idea that the mylonitic matrix deformed at very high strain rates and is related to the formation of pseudotachylite in nearby pull-apart structures (Fig. 4b).

Continued shearing at decreasing temperature and lithostatic pressure may have induced non-linear work-hardening (Fig. 10) due to the exponential (Arrhenius) dependence of viscous creep strength on temperature (Weertman 1970). However, the rock is inferred to have weakened again when serpentine and chlorite nucleated and grew at the expense of the high-grade minerals (Fig. 10). Laboratory experiments indicate that serpentinite (Raleigh & Paterson 1965) has a laboratory strength several orders of magnitude less than the strengths of olivine and pyroxenes at comparable homologous temperatures and strain rates (Fig. 1 in Brodie & Rutter 1987*b*). Below about 500–600 °C, olivine aggregates undergo cataclasis, even at low strain rates and very fine grain sizes (Handy 1989; Handy & Zingg 1991), whereas sheet silicates deform by dislocation glide and/or creep parallel to their 001 surfaces (e.g. biotite: Wilson & Bell 1979; Kronenberg *et al.* 1990) at natural strain rates and temperatures down to 150–250 °C (Lin 1997). On the crustal scale, the prime agent of weakening is inferred to have been the interconnection of the type 2 shear zones to form an anastomozing network subparallel to the extensional shearing plane. This is consistent with the observation above that most displacement was accommodated by type 2 shear zones.

The scenario above for the shear zones of the Balmuccia ultramafic body obviously does not preclude other strain localization mechanisms previously proposed for ultramafic rocks in the Ivrea–Verbano Zone. Indeed, a localization mechanism proposed for all occurrences so far is a strain-induced switch in deformation mechanisms for olivine from dislocation creep to viscous granular flow accommodated by grain boundary diffusion (Premosello body, Rutter & Brodie 1988; Balmuccia body, Jin *et al.* 1998) or syntectonic reaction (Finero body, Handy 1989). Yet, in all these studies pre- to syn-mylonitic fracturing and cataclasis have been either overlooked or attributed solely to late-mylonitic or post-mylonitic deformation under hydrous, sub-greenschist facies conditions.

Our study underscores the importance of combined fracturing and syntectonic metamorphism as a most effective agent of strain localization in the upper mantle (Vissers *et al.* 1997). Moreover, it confirms the predictions of Drury *et al.* (1991) that hydrous brittle deformation in the upper mantle localizes at temperatures less than 900 °C. We note that the weakening associated with coeval fracturing and heterogeneous nucleation of transiently fine-grained reaction products is stress- and strain-induced and occurs within an initially very strong lithology. Induced weakening therefore differs fundamentally from weakening which results from the localization of mylonitic deformation within pre-existing weak lithologies (Ranalli & Murphy 1987) or inherited structures, as proposed for natural mylonites (granitic mylonites: Segall & Pollard 1983, Segall & Simpson 1986; gabbro: Stünitz 1993; abyssal peridotite: Jaroslow *et al.* 1996) and as demonstrated in experiments (Tullis *et al.* 1990).

Despite their importance for localizing deformation in the upper mantle, type 1 shear zones are rarely preserved in naturally deformed peridotites. Due to overprinting during retrograde stage 2 deformation, the initial high temperature assemblages and microstructures of the type 1 shear zones are only preserved in narrow, 'arrested' zones of minor displacement at the margins and near the ends of the type 2 shear zones (Fig. 3b).

Implications for lithospheric extensional faulting and weakening

Induced weakening of the upper mantle potentially determines the large-scale structure and rheology of continental margins. The most striking consequence of the two-stage strength evolution outlined above is that the lithosphere may have weakened most where extrapolated experimental flow laws (e.g. as employed by Ranalli & Murphy 1987) indicate that it was initially strongest, *viz.*, at the top of the lithospheric mantle. Fault rocks formed there evidently have microstructures and rheologies which changed with strain during exhumation. To assess these effects on the lithospheric scale, we briefly review Early Mesozoic extensional structures in the Ivrea–Verbano crustal section, as shown in map view in Figure 1a and in the restored section in Figure 11.

Previous studies in the Ivrea–Verbano Zone have shown that rift-related attenuation of the lithosphere was accommodated primarily in quartz-rich granitoid and pelitic rocks within the Pogallo Shear Zone (Handy 1987; Handy & Zingg 1991), at the base of the intermediate crust (Fig. 11). The PSZ comprises retrograde amphibolite to greenschist facies mylonites, with cataclasites at its upper limit marking the Early Mesozoic, viscous-to-brittle transition in the crust (Handy 1987). Related mylonitic shearing also overprinted granulite facies metabasites and metasediments within the former lower crust in the Ivrea–Verbano Zone (Brodie & Rutter 1987*a*; Zingg *et al.* 1990). Comparison of pre- and post-rift crustal thicknesses from thermobarometric data at the northeastern end of the Ivrea–Verbano Zone indicate that the PSZ and lower crustal shear zones together

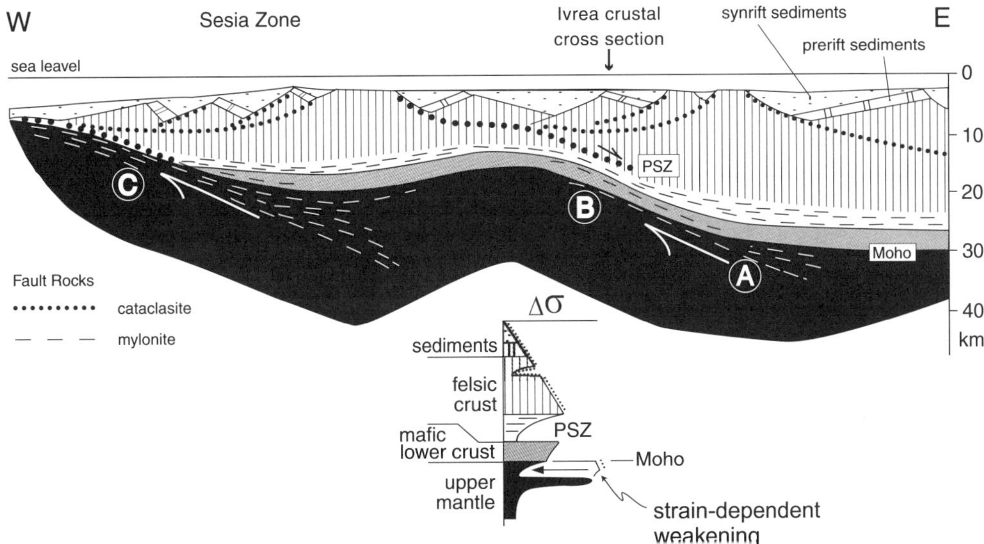

Fig. 11. Schematic cross section through the Apulian margin of Tethys at the end of Early Mesozoic rifting (after Handy & Zingg 1991; Schmid 1993). A, location for nucleation of type 1 shear zones; B, location for activity of type 2 shear zones; C, trans-lithospheric shear zone exhuming subcontinental mantle to the ocean floor. Approximate location of present Sesia Zone is labelled. Geometry of rift basins adopted from the references above. PSZ, Pogallo Shear Zone, as in Fig. 1a. Strength versus depth diagram below shows relative strength of lithospheric layers beneath the downward arrow, corresponding to the Ivrea crustal section. Note the horizontal, leftward arrow just below the Moho, indicating strain weakening described in the text.

excised a total of 20–25 km of crust in Early Mesozoic time (Handy *et al.* 1999).

Regarded on the lithospheric scale, these extensional structures are interpreted by some authors to be part of an asymmetrical rift system, with the lower crust and upper mantle of the Apulian margin, represented here by the Ivrea–Verbano Zone exposed in the lower plate of a uniform-sense master fault dipping to the west beneath the opposite European margin (Lemoine *et al.* 1987; Stampfli & Marthaler 1990; Vissers *et al.* 1991; Favre & Stampfli 1992; Froitzheim & Manatschal 1996). Unfortunately, this purported master fault is not exposed, either because it never existed or because Alpine overprinting of rift-related structures in the deep parts of the opposite European margin (all of which are located north and west of the Insubric Line in Fig. 1) was so complete. However, field studies clearly show that currently exposed Early Mesozoic shear zones were originally E-dipping (i.e. toward the rifted Apulian margin), both in the Ivrea–Verbano Zone and in the lower Austro-Alpine Margna-Sella units (Müntener & Hermann 2001). Together, these shear zones are inferred to have exhumed lower crustal and sub-continental, upper mantle rocks, as depicted in Figure 11 (Handy 1987; Handy & Zingg 1991). The Strona-Ceneri and Sesia Zones adjacent to the Ivrea–Verbano Zone (Fig. 1) formed, respectively, shallower and more distal parts of this lower plate margin (Fig. 11, Schmid 1993; Froitzheim & Manaschal 1996).

The limited displacement along the type 1 and 2 ultramafic shear zones in this paper suggests that they accommodated only a small proportion of the total extension within this extensional system. The type 1 shear zones are therefore small-displacement analogues for incipient extensional detachment faults in the uppermost mantle (A in Fig. 11), some of which are inferred to have connected upwards through the crust to faults bounding rift basins.

The critical growth of dilatant shear cracks and spontaneous nucleation of weak, lower pressure reaction products just prior to and/or during mylonitic shearing in the type 1 shear zones suggests that deformation of the upper mantle rocks may have begun after they had already entered the low pressure stability field for plagioclase lherzolite. Initial exhumation of the mantle must therefore have occurred within the overlying crust, along the Pogallo Shear Zone. The juxtaposition of hot mantle rock with cooler crustal rocks is reflected in the contrasting P–T paths for the mantle and crustal rocks in the Ivrea–Verbano Zone. While the exhuming mantle rocks underwent pronounced cooling (Fig. 9), the overlying crustal rocks were initially heated by these relatively hot mantle rocks in the footwall (Fig. 3c in Handy *et al.* 1999).

During this initial stage of exhumation, the upper mantle was stronger than the overlying crust (initial strength profile for the upper mantle in Fig. 11) and is inferred to have acted as a stress guide. Differential stress within this layer increased until it attained the fracture strength of upper mantle rock. Fracturing allowed the ingress of fluids and facilitated the phase changes which led to pronounced weakening in the type 1 shear zones. With continued stretching, a massive influx of fluids and the coalescence of serpentinitic, type 2 shear zones (B in Fig. 11) effected further weakening of the upper mantle. At this stage, the strength of the upper mantle is inferred to have dropped below that of the lower crust, as indicated by the arrow just below the Moho on the strength versus depth diagram in Figure 11.

The scenario above has several implications for the evolution of rifted continental margins. First, the lower crust is ultimately the strongest rather than the weakest layer and exceeds the strength of the upper mantle just beneath the Moho (Handy 1989). This is at odds with many models that incorporate a weak lower crust and strong upper mantle (e.g. Chen & Molnar 1983; McKenzie *et al.* 2000). However, a recent reassessment of seismological data from the continental lithosphere indicated that earthquakes are rare in the lithospheric mantle and usually occur in the crust (Maggi *et al.* 2000), lending support to the idea that, at least in some places, the lower crust is stronger than the upper mantle. Second, large extensional faults probably root within the top of the lithospheric mantle, rather than within the lower crust as previously proposed (e.g. Reston 1990). Lithospheric attenuation in rifted margins is therefore inferred to involve progressive delamination (see Fig. 3b of Lister *et al.* 1991) along layer-parallel weak zones (Brun & Beslier 1996) that nucleated within initially strong layers. The linkage of such strong-then-weak layers within the lithosphere leads to the formation of noncoaxial, trans-lithospheric shear zones with different rheologies in different depth-intervals. Noncoaxial extensional faults that transect the lithosphere are therefore believed to develop towards the end of rifting, rather than at the beginning as proposed by Wernicke (1985). Third, the formation of trans-lithospheric shear zones with very weak fault rocks in the distal parts of rifted margins (C in Fig. 11) is predicted to reduce the overall strength of the attenuating crust and therefore to accelerate rifting. Indeed, Vissers *et al.* (1995) have pointed out that the development of one or more through-going extensional shear zones may be responsible for increased spreading velocity in the Liguro-Piemontese domain from pre-middle Jurassic values of 0.5 cm/a to 2 cm/a from the middle Jurassic breakup onwards (Savostin *et al.* 1986). Finally, trans-lithospheric shear zones comprising weak, serpentinic mylonites and cataclasites in their shallowest segments (C in Fig. 11) may be responsible for the exhumation of subcontinental lithospheric mantle in continent–ocean transitional domains as observed, for example, in submarine surveys of the non-volcanic Galicia margin off the coast of Spain (Boillot *et al.* 1995).

Inclined extensional shear zones like those depicted in Figure 11 are not necessarily diagnostic of uniform sense simple shear (Wernicke 1985), but can be interpreted as localized non-coaxial shear of distal parts of the rifted continental margin within an overall regime of lithospheric scale pure shear (Brun & Beslier 1996). This scenario was first proposed for the latter stages of rifting of the lower plate Apulian margin by Handy (1987) and contrasts with other reconstructions of the Apulian margin in which lithospheric attenuation was accommodated solely by simple shear along uniform-sense master faults dipping either to the east (Trommsdorff *et al.* 1993) or west (e.g. Froitzheim & Manatschal 1996). In scaled analogue models of rheologically stratified, continental lithosphere subjected to vertical shortening and lateral, pure shear extension, Brun & Beslier (1996) showed that rheological instabilities within an initially stiff upper mantle layer induced asymmetrical rift geometries, such that a conjugate set of non-coaxial, extensional shear zones developed at the base of the lithosphere. Inserting weaker material at the top of the lithospheric mantle accentuated this asymmetry, favouring the further growth of one of these non-coaxial shear zones to accommodate rapid extensional exhumation of subcontinental mantle rock in its footwall, as modelled by Callot *et al.* (2001). The strikingly similar geometries of the rifted margins in the scaled models with the reconstructed Tethyan passive margins suggests that strain-dependent weakening at the top of the lithospheric mantle may well have facilitated extensional exhumation of the subcontinental mantle.

Conclusions

The ultramafic shear zones within what was originally the deepest part of the Ivrea–Verbano Zone (northern Italy) formed at or near the Moho during Early Mesozoic rifting of the non-volcanic, Apulian continental margin. They accommodated only small displacements and are therefore regarded as examples of incipient extensional detachment in the subcontinental mantle.

Two types of shear zones with different mineral parageneses and deformational microstructures formed during successive stages of extensional exhumation of the subcontinental mantle. (1) Shear zones at moderate to high angles to the pre-existing foliation in the host spinel lherzolite involved strain localization by initial fracturing transitional to viscous granular flow under retrograde, high temperature conditions. (2) Very long, narrow cataclastic shear zones coated with serpentine and chlorite accommodated most of the extensional strain under hydrous, low temperature (greenschist facies) conditions. These shear zones are oriented at low angles to the pre-existing foliation.

In type 1 shear zones, brittle–viscous shearing coincided with pronounced decompression, as inferred from the syntectonic replacement of a high-pressure assemblage (ol–cpx–opx–sp) by an hydrous lower pressure (ol–plag–hbl) assemblage. Phase equilibria constrain this reaction to have occurred at 750–1000 °C at pressures of 400–600 MPa. Syntectonic decompression is consistent with widespread evidence in the Ivrea–Verbano Zone for marked cooling and exhumation of the Apulian continental margin in Early Mesozoic time. The dominant deformation mechanism in the ultrafine-grained reaction products is inferred to have been viscous grain boundary sliding, probably accommodated by diffusional mass transfer along the grain boundaries.

The strain-dependent changes in mineralogy and deformation mechanisms above are believed to have decreased ultramafic rock strength by at least an order of magnitude from its peak value of about 100–200 MPa just after the onset of fracturing. These rocks weakened even further once type 2 shear zones containing serpentine and chlorite coalesced subparallel to the bulk extensional shearing plane.

Our findings suggest that extensional detachments nucleate as cracks at or near the top of the, initially hard, upper mantle. There, ultramafic rocks weaken with strain as grain size is reduced by cataclasis and heterogeneous nucleation of very fine-grained reaction products. The upper mantle therefore evolves into a low viscosity detachment layer sandwiched between stronger mafic lower crust above and mantle below. In the case of the Apulian continental margin, the interconnection of several such detachment layers to form large, non-coaxial extensional fault zones may have caused the observed Mid-Jurassic increase in rifting rate between Europe and Apulia. Such weak faults zones may also have accommodated extensional exhumation of subcontinental Apulian mantle within the continent-ocean transition zone.

The very constructive reviews of F. Gueydan and K. Kanagawa as well as editorial comments by H. de Bresser significantly improved the manuscript. Thanks also go to J. Babist for field assistance, and to B. Ernst and M. Grundmann for helping to prepare the figures. We acknowledge the support of the German Science Foundation (grant Ha 2403/5-1 to MH) and the Swiss National Science Foundation (grants 2000-055420.98/1 and 2100-057092.99/1 to HS).

References

BOILLOT, G., AGRINIER, P. *ET AL.* 1995. A lithospheric syn-rift shear zone at the ocean-continent transition: preliminary results of the GALINAUTE II cruise (Nautile dives on the Galicia Bank, Spain). *C.R. Académie des Sciences*, **321**, II a, 1171–1178.

BOUDIER, F., JACKSON, M. & NICOLAS, A. 1984. Structural study of the Balmuccia massif (Western Alps): a transition from mantle to lower crust. *Geologie en Mijnbouw*, **63**, 179–188.

BOULLIER, A. M. & GUEGUEN, Y. 1975. SP-Mylonites: origin of some mylonites by superplastic flow. *Contributions to Mineralogy and Petrology*, **50**, 93–104.

BRODIE, K. H. 1980. Variations in mineral chemistry across a shear zone in phlogopite peridotite. *Journal of Structural Geology*, **2**, 265–272.

BRODIE, K. H. & RUTTER, E. H. 1985. On the Relationship between Deformation and Metamorphism with Special Reference to the Behavior of Basic Rocks. *In*: THOMPSON, A. B. & RUBIE, D. (eds) *Kinematics, Textures and Deformation.* Advances in Physical Geochemistry, Springer Verlag, 138–179.

BRODIE, K. H. & RUTTER, E. H. 1987*a*. Deep crustal extensional faulting in the Ivrea Zone of Northern Italy. *Tectonophysics*, **140**, 183–212.

BRODIE, K. H. & RUTTER, E. H. 1987b. The role of transiently fine-grained reaction products in syntectonic metamorphism: natural and experimental examples. *Canadian Journal of Earth Sciences*, **24**, 556–564.

BRUN, J.-P. & BESLIER, M. O. 1996. Mantle exhumation of passive margins. *Earth and Planetary Science Letters*, **142**, 161–173.

BUCHER, K. & FREY, M. 1994. *Petrogenesis of Metamorphic Rocks*. Springer-Verlag, Berlin.

CALLOT, J.-P., GRIGNE, C., GEOFFREY, L. & BRUN, J.-P. 2001. Development of volcanic passive margins: Two dimensional laboratory models. *Tectonics*, **20**, 148–159.

CHAPMAN, D. S. 1986. Thermal gradients in the continental crust. *In*: DAWSON, J. B., CARSWELL, D. A., HALL, J. & WEDEPOHL, K. H. (eds) *The nature of the lower continental crust*. Geological Society, London, Special Publications, **24**, 51–62.

CHEN, W.-P. & MOLNAR, P. 1983. Focal depths of intracontinental earthquakes and their correlations with heat flow and tectonic age. *Seismological Research Lettters*, **59**, 263–272.

DE BRESSER, J. H. P., TER HEEGE, J. H. & SPIERS, C. J. 2001. Grain size reduction by dynamic recrystallization: can it result in major rheological weakening? *International Journal of Earth Sciences*, **90**, 28–45.

DIXON, J. & WILLIAMS, G. 1983. Reaction softening in mylonites from the Arnaboll thrust, Sutherland. *Scottish Journal of Geology*, **19**, 157–168.

DRURY, M. R., VISSERS, R. L. M., HOOGERDUIJN STRATING, E. H. & VAN DER WAL, D. 1991. Shear localisation in upper mantle peridotites. *Pure and Applied Geophysics*, **137**, 439–460.

EVANS, J. P. 1991. Textures, deformation mechanisms and the role of fluid in the cataclastic deformation of granitic rocks. *In*: KNIPE, R. J. & RUTTER, E. H. (eds) *Deformation Mechanisms, Rheology, and Tectonics*. Geological Society, London, Special Publications, **54**, 29–41.

FAVRE, P. & STAMPFLI, G. M. 1992. From rifting to passive margin: examples of the Red Sea, Central Atlantic and Alpine Tethys. *Tectonophysics*, **215**, 69–97.

FITZGERALD, J. D. & STÜNITZ, H. 1993. Deformation of granitoids at low metamorphic grade. I: reactions and grain size reduction. *Tectonophysics*, **221**, 269–297.

FROITZHEIM, N. & MANATSCHAL, G. 1996. Kinematics of Jurassic rifting, mantle exhumation, and passive-margin formation in the Austroalpine and Penninic nappes (eastern Switzerland). *Geological Society of America Bulletin*, **108**, 1120–1133.

FURUSHO, M. & KANAGAWA, K. 1999. Transformation-induced strain localization in a lherzolite mylonite from the Hidaka metamorphic belt of central Hokkaido, Japan. *Tectonophysics*, **313**, 411–432.

GANDHI, C. & ASHBY, M. F. 1979. Fracture-mechanisms maps for materials which cleave: F.C.C., B.C.C. and H.C.P. metals and ceramics. Overview No. 5. *Acta Metallurgica*, **27**, 1565–1602.

GARUTI, G. & FRIOLO, R. 1978/79. Textural features and olivine fabrics of peridotites from the Ivrea–Verbano Zone (Italian western Alps). *Memorie di Scienze Geologiche*, **33**, 111–125.

GASPARIK, T. 1987. Orthopyroxene thermobarometry in simple and complex systems. *Contributions to Mineralogy and Petrology*, **96**, 357–370.

GROCOTT, J. & WATTERSON, J. 1980. Strain profile of a boundary within a large ductile shear zone. *Journal of Structural Geology*, **2**, 111–117.

HANDY, M. R. 1987. The structure, age, and kinematics of the Pogallo fault zone, Southern Alps, northwestern Italy. *Eclogae geologicae Helvetiae*, **80**, 593–632.

HANDY, M. R. 1989. Deformation regimes and the rheological evolution of fault zones in the lithosphere: the effects of pressure, temperature, grain size, and time. *Tectonophysics*, **163**, 119–152.

HANDY, M. R. 1994. The energetics of steady state heterogeneous shear in mylonitic rock, *Materials Science and Engineering*, **A175**, 261—272.

HANDY, M. R. & ZINGG, A. 1991. The tectonic and rheologic evolution of the Ivrea Crustal Cross Section (Southern Alps of northwestern Italy and southern Switzerland). *Geological Society of America Bulletin*, **103**, 236–253.

HANDY, M. R., FRANZ, L., HELLER, F., JANOTT, B. & ZURBRIGGEN, R. 1999. Multistage accretion and exhumation of continental (Ivrea crustal section, Italy and Switzerland). *Tectonics*, **18**, 1154–1177.

HOBBS, B. E., MUHLHAUS, H.-B. & ORD, A. 1990. Instability, softening and localization of deformation. *In*: KNIPE, R. J. & RUTTER, E. H. (eds) *Deformation Mechanisms, Rheology and Tectonics*. Geological Society, London, Special Publications, **45**, 143–166.

JAROSLOW, G. E., HIRTH, G. & DICK, H. J. B. 1996. Abyssal peridotite mylonites: implications for grain-size sensitive flow and strain localization in the oceanic lithosphere. *Tectonophysics*, **256**, 17–37.

JIN, D., KARATO, S.-I. & OBATA, M. 1998. Mechanisms of shear localization in the continental lithosphere: inference from the deformation microstructures of peridotites from the Ivrea zone, northwestern Italy. *Journal of Structural Geology*, **20**, 195–209.

KARSONS, J. A. 1991. Accommodation zones and transfer faults: integral components of Mid-Atlantic Ridge extensional systems. *In*: PETERS, T., NICOLAS, A. & COLEMAN, R. J. (eds) *Ophiolite Genesis and Evolution of Oceanic Lithosphere*. Ministry of Petroleum and Minerals, Sultanate of Oman, 21–37.

KEEN, C. E., MCLEAN, B. C. & KAY, W. A. 1991. A deep seismic reflection profile across the Nova Scotia continental margin offshore eastern Canada. *Canadian Journal of Earth Sciiences*, **28**, 1112–1120.

KIRBY, S. H. 1985. Rock mechanics observations pertinent to the rheology of the continental lithosphere and the localization of strain along shear zones. *Tectonophysics*, **118**, 1–27.

KRONENBERG, A. K., SEGALL, P. & WOLF, G. H. 1990. Hydrolytic Weakening and Penetrative Deformation Within a Natural Shear Zone. *In*: DUBA, A. G., DURHAM, W. B., HANDIN, J. W. & WANG, H. F. (eds) *The brittle-ductile transition in rocks, the Heard volume*, AGU Geophysical Monograph, **56**, 21–36.

KRUHL, J. H. & VOLL, G. 1978/79. Deformation and Metamorphism of the Western Finero Complex. *Memorie di Scienze Geologiche*, **33**, 95–109.

KRUSE, R. & STÜNITZ, H. 1999. Deformation mechanisms and phase distribution in mafic, high-temperature mylonites from the Jotun Nappe, southern Norway. *Tectonophysics*, **303**, 223–250.

LEMOINE, M., TRICART, P. & BOILLOT, G. 1987. Ultramafic and gabbroic ocean floor of the Ligurian Tethys (Alps, Corsica, Apennines): in search of a genetic model. *Geology*, **15**, 622–625.

LIN, A. 1997. Ductile deformation of biotite in foliated cataclasite, Iida-Matsukawa fault, central Japan. *Journal of Asian Earth Sciences*, **15**, 407–411.

LISTER, G. S., ETHERIDGE, M. A. & SYMONDS, P. A. 1991. Detachment models for the formation of passive continental margins. *Tectonics*, **10**, 1038–1064.

MAGGI, A., JACKSON, J. A., MCKENZIE, D. & PRIESTLEY, K. 2000. Earthquake focal depths, effective elastic thickness, and the strength of the continental lithosphere. *Geology*, **28**, 495–498.

MCGEARY, S. & WARNER, M. 1985. Seismic profiling of the lower continental lithosphere. *Nature*, **317**, 795–797.

MCKENZIE, D., NIMMO, F. & JACKSON, J. A. 2000. Characteristics and consequences of flow in the lower crust. *Journal of Geophysical Research*, **105**, 11029–11046.

MITRA, G. 1978. Ductile deformation zones and mylonites: the mechanical processes involved in the deformation of crystalline basement rocks. *American Journal of Science*, **278**, 1057–1084.

MÜNTENER, O. & HERMANN, J. 2001. The role of lower crust and continental upper mantle during formation of non-volcanic passive margins: evidence from the Alps. *In*: WILSON, R. C. L., WHITMARSH, R. B., TAYLOR, B. & FROITZHEIM, N. (eds) *Non-volcanic rifting of continental margins: a comparison of evidence from land and sea*. Geological Society, London, Special Publications, **187**, 267–288.

NEWMAN, J., LAMB, W. M., DRURY, M. R. & VISSERS, R. L. M. 1999. Deformation processes in a peridotite shear zone: reaction-softening by an H_2O-deficient, continuous net transfer reaction. *In*: SCHMID, S. M., HEILBRONNER, R. & STÜNITZ, H. (eds) *Deformation Mechanisms in Nature and Experiment*. *Tectonophysics*, **303**, 193–222.

NIIDA, K. & GREEN, D. H. 1999. Stabilty and chemical composition of pargasitic amphibole in MORB pyrolite under upper mantle conditions. *Contributions to Mineralogy and Petrology*, **135**, 18–40.

OBATA, M. 1976. The solubility of Al_2O_3 in orthopyroxenes in spinel and plagioclase peridotites and spinel pyroxenite. *American Mineralogist*, **61**, 804–816.

OBATA, M. & KARATO, S. 1995. The solubility of Al_2O_3 ultramafic pseudotachylite from the Balmuccia peridotite, Ivrea–Verbano zone, northern Italy. *Tectonophysics*, **242**, 313–328.

PATERSON, M. S. 1987. Problems in the extrapolation of laboratory rheological data. *Tectonophysics*, **133**, 33–43.

PATERSON, M. S. & OLGAARD, D. L. 2000. Rock deformation to large strains in torsion. *Journal of Structural Geology*, **22**, 1341–1358.

POIRIER, J. P. & NICOLAS, A. 1975. Deformation-induced recrystallization due to progressive misorientation of subgrains, with special reference to mantle peridotites. *Journal of Geology*, **83**, 707–720.

QUICK, J. E., SINIGOI, S. & MAYER, A. 1994. Emplacement dynamics of a large mafic intrusion in the lower crust, Ivrea–Verbano, northern Italy. *Journal of Geophysical Research*, **99**, 21559–21573.

RALEIGH, C. B. & PATERSON, M. S. 1965. Experimental deformation of serpentinite and its tectonic implications. *Journal of Geophysical Research*, **70**, 3965–3985.

RANALLI, G. & MURPHY, D. C. 1987. Rheological stratification of the lithosphere. *Tectonophysics*, **132**, 281–296.

RESTON, T. 1987. Spatial interference, reflection character and the structure of the lower crust under extension – Results from 2-D seismic modelling. *Annales Geophysicae*, **5B**, 339–348.

RESTON, T. 1990. Shear in the lower crust during extension: not so pure and simple. *Tectonophysics*, **173**, 175–183.

RIVALENTI, G., GARUTI, G. & ROSSI, A. 1975. The origin of the Ivrea–Verbano basic formation (Western Italian Alps). *Bolletino Societa Geologica Italiana*, **94**, 1149–1186.

RIVALENTI, G., GARUTI, G., ROSSI, A., SIENA, F. & SINIGOI, S. 1981. Existence of different peridotite types and of a layered igneous complex in the Ivrea Zone of the Western Alps. *Journal of Petrology*, **22**, 127–153.

RIVALENTI, G., ROSSI, A., SIENA, F. & SINIGOI, S. 1984. The Layered Series of the Ivrea- Verbano Igneous Complex, Western Alps, Italy. *Tschermaks Mineralogische und Petrographische Mitteilungen*, **33**, 77–99.

RUTTER, E. H. & BRODIE, K. H. 1988. The role of tectonic grainsize reduction in the rheological stratification of the lithosphere. *Geologische Rundschau*, **77**, 295–308.

SAVOSTIN, L. A., SIBUET, J.-C., ZONENSHAIN, L. P., LE PICHON, X. & ROULET, M.-J. 1986. Kinematic evolution of the Tethys belt from the Atlantic Ocean to the Pamirs since the Triassic. *Tectonophysics*, **123**, 1–35.

SCHMID, S. M. 1993. Ivrea Zone and Adjacent Southern Alpine Basement. *In*: VON RAUMER, J. F. & NEUBAUER, F. (eds) *Pre-Mesozoic Geology in the Alps*. Springer Verlag, Berlin, 567–584.

SCHMID, S. M. & KISSLING, E. 2000. The arc of the western Alps in the light of geophysical data on deep crustal structure. *Tectonics*, **19**, 62–85.

SCHMID, S. M., AEBLI, H. R., HELLER, F. & ZINGG, A. 1989. The role of the Periadriatic Line in the tectonic evolution of the Alps. *In*: COWARD, M. P., DIETRICH, D. & PARK, R. (eds) *Alpine Tectonics*. Geological Society, London, Special Publications, **45**, 153–171.

SCHMID, S. M., PANOZZO, R. & BAUER, S. 1987. Simple shear experiments on calcite rocks: rheology and microfabrics. *Journal of Structural Geology*, **9**, 747–778.

SEGALL, P. & POLLARD, D. D. 1983. Nucleation and grwoth of strike-slip faults in granite. *Journal of Geophysical Research*, **88**, 555–568.

SEGALL, P. & SIMPSON, C. 1986. Nucleation of ductile shear zones on dilatant fractures. *Geology*, **14**, 56–59.

SHERVAIS, J. W. 1979. Thermal emplacement model for the Alpine Lherzolite massif at Balmuccia, Italy. *Journal of Petrology*, **20**, 795–820.

SINIGOI, S., QUICK, J., CLEMENS-KNOTT, D., MAYER, A. & DEMARCHI, G. 1994. Chemical evolution of a large mafic intrusion in the lower crust, Ivrea–Verbano Zone. *Journal of Geophysical Research*, **99**, 21575–21590.

SKROTZKI, W., WEDEL, A., WEBER, K. & MÜLLER, W. F. 1990. Microstructure and texture in lherzolites of the Balmuccia massif and their significance regarding the thermomechanical history. *Tectonophysics*, **179**, 227–251.

STAMPFLI, G. M. & MARTHALER, M. 1990. Divergent and convergent margins in the North-Western Alps confrontation to actualistic models. *Geodinamica Acta*, **3**, 159–184.

STECK, A. & TIÈCHE, J.-C. 1976. Carte geologique de l'antiforme peridotique de Finero avec des observations sur les phases de deformation et de recristallisation. *Bulletin de Minéralogie et Pétrographie de Suisse*, **56**, 501–512.

STÜNITZ, H. 1993. Transition from fracturing to viscous flow in a naturally deformed metagabbro. *In*: BOLAND, J. N. & FITZGERALD, J. D. (eds) *Defects and Processes in the Solid State: Geoscience Applications, the McLaren Volume*, Elsevier, Amsterdam, 121–150.

STÜNITZ, H. & TULLIS, J. 2001. Weakening and strain localization produced by syndeformational reaction of plagioclase. *International Journal of Earth Sciences*, **90**, 136–148.

STUNITZ, H., TULLIS, J. & YUND, R. 1999. The relation between deformation and reaction for experimentally deformed olivine + plagioclase. *EOS Transactions*, **80**, 1021.

TECHMER, K. S., AHRENDT, H. & WEBER, K. 1993. The development of pseudotachylite in the Ivrea–Verbano zone of the north Italian Alps. *Tectonophysics*, **204**, 307–322.

TROMMSDORFF, V. 1983. Metamorphose magensiumreicher Gesteine: Kritischer Vergleich von Natur, Experiment und thermodynamischer. *Datenbasis. Fortschr. Mineral.*, **61**, 283–308.

TROMMSDORFF, V. & EVANS, B. W. 1974. Alpine metamorphism of peridotitic rocks. *Schweizerische Mineralogische und Petrographische Mitteilungen*, **54**, 333–352.

TROMMSDORFF, V., PICCARDO, G. B. & MONTRASIO, A. 1993. From magmatism through metamorphism to sea floor emplacement of subcontinental Adria lithosphere during pre-Alpine rifting (Malenco, Italy). *Schweizerische Mineralogische und Petrographische Mitteilungen*, **73**, 191–203.

TULLIS, J. & YUND, R. A. 1985. Dynamic recrystallization of feldspar: a mechanism for ductile shear zone formation. *Geology*, **13**, 238–241.

TULLIS, J., DELL'ANGELLO, L. & YUND, R. A. 1990. Ductile shear zones from brittle precursors in feldspathic rocks: The role of dynamic recrystallization. *In*: DUBA, A. G., DURHAM, W. B., HANDIN, J. W. & WANG, H. F. (eds) *The Brittle-ductile Transition in Rocks, the Heard Volume*. Geophysical Monograph, **56**, 67–82.

TUSCHOLKE, B. E. & LIN, J. 1994. A geological model for the structure of ridge segments in slow-spreading oceanic crust. *Journal of Geophysical Research*, **99**, 11937–11958.

VAN DER WAL, D., CHOPRA, P. N., DRURY, M. R. & FITZGERALD, J. D. 1993. Relationships between dynamically recrystallized grain size and stress in experimentally deformed olivine rocks. *Geophysical Research Letters*, **20**, 1479–1482.

VAVRA, G., SCHMID, R. & GEBAUER, D. 1999. Internal morphology, habit and U-Th-Pb microanalysis of amphibolite-to-granulite facies zircons: geochronology of the Ivrea Zone (Southern Alps). *Contributions to Mineralogy and Petrology*, **134**, 380–404.

VISSERS, R. L. M., DRURY, M. R., HOOGERDUIJN STRATING, E. H., SPIERS, E. H. & VAN DER WAL, D. 1995. Mantle shear zones and their effect on lithosphere strength during continental break-up. *Tectonophysics*, **249**, 155–171.

VISSERS, R. L. M., DRURY, M. R., HOOGERDUIJN STRATING, E. H. & VAN DER WAL, D. 1991. Shear zones in the upper mantle: A case study in an Alpine lherzolite massif. *Geology*, **19**, 990–993.

VISSERS, R. L. M., DRURY, M. R., NEWMAN, J. & FLIERVOET, T. F. 1997. Mylonitic deformation in upper mantle peridotites of the North Pyrenean Zone (France): implications of strength and strain localization in the lithosphere. *Tectonophysics*, **279**, 303–325.

WALLACE, M. E. & GREEN, D. H. 1991. The effect of bulk rock composition on the stability of amphibole in the upper mantle: implications for the solidus positions and mantle metasomatism. *Mineralogy and Petrology*, **44**, 1–19.

WALTER, M. J. & PRESNALL, D. C. 1994. Melting behaviour of simplified lherzolite in the system $CaO-MgO-Al_2O_3-SiO_2-Na_2O_3$ from 7 to 35 kb. *Journal of Petrology*, **35**, 329–359.

WEERTMAN, J. 1970. The creep strength of the Earth's mantle. *Reviews of Geophysics and Space Physics*, **8**, 145–168.

WERNICKE, B. 1985. Uniform-sense normal simple shear of the continental lithosphere. *Canadian Journal of Earth Sciences*, **22**, 108–125.

WHITE, S. H. & KNIPE, R. J. 1978. Transformation- and reaction-enhanced ductility in rocks. *Journal of the Geological Society, London*, **135**, 513–516.

WIBBERLEY, C. 1999. Are feldspar-to-mica reactions necessarily reaction-softening processes? *Journal of Structural Geology*, **21**, 1219–1227.

WILSON, C. J. L. & BELL, I. A. 1979. Deformation of biotite and muscovite: optical microstructure. *Tectonophysics*, **58**, 179–200.

WINTSCH, R. P., CHRISTOFFERSEN, R. & KRONENBERG, A. K. 1995. Fluid-rock reaction weakening of fault zones. *Journal of Geophysical Research*, **100**, 13021–13032.

ZEUCH, D. H. 1982. Ductile faulting, dynamic recrystallization and grain-size sensitive flow of olivine. *Tectonophysics*, **83**, 293–308.

ZEUCH, D. H. 1983. On the inter-relationship between grain-size sensitive creep and dynamic recrystallization of olivine. *Tectonophysics*, **93**, 151–168.

ZINGG, A., HANDY, M. R., HUNZIKER, J. C. & SCHMID, S. M. 1990. Tectonometamorphic history of the Ivrea Zone and its relation to the crustal evolution of the Southern Alps. *Tectonophysics*, **182**, 169–192.

Index

Page numbers in italic, e.g. *173*, refer to figures. Page numbers in bold, e.g. **241**, signify entries in tables.